T0256316

Topographic Laser Ranging and Scanning

Topographic Laser Ranging and Scanning

Principles and Processing, Second Edition

Edited by
Jie Shan
Charles K. Toth

CRC Press
Taylor & Francis Group
Boca Raton London New York

CRC Press is an imprint of the
Taylor & Francis Group, an **informa** business

CRC Press
Taylor & Francis Group
6000 Broken Sound Parkway NW, Suite 300
Boca Raton, FL 33487-2742

First issued in paperback 2022

© 2018 by Taylor & Francis Group, LLC
CRC Press is an imprint of Taylor & Francis Group, an Informa business

No claim to original U.S. Government works

ISBN 13: 978-1-4987-7227-3 (hbk)
ISBN 13: 978-1-03-247629-2 (pbk)
ISBN 13: 978-1-315-15438-1 (ebk)

DOI: 10.1201/9781315154381

This book contains information obtained from authentic and highly regarded sources. Reasonable efforts have been made to publish reliable data and information, but the author and publisher cannot assume responsibility for the validity of all materials or the consequences of their use. The authors and publishers have attempted to trace the copyright holders of all material reproduced in this publication and apologize to copyright holders if permission to publish in this form has not been obtained. If any copyright material has not been acknowledged please write and let us know so we may rectify in any future reprint.

Except as permitted under U.S. Copyright Law, no part of this book may be reprinted, reproduced, transmitted, or utilized in any form by any electronic, mechanical, or other means, now known or hereafter invented, including photocopying, microfilming, and recording, or in any information storage or retrieval system, without written permission from the publishers.

For permission to photocopy or use material electronically from this work, please access www.copyright.com (http://www.copyright.com/) or contact the Copyright Clearance Center, Inc. (CCC), 222 Rosewood Drive, Danvers, MA 01923, 978-750-8400. CCC is a not-for-profit organization that provides licenses and registration for a variety of users. For organizations that have been granted a photocopy license by the CCC, a separate system of payment has been arranged.

Trademark Notice: Product or corporate names may be trademarks or registered trademarks, and are used only for identification and explanation without intent to infringe.

Publisher's Note
The publisher has gone to great lengths to ensure the quality of this reprint but points out that some imperfections in the original copies may be apparent.

Library of Congress Cataloging-in-Publication Data

Names: Shan, Jie, author. | Toth, Charles K., author.
Title: Topographic laser ranging and scanning : principles and processing, / Jie Shan and Charles K. Toth.
Description: Second edition. | Boca Raton : Taylor & Francis, CRC Press, 2018. | Includes bibliographical references and index.
Identifiers: LCCN 2017025409 | ISBN 9781498772273 (hardback : alk. paper)
Subjects: LCSH: Lasers in surveying. | Topographical surveying. | Aerial photogrammetry.
Classification: LCC TA579 .T66 2018 | DDC 526.9/8--dc23
LC record available at https://lccn.loc.gov/2017025409

**Visit the Taylor & Francis Web site at
http://www.taylorandfrancis.com**

**and the CRC Press Web site at
http://www.crcpress.com**

Contents

Preface to the Second Edition

There were two main reasons that motivated our decision to publish this 2nd edition of *Topographic Laser Ranging and Scanning*. First, we wanted to incorporate the large array of new laser ranging and scanning technologies that have become available for various scientific and engineering applications since the 1st edition's release in 2009. Examples of these technologies include various small, low-cost, and lightweight Light Detection and Ranging (LiDAR) sensors that are suitable for handheld and Unmanned Aerial Vehicle (UAV) applications; multispectral LiDAR sensors that collect spectral and ranging information simultaneously; and single photon LiDAR and Geiger-mode LiDAR, which became a reality since the 1st edition was published and now are being evaluated for civilian use. In the past eight years, we have also witnessed profound developments in LiDAR data processing techniques, largely due to the closer engagement between the remote sensing and computer vision communities. As a result, more sophisticated processing methodologies were developed and then implemented and utilized in mainstream production. The other reason that motivated us to produce the 2nd edition was that its predecessor is very popular and remains at the top of the best sellers in its category. In the past eight years or so, thousands of print and electronic copies have been sold and downloaded. In addition, numerous LiDAR-related courses are utilizing the 1st edition as their course text or reference, and it is among the most referenced materials for faculty, research scientists, and students. The remainder of this preface presents updated content information, summarizes and reflects on the changes and innovations.

Similar to the 1st edition, the 2nd edition is comprised of 19 chapters. A few chapters, which primarily focus on fundamental theories, were not changed or updated at the discretion of the authors, including Chapter 5, "Pulsed Laser Altimeter Ranging Techniques and Implications for Terrain Mapping"; Chapter 6, "Georeferencing Component of LiDAR Systems"; and Chapter 14, "Feature Extraction from LiDAR Data in Urban Areas."

All the other chapters were subject to updates and revisions to varying extents. Largely following the same structure as the 1st edition, Chapter 1 serves as an "Introduction to Laser Ranging, Profiling, and Scanning" with refinements and revision in the content as well. Chapters 2 and 3 are in reverse order of the 1st edition. Chapter 2 presents an updated and enhanced discussion on "Terrestrial Laser Scanners." Although Chapter 3 is still focused on "Airborne and Spaceborne Laser Profilers and Scanners" (Chapter 2 in the 1st edition), a number of new sensor systems are introduced with illustrations. In particular, recently introduced UAV-based LiDAR systems were added as new content. The text was revised in Chapter 4, but airborne LiDAR systems and calibration are still its focus. Chapter 7 enhances the discussions on the principles and fundamentals beyond small-footprint pulsed laser systems by revising the text and including new application examples. Chapter 8 still has a discussion on "Strip Adjustment", in which error sources and selection of overlapping areas are highlighted and surface modeling and representations are regrouped. The scope of Chapter 9 is unchanged (i.e., the quality control and quality assurance aspects of LiDAR mapping); however, the text and illustrations have been modified. Modifications were made in Chapter 10 to reflect a number of new developments in LiDAR data management, such as the latest LiDAR point cloud standard (LAS 1.4) and cloud hosting services (e.g., Microsoft Azure and Amazon Web Services). Although its focus remains on LiDAR data filtering, Chapter 11 presents revised discussions and a number of new examples and illustrations for digital terrain model (DTM) generation. Chapter 12 extends from airborne platforms to terrestrial and mobile ones to reflect the state of the art of forest inventory management. Following the same structure as the 1st edition, Chapter 13 discusses the integration of LiDAR and image data with the purpose of their precise registration and orthoimage production. Chapter 15 presents a new development of "Global Solutions to Building Segmentation and Reconstruction" from point clouds. Although the procedural descriptions of the previous edition were not changed in Chapter 16, a theoretical framework of context-based classification for building and road extraction was added. Chapter 17 presents a new development in progressive reconstruction of 3D building roof tops over time using multisensor data. Chapter 18 includes an updated workflow and revised discussions for building extraction and reconstruction from airborne LiDAR point clouds. Finally, Chapter 19 was revised to reflect the current test results of professional organizations.

The general structure of the 2nd edition is unchanged from the 1st edition and can be grouped into four topic areas. Part I, LiDAR Systems and Fundamentals (Chapters 1 through 3) presents an introduction and summary of various LiDAR systems and their fundamentals. Part II, Operational Principles and LiDAR Systems (Chapters 4 through 7) addresses the operational principles of the different components and ranging methods of LiDAR systems. Part III, Geometric Processing of LiDAR Data (Chapters 8 through 10) discusses the subsequent geometric processing of LiDAR data, in particular, with respect to quality, accuracy, and standards. Part IV, Information Extraction from LiDAR Data (Chapters 11 through 19) addresses the theories and practices of information extraction from LiDAR data, including terrain surface generation, forest inventory, orthoimage generation, building reconstruction, and road extraction.

The composition and publication of this 2nd edition was an endeavor that lasted almost two years after an early survey of the contributors revealed that the majority of the chapters needed certain updates. We, the editors, are very grateful to all the authors that agreed to include their most recent developments and carefully prepared their revised or new write-ups. Without them, it would not have been possible to reflect the state of the art in LiDAR remote sensing. Moreover, we are also highly indebted to the CRC staff for being patient and persistent in helping us and the authors to complete this book project.

Jie Shan
Purdue University

Charles K. Toth
The Ohio State University, May 2017

Preface to the First Edition

Light Detection and Ranging (LiDAR) is probably the most significant technology introduced in mainstream topographic mapping in the past decade. The main advantage of this technique is that it provides a direct method for 3D data collection. Furthermore, it is highly accurate because of the millimeter- and centimeter-level laser ranging accuracy and precise sensor platform orientation supported by an integrated position and orientation system (POS). Unlike the traditional photogrammetric methods, LiDAR directly collects an accurately georeferenced set of dense point clouds, which can be almost directly used in basic applications. However, the full exploitation of LiDAR's potentials and capabilities challenges for new data processing methods that are fundamentally different from the ones used in traditional photogrammetry. Over the past decade, there have been many significant developments in this field, mainly resulting from multidisciplinary research, including computer vision, computer graphics, electrical engineering, and photogrammetry. Consequently, the conventional image-based photogrammetry and vision are gradually adapting to a new subject, which is primarily concerned with point clouds data collection, calibration, registration, and information extraction.

Although there have been many studies and applications since the introduction of LiDAR, a comprehensive compilation is still missing that (1) describes the basic principles and fundamentals in laser ranging and scanning; (2) reflects the state of the art of laser scanning technologies, systems, and data collection methods; and (3) presents the data processing methods and recent developments reported in different subject areas, which is probably the most challenging task. It is not an overstatement that such a collection has been long overdue. The primary objective of this book is to meet these needs and present a manual to the LiDAR research, practice, and education communities.

The chapters in this book consist of four major parts, each addressing different topic areas in LiDAR technology and data processing. Part I, Chapters 1 through 3, primarily presents a brief introduction and a comprehensive summary of LiDAR systems. After a concise introduction to the laser fundamentals, Chapter 1 describes the two laser ranging methods (timed pulse method and phase comparison method) and the principles of laser profiling and scanning. The issues and properties of laser ranging techniques, such as safety concerns, returned power, beam divergence, and reflectivity, are also discussed. Chapters 2 and 3 present a rather comprehensive description of the airborne and spaceborne laser devices (profilers and scanners) and terrestrial scanners. Each of these two chapters covers a brief historical review, the taxonomy, and a thorough discussion on the components, working principles, and main specifications of the LiDAR equipment.

Part II addresses the operational principles of different units and ranging methods in a LiDAR system. Chapter 4 discusses the working principles of LiDAR systems and supporting devices and their synchronization. The theory and practices for georegistration and calibration of the boresight are also presented, along with testing examples. Finally, a discussion is offered on the working steps required to produce a digital terrain model (DTM), including flight planning, equipment installation, and data preprocessing. In order to most effectively use the products generated from LiDAR systems, Chapter 5 discusses the operational principles of LiDAR systems and the resulting effects on the acquired data. Laser ranging methods, including waveform recording and single-photon detection approaches, are examined. The relationships between ranging methods, the instrument parameters, and the character of the resulting elevation data are also considered, with emphasis on the measurement of ground topography beneath vegetation cover and characterization of forest canopy structure. Chapter 6 starts with an introduction to direct georeferencing technology. The discussion is focused on the analytics of the inertial navigation system (INS), which is important for achieving quality airborne LiDAR data. The principles for the integration of INS, global positioning system (GPS), and LiDAR are also presented. Chapter 7 is about small-footprint pulsed laser systems with emphasis on full-waveform analysis. Addressed with details in this chapter are approaches for designing a laser system, modeling the spatial and temporal properties of the emitted laser pulses, detecting return pulses, and deriving attributes from the waveform.

The theme of Part III relates to the subsequent geometric processing of LiDAR data, in particular, with respect to quality, accuracy, and standards. Chapter 8 starts with a review of the underlying theory for surface representation. The factors that affect strip adjustment to form one integrated consistent dataset are studied. The discussion covers both data-driven and sensor calibration-based methods. Chapter 9 identifies the unique properties for quality control and calibration of a LiDAR system. Its error sources and their impact on the resulting surface are identified and discussed. Such a study yields several recommended procedures for quality assurance and control of the LiDAR systems and their derived data. Chapter 10 addresses the data format, organization, storage, and standards, and their effect on production from a practical point of view.

Part IV is about information extraction from the LiDAR data. As the starting chapter of this topic, Chapter 11 focuses on the extraction of the terrain surface from either the original point cloud or gridded data. Emphasis is placed on the principles of different filters for ground separation and their performance. The role of structural lines in the terrain model generation is discussed to ensure its fidelity and quality. Chapter 12 presents a broad overview of laser-based forest inventory. This includes user requirements, laser-scattering process in forests, canopy height retrieval, and the corresponding accuracy. Besides tree or stand parameter retrieval, the change detection possibilities using multitemporal laser surveys are also discussed. Chapter 13 introduces algorithms for a multiprimitive and multisensor triangulation environment to assure the precise alignment of the involved LiDAR and imagery data. Such alignment leads to the straightforward production of orthophotos, for which several methodologies using LiDAR and photogrammetric data of urban areas are outlined. Chapters 14 through 18 focus on feature extraction in urban environments. Such a dynamic topic attracts many studies from different angles in different disciplines. These methods include random sample consensus (RANSAC) and Hough transform for model detection and estimation (Chapter 14), clustering approach for segmentation and reconstruction (Chapters 15, 16, and 18), combined use of multiple datasets (Chapters 16 and 17), region growing (Chapters 16 and 18), graph partition (Chapters 14 and 17), and rule-based approaches (Chapter 16). Although each of the above chapters has independent quality evaluation, Chapter 19 presents a comprehensive summary of building extraction qualities based on an organized joint effort of different participants. In summary, through the above independent yet related structure of the book contents, we present the audience a broad discussion on different aspects of laser topographic mapping—principles, systems, operation, geometric processing, and information extraction.

This book is targeted to a variety of audience and is in particular expected to be used as a reference book for senior undergraduate and graduate students, majoring or working in diverse disciplines, such as geomatics, geodesy, natural resources, urban planning, computer vision, and computer graphics. It can also be used by researchers who are interested in developing new methods and need in-depth knowledge of laser scanning and data processing. Other professionals may gain the same from the broad topics addressed in this book.

Despite the remarkable achievements in LiDAR technology and data processing, it is still a relatively young subject and would likely change its course rather quickly in the near future. We expect this book can serve as a durable introduction to the fundamentals, principles, basic problems, and issues in LiDAR technology for topographic applications. It also provides a rather comprehensive and sophisticated coverage on various data processing techniques, which we believe will sustain and ultimately lead to new developments while the technology is advancing.

Finally, we would thank all of the contributors and technical staff members at the publisher, without whom this book would not have been possible. It is greatly appreciated that the leaders in their area of expertise agreed to share their recent work and carefully prepared the manuscripts. The publisher's staff members helped the editors and authors throughout the entire publication period. It is our great pleasure to be working with all of them to present this book to the broad LiDAR community and beyond.

Jie Shan
Purdue University

Charles K. Toth
The Ohio State University

Editors

Jie Shan is currently a professor with the Lyles School of Civil Engineering, Purdue University, West Lafayette, Indiana. He earned his PhD in photogrammetry and remote sensing from Wuhan University, Wuhan, China. He has been a faculty member at universities in China and Sweden, and a research fellow in Germany. His areas of interests include sensor geometry and positioning, object extraction and reconstruction from images and point clouds, urban remote sensing, automated digital mapping, and pattern recognition of spatial, temporal, and semantic data. He has authored/coauthored more than 200 scientific publications, and is a recipient of multiple best paper awards, including the Talbert Abrams Grand Award and the Environmental Systems Research Institute Award. He is an elected American Society for Photogrammetry and Remote Sensing (ASPRS) fellow and a senior member of IEEE.

Charles K. Toth is a research professor in the Department of Civil, Environmental and Geodetic Engineering, The Ohio State University, Columbus, Ohio. He earned his MSc degree in electrical engineering and a PhD in electrical engineering and geoinformation sciences from the Technical University of Budapest, Budapest, Hungary. His research interest and expertise cover broad areas of spatial information sciences and systems, including photogrammetry, multisensor geospatial data acquisition systems, LiDAR, high-resolution imaging, surface extraction, modeling, integrating and calibrating multisensor systems, georeferencing and navigation, 2D/3D signal processing, and mobile mapping technologies. He has published more than 300 peer-reviewed journal and proceeding papers, and is the recipient of numerous awards, including the 2009 APSRS Photogrammetric Award, United States Geospatial Intelligence Foundation (USGIF) Academic Achievement Award, 2015, ISPRS Schwidefsky Medal Winner 2016, several Lumley Research Awards from The Ohio State University (OSU), and various best papers awards.

Contributors

Frédéric Bretar
Consulate General of France
Hong Kong

Shu-Ching Chen
School of Computing and Information Sciences
Florida International University
Miami, Florida

Simon Clode
Veris Australia Pty Ltd.
Osborne Park, Western Australia, Australia

Naser El-Sheimy
Department of Geomatics Engineering
The University of Calgary
Calgary, Alberta, Canada

Lewis Graham
GeoCue Group Inc.
Madison, Alabama

Eberhard Gülch
Faculty of Geomatics
Computer Science and Mathematics
Stuttgart University of Applied Sciences
Stuttgart, Germany

Ayman Habib
Lyles School of Civil Engineering
Purdue University
West Lafayette, Indiana

David Harding
Goddard Space Flight Center
National Aeronautics and Space Administration
Greenbelt, Maryland

Markus Holopainen
Department of Forest Sciences
University of Helsinki
Helsinki, Finland

Hannu Hyyppä
Department of Built Environment
Aalto University
Espoo, Finland

Juha Hyyppä
Centre of Excellence in Laser Scanning
    Research
Finnish Geospatial Research Institute
Masala, Finland

Anttoni Jaakkola
Centre of Excellence in Laser Scanning
    Research
Finnish Geospatial Research Institute
Masala, Finland

Wanshou Jiang
State Key Laboratory of Information
    Engineering in Surveying, Mapping and
    Remote Sensing
Wuhan University
Wuhan, China

Jaewook Jung
Department of Earth and Space Science and
    Engineering
York University
Toronto, Ontario, Canada

Boris Jutzi
Institute of Photogrammetry and Remote
    Sensing
Karlsruhe Institute of Technology
Karlsruhe, Germany

Harri Kaartinen
Centre of Excellence in Laser Scanning
    Research
Finnish Geospatial Research Institute
Masala, Finland

Zoltan Koppanyi
Department of Civil, Environmental &
    Geodetic Engineering
The Ohio State University

Antero Kukko
Centre of Excellence in Laser Scanning
    Research
Finnish Geospatial Research Institute
Masala, Finland

Xinlian Liang
Centre of Excellence in Laser Scanning
  Research
Finnish Geospatial Research Institute
Masala, Finland

Gottfried Mandlburger
Department of Geodesy and Geoinformation
TU Wien
Vienna, Austria

Gordon Petrie
School of Geographical and Earth Sciences
University of Glasgow
Glasgow, Scotland, UK

Norbert Pfeifer
Department of Geodesy and Geoinformation
TU Wien
Vienna, Austria

Franz Rottensteiner
Institute of Photogrammetry and
  GeoInformation
Leibniz University Hannover
Hannover, Germany

Gunho Sohn
Department of Earth and Space Science and
  Engineering
York University
Toronto, Ontario, Canada

Uwe Stilla
Photogrammetry and Remote Sensing
Technical University of Munich
Munich, Germany

Mikko Vastaranta
Department of Forest Sciences
University of Helsinki
Helsinki, Finland

Yunsheng Wang
Centre of Excellence in Laser Scanning
  Research
Finnish Geospatial Research Institute
Masala, Finland

Aloysius Wehr
Institute of Navigation
University of Stuttgart
Stuttgart, Germany

Jianhua Yan
Amazon.com, Inc.
Seattle, Washington

Jixing Yan
School of Highways
Chang'an University
Xi'an, China

Xiaowei Yu
Centre of Excellence in Laser Scanning
  Research
Finnish Geospatial Research Institute
Masala, Finland

Keqi Zhang
Department of Earth and Environment
International Hurricane Research Center
Florida International University
Miami, Florida

1 Introduction to Laser Ranging, Profiling, and Scanning

Gordon Petrie and Charles K. Toth

CONTENTS

1.1 INTRODUCTION

Topographic laser profiling and scanning systems have been the subject of phenomenal developments in recent years and, without doubt, they have become the most important geospatial data acquisition technology that has been introduced since the last millennium. Installed on both airborne and land-based platforms, these systems can collect explicit 3D data in large volumes at an unprecedented accuracy. Furthermore, the complexity of the required processing of the measured laser data is relatively modest, which has further fueled the rapid proliferation of this technology to a variety of applications. Although the invention of laser goes back to the early 1960s, the lack of various supporting technologies prevented the exploitation of this device in the mapping field for several decades. The introduction of direct georeferencing technology in the mid-1990s and the general advancements in computer technology have been the key enabling technologies for developing laser profiling and scanning systems for use in the topographic mapping field that are commercially viable. This introductory chapter provides a brief historical background on the development path and the fundamentals of laser technology before introducing the basic principles and techniques of topographic laser ranging, profiling, and scanning.

The present widespread use of laser profiling and scanning systems for topographic applications only really began in the mid-1990s after a long prior period of research and the development of the appropriate technology and instrumentation. In this respect, it should be mentioned that NASA played a large part in pioneering and developing the requisite technology through its activities in Arctic topographic mapping from the 1960s onwards. However, prior to these developments, lasers had been used extensively for numerous other applications within the field of surveying—especially within engineering surveying. Indeed, quite soon after the invention of the laser in its various original forms—the solid-state laser (in 1960), the gas laser (in 1961), and the semiconductor laser (in 1962)—the device was adopted by professional land surveyors and civil engineers. Various types of laser-based surveying instruments were devised and soon began to be used in their field survey operations. These early applications of lasers included the alignment operations and deformation measurements being made in tunnels and shafts and on bridges. Another early use was the incorporation of lasers in both manually operated and automatic (self-leveling) laser levels, again principally for engineering survey applications.

1.2   TERRESTRIAL APPLICATIONS

Turning next to laser ranging, profiling, and scanning, which are the main subjects of the present volume, lasers started to be used by surveyors for distance or range measurements in the mid- to late-1960s. These measurements were made using instruments that were based either on phase comparison methods or on pulse echo techniques. The latter included the powerful solid-state laser rangers that were used for military ranging applications such as gunnery and tracking. On the field surveying side, from the 1970s onwards, lasers started to replace the tungsten or mercury vapor lamps that had been used in early types of electronic distance measuring instruments. Initially, these laser-based electronic distance measuring instruments were mainly used as stand-alone devices measuring the distances required for control surveys or geodetic networks using trilateration or traversing methods. The angles required for these operations were measured separately using theodolites. Later, these two types of instruments were merged with the laser-based ranging technology being incorporated into total stations, which were also capable of making precise angular measurements using optoelectronic encoders. These total stations allowed topographic surveys to be undertaken by surveyors with field survey assistants setting up pole-mounted reflectors at the successive positions required for the construction of the topographic map or terrain model. This type of operation is often referred to as electronic tacheometry. With the development of very small and powerful (yet eye-safe) lasers, reflectorless distance measurements then became possible. These allowed manually operated ground-based profiling devices based on laser rangefinders to be developed, initially for use in quarries and open-cast pits and in tunnels (Petrie, 1990). Given all this prior development of lasers for numerous different field surveying applications, it was a natural and logical development for scanning mechanisms to be added to these laser rangefinders and profilers. This has culminated in the development of the present types of terrestrial laser scanners and profilers that are now being used widely for topographic mapping applications, either from stationary positions when mounted on tripods or in a mobile mode when mounted on vehicular platforms or used as hand-portable scanners.

1.3   AIRBORNE AND SPACEBORNE APPLICATIONS

Turning next to airborne platforms, laser altimeters that could be used to measure continuous profiles of the terrain from aircraft had been devised and flown as early as 1965. One early example was based on the use of a gas laser (Miller, 1965), whereas another was based on a semiconductor laser (Shepherd, 1965). As with ground laser ranging devices, at first, both the phase comparison and pulse echo methods of measuring the distance from the airborne platform to the ground were used for the purpose. A steady development of airborne laser profilers then took place throughout the 1970s and 1980s. However, the limitation in this technique is that it can only acquire elevation data directly below the aircraft along a single line crossing the terrain during an individual flight. Thus, if a large area of terrain had to be covered, as required for topographic mapping, a very large number of flight lines had to be flown to give complete coverage of the ground. So, in practice, airborne laser profiling was only applicable to those surveys being conducted along individual lines or quite narrow corridors—for example, when carrying out geophysical surveys—in which the method could be used in conjunction with other instruments such as gravimeters or magnetometers that had similar linear measuring characteristics. However, once suitable scanning mechanisms had been devised during the early 1990s, airborne laser scanning developed very rapidly and has come into widespread use. The introduction of direct georeferencing as an enabling technology was essential for the adoption of airborne laser scanners, since, unlike in photogrammetry, there is no viable method to reconstruct the sensor trajectory solely from the laser range measurements and thus to determine the ground coordinates of the laser points. The GPS constellation had been completed by the early/mid-1990s, and with the commercial availability of medium/high-performance Inertial Measurement Units (IMUs) by the mid-1990s, integrated GPS/IMU georeferencing systems were able to deliver airborne platform position and attitude data at an accuracy of 4–7 cm and 20–60 arc-seconds, respectively.

On the spaceborne side, given the huge distances of hundreds of kilometers over which the range has to be measured from an Earth-orbiting satellite—100 times greater than those being measured from an airborne platform—very powerful lasers have had to be used. Besides which, the time-of-flight (TOF) of the laser pulse is much greater, so the rate at which measurements can be made is reduced, unless a multipulse technique is used. Furthermore, the platform speed is very high at 29,000 km/h, which again is 100 times greater than that of a survey aircraft. Thus far, these demanding operational characteristics have limited spaceborne missions using laser ranging instruments to the acquisition of topographic data using profiling rather than scanning techniques.

1.4 BASIC PRINCIPLES OF LASER RANGING, PROFILING, AND SCANNING

A very brief introduction to the basic principles of laser ranging, profiling, and scanning will be provided here before going on to the remainder of the current chapter, which is concerned with laser fundamentals and laser ranging.

1.4.1 LASER RANGING

All laser ranging, profiling, and scanning operations are based on the use of some type of laser-based ranging instrument—usually described as a laser ranger or laser rangefinder—that can measure distance to a high degree of accuracy. As will be discussed in more detail later in the current chapter, this measurement of distance or range, which is always based on the precise measurement of time, can be carried out using one of the two main methods.

1. The first of these involves the accurate measurement of the TOF of a very short but intense pulse of laser radiation to travel from the laser ranger to the object being measured and to return to the instrument after having been reflected from the object—hence the use of the term *pulse echo* mentioned earlier. Thus, the laser ranging instrument measures the precise time interval that has elapsed between the pulse being emitted by the laser ranger located at point A and its return after reflection from a ground object located at point B (Figure 1.1).

$$R = \frac{\upsilon \cdot t}{2} \qquad (1.1)$$

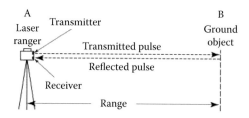

FIGURE 1.1 Basic operation of a laser rangefinder that is using the timed pulse or TOF method.

where:
    R is the slant distance or range
    υ is the speed of electromagnetic radiation, which is a known value
    t is the measured time interval

From this, the following simple relationship can be derived:

$$\Delta R = \frac{\Delta\upsilon \cdot t}{2} + \frac{\upsilon \cdot \Delta t}{2} \tag{1.2}$$

where:
    ΔR is the range precision
    $\Delta\upsilon$ is the velocity precision
    Δt is the corresponding precision value of the time measurement

Since the speed of light is very accurately known, in practice, the range precision or resolution is determined by the precision of the time measurement.

2. In the second (alternative) method, the laser transmits a continuous beam of laser radiation instead of a pulse. In this case, the range value is derived by comparing the transmitted and received versions of the sinusoidal wave pattern of this emitted beam and measuring the phase difference between them (Figure 1.2a). Since the wavelength (λ) of the carrier signal of the emitted beam of laser radiation is quite short—typically around 1 µm— and there is no need for such a measuring accuracy in topographic mapping applications, a modulation signal in the form of a measuring wave pattern is superimposed on the carrier signal, and its phase difference can be measured more precisely. Thus, the amplitude (or intensity) of the laser radiation will be modulated by a sinusoidal signal, which has a period T_m and wavelength λ_m. The measurement of the slant distance R is then carried out through the accurate measurement of the phase difference (or the phase angle, φ) between the emitted signal at point A and the signal received at the instrument after its reflection either from the ground itself or from an object that is present on the ground at point B. This phase measurement is usually carried out using a digital pulse counting technique. This gives the fractional part of the total distance ($\Delta\lambda$) (Figure 1.2b). By changing the modulation pattern, the integer number of wavelengths (M) can be determined and added to the fractional values to give the final slant range (R).

$$R = \frac{(M\lambda + \Delta\lambda)}{2} \tag{1.3}$$

where:
    M is the integer number of wavelengths
    λ is the known value of the wavelength
    $\Delta\lambda$ is the fractional part of the wavelength $= (\varphi/2\pi) \cdot \lambda$, where φ is the phase angle

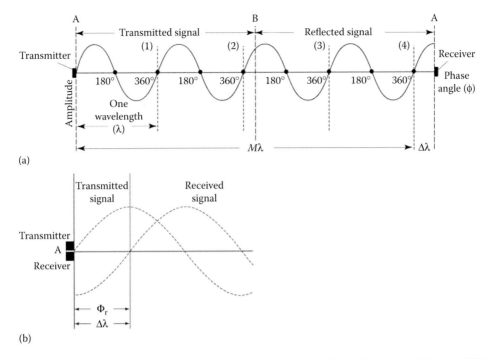

(a)

(b)

FIGURE 1.2 (a) Phase comparison is carried out between the transmitted and reflected signals from a CW laser and (b) the actual phase comparison between the two signals takes place at the laser rangefinder located at A.

1.4.2 LASER PROFILING

The use of a reflectorless laser rangefinder to measure the distances to a series of closely spaced points located adjacent to one another along a line on the terrain results in a 2D vertical profile or vertical cross section of the ground showing the successive elevations of the ground along that line.

1. In the case of a terrestrial or ground-based laser ranger, the measurement of the terrain profile is executed in a series of steps with the successive measured distances (slant ranges) and vertical angles (V) to each sampled point being recorded and stored digitally (Figure 1.3a). The profile of the terrain along the measured line can then be derived from this measured data by computation using the following quite simple relationships (Figure 1.3b):

$$D = R \cos V \tag{1.4}$$

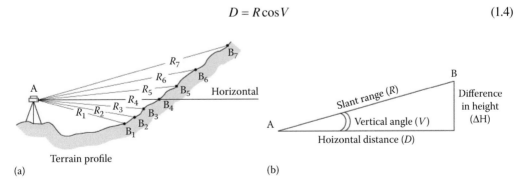

(a)

(b)

FIGURE 1.3 (a) Measurement of slant ranges (R) and vertical angles by a rangefinder located at A to a series of successive points located along a line on the ground to form an elevation profile. (b) The measured slant ranges (R) and vertical angles (V) are used to compute the horizontal distances and differences in height between the rangefinder at A and each of the ground objects at B.

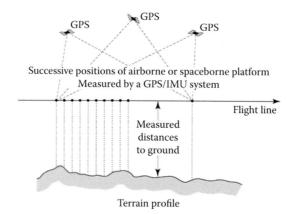

FIGURE 1.4 Profile being measured along a line on the terrain from an airborne or spaceborne platform using a laser altimeter.

> where:
>> D is the horizontal distance
>> R is the measured slant range
>> V is the measured vertical angle

$$\Delta H = R \sin V \tag{1.5}$$

where ΔH is the difference in height between the laser ranger and the ground point being measured.

2. In the case of a simple laser profiler that has been mounted on an airborne or spaceborne platform, the laser rangefinder, which, in this context, is often called a laser altimeter, is pointed vertically toward the ground to allow a rapid series of measurements of the distances to the ground from the successive positions of the moving platform. The measurements of the vertical distances from the platform to a series of adjacent points along the ground track are made possible through the forward motion of the airborne or spaceborne platform. If the positions and altitudes of the platform at these successive positions in the air or in space are known or can be determined, for example using a GPS/IMU system (or a star tracker in the spaceborne case), then the corresponding ranges measured at these points will allow their ground elevation values to be determined. Consequently, these allow the terrain elevation profile along the flight line to be constructed (Figure 1.4).

1.4.3 Laser Scanning

With the addition of a scanning mechanism, for example, utilizing a rotating mirror or prism, the laser ranging instrument is upgraded from being a 2D profiler to becoming a 3D scanner that can measure and map the topographic features of an area in detail instead of simply determining elevation values along a line in the terrain—as is done with a laser profiler.

1. With regard to a terrestrial or ground-based laser scanner instrument, the position of the platform is fixed; therefore, motion in two directions is needed to scan an area of the terrain. Thus, besides the vertical motion given by the rotating mirror or prism, the addition of a controlled (and measured) motion in the azimuth direction—usually implemented

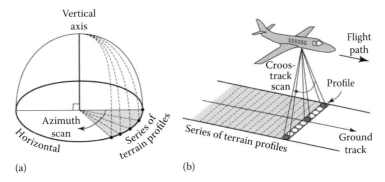

FIGURE 1.5 (a) Coverage of an area of terrain from a fixed ground station is achieved through the measurement of a series of elevation profiles adjacent to one another, the position of each new profile being defined by an angular step in the azimuth direction. (b) The coverage of a swath of terrain being carried out from an airborne platform is achieved through a series of elevation profiles measured in the cross-track direction. Each individual terrain profile is being measured by the laser rangefinder using an optical-mechanical scanning mechanism to point to each successive position that is required along that profile.

through the use of a motor drive—allows the measurement of a series of elevation profiles around the vertical axis of the laser ranger. This provides the position and elevation data that will allow a 3D model of the terrain and the objects present on it to be formed for the area around the position occupied by the laser scanner (Figure 1.5a).

2. In the case of airborne and spaceborne instruments, the area scanning is achieved by a series of profile measurements in the direction perpendicular to the flight line while the forward motion of the platform provides the second dimension. The addition of a scanning mechanism employing a reflective mirror or prism, angular rotation values of which can be continuously and precisely measured using an angular encoder, enables the additional elevation profiles of the terrain to be measured by the laser ranger in the lateral or cross-track direction, so supplementing the longitudinal profile being measured in the along-track direction of the flight line. Through a series of these profiles measured in the cross-track direction, the positions and elevations of a mesh of points, also called a Light Detection and Ranging (LiDAR) point cloud, is generated covering a swath or area of the terrain instead of the set of elevation values along the ground track of the flight line that is produced by the airborne or spaceborne profiler (Figure 1.5b).

1.5 LASER FUNDAMENTALS

The word laser is an acronym for light amplification by stimulated emission of radiation. Essentially a laser is an optical device that, when activated by an external energy source, produces and emits a beam or pulse of monochromatic radiation in which all the waves are coherent and in phase. The emitted radiation is also highly collimated and directional in the sense that it is emitted as a narrow beam or pulse in a specific direction.

In general terms, lasers are usually classified according to the type of material that is being used as the radiation source. The most common types are gas lasers, solid-state lasers, and semiconductor lasers. Other types that are used less frequently are liquid lasers, dye lasers, excimer lasers, and so on. For the laser ranging, profiling, and scanning that is being carried out for topographic mapping purposes, in which very high energy levels are required to perform distance measurements often over long ranges, only certain types of solid-state and semiconductor lasers have the very specific characteristics—high intensity combined with a high degree of collimation—that are necessary to carry out these operations.

1.5.1 LASER COMPONENTS

All lasers comprise three main elements:

1. The first of these elements comprises the active material of the laser that contains atoms, the electrons of which may be excited and raised to a much higher (metastable) energy level by an energy source. Examples of the materials that are being used extensively in the laser-based instruments that have been developed for the ranging, profiling, and scanning operations being carried out for topographic applications include a solid-state crystalline material such as neodymium-doped yttrium aluminum garnet (Nd:YAG). Lasers based on such crystalline materials are often described as being bulk lasers. Those based on the use of optical fibers as the solid-state material are termed fiber lasers. Semiconductor materials such as gallium arsenide (GaAs) are also used widely as the basis for the lasers that are used in certain types of laser rangefinders and scanners.

2. The second element that is present in every laser is an energy source. This provides the energy to start up and continue the lasing action. The continuous provision of energy to the laser is usually described as *pumping* the laser. Examples of suitable energy sources include optical sources such as a high-intensity discharge lamp or a laser diode, both of which are used with solid-state bulk lasers. Solid-state fiber lasers almost always utilize laser diodes as the pumping source. Alternatively, an electrical power unit producing a current that passes directly through the active material may be used as the energy source in the case of semiconductor lasers.

3. The third element is the provision of two mirrors: one that is fully reflective, reflecting 100% of the incident laser radiation; and the other that is semireflective (i.e., partly transmissive). Again these are integral components or features of every laser.

1.5.2 SOLID-STATE MATERIALS

The most frequently used Nd:YAG solid-state crystalline material that has already been mentioned earlier usually takes the form of a cylindrical rod. The growing and manufacture of the YAG laser rod is a slow and quite complex procedure, so the cost of the rod is comparatively high. The basic YAG material is transparent and colorless. When doped with a very small amount (1%) of Nd, the crystal takes on a light blue color. The Nd atoms act as the actual lasant. An alternative matrix material to the YAG crystalline material is glass. Indeed neodymium-doped glass has been used in a number of military rangefinders. Glass is a relatively inexpensive material and can easily be worked to give the desired shape (rod or disk) and dimensions. However, the problem with using glass as the matrix is that it has poor thermal conductivity and a low capability to dissipate heat. Thus, although the neodymium-doped glass laser can produce high-energy pulses, the repetition rate has to be kept low to keep the heat level low. So it requires cooling if operated at a high-power level and/or at a high pulse repetition rate, which will be the case with the airborne laser profiling and scanning carried out for topographic applications, though not in military rangefinders. By contrast, when the Nd:YAG laser is operated in a pulsed mode, very high pulsing rates can be achieved while still providing a very high average output power of 1 kW or more—since the thermal conductivity of the YAG crystal has a much higher rate than that of glass. The end faces of the Nd:YAG rod are made accurately parallel to one another with one of them silvered to make it fully reflective, whereas the other is semisilvered and partly reflective. However, more commonly, separate mirrors are utilized to provide the required reflections.

The Nd:YAG laser has been adopted quite widely as the basis of the laser rangefinders used in airborne laser scanners. A less commonly used solid-state laser is that employing yttrium lithium fluoride (YLF) as the matrix material. Again this is doped using a small amount of Nd to form the Nd:YLF laser. This has been used for example in NASA's RAster SCanning Airborne Laser (RASCAL) airborne laser scanner. Yet another material that has been used in NASA-built and NASA-developed airborne scanners is neodymium-doped yttrium vanadate (Nd:YVO$_4$). Compared with the Nd:YAG laser, the so-called vanadate laser has the advantage that it allows very high pulse repetition rates; on the other hand, it does not allow as high a pulse energy with Q-switching as the Nd:YAG laser. The Terrapoint Airborne LiDAR Terrain Mapping System (ALTMS) airborne laser scanners, which were developed originally with the assistance of NASA, all utilized these *vanadate* lasers.

Another highly miniaturized form of this type of solid-state laser that is being used in certain types of airborne laser scanner is the so-called microchip laser (Figure 1.6). This device is produced by polishing the two sides of a very thin wafer (an etalon) of an appropriate solid-state material (such as Nd:YAG or Nd:YVO$_4$) so that the sides are completely flat and are aligned very precisely parallel to each other (Zayhowski, 1990). The width of the wafer corresponds to the length of the laser cavity. The two polished surfaces are then dielectrically coated with a suitable material (e.g., magnesium fluoride) to form the mirrors of the cavity. The laser is then pumped, typically using a GaAlAs (gallium aluminum arsenide) diode laser, either coupled directly or using an optical fiber to transmit the radiation into the laser material. The resulting microchip lasers, complete with a pump diode and a suitable device to implement Q-switching to provide greater power, are dimensionally very small and quite rugged. Furthermore, there are no optical components within the microchip laser that can become displaced or misaligned. As will be seen later, microchip laser transmitters are being used in the laser rangefinders that are being employed in NASA's recently developed SIMPL and ALISTS airborne LiDAR systems.

In recent years, there has been an increasing use of fiber lasers, especially in airborne laser scanners. Fiber lasers utilize specially manufactured optical fibers, which are long flexible transparent fibers made of glass or plastic with a diameter slightly greater than that of a human hair. These optical fibers can be quite long, but they can be coiled up to form a very compact laser that is pumped by a relatively inexpensive diode laser (or a bank of such lasers) and uses fiber Bragg gratings on the ends of the fiber as internal reflectors (Figure 1.7a). These optical fibers also exhibit excellent thermal dissipation when used in a high-powered lasing operation, because of the fiber's high surface area to volume ratio that allows efficient cooling of the material. The transparent core of the optical fiber is surrounded by a transparent cladding material with a lower refractive index (Figure 1.7b). This ensures that the light or radiation is kept within the core by total internal reflection. For use in a fiber laser, the optical fibers are again doped with a rare earth such as erbium, neodymium, or ytterbium. Fiber lasers using erbium as the active lasant material have become particularly important in recent years, especially for use in those airborne laser scanners that are designed specifically to be operated from low altitudes (flying heights). This relates to the characteristic that they emit their pulses at wavelengths around $\lambda = 1550$ nm in the short wavelength infrared (SWIR) part of the spectrum—with wavelengths above $\lambda = 1400$ nm being reckoned to be *eye-safe*.

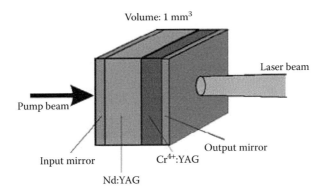

FIGURE 1.6 Diagram of a Q-switched microchip laser. The laser cavity comprises a one to two cubic millimeter Nd:YAG crystal bonded to a thin layer of Cr^{4+}-doped YAG material that acts as the saturable absorber (for Q-switching). The two dielectrically coated mirrors are located parallel to each other with one on each side of the thin wafer of the laser material. (Courtesy of ATlaser, Florence, Italy.)

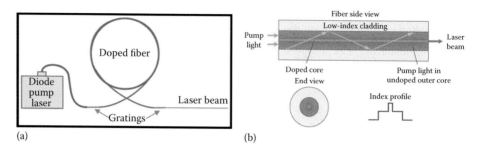

(a)   (b)

FIGURE 1.7 (a) Diagram showing the basic arrangement of a diode-pumped fiber laser. (From Optronics Research Centre, University of Southampton, UK) (b) The structure of a fiber laser comprises (1) a doped inner core, which acts as the laser; (2) an undoped outer core through which the light or infrared radiation from the pump laser diodes may be channeled; and (3) the outer cladding of the fiber. (Courtesy of Laser Focus World.)

1.5.3 LASER ACTION—SOLID-STATE MATERIALS

Although the laser action can be activated and maintained using a xenon lamp or krypton discharge tube (Figure 1.8), for most laser rangefinders, the Nd-doped bulk lasers described earlier are pumped continuously using a semiconductor diode or a diode array (Figure 1.9). The minimum pumping power that is required to begin the lasing action is called the lasing threshold. When the

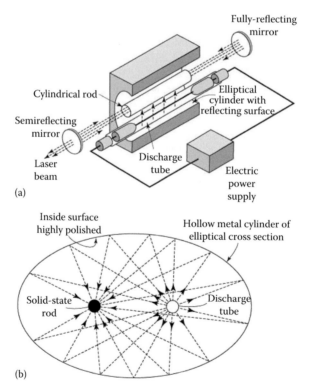

FIGURE 1.8 (a) Construction of a solid-state laser with the Nd-doped cylindrical rod placed at one focal point of a hollow metal cylinder with an elliptical cross section, the inside surface of which is highly polished; a high-energy discharge tube is placed at the other focal point to provide the energy needed to start up and maintain the lasing action. (b) The discharge tube is placed at one of the focal points of the elliptical cylinder, in which case, all the emitted light will be reflected by the polished surface to pass into the solid-state rod placed at the other focal point.

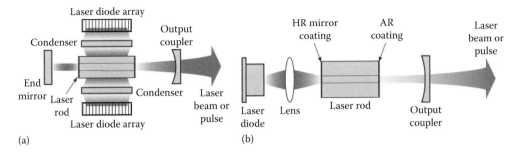

FIGURE 1.9 Different designs of Nd-doped lasers pumped by laser diodes (a) in the form of banks of diodes located above and below the laser rod and (b) using a single diode aligned with the central axis of the laser rod.

pumping action begins, it energizes a large number of the Nd atoms into an excited state, in which their electrons are raised from their normal ground state to a much higher but unstable energy level. As the electrons cascade down from this high-energy state back toward their normal ground state, they pass through an intermediate energy level called the metastable state. By continuing to pump energy into the active material, most of the electrons remain in this metastable state and a so-called *population inversion* is achieved. While they are in this metastable state, the electrons may spontaneously emit energy in the form of photons—an action called *spontaneous emission* (Figure 1.10). Other adjacent atoms are then stimulated to emit photons with the same frequency and phase—described as *stimulated emission*. Thus, amplification of the radiation is produced by stimulating coherent emissions from the other excited atoms.

For those photons that travel along the axis of the rod or cylinder, when they reach the fully reflective end or mirror, they are reflected back along their paths to the other end where they are reflected back again. In the course of their repeated forward and backward paths along the cylinder, they are joined by other photons emitted from other excited atoms, electrons of which have reached the metastable state (Figure 1.10). The new photons are precisely in phase with the original or primary photons. The overall effect may be envisaged or depicted as comprising an in-phase tidal

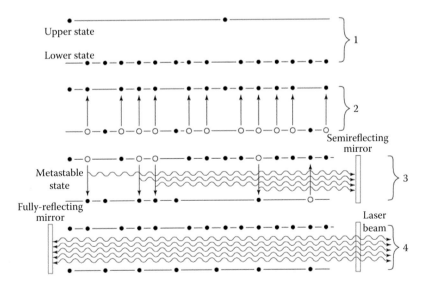

FIGURE 1.10 Simplified representation of the lasing action: (1) showing the situation prior to the application of the energy source; (2) situation with the application of energy from the external source; (3) *spontaneous emission* takes place with photons being emitted parallel to the main axis of the cylindrically shaped laser rod; and (4) after further *stimulated emission* and reflections to and from the mirrors, the laser beam or pulse is emitted through the semireflective (or partially transmitting) mirror.

wave of photons sweeping up and down the axis of the cylinder or rod, increasing in amplitude and intensity as more of the Nd atoms are stimulated into emitting photons by the passing wave. The beam of light (or infrared radiation) builds up very rapidly in intensity until it emerges from the semireflective (partly transmissive) mirror either as a continuous beam or as a short, intense, and highly directional pulse of light or infrared laser radiation. In the case of the Nd:YAG and Nd:YVO$_4$ lasers, the radiation is emitted at the wavelength (λ) of 1064 nm; other important emission wavelengths for the Nd:YVO$_4$ (vanadate) lasers are 914 and 1342 nm. With the Nd:YLF laser, the laser radiation is emitted at the wavelength (λ) of 1046 nm.

Through the use of frequency doubling techniques, which halve the emitted wavelengths, the various lasers mentioned earlier can also be adapted to emit their radiation at wavelengths of 532 nm (Nd:YAG and N:YVO$_4$) and 523 nm (Nd:YLF), respectively, in the green part of the electromagnetic spectrum. This is achieved by passing the pulse through a block of a special nonlinear crystalline material such as potassium dihydrogen phosphate (KDP) or potassium titanyl phosphate (KTP), which is transparent and has the specific crystalline structure and birefringent properties to achieve the half-wavelength that is required. The shorter wavelength laser light, for example, that emitted at $\lambda = 532$ nm, is used primarily for depth measurement in bathymetric mapping, as it allows for better penetration into water.

When operating in its pulsed mode, the power of the solid-state laser (including fiber lasers) can be built up still further through the use of a technique called Q-switching. This employs a special type of shutter that delays the release of the energy stored in the pulse until it reaches a very high power level (Figure 1.11). The shutter is then opened to allow the laser emission to take place in the form of a very intense pulse of coherent radiation. The Q-switch can take the active form of a Kerr cell with switchable polarizing filters. Alternatively, it may take the more passive form of a light-absorbing dye, which is bleached when the energy reaches a suitably high level, so releasing the pulse. Q-switching is a necessity when a laser rangefinder, profiler, or scanner has to operate over distances of hundreds of kilometers; for example, when it is mounted on a spaceborne platform. Typically Q-switched bulk lasers can produce pulses of intense energy over a time range of 10–250 ns. Analyzing the operation in more general terms, the Q-switching of bulk lasers leads to lower pulse repetition rates and longer pulse durations, as well as much higher pulse energies. Passively Q-switched microchip lasers generate pulses of much shorter duration and higher repetition rates, albeit with lower energies.

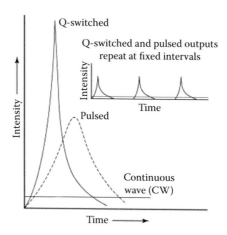

FIGURE 1.11 Laser output modes—CW, pulsed, and Q-switched. (1) CW lasers have low intensity outputs and consequently only require a comparatively small energy input for their operation; (2) pulsed lasers emit pulses with much higher energy levels; whereas, finally; (3) still higher energies are emitted with Q-switched lasers. (From Price, W.F. and Uren, J., *Laser Surveying*, Van Nostrand Reinhold, 1990. With Permission.)

1.5.4 SEMICONDUCTOR MATERIALS

Those lasers that are based on semiconductor materials all exhibit the electrical properties that are characteristic of electrical diodes. Thus, they are often referred to as diode lasers as an alternative to the description semiconductor laser. As noted previously, GaAs is the main semiconductor material that is used as the basis for the lasers utilised in rangefinders and scanners. Under suitable conditions, they will emit an intense beam or pulse of infrared radiation. The basic structure of a simple semiconductor laser is shown in Figure 1.12. It comprises various layers of the semiconductor material with metal conductors attached to the top and the bottom of the diode to allow electrical power to pass into the material. The upper semiconductor layer contains impurities that cause a deficiency in their electrons, leaving positive *holes*. This material is therefore called a p (= positive)-type semiconductor. The lower layer has impurities that result in an excess of electrons with negative charges. So this material is said to be an n (= negative)-type semiconductor. The active element of this type of laser is an additional layer called the p–n junction that is sandwiched between the p-type and the n-type semiconductor layers (Figure 1.12).

1.5.5 LASER ACTION—SEMICONDUCTOR MATERIALS

When stimulated by an electric current, electrons enter the p–n junction from the n-side and holes from the p-side. The two mix and again a population inversion takes place. In a simplified description, the electrons and holes move in the desired direction through the application of an electric voltage across the junction. Therefore, this type of laser is also referred to as an injection laser as the electrons and holes are injected into the junction by the applied voltage. The actual voltage is produced through the connection of a suitable electric power source to the two electrodes affixed to the top and bottom surfaces of the material. Thus, in this case, the pumping action is purely electrical in character. As described earlier, atoms within the active material of the p–n junction are excited to higher energy states and in falling back to a lower state, they exhibit electroluminescence and emit photons. The photons traveling along the junction region stimulate further photon emission and so produce still more radiation. The actual reflective lasing action can be produced by simply polishing one end face of the material so that it acts as a semireflective mirror. For the other (fully reflective) end, a metal film of gold—which is a good reflector at infrared wavelengths—is coated on to the end of the device, ensuring 100% reflection. The wavelength (λ) of the radiation being emitted from a GaAs semiconductor laser is normally around 835–840 nm in the near-infrared part of the electromagnetic spectrum. A laser based on a similar material—GaAlAs—produces its radiation over the much greater wavelength range of 670–830 nm, in which the actual wavelength value is being controlled by suitable variations of the bias current.

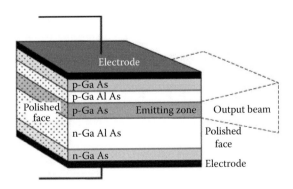

FIGURE 1.12 Basic construction of a semiconductor laser in which the active element is a p–n junction sandwiched between a p-type and an n-type semiconductor layer.

Physically, the GaAs semiconductor laser is quite small. However, it is a very efficient device, since up to half of all the electrons injected into the active GaAs material produce photons that are output in the form of laser radiation. Besides this high efficiency, it is very easy to control the output power through small changes to the supply current or voltage. Semiconductors are produced in very large quantities for other (nonsurveying) applications, so they are relatively inexpensive, very reliable, and have a long life with low-energy consumption. Negative points are that the output from a semiconductor laser has a rather poor spectral purity—which is not too important a characteristic in ranging applications—and has a rather wide divergent angle. When operated at the temperatures normally encountered in the field, the semiconductor material can reach quite high temperatures to generate suitable output power—for example, when operated in a pulsed mode over longer ranges. On the other hand, semiconductor diode lasers are often used in a continuous (nonpulsed) mode over shorter distances. In which case, it is easier to keep the power dissipation and heat levels at an acceptable level within the device. With continuous wave (CW) operation of this type of laser, the phase measurement technique allows range measurements to be made at a much higher rate than with the pulse-ranging technique.

1.6 LASER RANGING

The laser ranger or rangefinder is the device or instrument that is used to measure the slant range or distance to a reflective object such as the ground, a building, or a tree—that is often referred to in rather abstract terms as a *non-cooperative target*. Besides being used purely as a distance measuring device for surveying, military, or even sporting purposes, the laser rangefinder forms the basis of all laser profilers and scanners. As mentioned in the introduction to this chapter, there are two measurement technologies and methodologies that are in widespread use for topographic applications: (1) the TOF or timed pulse or pulse echo method; and (2) the multiple-frequency phase comparison or phase shift method for CW operation using amplitude modulation (AM). A third measuring methodology, called here the heterodyne method, is also based on a CW operation, but using frequency modulation (FM) instead of the AM used in the phase comparison method. At present (in 2016), it is in the research and development stage rather than in production service.

1.6.1 Timed Pulse Method

As discussed in the introductory section, the TOF method that is used to determine the slant range from the measuring instrument to an object involves the very accurate measurement of the time required for a very short pulse of laser radiation to travel to the target and back. The method has been used to measure distances of a few tens or hundreds of meters in the case of terrestrial or ground-based rangefinders, profilers, and scanners. Obviously, the distances that need to be measured from an airborne platform will be much greater—in the order of hundreds of meters to several kilometers—whereas the ranging devices mounted on spaceborne platforms such as the space shuttle and satellites will be measuring distances of several hundred kilometers.

The design of a generic type of laser ranger or rangefinder using the timed pulse method is presented in Figure 1.13. The actual design and the specific components that will be used in a particular instrument will depend on the particular application, for example, whether it is being operated from a ground-based, airborne, or spaceborne platform. The high-energy laser will have a lens placed in front of it to expand and collimate the laser pulse so that it has a minimum divergence. When a pulse is emitted by the laser, a tiny part of its energy is diverted by a beam splitter on to a photodiode whose signal or start pulse triggers the timing device (Figure 1.14); the triggering is usually accomplished by simple thresholding. In practice, this timer is a high-speed counter that is controlled by a very stable oscillator. To achieve a 1 cm ranging resolution, the timer should be able to measure a 66 ps interval, which would require a clock rate of about 15 GHz.

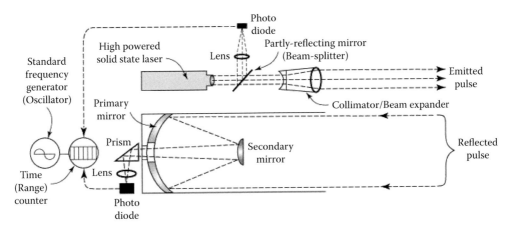

FIGURE 1.13 Layout of the main components of a pulse-type laser rangefinder and their interrelationship.

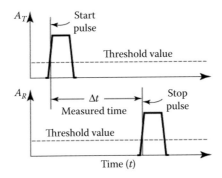

FIGURE 1.14 Diagrammatic representation of the action of the start and stop pulses used to control the time counter.

After its reflection from the ground object, the returning pulse that is being measured is picked up by the receiving lens or mirror optics. With the long ranges of hundreds of kilometers that need to be measured from a spaceborne platform, large diameter mirror optics are often required to receive and detect the quite faint signals that are returned over these huge distances. Besides, even at shorter ranges, when the emitted laser pulse reaches the object being measured, it will have diverged and spread. In addition, it will have been scattered in various directions by tiny particles that are present in the atmosphere, both on its outward path and on its return path to the ranger. So even at short ranges, there is a need to amplify the weak return signal electrically.

Whether large or small in terms of its physical size, the receiving lens or mirror optical system will focus the returning (reflected) pulse on to another very sensitive photodiode that acts as a detector. Typical of those in use in laser rangefinders are silicon avalanche photodiodes (Si APDs) or indium gallium arsenide PIN (positive–intrinsic–negative) semiconductor photodiodes. These convert the received power of the reflected pulse into an output voltage. Different types or models of these devices are used depending on the wavelength of the laser radiation being used. Besides which, narrowband optical filters will usually be placed over the photodiode to cut down or eliminate the effects of sunlight or other sources of optical noise that may cause spurious signals. In the simplest type of rangefinder, when the reflected pulse is received, the diode delivers a stop pulse to the time counter (Figure 1.14). Usually, the specific point on the leading edge, in which the voltage generated by the diode reaches a predetermined threshold value, is used to start and stop the time counter. It should be noted that in real state-of-the-art systems much more sophisticated triggering is implemented. The time-to-digital converter process is rather simple as the measured time in terms of timer counts is converted to the corresponding range or distance value by using the known speed of light. If a series of reflections are being returned from a particular object such as a tree or an electric pylon from a single pulse, then the time (epoch) of each of the returned pulses—for example, the first and last pulses—will be measured. Some systems can measure and record up to four separate reflections. Alternatively, the complete waveform or shape of the whole of the return signal can be measured, digitized, and recorded on a time base for later analysis (Figure 1.15), as will be discussed in more detail in subsequent chapters—especially in Chapter 5.

Quite another approach that has been adopted by NASA and defense agencies in the United States, though not so far (in 2016) by the commercial mainstream system suppliers of laser scanners, is to utilize photon counting detectors. There are a number of different types including photomultiplier tubes (PMTs) and certain specific forms of Si APD detector. The choice depends partly on the wavelength (λ) of the laser being used in the rangefinder, since PMTs can only operate in the visible part of the electromagnetic spectrum. Another factor to be considered is the detector dead-time, which is the time needed by the device to recover after detection of a single photon. PMT detectors have very short dead-times, whereas Si APD detectors have longer dead-times. A further development of photon counting detectors has been the use of arrays of these detectors operated in the so-called Geiger mode (Abdullah, 2012, 2016). This involves the use of a rangefinder equipped with a relatively low-energy laser (of a few micro joules) that emits short (subnanosecond) pulses at a very high rate. Each emitted pulse is split by a diffractive optical beam splitter into a set of subpulses that illuminate a small area of the ground surface (Figure 1.16), the size of which is dependent on flying altitude and the laser aperture. The reflections from this set of subpulses hitting the ground are focused by a suitable set of receiving optics on to a corresponding array of single-photon detectors. Arrays of 32 × 128 photon detectors have been used in the military Airborne LiDAR Imaging Research Testbed system (Fried, 2015) and the civilian Harris IntelliEarth systems (Clifton et al., 2015) that employ this technology.

The actual length (or width) of the pulse that is emitted by the laser is another important characteristic of any laser system, as it has a major impact on how multiple returns can be detected or differentiated. If the laser pulse lasts 10 ns (where 1 ns is 10^{-9} s) then the pulse length will be 3 m at the speed of light (v). Thus, the measuring resolution of a ranging device utilizing timed pulses is governed, in the first place, by the length of the emitted pulse and the degree to which either the center or the leading or the trailing edge of the return pulse can be determined by the conventional type of detector. In general, this can be a difficult task to perform as the return pulse will often be quite distorted after its reflection from the ground targets. Since the speed of electromagnetic energy (v) is 299,792,458 m/s, a resolution of 1 m in a single measured range requires the timing of the pulse to be accurate to 1/300,000,000 (3×10^{-9}) of a second. A resolution of 1 cm in the measured range requires a time measurement of 3×10^{-11} s, whereas a 1 mm resolution requires 3×10^{-12} s, equal to 3 ps (picoseconds). Since most pulsed systems currently use a 1 GHz sampling rate, a backscattered

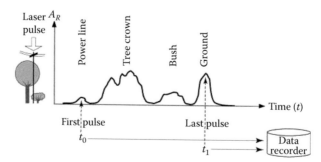

FIGURE 1.15 Showing the shape of the complete waveform of the returned (reflected) pulse that can be used for further analysis. (From Brenner, C., Aerial laser scanning. International summer school on "Digital Recording & 3D Modeling," Aghios Nikulaos, Crete, Greece, April 24–29, 2006. With Permission.)

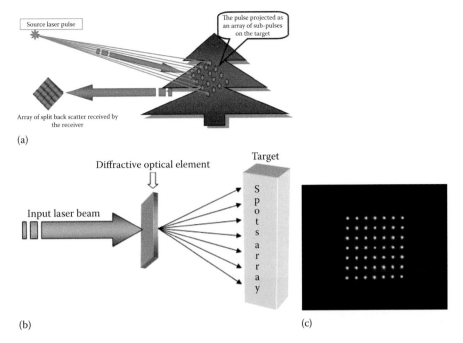

FIGURE 1.16 (a) Showing the principle of a laser scanner operating in Geiger-mode. The emitted pulse is divided into subpulses that travel onwards and strike the ground. The backscattered laser subpulses are then reflected back to the receiver optics and are then directed toward a corresponding array of segmented detectors. (From Abdullah, Q.A., *Photogramm. Eng. Remote Sens.*, 82, 307–312, 2016. With Permission.) (b) Showing the splitting of the emitted pulse into subpulses using a diffractive beam splitter; the subpulses are all directed towards the ground target. (From Abdullah, Q.A., *Photogramm. Eng. Remote Sens.*, 82, 307–312, 2016. With Permission.) (c) The diffractive beam splitter is used to split a single laser pulse into several subpulses, in this case, having a 7 × 7 matrix pattern. Each subpulse has the characteristics of the original pulse, except for its power and angle of propagation. (Courtesy of HOLO/OR.)

laser pulse, waveform, is sampled at 1 ns, which is 30 cm in distance. Note that because the pulse travels twice the range being measured, see Equation 1.1, the 1 ns equals 15 cm range resolution. As mentioned earlier, a timing device based on a very accurate quartz-stabilized oscillator is required to carry out the measurement of the elapsed time (or round trip transit time) to the required level. In fact, the quality of the timer, including the start/stop triggering components, determines the achievable accuracy of the laser ranging instrument.

When short distances are being measured, as is the case with a terrestrial or ground-based ranging instrument, it is possible to make a relatively large number of measurements over a short period of time, that is, the pulse repetition frequency (PRF) can be high. Thus, for example, if a distance of 100 m has to be measured with a ground-based laser ranger, then the travel time over the 200 m total distance out and back is 200/300,000,000 = 0.67 μs. With longer distances, such as those required to be measured from airborne and spaceborne platforms, a longer time interval will elapse between the emission and the reception of individual pulses. For a distance of 1000 m that is typical for an airborne laser rangefinder, the elapsed time will be 6.7 μs, in which case, the maximum PRF will be 1/6.7 μs, which is 150 kHz. Thus, in general terms, a longer time interval between successive pulses will usually be required when longer distances need to be measured. However, this remark needs to be tempered by the introduction of the technique of having multiple laser pulses in the air simultaneously (Figure 1.17) (Toth, 2004). This gives the capability for a laser ranger to emit a new pulse without waiting for the reflection from the previous pulse being received at the instrument. Thus, more than one measuring cycle can be taking place at any moment of time. Note that the concept of using multiple pulses was introduced in RAdio Detection And Ranging (RADAR) systems much earlier (Edde, 1992).

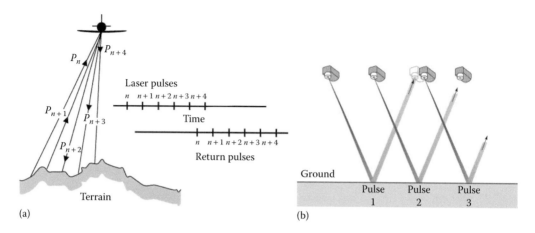

FIGURE 1.17 (a) Pulse interleaving concept for a five-time frequency rate is being applied in the cross-track direction from an airborne laser scanner. (From Toth, C.K., Future trends in LiDAR, *ASPRS Annual Conference*, Denver, CO, May 23–28, Paper No. 232, 2004. With Permission.) (b) multiple pulses through space and in the air are being emitted in the along-track direction to measure a longitudinal profile from a satellite.

In the multiple pulse case, the limit of the number of pulses is primarily imposed by the laser source; in other words, how frequently can a pulse be emitted by the source. Note that systems from the early 2000s could increase their PRF at the price of reducing the energy of the emitted pulse and, in this way, the ranging accuracy was decreased due to the less favorable signal-to-noise ratio. Current laser supplies are much more powerful, so this phenomenon does not impact current state-of-the-art systems. The fact that there are multiple pulses traveling back and forth, however, requires additional processing to resolve the ambiguity (to properly pair an emitted pulse to a received one). Since the number of pulses is low and the object distance is not changing too much between two pulses, this task is almost trivial and requires limited effort.

1.6.2 PHASE COMPARISON METHOD

The phase comparison method is used with those rangefinders in which the laser radiation is emitted as a continuous beam—often referred to as a CW laser—instead of discrete pulses. In practice, due to the limited power of the CW laser, there are very few laser rangefinders, profilers, or scanners of this type in actual operation from airborne or spaceborne platforms as compared with those rangers using the timed pulse or pulse echo technique. By contrast, the phase comparison technique is much used in short-range terrestrial or ground-based laser scanners in which the distances that need to be measured are often much shorter—typically <100 m. In those devices using the CW approach, the transmitted beam comprises a basic (laser) carrier signal on which a modulation signal has been superimposed for the purpose of measurement (Figure 1.18). This modulation signal or measuring wave is derived from and is held at a constant value using a stable frequency oscillator. Thus, the measuring wave is used to control the amplitude of the carrier wave—a process known as AM.

The reflection of the signal for a CW laser is similar to that for a pulsed one, except it is continuous. The transmitted beam then strikes the object being measured, and a small part of the signal is reflected back along the identical path to the instrument, in which it is detected using a silicon photodiode. The weakened reflected signal is then amplified and subjected to demodulation—which is the separation of the measuring and carrier waves. Next, the reflected signal is compared with the signal of the original transmitted beam (or reference signal). As mentioned previously, the difference in phase (or phase angle, φ) between the two signals is then measured and represents the fractional part ($\Delta\lambda$) of the total slant range that is required (Figure 1.2). The integer number of wavelengths, $M\lambda$, cannot be derived from this single measurement; the task of determining M is the so-called ambiguity resolution problem. There are several ways to solve the ambiguity, depending on using

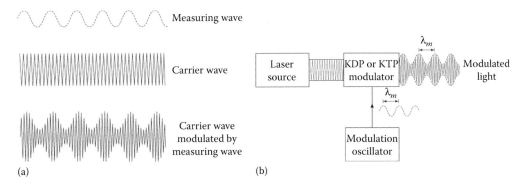

FIGURE 1.18 (a) Amplitude modulation with the measuring wave being used to vary the amplitude of the carrier wave and (b) amplitude modulation of the emitted beam is performed by passing it through a potassium dihydrogen phosphate or potassium titanyl phosphate crystal to which a voltage has been applied. (From Price, W.F. and Uren, J., *Laser Surveying*, Van Nostrand Reinhold, London, UK, 1990. With Permission.)

various modulation frequencies, or simply following the range changes, and so on. The most widely applied method to resolve the ambiguity is to make a number of changes to the wavelength (λ) and therefore to the frequency (f) of the emitted beam very rapidly in succession—hence the description of this being a multifrequency phase measuring system. The required changes in wavelength are carried out automatically and very rapidly within the instrument without any human interaction. Thus, there is no need for this to be done manually in modern instruments. The resulting measurements of the series of phase angles φ lead to a set of simple equations having the general form:

$$R = \frac{M_1 \cdot \lambda_1}{2} + \left(\frac{\varphi_1}{2\pi}\right) \cdot \left(\frac{\lambda_1}{2}\right)$$

$$R = \frac{M_2 \cdot \lambda_2}{2} + \left(\frac{\varphi_2}{2\pi}\right) \cdot \left(\frac{\lambda_2}{2}\right) \tag{1.6}$$

$$R = \frac{M_n \cdot \lambda_n}{2} + \left(\frac{\varphi_n}{2\pi}\right) \cdot \left(\frac{\lambda_n}{2}\right)$$

where n is the number of different frequencies that are being employed. Solving the set of simultaneous equations will give the final value of the slant range (R).

It is important to note that, when using a modulated laser beam for ranging, the characteristics of the modulation signal, including the wavelength and phase measurement resolution, will determine the ranging precision of the system. When this type of multifrequency CW laser ranging system is being used, the lowest frequency associated with the longest wavelength, λ_{long}, will determine the longest slant range, R_{max}, that can be measured. By the same token, the highest frequency with the shortest wavelength, λ_{short}, will determine the range resolution and therefore the accuracy with which the slant range can be measured by such a device or system. For example, a 10 MHz modulation signal has a wavelength of 30 m, which with a 1° phase resolution could result in a precision of about 8 cm (30/360).

1.6.3 HETERODYNE METHOD

Another CW laser ranging method is based on the FM of the laser signal. In an analogous manner to that already described earlier for AM (Figure 1.18a), a measuring or modulation wave is superimposed to control and change the frequency of the carrier wave (Figure 1.19a) with the amplitude remaining fixed. There are many ways to modulate the laser signal frequency around

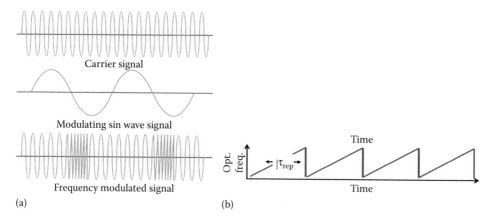

(a)   (b)

FIGURE 1.19 (a) Frequency modulation (FM) with the measuring or modulating wave being used to vary the frequency of the carrier wave; note that the amplitude value remains fixed. (b) The laser signal frequency when plotted against time to illustrate the *saw tooth* type of FM signal.

a base frequency. Figure 1.19b shows a typical *saw tooth* arrangement in which the laser signal frequency is linearly increasing in time and then drops back to an initial base value after a given time period (repetition time).

In the current implementation of this type of frequency-modulated carrier-wave signal for ranging (Barber et al., 2014, Mateo and Barber, 2015), a chirped laser generates the appropriate FM signal. The heterodyne technique, which can be used to shift and compare one frequency range against another, is then utilized for the actual measurement of the range. The signal from the chirped laser is first split into two (Figure 1.20a). The first part (Tx) is transmitted toward the object whose range is being measured. The second part, labeled LO (= local oscillator), is retained as a reference signal. The reflected return signal (Rx) from the object is then combined and compared with the reference LO signal (Figure 1.20b).

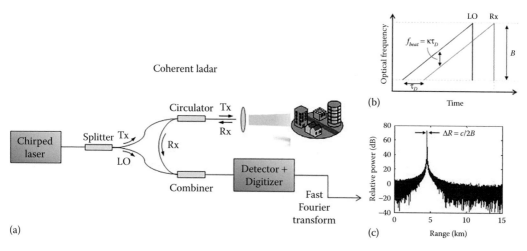

(a)   (c)

FIGURE 1.20 (a) Diagram of a frequency-modulated carrier-wave LiDAR system including (1) the splitter producing the Tx and LO signals, (2) the circulator which transmits the ranging signal (Tx) and receives the reflected signal (Rx) from the target, and (3) the combiner which combines and mixes the Rx and LO signals. (b) Showing the comparison of the Rx and LO signals. (at upper right); note that the range delay is encoded in the RF spectrum of the detected signal. (c) Diagram of the power spectrum after the fast Fourier transform computation showing power plotted against range; the peak (maximum) value in the power spectrum provides the range (at lower right). (From Bridger Photonics, Bridger photonics distance measurement technology, Bridger Photonics White Paper, 2pp, 2015. With Permission.)

The value of the range that is obtained is based on spectral analysis and requires substantial computation. The main steps of the signal processing are shown in Figure 1.20b and c. The reference LO and the received Rx signals are combined and the power spectrum is then computed using a fast Fourier transform. The maximum value in the power spectrum is correlated to the range. Since the fast Fourier transform can be implemented in hardware, for example using field-programmable gate array technology, the computation time is acceptable for normal mapping applications.

This technique is significantly more complex when compared with the AM method employing phase comparison that was described in the previous section. Consequently, at the present time, it is mainly in the research and development stage (Massaro et al., 2014, Bridger Photonics, 2015). In a similar manner to broadcast radio, there are certain clear advantages of the FM approach over the AM. Most importantly, additive signal noise has no impact on the range computation, which is not the case for the AM method. One practical aspect of this characteristic is the better penetration capability that is offered by frequency-modulated carrier-wave systems. Moreover, as with radar, Doppler measurements can easily be made available, so object tracking can be directly supported; for example, the leaves of vegetation can be detected, since they are never totally static.

1.6.4 POWER OUTPUT

Usually, the output power as it relates to a laser is expressed in watts (W) or milliwatts (mW), where 1 W (the SI unit of power) = 1 J/s. The joule is the SI unit of energy or capacity for doing work, equal to the work done when a current of 1 A is passed through a resistance of 1 Ω for 1 s. The output for pulsed lasers is often expressed as the radiant exposure, which is the concentration of the laser energy on a given area, expressed in terms of joules per square centimeter (J/cm^2).

Very different output power levels are encountered depending on the type of laser—pulsed or CW—that is being considered. Thus, at one end of the power spectrum are the terrestrial or ground-based laser scanners using CW lasers—such as those manufactured by Zoller & Fröhlich and Faro—that measure over short ranges (up to 70 m). By using the phase measuring technique, they employ CW semiconductor lasers with a relatively low power—between 10 and 20 mW for slant ranges up to 25 m and between 20 and 40 mW for distances up to 50–70 m. By contrast, the pulse-type lasers being used in airborne laser scanners measuring ranges of a few kilometers typically operate at peak power levels of 1–2 kW or even more (Flood, 2001b). In the latter case, if the pulse duration (t_p) is 10 ns, then the energy (E) generated per pulse is $E = P_{peak} \cdot t_p = 20\ \mu J$. The average power for a pulse repetition rate of 50 kHz will be $P = E\ F = 0.000020 \times 50,000 = 1$ W.

The actual measurement of the laser power in a laboratory or a manufacturing facility is normally carried out using instruments based on one or other of two different types of detector: quantum detectors and thermopile detectors. The quantum detector is based on a semiconductor material such as indium gallium arsenide and measures laser power by counting photons. It converts the incoming photons into charge carriers (electrons and holes), which are then summed as a voltage or current using an amplifying circuit. By contrast, the thermopile detector acts essentially as a calorimeter. The incident radiation heats the device and a circuit measures the heat differential between the detector and an attached heat sink. Both detectors provide their final measurements to the user in units of watts. The quantum detector is more sensitive at low power levels—at the microwatt level or below. Both types of instruments can be used at higher power levels, but the thermopile detector is best if very high power levels need to be measured as the quantum detector can suffer from saturation effects (and possible damage) at these levels.

1.6.5 POWER AND SAFETY CONCERNS

Given the fact that many of the devices used in laser ranging, profiling, and scanning utilize powerful pulse-type lasers, the matter of safety and of safety standards is a matter of prime importance, especially during the actual use of these instruments in the field and in the air. For most of the world, the applicable safety standards are those set by the International Electro-technical Commission (IEC). These are known as the International IEC 60825 Standards. The exception to the adoption of this standard is the United States, which, so far, has not adopted this standard. Instead, it has its own standards set by the Center for Devices & Radiological Health (CDRH) and administered by the Food & Drug Administration. The American standard is ANSI Z136.1, whereas the manufacturer's standard is called CDRH 21 CFR, parts 1040.10 and 1040.11. Within Europe—where many of the manufacturers of laser rangefinders, profilers, and scanners are located—the IEC standard has been adopted and is known as the EN 60825 Standard. Each European country has its own version of this standard; for example, in the United Kingdom, it is known as the British Standard BS EN 60825. Recently the CDRH in the United States has decided to harmonize its standards with the IEC 60825 standard. For an introductory overview of the current situation regarding laser classification on the basis of laser hazards and safety, reference may be made to the following web page on Wikipedia:-http://en.wikipedia.org/wiki/Laser_Safety.

1.6.6 LASER HAZARD CLASSIFICATION

Of most importance to the present discussion is that both the International and U.S. standards divide lasers into more or less the same four broad categories on the basis of their hazards, especially concerning the risk of them causing damage to human eyes or skin. A very brief outline is given in the following; for detailed information, reference should be made to the appropriate documentation issued by the relevant standards authorities.

1. Class 1: These comprise those lasers and laser-based systems that do not emit radiation at known hazard levels and are exempt from radiation hazard controls during their operation in the field, though not during servicing. Thus, the CW lasers used in ground-based laser scanners such as the Z + F and Faro scanners mentioned earlier operate with Class 1 lasers.
2. Class 1M: A Class 1M laser is safe for all conditions of use, except when passed through a magnifying optical device such as a microscope or a telescope. Normally Class 1M lasers produce large-diameter beams or beams that are divergent. If, however, focusing optics are used to narrow the beam, then the hazard level will be increased, in which case, the Class will need to be changed.
3. Class 2: These are low-power visible light lasers that emit their radiation above Class 1 levels, but at a radiant power not above 1 mW. The concept is that the human aversion to bright light (the so-called blink reflex) will protect a person. Only limited controls are specified regarding their operation.
4. Class 2M: Again this class of laser is safe because of the blink reflex, as long as it is not viewed through an optical instrument such as a microscope or telescope. As with the Class 1M laser, this applies mainly to laser beams having a large diameter or large divergence, for which the amount of light passing through the eye pupil cannot exceed the limits set for Class 2.
5. Class 3R: These comprise lasers with intermediate power levels—typically CW lasers in the range 1–5 mW that are only hazardous for intrabeam viewing. So only certain limited controls are specified for their operation. However, for pulsed lasers falling into this particular class, quite other limits and precautions will apply.
6. Class 3B: These comprise lasers with moderate powered emissions—for example, CW lasers operating in the range 5–500 mW, or pulsed lasers with outputs up to 10 J/cm². A Class 3B laser is hazardous if the beam or pulse is viewed directly, but diffuse reflections are not harmful. There are a set of specific controls recommended for their operation. In particular, Class 3B lasers must be equipped with a key switch and a safety interlock. Furthermore, protective eyewear is typically required if any direct viewing of the laser beam is contemplated.

7. Class 4: This category comprises those lasers with power levels above those specified for Class 3 lasers. They are hazardous to view under any circumstances; besides which, they are a skin hazard; and potentially a fire hazard. Class 4 lasers must be equipped with a key switch and a safety interlock. Most lasers that are being used for industrial, scientific, military, and medical purposes fall into this category. In the context of the present discussion, many airborne laser scanners, including the ALTM series from the leading supplier, Teledyne Optech, use Class 4 lasers (Flood, 2001a). So the appropriate safety measures must be followed both during their manufacture and servicing and also during their operation in the field or in the air.

It is important to note also that certain lasers emitting radiation in the short wavelength infrared (SWIR) part of the electromagnetic spectrum—at wavelengths greater than 1400 nm are labeled as being *eye-safe*. An example is the Teledyne Optech ILRIS-3D terrestrial laser scanner, which uses a laser rangefinder that emits its radiation at $\lambda = 1550$ nm. This designation arises from the fact that the water content of the cornea of the human eye absorbs radiation at these wavelengths. However, this label can be misleading as it really applies only to low-powered CW beams. Any high-powered Q-switched pulsed laser operating at these wavelengths can still cause severe damage to the eye of an observer.

1.6.7 Beam Divergence

No matter how well collimated the laser beam or pulse is when it leaves the ranging instrument, it will have spread to illuminate a circular or elliptical area when it reaches the ground or the object on the ground. For a given angular spread of the beam, the greater the range, the larger the diameter of the area that is being covered on the ground or at the object. Thus, if the ground is irregular in shape or elevation, the return signal will be the average of the mixture of reflections occurring within the circular or elliptical area illuminated by the incident laser radiation.

For a terrestrial or ground-based laser rangefinder, the geometry of the beam traveling between the rangefinder and the target is shown in Figure 1.21. The area (A) covered by the diverging beam when it reaches the target is equal to π ($\theta/2R + d$)/2, where θ is the angle of divergence in radians; d is the diameter of the aperture; and R is the range.

Brenner (2006) gives further information about this matter. If the beam divergence is γ, then the theoretical limit of the divergence caused by diffraction is $\gamma \geq 2.44\ \lambda/d$, where d is the diameter of the aperture of the laser. In this case, using a pulsed laser emitting its radiation at $\lambda = 1064$ nm and having an aperture of 10 cm, the value of γ will be 0.026 mrad. Brenner quotes the typical values of γ for an airborne laser scanner as being in the range 0.3–2 mrad. The Teledyne Optech

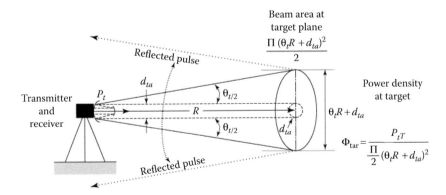

FIGURE 1.21 Spread of the beam or pulse that has been emitted by the laser rangefinder and its divergence on its path to the ground object or target and on its return from the object or target.

ALTM 3100 and the Leica Geosystems ALS50-II airborne laser scanners both have beam divergences at the smaller end (0.3 mrad) of this range. The *RIEGL* LMS-Q560 laser scanner and the IGI LiteMapper and Trimble Harrier airborne laser scanners that use the *RIEGL* device as the laser engine in their systems have a beam divergence of 0.5 mrad. At a typical flying height of 1000 m, the laser footprint is about 30 and 50 cm for the beam divergences of 0.3 and 0.5 mrad, respectively.

1.6.8 REFLECTIVITY

The reflectance or reflectivity of a ground target is another important matter to be considered in terms of the ranging performance of laser profiling and scanning systems. Reflectance is formally defined as the ratio of the incident radiation on a particular surface to the reflected radiation from that surface. Obviously, if the strength of the reflected signals that are received by the ranger is very weak, the range over which the required measurements can be made will be reduced as no detectable return signal will reach the receiver. The backscattering properties of the particular piece of ground or object being measured are therefore matters of considerable importance. For a diffusively reflective target such as a building, rock, or tree trunk—also called a hard surface—the reflected radiation can be envisaged (or idealized) as having been scattered into a hemispherical pattern with the maximum reflection taking place perpendicular to the target plane and the intensity diminishing rapidly to each side (Figure 1.22).

Furthermore, the reflective properties of the target will vary according to the wavelength of the radiation being emitted by the laser. In practice, it is often difficult to obtain information on these properties, as much of the published information on the subject of reflectivity and backscattering comes from remote sensing. In many of these reports on reflectivity, the incident radiation comes from the Sun and is incoherent in nature. Thus, the reflectivity may be very different when the radiation comes from a coherent source such as a laser emitting its radiation at a very specific wavelength. Wehr and Lohr (1999) have published a table based on information provided by the *RIEGL* company that manufactures both airborne and terrestrial laser rangefinders, profilers, and scanners. This shows the typical reflectivity of various diffuse reflectors for laser radiation with a wavelength (λ) of 900 nm (Table 1.1).

Further information provided to the same two authors (Wehr and Lohr, 1999) by the *RIEGL* company with regard to the variation in the maximum range with target reflectivity is given in Figure 1.23.

Still further information on the effects of variations in surface reflection comes from various manufacturers of ground-based laser scanners, particularly with regard to their effects on the range over which measurements can be made. Thus, the *RIEGL* company quoted a maximum measuring range for its LMS-Z210i pulse-type terrestrial laser scanner of 350 m for natural targets having a reflectivity ≥50%. However, this was reduced to 150 m for natural targets having a reflectivity of ≥10%. Further information on this matter has been provided by the i-site company from Australia in respect of its 4400LS terrestrial laser scanner. The company quotes a maximum range of 150 m for measurements made with this instrument to a surface of black coal (with 5%–10% reflectivity);

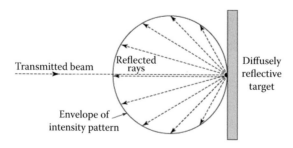

FIGURE 1.22 Reflectivity of a diffuse target.

TABLE 1.1

Typical Reflectivity of Various Reflecting Materials at a Wavelength of 900 nm

Material	Reflectivity (%)
Lumber (pine, clean, dry)	94
Snow	80–90
White masonry	85
Limestone, clay	Up to 75
Deciduous trees	Typical 60
Coniferous trees	Typical 30
Carbonate sand (dry)	57
Carbonate sand (wet)	41
Beach sand; bare areas in desert	Typical 50
Rough wood pallet (clean)	25
Concrete, smooth	24
Asphalt with pebbles	17
Lava	8
Black neoprene (synthetic rubber)	5

Source: Wehr, A. and Lohr, U., *ISPRS J. Photogram. Rem. Sens.,* 54, 68, 1999. With Permission.

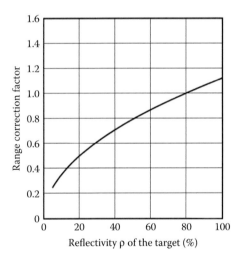

FIGURE 1.23 Correction factors for maximum range depending on the reflectivity of diffuse targets. (From Wehr, A. and Lohr, U., *ISPRS J. Photogram. Rem. Sens.*, 54, 68, 1999. With Permission.)

600 m for measurements made to rock or concrete surfaces (with 40%–50% reflectivity); and up to 700 m for surfaces having a still higher reflectivity. Both the *RIEGL* and the i-site instruments use the pulse-based ranging technique with their lasers emitting their radiation at the wavelength (λ) of 905 nm.

Turning next to those terrestrial short-range high-speed laser scanners that utilize CW lasers and the phase-based measuring technique, Leica Geosystems has provided information in its literature on the effects of reflectivity on the precision with which different surfaces can be modeled using its HDS4500 scanner. This utilizes a CW semiconductor laser emitting its radiation at a wavelength (λ) of 780 nm. At a shorter range of 10 m, the precision is quoted as ≤1.6 mm for dark gray surfaces

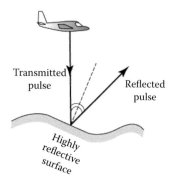

FIGURE 1.24 Highly reflective surface on the terrain that is not at right angles to the incident laser pulse or beam will reflect its radiation off to the side and will not return a signal to the laser rangefinder.

with 20% reflectivity, but ≤1.0 mm for white surfaces with 100% reflectivity. At a longer range of 25 m, the respective values of precision were ≤4.4 mm for the dark gray surface and ≤1.8 mm for the white surface.

Another important factor that governs the reflectivity of laser radiation is the angle that the ground or object makes with the incident pulse or beam (Figure 1.24). With an airborne laser scanner illuminating an area of smooth sloping terrain having a high (specular) reflectance, a very large part of the incident radiation from the scanner will be reflected off to the side and will not be returned to the laser scanner. In the case of a forest canopy, the laser pulse emitted from the airborne scanner in the vertical or nadir direction will be able to penetrate the spaces in the canopy and produce a reflection from the ground below the canopy. However, the ability to penetrate the canopy and reach the ground will decrease greatly with increasing scan angles away from the nadir direction.

1.6.9 POWER RECEIVED AFTER REFLECTANCE

Besides the loss of power in the emitted pulse through its spread caused by beam divergence and the limited reflectances from the ground and its objects, the radiation will also be subjected to scattering by the particles of dust and water droplets that are present in the air between the rangefinder and the target. Furthermore, it can be affected by scintillation effects in the atmosphere through which it is traveling. These effects will also be present during its return path after reflection. This results in a further diminution of the strength of the signal that will be returned to the measuring instrument. Indeed the power that will be reflected and received back at the rangefinder, profiler, or scanner will be a tiny fraction of that originally emitted by its transmitter.

If P_T is the power of the pulse transmitted by the airborne scanner, then the following expressions can be derived (Brenner, 2006):

1. Power that has been transmitted is P_T.
2. Power received at the ground will be $M \cdot P_T$ where M is the transmission factor through the atmosphere.
3. Power that is reflected from the ground (assuming a Lambertian surface) will be $(\psi/2\pi) \cdot \rho \cdot M \cdot P_T$.
4. The power received after reflectance from the ground and the return path through the atmosphere will be $P_R = A/(2\pi R^2) \cdot M \cdot \rho \cdot M \cdot P_T = \rho \cdot \{(M^2 \cdot A)/2\pi R^2\} \cdot P_T$ (Figure 1.25).

For example, if $P_T = 2000$ W, atmospheric transmission $(M) = 0.8$, $A = 100$ cm^2, where D (aperture) = 10 cm, H (Flying Height) = 1 km, and reflectivity $(\rho) = 0.5$, then $P = 4.10^{-10}$. $P_T = 800$ nW.

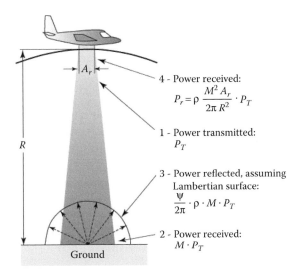

4 - Power received:
$$P_r = \rho \, \frac{M^2 A_r}{2\pi R^2} \cdot P_T$$

1 - Power transmitted:
$$P_T$$

3 - Power reflected, assuming Lambertian surface:
$$\frac{\psi}{2\pi} \cdot \rho \cdot M \cdot P_T$$

2 - Power received:
$$M \cdot P_T$$

FIGURE 1.25   Power transmitted, reflected, and received is shown graphically for an airborne laser scanner. (From Brenner, C., Aerial laser scanning. International summer school on "Digital Recording & 3D Modeling," Aghios Nikulaos, Crete, Greece, 2006, April 24–29. With Permission.)

Thus, in terms of its power, only the tiniest fraction of the power of the emitted pulse is being received back at the laser rangefinder, profiler, or scanner after its travel to and from the ground target. As already mentioned, the received signal needs to be amplified and any accumulated noise needs to be filtered out before the final range value can be determined from the elapsed time.

1.7   CONCLUSION

The current chapter has provided an introduction to lasers and laser ranging in the very specific context of their application to those rangefinders that form the basis of the laser profilers and scanners that are now being used extensively in topographic surveying, mapping, and modeling. Based on this introduction, the two chapters that follow will describe the laser profilers and scanners that are currently in use in the terrestrial or ground-based applications of the technology (Chapter 2) and those being operated from airborne and spaceborne platforms (Chapter 3). The authors are very grateful to Mike Shand who has provided most of the drawings that have been included in this chapter.

REFERENCES

Abdullah, Q.A., 2012. Mapping matters: Geiger mode lidar. *Photogrammetric Engineering and Remote Sensing,* 78 (7): 664–665.

Abdullah, Q.A., 2016. A star is born: the state of new lidar technologies. *Photogrammetric Engineering and Remote Sensing,* 82 (5): 307–312.

Barber, Z., Erkmen, B., Dahl, J., Sharpe, T., 2014. Photon-limited information in high resolution laser ranging. Final Report No. 61101-PH-DRP.2, U.S. Army Research Office, Research Triangle Park, Durham, NC. 46pp.

Brenner, C., 2006. Aerial laser scanning. International summer school on "Digital Recording & 3D Modeling," Aghios Nikulaos, Crete, Greece, April 24–26.

Bridger Photonics, 2015. Bridger photonics distance measurement technology. Bridger Photonics White Paper. 2pp.

Clifton, W.E., Steele, B., Nelson, G., Truscott, A., Itzler, M., Entwisle, M., 2015. Medium altitude airborne Geiger-mode mapping lidar system. *Laser Radar Technology and Applications XX, Proceedings of the SPIE,* Vol. 9465, pp. 946506-1-8.

Edde, B., 1992. *RADAR; Principles, Technology, Applications*. Prentice Hall, Upper Saddle River, NJ, p. 816.

Flood, M., 2001a. Eye safety concerns in airborne Lidar mapping. http://www.asprs.org/society/comittees/lidar/Downloads/Flood%20–%20Eye%20safety.zip.

Flood, M., 2001b. Laser altimetry: From science to commercial Lidar mapping. *Photogrammetric Engineering and Remote Sensing*, 67(11): 1209–1217.

Fried, D.G., 2015. Fast, cost-efficient airborne 3D imaging with Geiger-mode detector arrays. Presented Paper, International Lidar Mapping Forum (ILMF 2015), Denver, CO. February 23, 2015. 27 pp.

Massaro, R. D., Anderson, J.E., Nelson, J. D., Edwards, J. D., 2014. A comparative study between frequency-modulated continuous wave ladar and linear mode lidar. *International Archives of the Photogrammetry, Remote Sensing and Spatial Information Sciences*, 40: 233–239.

Mateo, A.B. and Barber, Z.W., 2015. Precision and accuracy testing of FMCW ladar-based length metrology. *Applied Optics*, 54: 6019–6024.

Miller, B., 1965. Laser altimeter may aid photo mapping. *Aviation Week & Space Technology*, March 29, 1965, 4 pp.

Petrie, G., 1990. Laser-based surveying instrumentation & methods (Chapter 2). In Kennie, T.J.M. and Petrie, G. (Eds.) *Engineering Surveying Technology*, New York: Blackie, Glasgow and London and Halsted Press, pp. 48–83.

Price, W.F. and Uren, J., 1990. *Laser Surveying*. Van Nostrand Reinhold (International). 256 pp.

Shepherd, E.C., 1965. Laser to watch height. *New Scientist*, April 1, 1965, 33 pp.

Toth, C.K., 2004. Future trends in LiDAR. *ASPRS Annual Conference*, Denver, CO, May 23–28, 2004. Paper No. 232, 9 pp.

Wehr, A. and Lohr, U., 1999. Airborne laser scanning—an introduction and overview. *ISPRS Journal of Photogrammetry and Remote Sensing*, 54(2/3): 68–82.

Zayhowski, J.J., 1990, Microchip Lasers. *The Lincoln Laboratory Journal*, 3(3): 427–446.

FURTHER READING

Baltsavias, E.P., 1999a. Airborne laser scanning: Existing systems, firms, & other resources. *ISPRS Journal of Photogrammetry and Remote Sensing*, 54(2–3): 164–198.

Baltsavias, E.P., 1999b. Airborne laser scanning: Basic relations & formulas. *ISPRS Journal of Photogrammetry & Remote Sensing*, 54(2–3): 199–214.

Jensen, H. and Ruddock, K.A., 1965. Applications of a Laser Profiler to Photogrammetric Problems. Brochure published by Aero Service Corporation, Philadelphia, PA, 22 pp.

Thiel, K.-H. and Wehr, A., 2004 *Performance Capabilities of Laser Scanners—An Overview & Measurement Principle Analysis*. Proceedings of ISPRS WG 8/2 Workshop—"Laser Scanners for Forest & Landscape Assessment," Freiberg, Germany, October 3–6.

2 Terrestrial Laser Scanners

Gordon Petrie and Charles K. Toth

CONTENTS

2.1 INTRODUCTION

As discussed in the introduction to Chapter 1, there has been a widespread use of lasers in land and engineering surveying for the last 30 years. This can be seen in the incorporation of lasers into standard surveying instruments such as total stations and the many laser rangefinders, profilers, levels, and alignment devices that are in daily use within these fields of activity. Given this background, it would appear to be a natural development for scanning and profiling mechanisms to be added to the total stations that were already equipped with laser rangefinders and angular encoders and for a terrestrial or ground-based laser scanner to be developed for use in field surveying. In classical field surveys, individual points are being measured on very specific ground features (usually marked by a reflective prism) to a high degree of accuracy by a field surveyor using a total station—as required for large-scale cadastral, engineering, and topographic maps. By contrast, the use of a laser scanner will allow the automated measurement and location of hundreds or thousands of nonspecific points (without the use of reflective prisms) in the area surrounding the position in which the instrument had been set up, all within a very short time-frame (Lichti et al., 2002). Besides this favorable technological background against which terrestrial laser scanner development could have been expected to take place, there was also no requirement for the supporting georeferencing technology of integrated GPS/IMU systems that were essential for the development of airborne profiling and scanning. Indeed, the locations of the ground stations where a terrestrial laser scanner has been set up on a tripod can easily be carried out using the traversing and resection methods that are used as standard practice by land surveyors.

However, for whatever reasons, the development of terrestrial or ground-based laser scanners has tended to lag behind that of airborne laser scanners. As Pfeifer and Briese (2007) pointed out, terrestrial laser scanning (TLS) matured and has been accepted by the overall geoinformatics industry rather later than its airborne equivalents. It may be that this was partly the result of the fact that a large segment of the professional surveying community comprises individual practitioners or small partnerships who operate purely on a local basis. Many of these do not possess the financial resources needed to acquire laser scanners costing much more than the total stations, GPS receivers, theodolites, and surveyors' levels that are the traditional forms of instrumentation being used in land surveying. In fact, in many cases, the first users of terrestrial laser scanners came from the surveying departments of large airborne mapping companies who had already been using airborne laser scanners and thus had a good understanding of the underlying technology. However, now that terrestrial laser scanners have become sufficiently developed and have been accepted by the surveying profession, they have proliferated quite quickly and in large numbers, even though the investment is by surveying standards considerable—although it is still only around one-tenth of that required for the purchase of an airborne laser scanner.

In this particular context, it is also interesting to note that, when it did take place, much of the early development of terrestrial laser scanners, especially over short ranges, was carried out by companies that were active in the manufacture and supply of measuring instruments and equipment for use in industrial metrology rather than the mainstream suppliers of surveying instruments—though the latter were quick to adopt the technology once it had been fully developed. Another interesting consequence of the developments that have come from the metrology side has been the introduction and widespread adoption of phase measuring techniques for the laser range-finders that are used in short-range terrestrial laser scanners. This is in complete contrast to the situation in airborne laser scanning (ALS) in which the pulse-ranging technique has been totally dominant. Moreover, pulse-ranging has been adopted in those terrestrial laser scanners that are being used over the medium and long ranges (LRs) that are similar to those that are commonly encountered in ALS.

Turning next to the vehicle-mounted laser scanners and profilers that form parts of certain mobile mapping systems (MMSs), the technology has, in many respects, more similarities to that of the ALS systems that are being deployed on Unmanned Aerial Vehicle (UAV)/Unmanned Aerial System (UAS) platforms than that used in terrestrial laser scanners. In particular, the requirements for the continuous georeferencing of the moving vehicle have resulted in the use of integrated GPS/ or GNSS/IMU systems as essential components of the overall MMS. However, a special problem with these systems concerns their application to the mapping of built-up urban areas where the interruptions in the GPS or GNSS signals caused by high buildings result in poor satellite configurations or even a complete loss of signal (Bohm and Haala, 2005). Besides which, the reflections of signals from nearby buildings can give rise to multipath effects that can reduce the positional accuracy of the GPS or GNSS measurements. These deficiencies in the GPS or GNSS measurements have resulted, in most cases, in their supplementation by distance measuring devices (odometers) that are attached to the wheels of the vehicles on which the laser scanners have been mounted. In this context, it should also be noted that many terrestrial laser scanners are mounted or installed on vehicles to improve their production efficiency—basically, to allow for their faster deployment and more rapid transport in the field and, in general, to provide more flexibility. In many cases, the laser scanner is attached to a mast that can be erected to higher positions, providing a better observation potential as compared with simple tripod-mounted installations. However, these vehicular-mounted *stop-and-go* systems do not fall into the category of MMSs that have a continuous movement of the supporting platform as they carry out their measurements.

Terrestrial laser scanners, whether tripod- or vehicle-mounted, do have the great advantage of providing the details of building facades as required for the production of realistic 3D city models. In this respect, the data derived from terrestrial laser scanners complement the data acquired using airborne stereo-photogrammetric and laser scanning methods. The latter can produce an overall 3D city model at roof or ground level, but it will lack detailed information about the building facades. Equally, the terrestrial laser methodology can produce the facades along the narrow corridor being traveled by the mapping vehicle, but it does not provide information about the areas hidden by the buildings, woods, and so on that can be covered from the air. Thus, a combination of the two methods (airborne [in particular, that utilizing UAV platforms] and terrestrial) is becoming increasingly common for the production of these 3D city models (Bohm and Haala, 2005).

As noted earlier, the terrestrial laser scanners that are being utilized for topographic mapping and modeling operations can be operated either from a static (or stationary) position—for example, being mounted on a tripod over a ground mark—or they can be operated from a dynamic (or moving) platform such as a van, truck, or railcar. Quite different instrumental or system designs will result depending on which of these two main operational modes has been adopted for the acquisition of the terrestrial laser scan data for mapping operations (Ingensand, 2006). Thus, the coverage of terrestrial laser scanner technology that will be given in this chapter has been organized to fall into these two distinct categories—those covering static and dynamic laser scanners, respectively.

2.2   STATIC TERRESTRIAL LASER SCANNERS

The types of laser scanner that will be considered in this section of the chapter are those that carry out the 3D measurements of the topographic features that are present on the ground in the area around the fixed (static) position that has been occupied by the instrument. They do this through the simultaneous measurement of (1) slant range by a laser rangefinder and (2) the two associated angles in the horizontal and vertical planes passing through the center of the instrument by angular encoders (Figure 2.1). In most cases, prior to the actual scanning process, the angular increments in both directions, comprising the azimuth and vertical rotations, can be set by the user. Typically, the angular step sizes are set to identical values; in which case, the laser scanner provides an equal spatial sampling in an instrument-centered polar coordinate system.

These simultaneous measurements of distance and angle are carried out in a highly automated manner using a predetermined scan pattern, usually at a measuring rate of 1000 Hz or more—indeed up to 1 million points per second in some current instruments. As discussed in Chapter 1, the distance measurements that are made by the laser rangefinder will utilize either the pulse ranging or the phase difference measuring technique. However, in this introductory discussion, it should be mentioned that there also exist a number of very short-range laser scanners, which are based on other measuring principles and which operate over ranges of a few meters, often to accuracy values of a fraction of a millimeter. A comprehensive review of the principles and technologies utilized in this ultra-high accuracy (HA), very short-range scanners is given in Blais (2004). These instruments are much used in metrology, industrial applications, and reverse engineering; in body scanning and medical research; and in the recording of cultural objects by museum staff and archaeologists. A representative example of such an instrument—one among very many—is the Konica Minolta VIVID laser scanner. This operates on the basis of optical triangulation with the target or object being scanned with laser stripes whose reflected images are being recorded simultaneously by a digital camera. The maximum measuring range of such an instrument is only 2.5 m. Over the last few years, quite a number of hand-portable range measuring and scanning devices such as the DotProduct DPI-7, Mantis Vision MVC-F5 SR, Artec Spider, Creaform HandyScan, Microsoft Kinect, and so on have appeared in the market. Again, these devices are concerned with measurements over very short ranges, usually <5 m, and mostly they utilize structured light technology for applications such as reverse engineering, medical applications, the measurement of cultural objects, the measurement of small rooms for indoor mapping purposes, and so on. These very short range laser scanners (and other similar scanner-less devices) will not be considered here, in which the emphasis is exclusively (or mainly) concerned with topographic applications.

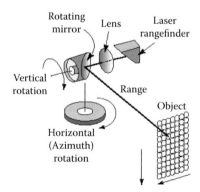

FIGURE 2.1   The range and the horizontal and vertical angular rotations that are measured toward objects in the terrain using a terrestrial laser scanner. (Drawn by M. Shand.)

2.2.1 OVERALL CLASSIFICATION

The primary classification that has been adopted widely in the published literature on static terrestrial or ground-based 3D laser scanners differentiates between those instruments that utilize the pulse ranging or time-of-flight (TOF) measuring principle and those that employ the phase measuring technique—as set out in Chapter 1 (Fröhlich and Mettenleiter, 2004). Within the specific context of static terrestrial laser scanners, it can be said that the phase difference method produces a series of successive range measurements at a very high rate and to a high degree of accuracy, but usually over distances of some tens of meters—although this has been increased to well over 100 m in some recent instruments. Whereas the pulse-based TOF method allows much longer distances of hundreds of meters (or, in some cases, some thousands of meters) to be measured, albeit often at a reduced rate and a somewhat lower (though quite acceptable) accuracy as compared with those of the phase measurement method. However, recently, the lower rate of the TOF method has been increased to a similar level to that of phased-based methods—using the multipulse technique in some instruments.

However, this simple classification that is based solely on the measuring technique (phase or pulse) that is used to measure distance does not take account of the angular scanning actions that are used to ensure the required coverage of the ground and of the objects that are present on it. Therefore, it is essential to have a secondary classification that will account for both the scanning mechanism and the pattern or coverage that this produces over the ground. The classification that has been adopted here is that introduced by Staiger (2003)—which differentiates between three types of static 3D terrestrial or ground-based laser scanners—(1) panoramic-type scanners, (2) hybrid scanners, and (3) camera-type scanners (Figure 2.2).

1. Within the first of these three categories, panoramic-type 3D scanners carry out distance and angular measurements in a systematic pattern that gives a full 360° angular coverage within the horizontal plane passing through the instrument's center and typically a minimum 180° coverage in the vertical plane lying at right angles to the horizontal plane—thus giving hemispheric coverage. However, a still greater vertical field-of-view (FOV) of 270° or more is not uncommon—which means that a substantial coverage of the ground lying below the instrument's horizontal plane can be achieved. Indeed, the only gap or void in the coverage of a full sphere on a number of instruments is that produced by the base of the scanner instrument and its supporting tripod. Although this panoramic scanning pattern is very useful in the context of topographic mapping, it is even more desirable, indeed often obligatory, in the measurement of complex industrial facilities, deep quarries and

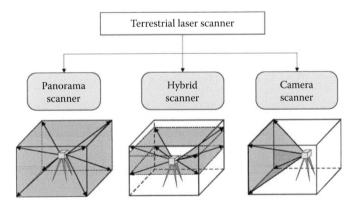

FIGURE 2.2 The classification of 3D terrestrial laser scanners based on their respective scanning mechanisms and coverages. (From Staiger, R., Terrestrial laser scanning—Technology, systems, and applications. Second FIG Regional Conference, Marrakech, MO, December 2–5, 2003; redrawn by M. Shand. With Permission.)

open-cast mines, and the facades of buildings within urban areas—or even indoors in large halls, churches, rooms, and others.

2. The instruments falling within the second category of hybrid 3D scanners are those in which the scanning action is unrestricted around one rotation axis—usually the full 360° horizontal scanning movement in the azimuth direction that is produced by a rotation of the instrument around its vertical axis. However, the vertical angular scan movement in elevation around the horizontal axis of this type of scanner instrument is restricted or limited—typically to 50°–60°. This reflects the situation that is commonly encountered in medium- and long-range 3D laser scanning that is being carried out for topographic mapping purposes in which there is no requirement to measure objects overhead or at steep vertical angles, as will be needed within deep quarries or urban areas.

3. The camera-type 3D laser scanners that make up the third category carry out their distance and angular measurements over a much more limited angular range and often within a quite specific FOV. Typical might be the systematic scanning of the surrounding area over an angular field of 40° × 40° in much the same manner as a photogrammetric camera—at least in terms of its angular coverage, though obviously not in terms of the actual measurements that are being made and recorded. This type of configuration is well suited to the survey of a specific site or a building facade.

As with airborne laser scanners, a third or tertiary classification can be envisaged, based on the range or distance over which the static terrestrial or ground-based 3D laser scanners can be used.

1. The first group that can be distinguished comprises those laser scanners that are limited to short ranges up to 150 m—indeed some are limited to distances of 30–60 m. They mostly comprise those instruments that employ the phase measuring principle for distance measurement using the laser rangefinder—although, as will be seen later, there are instruments that employ pulse (TOF) ranging over these short ranges. Usually, the limitations in range of these instruments are offset by the very high accuracies that they achieve in distance measurement—often to a few millimeters.

2. A second group, almost entirely based on pulse ranging using TOF measurements of distance, can measure over medium ranges with maximum values from 150 to 450 m, often at a somewhat reduced accuracy as compared with the first group.

3. A third long-range group, again using the pulse ranging technique, can measure still longer distances—up to some kilometers in the case of certain of the laser scanner instruments manufactured by Teledyne Optech, *RIEGL*, and Maptek. However, the gain in range is normally accompanied by a reduction in the accuracy of the measured distances and in the pulse repetition rate—though this is still appropriate to the applications concerned and is very acceptable to the users of these instruments.

2.2.2 SHORT-RANGE LASER SCANNERS USING PHASE MEASUREMENT

Of the main measuring principles outlined earlier—phase measurement and pulse ranging—the former is almost exclusively in use in those ground-based laser scanners that measure over short ranges—typically with maximum ranges from 50 to 150 m. It will be seen that all of the instruments that have been included in this group fall into the category of panoramic scanners. This enables them to be used indoors within buildings as well as outdoors in which their operational characteristics make them well suited for use in the surveys and mapping of urban areas with high buildings where the measurement of short distances and high vertical angles is required. The factories of two of the most prominent manufacturers of this type of scanner—Zoller + Fröhlich (Z+F) and Faro—are both located in Germany, though the Faro company itself has its headquarters in North America. A third supplier is Basis Software Inc. which is based in the United States.

2.2.2.1 Zoller + Fröhlich

The current Z+F ground-based 3D laser scanner is called the Imager 5010; previous models from this manufacturer that have been widely used by surveyors were the Imager 5003 (Mettenleiter et al., 2003) and the Imager 5006. All of these instruments have a similar design, construction, and operation, but with greatly improved performance characteristics with each new successive model in the series. The instruments all feature a servo motor and an angular encoder that measures and implements the angular scan movement in the horizontal plane (in azimuth) around the instrument's vertical axis through a rotation of the upper part of the instrument against its fixed base—which is normally mounted on a tripod so that it can be set accurately over a ground mark. The scan movement in the vertical direction is implemented using a lightweight, fast-rotating mirror that has been placed on the horizontal (trunnion) axis of the instrument, which is normally supported on two vertical standards or pillars. The horizontal rotation in azimuth covers the full circle of 360°, whereas the rotational movement of the mirror in these Imager scanners allows scan angles of 310° within the vertical plane (in the Imager 5006 model) and 320° (in the current 5010 model), respectively. The manufacturer's claimed accuracy in both horizontal and vertical angular measurement is ±0.007°—which is equivalent to ±6 mm accuracy at the measured points in both directions in the plane that is perpendicular to the laser direction at 50 m object distance. The maximum scan rate in the vertical (elevation) plane is 50 Hz; however, a more typical rate in actual operational use is 25 Hz.

The LARA laser rangefinder that was used in the original Imager 5003 and 5006 scanner instruments employed a Class 3R continuous wave (CW) laser that operated in the near infrared part of the spectrum at $\lambda = 780$ nm. In the earliest Imager 5003 model, the rangefinder was available in two alternative versions giving maximum possible ranges (or ambiguity limits) of 25.2 and 53.5 m, respectively (Mettenleiter et al., 2000). A detailed investigation into the accuracies of the distance and angular measurements of the Imager 5003 was made by Schulz and Ingensand (2004) of the ETH, Zürich. The later Imager 5006 model had an improved maximum range (or ambiguity limit) of 79 m. The accuracy in distance quoted by the manufacturer was ±6.5 mm over a range of 25 m. This instrument continues to be sold currently (in 2016) as the Imager 5006h model, being marketed as a lower cost, entry-level scanner for first-time users. Another variant is the Imager 5006EX model (Figure 2.3a), part of which is encased in a special protective jacket and is designed to be used in areas such as mines and chemical plants in which there is a potentially explosive environment. This specialist Imager 5006EX scanner model is also still available for sale in 2016.

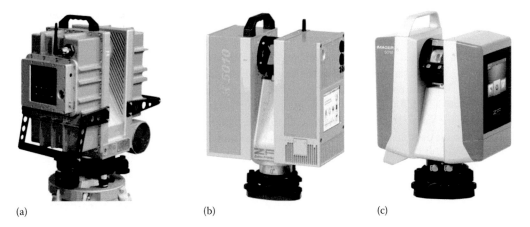

(a)   (b)   (c)

FIGURE 2.3 (a) The Z+F Imager 5006EX phase-based laser scanner that has been developed in cooperation with DMT GmbH for use in potentially explosive environments. (b) The current Z+F Imager 5010 high-speed phase-based 3D laser scanner. (c) The recently announced Imager 5016 laser scanner. (Courtesy of Zoller + Fröhlich, Wangen im Allgäu, Germany. With Permission.)

In the current Imager 5010 model, a Class 1 laser operating at $\lambda = 1500$ nm is being used in the improved rangefinder design (Figure 2.3b). With the longer laser wavelength and the new rangefinder design, a maximum range (or ambiguity limit) of 187.3 m is possible, while the maximum possible measurement rate has been raised from 500,000 measurements per second in the original models to 1 million in the Imager 5010 model—although, in normal operation, the rate is likely to be somewhat lower. The instrument also features a dual-axis compensator that corrects the horizontal angle readings when the instrument is not truly vertical and the telescope is inclined above or below the horizontal plane—which is also a common feature in total stations. A variant is the Imager 5010C which has an integrated digital color camera that implements Z+F's high dynamic range (HDR) technology and aims to produce a better color balance across the whole of each frame image. Another variant is the Imager 5010X model that is equipped with a GPS or GNSS receiver and an indoor navigation system (including a gyroscope, an accelerometer, a compass, and a barometer) for operation when GPS signals are not available.

In October 2016, Z+F announced that it would be introducing a still newer model in the form of its Imager 5016 instrument (Figure 2.3c). This will have a new and substantially different ergonomic design that will make it smaller and lighter than the Imager 5010 model, weighing 5.6 kg against the 9.8 kg of the latter. Besides which, the new instrument will again allow a maximum measuring rate of 1 million points per second, but it will also have a much greater range with a maximum value of 360 m instead of the 187.3 m that is the maximum value in the Imager 5010 series of instruments. The new Imager 5016 model will also include both the integrated HDR camera and the positioning system that feature in the Imager 5010C and 5010X models. The new HDR camera will be equipped with integrated LED spots for illumination purposes, which will add additional flexibility while imaging during the scanning operations. This feature should ensure that additional external lighting sources will not normally be required when capturing images in dark environments.

The previous Z+F Imager 5003 instrument was also marketed and sold very widely for mapping applications by Leica Geosystems as its HDS4500 laser scanner, where HDS is an acronym for *high definition surveying*. Similarly, the later Z+F Imager 5006 instrument was sold by Leica Geosystems as its HDS6000 scanner and the current Imager 5010 model as the HDS7000, both with the same product performance as the corresponding Z+F versions of the instruments. However, this arrangement was terminated with the introduction of the short-range, pulse-ranging ScanStation laser scanners by Leica Geosystems (that will be described later). Thus, the Leica company no longer markets and sells these rebadged versions of the Z+F Imager scanner instruments. Another company that has a similar arrangement of selling rebadged Z+F phase-based scanners is the Japanese company, TI Asahi, which, since 2013, has marketed the 5010 and 5010C models as the Pentax S-3180 and S-3180V 3D scanning systems, respectively, while the older entry-level Imager 5006h scanner is marketed as the S-3080 3D scanning system. The Italian company, Stonex, has also offered a rebadged version of the Imager 5010 as its X9 model in its range of laser scanners.

In parallel with its range of 3D laser scanners, Z+F has also developed a corresponding range of 2D profilers. Initially, each of these instruments were only slightly modified versions of the different Imager models that were constrained to operate in an individual vertical 2D plane (1) without the need for the motor to provide the rapid change in the azimuth direction required to implement successive scans over the whole of the 360° horizontal range and (2) without the horizontal angular measuring capabilities. Thus, for example, the 2D Profiler 5010 was derived in this way from the Imager 5010 3D laser scanner. However, Z+F has also designed and built a number of custom-made profilers for different customers with specialist requirements. Examples of these instruments built for rail mapping applications include the Profiler 6000-300 (Fröhlich and Mettenleiter, 2004) (Figure 2.4a) and the Profiler 6007 Duo (Figure 2.4b), the latter featuring dual laser scanner/profilers. More recently, Z+F has built a new Profiler 9012 instrument (Figure 2.5a), which is substantially different in its design and construction to the current Imager 5010 scanners. This instrument has a scanning head mounted on the side of the instrument that allows a full unobstructed 360° rotation in the vertical plane instead of the 320° that is possible in the current Imager 5010 model and in

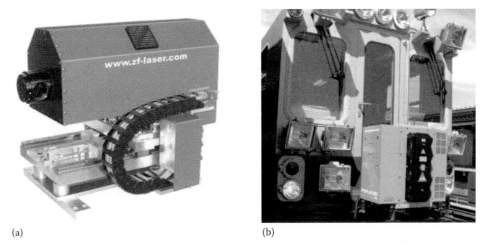

(a)   (b)

FIGURE 2.4   (a) The Z+F Profiler 6000-300 built for rail mapping. (b) A Z+F Profiler 6007 Duo system mounted on the front of a train in New York to carry out surveys of the railroad track and its fittings. (Courtesy of Zoller + Fröhlich. With Permission.)

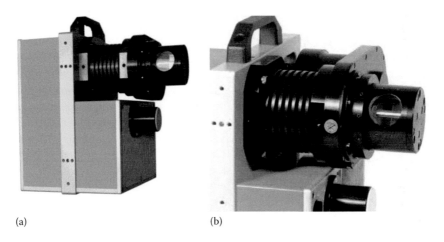

(a)   (b)

FIGURE 2.5   (a) The Z+F Profiler 9012 instrument. (b) A close-up photo of the Profiler 9012's scanning head, which is mounted clear of the main body of the instrument to allow an unobstructed 2D profile scan over a full 360° rotation. (Courtesy of Zoller + Fröhlich. With Permission.)

the previous Profiler 5010 (Figure 2.5b). This has proven to be popular for installation in several different MMSs (as will be discussed later). The Profiler 9012 has a similar performance to that of the Imager 5010 scanner, including a maximum data acquisition rate of 1 million points per second and a maximum scan speed of 200 Hz (12,000 rpm). However, its maximum range (or ambiguity limit) is 119 m instead of the 187.3 m of the Imager 5010. Recently introduced variants are the Profiler 9012A, which is optimized for close-range operations (up to 5 m) and the Profiler 9012 M/A, which combines a marker mode with high-precision close-range surveying.

2.2.2.2   Faro

The Faro company, which is mainly based in North America with its headquarters in Florida, is involved in the manufacture of a wide range of portable computer-based measuring instruments, including laser trackers, gauges, and measuring arms, for use in industrial plants and factories.

However, its range of terrestrial or ground-based laser scanners was developed originally by the IQvolution company, based in Stuttgart, Germany, which Faro purchased in 2005. The IQvolution 3D laser scanner was called the IQsun 880. This instrument was renamed as the LS 880 after the take over by Faro (Mechelke et al., 2007) after which, two new shorter range models—the LS 420 and LS 840—have been introduced to the market.

The basic design, construction, and operation of all three models of this panoramic-type scanner were very similar. The laser rangefinder was mounted in the horizontal plane and aligned with the horizontal (trunnion) axis of the instrument. The output beam from the laser passed into the center of a continuously rotating (motor-driven) mirror that deflected it through a fixed angle of 90° to produce a vertical profile scan of the laser beam giving an angular coverage of 320° in the vertical plane. The 360° azimuth scan was implemented using a motor whose power was derived from a 24 V DC battery pack. These angular motions placed all three models in the category of panoramic scanners. All three models used a Class 3R CW semiconductor laser operating at $\lambda = 785$ nm in the near infrared part of the spectrum as the basis of their rangefinders. The LS 420 (Figure 2.7a), LS 840, and LS 880 models allowed maximum ranges of 20, 40, and 76 m to be measured, respectively. With regard to the phase measuring technique used to measure distances, taking the LS 880 model as an example, the output laser beam was split and the amplitude was modulated to operate at three different wavelengths—76, 9.6, and 1.2 m—as shown in Figure 2.6. This allowed the measured range to be determined to ±0.6 mm in terms of its resolution value. The claimed accuracy of the measured ranges was ±3 mm at a distance of 10 m. At the time of their introduction, the obvious advantage of using the phase measurement technique over the TOF or pulse ranging method was its speed of measurement. In the case of the LS 880 model, the laser rangefinder could measure distances at rates up to 120,000 points per second. Improved versions of these scanners with upgraded electronics and better positional accuracy were introduced in March 2008 at the SPAR 2008 Conference under the name Photon and were called the Photon 20 (with 20 m range), Photon 80 (with 76 m range), and Photon 120 (with a 153 m ambiguity interval [wavelength] and a maximum range of 120 m). In the case of the last of these three models, the maximum speed of measurement was increased to 976,000 points per second.

In 2010, Faro introduced its Focus 3D range (Figure 2.7b). Initially, this comprised two models— the Model 20 (with 20 m range) and the Model 120 (with 120 m maximum range). The maximum speed of measurement was 976,000 points per second for both models. Although the performance values were not too different than those of the previous models, the instruments were much more compact, much lighter in weight, and utilized a rangefinder with a Class 3R laser having a wavelength of $\lambda = 905$ nm in the near-infrared part of the spectrum. They also featured an integrated color camera producing 70 Megapixel images. As a result of these various design changes, the instruments were not only enhanced in terms of their performance but were also considerably lower in cost than their

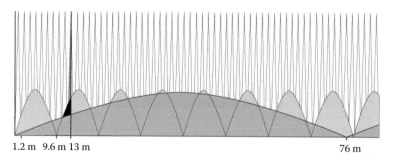

1.2 m 9.6 m 13 m   76 m

FIGURE 2.6 The measurement scheme used in the Faro LS 880 laser scanners showing the three different frequencies corresponding to distances of 76, 9.6, and 1.2 m, respectively, at which the phase differences are measured. An example is shown for a measured distance of 13 m. (Courtesy of Faro, Korntal-Munchingen, Germany. With Permission. Redrawn by M. Shand.)

(a) (b) (c)

FIGURE 2.7 (a) The earlier Faro LS 420, (b) the later Faro Focus 3D, and (c) the latest Faro Focus S laser scanner, all of which measure their ranges employing phase differences. (Courtesy of Faro. With Permission.)

predecessors. One result of this development was that, between 2012 and 2015, under an Original Equipment Manufacturer (OEM) agreement, the Faro Focus 3D scanners were also sold by Trimble, rebadged as its TX5 scanner. A similar agreement was also made at the same time with Topcon* that allowed Topcon to sell the Focus 3D scanners as a distributor in Europe, the Middle East, and Africa.

As a result of a series of further developments that have been implemented since 2013, the Faro Focus 3D laser scanner is currently (in 2016) available in three different forms—the X30 model (with a maximum range of 30 m), the X130 model (with a range of 130 m), and the X 330 model (with a range of 330 m). They all utilize a rangefinder equipped with a Class 1 laser having a wavelength of $\lambda = 1550$ nm. Both of the longer range models also incorporate a GPS receiver to provide the accurate location of the scanner, supplemented by a dual axis compensator, a compass, and a height sensor. The short range X30 model also incorporates the last three devices, but not the integrated GPS receiver, since it is very often employed on interior scanning applications within buildings.

In October 2016, Faro introduced its Focus S laser scanner range (Figure 2.7c), which consists of two models, the Focus S150 and the S350. The former has a measuring range from 0.6 to 150 m, whereas the corresponding figures for the latter are an extended measuring range from 0.6 to 350 m. For both models, the claimed accuracy of the measured distances is ±1 mm. The Focus S scanners also include a substantial upgrade to the instruments' sealed construction, bringing the scanners into the Class 54 IP (Ingress Protection) rating, as defined in the international standard IEC 60529. This means that the scanner provides protection against dirt, dust, and various types of unwanted water, such as rain and splashes. Thus the design is intended to ensure the instrument's use in difficult environments, such as dusty or humid work sites. The two new Focus S scanners also include accessory bays. These are ports on the instruments that surveyors can use to attach accessories to the scanner.

2.2.2.3 Basis Software Inc.

This American company is based in Redmond, Washington. Its main 3D laser scanner product that can be used in topographic applications is the Surphaser, which again is a panoramic-type phase-based scanner. The company also manufactures short-range laser scanners for industrial use that will not be discussed here. The Surphaser was originally available in two models—the 25HS (Hemispheric Scanner) that had a maximum range of 22.5 m, and the 25HSX that had an extended range of 38.5 m. However, the latter model was developed further into various versions or configurations with extended ranges of 46 m (version SR), 70 m (IR_X), and 140 m (ER_XS). The basic design and construction of all of these Surphaser 25HSX instruments is somewhat similar to that

---

* Topcon is a registered trademark of Topcon Corporation.

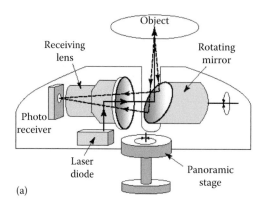

(a)

(b)

FIGURE 2.8 (a) Diagram showing the design and the main features of the Surphaser series of laser scanners. (b) The more compact ultrashort-range Surphaser 3D scanner Model 75USR is at the front of this photo, whereas the larger and longer range Surphaser 105HSX model is at the rear. (Courtesy of Basis Software, Redmond, Washington; Diagram (a) redrawn by M. Shand. With Permission.)

of the Faro scanners described previously in which they both employ a laser diode rangefinder that emits its continuous beam of laser radiation along the instrument's horizontal axis on to a continuously rotating mirror located at the end of a motor-driven shaft (Lichti et al., 2007). This turns the laser beam through a right angle to produce the required scanning motion in the vertical plane giving an angular coverage of 270° (Figure 2.8a). The receiving lens then focuses the reflected radiation from the object on to the photodiode that acts as the receiver, thus allowing the continuous measurement of the phase differences and the intensity values. The maximum rate of range measurements of the laser rangefinder varies according to the model or version between 190,000 to 1200,000 points per second, though often, in practice, a lower rate will be used. The semiconductor laser diode that is used as the basis of the Class 3R laser employed in the 25HSX instruments emits its radiation at the wavelength (λ) of 685 nm on the red edge of the visible part of the spectrum with a power of 15 mW. The 360° angular scan in the horizontal plane is implemented through the rotation of the upper part of the instrument that is driven in azimuth by a stepping motor and gearbox located in the lower (fixed) part of the instrument which can be fitted into a standard Wild/Leica tribrach. The instrument's power requirements are supplied by a standard 18/24 V DC battery. The measured data are transferred via an appropriate interface to be recorded on a laptop computer.

In November 2007, a version of the Surphaser was introduced by Trimble as the FX Scanner, the instrument being supplied by Basis Software to Trimble under an OEM agreement. The FX scanner was used in conjunction with Trimble's FX controller software, while Trimble's Scene Manager software was used to locate the positions where the scanner had been set up for its scanning and measuring operations. The measured data could then be used in Trimble's LASERGen suite of application software. However, this arrangement has since been discontinued, and the FX instrument is no longer offered by Trimble. The current versions of the Surphaser that are on offer by Basis Software are (1) the Model 100HSX that has a similar specification and uses a similar Class 3R laser with a wavelength (λ) of 685 nm to that of the 25HSX model that has been described earlier, (2) the Model 105HSX that utilizes a Class 1 laser with a wavelength (λ) of 1550 nm and an extended ambiguity range of 180 m, and (3) the Model 75USR (=ultrashort range) (Figure 2.8b) that is much more compact and lower in weight than the other models and, as its title suggests, is optimized for very short range scanning operations. It also uses a laser with a wavelength (λ) of 685 nm like the Model 100HSX model.

2.2.2.4 Leica Geosystems

As noted previously, Leica Geosystems sold substantial numbers of phase-based laser scanners in the form of their HDS 4500, HDS 6000, and HDS 7000 models, all of which were rebadged versions

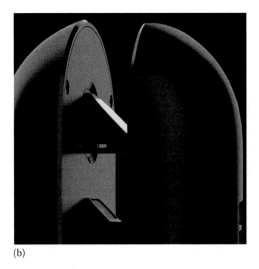

(a)   (b)

FIGURE 2.9 (a) A Leica BLK360 phase-based laser scanner. (b) Showing the high-speed scanning mirror of the BLK360 scanner. (Courtesy of Leica Geosystems, St. Gallen, Switzerland. With Permission.)

of models in the Z+F Imager range of scanners. However, in November 2016, the Leica company reentered this particular area with the introduction of its BLK360 scanner instrument (Figure 2.9) in partnership with Autodesk, the major supplier of software for use in engineering, architecture, and construction. The BLK360 instrument is physically very small, having a diameter of 4 in (10 cm) and a height of 6.5 in (16 cm) and weighs only 1 kg. The scanner instrument has a rangefinder that can operate at a maximum rate of 360,000 measurements per second over ranges up to 60 m with a claimed accuracy of ±4 mm. Simultaneously with its laser scanning, the BLK360 instrument also collects image data with an internal camera operating in a HDR capture mode in conjunction with a thermal imager. The scanner operates in a panoramic mode having a 360° scan in azimuth, with a single complete scan taking 3 min. The transfer of the measured data is carried out using a wireless connection to a hand-held iPad computer. Autodesk's contribution to the overall system is a Mobile edition of its ReCap 360 Pro software that runs on the iPad computer and provides the controlling software for the operation of the BLK360 instrument. The operation of the ReCap 360 Pro software also allows the BLK360 scans to be registered in the field in real time.

2.2.3 Medium-Range Laser Scanners Using Pulse Ranging

As defined earlier, this group of 3D laser scanners can measure distances over medium ranges with maximum values lying between 150 and 450 m. Although several of the manufacturers of short-range 3D laser scanners, such as Faro and Basic Software, have strong interests in metrology—and indeed they manufacture and sell various other mensuration products that fall into that area—the situation is quite different with the manufacturers of medium-range 3D laser scanners. In this area, several of the principal system suppliers of these 3D scanner instruments, such as Leica, Trimble, and Topcon, are also major manufacturers of surveying instrumentation such as GPS receivers, total stations, and laser levels. The medium-range ground-based 3D laser scanners that are manufactured and supplied by each of these companies are all based on the pulse ranging technique.

2.2.3.1 Leica Geosystems

Leica entered this field in 2001 through its purchase of the pioneering Cyra Technologies Inc. company based in California. The rangefinder of the original Cyrax 2400 model used a diode-pumped, passively Q-switched, Nd:YAG microchip laser, whose output was frequency doubled to operate at

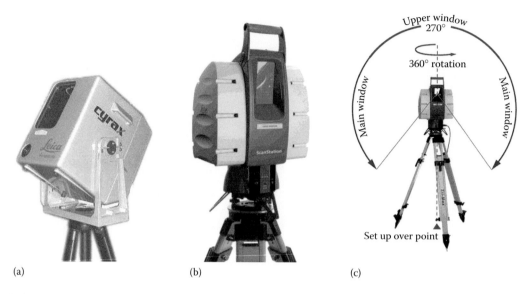

(a)   (b)   (c)

FIGURE 2.10 (a) An early Leica Cyrax camera-type laser scanner. (b) The Leica Geosystems ScanStation laser scanner with (c) its twin scan windows that together provided *panoramic* coverage of the surrounding area that was also utilized in the previous HDS3000 scanner. (Courtesy of Leica Geosystems, St. Gallen, Switzerland. With Permission.)

$\lambda = 532$ nm in the visible (green) part of the spectrum. This allowed a maximum speed of measurement of 800 points per second with a maximum range of 100 m. The Cyrax 2400 was a camera-type laser scanner that could scan a $40° \times 40°$ window using a twin mirror optical scanning system and could achieve a positional accuracy of ± 6 mm for scan data located 50 m from the scanner (Figure 2.10a). The main body of the instrument sat in a simple nonmotorized pan-and-tilt mount that allowed it to be pointed manually in steps over an angular range of 360° in azimuth and 195° in the vertical plane. It was followed by the improved Cyrax 2500 model that was later renamed as the Leica Geosystems HDS2500 scanner (Sternberg et al., 2004). In 2004, the Cyrax models were replaced by the Leica Geosystems HDS3000 scanner that had a very different design and specification. In particular, the camera-type layout of the previous Cyrax instruments was replaced by a dual-window design that, in total, gave a fully panoramic coverage of 360° in azimuth and 270° in the vertical plane (Figure 2.10c) The HDS3000 scanner used servo motors, both to rotate the scan mirror in the vertical plane and for the horizontal azimuth drive, resulting in a much higher scan rate than its Cyrax predecessors. The rangefinder that was used in the HDS3000 was based on a Class 3R microchip laser, again operating in the green part of the spectrum at $\lambda = 532$ nm, providing a scan rate of up to 1800 points per second and a maximum specified operating range of 100 m.

In 2006, the HDS3000 scanner was superseded by the ScanStation (Figure 2.10b). This retained the overall design and construction of the HDS3000 panoramic-type instrument. Moreover, it had a much greater maximum operational range of 300 m, a maximum measuring rate of 4000 points per second, and incorporated a number of additional features such as a dual-axis compensator to allow conventional surveying operations such as resection and traversing to be carried out using the instrument for the required angular measurements. In July 2007, the next model in the series, called the ScanStation 2, was introduced. Superficially, this had the same overall design and appearance of the HDS3000 instrument and the original ScanStation. However, the ScanStation 2 instrument featured a completely new laser rangefinder utilizing a Class 3R passive Q-switched microchip laser, again operating at $\lambda = 532$ nm in the green part of the spectrum. Moreover, its pulse repetition frequency (PRF) was 50 kHz—12 times faster than the speed that could be achieved in the previous

ScanStation model. This was combined with more advanced timing electronics, comprising a new time-to-digital converter and a high sensitivity receiver. All of which allowed the instrument to carry out its range measurements at a very much higher speed—with a maximum rate of 50,000 points per second—while still retaining the same high positional accuracy and the maximum range of 300 m with a reflectivity of 90% that had been achieved in the previous ScanStation model. With a ten-fold improvement in its measuring rate, the ScanStation 2 really marked a huge advance in performance as compared with its predecessors. The user interface and instrument controller was a laptop or tablet PC. With the introduction of this instrument, the in-house production of the ScanStation laser scanner instruments was moved from the Cyra factory in San Ramon, California to the main Leica Geosystems manufacturing plant located in Heerbrugg, Switzerland.

In September 2009, Leica Geosystems introduced the first of its "C" series of ScanStation 3D laser scanners in the form of its C10 model (Figure 2.11a). Although this instrument retained the ScanStation name, in fact, it underwent a very substantial redesign and upgrade as compared with the ScanStation 2. In particular, it featured a built-in touch screen color graphical display with an onboard computer/controller, an integrated 80 GB data storage and an internal battery, all of which eliminated the need for a separate laptop PC to act as the instrument's control device. Data transfer to an external computer could be carried out using built-in USB and Ethernet W-LAN (wireless) ports. New firmware allowed the onboard panel display to be repeated on an Apple iPad tablet computer, allowing it to be used as a remote control device, if required. The laser rangefinder was, once again, based on a Class 3R microchip laser emitting its pulses at $\lambda = 532$ nm. The dual-window design of the previous ScanStation instruments was replaced by a continuously rotating mirror spinning in the vertical plane, which meant that it no longer required two scans to capture a *full dome* FOV. Thus, the ScanStation C10 instrument could be operated to scan in full panoramic mode (Figure 2.11b) with a 360° (horizontal) × 270° (vertical) angular coverage or in a restricted camera mode using an oscillating scan motion over a specified window with selectable angular coverage.

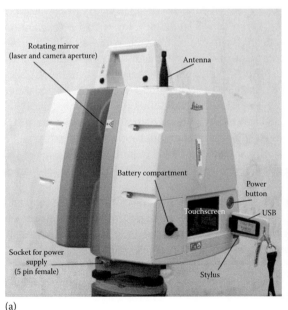

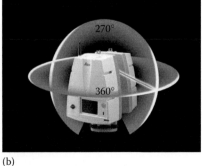

(a)   (b)

FIGURE 2.11   (a) The Leica Geosystems ScanStation C10 laser scanner showing its main external components. (b) Showing the *panoramic* coverage of the C10 model using a single rotating mirror. (Courtesy of Leica Geosystems, St. Gallen, Switzerland. With Permission.)

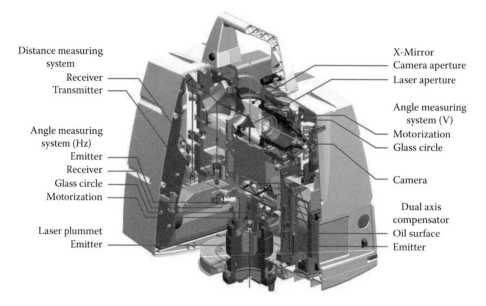

Distance measuring system
Receiver
Transmitter
Angle measuring system (Hz)
Emitter
Receiver
Glass circle
Motorization
Laser plummet
Emitter

X-Mirror
Camera aperture
Laser aperture
Angle measuring system (V)
Motorization
Glass circle
Camera
Dual axis compensator
Oil surface
Emitter

FIGURE 2.12 A cut-away diagram showing the internal design and construction and the major components of the Leica Geosystems C10 ScanStation laser scanner. (Courtesy of Leica Geosystems, St. Gallen, Switzerland. With Permission.)

The instrument was also fitted with a video camera that has a zoom capability and implements real-time video streaming of scanned targets (Figure 2.12). This same video camera also serves as an internal small-format (4 Megapixel) digital frame camera that can acquire high-quality color images of the objects that are being scanned and align the captured images with the points that have been scanned. Finally, within the ScanStation "C" series, in June 2011, Leica Geosystems introduced its ScanStation C5 instrument (Petrie 2012). This was a substantially lower cost instrument than the C10 model with a correspondingly lower performance. Compared with its C10 *big brother*, it still retained the same basic construction, measuring principle, and scan movements, together with the integrated onboard controller, color display, storage, and battery. However, the C5 model has a much more restricted measuring range (distance) of 35 m (v. 300 m for the C10) and a maximum measurement rate of 25,000 points per second (v. 50,000 points per second for the C10). It also lacks the dual-axis level compensator and the high-resolution digital frame camera of the C10 model. However, all these restrictions and missing features are available as optional extras and can be added later to the basic C5 instrument to bring it up to the C10 standard. According to the Leica Geosystems Website, the two "C" series scanners (C5 and C10) are still currently (in 2016) being offered by the company.

In 2012, Leica introduced the first in its "P" series of ScanStation instruments in the shape of its P20 model. This had the same external design and shape as the "C" series instruments and the same panoramic 360° (horizontal) × 270° (vertical) angular coverage. However, it utilized waveform digitizing technology, and it provided a very large increase in its range measuring rate with a maximum value of 1,000,000 points per second. To achieve this speed, a very different rangefinder was utilized, based on a Class 1 laser emitting its radiation at the wavelengths (λ) of 808 nm/ 658 nm in the near infrared and red parts of the spectrum, respectively, instead of the green wavelength lasers that had been used in the previous ScanStation models. However, the P20 model had the more restricted maximum range of 120 m as compared with the C10 model. The P20 also featured a 5 Megapixel internal camera operating in a HDR image capture mode, which involved the scanner taking multiple images at each image position together with the matching scan data (Walsh 2015a, b). The initial P20 model combined these features along with high-accuracy angular measurements and

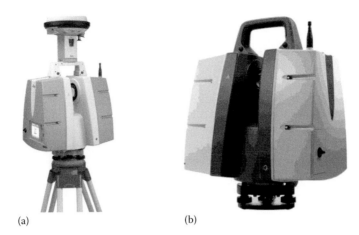

(a)   (b)

FIGURE 2.13 (a) A Leica Geosystems ScanStation P16 laser scanner with a GPS positioning receiver mounted on its carrying handle. (b) An example of a Leica Geosystems ScanStation P40 laser scanner. (Courtesy of Leica Geosystems, St. Gallen, Switzerland. With Permission.)

survey-grade tilt compensation. Since then, a number of other instruments have been added to the "P" series.

The first of these new instruments were the P15 and P16 models (Figure 2.13a) that were introduced in 2014 and 2015, respectively. They are broadly similar to the P20 model in terms of their construction, angular coverage (360° × 270°), and measuring performance (up to 1 million measured points per second). However, both models are designed specifically for very short range operations with a maximum range of 40 m. Within this context, it was noted earlier that Leica was no longer offering the phase-based HDS7000 short-range scanner following the introduction of these instruments. The main distinguishing feature between the P15 and P16 models is that their rangefinders operate at different wavelengths—using an 808 nm/658 nm combination in the case of the former instrument and a 1550 nm/658 nm combination in the case of the latter. Following on from these instruments, the P30 and P40 models (Figure 2.13b) with considerably increased maximum range measuring capabilities were introduced late in 2015. They both follow the P16 model in their use of the 1550 nm/658 nm combination of wavelengths in their rangefinders. The distinguishing feature between the two newer models is their maximum range capabilities—which is 120 m in the case of the P30 and 270 m in the case of the P40. In summary, the current range of the "P" series of ScanStation 3D laser scanners, comprising the P16, P30 and P40 models, all operate with an eye-safe laser having a 1550 nm/658 nm wavelength combination and all three models utilize a 4 Megapixel camera, besides which, they share very similar operating characteristics and performances. They are chiefly distinguished from one another by their respective maximum ranging capabilities (of 40, 120, and 270 m, respectively).

2.2.3.2 Trimble

Trimble is another of the major suppliers of surveying instrumentation that has entered the field of terrestrial or ground-based 3D laser scanning through the acquisition of a much smaller company that specialized in this area. It did so in September 2003 through its purchase of the Mensi company based in Fontenay-sous-Bois in France—which had been one of the pioneering developers of laser measurement technology that was used principally in metrology applications that were being carried out within industrial facilities. In 2001, the Mensi company had introduced a longer range 3D laser scanner in the form of its GS 100 instrument, based on pulse ranging. This was an early example of a hybrid type of scanner and used a rangefinder based on the use of a Class 2 laser operating at $\lambda = 532$ nm in the green part of the spectrum. A second model in the series, called the GS 200,

(a)   (b)

FIGURE 2.14   (a) The Trimble GX 3D laser scanner. (b) The Callidus CPW 8800 scanner (later the Trimble CX scanner) that utilized a combination of pulsed (TOF) and phase-based laser ranging technologies. (Courtesy of Trimble, Sunnyvale, CA. With Permission.)

followed soon after (Kersten et al., 2005), which allowed data to be captured at ranges up to 350 m. Following the takeover, the later models in the series were called the Trimble GX 3D laser scanners (Figure 2.14a). In March 2007, the production of the GX instruments was shifted from France to the former Spectra Precision factory in Danderyd, a suburb of Stockholm, the Swedish capital city.

In 2009, Trimble also acquired Callidus Precision Systems GmbH of Halle, Germany, which manufactured a short-range CP 3200 laser scanner based on pulse ranging and an unusual hybrid scanner called the CPW 8000 (Figure 2.14b) that combined the two main mensuration methods employed in laser scanners—those of TOF pulse ranging and CW phase measurement—in a single instrument. After the takeover, this latter instrument was sold as the CX laser scanner by Trimble.

In 2013, Trimble discontinued both the CX and GX scanners and introduced a completely new and very different scanner design in the shape of its TX8 model (Figure 2.15), which is manufactured at its Swedish factory. This scanner features TOF pulse measurement technology using a rangefinder equipped with a Class 1 laser having a wavelength of 1.5 μm. This provides a maximum

FIGURE 2.15   The Trimble TX8 laser scanner. (Courtesy of Trimble, Sunnyvale, CA. With Permission.)

standard range of 120 m on highly reflective targets and 100 m even on poorly reflective objects. However, this maximum distance can be extended to 340 m through the installation of an optional upgrade package. The maximum measuring rate of the rangefinder is 1 million points per second. The scanning arrangement comprises a conventional horizontal rotation of 360° in azimuth and a rotating mirror providing a vertical scan coverage of 317°. Like many of its competitors, the instrument also features dual axis compensation for angular measurements and has a color touchscreen display. An external camera needs to be fitted if high-resolution imaging is required.

2.2.3.3 Topcon

Another of the major manufacturers of surveying instrumentation, Topcon, entered the field of terrestrial 3D laser scanners early in 2008 with the introduction of its GLS-1000 model (Figure 2.16a). This was a hybrid-type instrument with a vertical angular coverage of 70°, besides the normal horizontal rotation in azimuth of 360°. The instrument's pulse-based rangefinder featured a Class 1 laser that had a maximum range of 330 m to objects having a high (90%) reflectance and 150 m to those objects having a low reflectance (18%). The maximum rate of measurement was 3000 points per second. The GLS1000 scanner also featured an integral camera producing 2 Megapixel frame images and an integrated control panel located on the side of the instrument. The instrument also featured a dual-axis compensator for the correction of the measured angles. The data measured by the scanner were recorded on a removable storage card. This instrument was soon upgraded with the introduction of the GLS-1500 model that had the same basic construction and a very similar specification to that of the GLS-1000 model but possessed the vastly improved measuring rate of 30,000 points per second.

In 2013, Topcon introduced its GLS-2000 model (Figure 2.16b). This has a maximum range of 350 m when used in its standard mode and 210 m when operated in its low-powered mode. These distances are achieved using a rangefinder that can be switched between the Class 3R laser utilized in the standard mode and a Class 1M laser with lower power that is used to alleviate eye safety concerns in populated areas. The maximum measuring rate is 120,000 points per second. Again, the instrument features a dual-axis tilt sensor for the correction of those angles measured while the instrument is slightly disleveled. As is now common, the GLS-2000 is also fitted with a color touch screen display for control purposes. The instrument is also equipped with dual 5 Megapixel cameras. One of the cameras is equipped with very wide angle (170°) lens that is used to obtain images while scanning at high speed, whereas the second camera is equipped with a narrow-angle (8.9°) telephoto lens that is mounted coaxial with the instrument's measuring axis.

(a)   (b)

FIGURE 2.16 (a) The Topcon GLS1000 terrestrial laser scanner and (b) the GLS2000 laser scanner. (Courtesy of Topcon, Tokyo, Japan. With Permission.)

2.2.3.4 Dr. Clauss Bild- and Datentechnik GmbH

This German company is best known for its manufacture of (1) panoramic line-scanner imagers and tilt-and-pan heads for use with these imaging devices and (2) of large turntables for industrial use. However, recently it has entered the field of 3D laser scanning. Its RODEON Scan product (Figure 2.17a) allows the simultaneous capture of laser scan and image data in a panoramic-type coverage mode with 360° horizontal × 330° (−75° to +90°) vertical angular coverage. The vertical scanning resolution is 0.125°, whereas that of the horizontal scan is a minimum of 0.015°. The range-finder employs a Class 1 laser that has a maximum range of 250 m. The instrument is also equipped with twin integrated flash bulbs that allow scanning and imaging to be undertaken without having to be dependent on ambient light. A second instrument is the RODEON Smartscan (Figure 2.17b) that comprises only the laser scanner. This utilizes a similar rangefinder to that used in the RODEON Scan providing a 250 m maximum range with a maximum scan rate of 14,000 measurements per second.

2.2.3.5 Stonex

This Italian company produces a wide range of surveying instruments such as levels, theodolites, total stations, and GPS receivers. Originally, it entered the laser scanning field by offering its X9 scanner, which was a rebadged Z+F 5010 short-range phase-based scanner. In 2013, it introduced a new laser scanner, the X300 instrument (Figure 2.18), which has been designed and manufactured in Italy and employs the pulse ranging technique. It is a hybrid-type scanner with its base rotating horizontally through 360° in azimuth and a mirror rotating vertically through 90° (from −25° to

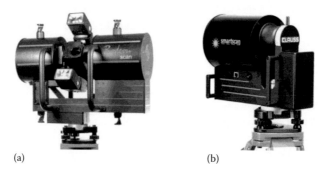

(a)   (b)

FIGURE 2.17 (a) The RODEON Scan instrument that combines both a high-resolution panoramic camera and a laser scanner and (b) the RODEON Smartscan laser scanner, both manufactured by the Dr. Clauss company. (Courtesy of Dr. Clauss. Zwoenitz, Germany. With Permission.)

FIGURE 2.18 Stonex X300 laser scanner. (Courtesy of Stonex, Lissone, Italy. With Permission.)

+65°). The rangefinder features a Class 1M laser emitting its pulses at the wavelength (λ) of 905 nm at a maximum speed of 40,000 pulses per second. The maximum range is 300 m with 100% (white) reflective targets, whereas the quoted range accuracy is ±6 mm at a distance of 50 m and <40 mm at 300 m. The X300 scanner is fitted internally with twin cameras located at the base of the instrument, with each camera generating a 5 Megapixel image. The instrument is also fitted with a dual-axis compensator for the correction of the measured angles if the instrument is slightly disleveled. In 2015, Stonex introduced its X300L model with a similar specification, except that the maximum range was reduced to 180 m with 95% reflectivity. As an option, the X300L instrument could be purchased from Stonex at a reduced cost on a pay per use basis. Both models are delivered with versions of the Stonex Reconstructor software package that has been developed in collaboration with the Gexcel software house. This particular package has been built on the basis of the 3D Reconstructor software that was produced originally by the Joint Research Centre of the European Commission for the processing of the measured 3D data obtained by laser scanners.

A rebadged version of the Stonex X300 laser scanner is offered by the GeoMax company, which is based in Switzerland and is part of the Hexagon organization, which also owns Leica Geosystems and Intergraph. This particular version of the instrument is called the SPS Zoom 300 laser scanner. It has the same technical specification as the Stonex X300 scanner, but it is supplied instead with the basic module of the GeoMax X-PAD MPS (Multi Positioning Software)—which can be supplemented by other software modules that can be purchased by the user from GeoMax as needed.

2.2.3.6 Smart Max Geosystems

This Chinese company, which is a subsidiary of the Hefei Simai Surveying Instrument Company, is another recent entrant into the manufacture and supply of terrestrial laser scanners. Two ranges of laser scanner—the VS series and the SC series—are on offer from this supplier. The VS series comprise three models—the VS100, VS200, and VS1000—all with a similar though not identical design, construction, and operation, and all operating as hybrid-type scanners with a coverage of 360° horizontal × 90° vertical angular coverage. All three models feature rangefinders based on the use of a Class 1 laser operating at λ = 905 nm. However, the power and the scan rates differ. The VS100 model has a maximum range of 380 m on targets having a 90% reflection and 100 m with targets having an 18% reflectivity and can perform measurements at a maximum rate of 24,000 points per second. With the VS200 model, the maximum range is extended to 500 m at 90% reflectivity and 200 m with targets having an 18% reflection; however, the maximum measuring rate is then reduced to 10,000 points per second. Finally with the VS1000 model (Figure 2.19a), the maximum

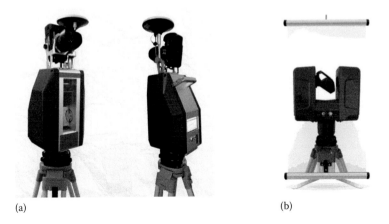

(a)   (b)

FIGURE 2.19 (a) The VS1000 3D laser scanner with a digital frame camera and a GPS receiver mounted on top of the instrument, as viewed from the front (at left) and the back (at right) of the instrument. (b) An example of the SC500 3D laser scanner. (Courtesy of Smart Max Geosystems, Hong Kong. With Permission.)

range is 1200 m with 90% reflection and 600 m at 18% reflectivity, but with the maximum measuring rate also increased to 36,000 points per second. The claimed accuracy of its angular measurements is ±5 s. The VS1000 model also features a touch screen display, external remote control from a tablet PC or smart phone via a wireless connection, built-in GPS positioning, and tilt compensation for measured angular values. The instrument can be fitted externally with a Canon 5D Mk II digital frame camera providing 21 Megapixel images.

The other models offered by the company are the SC70 and SC500. Similar to the VS model, they both utilize Class 1 lasers, but obviously each has a quite different power value to the other. Thus, the SC70 has a maximum range of 380 m with a 90% reflection but 70 m with 18% reflection from the target, whereas the maximum rate of measurement is quoted as 2000 points per second. By contrast, the SC500 (Figure 2.19b) is a much longer range model, the maximum ranges being quoted a 3000 and 500 m, respectively, at the 90% and 18% reflection values. The SC500 model can also measure at much higher speeds with a quoted maximum rate of 36,000 points per second. The SC scanners differ from the VS instruments in providing a panoramic scanning coverage of 360° × 300° (instead of a hybrid configuration) and incorporating an internal video camera to provide accompanying imagery.

2.2.3.7 Hi-Target

Hi-Target is another Chinese manufacturer and supplier of surveying instruments with a range of products that includes total stations, GNSS receivers and marine positioning systems, and echosounders. In 2015, the company decided to enter the 3D laser scanning market with its HDS450 hybrid-type laser scanner (Figure 2.20a). The instrument's rangefinder uses a Class 1 laser operating at the wavelength $\lambda = 1545$ nm with a maximum range of 450 m to targets having a 90% reflectivity and 300 m with those having a 60% reflectivity. The maximum data acquisition rate of the HS450 is 300,000 points per second, while the stated accuracy is ±1 cm at a range of 100 m. The vertical FOV of this hybrid-type instrument is 100°, covering the range between −40° and +60° relative to the horizontal plane, while the maximum scanning speed is 150 lines per second. The instrument also features a bi-axial compensator and can be fitted with an external camera (Figure 2.20b).

(a)   (b)

FIGURE 2.20   (a) The Hi-Target HS450 laser scanner, as seen from the rear of the instrument. (b) A Hi-Target HS450 scanner, as seen from the front of the instrument and fitted with an external camera. (Courtesy of Hi-Target, Guangzhou, China. With Permission.)

2.2.4   LONG-RANGE LASER SCANNERS USING PULSE RANGING

Within this group of laser scanners, the maximum measuring range to highly reflective targets has been raised to more than 500 m; indeed, in some cases, it can be very substantially more—up to several kilometers. All of the instruments use pulse-ranging based on the TOF measuring principle. It will be noted that two of the principal suppliers of these long-range scanners—Teledyne Optech and *RIEGL*—are also the manufacturers of airborne laser scanners that measure over similar distances in terms of the flying heights at which they are operated. Within this context, it should be noted that the pulse travel time for a single pulse is about 3.3 μs for the 500 m range, so the pulse rate is limited to 300 kHz, unless multipulse technology is used.

2.2.4.1   Teledyne Optech

The Optech company has been based in the Toronto area of Canada for 40 years. In 2015, it was acquired by Teledyne Technologies, a large American aerospace and defense electronics corporation based in California. Optech's original activities encompassed laser ranging and scanning devices operating in many different environments—spaceborne, airborne, and terrestrial. Within this context, the intelligent laser range imaging system (ILRIS) was developed originally by Optech for the Canadian Space Agency in the late 1990s, as a combined ranging and imaging device for use onboard spacecraft. In its original form, as the ILRIS-100, it used pulse ranging, operating in both the visible (green) and infrared parts of the spectrum to scan a scene and produce simultaneous range, angular and intensity data at ranges up to 500 m. When processed, these data were combined to form a range and intensity dataset that could be produced either in gray scale or color-coded form based on optical images that were 2 k × 2 k pixels (4 MB) in size.

In June 2000, Optech introduced its tripod-mounted ILRIS-3D version of the instrument that it had developed specifically for TLS (Figure 2.21a). It was intended to be used on topographic and open-cast mining applications on the one hand, and for industrial applications, especially the measurement and modeling of industrial plants and facilities, on the other. Thus, the instrument was designed and constructed with long-range capabilities from the outset. Using a Class 1 laser rangefinder emitting its infrared radiation at λ = 1550 nm, it had a maximum range of 800 m with a range resolution of 1 cm, even with a target having only 20% reflectance and a range of 350 m with a very low-reflectance (4%) target such as a coal stockpile. A measuring rate of 2000 points

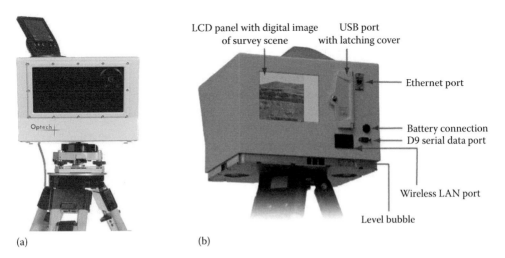

(a)   (b)

FIGURE 2.21   (a) An early model of the Optech ILRIS-3D laser scanner, as seen from the front. (b) The ILRIS-3D laser scanner as seen from the rear. (Courtesy of Teledyne Optech, Vaughan, ON. With Permission.)

per second could be utilized while scanning over a 40° × 40° FOV using two internal deflection mirrors, producing a camera-type configuration. The main case of the ILRIS-3D instrument
with its distinctive box shape also contained a bore-sighted small-format digital camera giving
a 640 × 480 pixel image. This camera also included an LCD viewfinder fitted to the back of the
instrument that could be used both for set-up purposes and for the display of the captured data
(Figure 2.21b). Options included a differential GPS unit and an attitude measurement system.

In 2004, a substantially upgraded model of the ILRIS-3D instrument was introduced. This
provided an increased range (beyond 1000 m to highly reflective targets), an improved accuracy, an integrated Complementary Metal-Oxide-Semiconductor (CMOS)-based camera giving
a 6 Megapixel image, and an integrated handle for carrying purposes. As noted previously, the
ILRIS-3D was a camera-type instrument with a fixed FOV of 40° × 40°. To cover a much larger
area, Optech also introduced the ILRIS-36D version of the instrument, the first examples of which
were shipped in May 2005. This instrument was equipped with a motorized pan-and-tilt base that
allowed the scanner to cover a 360° × 230° FOV (Figure 2.22a). For this to be implemented, the
motorized base unit moved the scanner unit of the ILRIS-3D with its 40° × 40° FOV in a series of
steps that are measured by angular encoders. Each 40° × 40° scan patch or window overlapped on
its neighbors by 5°. However, a further development in the form of a new profiling feature eliminated the need for these overlaps using multiple scan windows. Another option that was also made
available was that of motion compensation (MC), which was designed to allow ILRS-3D scanners
to be used from a moving or dynamic platform such as a boat or a vehicle. This was based on the
use of an additional integrated GPS/IMU subsystem to provide the positional and orientation data
required for the MC.

The basic ILRIS-3D instrument with its distinctive box-style design continues to be developed.
Back in October 2006, a new optional feature was introduced in the form of the enhanced range (ER)
option. This increased the range of the instrument still further by 40% for use in open-cast mining
and large-area topographic surveys. In recent years, Optech has also introduced a high-density (HD)
version of the ILRIS with a smaller step size and a much higher maximum pulse repetition rate
of 10000 Hz, allowing that number of measured points per second. Another comparatively recent
development has been to introduce a LR version of the instrument (Figure 2.22b) that employs a

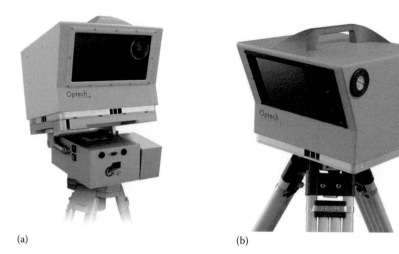

(a)   (b)

FIGURE 2.22   (a) An example of the Optech ILRIS-36D laser scanner with its motorized pan and tilt base.
(b) The Optech ILRIS-LR laser scanner. (Courtesy of Teledyne Optech, Vaughan, ON.)

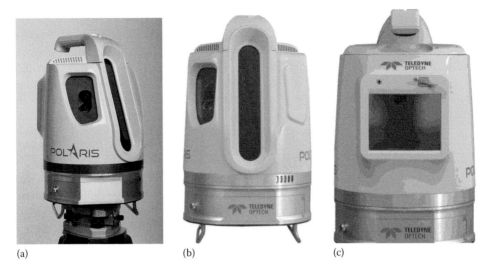

(a)   (b)   (c)

FIGURE 2.23 The new Polaris TLS as seen from (a) the side of the instrument with the camera window at left, (b) the front of the instrument with the camera window at left and the scanner window in the middle, and (c) the rear of the instrument showing the touchscreen color LCD display. (Courtesy of Teledyne Optech, Vaughan, ON. With Permission.)

rangefinder with a higher powered Class 3R laser having a wavelength (λ) of 1064 nm instead of the $\lambda = 1535$ nm wavelength that is being used in the current HD and ER models. As a result, the LR model can achieve a maximum range of 3000 m with targets having an 80% reflectivity compared with the values of 1250 m (HD) and 1800 m (ER), respectively, for the maximum ranges with 80% reflectivity that can be measured by the other two current models. With a 10% reflectivity, the respective maximum ranges are 400 m (HD), 650 m (ER), and 1330 m (LR). The 36D and MC options can still be applied to all three of these current models.

For the processing of the measured ILRIS-3D data, Optech has utilized the Quick Terrain Modeler from the Applied Imagery company based in Maryland, and modules from the Polyworks software package developed by another Canadian company, Innov-Metric Software from Quebec. However currently Optech is offering its own ILRIS Scan software, which is an all-in-one scanning, viewing, and processing software package with various options that it has developed in cooperation with the Gexcel company from Italy.

In October 2016, Optech introduced a completely new line of laser scanners called the Polaris TLS (Terrestrial Laser Scanner) (Figure 2.23) that will start to be delivered to customers early in 2017. These scanners will feature a panoramic type of operation covering a full 360° in azimuth and providing 120° angular coverage in the vertical plane instead of the camera-type operation of the ILRIS series with its restricted angular coverage. There will be three different models in the Polaris series—named TLS-250, TLS-750, and TLS-1600,respectively. The rangefinder that is used in each of these models uses a Class 1M laser emitting its pulses at $\lambda = 1535$ nm with a range resolution of 1 mm. The TLS-250 model will have a maximum range of 250 m to targets with a 90% reflectivity and has a peak pulse repetition rate of 500 kHz at short distances. The TLS-750 model has a similar design, but it also features two programmable data collection rates (200 and 500 kHz) and has a maximum range capability that has been increased to 750 m at the lower of these two rates, together with options such as external cameras and a GPS receiver. The third model, the TLS-1600, has a choice of three possible collection rates—50, 200, and 500 kHz with a maximum range of 1600 m using the lowest of these three rates.

2.2.4.2   *RIEGL*

This company, based in Horn, Austria, began its activities in this particular field by building a number of long-range hybrid-type 3D laser scanners based on pulse ranging, which were entitled the LMS-Zxxx (laser measurement systems) series (Figure 2.24a). This comprised four different models—the LMS-Z210ii, LMS-Z390i, LMS-Z420i, and LMS-Z620—having maximum measuring ranges of 400, 650, 1000, and 2000 m, respectively, with objects having a reflectance of 80%. They all had similar design characteristics (Figure 2.24b) and utilized the same type of upward-pointing laser engine with a continuously rotating optical polygon placed in front of the laser range-finder to produce the basic optical scan within the vertical plane. The rangefinders used a Class 1 laser operating in the near infrared part of the spectrum, either at λ = 905 or λ = 1550 nm. The maximum measuring rates that could be achieved using pulse ranging with these LMS-Zxxx models lay between 8000 and 12,000 points per second, although lower rates were often used in actual practice. The four LMS-Zxxx models could all be fitted optionally with a mount, with two vertical posts that provided an additional horizontal rotation axis that allows the scanner engine as a whole to be inclined at an oblique angle to the vertical axis (Figure 2.24a). The various models could also be equipped optionally with a small-format charge-coupled device (CCD) digital frame camera that sat on a mount on top of the main scanner engine (Figure 2.24a and b).

Starting in 2008/2009 with the VZ-400 model, over the years since then, *RIEGL* has introduced a new series of terrestrial 3D laser scanners, called the VZ-xxxx series, which forms part of its V-Line range of scanner products (Figure 2.25). As with the LMS-Zxxx series, the individual models in the VZ-xxxx series—the VZ-400, VZ-1000, VZ-2000, VZ-4000,and VZ-6000— are all of a similar, though not identical, design, and featuring ever greater ranges as the model numbers increase. The lasers that have been used as the basis for the rangefinders used in these scanners all fall into the Class 1 category, except for the longest range VZ-6000 instrument,

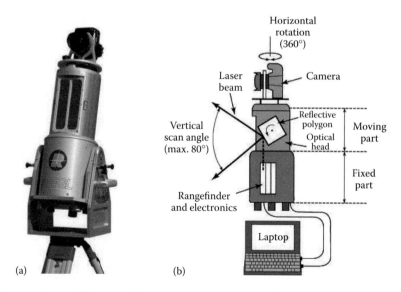

(a)   (b)

FIGURE 2.24   (a) This *RIEGL* LMS-Z390i laser scanner is shown with an additional mount having a horizontal axis around which the main part of the instrument can be rotated and operated at an oblique angle to the vertical—if that is required. This particular example is also fitted with a calibrated small-format digital frame camera producing colored images. (b) Diagram showing the general layout of the *RIEGL* LMS-Zxxx series of terrestrial laser scanners. (Courtesy of *RIEGL*, Horn, Austria; redrawn by M. Shand. With Permission.)

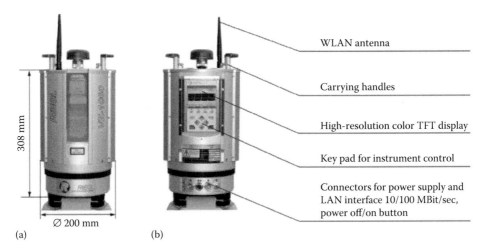

308 mm

Ø 200 mm

(a)   (b)

WLAN antenna

Carrying handles

High-resolution color TFT display

Key pad for instrument control

Connectors for power supply and
LAN interface 10/100 MBit/sec,
power off/on button

FIGURE 2.25 Showing the main external components of the *RIEGL* VZ-1000 laser scanner as seen from the front (a) and the rear (b) of the instrument.

which utilizes a Class 3R laser. The respective maximum ranges (achieved on targets having a 90% reflectivity) are 600 m (VZ-400), 1400 m (VZ-1000), 2050 m (VZ-2000), 4000 m (VZ-4000), and 6000 m (VZ-6000). However, these ranges will fall to roughly half of these values when targets having a reflectivity of 20% are encountered. With regard to the maximum laser pulse repetition rates, these vary from 100 kHz in the earliest (VZ-400) model to 300 kHz in the case of the later (VZ-1000, VZ-4000, and VZ-6000) models to 1 MHz in the case of the VZ-2000 model that utilizes *RIEGL*'s multiple pulse technology that is also employed in its airborne laser scanners. The V-line of scanners also makes use of echo digitization and online waveform analysis during their range measurements, the whole process being carried out in a real time within the scanner hardware. The quoted accuracies of the range measurements vary with distance from ±5 mm (with the VZ-400), to ±8 mm (with the VZ-1000 and VZ-2000), and to ±15 mm (with the VZ-4000 and VZ-6000).

Since the VZ scanners all fall into the hybrid classification, the scan angle range in the vertical plane is restricted to 100° (from +60° to −40°) with the shorter range VZ-400, VZ-1000, and VZ-2000 models and 60° (from +30° to −30°) in the longer range VZ-4000 and VZ-6000 models. The actual scan speed can be varied from 3 lines per second to 120 lines per second for the two earliest models, and to 240 lines per second for the later models. The resolution of angular measurement in the VZ-xxxx scanners is quoted as being ±1.8 arc-seconds. It should be noted that all the VZ-xxxx models are also fitted with an integrated GPS receiver with an L1 antenna and have integrated inclination sensors for use with the angular measurements being made during a vertical set-up of the scanner. With regard to imaging capabilities, both of the longest range models (VZ-4000 and VZ-6000) have built-in digital frame cameras producing 5 Megapixel images, whereas the other shorter range VZ scanner models all have a mount with fittings on the top of the instrument to accommodate a suitable camera externally (Figure 2.26a).

Besides the LMS-Zxxx and VZ-xxxx series of 3D laser scanners, over the last decade, *RIEGL* has also manufactured a series of long-range laser profile measuring (LPM) systems. In 2008, it introduced the latest model in the series, which is called the LPM-321 profiler. This has a very different mechanical and optical design to those of the LMS-Zxxx and VZ-xxxx laser scanners, with the LPM-321 laser rangefinder unit mounted on the horizontal (trunnion) axis

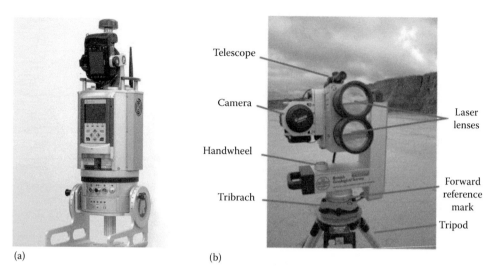

(a)   (b)

FIGURE 2.26   (a) A *RIEGL* VZ-1000 laser scanner with a digital frame camera and a GPS receiver mounted on top of the main body of the instrument—which is itself mounted on a tilt mechanism that allows the whole system to be tilted at an oblique angle when required. (Courtesy of *RIEGL*, Horn, Austria. With Permission.) (b) Showing the main features of a *RIEGL* LPM-321 profiler that is being operated by the British Geological Survey (BGS) on a glaciological survey. (Reproduced with permission of the British Geological Survey © NERC 2017.)

of the instrument, supported by a single vertical pillar (Figure 2.26b). The rangefinder of the LPM-321 can measure ranges up to a maximum distance of 6000 m to targets with 80% reflectivity (without the use of a reflector) and can be operated to measure profiles either manually or in an automated mode. It also offers a full waveform digitizing capability in the same manner as in *RIEGL*'s airborne laser scanners and in the VZ-xxxx series of terrestrial laser scanners described previously. In its automated mode of operation, the LPM-321 can measure up to 1000 points per second. The instrument has a sighting telescope with up to 20× magnification that sits on top of the rangefinder and can have a calibrated digital camera fitted to the side of the rangefinder (Figure 2.26b).

2.2.4.3   Maptek

The i-Site company was founded as a subsidiary of the Maptek company based in Adelaide in South Australia. Maptek is the developer of the Vulcan 3D geological modeling and mine-planning software that is used extensively in the mining industry world-wide. The i-Site company was formed specifically to support and develop Maptek's early interest and activity in laser scanning. The Maptek Studio software for use with ground-based laser scanned data was an early product from the company, often sold to users together with *RIEGL* terrestrial 3D laser scanners as a bundled package.

However, after two or three years of development, in 2004, Maptek brought out its own Model 4400 3D laser scanner (Figure 2.27a). This was a hybrid-type scanner having a motorized rotation of 360° in azimuth and providing 80° angular coverage in the vertical plane. The stated angular accuracy was ±0.04°. The instrument's rangefinder was based on a Class 3R laser that emitted its pulses at λ = 905 nm with a power of 10 mW and measured its ranges at a maximum rate of 4400 points per second. The maximum range with highly reflective surfaces was 700 m.

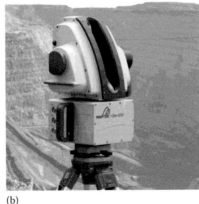

(a)   (b)

FIGURE 2.27 (a) An example of the earlier Maptek model 4400 laser scanner. (b) An example of the current Maptek model 8200 laser scanner. (Courtesy of Maptek, Adelaide, South Australia. With Permission.)

The claimed accuracy in range was ±5 cm at a range of 100 m. Furthermore, the Maptek 4400 instrument also incorporated a digital panoramic line scanner equipped with a Nikon $f = 20$ mm lens that produced a 37 Megapixel linescan image that was acquired concurrently during the laser scanning/ranging operation. The instrument also incorporated a viewing telescope with 14× magnification that could be used for alignment and back-sighting operations. Slightly different versions of the instrument were developed for specific applications: (1) In 2006, the 4400LR model was introduced, designed for use over the longer ranges encountered in topographic and open-cast mining applications and (2) in 2007, the 4400CR model came on the market for use in the close ranges encountered in police and forensic applications and in underground mining.

Starting in 2010, Maptek has introduced its 8000 series of scanners to replace the earlier 4400 series. The 8800 model appeared in 2010, the 8400 in 2011, the 8810 in 2012, the 8200 in 2013, and the 8820 in 2014 (Figure 2.27b). The two initial models were designed specifically for use over the longer ranges encountered in topographic and geological mapping and large open-pit surveys in the case of the 8800 model, and for shorter ranges for operations in indoor sheds and silos, stockpile surveys and underground mining in the case of the 8400 model. Thus, the 8800 model could measure a maximum range of 1400 m with an accuracy of ±1 cm to objects with 80% reflectivity, whereas the 8400 could measure a maximum range of 700 m with an accuracy of ±2 cm to objects having the same reflectivity. Both models used a rangefinder based on the use of a Class 1 erbium laser that emitted its pulses at the wavelength of $\lambda = 1545$ nm and had a maximum data acquisition rate of 8800 points per second. Both models are hybrid-type scanners with a 360° coverage in azimuth and an 80° angular coverage in the vertical plane. However, the 8800 model had a digital imager similar to that of the earlier 4400 model, and it also possessed a back-sight telescope for surveying operations, whereas these components were not fitted to the 8400 model. The later 8810 and 8820 models have a longer range, significantly improved acquisition speeds (with a maximum rate of 80 kHz in the case of the 8820 model), and an improved range accuracy over the initial 8800 model. Both models also incorporate an internal GPS receiver and compass. The 8200 model is a lighter weight and more portable model within the 8000 series for use in the surveys of tunnels, cavities, and slopes underground as well as silos and small stockpiles during surface mining operations. For these operations, it has a minimum range of 1 m and an increased angular range of 125° (vertical) × 360° (horizontal).

In 2009, Leica Geosystems announced that a rebadged version of the Maptek 4400 model and its accompanying software would be offered to its customers as its HDS4400 model. When the Maptek 8800 and 8400 models were introduced during the 2010/11 period, Leica also sold these as its HDS8800 and HDS8400 models and then the HDS8810 model later after its introduction by Maptek in 2012 (Petrie, 2012). However, at the present time (in 2016), these various models no longer appear on the Leica Geosystems Website and catalog.

2.2.5 Total Stations with Scanning Capabilities

Total stations are in widespread use in land and engineering surveying practice. Basically, they comprise an electronic theodolite (measuring horizontal and vertical angles) that is integrated with a laser rangefinder (measuring distances) to establish the positions and elevations of specific points in the terrain. These instruments are usually operated manually by surveyors and engineers; most total stations are nonrobotic in their operation. However, some manufacturers have developed total stations further to incorporate a limited robotic capability that allows the instrument to undertake 3D laser scanning, albeit with a much reduced performance as compared with the purpose-designed and built laser scanners that have been described thus far.

2.2.5.1 Renishaw

Renishaw, which is a specialist manufacturer of metrology equipment in the United Kingdom, acquired Measurement Devices Ltd. (MDL) in 2013. The MDL company, which was based in Aberdeen and York within the United Kingdom, had been engaged in the manufacture of laser-based measuring systems since its foundation in 1983. One of its products was the Quarryman instrument that has been used extensively for many years in quarries and open-cast mines world-wide to measure profiles across rock faces employing laser ranging techniques. The production of this instrument has been continued by Renishaw. The Quarryman employs a rangefinder that is mounted on the instrument's horizontal (trunnion) axis supported by two pillars or standards in the manner of a theodolite telescope. The rangefinder can utilize one of two semiconductor lasers. The first is a Class 2M InGaAs laser diode emitting its pulses at $\lambda = 905$ nm in the near infrared part of the spectrum; the other is a more powerful Class 3R InGaAs laser diode, also emitting its pulses at $\lambda = 905$ nm. Thus, the Quarryman is available in either of two forms—(1) the Quarryman Pro with the less powerful laser and the (2) Quarryman Pro LR (=Long Range), in which the more power-ful of the two lasers is employed. The Pro version can measure a maximum range of 750 m to a moderately reflective target, whereas the Pro LR has a maximum range of 1200 m. The instrument can be operated manually as a total station using a telescope that is mounted on top of the range-finder. Alternatively, it can be operated in an automated (robotic) mode as a 3D laser scanner using its built-in stepper motors, in which case, it can measure objects at a rate of 250 points per second. The corresponding angles are measured using horizontal and vertical angular encoders. The full angular coverage provided by the instrument is 360° (horizontal) × 135° (from −45° to +90° within the vertical plane). However, specific areas can be measured through the prior definition of a rect-angle or polygon by the operator using the instrument's built-in numeric keyboard. The measured ranges and angular data are stored on a standard interchangeable flash card or transferred via a USB port to another suitable storage device. The instruments can use the company's Quarryman Viewer software for the processing of the measured data to form a point cloud and, from this, the data can be used to generate contours and sections and to compute volumes using various external software packages.

2.2.5.2 Trimble

In June 2007, Trimble introduced its VX Spatial Station (Anon, 2008). This instrument was constructed and operated on an entirely different concept to that of the company's CX, FX, GX, and TX series of scanners described earlier. Basically, it was operated as an automated (robotic) total station that also had an imaging and scanning capability. The instrument was equipped with a conventional surveyor's telescope rather than the scanning mirror that is used in a dedicated laser scanner. The total station capability of the instrument that was used for positioning and traversing provided a ± 1 s angular accuracy and a ± 3 mm distance accuracy with a maximum range of between 300 and 800 m without a prism and 3 km with a single prism. The 3D laser scanning operation was carried out using a rangefinder equipped with a Class 1 pulsed laser diode emitting its radiation at $\lambda = 905$ nm and measuring at a rate of 5–15 points per second with a scanning range of up to 150 m. A tiny onboard digital frame camera was fitted to the underside of the telescope with its optical axis parallel to that of the telescope. This calibrated camera produced HDTV images that were 2048×1536 pixels in size and could be recorded in JPEG format at a maximum rate of 5 FPS on the instrument's removable data collector unit. Obviously, the instrument was basically a total station that was designed to provide surveyors with 3D laser scanning and imaging capabilities over limited areas or specific objects such as buildings, rather than carrying out the systematic scanning of large areas or objects that the Trimble CX, FX, and GX series were designed specifically to perform.

Starting in 2010, Trimble has continued to offer an upgraded version of the VX instrument as part of what is now marketed as the S-series of total stations (Frazier, 2012). The S7 (Figure 2.28a) and S9 models in this series are basically similar in specification, construction, and operation to the VX Spatial Station; these two newer models differing mainly in their respective angular accuracies—± 1 s in the case of the S7 model; ± 0.5 s in the case of the S9 instrument. However, the maximum range that can be achieved with both models while scanning has been extended to 250 m, though the maximum scanning speed remained at 15 points per second. The frame size of the imaging camera in both instruments remains as in the earlier VX instrument, as does the maximum speed of imaging data capture.

However, in October 2016, Trimble introduced a completely new model in the series, called the SX10 Scanning Total Station (Figure 2.28b) that is manufactured in its Swedish factory. This instrument represents a large jump in performance as compared with the previous S-series models just described. In particular, the SX10 instrument carries out its measurements at ranges up

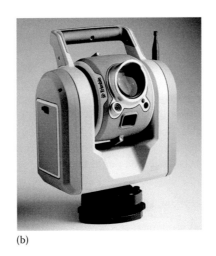

(a)   (b)   (c)

FIGURE 2.28 (a) The Trimble S7 total station, (b) the Trimble SX10 scanning total station (Courtesy of Trimble, Sunnyvale, CA. With Permission.), and (c) the Topcon IS-3 total station. (Courtesy of Topcon, Tokyo, Japan. With Permission.)

to a maximum of 600 meters at rates up to 26,000 points per second over a 360° by 300° FOV. The instrument can execute its scans with a spot size of 14 mm at a range of 100 m and produces each of its frame images with a 5 Megapixel size over the whole of its panoramic coverage.

2.2.5.3   Topcon

In 2007, Topcon also introduced a total station with a LR imaging and scanning capability in the shape of its IS Imaging Station (Billings, 2008). (N.B. The instrument was also sold as the Topcon GPT-9000-IS Imaging Station.) Similar to Trimble's VX instrument from the same period, it combined the angular measuring and ranging capabilities of a total station with the imaging produced by an integrated camera system, allied to a laser scanning capability. In fact, two cameras were fitted. These were equipped with wide-angle and 30× optical zoom lenses, respectively, each camera acquiring images that were 1.3 Megapixels in size at a rate of 10 frames per second. The former camera was mounted above the objective lens of the instrument's telescope, whereas the latter operated coaxially within the telescope. The IS instrument had a maximum range of 3000 m using a single prism and an angular measuring accuracy of ±1 s. The maximum scanning speed was 20 points per second, whereas the maximum scanning range was 250 m. The IS was also fitted internally with a radio that allowed wireless monitoring of the operation of the instrument from an external location over limited distances. The IS-2 model followed with only minor changes. In 2012, Topcon introduced a further upgraded version of the instrument called the IS-3 (Figure 2.28c) with a much extended WiFi range, improvements to its robotic tracking, and the introduction of user-defined scanning boundaries (Billings, 2012).

2.2.5.4   Leica Geosystems

In 2013, Leica Geosystems also entered this particular area of total stations with an imaging and 3D laser scanning capability with the introduction of its Nova MS50 MultiStation (Figure 2.29a) (Grimm, 2013; Grimm and Zogg, 2013). Similar to the products from the other system suppliers in this area, this instrument combined total station functionality with digital image acquisition and 3D laser scanning. The MS50 instrument uses a Class 3R laser as the basis of its rangefinder and employs Leica's Wave Form Digitizer technology that combines both the TOF and phase shift measurement methodologies (Maar and Zogg, 2014). The MS50 provides a ±1 s angular accuracy and a ±2 mm distance accuracy in combination with a maximum range of 2000 m without a prism and 10 km with a single prism. The maximum speed of measurement during scanning operations is 1000 points per second at ranges up to 300 m. However, if the scan speed is lowered, then the maximum range can be increased to 400 m at 250 Hz, 500 m at 62 Hz, and 1000 m at 1 Hz. The MS50 instrument also features two digital cameras, each producing 5 Megapixel images—the first being an overview camera mounted on top of the instrument's main telescope with its optical axis parallel to it, whereas the second is mounted coaxially along the optical axis of the main telescope and provides a 30× magnified image. Both cameras can operate at a maximum rate of imaging of 20 Hz. A novel feature of the MS50 is that it also provides GNSS connectivity. In 2015, Leica introduced its Nova MS60 MultiStation (Figure 2.29b), in which the main change has been the incorporation of the Leica Captivate 3D field software that generates 3D models and allows them to be displayed on the instrument's improved built-in display screen.

2.2.6   SUMMARY—STATIC TERRESTRIAL LASER SCANNERS

The field of static terrestrial 3D laser scanners has burgeoned since the first edition of this book was published. In particular, there have been enormous developments and advances in the design, construction, and performance of these static 3D laser scanners that have led to their widespread adoption and application by the land and engineering surveying professions. These developments have seen measuring rates improve by more than 50 times in some cases, while the maximum distances that can be measured have increased greatly in all three range classes—in the case of the long-range models to several kilometers. The instruments have also been fitted with advanced electronics,

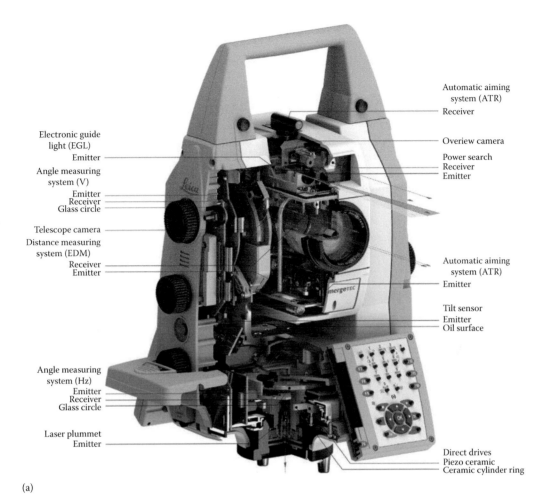

Automatic aiming
system (ATR)
Receiver

Electronic guide
light (EGL)
Emitter

Overiew camera

Power search
Receiver
Emitter

Angle measuring
system (V)
Emitter
Receiver
Glass circle

Telescope camera
Distance measuring
system (EDM)
Receiver
Emitter

Automatic aiming
system (ATR)
Emitter

Tilt sensor
Emitter
Oil surface

Angle measuring
system (Hz)
Emitter
Receiver
Glass circle

Laser plummet
Emitter

Direct drives
Piezo ceramic
Ceramic cylinder ring

(a)

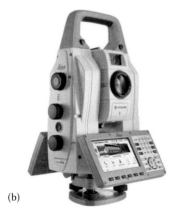

(b)

FIGURE 2.29 (a) A cut-away diagram showing the internal design and construction of the Leica Geosystems Nova MS50 MultiStation total station with its laser scanning and imaging capabilities. (b) An example of the Leica MS60 MultiStation. (Courtesy of Leica Geosystems, St. Gallen, Switzerland. With Permission.)

considerable computing power, and firmware that make their operational use much easier, yet much more sophisticated than before. It is worth noting too the growth in the number of suppliers of total stations that now have a built-in 3D laser scanning capability.

2.3 DYNAMIC TERRESTRIAL LASER SCANNERS

The subject of vehicle-borne terrestrial laser scanners is an exciting one to explore. It forms only a part of a much wider subject area—namely that of conducting surveying and mapping operations from dynamic vehicular platforms. To begin with, this wider subject area of MMSs was concerned mainly with the imagery that had been acquired from these platforms using multiple video and digital cameras in combination with the data acquired concurrently for direct georeferencing purposes by integrated GPS/IMU units. This was then followed by the photogrammetric evaluation of the imagery on a highly automated basis to deal with the enormous number of small-format images that are collected during these operations. Initially, this activity was a pioneering research effort that was carried out in the late 1980s by the Center for Mapping of the Ohio State University using its GPSVan (Bossler and Toth, 1996; Toth and Grejner-Brzezinska, 2003, 2004). It was joined somewhat later by an independent but parallel project at the University of Calgary using its similar VISAT van. Other similar projects were undertaken by various European universities and institutions such as the EPFL, Lausanne (with its Photobus), and the Institut Cartogràfic de Catalunya, Barcelona (with its GeoVan). These various research efforts resulted in the establishment of a number of commercial companies carrying out mapping from vehicles using purely camera-based technology. These systems were concerned principally with the collection of geospatial data for road inventory purposes. There are still (in 2016) quite a number of companies that carry out their survey and mapping operations from vehicles using solely video or digital imagery and subsequent photogrammetric processing.

The development and use of laser scanners on vehicular platforms, either instead of or in conjunction with cameras, have taken place somewhat later. An early research effort was the vehicle-borne laser mapping system of the Centre for Spatial Information Science of the University of Tokyo (Inaba et al., 1999, Manandhar and Shibasaki, 2001, 2003; Zhao and Shibasaki, 2003, 2005). Shortly afterwards, the camera-based GeoVan of the Institut Cartogràfic de Catalunya, Barcelona, was transformed into the Institute's Geomobil project (Talaya et al., 2004 a, b) that featured both multiple cameras and laser scanners for data collection purposes. A year or two later came the introduction of vehicular-based systems using laser scanners that could operate on a commercial basis. Now (in 2016) vehicular-based MMSs featuring laser scanners and profilers are available for purchase from a wide range of system suppliers. These are used routinely by numerous service providers who provide a commonly available service—at least in the more advanced industrialized countries.

In the introduction to this section, the use of laser scanners and profilers on vehicles for nontopographic purposes should also be mentioned—in particular, their use in vehicle navigation and collision avoidance systems. If these come to be mass produced and used widely, then undoubtedly this will have a strong influence, especially with regard to the costs of those laser scanners and profilers that can be mounted on vehicles (Toth, 2009). A further mention should also be made of the influence of the Defense Advanced Research Projects Agency (DARPA) Grand Challenge (DGC) and DARPA Urban Challenge (DUC) contests that hastened the development of these navigation and collision avoidance technologies for use in unmanned vehicles. There was a striking difference between the imaging sensor configurations that were used in the first DGC held in 2004 and those used in the later DUC held in 2007. In 2004, digital cameras represented the primary sensors to observe the object space around the vehicles, and stereo-photogrammetric techniques were widely used for object space reconstruction. Laser profilers, predominantly SICK models, were only used to supplement the image sensing capabilities. The situation was reversed by the 2005 DGC, when the winner, the Stanley vehicle from Stanford University, used five SICK roof-mounted laser profilers to map the area in the front of the vehicle, and cameras were only used at high speed to look ahead for objects that were further away than 40 m.

Recognizing the potential of laser scanning, various manufacturers developed dedicated laser scanners for the 2007 DUC. The Ibeo system provided laser profiling capabilities in four planes, thus replacing four of the SICK units. Then, more importantly, the introduction of the Velodyne laser scanner, which was used by the winner, the Boss, a fully autonomous Chevy Tahoe vehicle from Carnegie Mellon University, and by most of the top ten participants, represented a major technological breakthrough in 2007 (Ozguner et al., 2007). Since then, Velodyne's technology has been adopted quite widely in MMSs, along with other purpose-built laser scanners and profilers—as will be discussed in the next section.

A few of these mobile systems rely wholly on the use of laser scanners to collect the relevant data for mapping purposes. This is referred to as mobile laser scanning in much the same way as TLS and ALS. However, most MMSs combine and integrate the operation of their laser scanners and profilers with that of digital frame cameras in much the same way as is done with airborne mapping systems that employ the two technologies in combination. Indeed, the similarities run even closer between the two in which both of these systems require the further integration of a GPS or GNSS/IMU position and orientation subsystem, since, without accurate georeferencing, no point cloud of accurately positioned elevations can be generated from the measured laser scan data produced by the system, regardless of whether it is mounted on a land vehicle or an airborne platform.

2.3.1 3D LASER SCANNERS FOR MOBILE MAPPING SYSTEMS

The 3D terrestrial laser scanners that have been described in Section 2.2 of this chapter have played only a small part in the development of laser-based MMSs. Indeed, the 3D laser scanners that are used for surveying work can only be used for static measurements when operated in their native 3D operational mode that involves a 360° rotation in azimuth. Of course, these instruments can be mounted on a vehicle and transported from one position to another in a so-called stop-and-go mode of operation. However, in such a mode, the actual scanning operation still needs to be carried out at a single fixed location with the 360° azimuthal rotation in operation, which often takes several minutes. However, a very small number of terrestrial 3D laser scanners from Faro (Figure 2.30), Z+F, and Leica Geosystems (Figure 2.31) have been operated while in motion on mobile mapping vehicles, but with their horizontal (azimuth) angular movements disabled or locked—which effectively makes them into 2D laser profilers.

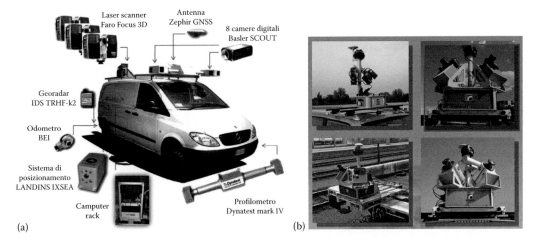

FIGURE 2.30 (a) Diagram showing the components of the Siteco Road Scanner mobile mapping system; note the use of Faro Focus 3D terrestrial laser scanners. (b) Photos of the three Faro Focus 3D laser scanners being deployed and operated as 2D profilers on the Siteco Road Scanner system. (Courtesy of Siteco, Bologna, Italy. With Permission.)

FIGURE 2.31   (a) An annotated photo showing the principal components of the ABA mobile mapping system; note the use of three Leica HDS6000 phase-based 3D terrestrial laser scanners as 2D profilers. (Courtesy of ABA Surveying, Knaphill, UK. With Permission.) (b) A Leica Geosystems Pegasus One mobile mapping system featuring an HDS7000 phase-based terrestrial 3D laser scanner, again being utilized as a 2D laser profiler. (Courtesy of Leica Geosystems, St. Gallen, Switzerland. With Permission.)

2.3.2 2D LASER PROFILERS FOR MOBILE MAPPING SYSTEMS

Thus, the main emphasis in mobile mapping operations is on the use of dedicated and purpose-built 2D laser profilers that can very rapidly acquire a large number of range or elevation profiles comprising the distance and angular values measured within a single 2D plane (Petrie, 2010). These profile measurements are carried out using the 2D laser profiler instrument to measure the required distances and angles simultaneously within a series of successive parallel planes intersecting the corridors comprising the road surfaces, pavements, *street furniture*, buildings, and vegetation that are located adjacent to the roads or streets along which the mobile mapping vehicles are being driven. Indeed, in many respects, the 2D laser scanner/profilers that are mounted on mobile mapping vehicles are, in principle, quite similar to the laser scanners that are being used in ALS (that will be discussed in Chapter 3)—except that they are often being operated over distances of a few tens of meters, instead of the several hundreds or thousands of meters that are encountered in ALS. As with ALS, the third dimension to the captured 2D profile data is being created by the forward movement of the vehicular platform on which the 2D laser profiler is mounted. At the same time, the location of each new range profile is being measured continuously (and very accurately) using an integrated suite of positioning devices—comprising a GPS or GNSS receiver, an IMU, and an odometer or distance measuring instrument—as the vehicle travels forward.

It is also worth noting that quite different scan patterns will result over the ground depending on (1) how the 2D laser profilers are mounted on the vehicle—for example, with either a horizontal, vertical, or inclined spinning axis—and (2) the speed of rotation of the profiler and the velocity of the vehicular platform on which it is mounted. A detailed account and analysis for one particular laser scanning device is presented in the paper given by Elseberg et al., 2013. As noted earlier, many MMSs operate a 2D *full-circle* laser profiler spinning around a horizontal axis and use the forward motion of the vehicle to provide 3D data. This results in a corkscrew pattern if viewed, for example, on the surface of a tunnel. When this scanning pattern intersects with a planar surface such as the ground, then the measured laser points are distributed over the surface in a linear pattern.

The 2D laser profilers that have probably been the most commonly used in mobile mapping are those manufactured by the SICK company, which is based in Waldkirch, Germany. Certain models in the SICK LMS series of 2D laser profilers are designed specifically for outdoor use, the LMS 291 model being that mainly used in mobile mapping (Figure 2.32). This profiler combines (1) a low-powered rapid-firing laser rangefinder using the TOF or pulse-based distance measuring principle, with (2) a rotating mirror whose angular directions are also being measured continuously using an angular encoder. The laser rangefinder utilizes a semiconductor diode laser that emits its radiation in the near infrared part of the spectrum. The rangefinder also has a photo-diode receiver that can receive up to three returning echoes (reflections) from a single transmitted pulse. The angular scanning/profiling motion is implemented using a continuous rotating mirror that has a maximum scan frequency of 75 Hz and is equipped with a rotary (angular) encoder. By using this technology, the LMS 291 generates a fan-shaped scanning angle of 180° within its 2D scanning plane and can measure ranges of up to 80 m to objects having a reflectivity of 70%, 60 m to objects (such as a wooden house) with a reflectivity of 40%, and 30 m to objects with 10% reflectivity. The measuring resolution in range of the LMS 291 model is stated to be 1 cm, whereas the quoted accuracy is ±6 cm. An additional Laser Measurement Interface (LMI) controller can be supplied to control the operation of multiple LMS scanners. As will be seen later in this chapter, these units from SICK have been used widely in the MMSs built, for example, by Topcon, Mitsubishi, and Google.

Still longer range 2D laser profilers are available from LASE GmbH, which is another company in the SICK Group that is based in Wesel, Germany. The LASE LD-LRS laser scanner can measure ranges up to 250 m with suitable highly reflective objects, 110 m to objects with 20% reflectivity, and 80 m to objects with 10% reflectivity, and has a 300° scanning angle. Yet another company in the SICK Group, Ibeo, based in Hamburg, Germany, also offer multilayer laser scanners and profilers that are suitable for mapping purposes (Schulz et al., 2007). These Ibeo instruments will be discussed in more detail in the UAV section of Chapter 3. Besides, the SICK scanners, there are several other TOF 2D laser profilers that are used quite widely in mobile mapping operations. These are made by the specialist manufacturers and suppliers of the terrestrial 3D laser scanners described previously, such as *RIEGL*, Optech, and MDL/Renishaw. In general terms, the specialized 2D profiler units from these companies provide greater ranges, faster speeds, and higher measuring accuracies than those provided by the SICK laser scanners discussed earlier.

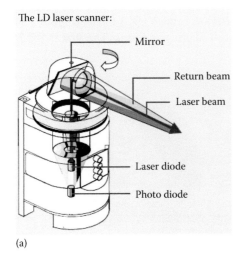

(a)   (b)

FIGURE 2.32 (a) Diagram showing the construction of a SICK 2D laser profiler. (b) A photo of the SICK LMS 111 model that is widely used in mobile mapping systems. (Courtesy of SICK, Inc., Waldkirch, Germany. With Permission.)

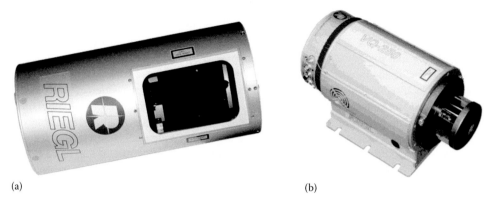

(a)　　　　　　　　　　　　　　　　　　(b)

FIGURE 2.33 (a) An example of a *RIEGL* LMS-Q120 2D profiler with its viewing window having an 80° angle over which it can scan and measure ranges to objects. (b) The *RIEGL* VQ-250 profiler with its spinning head providing a 360° *full circle* scan angle over which ranges can be measured. (Courtesy of *RIEGL*, Horn, Austria. With Permission.)

The *RIEGL* LMS-Q120 laser profiler (Figure 2.33a) with its continuously rotating polygon mirror has been used in a number of early MMSs. It has a PRF of 30 kHz, a range of 150 m to targets with an 80% reflectivity, a ranging accuracy of ±25 mm, a scanning angle of 80° within the 2D plane in which it is scanning, and can be operated at scan rates up to 100 Hz. Since then, *RIEGL* has produced a number of more advanced models having an improved performance as part of its *V-Line* of products. The company's VQ-180 model, first introduced in 2010 and still being produced in 2016, offers still higher PRF values (up to 200 kHz) and scan rates (up to 120 Hz), and it also has a larger scan angle (of 100 degrees). In 2012, *RIEGL* also introduced a more powerful VQ-250 model (Figure 2.33b) that is designed specifically for use in mobile mapping. It provides a *full circle* 360° scan within its 2D scanning plane and can measure ranges up to 200 m (with 80% reflectivity) with PRF values up to 300 kHz and scan rates of 100 Hz, while still maintaining an accuracy of ±10 mm. This was followed by the VQ-450 model, again having a 360° *full circle* capability, but with a rangefinder based on a Class 1 laser giving a still longer maximum range of 800 m (with 80% reflectivity) and PRF values of up to 550 kHz and scan rates of 200 Hz. A derivative of this instrument is the VMQ-450 system, which integrates a single VQ-450 profiler unit with a GNSS/IMU subsystem and a system control unit. Another single 2D profiler from *RIEGL* is the VUX-1 model (Figure 2.34b) that was introduced in 2014. It has been produced in two alternative forms. The first is the HA

(a)　　　　　　　　　　　　　　　　　　(b)

FIGURE 2.34 (a) The *RIEGL* VMQ-1HA system based on the VUX-1HA laser profiler and integrated with a GNSS sub-system and a system control unit. (b) The *RIEGL* VUX-1HA 2D laser profiler. (Courtesy of *RIEGL*, Horn, Austria. With Permission.)

version with a 400 m maximum range (with 80% reflectivity), a maximum PRF of 1 million points per second, and a maximum scan speed of 250 scans per second. The second model is the LR version with a maximum range of 1350 m (with 60% reflectivity), a maximum PRF of 750,000 points per second, and a maximum scan speed of 200 scans per second. As with the VQ-450, *RIEGL* has developed a derivative of the VUX-1 called the VMQ-1HA system (Figure 2.34a) that integrates the VUX-1HA model with a set of up to four cameras, together with a system control unit.

RIEGL has also produced a number of VMX models utilizing multiple units of these 2D laser profilers for mobile mapping purposes. Thus, the VMX-250 system integrates two of the VQ-250 profilers along with a GNSS receiver and an IMU, all housed within a suitable protective cover and mounted on a roof platform that could also accommodate up to six small-format digital frame cameras. The VMX-450 system has a similar layout and components but is based on two of the VQ-450 2D profilers, whereas the VMX-1HA system (Figure 2.35) also has a similar layout based on two VUX-1HA profilers. As will be seen later in this chapter, these numerous individual profilers and systems from *RIEGL* have been widely utilized by system suppliers such as 3D Laser Mapping and Trimble.

By contrast, Teledyne Optech does not sell its in-house-built laser scanners and profilers as separate (OEM) products to external system integrators and suppliers in the manner of SICK and *RIEGL*. Instead, it manufactures its 2D laser profilers (Figure 2.36a) to form part of its own LYNX MMSs. The laser profilers that have been used in its LYNX V200 system from 2009 employing a rangefinder based on a Class 1 laser provided a 360° *full circle* profile scan, a maximum range of 200 m, a PRF of up to 200 kHz, a scan rate of up to 200 Hz, and a range accuracy of circa ±10 mm. In the current Lynx SG models, a similar 360° *full circle* profiler is used with the maximum range increased to 250 m, (at 10% reflectivity), the maximum PRF to 1200 kHz, and the maximum scan rate to 600 scans per second.

The MDL (now Renishaw) company also manufactures its own 2D laser profilers (Figure 2.36b) with a 360° *full circle* capability, which it calls its scanning laser modules (SLM). These are available and can be supplied to other system suppliers on an OEM basis, besides being fitted to Renishaw's own range of Dynascan MMSs. The company offers two different modules the SLM-150 (with a maximum range of 150 m) and the SLM-500 (with a maximum range of 500 m). Both modules utilize a Class 1 laser and produce a ranging accuracy of ±10 cm, an angular accuracy of ±0.02°, and a measuring rate of 36,000 points per second.

As mentioned earlier, initially, the Velodyne company developed its multiple spinning laser scanner/profiler systems for use in the DARPA DGC and DUC contests that promoted the development of unmanned vehicles. The original Velodyne HDL-64E High Definition LiDAR (Figure 2.37)

   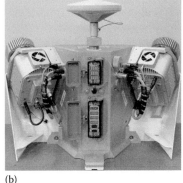

(a)   (b)

FIGURE 2.35 (a) The *RIEGL* VMX-1HA mobile mapping system featuring two of the VUX-1HA laser profilers, mounted with inclined spinning axes; three of the system's small-format cameras can also be seen, together with the antenna of the GNSS receiver. (b) The VMX-1HA system with the two VUX-1HA profilers and the GNSS antenna as viewed from the rear. (Courtesy of *RIEGL*, Horn, Austria. With Permission.)

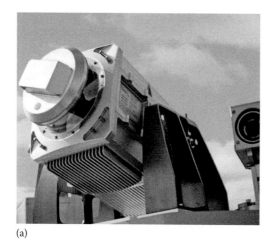

(a)   (b)

FIGURE 2.36 (a) A 360° *full circle* 2D laser profiler as fitted to the Optech Lynx mobile mapping system. (Courtesy of Optech, Vaughan, ON. With Permission.) (b) A Renishaw 360° *full circle* SLM 2D laser profiler as fitted to Renishaw's Dynascan mobile mapping system. (Courtesy of Renishaw, Wotton-under-Edge, UK. With Permission.)

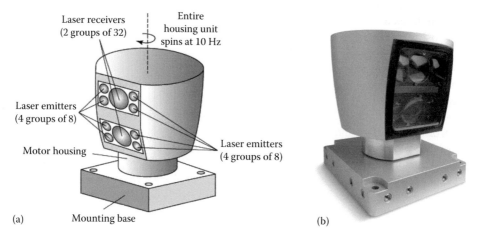

(a)   (b)

FIGURE 2.37 (a) Diagram showing the main external features of the Velodyne HDL-64E spinning laser scanner. (Drawn by M. Shand.) (b) The Velodyne HDL-64E laser scanner. (Courtesy of Velodyne, San Jose, CA. With Permission.)

was based on the simultaneous use of 64 individual laser rangefinder units spinning around a common axis and covering a 26.8° vertical spread with a series of 2D profiles, thus eliminating the need for any vertical mechanical motion. The system features high horizontal rotation rates of the laser rangefinders spinning around the vertical axis of the unit, at rates up to 15 Hz, with an angular resolution of 0.09°. The Class 1 laser operates at the wavelength (λ) of 905 nm with a 10 ns pulse width. The ranging accuracy is claimed to be less than ±5 cm for measured ranges of 50 and 120 m with reflectivities of 10% and 80%, respectively. The total data collection rate is more than 1 million points per second. Since then, system suppliers, such as Topcon, have made extensive use of the HDL-64E laser scanner/profilers and have integrated them with the digital cameras, GNSS/IMU subsystems, and control units, which are the principal building blocks of the vehicle-based MMSs that have been developed for topographic applications.

In 2010, Velodyne introduced a smaller and lower cost model, named the HDL-32E. This operates on the same basic principle as the HDL-64E model, having a set of laser rangefinders that spin

FIGURE 2.38 Showing the comparative sizes of the HDL-64E, HDL-32E, and VLP-16 multilaser spinning profilers. (Courtesy of Velodyne, San Jose, CA. With Permission.)

around the vertical axis of the instrument at 10 Hz. However, it uses a bank of 32 individual laser ranging units instead of the 64 used in the HDL-64E model and its vertical FOV stretches from +10.67° to −30.67°. Similar to the HDL-32E model, it uses Class 1 lasers as the basis for its range-finders and these have a maximum range of 80 to 100 m. The maximum rate of measurement is 700,000 points per second, whereas the claimed accuracy in range is ±2 cm. The HDL-32E instrument is much smaller (Figure 2.38) and lighter, weighing just under 2 kg, compared with the 15 kg for the earlier HDL-64E model. In 2014, Velodyne introduced a further model in the series, called the VLP-16 or *Puck* (Figure 2.38), which uses only 16 laser ranging units. The performance characteristics of the VLP-16 model include a maximum range of 100 m, a PRF of 300,000 points per second, a range accuracy of ±2 cm, and a vertical FOV from +15° to −15°. As Figure 2.36 shows, it is still smaller than the HDL-32E model, and it has a very low weight of only 830 g. A still lighter weight version of this device, called the *Puck LITE*, has just been introduced in February 2016. This has the same performance characteristics as the *Puck* model but weighs only 590 g. The various Velodyne models with their simultaneous measurements of multiple 2D profiles have been used widely for mobile mapping purposes, including Topcon and 3D Laser Mapping among the system suppliers and Navteq and Apple among service providers operating large fleets of mapping vehicles.

2.3.3 COMMERCIAL SYSTEM SUPPLIERS

The 2D laser profilers discussed previously are then combined together with GNSS/IMU subsystems and control hardware and software to form complete MMSs (Figure 2.39a) by various system suppliers, who then sell them to numerous service providers.

2.3.3.1 Teledyne Optech

As already mentioned earlier, Optech developed a special MC version of its ILRIS-3D laser scanner for use on dynamic or moving platforms. This system was used, for example, by the Sineco company in Italy for surveys of an open cast mine and for road surveys conducted from a moving van (Zampa and Conforti, 2008). Besides the ILRIS-3D laser scanner unit that generated the range, intensity, angle, and time data, the overall system included an Applanix POS/LV 420 GPS/IMU subsystem that provided the position and orientation data against time. The Polyworks software package was used by Sineco to carry out the subsequent processing of all the captured data.

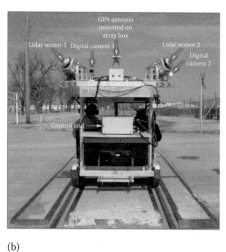

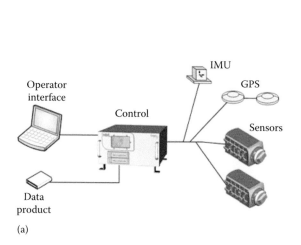

IMU

Operator
interface

GPS

Control

Sensors

Data
product

(a)   (b)

FIGURE 2.39   (a) Diagram of a control set-up for a mobile mapping system showing the main components of such a system. (b) Showing the actual configuration of the control unit and the various sensors depicted in diagram (a) that has been adopted with an Optech Lynx mobile mapping system—including its 2D laser profilers, digital cameras, and GPS receiver—that has been mounted on a railroad speeder (track maintenance car) in Oklahoma. (Courtesy of Teledyne Optech, Vaughan, ON. With Permission.)

In 2007, Optech released a completely new product, called the Lynx Mobile Mapper (Figure 2.39b). This was a purpose-built spinning 2D laser profiling system designed specifically for attachment to standard vehicle roof racks with mounts for two laser 2D profilers and two (optional) calibrated frame cameras as well as the system IMU and GPS antenna. The third dimension was provided by the vehicle motion. Although the scanner/profilers used in the Lynx system also utilized a Class 1 laser as the basis for their laser rangefinders, they were very different units to those used in the camera-type ILRIS-MC scanner, having a maximum range of 100 m, a 360° *full circle* angular coverage within the 2D plane of operation, a pulse measuring rate of 100 kHz, and a scan rate of 9000 rpm (150 Hz). The system control unit with its embedded navigation solution was based on the Applanix POS/LV 420 subsystem and could control up to four 2D laser profilers simultaneously using the laptop computer attached to the unit. The Applanix POSPac software was used to process the POS/LV data, whereas Optech supplied its own Lynx-Survey and Lynx-Process software for final postprocessing.

Since its introduction in 2007, Optech has continued to develop the Lynx mobile mapper system. The first of these developments was to offer two separate models—the V100 and V200 systems. The latter gave an increased range (200 v. 100 m), a higher PRF (200 kHz v. 100 kHz), and a higher scan rate (200 Hz v. 150 Hz) than the former, which was essentially the Lynx in its original form. Currently (in 2016), three different Lynx systems are on offer—the MG, SG, and SG-S systems, in which the letters MG = Mapping Grade and SG = Survey Grade in terms of the accuracy of the captured data. This is claimed to be ±20 cm in the case of the MG system (Figure 2.40a) and ±5 cm with the SG system. All three systems share a similar rangefinder specification of a Class 1 laser with a maximum range of 250 m with a 10% reflectivity, a range precision of 5 mm, and the ability to measure four return pulses simultaneously. However, the SG system (Figure 2.40b) features twin laser profilers instead of the single profiler of the MG and SG-S systems. Furthermore, the laser range measuring rates differ with maximum rates of 500 kHz (MG), 600 kHz (SG-S), and 1200 kHz (SG), respectively, as do the respective maximum scan frequency rates of 200 Hz (MG), 300 Hz (SG-S), and 600 Hz (SG). The greater accuracy of the SG systems is stated to be the result of the superior data processing and rectification that is achievable using Optech's LMS Pro software as well as the improvements in the hardware performance. It should also be stated that although all

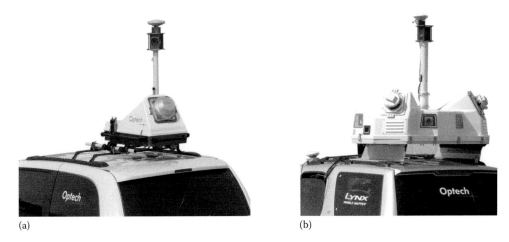

(a) (b)

FIGURE 2.40 (a) A Lynx MG mobile mapping system featuring a single laser 2D scanner/profiler; a Ladybug multiple camera unit; and a GPS antenna, all mounted on the roof of the vehicle. (b) A Lynx SG system featuring twin laser 2D scanner/profilers, a Ladybug multiple camera unit, four 5 Megapixel digital frame cameras, and a GPS antenna; the two 2D laser profilers and the four 5 Megapixel cameras are all housed in a protective housing and mounted on a vehicle roof rack. (Courtesy of Optech, Vaughan, ON. With Permission.)

three systems also feature a Point Grey Research Ladybug3 or Ladybug5 multiple camera unit, the SG model can also be fitted with up to four additional 5 Megapixel digital frame cameras. All of the Lynx systems appear to utilize Applanix POS/LV GNSS/IMU subsystems and POSPac software as standard.

Another new product within this particular application area was introduced by Teledyne Optech in October 2016 in the form of its lightweight Maverick MMS (Figure 2.41). This system can be mounted on a wide variety of platforms, including road and rail vehicles and backpacks. It combines (1) a Velodyne HDL-32 laser scanner/profiler, with (2) a Point Grey Ladybug5 unit comprising six digital frame cameras, each generating a 5 Megapixel image, resulting in 360° horizontal imagery coverage, and (3) an integrated NovAtel SPAN position and orientation subsystem. The system is delivered with the Distillery software from Mandli Communications that carries out the postprocessing of the captured laser scan, image, and GPS/IMU data.

FIGURE 2.41 An Optech Maverick mobile mapping system showing the Velodyne HDL-32 laser scanner/profiler at left and the NovAtel GPS receiver with its white cap at right, with the Point Grey Ladybug5 multiple camera unit placed between them. (Courtesy of Teledyne Optech, Vaughan, ON. With Permission.)

2.3.3.2 3D Laser Mapping

This company, which is based in Bingham, a small town near the city of Nottingham in the United Kingdom, has developed its portable StreetMapper system specifically for use on moving vehicles. It has done this in close collaboration with the German systems supplier, Integrated Geospatial Innovations (IGI), which also produces the LiteMapper ALS system (Hunter et al., 2006; Kremer and Hunter, 2007). For the StreetMapper system, IGI supplied its TERRAcontrol GPS/IMU system, which is derived from its airborne AEROcontrol unit, together with its own hardware and software solutions for the control of the 2D laser profilers and data storage. The control unit was housed in a cabinet that was mounted inside the vehicle. IGI also contributed its TERRAoffice software (derived from its AEROoffice package) for the processing of the IMU data, whereas the differential GPS data were processed using the GrafNav package from the Waypoint division of NovAtel based in Canada. The TerraScan/TerraModeler/TerraMatch suite of programs from Terrasolid in Finland was utilized for the processing of the laser scan data and its transformation into the final 3D model data. From the start of this development in 2004/5, the multiple 2D laser profilers were supplied by *RIEGL*, between two and four of its LMS-Q120 units (with their 150 m range) being fitted on to a roof rack together with the IMU and the GPS antenna (Figure 2.42a). Either video or digital still cameras were added to generate high quality images that would complement the laser profile data. Touch screen displays installed on the dashboard of the vehicle were used for the display of the captured data.

Two further generations of the StreetMapper appeared first in 2010 and then in 2012, respectively, still using *RIEGL* 2D scanner/profilers. These made use of the *RIEGL* VQ-250 or VQ-450 profilers or the twin-headed VMX-250 profiler, together with various covered housings or pods (Figure 2.42b) in the StreetMapper 360 version of the system that covers a 360° FOV. In 2015, a fourth generation of the system, named StreetMapper IV (Figure 2.44a), was introduced, which is more compact, lighter in weight, and more portable than the previous models in the series. This StreetMapper IV system can be supplied in any one of three different versions, labeled (1) Single GIS Grade, (2) Single Survey Grade, and (3) Dual Survey Grade, the single and dual referring to the number of *RIEGL* VUX 2D scanner/profilers that are being used as the basis of the system. Complementing these laser profilers are the six digital frame cameras in a Point Grey Ladybug3 unit that produce 2 Megapixel images with the GIS Grade version, and 5 Megapixel images from a

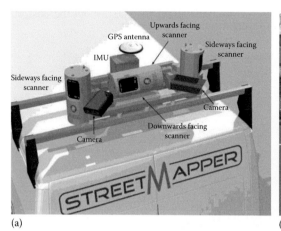

(a)   (b)

FIGURE 2.42 (a) This van is equipped with a StreetMapper system from 3D Laser Mapping. Its roof rack supports four *RIEGL* laser scanner/profilers—two sideways-facing; one upwards-facing and one downwards-facing—and two small-format digital cameras; together with the IMU and GPS antenna of the position and orientation system that has been supplied by IGI. (b) A StreetMapper system with twin laser profilers and multiple cameras encased in a protective pod that has been mounted on the roof of a survey vehicle in Japan. (Courtesy of 3D Laser Mapping, Nottingham, UK. With Permission.)

(a)   (b)

FIGURE 2.43 (a) This RailMapper system contained in a protective pod is mounted on the roof of a DB (Deutsche Bahn) railway maintenance railcar. (b) A RailMapper vehicle equipped with twin Z+F Profiler 9012 profilers as seen from the rear of the vehicle. Note the rerailer equipment that keeps the wheels of the pick-up vehicle on the railway track. (Courtesy of IGI, Kreuztal, Germany. With Permission.)

six camera Ladybug5 unit with the two Survey Grade versions. Again, the system features a high-end GNSS/IMU subsystem based on the use of fiber-optic gyros supplied by IGI. The software for the processing of the data acquired by the GIS Grade System is supplied by the Belgian company Orbit Geospatial Technologies, whereas that used with the Survey Grade systems continues to be supplied by Terrasolid. In parallel with the development of the StreetMapper, the two companies have also developed the very similar RailMapper system (Figure 2.43a) for surveys concerned with railroads and tramway systems, including surveys of tunnels. Initially, *RIEGL* 2D scanner/profilers were used; however, the current version of the RailMapper system utilizes the Z+F Profiler 9012 as its laser measuring unit (Figure 2.43b).

(a)   (b)

FIGURE 2.44 (a) A StreetMapperIV system from 3D Laser Mapping showing the rotating head of the single RIEGL VUX 2D laser profiler and the GNSS antenna protruding above and external to the protective hood of the overall system. (b) A 3D Laser Mapping ROBIN system showing the RIEGL VUX laser profiler, twin GNSS antennae and a Nikon small-format digital frame camera. (Courtesy of 3D Laser Mapping, Nottingham, UK. With Permission.)

Besides the various models of the StreetMapper system, in 2013, 3D Laser Mapping and IGI also introduced an alternative more compact and lower cost GIS or mapping grade product called the V-Mapper system. This utilizes the Velodyne HDL-32E unit generating its multiple array of 2D profiles with its more limited maximum range of 100 m range (as compared with the 400 m of the StreetMapper) in combination with a multiple small-format camera unit and a GNSS/IMU positioning and orientation subsystem. The two collaborating companies have also introduced another more precise product, again with a maximum range of around 100 m, in the form of the Z-Mapper that utilizes the phase-based Z+F Profiler 9012 as its laser measuring unit. Still another new MMS has just been introduced by the two partners (in May 2016) in the form of the ROBIN system (Figure 2.44b). This is an integrated system that is based on the *RIEGL* VUX range of 2D laser profilers and includes both a 12 and a 30 Megapixel camera, two GNSS antennae, a survey grade IMU, a touch-screen control unit, and software for data capture and postprocessing. It also features a three-in-one mounting system that allows it to be used either as a backpack system, a vehicle-mounted system or a UAV-mounted system.

2.3.3.3 Topcon

Topcon Positioning Systems introduced its MMS—called the *IP-S2 Integrated Positioning System* (Figure 2.45a)—to the market in the spring of 2009. However, when the company announced during this introduction that "more than 400 units are currently in use world-wide," it was only too obvious that it had indeed been supplying these systems for some time—mainly, it seemed, to Google Inc. to partially equip its StreetView mobile mapping vehicles. The IP-S2 system (Talend, 2010) included a Topcon dual-frequency 40 channel GNSS receiver operating at 20 Hz, which was coupled to a Honeywell HG1700 tactical-grade IMU based on a ring laser gyro that was operating at 100 Hz. The resulting Differential Global Positioning System (DGPS)/IMU positional data was supplemented by the data generated by a wheel-mounted odometer with an angular encoder operating at 30 Hz to complete the overall positioning capability for the IP-S2 system. Besides these positioning devices, the imaging and laser scanning/profiling capabilities of the IP-S2 were

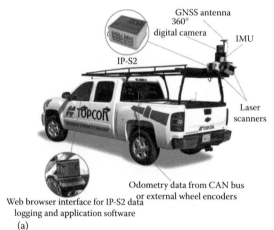

(a)   (b)

FIGURE 2.45   (a) Showing the main components of a Topcon IP-S2 system mounted on a pick-up truck. This particular version uses three SICK laser profiler units. (b) The Topcon IP-S2 HD version utilizing a Velodyne HDL-64E scanner/profiler mounted at an angle from the vertical to measure the ground and surrounding features. Note the very different support frame and the special mount above the Velodyne unit to ensure that the multicamera unit and the GPS antenna are both set in a vertical orientation. (Courtesy of Topcon, Tokyo, Japan. With Permission.)

based on well-known units that were also available off-the-shelf. They included the Point Grey Ladybug multicamera unit from Point Grey Research that carried out the 360° panoramic imaging with frame rates of up to 15 frames per second. The laser profiling that was carried out using the standard configuration of the IP-S2 was provided by three SICK LMS 291 2D profilers operating at 75 Hz. One of these laser profilers pointed directly forwards (or backwards) towards the road in front of (or behind) the vehicle, whereas the other two scanners pointed to each side to provide a continuous series of range or elevation profiles within the vertical plane. All of these imaging and scanning devices sent their data to a central system control box built by Topcon that then passed it via a high-speed FireWire-B (IEEE1394-B) link to the PC that was mounted in the vehicle for the recording and processing of the data. An LCD display screen allowed the vehicle's crew to monitor the connectivity and operation of all the various positioning, laser scanning, and frame imaging devices.

Following on from the introduction of the IP-S2 system, Topcon soon produced other versions of the basic system. With the IP-S2 Lite model, the laser profilers were omitted altogether, so the data collection comprised only the imagery acquired by the multiple camera unit. However, at the other end of the scale, Topcon also produced a high-density version of the system, labeled IP-S2 HD (Figure 2.45b), in which the SICK laser profilers were replaced by a single Velodyne HDL-64E scanner/profiler with its 64 spinning laser rangefinders that were inclined at a considerable angle from the vertical to measure the ground and the road surface that was being covered by the mapping vehicle. To accommodate this device, the supporting mast was replaced by a much sturdier structure. The position and orientation subsystem in both the Lite and HD versions were also changed to utilize a MEMS-based IMU. In 2015, Topcon introduced yet another new model in the series in the shape of its IP-S3 HD1 system (Figure 2.46a). This again utilizes a Velodyne scanner/profiler, but in this case, it is the smaller HDL-32E model (Figure 2.46b) with its 32 spinning lasers that is being used instead of the larger and more expensive HDL-64E model that was used in the IP-S2 HD system.

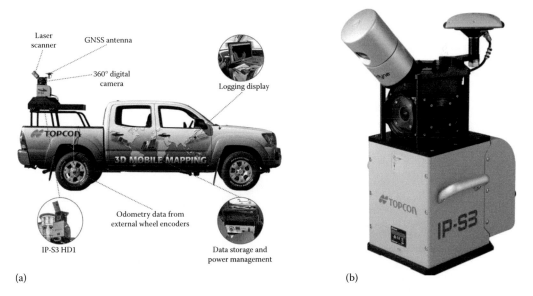

(a)   (b)

FIGURE 2.46 (a) Showing the main components of a Topcon IP-S3 system mounted on a pick-up truck. (b) A close-up photo showing the main control box of the IP-S3 system, with the multiple-camera unit mounted on top of the box, and with the Velodyne HDL-32E scanner/profiler and the GNSS antenna both mounted on top of the camera unit. (Courtesy of Topcon, Tokyo, Japan. With Permission.)

2.3.3.4   Renishaw

As noted earlier, Renishaw acquired the MDL company in 2013. The latter had already introduced its Dynascan MMS prior to the acquisition. The system was available in different forms with either single, dual, or triple profiling heads (Figure 2.47) employing its own SLM profiler heads having a 360° rotation to define a 2D scanning plane. These *full-circle* profilers could be arranged to produce their scans either in a vertical plane or inclined in a plane at 45° to the vertical (Figure 2.48), or in a combination of these planes. These profilers were then mounted on the compact box containing the system's control electronics, which also had a GPS antenna mounted on it. Besides the choice regarding the number of profiler heads that could be fitted to the system, there is also a choice in the accuracy of the systems that could be supplied, with either mapping (M) or survey (S) grade systems

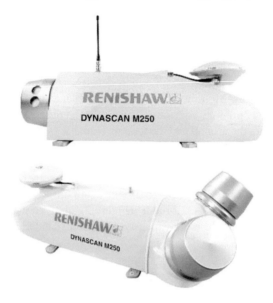

FIGURE 2.47   Single- and dual-headed Renishaw Dynascan M250 photo. (Courtesy of Renishaw, Wotton-under-Edge, UK. With Permission.)

FIGURE 2.48   Vehicle-mounted Dynascan photo. (Courtesy of Renishaw, Wotton-under-Edge, UK. With Permission.)

being available. Taking the S250 version as an example, the twin scanner/profilers use rangefinders utilizing Class 1 InGaAs laser diodes having a wavelength of $\lambda = 905$ nm. These have a maximum pulse repetition rate of 36,000 pulses per second and a maximum range of 250 m, as required for surveys of open-cast mines and quarries in which MDL had already established a considerable user base through its Quarryman product. These Dynascan units have also been mounted on various marine craft to carry out surveys of harbors and other coastal features as well as being used for vehicle-based surveys and mapping.

2.3.3.5 Trimble

The MMSs that are being supplied by Trimble were developed originally by the Geo-3D company, which was based at Brossard, near Montreal in Canada and was acquired by Trimble in January 2008. Its main product had been its Trident-3D MMS. This had been fitted to a variety of different vehicles and in a number of different configurations as specified by the customers. Digital video and still frame cameras from various suppliers had been fitted—including, in one case, the use of a Redlake multispectral camera—while the 2D laser profilers that had been used were supplied by SICK and by *RIEGL*. The DGPS/IMU subsystems that were used for geopositioning were the POS LV units supplied by Applanix (another Trimble company) and included Trimble GPS receivers. The system controller and rack-mounted computers that formed parts of the overall system were built by Geo-3D, which also supplied the distance measuring instrument. Various display options were also offered by Geo-3D. The final version of this series of vehicle-based mapping system was then called the Trimble Cougar system.

The direct descendent of this Cougar system is the current Trimble MX8 MMS (Figure 2.49a). In its standard form, it is fitted with a *RIEGL* VQ-250 or VQ-450 twin-headed scanner/profiler unit. Supplementing these laser-based units are typically (1) a multicamera unit comprising three forward-facing and one rear-facing digital frame camera, which is designed to record panoramic imagery of the corridor along which the mapping vehicle is passing and (2) an additional rear-facing three digital frame camera unit producing imagery of the road surface for those users with this particular requirement. Each individual camera produces 5 Megapixel images. The MX8 system also includes an Applanix POS LV-420 or LV-520 GNSS/IMU subsystem for positioning and orientation purposes. These various units and subsystems, together with their control electronics, are housed in a large protective pod that is mounted at the rear of the van on a roof rack that is carrying the overall MX8 system. Besides the MX8 system, Trimble also offers its MX2 system that is based on the MDL/ Renishaw Dynascan system that has already been described in the previous section (Figure 2.49b).

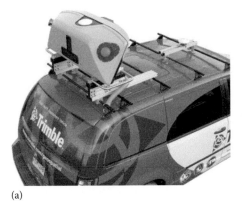

(a)   (b)

FIGURE 2.49 (a) An example of a Trimble MX8 mapping system in its protective pod mounted on top of its mapping van. (b) A Trimble MX2 system mounted on top of the mapping van. (Courtesy of Trimble, Sunnyvale, CA. With Permission.)

2.3.3.6 Mitsubishi

Another MMS supplier is the Mitsubishi Electric Corporation from Japan, the IT Space Solutions Division of which introduced its MMS at the Intergeo trade fair held in Karlsruhe, Germany, in September 2009. This product had been developed jointly by staff members of Waseda University in Tokyo in collaboration with Mitsubishi from 2006 onwards. The system was offered in three different versions: (1) The most basic version was the MMS-A, which had three roof-mounted GNSS receivers arranged in a triangular pattern for positioning and orientation purposes; an IMU; an odometer; and a sensor control box. This version was offered mainly as a vehicle positioning device, with the choice of cameras and laser scanners and their integration being left to the customer. (2) The second version was the MMS-S that was offered with two video cameras and two 2D laser profilers in addition to the positioning devices included in the basic MMS-A version. (3) Finally, the MMS-X version was offered with multiple (up to 6) cameras and (up to 4) laser scanners, again in addition to the positioning instrumentation included in the MMS-A version. In the literature that accompanied this introduction, the supplier of the dual-frequency GNSS receivers was stated to be Trimble; the IMU was from Crossbow, using a FOG gyro supplied by Japan Aviation Electronics; the frame cameras were supplied by IMPERX from the United States; while the laser profilers were the ubiquitous LMS 291 model from SICK.

Since then, the MMS system has been developed still further. Currently (in 2016), three different versions of the system are again being offered commercially. These are labeled the MMS-X, the MMS-X320R (Figure 2.50a), and the MMS-K320, respectively. All three versions are fitted with digital frame cameras capturing 5 Megapixel images at a maximum rate of 10 images per second for a single camera. With regard to the 2D laser profilers, each system deploys two SICK instruments, each with a 180° coverage and having a maximum quoted range of 65 m and measuring a maximum rate of 27,100 points per second. However, the MMS-X versions can be supplied with four of these SICK laser profilers. The MMS-X320R version also features an additional *RIEGL* VQ-250 laser profiler with its 360° coverage that operates at a maximum acquisition rate of 300,000 points per second over ranges up to a maximum of 200 m (Figure 2.50b).

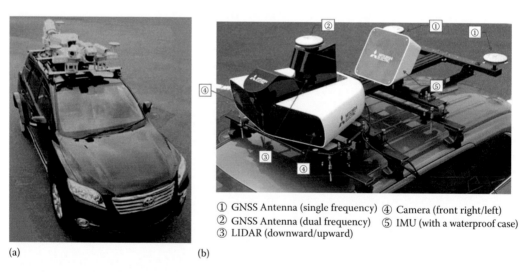

① GNSS Antenna (single frequency) ④ Camera (front right/left)
② GNSS Antenna (dual frequency) ⑤ IMU (with a waterproof case)
③ LIDAR (downward/upward)

(a) (b)

FIGURE 2.50 (a) An example of the Mitsubishi MMS-G220 mobile mapping system. (b) Showing the roof unit of detachable MMS-G220 system in more detail. Note the three GPS antennae for positioning and orientation purposes, the three 5 Megapixel cameras (two pointing forwards, the third to the rear), the two SICK 2D laser profilers pointing forward (one upwards, the other downwards), and the *RIEGL* VQ-250 laser profiler at the rear of the supporting platform. The system control box lies immediately forward of the *RIEGL* profiler. (Courtesy of Mitsubishi Electric, Tokyo, Japan. With Permission.)

2.3.3.7   Leica Geosystems

Leica Geosystems was a relatively late entrant to the mobile mapping field, introducing its Pegasus One system (Figure 2.51) to the market in June 2013 (Petrie, 2014; Duffy, 2014). The system had been developed in cooperation with the Geosoft software company based in Italy, which was in fact acquired by Leica Geosystems at the same time as the Pegasus One was introduced. Basically, the system was fitted with six 2MB cameras that were positioned to capture digital frame imagery within a full 360° (horizontal) × 70° (vertical) FOV. The system also employed NovAtel's SPAN GNSS/IMU subsystem to satisfy its positioning and orientation requirements. For the laser scanning/ profiling components of the overall system, Leica offered alternative solutions. For those customers who had already invested in a static terrestrial 3D laser scanner, they could utilize, for example, their Leica HDS7000 or ScanStation P20 instrument for the purpose. The terrestrial scanner instrument could then be placed on its side with its horizontal (azimuthal) rotation movement locked so that it scanned purely within a fixed 2D plane, so acting as a laser profiler. For those customers requiring a dedicated mapping system, the Pegasus One system was fitted with a Z+F Profiler 9012 that has a maximum scanning/profiling speed of 200 Hz and can measure at a maximum rate of 1.1 million points per second. In 2014, Leica introduced a new revised version of the system, called the Pegasus Two. Again, this can be fitted either with an existing scanner such as the ScanStation P20 (Figure 2.52a) or a dedicated 2D profiler such as the Z+F Profiler 9012 (Figure 2.52b). The Pegasus Two can also be fitted with up to eight cameras, each generating 4 Megapixel images. The new system also utilizes a NovAtel SPAN GNSS/IMU subsystem and offers the same choice of laser scanner/profiler instruments as the Pegasus One system. However, the new system is mounted in a more compact and convenient housing.

In June 2015, the Pegasus Backpack system (Figure 2.53) was introduced. It includes five 4 Megapixel digital frame cameras that are mounted to provide maximum coverage around the surveyor and two Velodyne VLP-16 Puck laser scanners that can measure a maximum of 600,000 points per second at ranges up to 50 m. The positioning and orientation subsystem comprises a NovAtel OEM 638 GNSS receiver and an accompanying IMU. These various components, together with the system control unit and the batteries that are required to power the

FIGURE 2.51   This Pegasus One system has been fitted with a Leica HDS7000 terrestrial laser scanner with a locked azimuth rotation, so that it acts as a 2D laser profiler. (Courtesy of Leica/Geosoft, Pordenone, Italy. With Permission.)

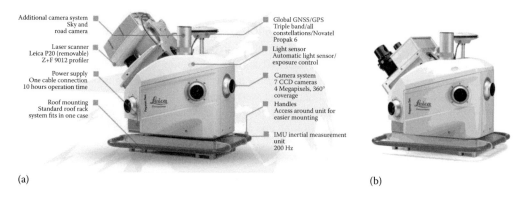

Additional camera system
Sky and
road camera

Laser scanner
Leica P20 (removable)
Z+F 9012 profiler

Power supply
One cable connection
10 hours operation time

Roof mounting
Standard roof rack
system fits in one case

Global GNSS/GPS
Triple band/all
constellations/Novatel
Propak 6

Light sensor
Automatic light sensor/
exposure control

Camera system
7 CCD cameras
4 Megapixels, 360°
coverage

Handles
Access around unit for
easier mounting

IMU inertial measurement
unit
200 Hz

(a)   (b)

FIGURE 2.52 (a) An annotated photo showing the various components of a Pegasus Two mobile mapping system that has been fitted with a Leica ScanStation P20 terrestrial 3D laser scanner to act as a 2D profiler. (b) A Leica Pegasus Two system fitted with a phase-based Z+F Profiler 9012 instrument. (Courtesy of Leica Geosystems, St. Gallen, Switzerland. With Permission.)

FIGURE 2.53 A Pegasus Backpack mobile mapping system. Note the twin Velodyne Puck profilers, one mounted horizontally, the other vertically; the GNSS receiver; and the ports of three of the small-format digital frame cameras. (Courtesy of Leica Geosystems, St. Gallen, Switzerland. With Permission.)

equipment, are all mounted on the lightweight carbon-fiber frame of the backpack. The overall system is controlled by a tablet device that is carried by the operator.

2.3.3.8 Hi-Target

As noted previously, the Hi-Target company is a Chinese surveying instrument supplier that manufactures total stations, GNSS receivers and marine positioning systems, and echo-sounders. Through its Haida Cloud subsidiary company, it announced its entry into the mobile mapping market in 2015 via its iScan range of products. In somewhat the same manner as the Renishaw Dynascan systems described earlier, the iScan system from Hi-Target is available in three different forms with either single, dual, or triple 2D profiling heads (Figure 2.54). These are labeled the iScan-V, iScan-C, and iScan-3, respectively. An example of the last (triple-headed) of these models is shown in Figure 2.55a. In addition to which, an iScan-P (= Portable) system is available that is mounted on a specially built framed backpack that is carried on the back of the surveyor (Figure 2.55b).

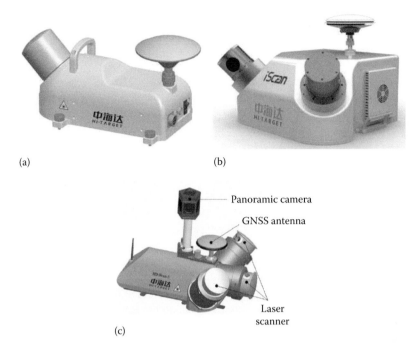

(a) (b)

Panoramic camera

GNSS antenna

Laser
scanner

(c)

FIGURE 2.54 Showing the various forms of the iScan system equipped with (a) single, (b) dual, and (c) triple 2D profiling heads. (Courtesy of Hi-Target, Guangzhou, China. With Permission.)

(a) (b)

FIGURE 2.55 (a) A mobile mapping vehicle with a roof-mounted iScan system equipped with triple 2D profiling heads, a GNSS receiver, and a multicamera imaging system. (b) A backpack version of the iScan system. (Courtesy of Hi-Target, Guangzhou, China. With Permission.)

2.3.4 CUSTOM-BUILT AND IN-HOUSE OPERATED SYSTEMS

There are a number of very large service providers that operate large numbers of MMSs—in some cases, numbering hundreds of vehicles. The sheer number of these systems has meant that these large organizations have designed and constructed their own systems in-house, often using the laser scanning and profiling instruments and components that have been outlined previously. The MMSs that have been constructed and operated by some of these organizations will be discussed in the following sections.

2.3.4.1 Tele Atlas and TomTom

The Tele Atlas company, which is based in Ghent, Belgium, has been a leading supplier of digital road map data for use in vehicle navigation and location-based systems and in the production of road maps and atlases for many years. In 2008, Tele Atlas was acquired by the Dutch TomTom organization that produces a variety of vehicle navigation and mapping products. For well over a decade, Tele Atlas operated fleets of vans that continually acquired data for the revision of its digital map database. For this purpose, Tele Atlas had a fleet of 22 camper vans (in an eye-catching orange color!) operating throughout Europe. Each van was equipped with six digital cameras, a GPS receiver that used the Fugro OmniSTAR positioning service, a gyro unit, and a distance measuring device that was attached to the rear wheel of the van for use when the GPS signals were lost. The resulting data were processed and analyzed in data centers that were located in Lodz, Poland, and Pune, India. In North America, a fleet of smaller vans was used, based in Lebanon, New Hampshire. Each van was equipped with a roof rack containing the cameras, GPS antenna, and others as before. However, each of the systems used in North America also featured twin SICK 2D laser profilers that were pointed to the side of the van to continuously collect street-level range data as the van traveled forward.

Since its takeover of Tele Atlas, TomTom has continued to operate fleets of mobile mapping vans equipped with a mixture of digital cameras and laser profilers. Typical of the current vehicles are the vehicles shown in Figure 2.56, which are equipped with a multiple camera unit and SICK and Velodyne laser profilers. Since 2012, TomTom has been one of the main suppliers of mapping data to Apple for use in its map products. From 2015 onwards, TomTom has also been a major supplier of map data to the Uber transportation networking company.

2.3.4.2 Google

Google's Street View imagery is a special feature of the well-known Google Maps and Google Earth services that can be accessed via the Internet. The Street View software provides access to the panoramic images that have been acquired at intervals of 10 to 20 m along the streets of many cities within the more highly developed countries of the world—in the United States, Western Europe, Japan, and Australasia. The service was first introduced with coverage of a few cities in the United States in May 2007. The American coverage has been continually extended since then. Just over a year later, in July 2008, Street View was introduced to Europe, first of all in France. Later that year, coverage of certain cities in Spain, Italy, the Netherlands, and the United Kingdom began. Since then, work has continued intensively and on a massive scale to extend the coverage to even more cities and to expand the coverage of the streets within each city that is being covered. The numbers and types of mobile mapping vehicles that have been used

(a)   (b)

FIGURE 2.56 (a) A fleet of TomTom mobile mapping vehicles that are based in Poland. (b) Showing in close-up the mobile mapping system featuring a multiple camera unit, SICK 2D laser profilers, and a GNSS receiver antenna that is fitted to the roof of each of the TomTom mapping vehicles. (Courtesy of TomTom, Amsterdam, the Netherlands. With Permission.)

to acquire Street View imagery in different countries has varied considerably from country to country. At the start of the program, the imaging technology that was used also varied considerably.

Initially, in 2007, much of the imagery of the first batch of cities that were covered in the United States had been collected by the Immersive Media company on contract using its distinctive Dodeca multiple camera systems. However, since 2008, Google has been collecting the required image, laser profile and positional data using its own vehicles, steadily expanding its fleets of mobile mapping cars for the purpose. In the United States, Australasia, and Japan, the cars were at first equipped with Point Grey Research Ladybug multicamera units attached to a mast mounted on the roof of the car. However, since then, these have been replaced by various multicamera systems that are unique to Google. These cameras are mounted on a sturdy mast that is itself attached to a roof rack that is fitted to the roof of the car. According to Anguelov et al., 2010, three successive generations of multicamera systems have been deployed: (1) The R2 system comprised a ring of eight 11 Megapixel CCD cameras; (2) the R5 system utilized a ring of eight CMOS cameras with a ninth camera equipped with a fisheye lens placed on top of the main ring and pointing vertically upwards to photograph the upper levels of high buildings (Figure 2.57a and b); and (3) the R7 system that uses 15 of the same cameras employed in the R5 series to provide comprehensive panoramic coverage without the need for a fisheye camera (Figure 2.57c). The mast and camera system can be folded down on to the roof rack when not in use. Each car is equipped with a combined DGPS/IMU system that has been supplied by Topcon, together with a wheel-mounted odometer that, in conjunction with the IMU, can help establish position wherever GPS coverage is poor or has been lost in tunnels or within high-rise urban areas. All the Google Street View cars also feature a pair of SICK LMS 291 2D laser profilers that are attached to the mast and continuously measure a series of range or elevation profiles to either side of the mapping vehicle. A third SICK LMS291 profiler measures the road surface in front of the vehicle. Besides the car-based MMSs (Figure 2.58a), Google has also introduced a number of pedal-powered tricycles (trikes) (Figure 2.58b) that are equipped with a similar set of cameras, 2D laser profilers, and positioning equipment to those used on the Google mapping vehicles. These are being used for data collection in areas such as pedestrian precincts and public parks and along cycle tracks on which cars cannot be operated. The article by Anguelov et al. (2010) quoted previously also gives details of the use that is made of the 2D laser profile data, which is used to form a 3D model or mesh on to which the photographic images can be fitted. Information on the further processing carried out by Google on this captured data is given in the article by Madrigal 2012.

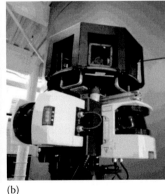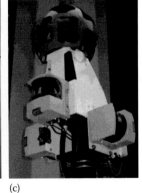

(a)   (b)   (c)

FIGURE 2.57 (a) Showing the various components at the top of the mast mounted on the roof of a Google Street View mapping car, including an R5 multicamera unit with its fisheye lens at the top, the Topcon control box, and the three SICK LMS291 2D laser profilers. (b) A close-up photograph of the R5 camera and two of the SICK LMS291 laser profilers, the one measuring a 2D vertical profile, the other a 2D horizontal profile. (c) The later R7 camera unit containing 15 individual cameras and the three accompanying SICK LMS291 2D laser profilers. (Courtesy of Google, Mountain View, California. With Permission.)

(a)   (b)

FIGURE 2.58   (a) A Google Street View car with the mast-mounted system as shown earlier in Figure 2.57c. (b) A pedal-powered tricycle that has been equipped with a similar set of imaging, laser profiling, and positioning devices as the Google car, as seen at Warwick Castle in England. (Courtesy of Google. With Permission.)

2.3.4.3   Navteq and Nokia

Navteq was a large American mapping company with its headquarters in Chicago and a large map data production center located in Fargo, North Dakota, that was supplemented and supported by a network of smaller national and regional offices located worldwide. In December 2007, the company was purchased by the Finnish Nokia organization, which is a major supplier of telecom networks and infrastructure on a worldwide scale. Nokia also provided its Ovi Maps product (previously called Nokia Maps), which could be downloaded free by those customers who had purchased the company's smartphones that were equipped with a suitable processor, display screen, and operating system. However, besides supplying digital map data to Nokia for incorporation in these products, Navteq also provided digital map databases for the navigation systems that were being installed in the cars that are being built by several different manufacturers. Besides which, the company also supplied digital map data for use in portable GPS sets and in the Internet-based map applications that are being provided by Microsoft (Bing Maps) and Yahoo (Yahoo Live Maps). In 2012, Nokia rebranded its mapping and location services under the single title *Here*. In November of that year, it also acquired the mobile mapping service of the Earthmine company that was based in California. In December 2015, Nokia sold the entire *Here* business to a consortium of very large German car manufacturing companies comprising Audi, BMW, and Daimler (Mercedes).

Originally, the revision of the Navteq map databases of road networks had been carried in a relatively simple manner using survey cars with a crew of two. These cars were equipped solely with a roof-mounted GPS receiver and a laptop computer installed inside the car that had been loaded with the map database for the local area that was being surveyed. This database was revised manually and visually by the crew as they drove around the area that needed revision. However, by 2006, a digital video camera had been installed in many of the Navteq survey cars to provide a video record of each survey trip. In 2008, a new fleet of mobile mapping vehicles was introduced by Navteq. Each vehicle was equipped with an array of six or eight digital video cameras and an Applanix POS LV GPS/IMU subsystem based on a Trimble GPS receiver. It was not until 2010 that Navteq started to utilize 2D laser profilers on its mapping vehicles, first of all with its adoption of the Topcon IP-S2 HD system that was based on the Velodyne HDL-64E unit, as described earlier (Figure 2.59a). With the purchase of Earthmine in 2012, Nokia also acquired that company's digital camera technology that was licensed from Caltech-JPL. These cameras continued to be used in a modified (non-stereo) configuration, supplemented by a Velodyne HDL-32E unit to provide the 3D data that had formerly been produced by the stereo-imagery collected by Earthmine (Figure 2.59b).

(a)   (b)

FIGURE 2.59 Showing two alternative mobile mapping systems used by HERE Technologies' mapping vehicles (a) using a Point Grey Ladybug multi-camera unit in combination with a Velodyne HDL-64E spinning laser unit; and (b) using an Earthmine four-camera unit in combination with a Velodyne HDL-32E spinning laser unit. (From HERE Technologies.)

2.3.4.4 Apple

Apple Inc. announced in 2015 that it was operating a fleet of vehicles that were collecting data for its *Apple Maps* web mapping service to supplement the map and image data that it was already sourcing from other established providers such as TomTom and DigitalGlobe. Reports suggest that the main coverage is in the United States with some limited areas in Europe (in France, the United Kingdom, Ireland, Italy, and Sweden) also being covered. It would appear that two different systems have been developed for the purpose. The first type (Figure 2.60a) of which numerous photographs have been posted on the Web, has a roof-mounted platform equipped with digital cameras located at each of the four corners of the platform. These are supplemented by two Velodyne HDL-32E laser scanner/profilers, one pointing forward, the other backwards from the front and the rear of the roof-platform, respectively. The system also features twin GPS or GNSS receiver antennae. The second, more recent, system (Figure 2.60b) has a quite different configuration employing four of the Velodyne HDL-32E units, with one located at each of the four corners of the roof platform. The imaging cameras have been displaced and are now located midway between each pair of laser units, with a single camera pointing in the forward and backward directions and an unusual arrangement of double cameras pointing in diverging directions to the side of the mapping vehicle.

(a)   (b)

FIGURE 2.60 Mapping vehicles from *Apple Maps* with different configurations of digital cameras and Velodyne laser profilers. (a) This example has each of the four camera units mounted at the corners of the roof-top platform and has one of the two Velodyne laser units pointing forwards and the other pointing backwards. (b) This other example has each of the four Velodyne laser units located at the corners of the roof-top platform with the camera units located midway along each of the four sides on the roof of the vehicle. (Enterprise Garage, Los Altos, California. With permission.)

2.3.5 SUMMARY—DYNAMIC TERRESTRIAL LASER SCANNERS

When the first edition of this book was written, mapping systems that were capable of being operated from moving vehicles were only commercially available off-the-shelf from one or two suppliers. Indeed, the commercial operation of mobile mapping vehicles by service providers using laser scanners and profilers was in its infancy. The change that has occurred since then has been dramatic. Numerous systems are now available from different suppliers employing differing numbers of laser units in a wide variety of configurations. Furthermore, large fleets of mobile mapping vehicles are in operation by service providers in many parts of the world. Most of the systems that are being operated combine the data from laser profiling with the images produced by digital frame cameras. However, looking at the current situation overall, the differences in the performance of the different systems and the accuracy of the resulting data have resulted in two distinct categories of systems being created. (1) Into the first category fall those systems that are being used for the acquisition of the digital images and coordinate data that are required primarily for street-level image display purposes, for vehicle navigation, and for cartographic mapping applications. (2) Into the second category fall those systems that are being used to collect the more accurate 3D coordinate data that can be used to describe very accurately the road and rail infrastructure that is needed for engineering, maintenance, and management purposes. However, within this whole context of vehicle-based MMSs, it is important to remember that the methodology is limited to the surveying and mapping of quite narrow corridors of terrain. The surveying and mapping of large areas of topography require the use of the airborne systems that are the subject of the next chapter of this book.

REFERENCES AND FURTHER READING

STATIC TERRESTRIAL LASER SCANNERS

Anon, 2008. The Trimble VX Spatial Station. *The American Surveyor*, 5(9): 10.
Billings, S., 2008. The Topcon IS Imaging Station. *The American Surveyor*, 5(11): 61–66.
Billings, S., 2012. IS-3 Imaging Station. *The American Surveyor*, 9(6): 35–38.
Blais, F., 2004. Review of 20 years of range sensor development. *Journal of Electronic Imaging*, 13(1): 231–240.
Bohm, J. and Haala, N., 2005. Efficient integration of aerial and terrestrial laser data for virtual city modelling using lasermaps. Proceedings ISPRS WG III/3, III/4, V/3 Workshop Laser Scanning 2005, Enschede, the Netherlands, September 12–14, 2005, pp. 192–197.
Fröhlich, C. and Mettenleiter, M., 2004. Terrestrial laser scanning—New perspectives in 3D surveying. Laserscanners for forest and landscape assessment. *International Archives of Photogrammetry and Remote Sensing*, 36(Part. 8, W2): 7–13.
Frazier, T.J., 2012. Hardware review: Trimble VX Spatial Station. *Professional Surveyor*, 32(8).
Grimm, D.E., 2013. Leica Nova MS50: The world's first multi station. *GEOInformatics*, 16(7): 22–26.
Grimm, D.E. and Zogg, H-M., 2013. Leica Nova MS50. Leica Geosystems White Paper, 12 pp.
Ingensand, H., 2006. Metrological aspects in terrestrial laser-scanning technology. *Third IAC/12th FIG Symposium*, Baden, Germany, May 22–24, 2006, 10 pp.
Kersten, T., Sternberg, H., and Mechelke, K., 2005. Investigations into the accuracy behaviour of the terrestrial laser scanning system Trimble GS100. In Gruen, A. and Kahmen, H., (Eds.) *Optical 3D Measurement Techniques VII*, Vol. 1, pp. 122–131.
Lichti, D., Gordon, S., and Stewart, M., 2002. Ground-based laser scanners: Operation, systems and applications. *Geomatica*, 56(1): 21–33.
Lichti, D., Brustle, S., and Franke, J., 2007. Self calibration and analysis of the Surphaser 25HS 3D scanner. Strategic Integration of Surveying Services, FIG Working Week 2007, Hong Kong SAR, China, May 13–17, 2007, 13 pp.
Maar, H. and Zogg, H-M., 2014. WFD—Wave form digitizer technology. Leica Geosystems White Paper, 12 pp.
Mechelke, K., Kersten, T., and Lindstaedt, M., 2007. Comparative investigations into the accuracy behaviour of the new generation of terrestrial laser scanning systems. In Gruen, A. and Kahmen, H., (Eds.) *Optical 3-D Measurement Techniques VIII*, Vol. I, Zurich, Switzerland, July 9–12, 2007, pp. 319–327.
Mettenleiter, M., Hartl, F., and Fröhlich, C., 2000. Imaging laser radar for 3-D modelling of real world environments. *International Conference on OPTO/IRS2/MTT*, Erfurt, Germany, May 11, 2000, 5 pp.

Mettenleiter, M., Härtl, F., Heinz, I., Neumann, B., Hildebrand, A., Abmayr, T., Fröhlich, C., 2003. 3D Laser scanning for engineering and architectural heritage conservation. *Proceedings of the XIXth International Symposium CIPA 2003 "New Perspectives To Save Cultural Heritage,"* Antalya, Turkey, pp. 484–489.

Petrie, G., 2012. Technology developments in terrestrial laser scanners: With specific reference to the Leica Geosystems HDS and ScanStation product ranges. *GEOInformatics*, 15(5): 20–26.

Pfeifer, N. and Briese, C., 2007. Geometrical aspects of airborne laser scanning and terrestrial laser scanning. *International Archives of Photogrammetry and Remote Sensing*, 36(Part 3, W52): 311–319.

Schulz, T. and Ingensand, H., 2004. Influencing variables, precision and accuracy of terrestrial laser scanners. INGEO 2004 and FIG Regional Central and Eastern European Conference on Engineering Surveying, Bratislava, Slovakia, November 11–13, 2004, 8 pp.

Staiger, R., 2003. Terrestrial laser scanning—Technology, systems and applications. Second FIG Regional Conference, Marrakech, Morocco, December 2–5, 2003, 10 pp.

Sternberg, H., Kersten, T., Jahn, I., and Kinzel, R., 2004. Terrestrial 3D laser scanning—Data acquisition and object modelling for industrial as-built documentation and architectural applications. *International Archives of Photogrammetry, Remote Sensing and Spatial Information Sciences*, 35(Commission VII, Part B2): 942–947.

Walsh, G., 2015a. Leica ScanStation P-Series–Details that matter. Leica Geosystems White Paper, 20 pp.

Walsh, G., 2015b. Tilt Compensation for Laser Scanners. Leica Geosystems White Paper, 16 pp.

DYNAMIC TERRESTRIAL LASER SCANNERS

Anguelov, D., Dulong, C., Filip, D., Frueh, C., Lafon, S., Lyon, R., Ogale, A., Vincent, L., and Weaver, J., 2010. Google Street View: capturing the world at street level. *Computer*, 43(6): 32–38.

Bossler, J. and Toth, C., 1996. Feature positioning accuracy in mobile mapping: Results obtained by the GPSVan™. *International Archives of Photogrammetry and Remote Sensing*, ISPRS Comm. IV, 31(Part B4): 139–142.

Duffy, L., 2014. Can one mobile mapping system do it all? *Lidar News Magazine*, 4(4): 38–42.

Elseberg, J., Borrmann, D., and Nuchter, A., 2013. A study of scan patterns for mobile mapping. *International Archives of Photogrammetry, Remote Sensing and Spatial Information Sciences*, Xl-7/W2: 75–80.

Hunter, G., Cox, C., and Kremer, J., 2006. Development of a commercial laser scanning mobile mapping system—StreetMapper. Second International Workshop, The Future of Remote Sensing, Antwerp, Belgium, October 17–18, 2006, 4 pp.

Inaba, K., Manandhar, D., and Shibasaki, R., 1999. Calibration of a vehicle-based laser/CCD sensor system for urban 3D mapping. *Proceedings 20th Asian Conference on Remote Sensing (ACRS)*, Hong Kong, China, 7 pp.

Kremer, J. and Hunter, G., 2007. Performance of the StreetMapper mobile LiDAR mapping system in "real world" projects. In Fritsch, D., (Ed.) *Photogrammetric Week '07*, pp. 215–225.

Madrigal, A., 2012. How Google builds its maps and what it means for the future of everything. *The Atlantic*, September 2012 issue, 7 pp.

Manandhar, D. and Shibasaki, R., 2000. Geo-referencing of multi-range data for vehicle-borne laser mapping system (VLMS). *Proceedings 21st Asian Conference on Remote Sensing (ACRS)*, Taipei, Taiwan, pp. 974–979.

Manandhar, D. and Shibasaki, R., 2003. Accuracy assessment of mobile mapping system. *Proceedings 24th Asian Conference on Remote Sensing (ACRS)*, Busan, South Korea, 3 pp.

Ozguner, U., Redmill, K., Toth, C., and Grejner-Brzezinska, D., 2007. Navigating these mean streets: Real-time Mapping in Autonomous Vehicles. *GPS World*, 18(10): 32–37.

Petrie, G., 2010. Mobile mapping systems: An introduction to the technology. *GEOInformatics*, 13(1): 32–43.

Petrie, G., 2014. SPAR Europe and European lidar mapping forum: Review of the exhibition at the joint meeting held in Amsterdam. *GEOInformatics*, 17(1): 36–42.

Schulz, R., Justus, W., and Kohler, M., 2007. Mapping Based Laserscanner ACC, Presented Paper, 11th International Forum on Advanced Microsystems for Automotive Applications, May 9 and 10, 2007, 11 pp.

Talaya, J., Bosch, E., Alamus, R., Serra, A., and Baron, A., 2004a. Geovan: The mobile mapping system from the ICC. *Proceedings 4th International Symposium on Mobile Mapping Technology*, Kunming, China, March 29–31, 2004, 7 pp.

Talaya, J., Alamus, R., Bosch, E., Serra, A., Kornus, W., and Baron, A., 2004b. Integration of terrestrial laser scanner with GPS/IMU orientation sensors. *International Archives of Photogrammetry and Remote Sensing*, ISPRS Comm. V, Vol. 35, Part B5, 6 pp.

Talend, D., 2010. Comprehensive collection. *The American Surveyor*, 7(6): 52–56.

Toth, C., 2009. R&D of mobile lidar mapping and future trends. *Proceedings of the American Society of Photogrammetry and Remote Sensing*, Baltimore, MD, 7 pp.

Toth, C. and Grejner-Brzezinska, D., 2003. Driving the line: Multi-sensor monitoring for mobile mapping. *GPS World*, 14(3): 16–22.

Toth, C. and Grejner-Brzezinska, D., 2004. Redefining the paradigm of modern mobile mapping: An automated high-precision road centerline mapping system. *Photogrammetric Engineering and Remote Sensing*, 70(6): 685–694.

Zampa, F. and Conforti, D., 2009. Mapping with mobile lidar. *GIM International*, 23(4): 35–37.

Zhao, H. and Shibasaki, R., 2003. A vehicle-borne urban 3D acquisition system using single-row laser range scanners. *IEEE Transactions, SMC Part B: Cybernetics*, 33(4): 658–666.

Zhao, H. and Shibasaki, R., 2005. Updating a digital geographic database using vehicle-borne laser scanners and line cameras. *Photogrammetric Engineering and Remote Sensing*, 71(4): 415–424.

3 Airborne and Spaceborne Laser Profilers and Scanners

Gordon Petrie and Charles K. Toth

CONTENTS

3.1 INTRODUCTION

As already outlined in the introductory section of Chapter 1, the direct measurement of terrain elevations from an airborne platform using laser rangefinders began with the laser profiling technique. This involved the measurement of successive ranges continuously from an aircraft vertically downwards toward the terrain along the ground track of the aircraft. Essentially, these measured ranges gave the successive values of the ground clearance between the aircraft and the terrain. If the corresponding values of the successive positions and altitude of the aircraft in the air from which the laser ranges were made could be measured accurately, for example, using an Electronic Distance Measuring (EDM) system, then, this allowed the determination of the profile of the terrain along the ground track. The initial development of this airborne profiling technology and methodology was an important first step that led eventually to the development of the airborne laser scanners that are now in widespread use for topographic mapping purposes. In the case of spaceborne platforms, given the great distances (flying heights) and very high speeds (29,000 km/h) at which these platforms orbit the Earth, until now, only laser profilers can be operated from them. Laser scanners cannot be used from spacecraft as yet—given the present limitations in the laser ranging technology that is currently available.

Valuable though the terrain profiles were (and are), they have obvious limitations when large areas of terrain have to be measured, either to form a terrain elevation model or to form part of a mapping project. However, once suitable scanning mechanisms had been devised successfully in the early to mid-1990s, airborne laser scanners were then designed and built, and they quite quickly started to come into operational use for the determination of terrain elevation values over large swaths of terrain. The practical implementation of the airborne laser scanning technique was also made possible through the successful concurrent development of integrated GPS/inertial measurement unit (IMU) systems that could determine the successive positions, altitudes, and attitudes of the scanner systems while they carried out their scanning/ranging measurements. After their introduction in the mid-1990s, airborne laser scanners were at first adopted gradually and rather cautiously. However, since 2002, as the technology has improved and experience was gained, they have been brought into service in ever increasing numbers—in spite of their considerable cost of purchase. Now, they are a major element in the current topographic mapping scene and in coastal bathymetric mapping and charting. Since the first edition of this book was published, the technology of mainstream airborne laser scanners has developed apace allowing ever more rapid and denser acquisition of terrain elevation and bathymetric depth data. At the same time, a new and somewhat different technology has been developed that allows laser scanning to be carried from very lightweight Unmanned Aerial Vehicle (UAV) aircraft. As will be seen, much of this latter development is based on the instrumentation that had already been developed for terrestrial laser scanning and for vehicle-based mobile mapping and navigation purposes.

The objective of this chapter is to present to the reader an account of the technology of airborne and spaceborne laser profilers and scanners as it exists in 2016. The arrangement of the chapter will be to first cover airborne and spaceborne laser profilers in Sections 3.2 and 3.3, after which, airborne laser scanners will be dealt with in much greater detail and at greater length starting in Section 3.4— as befits their relative importance in carrying out mapping operations. The coverage of airborne laser scanners will be subdivided into three main parts—(1) those used for topographic mapping from manned aircraft covered in Section 3.5, (2) those that are being utilized on unmanned aircraft in Section 3.6, and (3) those employed in bathymetric mapping and charting operations in Section 3.7.

3.2 AIRBORNE LASER PROFILERS

Soon after the first lasers had been constructed and demonstrated, they began to be used for range-finding purposes. In particular, laser altimeters were devised for use in aircraft—for example, in the United Kingdom, an experimental airborne laser altimeter using a Gallium Arsenide (GaAs) semiconductor laser was demonstrated as early as 1965 (Shepherd, 1965). Shortly after that, the first airborne laser profiler was introduced for use in commercial topographic mapping operations in the United States (Miller, 1965; Jensen and Ruddock, 1965). The instrument had been developed jointly by the Spectra Physics Company—which had built the laser—and the Aero Service Corporation— then a major aerial survey and mapping company that was owned by Litton Industries. The range-finder part of the system was based on a helium–neon gas laser operating as a continuous wave device at the wavelength of $\lambda = 632.8$ nm. By the mid-1970s, the situation had changed entirely, and pulsed time-of-flight (TOF) laser rangefinders based on both Nd:YAG solid-state lasers and GaAs semiconductor lasers had been developed, which could be used to form the basis of airborne laser profiling systems. For example, the Avco Everett company from Everett, Massachusetts, produced its Avco Airborne Laser Mapping System in 1979 (McDonough et al., 1979). Another slightly later example from the 1980s was the PRAM III laser profiler that was built by Associated Controls & Communications (later Dynatech Scientific) of Salem, Massachusetts. The laser rangefinder used in this profiler was based on a GaAs semiconductor diode laser operating in a pulsed mode at $\lambda = 904$ nm. Yet another well-known airborne profiling system from the 1980s was the Model 501 SX built by the Optech (later the Teledyne Optech) company based in Toronto, Canada, which is now such a major force as a system supplier of airborne laser scanners. The Model 501 SX profiler used a GaAs semiconductor diode laser operating at the wavelength (λ) of 904 nm with a 15 ns pulse width and measuring rates up to 2 kHz. Since, at that time, the GPS system was not fully operational, various types of microwave ranging systems were utilized by different operators to provide the continuous accurate positioning of the PRAM III and Model 501 SX profiling systems in the air.

It may well seem that, given the extra coverage of the ground provided by the cross-track scanning method, the airborne laser profiler would by now have disappeared. However, a very small number have remained in operational service until very recently, being used typically by scientific organizations for the mapping of barely changing surfaces such as water bodies, ice-covered terrain, and flat land areas. A notable example has been the Geodesy and Geodynamics Laboratory Airborne Laser Profiler of the Geodesy and Geodynamics Laboratory of the Eidgenössische Technische Hochschule (ETH), Zürich (Geiger et al., 2007). This system was based on a Teledyne Optech Model 501 profiler complemented by a position and attitude measurement subsystem provided by four GPS receivers (Figure 3.1). Forest research is another area where the use of airborne profilers has continued till very recently. Indeed, a system—called the Portable Airborne Laser System—was developed over a decade ago specifically for this particular application by NASA's Goddard Space Flight Center (GSFC) using low-cost commercial-off-the-shelf (COTS) components (Nelson et al., 2003). This system was based

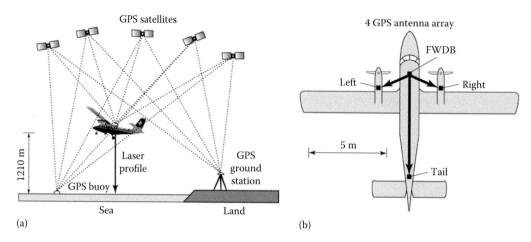

FIGURE 3.1   (a) Shows the overall concept of the airborne profiling being carried out by the Geodesy and Geodynamics Laboratory airborne laser profiler of the ETH, Zürich. (b) Shows the arrangement of the four GPS receivers that are mounted on the ETH Twin Otter aircraft to provide positional and attitude data. (Courtesy of Institute of Geodesy and Photogrammetry, ETH, Zürich; redrawn by M. Shand. With Permission.)

on a *RIEGL* LD90-3800-VHS laser rangefinder with a 20 ns wide laser pulse and having a pulse repetition rate of 2 kHz. Another interesting development was the incorporation of an airborne laser profiler into the Geographic Synthetic Aperture Radar (GeoSAR) synthetic aperture radar system operated by the Fugro EarthData survey and mapping company. The GeoSAR features a dual-sided synthetic aperture radar (SAR), comprising two X-band and two P-band systems, collecting IfSAR image data on either side of the aircraft's flight line. This was supplemented by the nadir-pointing laser profiler that provided additional high-accuracy elevation data of the ground directly under the aircraft's flight path. These data could then be used for sensor calibration or for quality control purposes by providing reference data for neighboring strips.

3.3   SPACEBORNE LASER PROFILERS

A number of microwave (radar) profilers have been mounted in satellites and used very successfully to measure sea surface topography over large areas of the world's oceans. They include the well-known TOPEX/Poseidon and Jason-1 missions operated jointly by NASA and the French CNES space agency and the various radar altimeters operated by the European Space Agency on board the ERS-1 and -2 and Envisat satellites. These microwave profilers work less well over areas of land and ice that are characterized by large variations in elevation. This is due in large part to the comparatively large angular beam widths of these microwave radar instruments and the resulting large area footprints on the ground. Thus, a substantial group of scientists concerned with research studies of large ice sheets, such as those covering Greenland and Antarctica, campaigned for the development and use of laser altimeters (profilers) from satellites (Figure 3.2), since they appeared to offer much narrower angular beams and smaller ground footprints over the ice-covered terrain than those produced by microwave (radar) profilers. In turn, this would lead to much finer resolution (from the smaller footprints) and higher accuracies in the resulting elevation data. Besides which, the recording of the returned waveform would allow studies of surface roughness to be undertaken. Other scientists concerned with research into desert topography and the movement of sand-based geomorphological features were also interested in the improved elevation data that could be generated by spaceborne laser profilers. Furthermore, atmospheric scientists also pressed to have a laser profiler operated from space to measure cloud structures and aerosol layers. NASA responded to these various pressures from the scientific community by conducting a series of experimental missions using laser profilers mounted on various Space Shuttle spacecraft. This led ultimately to

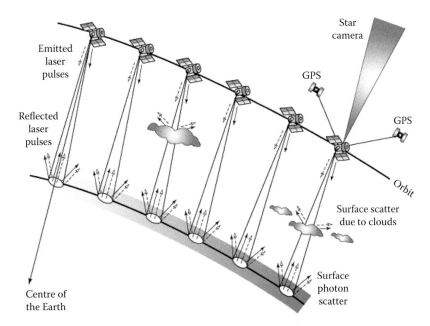

FIGURE 3.2 Diagram showing the operational mode of a spaceborne laser profiler with the transmitted pulses being reflected both from the ground and from the cloud cover. (Courtesy of NASA, Washington, DC; redrawn by M. Shand. With Permission.)

the development and deployment of a dedicated Ice, Cloud, and land Elevation Satellite (ICESat) equipped with a specially designed laser profiler, as will be discussed later.

3.3.1 LiDAR In-Space Technology Experiment

The characteristics of a laser profiler suitable for operation from space needed of course to be substantially different to those of an airborne profiler. In particular, the operational height over which the distances had to be measured would be in the order of 250–800 km instead of the few kilometers of an aircraft flying height, whereas the speed over the ground would be 29,000 km/h instead of the 200–300 km/h of survey aircraft (Figure 3.2). NASA's first experimental mission was the LiDAR In-space Technology Experiment (LITE) that was mounted on board the Shuttle flight STS-64 in 1994. This used large diameter (1 m) optics that were very heavy, together with a very powerful flashtube-pumped Nd:YAG laser that required considerable electrical power (3 kW)—all of which could be coped with by the Shuttle spacecraft. In the event, the mission concentrated mainly on satisfying the requirements of atmospheric, weather, and climatic research scientists through the measurement of cloud height and aerosol and dust layers rather than the Earth's topography.

3.3.2 Shuttle Laser Altimeter

However, the two follow-up missions that were accommodated on Shuttle flights in 1996 and 1997 were much more oriented toward topographic applications, including the measurement of Earth surface relief and vegetation canopies. For these applications, both missions used the Shuttle Laser Altimeter (SLA). This profiler employed a much smaller, lighter, and less power-hungry laser rangefinder based on a Q-switched diode-pumped Nd:YAG laser (with $\lambda = 1{,}064$ nm) and a much smaller (38 cm) diameter optical telescope. The pulse energy of this laser was 40 mJ, and the pulse repetition rate was 10 Hz. The first of these two missions—called SLA-01—was flown in January 1996 on Shuttle flight STS-72 and lasted 10 days. A major problem with the SLA profiler that was experienced during this first mission was the deep saturation of the detector and its electronics arising from the strong levels of reflection that occurred over deserts and other flat surfaces. However, notwithstanding these difficulties, the SLA-01 mission was judged to be successful.

Based on the experience gained during the SLA-01 mission, the profiler was modified extensively for the SLA-02 mission that was flown on the Shuttle STS-85 flight that took place over a 12 day period in August 1997 (Carabajal et al., 1999). In particular, the laser rangefinder was fitted with a variable gain-state amplifier that allowed the signal intensity that was passed to the detector to be controlled from the mission ground control station to prevent the detector becoming saturated. The pulses from the SLA profiler generated 100 m footprints of the ground at a rate of 10 pulses per second (10 Hz) giving a spacing of 700 m between successive footprints. With the Shuttle having an orbital inclination of 57°, much more of the Earth's surface was covered by the SLA-02 mission than was achieved with the previous SLA-01 mission that had an inclination of 28.45° (Harding et al., 1999).

3.3.3 Geoscience Laser Altimeter System

Based on the experience gained from the two Shuttle SLA missions, the Geoscience Laser Altimeter System (GLAS) was developed for deployment on the dedicated ICESat satellite that was launched in January 2003 (Figure 3.3a). Changes in the elevations of regional ice sheets were to be determined by comparison of the successive sets of elevation profile data measured using the GLAS instrument. The GLAS profiler was equipped with three separate laser rangefinders that were fitted to a common optical bench. These rangefinders were designed to be operated independently so that, if one failed, another could then be used to replace it. In the event, this foresight proved to be invaluable since the first of the three rangefinders did in fact fail in March 2003 after only 38 days of operation. So the other two were then brought into operation in turn to replace it. They were operated for one month or so at a time every three to six months to extend the time series of measurements over the ice sheets. The third laser failed finally in October 2009.

Each of the GLAS rangefinders used a powerful Q-switched diode-pumped Nd:YAG laser with a pulse width of 6 ns that operated at the dual wavelengths of $\lambda = 1,064$ and $\lambda = 532$ nm in the near infrared (NIR) and green parts of the spectrum, respectively (Abshire et al., 2005). These lasers

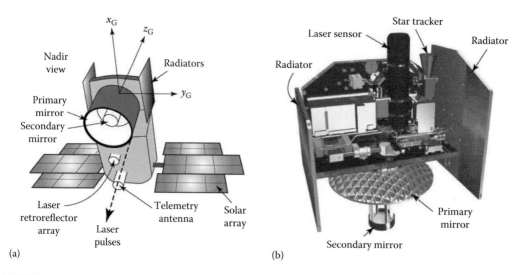

(a)   (b)

FIGURE 3.3 (a) The diagram provides a nadir view of the ICESat, showing the main features of the satellite and its GLAS laser profiler. (Courtesy of NASA, Washington, DC; simplified and redrawn by M. Shand. With Permission.) (b) A CAD drawing of the main body of ICESat when viewed from the side and showing the main elements of the GLAS laser profiler. (Courtesy of NASA, Washington, DC; additional annotation by M. Shand. With Permission.)

were operated with a pulse repetition frequency (PRF) of 40 Hz and produced pulse energies of 74 and 36 mJ, respectively, at the two different wavelengths. The lightweight beryllium mirror of the receiver telescope had a diameter of 1 m. The pulses emitted from the laser rangefinder at its orbital height of 600 km illuminated a ground area that was 70 m in diameter, spaced at 170 m intervals along the orbital track (or profile line) over the Earth's surface (Figure 3.2). The positional data were generated using onboard GPS receivers carrying out differential measurements in conjunction with an extensive network of GPS receivers located at base stations on the ground. These measurements of the satellite's position were supplemented by the measurements made by a network of satellite laser rangers located at base stations on the ground. The attitude of the satellite was measured continually using star tracking cameras mounted on the upper (zenith) side of the ICESat that constituted its Stellar Reference System (Figure 3.3b).

3.3.4 ADVANCED TOPOGRAPHIC LASER ALTIMETER SYSTEM

Currently, the Advanced Topographic Laser Altimeter System (ATLAS) profiler system is under construction by NASA's GSFC with a view to being deployed on the ICESat-2 satellite that currently is planned to be launched in 2018. The laser rangefinder will emit its pulses at the wavelength (λ) of 532 nm in the green part of the spectrum at the much higher rate of 10 kHz compared with the 40 Hz that was used in the GLAS system mounted on ICESat. It also utilizes a micropulse laser with a much lower pulse energy of 2 mJ and a 1.5 ns pulse width as compared with the corresponding values that had been used in the GLAS system. Each emitted pulse will be passed first through optical elements that steer and shape the beam and will then pass through a Diffractive Optical Element (acting as a beam splitter) that splits the pulse into six individual subpulses. These will be arranged as three pairs (Figure 3.4) with a gap of 3.3 km between each pair on the ground from the planned orbital height of 500 km. There will be a 90 m (295 ft) spacing of the profiles within each pair at the

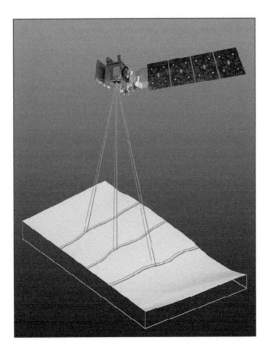

FIGURE 3.4   Diagram showing the ground coverage of the three pairs of profiles of the ATLAS spaceborne profiler as proposed for use in the NASA ICESat-2 mission. Each pair of profiles has a 90 m gap at ground level between the two individual profile lines making up the pair. There will be a gap of 3.3 km between each pair and its neighboring pair. (Courtesy of NASA, Washington, DC. With Permission.)

ground level. This is intended to allow the determination of the ground surface slope between the pair. The much higher 10 kHz pulse repetition rate will allow the ATLAS system to measure an elevation value every 2.3 ft (70 cm) along the satellite ground track. The actual spot diameter covered on the Earth's surface will be 10 m. Thus, there will be considerable overlap between the individual pulse footprints. With the comparatively low-powered micropulse laser that is being used, only a relatively few photons will be received from each reflected pulse, so, on the receiver side, a single photon counting technique will be employed using photomultiplier tube detector arrays.

3.4   AIRBORNE LASER SCANNERS

The overall concept of the airborne laser scanner is that (1) the position, height, and attitude of the airborne platform (together with that of the scanner that is mounted on it) are being measured continuously in-flight by an onboard GPS/IMU or GNSS/IMU unit with specific reference to a nearby GPS ground base station or a wide-area correction service such as OmniSTAR. (2) Simultaneously, a dense series of ranges and the corresponding scan angles from the platform to the ground are being measured very rapidly across the terrain in the cross-track direction by the laser rangefinder and by the angular encoder that is attached to the scanning mechanism. (3) Combining these two sets of measurements results in the determination of a line of elevation values at known positions (with X, Y, and Z coordinates) forming a profile across the terrain in the cross-track direction (Figure 3.5). The successive series of these measured profiles that are acquired in parallel as the airborne platform flies forward form a digital terrain model or 3D point cloud of the terrain area that has been scanned. Besides the measurement of the slant range values using a very precise clock, the scanner detectors will also measure the intensity (or energy) value of the returned pulse. However, this latter information is often very *noisy* and can be difficult to

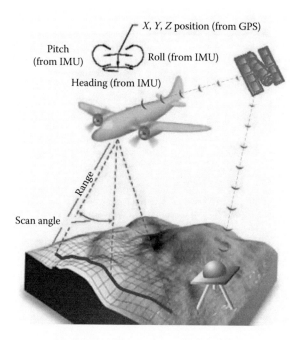

FIGURE 3.5   Diagram showing the overall principle of airborne laser scanning, with (1) the position, height, and attitude of the scanner being measured by the GNSS/IMU subsystem on board the aircraft, while (2) the slant range and scan angle values between the aircraft and the ground are being measured simultaneously by the laser rangefinder and the angular encoder attached to the scanning mechanism, resulting (3) in a profile of measured ground elevation values in the cross-track direction. (Courtesy of Leica Geosystems, St. Gallen, Switzerland; redrawn by M. Shand. With Permission.)

interpret or to utilize. Up till now, the intensity values appear to be of limited interest to most users of airborne laser scan data, whose interest and attention is usually focused on the positional and elevation data that are provided by the laser scanner for topographic mapping purposes.

3.4.1 System Configuration

A typical airborne laser scanner system comprises the following main components:

1. The basic laser ranging unit, including its transmitter and receiver optics that has already been discussed in Chapter 1.
2. An optical scanning mechanism such as a rotating mirror (together with its angular encoder) that is used to carry out the scanning of the terrain in the cross-track direction.
3. An electronics unit that comprises various hardware elements that provide many of the control and processing functions of the overall system.
4. A positioning and orientation subsystem comprising an integrated differential GPS or GNSS/IMU unit forms an essential element of the total airborne laser scanning system—without which it cannot be operated in any really practical or useful manner.
5. The software that will be used to control and coordinate the operation of each of the main elements of the system and to carry out the recording, storage, and preliminary processing of the measured data collected in-flight is a further essential component of the overall system. Furthermore, substantial processing is needed postflight to create the LiDAR point cloud.
6. An imaging device such as a digital camera, video camera, or pushbroom line scanner often forms an integral part of the overall airborne laser scanning system.

The overall configuration and the relationship of the major hardware components or units within the overall system are given in Figure 3.6a.

3.4.2 System Components

Next, a more detailed discussion of the role that each of these system components or elements plays in the overall laser scanning system will be undertaken.

1. The technology and operation of laser ranging units has already been discussed in Chapter 1. This rangefinder unit used in the airborne laser scanner will include the actual laser; the transmitting and receiving optics; and the receiver with its detector, time counter, and digitizing unit. In practice, due to the relatively large distances over which the slant ranges to the ground have to be measured from an airborne platform, in most systems, powerful pulsed lasers are used for the required TOF measurements of range (Pfeiffer and Briese, 2007). However, low-powered lasers are now being employed in a few systems in conjunction with receivers that can detect and count the individual photons that are being reflected back from the ground.
2. An optical scanning mechanism is attached rigidly in front of the output end of the laser ranging unit. As the name suggests, it uses an optical element such as a rotating plane or polygon mirror or a fiber-optic linear array to send a stream of pulses of laser radiation at known angles and at high speed along a line crossing the terrain in the lateral or cross-track direction relative to the airborne platform's flight path. This allows the sequential measurements of a series of ranges and the corresponding angles to be made to successive points along this line allowing a profile of the ground elevations to be constructed along that line. The forward motion of the aircraft or helicopter on which the laser scanner is mounted allows a series of range measurements to be made along successive profile lines in the cross-track direction. The combination of the successive sets of range and angular measurements made in both the along-track and cross-track directions allows the elevations of the ground surface and its objects to be determined for a wide swath of the terrain.

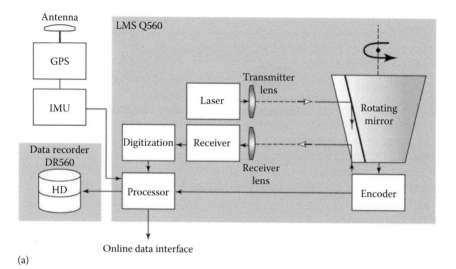

(a)

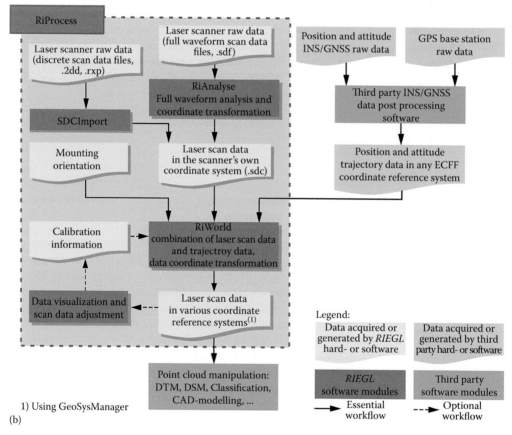

(b)

FIGURE 3.6 (a) This diagram shows the relationship between the main hardware components of an airborne laser scanner system—using the *RIEGL* LMS-Q560 as the model. (Courtesy of *RIEGL*, Horn, Austria; redrawn by M. Shand. With Permission.) (b) Diagrammatic representation of the data processing operations being carried out by a representative software package—the *RIEGL* RiPROCESS 560. (Courtesy of *RIEGL*, Horn, Austria; redrawn by M. Shand. With Permission.)

3. The electronics unit is normally computer based and is equipped with a display and an operator interface through which commands can be given to the system to execute specific actions. Besides the dedicated computer, certain electronic hardware components or cards may also be incorporated into the unit. In addition to controlling the operation of the laser rangefinder and the scanning mechanism, this unit controls the data recording system that collects and stores the measured time and waveform data from the laser rangefinder and the angular data from the scanning mechanism, together with the corresponding positional and attitude data from the GPS or GNSS receiver and the IMU. The storage of all of this data is normally carried out using the hard disk or solid state drive of the computer, although earlier systems often recorded and stored their data on digital tape drives.

4. As its title suggests, the position and orientation subsystem records various items of raw navigation data—including GPS or GNSS observations, linear and angular accelerations, and angular encoder values—that are measured continuously in-flight. After the mission has been completed, the accurate position, altitude, and attitude of the laser rangefinder and its attached scanning mechanism will be determined during a postflight computer-based data processing operation. The hardware side of the system comprises a closely integrated combination of a high quality GPS or GNSS receiver and an IMU. In many cases, the GPS receiver takes the form of a specially designed card (or cards), which is mounted on a rack within the electronics unit. However, stand-alone and/or additional receivers can also be used. The differential GPS processing requires ground reference GPS data that can be provided in any one of the four ways: (1) via a separate GNSS or GPS receiver that is being operated at a base station on the ground; (2) using a wide-area service such as OmniSTAR that employs satellite broadcast technology to provide corrections in real-time, based on a worldwide network of reference stations; (3) utilizing data provided by a local network of reference stations, a solution that is increasingly becoming popular in the industrialized world due to the widespread availability of Continuously Operating Reference Stations (CORS) networks; and (4) employing Precise Point Positioning technology that is based on the use of precise orbits and clock corrections for the satellites provided by IGS, based on worldwide network of reference GPS stations. If a wide-area service such as OmniSTAR is used, then either an OmniSTAR-ready GNSS or GPS receiver or an additional OmniSTAR L-band receiver is needed. Wide-area services allow for in-flight processing, whereas all the other solutions are based on postmission processing; note that, in some cases, the precise corrections may be available only after longer time periods, such as days or weeks.

5. Software is required for various purposes and for use in different stages of the overall airborne laser scanning operation. Initially, a package or module will be provided for mission planning and for the implementation of the planned flight lines by the pilot. This latter operation will be carried out in conjunction with the data from the GNSS or GPS receiver using a suitable cockpit display for the pilot who is flying the mission. This allows him (or her) to make the necessary adjustments to the actual path being flown by the aircraft to ensure that it is as close as possible to the planned flight line. A further module will allow the settings of various parameters of the laser rangefinder such as its scan rate, pulse rate, and scan angle to be controlled by the system operator. Additional software will also be provided for the collection and storage of all the data being measured by these different parts of the overall system. The synchronization and time-tagging of this data in-flight is normally provided by the GPS or GNSS receiver. Finally, software needs to be provided to execute any preliminary processing and display of the raw data that need to be carried out in-flight as well as that required for postflight processing (Figure 3.6b).

6. The most commonly used imaging device that will form an essential component of the airborne laser scanning system is a small- or medium-format digital frame camera. Usually, this camera will be attached rigidly to the same base plate or mount as the laser ranging unit, and its operation will be closely integrated with that of the rangefinder and the scanning mechanism.

3.4.3 System Solutions

It is appropriate next to discuss, in broad terms, the actual solutions containing these various components or subsystems that have been integrated into a system that is capable of being operated onboard airborne platforms, both for commercial service and for research purposes (Figure 3.7).

1. The first group comprises those systems that employ discrete pulse-based laser rangefinders that use a single or multiple optical-mechanical scanning profiler(s) to provide range and angular measurements in a 2D plane with the platform motion providing the third dimension. (i) Within this group, there are a substantial number of systems that simply employ a single laser rangefinder. Such devices have been built in large numbers by each of the main system suppliers—Teledyne Optech, Leica Geosystems, and *RIEGL*. (ii) The second group comprise those systems that utilize either two laser rangefinders or a single rangefinder generating dual streams of laser pulses, again being used in combination with an optical-mechanical scanning mechanism. Again, such solutions are available from Teledyne Optech, Leica Geosystems, and *RIEGL*. (iii) The third group comprises systems making use of multiple (more than two) rangefinders as the basis of the system in conjunction with an optical scanning mechanism. As will be seen later, these multiple rangefinders can be emitting their pulses at differing wavelengths resulting in a multiwavelength or multispectral laser scanner system.

2. The second group employs rangefinders generating a linear or area array of pulses that produce range and angular measurements in multiple 2D planes simultaneously, again with the platform motion imparting the third dimension to the measured dataset. (1) These arrays can be of a linear type—such as those manufactured by Velodyne, Ibeo, and

Airborne LiDAR sensor/system solutions

A. Single sensor systems (pulsed), opto-mechanical scanning profiler, platform motion provides 3rd dimension

❖ Single-sensor (Leica, Optech, *RIEGL*)

❖ Dual-sensor, two sensors sharing the same optical system (Leica, Optech, *RIEGL*)

❖ Multiple units (Optech, Selex)

B. Multiple array sensor systems (pulsed), multiple profiles, platform and/or spinning motion provides 3rd dimesion

❖ Linear arrays (Velodyne, SICK and Ibeo, Quanergy)

❖ Area arrays (Harris, Sigma Space)

FIGURE 3.7   Diagram showing, in general terms, the various solutions that have been devised for airborne laser scanning using different types of sensor systems (rangefinders).

Quanergy—that have already been described in the previous chapter (Chapter 2) concerned with terrestrial laser scanners. In some cases, the multiple pulses of the array can be generated from a single rangefinder whose emitted pulses are divided using a suitable optical device into a linear array of subpulses that cover a line on the ground—with the motion of the airborne platform providing a pushbroom coverage over the terrain. In the case of the Velodyne scanners, the 3D data are produced through a spinning action of the array as well as the platform motion. (2) Alternatively, the arrays can be of an area type—often described as focal plane arrays. Again, as will be seen later, often these area arrays comprise multiple subpulses that are generated from a single pulse emitted from an individual rangefinder with the subdivision of the original emitted pulse being effected by a suitable optical device.

3.4.4　CLASSIFICATION AND DESCRIPTIONS OF INDIVIDUAL SYSTEMS

Taking the broadest view, airborne laser scanner systems may also be divided into three quite distinct groups on the basis of their platforms (manned or unmanned) and their applications (topographic or bathymetric). These different groups will form the basis of the more detailed description of the individual systems that will follow.

1. The first of these groups comprises those systems that are mounted on manned aircraft and are designed solely for topographic mapping operations (i.e., the systems cannot carry out bathymetric mapping and charting). These topographic oriented systems will be covered in Section 3.5.
2. The second group comprises those systems that are operated on civilian UAVs—which are highly regulated; in particular, they are highly restricted with regard to the altitudes that they are allowed to fly at. This has impacted strongly on the design and the construction of the laser scanners that can be used on these aircraft. Those systems that are mounted on UAVs will be covered in Section 3.6.
3. The third group consists of those systems mounted on manned aircraft that are designed principally to carry out measurements of water depth in the bathymetric mapping operations undertaken for marine or lacustrine charting applications. However, it should be noted that these charting operations will often include some topographic mapping of the coast or shoreline areas. These bathymetric systems will be covered in Section 3.7.

3.5　AIRBORNE TOPOGRAPHIC LASER SCANNERS

As noted earlier, these systems comprise the classical types of airborne laser scanners that are designed primarily for topographic mapping purposes, including mapping carried for engineering, environmental, and scientific research applications. Primarily, they are deployed on conventional manned aircraft.

3.5.1　PRIMARY, SECONDARY, AND TERTIARY CLASSIFICATIONS

The basis for the *primary* classification of the airborne laser scanning systems used for topographic mapping purposes that has been adopted here is to divide them into three distinct categories (Petrie 2011a):

1. Scanner systems that have been manufactured in substantial numbers for commercial sale by the major system suppliers
2. Custom-built systems that have been built and operated in-house by service providers, usually in relatively small numbers

3. Airborne laser scanners that have been developed as technology demonstrators, princi-pally by NASA and, on the application side, have been used mainly for scientific research purposes

It is useful also to employ another *secondary* classification that is based on the different types of scanning mechanisms that are being utilized to scan the ground and measure its topography. The scanning action of a particular mechanism defines the pattern, the spacing, and the location of the points being measured both on the ground surface and on the objects that are present on it.

Besides this classification, it should be borne in mind that a further *tertiary* classification can also be envisaged and utilized on the basis of the range of flying heights over which the system can be used. In practice, this is based largely on the maximum distance or slant range that can be measured by the laser ranging unit. Thus, a certain number of airborne laser scanning systems with relatively short-range capabilities are designed primarily for operation from relatively low altitudes and at slow flying speeds, most often using a helicopter as the airborne platform. Often, such systems are used primarily to carry out the corridor mapping of linear features such as roads, railways, rivers, canals, power lines, and so on. At the other end of the operational range are those systems designed for commercial operation over a much greater range of flying heights at higher altitudes up to 6 km (20,000 ft.). These tend to be used in the topographic mapping of substantial areas of the Earth's terrain.

3.5.2 Scanning Mechanisms and Ground Measuring Patterns

With regard to the secondary classification outlined earlier, four main scanning mechanisms, each with its own distinctive ground measuring pattern, can be distinguished as follows:

1. Either a single mirror or a pair of oscillating plane mirrors is used in the systems that have been constructed by the two largest commercial suppliers of airborne laser scan-ning systems—Teledyne Optech with its Airborne Laser Terrain Mapper (ALTM) series and Leica Geosystems with its Aeroscan and ALS scanners. The precise angle that the mirror makes with the direction of the vertical is measured continuously using an angu-lar encoder. The use of this type of bidirectional scanning mechanism results either in a Z-shaped (saw-toothed) pattern or in a very similar sinusoidal pattern of points being measured on the ground (Figure 3.8).

2. An optical polygon that is continuously spinning in one direction providing a unidirec-tional scanning motion is used in the various systems such as the IGI LiteMapper, Trimble Harrier, and so on—that utilize the *RIEGL* laser ranging and scanning engines, besides the complete airborne laser scanning systems that are supplied by the *RIEGL* company (Figure 3.9a). The use of this type of scanning device results in a series of parallel lines of measured points being generated over the ground (Figure 3.9b). The constant rotational velocity of the optical polygon means that there is no repetitive acceleration or deceleration of the mirror. This provides for high frequencies such as the 400 Hz scan rate reached by the *RIEGL* LMS-Q1560i laser scanner. In turn, this offers better control over the spacing of the LiDAR points. The raster scanning airborne laser (RASCAL) scanner constructed and operated by NASA also uses a polygon mirror. However, this does not spin continu-ously in one direction. Instead, it oscillates in a unidirectional mode over a range of ± 16° from the nadir. Thus, it falls into the previous category.

3. A nutating mirror or prism producing an elliptical scan pattern—the so-called Palmer scan (Figure 3.10a)—over the ground is used in NASA's Airborne Topographic Mapper (ATM) and Airborne Oceanographic LiDAR (AOL) series of laser scanners; in the Coastal Zone Mapping & Imaging LiDAR (CZMIL) bathymetric scanner built by Teledyne Optech; in the Dragon Eye topographic and the Chiroptera and Hawk Eye bathymetric scanner systems

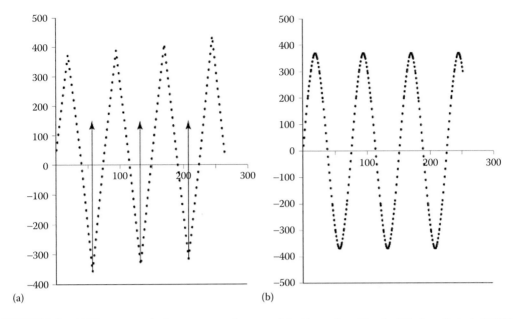

(a)   (b)

FIGURE 3.8 (a) The saw-toothed pattern over the ground that is produced by the Teledyne Optech ALTM series of laser scanners and (b) the sinusoidal pattern produced by the Leica Geosystems ALS laser scanners—in both cases, using oscillating mirrors as the scanning mechanisms. (Courtesy of Teledyne Optech, Vaughan, ON; redrawn by M. Shand. With Permission.)

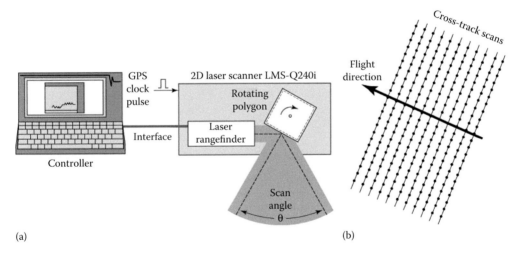

(a)   (b)

FIGURE 3.9 (a) Diagram showing the unidirectional rotating optical polygon that forms the scanning mechanism that is used in *RIEGL* laser scanning engines. (Courtesy of *RIEGL*, Horn, Austria; redrawn by M. Shand. With Permission.) (b) The resulting raster scanning pattern that covers the ground. (Drawn by M. Shand.)

built by AHAB ([Airborne Hydrography AB] which is now owned by Leica Geosystems); and in the later models in the Blom TopEye series of airborne laser scanners, also partly built by AHAB. A similar mechanism has been adopted by *RIEGL* in its VQ-880-G model that is again designed for bathymetric mapping operations. These mechanisms all produce a series of overlapping elliptical scans over the ground or water surfaces that are being mapped (Figure 3.10b).

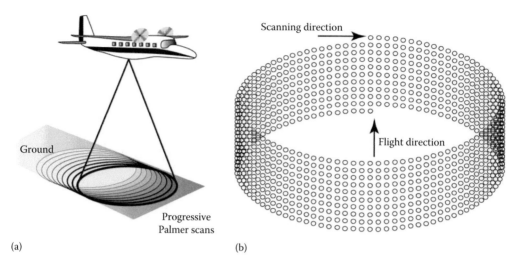

(a)   (b)

FIGURE 3.10   (a) Diagram showing the elliptical ground scanning pattern and coverage of an airborne laser scanning system that is utilizing progressive Palmer scans. (Drawn by M. Shand.) (b) LiDAR point distribution on the ground of an airborne laser scanner using progressive Palmer scans. (Drawn by M. Shand.)

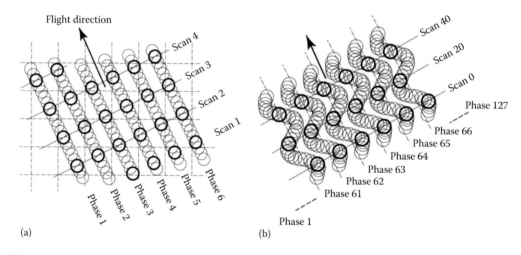

(a)   (b)

FIGURE 3.11   (a) The pattern of points measured over the ground by the TopoSys Falcon system was a series of lines parallel to the flight direction. (b) The use of a swing mirror caused systematic displacements of the laser scan pulses in the cross-track direction resulting in a *snake-like* pattern of measurements across the terrain. (Courtesy of Trimble; redrawn by M. Shand. With Permission.)

4. A pair of linear fiber-optic arrays was employed in the Falcon systems that were built and operated in-house by the German TopoSys company before its acquisition by Trimble. A pair of tilted mirrors driven by a motor distributed and collected the pulses being sent to and from the fixed set of arrays. This arrangement resulted in a series of scan lines or profiles that run parallel to the flight line as the measuring pattern covering the ground (Figure 3.11a). However, given the gaps between these measured profile lines, this basic pattern was modified somewhat in the later models in the Falcon series through the use of an additional swing mirror (Figure 3.11b). As will be seen later, the NASA Slope Imaging Multi-polarization Photon-counting LiDAR (SIMPL) and Multiple Altimeter Beam Experimental LiDAR (MABEL) research systems also employ fiber-optic arrays to transmit the laser pulses and to receive the reflections, albeit in a somewhat different manner, again resulting in a series of profiles of ground elevations being measured simultaneously along lines parallel to the flight line.

3.5.3 Mainstream Commercial System Suppliers

At the present time (in 2016), there are three main commercial suppliers of airborne laser scanner systems. The two principal suppliers in terms of volume of complete systems are (1) Optech International Inc., which has been renamed Teledyne Optech since Teledyne acquired full ownership of the company in 2015 and (2) Leica Geosystems, which, in 2013, acquired the Swedish AHAB company whose laser scanner products were first sold under the label Leica-AHAB, but are now, in 2016, being sold directly under the Leica Geosystems label or brand. The third supplier, *RIEGL*, also supplies complete systems. However, the company has also supplied quite a large number of laser scanning engines (comprising a laser rangefinder and a scanning mechanism) to a number of other system suppliers such as Trimble and IGI on an Original Equipment Manufacturer (OEM) basis. Besides which, *RIEGL* has also sold a number of these laser scanning engines to several service providers in North America. These companies have then added the other required components or subsystems that are available on a COTS basis to build up the final complete systems in-house.

3.5.3.1 Teledyne Optech

This company, based in Vaughan, Ontario in Canada, located just north of Toronto, is one of the largest commercial suppliers of airborne laser scanners. The first model in its ALTM series—the ALTM 1020—was introduced in 1993. For some years prior to this, the Optech company (as it then was) had been involved in a number of laser research and development projects, including the construction and supply of laser rangefinders and airborne laser profilers. Since its introduction, the basic design of the ALTM series has stayed much the same, but, with each new model in the series, the main elements of the design have been steadily improved and refined. The result of these developments has been a marked improvement in almost all of the operational characteristics of the ALTM series with a very substantial increase in performance and accuracy. A typical operational system showing its major components and operational parameters is shown in Figure 3.12.

The basic design of the ALTM series features a rangefinder employing a powerful Class 4 pulsed laser emitting its radiation at $\lambda = 1{,}047$ nm wavelength in the NIR part of the e–m spectrum. In late 2006, Optech introduced its multiple pulse technology that allows the firing of a second laser pulse by the rangefinder before the reflected signal from the previous pulse has been received and detected by the system (Toth, 2004). This has allowed the use of a much higher pulse repetition rate—in this case, to a maximum value of 550 kHz—to be reached in the latest ALTM Galaxy model (Figure 3.13). The ALTM scanners all employ oscillating plane mirrors as the mechanism for scanning the ground in the cross-track direction. This results in a saw-toothed pattern of measured points over the ground (Jenkins, 2006). The position and orientation subsystems that have been used throughout the ALTM series are the POS/AV models supplied by the Applanix company (now owned by Trimble), which is a near neighbor of Teledyne Optech, based at Richmond Hill in the Toronto area. A variety of dual-frequency GPS or GNSS receivers—from Trimble, Ashtech, NovAtel, and so on—that are suitable for use in aircraft have been utilized in the POS/AV systems supplied for incorporation in the different ALTM models. Similarly, with the IMUs, different units—from Litton (now Northrop Grumman), Honeywell, and Sagem—have been used, depending on the specified accuracy and the particular market into which the ALTM is being sold. In this last respect, licenses for the sale of American-manufactured IMUs from Litton and Honeywell will not be granted by the relevant U.S. government agencies—the State Department or the Department of Commerce—if the scanner is to be sold to certain countries. The accuracies offered by the different versions of the POS/AV system that have been used in conjunction with the ALTM scanners are summarized in Table 3.1.

The steadily improving performance of the different models of the mainstream ALTM scanners over the years can be seen, for instance, in the substantial rise in the PRF value—from 5 kHz in the initial ALTM 1020 model from 1993 to 10 kHz in the ALTM 1210 model and 25 kHz in the ALTM 1225 model from 1999. Since then, the frequency has risen first to 33 kHz in the ALTM

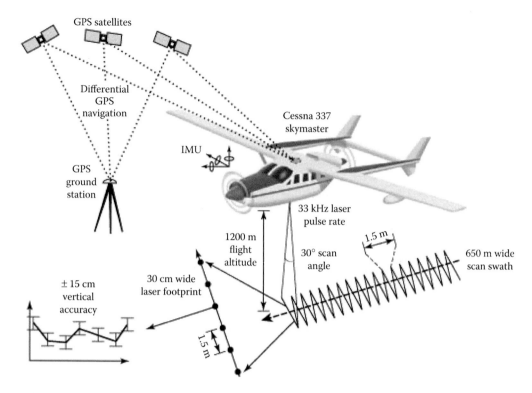

FIGURE 3.12 Diagram showing the operational concept of a Teledyne Optech ALTM airborne laser scanner, including the use of GPS in the aircraft and at the ground base station. Note the zigzag (or saw-toothed) pattern of the points that are being measured on the ground. (Courtesy of Florida International University, Miami, FL; redrawn by M. Shand. With Permission.)

FIGURE 3.13 The principal elements of a Teledyne Optech ALTM Galaxy laser scanning system with the laser scanner unit located at center and the case containing the control electronics at front left. Behind the control box unit is a small display screen that is used in the navigation of the aircraft in-flight, while at right there is a laptop computer that is used for flight management, control purposes, and data storage. (Courtesy of Teledyne Optech, Vaughan, ON. With Permission.)

2033 and 3033 models dating from the first three years of the twenty-first century to the 70 kHz of the ALTM 3070 model, the 100 kHz rate of the ALTM 3100 models, and the 167 kHz rate of the ALTM Gemini model with its multiple-pulse technology (Petrie, 2006). In 2014, Optech introduced the ALTM Galaxy model (Figure 3.13) in which the maximum PRF value could reach 550 kHz. Similarly, the maximum scan rates have risen from 28 Hz in the early ALTM models to 100 Hz in

TABLE 3.1
POS/AV Absolute Accuracy Specifications—RMSE Values

Model Number	210	310	410	510	610
Position (m)	0.05–0.3	0.05–0.3	0.05–0.3	0.05–0.3	0.05–0.3
Velocity (m/s)	0.01	0.075	0.005	0.005	0.005
Roll and pitch (°)	0.04	0.015	0.008	0.005	0.025
True heading (°)	0.08	0.035	0.015	0.008	0.005

Source: Optech and Applanix, Vaughan, ON.
Notes: The lower end POS/AV 210, 310, and 410 systems all use MEMS quartz gyros; the POS/AV 510 uses FOGs; whereas the POS/AV 610 uses ring laser gyros.

the current Galaxy model. In operational terms, the maximum operational altitude has also risen from 1,000 m with the ALTM 1020 model and 1,200 m in the ALTM 1210 and 1225 models to 4,700 m with the current ALTM Galaxy model (Young, 2015). Although the quoted range resolution of the rangefinder has remained throughout at ±1 cm, a big improvement has been achieved in terms of the actual point positioning accuracy that is achieved in object space. This is mainly due to better scanning performance and improved georeferencing performance. The resulting improvements are especially noticeable in the horizontal accuracy that has steadily improved from 1/1,000 of the flying height (H) in the initial ALTM 1020 model to 1/5500 H in the current ALTM Galaxy model. The elevation accuracy will also vary with flying height and the pulse repetition rate. The elevation accuracy values that can be achieved for the current ALTM Galaxy model, as quoted by Teledyne Optech, are given in Table 3.2.

Since 2008, Teledyne Optech has also offered its compact ALTM Orion models for use in lower altitude operations (Figure 3.14a). The initial Orion M200 model could operate over the range of 200–2500 m above ground level (AGL) to carry out reasonably wide area scanning from medium altitudes. The later Orion C200 model that was introduced in 2009 was designed specifically for low altitude operation, for example, for use in corridor surveys or engineering surveys at large scales, over an altitude range from 50 to 1,000 m. The higher altitude M200 model had a laser rangefinder operating at $\lambda = 1,064$ nm and had a pulse repetition rate of up to 200 kHz. The rangefinder in the C200 model operated at $\lambda = 1,541$ nm with an emphasis on eye safety, having regard to the very low altitudes at which it could potentially be operated. The M200 model utilized the higher grade Applanix POS AV 510 system as its georeferencing system, whereas the C200 model was fitted with a POS AV 410 unit as its GNSS/IMU subsystem. The current range of Orion scanners that is being

TABLE 3.2
Elevation Accuracy—RMSE Values

Laser Repetition Rate (kHz)	500 m Altitude (cm)	1000 m Altitude (cm)	2000 m Altitude (cm)	3000 m Altitude (cm)	4000 m Altitude (cm)
33	<5	<10	<15	<20	<25
50	<5	<10	<15	<20	N/A
70	<5	<10	<15	N/A	N/A
100	<10	<10	<15	N/A	N/A

Source: Optech, Vaughan, ON.
Note: The quoted accuracies do not include GPS errors.

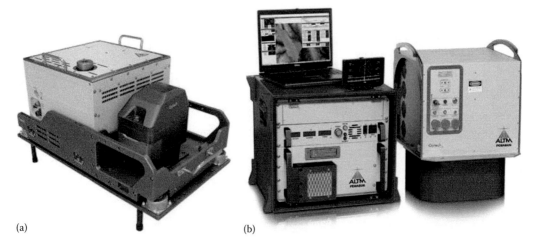

(a)   (b)

FIGURE 3.14   (a) A Teledyne Optech Orion laser scanner and a Teledyne Optech CS-10000 medium-format digital camera are mounted rigidly together on a base plate. (b) The Teledyne Optech Pegasus HD400 twin-laser system with the scanner unit at right. The control electronics cabinet is at left with a laptop computer and a system display monitor sitting on top of it. (Courtesy of Teledyne Optech, Vaughan, ON. With Permission.)

offered for sale in 2016 continues much as before with the upgraded C300 model that is designed for lower altitude operation and the M300 for medium altitudes. However, a third model has been added to the range in the shape of the H300 model that is designed to be operated at still higher altitudes up to 4,000 m.

The Pegasus scanners that were first introduced in 2010 represent another line of development in Teledyne Optech's ALTM family of airborne laser scanners (Smith and Sitar, 2010). Their design is based on the use of multiple laser rangefinders operating at $\lambda = 1,064$ nm, a single shared scanner mechanism, and a single shared Applanix POS/AV 510 GNSS/IMU subsystem. Operating in combination, a twin rangefinder version of the Pegasus scanner system could produce PRF values up to 400 kHz, giving rise to the initial commercial version that was called the Pegasus HD400 (Figure 3.14b). This particular configuration ensured its operation from medium altitudes (up to 2.5 km) and/or its production of a high density of elevation values on the ground. It was also possible to configure the instrument so that one stream of scanner pulses was nadir pointing, whereas the other pointed in a forward direction. This allowed better coverage of the vertical sides of cliffs and buildings for example. The current models are the Pegasus HA500 for high-altitude operation (up to an altitude of 5 km maximum above the terrain), and the Pegasus HD500 for operation at a lower altitude (2,500 m maximum above average ground level) and with its maximum PRF increased to 500 kHz, so allowing a still higher density of measured points over the terrain.

Yet another new line of development of Teledyne Optech's ALTM line of laser scanners is the Titan model that was first shown at the end of 2014 (Sitar, 2015). This is the first commercial multispectral or multiwavelength airborne laser scanner featuring three laser rangefinders emitting pulses simultaneously in the green ($\lambda = 532$ nm), near-infrared ($\lambda = 1,064$ nm), and short-wavelength infrared ($\lambda = 1,550$ nm) parts of the electromagnetic spectrum, together with a fully integrated medium-format (29 Megapixel) CS-6500 digital frame camera and an Applanix POS/AV AP50 subsystem as standard parts of the overall system (Figure 3.15). The green laser rangefinder is tilted forward by 7° to optimize its possible application to shallow-water bathymetric mapping. Besides which, either a Teledyne Optech CS-MS 1920 small-format (2 Megapixel) multispectral camera or a Teledyne Optech CS10000 80 Megapixel metric frame mapping camera is available as optional additions to a Titan system. The novelty of the basic system is of course that it allows mapping of the terrain at multiple wavelengths using active laser scanning techniques instead of the

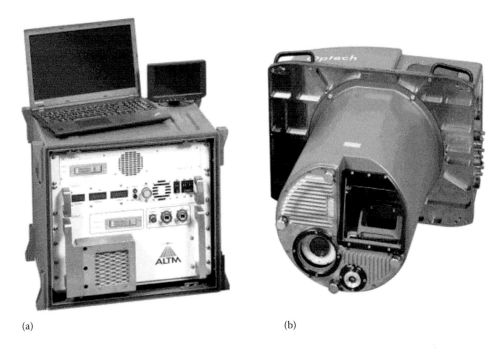

(a) (b)

FIGURE 3.15 (a) The electronics control unit with the laptop computer and display unit sitting on top of it. (b) The Teledyne Optech Titan scanner with its three laser rangefinders and built-in CS-6500 camera. (Courtesy of Teledyne Optech, Vaughan, ON. With Permission.)

purely passive imaging of the ground that is produced by a multispectral camera. Obviously, the resulting data from a Titan system will be of a wholly different type and character to that produced by the multispectral imaging camera. Furthermore, the system will produce environmental mapping of a unique kind on a 24 h (day/night) basis without the restriction to daylight hours that is, of course, a feature of the passive imaging of a conventional multispectral imaging camera.

Many of the earlier models in the ALTM scanner series were fitted with medium-format cameras as a supplementary imaging device or subsystem that could generate much higher quality images in terms of their resolution and texture as well as their color content than the rather *noisy* or grainy image produced by the intensity values that are returned by the laser pulses that are being reflected from the topographic features of the terrain. One of these digital frame cameras would be fitted rigidly to the mount or base plate of the scanner and was used to acquire imagery of the ground simultaneously during the operation of the laser scanner. Initially, these medium-format cameras were supplied either by Applanix or Rolleimetric. However, in 2010, Teledyne Optech decided that such cameras should be manufactured in-house. To implement this decision, it purchased first of all the Belgian supplier, DiMAC Systems, and then Geospatial Systems Inc. (GSI) based in Rochester, New York. The manufacture of these DiMAC and GSI cameras was then shifted to Teledyne Optech's manufacturing facility in Vaughan, Ontario. Currently, Teledyne Optech offers the range of cameras already mentioned previously. Indeed, the majority of Teledyne Optech's ALTM laser scanner systems are now supplied integrated with one or other of these frame camera models (Figure 3.14a).

3.5.3.2 Leica Geosystems

This major supplier of instrumentation and systems to the whole of the surveying and mapping industry entered the airborne laser scanning field in January 2001 through its purchase of the Azimuth Corporation based in Massachusetts in the United States. Azimuth was a small specialist company that originally supplied laser rangefinders that were incorporated into various custom-built airborne laser scanners such as the Remote Airborne Mapping System (RAMS) that was built in-house by the Enerquest (later Spectrum Mapping) company based in Denver, Colorado. However, in 1998, the Azimuth company started to build complete airborne scanner systems, under the brand name AeroScan. After the company's acquisition by Leica Geosystems in 2001, this system was rebranded and sold as the ALS40 in parallel with Leica's ADS40 pushbroom scanner producing optical linescan imagery. The original AeroScan and ALS40 systems used a large and powerful laser rangefinder emitting its pulses at the wavelength (λ) of 1,064 nm. Indeed, the special customized versions of the AeroScan built for the Fugro EarthData and 3001 aerial mapping companies in the United States could be operated at altitudes up to a maximum of 6 km. In 2003, a new and much more compact model called the ALS50 was introduced (Figure 3.16). In May 2006, an improved second generation model, called the ALS50-II, was introduced (Petrie, 2006). This featured multiple-pulse laser ranging technology (Roth and Thompson, 2008). With regard to the optical scanning mechanism, both the ALS40 and ALS50 models featured a bidirectional oscillating mirror that produced a sinusoidal pattern of measured points over the ground. With the advent of the ALS50-II model, the production of the ALS50 series was moved from Massachusetts to Leica Geosystem's main manufacturing plant located in Heerbrugg, Switzerland.

In 2007, Leica introduced its Corridor Mapper (CM) design, which, as the title suggests, was designed for corridor and other types of mapping at large scales from lower altitudes up to 1,000 m AGL. This was followed by the introduction of the ALS60 model at the ISPRS Congress in 2008. This represented a still further upgradation of the ALS50-II design with pulse repetition rates of up to 200 kHz employing multiple-pulse technology, while the maximum scan rate had increased to 100 Hz. In 2011, Leica introduced its ALS70 model (Figure 3.17) that offered a still higher performance employing its so-called *Point Density Multiplier* technology (Roth, 2011). As discussed earlier, this model generated two streams of laser pulses simultaneously using a single laser rangefinder and scanning mechanism in combination with a beam splitter and a multiple pulse operating regime. The system offered a maximum effective pulse rate of 500 kHz from a flying height of 1,000 m AGL and a 200 Hz scan rate using the dual stream of laser pulses. In fact, the ALS70 was later produced in three different forms, each optimized for a particular application. Thus, the Leica ALS70-CM

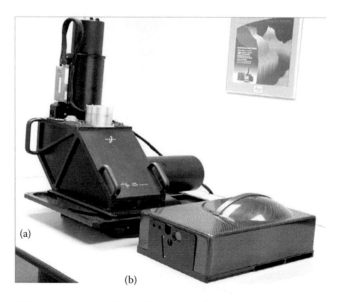

FIGURE 3.16 The difference in size (and weight) between the older Leica ALS40 laser scanner (a) and the newer ALS50 model (b) is quite striking. (Courtesy of Spencer B. Gross , Reno, Nevada. With Permission.)

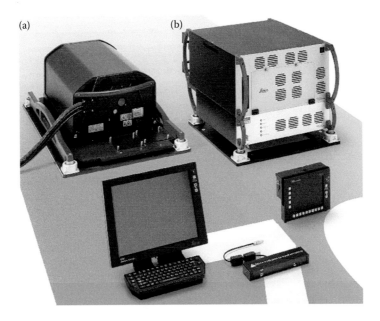

FIGURE 3.17 The Leica ALS70 airborne laser scanner is shown (a) together with its control electronics cabinet (b) and its aircraft-certified LCD displays (at front) that are used for flight management and system control purposes. (Courtesy of Leica Geosystems, St. Gallen, Switzerland. With Permission.)

variant was designed for city and corridor mapping applications at large scales in which it would be mounted in small aircraft or in a helicopter pod and flown from relatively low flying heights (up to 1,600 m). The Leica ALS70-HP model was designed for use in general-purpose medium-scale topographic mapping at the flying heights (up to 3,500 m) that are most widely used for such an application. Besides which, it could be operated over areas of greater terrain relief due to its greater ranging capability, thus allowing a higher maximum flying height. The third model was the Leica ALS70-HA, a high-altitude variant, allowing the highest flying heights (up to 5 km), for use in small-scale wide-area topographic mapping at a regional or national level, albeit with lower scan rates and a lower point density in the resulting terrain model. Another innovation that was introduced with the ALS70 was the possibility for the user to select different scan patterns—labeled as sine (sinusoidal), triangle (saw-toothed), and raster (Figure 3.18). The current model in the series is the ALS80 that was introduced in 2014. It is chiefly notable for the still further increases in the maximum measurement rate to 1000 kHz (1 MHz) from a flying height of 1000 m. Again, it is available in the three different variants (CM, HP, and HA) covering the same range of applications as the preceding ALS70 model.

Regarding the position and orientation subsystems, initially, the various models of airborne scanners in the ALS range produced by Leica Geosystems all used one or other of the models in the POS/AV range of GPS/IMU systems from Applanix. However, in 2004, the Applanix company was acquired by Trimble, one of Leica's main commercial rivals in the surveying instrument field. This led Leica Geosystems to purchase (in 2005) the small specialist Terramatics company based in Calgary, Canada. This company had already developed its IPAS (Inertial Position and Attitude System), and this was quickly adapted for use in the ALS50 airborne laser scanning system (and also in Leica's ADS40 airborne pushbroom line scanner). At the beginning of 2008, NovAtel Inc., another Canadian company based in Calgary—which is a major provider of GNSS components and integrates them with IMUs—was purchased by Hexagon, which has been the owner of Leica Geosystems since 2005. Thus, many NovAtel hardware components and software have been integrated into the appropriate Leica products and systems, including the current ALS80 system. As with the Applanix POS/AV system, a range of different IMUs with different capabilities can

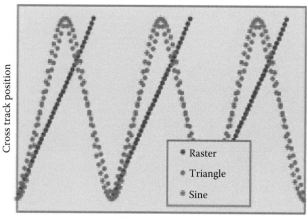

Cross track position

Along track position

* Raster
* Triangle
* Sine

FIGURE 3.18 The newer ALS70 and ALS80 airborne laser scanners offer a choice of sine, triangle, and raster scan patterns. (Courtesy of Leica Geosystems, St. Gallen, Switzerland. With Permission.)

TABLE 3.3
Specification and Accuracy Values for the Leica IPAS10 System

		NUS4	DUS5	NUS5	CUS6
Absolute	Position	0.05–0.3 m	0.05–0.3 m	0.05–0.3 m	0.05–0.3 m
Accuracy after	Velocity	0.005 m/s	0.005 m/s	0.005 m/s	0.005 m/s
Postprocessing	Roll and pitch	0.008°	0.005°	0.005°	0.0025°
RMS	Heading	0.015°	0.008°	0.008°	0.005°
Relative accuracy	Angular random noise	<0.05 deg/sqrt (h)	<0.01 deg/sqrt (h)	<0.01 deg/sqrt (h)	<0.01 deg/sqrt (h)
	Drift	<0.5 deg/h	<0.1 deg/h	<0.1 deg/h	<0.01 deg/h
IMU	High performance gyros	200 Hz FOG	200 Hz FOG	256 Hz Dry-tuned gyro	200 Hz Ring laser gyro
GPS receiver	Internal in IPAS10 control unit	12-channel dual frequency receiver (L1/L2) low noise, 20 Hz raw data, DGPS ready			

Source: Leica Geosystems, St. Gallen, Switzerland.
Notes: NUS4 is the iMAR FSAS unit; DUS5 is the Litton LN-200 unit; NUS5 is a Sagem unit; and CUS6 is the Honeywell MicroIRS unit.

be utilized in an IPAS system according to the user's needs. These IMUs are supplied either by Northrop Grumman (formerly Litton) and Honeywell in the United States or from the European suppliers iMAR (Germany) and Sagem (France). Since the georeferencing accounts for the largest part of the error budget of an airborne laser scanning system, the accuracy values specified for each of these units are summarized in Table 3.3.

A small-format digital frame camera with a 1,024 × 1,280 pixel format size was available as an optional item that could be fitted to the early ALS50 model. However, many users of the ALS scanner series preferred to use the medium-format digital frame cameras available from Applanix (DSS), Rollei (AIC), and Spectrum (NexVue), all of which were integrated by Leica to operate in conjunction with its ALS50 scanners (Roth, 2005). Then, in the summer of 2007, Leica started to

FIGURE 3.19 This Leica ALS70 laser scanner and its accompanying RCD30 digital frame camera have been placed together on a supporting base plate and mounted on a PAV80 gyro-controlled mount. (Courtesy of Leica Geosystems, St. Gallen, Switzerland. With Permission.)

supply its own RCD105 medium-format digital frame camera with a 39 Megapixel back produced by GSI based in Rochester, New York in the United States. However, GSI was acquired by Teledyne Optech in 2010, since when, the Leica ALS models have been offered with the RCD30 frame camera with a 60 or 80 Megapixel format that is produced in-house by Leica Geosystems as an optional addition to an ALS laser scanner system (Figure 3.19).

The AHAB company, which is located in Jonkoping, Sweden, was best known for its development of airborne bathymetric laser scanner systems. However, AHAB also developed the laser rangefinders and scanners that formed part of the upgrade of the original Blom TopEye laser scanner systems into the TopEye Mk. II systems. Following on from this, AHAB also developed a similar very compact topographic laser scanner system, called DragonEye, for sale to other customers. This system was designed to operate from lower altitudes (up to 1 km) over land surfaces. It was equipped with a laser rangefinder having maximum PRF values of 300 kHz at a flying height H = 200 m and 200 kHz at H = 500 m and had detectors that can record up to four return echoes per pulse. The DragonEye had a Palmer scan mechanism that generated an elliptical scanning pattern over the ground at scan rates up to 100 Hz. The GNSS/IMU subsystem that was used in the DragonEye was the iTraceRT-F200-E model manufactured by iMAR in Germany, which utilized a FOG (Fibre Optic Gyro)-based IMU. In October 2013, AHAB was acquired by the Hexagon company and became part of Leica Geosystems. Since then, the DragonEye II design (Figure 3.20) has been developed to feature twin (topographic) laser rangefinders operating at $\lambda = 1,047$ nm and set up to point obliquely forward and backward along the flight line. These allow the PRF value of the overall system to reach 1 MHz from low altitude, whereas the operational flying heights cover the range 300 to 1,600 m. The GNSS/IMU subsystem that is utilized in the DragonEye-II is a SPAN unit from NovAtel, which is of course yet another company within the Hexagon Group. It is also possible to replace one of the so-called *topographic heads* with a rangefinder operating at $\lambda = 532$ nm in the green part of the spectrum to provide a capability to carry out shallow-water bathymetric mapping.

3.5.3.3 *RIEGL*

This company is based in the town of Horn, Austria. For many years, it has been engaged in the manufacture of laser measuring instrumentation such as distance meters, speed meters, levels, altimeters, and anticollision devices for a wide range of industrial applications. Later, during the first decade of the twenty-first century, it developed a range of laser scanners for ground-based (terrestrial) surveying applications. With all this relevant background and experience, *RIEGL* also entered the field of airborne laser scanners in 2003. However, it did this in a very different manner to that of

FIGURE 3.20 The compact Leica Geosystems DragonEye II airborne laser scanner featuring twin laser rangefinders designed for low altitude operations. (Courtesy of Leica Geosystems, St. Gallen, Switzerland. With Permission.)

Teledyne Optech and Leica Geosystems discussed previously. Initially, the company developed a range of laser scanning engines, each comprising a laser rangefinder and a scanning mechanism together with the associated timing circuits and control electronics. It supplied these laser scanner engines on an OEM basis to a number of system suppliers who then added a suitable position and orientation subsystem, developed the appropriate software, and integrated all these components to form a complete airborne laser scanning system. After this initial effort, *RIEGL* then developed its first complete system, which was first shown publicly at the end of 2006. However, it has continued to build laser scanning engines as a major part of its airborne laser scanner business.

The *RIEGL* LMS-Qxxx line of laser scanning engines that are offered for use in various airborne systems come in two main flavors. The first of these is designed for relatively low-altitude applications—for example, for detailed corridor mapping or power line surveys, typically using a helicopter as the airborne platform. The second is designed for operation from much higher altitudes for large-area topographic surveys. The earliest models were the LMS-Q140 for low-altitude applications and the LMS-Q280 for use at higher altitudes. The next generation low-altitude scanner engine was the LMS-Q240 model, which could be operated from flying heights up to 450 m with targets having an 80% reflectivity. The corresponding higher altitude model was the LMS-Q560 that could be used at flying heights up to 1,500 m with targets having an 80% reflectivity. This latter engine could also be fitted with a unit carrying out the recording of the full intensity waveform of the signal that had been reflected from the ground (Ullrich and Pfennigbauer, 2011). The current models in this single rangefinder LMS-Qxxx series are the LMS-Q680i (Figure 3.21) and the LMS-Q780.

These last two *RIEGL* laser rangefinder units feature (1) a powerful Class 3R pulsed laser emitting its radiation at $\lambda = 1,550$ nm in the case of the LMS-Q680i model and (2) a Class 3B laser operating at $\lambda = 1,064$ nm on the LMS-Q780—in both cases, with pulsed repetition rates of up to 400 kHz. All of these different LMS-Qxxx engines utilize a scanning mechanism featuring a continuously rotating polygon block with a number of reflective surfaces giving a unidirectional scan of the terrain. This results in a raster scan pattern of measured points over the terrain (Figure 3.8b). The rotational speed of the reflective polygon is adjustable—in the case of the LMS-Q680i model to speeds between 10 and 200 Hz. The angular value of the rotating polygon block can be read out continuously to a resolution of 0.001°. Both the LMS-680i and LMS-Q780 models also feature *RIEGL*'s multiple-time-around technology that allows the simultaneous measurement of several pulses in the air.

FIGURE 3.21 *RIEGL*'s LMS-Q680i scanning engine that is designed for operation from lower altitudes. (Courtesy of *RIEGL*, Horn, Austria. With Permission.)

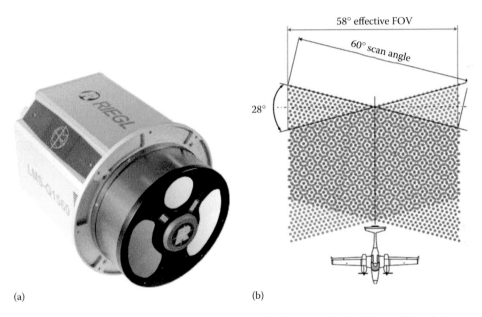

(a)   (b)

FIGURE 3.22 (a) *RIEGL*'s LMS-Q1560 laser scanner showing the two optical windows of its twin laser rangefinders and the window of the digital frame camera. (b) Showing the scanning pattern over the ground generated by the twin rangefinders of the LMS-Q1560 scanner. (Courtesy of *RIEGL*, Horn, Austria. With Permission.)

A notable development in the LMS-Qxxx series is the LMS-Q1560 model (Figure 3.22a) that was first introduced in 2013. This features the use of twin Class 3B laser rangefinders, each operating at the $\lambda = 1,064$ nm wavelength and with a maximum PRF of 400 kHz, as on the LMS-Q780 model. This pairing allows an overall maximum PRF of 800 kHz. The individual rangefinders can be rotated or tilted to point at different angles, including forward or backward, to produce different scanning patterns over the terrain (Figure 3.22b). The LMS-Q1560 scanner system is also equipped with an integrated GNSS/IMU subsystem and an 80 Megapixel digital frame camera, together with an optional capability to integrate a thermal intrared camera.

Another line of development from *RIEGL* was the introduction of its new *V-Line* series of compact and lightweight laser scanner engines in 2010. Initially, three models—the VQ-380, VQ-480i,

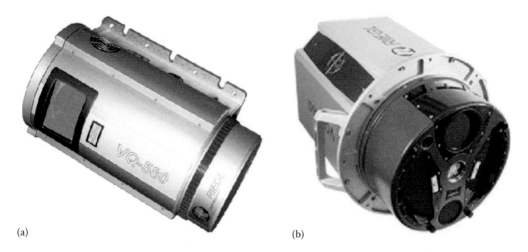

(a)   (b)

FIGURE 3.23 (a) The compact *RIEGL* VQ-580 laser scanning engine. (b) The twin-rangefinder *RIEGL* VQ-1560i scanner system designed for wide area coverage from high altitude flights. (Courtesy of *RIEGL*, Horn, Austria. With Permission.)

and VQ-580—appeared in this series (Figure 3.23a). A comparison of the data sheets for these three products reveals the following performance parameters: (1) the laser rangefinders of the VQ-380 and VQ-480 models operate at the wavelength $\lambda = 1,550$ nm, whereas that of the VQ-580 model operates at $\lambda = 1,064$ nm; (2) a maximum laser pulse repetition rate of 550 kHz can be achieved in the case of the V380 and 480i models, and 380 kHz in the case of the VQ-580; (3) a maximum scan rate of 120 Hz is possible for the VQ-380 and 150 Hz for the other two VQ models; and (4) a maximum (selectable) scan angle of 100° for the VQ-380 model and 60° for the other two VQ models. The maximum operating altitude for these VQ scanner engines is circa 500 m for the VQ-380, 1,050 m for the VQ-480i, and 1,200 m AGL for the VQ-580. The VQ-380 scanner is intended for lower altitude applications such as corridor mapping and power line inspection from light aircraft and helicopters, whereas the VQ-580 is said by *RIEGL* to be "especially designed to measure on snow and ice."

In October 2016, *RIEGL* introduced a further model in the VQ series in the shape of the VQ-1560i model (Figure 3.23b). This represented a powerful upgrade to the twin-rangefinder LMS-Q1560 model already described. The new system offers dual rangefinders that, in combination, provide a maximum pulse repetition rate of 2 MHz (2 × 1,000 kHz) and a maximum 1.33 million measurements per second over the ground and can be operated from a maximum flying height of 4.7 km AGL. The VQ-1560i model also features an improvement in *RIEGL*'s multiple-time-around technology, which enables the systems to handle multiple pulses in the air simultaneously without ranging ambiguities. *RIEGL*'s older LMS-Q1560 system could handle up to 10 pulses in the air simultaneously, whereas this new VQ-1560i model can handle a maximum total of 20. The VQ-1560i system also includes a fully integrated Applanix IMU and a GNSS receiver. Optional imagers that can be added to the VQ-1560i system include a Phase One medium-format (100 Megapixel) digital frame camera and an infrared camera.

Based on these various *RIEGL* laser scanning engines with their different characteristics and capabilities, a number of airborne laser scanning systems have been developed by various system suppliers—for example, IGI and Trimble—who offer their systems for sale to service providers. The *RIEGL* laser scanning engines also form the basis of a number of systems that have been built up in-house by various companies in Canada (e.g., LiDAR Services International) and the United States (e.g., Tuck Mapping Solutions), which use them to carry out their operations as service providers. It is very interesting to note the variety of different solutions that have been developed

(a)   (b)

FIGURE 3.24 (a) The *universal nose* of the Diamond DA42MPP aircraft into which the *RIEGL* LMS-Q680i laser scanner can be fitted and (b) showing an alternative design of nose cone that has been fitted to a DA42 MPP aircraft. (Courtesy of *RIEGL*, Horn, Austria. With Permission.)

using the same basic laser scanning engines from *RIEGL*. In total, they amount to a substantial part of *RIEGL*'s market for commercially supplied airborne laser scanning systems.

Besides supplying these various laser scanning engines discussed earlier, *RIEGL* also entered the market on its own account as a supplier of a complete airborne laser scanning system with the introduction of its LMS-S560 product. The version of this system that was supplied to the Diamond Airborne Sensing company featured two of *RIEGL*'s LMS-Q560 scanner engines, each able to operate with a PRF of up to 200 kHz, resulting in a 400 kHz combined system pulse rate. These two scanner engines were mounted together on a specially built carbon fiber reinforced plastic frame. This frame could also accommodate the IGI AEROcontrol GPS/IMU subsystem that was used for position and orientation purposes and a medium-format digital frame camera. One of the scanner engines sat in a separate tilt mount within the frame. This allowed it to be tilted through a range of ±45° to improve the data acquisition possibilities in mountainous areas. The whole system including the CFRP frame was enclosed in a very stiff and aerodynamically optimized CFRP cover to form a belly pod (=BP) that was fitted to the underside of the twin-engined Diamond DA42 multipurpose platform aircraft.

Another complete system offered by *RIEGL*, which is based on a special version of the LMS-Q680i scanner, is the NP680i (in which NP = Nose Pod). This includes a GNSS/IMU unit and a medium-format digital frame camera and is also designed to be fitted into the new *universal nose* of the Diamond DA42 Multipurpose Platform aircraft (Figure 3.24). Currently, *RIEGL* also offers its so-called CP (=Complete Platform) systems comprising any one of the LMS-Qxxx or VQ-xxx models that have been described earlier, together with an integrated GNSS/IMU subsystem. However, so far, this concept appears to have been implemented mainly with bathymetric systems (e.g., with the CP-820-GU) that will be described later in this chapter.

3.5.3.4 IGI

This company, based in Kreuztal, Germany, has developed its LiteMapper airborne scanner systems based on the *RIEGL* laser scanning engines discussed previously. To these basic rangefinder/scanning units, IGI adds its own CCNS (Computer Controlled Navigation System) and its AEROcontrol GPS/IMU subsystem. The latter uses IMUs based on FOGs that are constructed by the German Litef company, whereas the dual-frequency L1/L2 GPS receivers have been supplied by Ashtech or NovAtel. These component subsystems are integrated together with a purpose-built, computer-based LMcontrol unit for the control and operation of the scanner. To these hardware

FIGURE 3.25 A helicopter fitted with a specially built Helipod box that is mounted on an outrigger frame and contains an Integrated Geospatial Innovations (IGI) LiteMapper system. (Courtesy of IGI. With Permission.)

components of the overall system, IGI has added its own software modules developed in-house—WinMP (for use with CCNS), AEROoffice (for use with AEROcontrol), IGIplan (for mission planning), and LMtools (for in-flight data registration). The systems that resulted from the integration of all these components are the older LiteMapper 1400 and 2800 models for low- and high-altitude operations, respectively, based on *RIEGL*'s earlier LMS-Q140 and LMS-Q280 scanner engines and the later LiteMapper 2400 and 5600 models that are based on *RIEGL*'s later LMS-Q240 and LMS-Q560 scanner engines.

The current offerings in the LiteMapper series from IGI are the 2400 model (from before) for low-altitude operations and the 6800 and 7800 models based on the *RIEGL* LMS-Q680i and LMS-Q780 scanner engines that have been outlined previously. The LiteMapper models can all be supplied with a digital frame camera—in this case, from IGI's own range of single or multiple DigiCAM models based on Hasselblad cameras with digital backs producing frame images over a range from 40 to 80 Megapixels. IGI have supplied quite a number of LiteMapper systems mounted in a specially built Helipod (Figure 3.25) that has been certified for attachment to wide range of helicopters by the relevant airworthiness authorities.

3.5.3.5 Trimble

Trimble acquired TopoSys, which is another German company that is based in Biberach, located in the southern part of that country. TopoSys was best known for its Falcon airborne laser scanners based on the use of linear fiber-optic arrays that it operated in-house as a service supplier. However, the company also built and supplied systems to other service providers, based on the *RIEGL* LMS-Q240 and LMS-Q560 laser scanning engines. The two resulting products were marketed and sold as the Harrier 24 and 56 systems, respectively. The POS/AV position and orientation subsystems and the medium-format DSS digital frame cameras that were being utilized in these Harrier scanner systems were all supplied by Applanix—which was already owned by Trimble. The construction and sale of the Harrier systems was continued by Trimble after its acquisition of TopoSys, whereas the Falcon systems were discontinued. Currently, the Trimble company continues to base

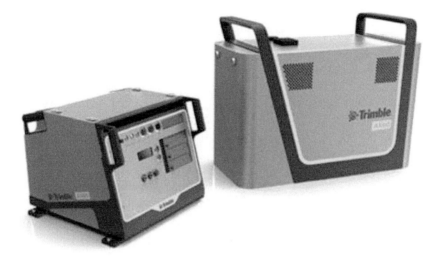

FIGURE 3.26   The Trimble AX60 airborne system is based on a *RIEGL* laser scanning engine, shown on the right of the picture. On the left is the control electronics unit. (Courtesy of Trimble, Sunnyvale, CA. With Permission.)

its airborne laser scanning systems on certain specific laser scanner engines supplied by *RIEGL*. These comprise the AX60 (Figure 3.26) and AX60i models that are based on the LMS-Q780 and LMS-Q680i engines, respectively, and the AX80 model that is based on the LMS-Q1560 engine with its dual laser rangefinders.

3.5.3.6   Helimap

Yet another European built airborne laser scanner that makes use of a *RIEGL* laser scanning engine is the Helimap system that was developed originally by the Photogrammetry and Topometry laboratories of the Federal Institute of Technology of Lausanne in Switzerland. This is a highly unusual hand-held laser scanning system that has been developed specifically for the mapping of areas of steep slopes in mountainous terrain that constitutes a natural hazard—such as avalanche slopes, rock falls, landslides, and glaciers—which are usually located in fairly inaccessible areas (Vallet and Skaloud, 2004, 2005). The system is operated from low altitudes from a helicopter. In its original form, the Helimap system comprised the following four main elements—a *RIEGL* LMS-Q140 laser scanner engine, a Javad dual-frequency L1/L2 GPS receiver, a Litton LN-200 IMU, and a Hasselblad medium-format digital frame camera equipped with a Kodak 16 Megapixel digital back. The measured data from the GPS receiver and the IMU were passed via a specially built interface-cum-logger to be recorded on a portable PC. The system was then upgraded to feature a *RIEGL* LMS-Q240 scanner, an IMU from iMAR, and a Hasselblad H1 camera equipped with an Imacon digital back generating 22 Megapixel images. All of these components were mounted rigidly together on a light and compact carbon-fiber and aluminum frame so that they maintained a constant stable relationship with one another. The frame with its instrumentation was hand-held by an operator seated on the side of the helicopter, who, if required, could point the whole system obliquely to the side to measure steep slopes (Figure 3.27). Although the original system was built and operated for scientific research purposes, several other Helimap systems have been built since then, employing different models of *RIEGL* scanners. These units have been operated commercially for the corridor mapping of roads, railways, power lines, pipelines, and so on by the Helimap System company based in Switzerland (Skaloud et al., 2005).

FIGURE 3.27   This example of a Helimap system is being operated from the side of an Alouette III helicopter to measure the steep sides of a valley. (Courtesy of Helimap Systems, Epalinges, Switzerland. With Permission.)

3.5.4   CUSTOM-BUILT AND IN-HOUSE OPERATED SYSTEMS

Besides the commercially available systems described earlier, a number of systems have been built in-house by service providers using COTS components bought from a wide variety of suppliers. In the case of two of the biggest of these service providers—Blom with its European-wide operations and Fugro with its world-wide coverage—their in-house produced systems have been built in some numbers. A third large service provider that has decided recently to undertake the construction of laser scanning systems for its own in-house use is the Harris Corporation in the United States. Most other systems built by service providers have often been produced as an individual one-of-a-kind system.

3.5.4.1   Blom

The origins of the Blom TopEye airborne laser scanning systems lie with the Saab Dynamics Company from Sweden, which had previously developed the laser scanner-based HawkEye airborne bathymetric mapping and charting system. Starting in 1993, Saab Combitech (another company in the Saab aerospace group), in combination with Osterman Helicopters, formed the Saab Survey Systems company and developed an airborne topographic laser scanning system, which it called TopEye. In 1999, Osterman Helicopters purchased all the shares in the partnership and renamed the company TopEye AB. In 2005, the Blom mapping company from Norway bought TopEye AB, which is still based in Gothenburg, Sweden. It provides airborne laser scanning services all over Europe under the Blom label, while still undertaking the development and maintenance of the TopEye scanners.

The original TopEye Mk I scanners that came into service in the mid-1990s used a laser range-finder with a maximum PRF of 8 kHz in combination with twin oscillating mirrors producing a Z-shaped pattern of measurements of the ground—similar to that produced by the Optech ALTM series. The system could record four returns from each transmitted laser pulse (Al-Bayari et al., 2002). Over the period 2003–2007, these instruments were radically rebuilt using components such as a new laser rangefinder supplied by the AHAB company. Thus, the resulting TopEye Mk II

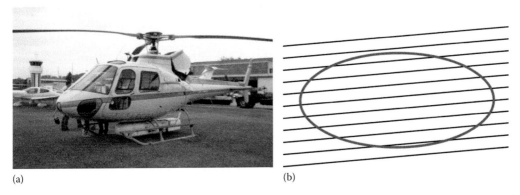

(a)   (b)

FIGURE 3.28   (a) The TopEye Mk. III laser scanner system is mounted in the box that is fitted to the underside of this helicopter. (b) Showing the respective ground patterns—elliptical and raster—that are being measured by the dual laser scanners of the Blom TopEye Mk. III system. (Courtesy of Blom. With Permission.)

scanners were equipped with new fiber-based laser rangefinders having a PRF of 50 kHz and nutating mirrors that produced a series of overlapping elliptical (Palmer scan) patterns of measured points over the ground (Figure 3.9). The rangefinder's receiver was equipped with avalanche photo diodes (APDs) that could record simultaneously both the first and last echoes and the full intensity waveform (with 128 samples) of the signal received after the reflection of the emitted pulse from the ground. The positioning and orientation subsystems that were used in these TopEye Mk II scanners were POS/AV units supplied by Applanix.

The latest model in the series is the TopEye Mark III (Figure 3.28a). This incorporates a proprietary dual-scan technology using (1) a laser rangefinder operating at 210 kHz in conjunction with a Palmer (elliptical) scan mechanism and (2) a *RIEGL* LMS-Q560 laser scanner with its linear/raster pattern of measurements over the ground. The reflected pulse echo data from each of the two scanners are again being captured as a full intensity waveform. Not only does this dual scanner arrangement provide higher density data over the ground (Figure 3.28b), but it also acquires its data at two quite different wavelengths—at $\lambda = 1,064$ nm (NIR) and $\lambda = 1,550$ nm (SWIR), respectively. All of these different models of TopEye laser scanners have been designed to be operated from helicopters flying at low altitudes. Seven of these TopEye scanners have been built, with five being operated by the Blom group and a further two with the Aerotec company in the United States.

3.5.4.2 Fugro

The title Fast Laser Imaging—Mapping and Profiling (FLI-MAP) is that given to a unique series of airborne laser scanning systems that have been produced in-house by the Fugro group of surveying and mapping companies. The systems have been operated principally by two of its subsidiaries—John E. Chance & Associates based in Lafayette, Louisiana in the United States and the Fugro Europe organization based in the Netherlands. They have also been operated by Fugro subsidiary companies in Australia and Canada. The development of the FLI-MAP airborne laser scanners began in 1993 with the first system entering commercial service in 1995. The initial systems, named FLI-MAP I and II, all date from the late 1990s. Since then, two newer systems—entitled FLI-MAP 400 and FLI-MAP 1000—have entered service.

The original FLI-MAP I system was based on the use of a rangefinder utilizing a Class 1 semiconductor laser that emitted its pulses at the wavelength $\lambda = 900$ nm and at a PRF of 8 kHz. The maximum operating altitude of the system was 200 m AGL. The scanning mechanism made use of a continuously rotating mirror giving a scan rate of 40 Hz. With the FLI-MAP II system, while the basic configuration remained the same, the system performance was upgraded substantially. The PRF of the laser rangefinder was increased to 10 kHz; the maximum altitude to 300 m; and the scan rate to 60 Hz. Both systems used a pair of video cameras, the one pointing vertically downwards, whereas the other was mounted obliquely in the forward pointing direction. These produced color S-VHS video images with a format size of 352 × 288 pixels.

The next model in the series was the FLI-MAP 400, which was substantially upgraded as compared with its predecessors. In particular, it uses a more powerful laser (which is still Class 1 eye-safe) and had the much higher maximum PRF of 150 kHz. These features have allowed the system to operate at higher altitudes—up to 350 m—which has enabled it to cover corridors having a much wider swath over the ground than the previous models. Although the earlier FLI-MAP I and II models only recorded a single (first) return from the reflected pulse from the ground, the FLI-MAP 400 allows the recording of four returns per emitted pulse. Furthermore, as well as the two video cameras that were used in the two earlier systems, the FLI-MAP 400 features twin small-format digital (still) frame cameras, each producing images of the ground that are 11 Megapixels in terms of their format size.

The position and orientation subsystems that have been used in these three FLI-MAP models is of particular interest. Each utilizes four Trimble 4000SE dual-frequency GPS receivers and a vertical gyro, all of which are mounted onboard the helicopter platform (Figure 3.29). In combination with additional GPS receivers used as base stations on the ground, they are used to determine the position, altitude, heading, pitch, and roll of the platform and its FLI-MAP scanner system. One of the four onboard GPS receivers is designated as the primary navigation receiver. After the postprocessing of its data in combination with the data collected by the ground-based stations, it generates the accurate X, Y, Z coordinates of its antenna. This primary navigation receiver also produces real-time positional values that are used as the basis for a flight navigation system that provides information to the pilot regarding deviations from the planned flight path and altitude using a system of LED light bars. The other three (secondary) GPS receivers that are mounted onboard the helicopter platform are used together with the primary GPS receiver and the vertical gyro to determine the heading, roll, and pitch of the platform and system every half second. For this purpose, the antennas of two of the GPS receivers are mounted on outrigger pylons attached to each side of the helicopter (Figure 3.29). This multiple GPS antennae-based attitude determination subsystem provides an adequate solution, since the flying height of the overall system is limited. Thus, the accuracy requirements for the attitude angles are less stringent as compared with those of a fixed-winged aircraft in which longer ranges need to be measured from the greater flying height.

All these three earlier models in the FLI-MAP scanner series have been mounted on helicopter platforms and have been used primarily for corridor surveys from low altitudes. However, the newest

FIGURE 3.29 A helicopter equipped with a FLI-MAP system, which is mounted inside the frame that is attached to the underside of the aircraft; note also the two outrigger pylons, each supporting a GPS antenna. (Courtesy of Fugro, Leidschendam, the Netherlands. With Permission.)

model in the series, called FLI-MAP 1000, is designed for use from higher altitudes and is mounted on a fixed-wing aircraft. With the FLI-MAP 1000 system, the PRF of the laser rangefinder has been increased to 375 kHz; the scan speed has been doubled; and a multiple-pulse-in-the-air technology has been implemented, all of these improvements allowing 250,000 elevation points to be acquired per second. An Applanix POS/AV subsystem allied to twin Trimble GPS receivers is used to give the positioning and orientation of the overall FLI-MAP 1000 system, while an Omnistar DGPS receiver is utilized to provide real-time navigation of the aircraft at an adequately precise level.

3.5.4.3 Harris Corporation

The LiDAR technology that has been developed by the Lincoln Laboratories of the Massachusetts Institute of Technology, largely with the support of U.S. defense agencies, resulted in the development of compact solid-state microchip lasers and high-efficiency Geiger-mode APD detector arrays that were capable of detecting a single photon. These developments resulted finally in the Airborne Ladar Imaging Research Testbed system that came to widespread notice in nonmilitary circles with its use to quickly create digital terrain models of extensive areas in and around the city of Port-au-Prince in Haiti following the devastating earthquake of January 2010 (Knowlton, 2011). Since then, the DARPA military research agency has facilitated the transfer of the Geiger-mode APD technology to Princeton Lightwave Inc. and Boeing Spectrolab to establish commercial sources of the Geiger-mode technology. In turn, this has led to the development of the first commercial systems utilizing the technology in the form of the Harris Corporation's LiDAR systems (Clifton et al., 2015). The IntelliEarth Geospatial Solutions division of this Corporation, which is a major American telecommunications and defense contractor, has built a number of LiDAR systems in-house for its own use as a commercial provider of services to external clients (Figure 3.30). These systems comprise (1) a rangefinder equipped with a Class 4 diode-pumped Nd:YAG solid state laser operating at the wavelength (λ) of 1,064 nm and having a very short pulse width of 550 ps and a PRF of 50 kHz; (2) a so-called holographic optical element that constitutes a diffractive optical system that splits the laser pulse into multiple subpulses and has a scanning subsystem to implement a Palmer-type scan with a 15° scan half-angle that is capable of rotational speeds in excess of 2,000 rpm; (3) a 128 × 32 pixel InP/InGaAsP Geiger-mode APD detector array that is capable of readout rates in excess of 100 kHz; and (4) the optical telescopes required for the transmission of the laser pulses and the reception of their

(a)

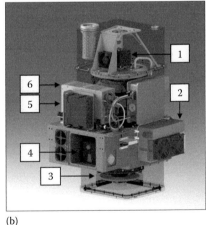
(b)

FIGURE 3.30 (a) A Harris IntelliEarth™ Geospatial Solutions Geiger-mode LiDAR system showing the holographic optical element of the Palmer scanner. (b) A cut-away drawing of the Harris LiDAR system. 1 = the Geiger-mode detector array; 2 = the laser rangefinder; 3 = the optical element of the Palmer scanner; 4 = the IMU; 5 = the GPS unit; 6 = the data acquisition subsystem. (Courtesy of Harris Corporation, Melbourne, FL. With Permission.)

reflections from the ground. The Harris system also utilizes multiple-pulse-in-the-air technology, while the position and orientation of the system are provided by a GPS/IMU subsystem.

3.5.4.4 Sigma Space Corporation

This company has also specialized in developing airborne laser scanners in which the ranging returns are being measured through the detection of single photons. In this case, the basic concept was implemented originally in the NASA Microlaser Altimeter from 2001. This was further developed by Sigma Space, first of all in the Leafcutter instrument that was funded by the USAF and then in the Mini-ATM LiDAR that was built for NASA (Figure 3.31). In both cases, the instruments were installed and flown in UAV aircraft. More recently, Sigma Space has developed its High Resolution Quantum LiDAR System (HRQLS) (Figure 3.32) and High Altitude LiDAR

(a)   (b)

FIGURE 3.31 (a) Shows the Mini Airborne Topographic Mapper (Mini-ATM) laser scanner that has been built for NASA by the Sigma Space Corporation. (b) Shows the L-3 Unmanned Systems Viking 300 fixed-wing UAV platform with the Mini-ATM shown in the left lower corner at the same scale as the UAV aircraft for comparative purposes. (Courtesy of NASA, Washington, DC. With Permission.)

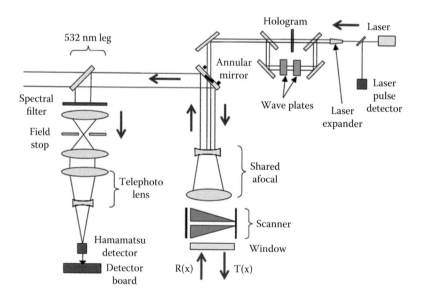

FIGURE 3.32 Schematic diagram of the optical component chain and flow of photons in the HRQLS instrument. Arrows show the movement of photons in the transmitted beam (Tx) and received energy (Rx). (Courtesy of Sigma Space Corporation, Lanham, MD, circa 2006. With Permission.)

(Degnan et al., 2013) and, most recently of all, it has built its HRQLS-2 system (Li et al., 2016; Sirota et al., 2016). With all three of these systems, each individual pulse being emitted by the laser rangefinder at the wavelength (λ) of 532 nm is split into 100 beamlets (subpulses) using a diffractive optical element and transmitted towards the terrain. The reflection from the terrain surface from each beamlet (subpulse) is then imaged on to the corresponding pixel of a 100 pixel detector array. In the case of the HRQLS instrument, the instrument emits its pulses with an output power of 1.7W and a pulse width of 700 ps at the rate of 25 kHz. These various models of the Sigma Space LiDARs are normally operated at flying heights AGL of 2.3 km (HRQLS), 3.8 km (HRQLS-2), and 7.6 km (High Altitude LiDAR), respectively. As one would expect, the use of the greater flying heights allows wider swaths and greater areal coverage to be achieved from a single flight, but with a reduced point density. Each system can be operated with either a dual-wedge or a single-wedge optical scanning device to produce a range of scan widths and (conical) scan patterns.

The Sigma Space HRQLS LiDAR has been flown on a number of projects, including that flown to carry out forest and terrain mapping for the University of Maryland, which has been reported in Swatentram et al. (2016). However, in February 2016, the Sigma Space company was acquired by Hexagon, which already owns Leica Geosystems, Intergraph and several other major suppliers of systems to the surveying and mapping industry. The expectation is that, through this acquisition, Sigma Space's LiDAR technology will become available to service providers within the industry for purchase on a commercial basis.

3.5.5 Research Systems

Besides the various systems described earlier that are being used commercially to provide airborne laser scanning services to customers, a number of interesting systems have been built by U.S. government research agencies, principally NASA, and used primarily for scientific research purposes.

3.5.5.1 NASA

NASA has long been in the forefront of the organizations developing airborne laser profilers and scanners over a very long period—since the mid-1970s. Much of its initial development took place within the AOL program that was sponsored jointly by NASA and the National Oceanic and Atmospheric Administration (NOAA). Essentially, this program acted as a test-bed both of the technology and of its applications. The AOL system was continually modified and upgraded over a 20 year period. On the one hand, research was conducted into the use of airborne lasers to carry out the measurement of elevations and depths for both topographic and bathymetric mapping applications. In parallel with this, extensive research was undertaken into the fluorescence induced in the plankton that is present in the surface layers of the ocean when illuminated by an airborne laser. These two parts of the research program were separated in 1994 when two new systems were introduced. The fluorescing system retained the AOL title, whereas the topographic mapping oriented system was given the title ATM.

3.5.5.2 Airborne Topographic Mapper

The ATM-I system was first flown in 1994, while the ATM-II was brought into service in 1996. The rangefinders of the two systems use a Spectra Physics Nd:YLF laser generating 7 ns wide pulses at the wavelength of $\lambda = 1,046$ nm. The output pulses have been frequency doubled to be emitted simultaneously at $\lambda = 523$ nm in the green part of the spectrum. The rangefinder can be operated at pulse repetition rates between 2 and 10 kHz. The scanning mechanism of the ATMs uses a nutating mirror generating an elliptical (Palmer) scan pattern over the ground with a scan rate of 20 Hz. From 1997 onwards, the ATM-II system has been equipped with an imaging device that records passive panchromatic image data following the same elliptical scanning path as the active laser scanner. The ATM-II has also been equipped with a dual-frequency GPS receiver allied to an IMU based on a Litton laser ring gyro to act as its position and orientation subsystem. The two ATM systems

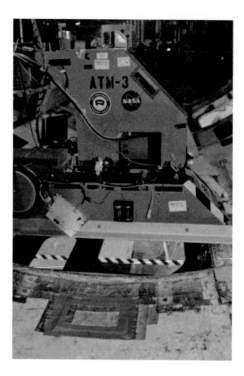

FIGURE 3.33 An ATM laser scanning system in use during Operation IceBridge. (Courtesy of NASA, Washington, DC. With Permission.)

have been used extensively by NASA for the mapping of coastal beaches on both the Atlantic and Pacific coasts of the United States and in the Gulf of Mexico. The systems have also been used for the surveys of large areas of ice sheets and glaciers in Greenland, Northern Canada, Svalbard (Spitzbergen), and Iceland. Indeed, since 2009, the two ATM systems have been the primary instruments used in Operation IceBridge (Figure 3.33). This program was instituted by NASA to provide images and LiDAR data of parts of Greenland and Antarctica during the period between the end of the ICESat satellite's operation in 2010 and the start of ICESat-2's operations, which, after several postponements, is now scheduled to begin in 2018. In connection with this program, the Applanix POS/AV 610 and 510 GPS/IMU subsystems deployed on the ATM systems have been upgraded to use Javad GPS units. The ATM systems have been operated from relatively low altitudes between 500 and 1,500 m AGL from a great variety of aircraft platforms during Operation IceBridge.

3.5.5.3 Raster Scanning Airborne Laser

In parallel with its ATM systems, NASA also developed its Raster Scanning Airborne Laser (RASCAL) system at the GSFC in 1995 for the airborne mapping of surface topography. Similar to the ATM systems, the scanner's rangefinder utilized a Spectra Physics Q-switched, diode-pumped, and frequency-doubled Nd:YLF laser to generate its pulses at 5 kHz, each of which is 6–10 ns long. However, the scanning mechanism was very different to that of the ATM, resulting in a series of parallel scans in the cross-track direction producing a grid or raster pattern of measurements over the terrain—hence the RASTER name of the system. Further details are given in the corresponding section of the first edition of this book.

3.5.5.4 Scanning LiDAR Imager of Canopies by Echo Recovery

The SLICER (Scanning LiDAR Imager of Canopies by Echo Recovery) was yet another airborne laser scanner that was developed by NASA in the mid-1990s. The system was based on the ATLAS, an earlier airborne laser profiler that had been built by NASA. The ATLAS was first modified into the SLICER through the installation of a more powerful laser and the addition of a waveform digitizer. The scanning mechanism that was used on the SLICER comprised a simple oscillating mirror that was rotated rapidly to the successive positions of a set of fixed scan angles using a computer-controlled galvanometer. Usually, five fixed angular positions were used for each cross-track scan, so only a very narrow scan width was achieved. Again, some further information about this scanner system is given in the first edition of this book.

3.5.5.5 Laser Vegetation Imaging Sensor

As its title of the system suggests, the Laser Vegetation Imaging Sensor (LVIS) was developed by NASA originally for the mapping of topography and of the height and structure of the vegetation growing on it, based on the experience that it had gained with the previous SLICER scanner (Blair et al., 1999). In particular, it has been designed to be operated from high altitudes for the coverage of a fairly wide swath over the terrain. Furthermore, it was unique in recording the active time history of both the emitted and reflected pulses. The LVIS rangefinder used a powerful Q-switched, diode-pumped Nd:YAG laser supplied by Cutting Edge Optronics (Figure 3.34a). This emitted pulses having a power of 5 mJ (permitting high-altitude operations) and a pulse length of 10 ns at the wavelength (λ) of 1,064 nm. In its earliest form, the pulse repetition rate of the laser could be varied over the range 100–500 Hz. In its later upgraded form, the rate was increased to 5 kHz. Furthermore, after the installation of a still more powerful laser, the flying height has been extended to 12 km (40,000 ft), being operated from a Sabreliner business jet aircraft equipped with a high-quality optical window. From this altitude, the swath width has been extended to 2 km. The ranging and the waveform recording are all carried out using a single detector, digitizer, and oscillator, the last acting as a clock or time base.

The scanning of the ground with its vegetation canopy is carried out in a stepped pattern with the galvanometer-driven mirror being stationary, when the laser emits its pulse, and the reflected energy is received from the ground (Figure 3.34b). This solution is necessary as the travel time of the pulse is rather long at high altitudes. The scanning mechanism uses a very stiff but lightweight mirror that remains flat during the severe accelerations and decelerations that are experienced during the rapid starts and stops of this stepped scanning action with the mirror remaining stationary during the measurement process. The position and orientation subsystem is similar to that of the SLICER system, using a Litton LTN-92 IMU in combination with two Ashtech Z-12 dual-frequency GPS receivers. Initially, the LVIS was regarded principally as an airborne prototype or simulator for the VCL spaceborne laser profiling mission. However, the Vegetation Canopy Lidar (VCL) mission was first delayed and then cancelled in 2002. After which, the LVIS was upgraded and repackaged to

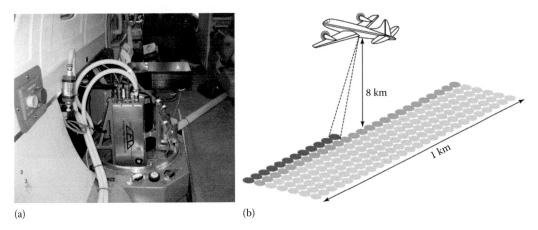

(a)   (b)

FIGURE 3.34 (a) The NASA LVIS laser scanner with its Cutting Edge Optronics rangefinder installed in a Wild camera mount. (b) Stepped pattern of the coverage of the ground and its vegetation canopy that is implemented by the LVIS laser scanner. (Courtesy of NASA, Washington, DC; redrawn by M. Shand. With Permission.)

fit within a standard Wild RC10 aerial camera mount (Figure 3.34a). Since then, the LVIS scanner has been used extensively for vegetation mapping both in the United States (in California and in the eastern states) and in Central America (in Costa Rica and Panama). Since 2009, LVIS has also been used to map large areas of sea ice and glacier zones from high altitudes operating over Greenland and Antarctica as part of NASA's Operation IceBridge. These data have also been used extensively to calibrate and validate the measurements made from the ICESat and CryoSat-2 satellites. In 2013, a smaller and more compact version of the LVIS instrument, known as LVIS-GH, was introduced for use on NASA's Global Hawk UAV aircraft onboard which can be operated at altitudes up to 60,000 ft (18 km). However, it has been flown from lower altitudes on other aircraft such as a Lockheed C-130 at 28,000 ft (8.5 km) during Operation IceBridge flights.

3.5.5.6 Slope Imaging Multipolarization Photon-Counting LiDAR

In recent years, NASA has developed a series of airborne laser scanners with a view to improving the technologies that are required to undertake certain specific spaceborne missions—in particular, its ICESat-2 and LiDAR Surface Topography (LIST) missions. The first of these scanners was its SIMPL instrument (Harding et al., 2010). Its main objective was to evaluate lasers that had low pulse energies and high pulse repetition rates allied to single photon counting in contrast to those using high pulse energies and low pulse rates allied to analogue receivers that had been used in the previous SLA and ICESat missions. Thus, the laser rangefinder used in SIMPL employed a solid-state microchip laser transmitter operating at 11 kHz and having a short 1 ns pulse width. The laser rangefinder generated its pulses at $\lambda = 1,064$ nm with frequency doubling used to generate pulses at $\lambda = 532$ nm. Each of these pulses was split into four separate plane-polarized subpulses that were emitted in parallel with one another, resulting in a very narrow pushbroom configuration of coverage over the ground (Figure 3.35). On the receiver side, Geiger-mode silicon APD (Si:APD) detectors

FIGURE 3.35 The basic concept of SIMPL by which the ground is covered in a pushbroom scanning mode using multiple fiber lasers, thus eliminating the requirement for optical scanning elements. The SIMPL development LiDAR instrument utilized only four adjacent pulses with a very narrow swath width to prove the basic concept. (Courtesy of NASA/GSFC, Greenbelt, MD. With Permission.)

employing single photon counting provided sixteen channels—four for each of the four subpulses. Each set of four channels detected the returning radiation with parallel and perpendicular polarization with respect to the transmitted polarization plane at both the $\lambda = 532$ nm and $\lambda = 1,064$ nm wavelengths. Successful test flights from a high-flying Learjet aircraft equipped with a high-quality optical window were carried out with SIMPL over both the ice and the land in the area of Lake Erie.

3.5.5.7 Multiple Altimeter Beam Experimental LiDAR

Next, the MABEL instrument was developed to meet the needs of the ICESat-2 project (McGill et al., 2013). Similar to SIMPL, MABEL uses a high repetition-rate, low pulse energy laser, and a photon-counting system featuring single-photon sensitive detectors. However, it was designed to be operated from very high altitudes—60,000 to 65,000 ft (18.2 to 19.8 km)—on-board either the NASA ER-2 (U-2) or its Global Hawk UAV aircraft. The laser repetition rate is variable from 5 to 25 kHz, although typical operations have used 10 kHz to mimic the on-orbit characteristics of the ICESat-2. The laser pulse length is 2 ns. If the laser is operated in 10 kHz mode, then a pulse is emitted every 2 cm along track, given the nominal 200 m/s speed of the ER-2 aircraft. However, since only a single laser with the wavelength $\lambda = 1,064$ nm is used, a transmitter fiber splitter box comprising a series of cascading beam splitters in combination with a potassium dihydrogen phosphate (KDP) crystal was developed to divide the output energy of each pulse into a series of eight output subpulses at $\lambda = 1,064$ nm and sixteen output subpulses at $\lambda = 532$ nm. These are fed into a fiber optic array comprising 215 individual fibers (107 for each wavelength, plus 1 central fiber) (Figure 3.36a). At any given moment of time, 16 of the 107 green ($\lambda = 532$ nm) wavelength fibers in its array can be coupled to the 16 corresponding ($\lambda = 532$ nm) transmitter fibers and 8 of the 107 NIR ($\lambda = 1,064$ nm) wavelength fibers can be coupled to the 8 corresponding ($\lambda = 1,064$ nm) transmitter fibers. The selection of a particular setting of the transmitter fibers thus defines the instrument's viewing geometry, which can be viewed as having a pushbroom configuration with some wide gaps in it, producing a series of parallel profile lines (Figure 3.36b). The receiver side has a similar set up of fibers with a lens that matches the transmitter fibers' lens. The viewing geometry of the MABEL can be changed by rearranging the connections between the receiver and transmitter fiber select boxes and the fiber arrays, as described earlier. Only the eight NIR ($\lambda = 1,064$ nm) receiver modules carry out single photon counting; the other sixteen green ($\lambda = 532$ nm) receiver modules have utilized photo-multiplier tube detectors. The GNSS/IMU subsystem employs a NovAtel model HG1700 IMU. The subsequent series of test and scientific flights using this system, which began in 2011 and have continued to be undertaken over large areas of land and sea ice since then, have shown that the technology that had been developed for the MABEL airborne instrument is indeed suitable and viable for the ICESat-2 spaceborne mission (Figure 3.37).

3.5.5.8 Airborne LIST Simulator

A still further development of the technology that is required for NASA's future spaceborne LiDAR missions has been implemented through the ALISTS (Airborne LIST Simulator) system. As the name suggests, this was oriented specifically towards the requirements of the proposed LIST mission, which would utilize a wide-swath nonscanning multibeam LiDAR to map the Earth's solid (land and ice) topography and vegetation on a global scale from space. This particular mission has been strongly recommended to NASA by the National Research Council of the U.S. National Academy of Sciences to implement vital Earth science applications as a result of its decadal survey that was published in 2007. The concept that was devised by NASA to satisfy this mission will involve the coverage of the ground areas with elevation measurements over a full 5 km wide swath from each pass of the LIST satellite orbiting at a height of 400 km. The nonscanning LiDAR would employ a full pushbroom configuration measuring 1,000 beam spots simultaneously in the cross-track direction, with each spot being 5 m in terms of its spatial resolution, so producing the required ground swath width of 5 km.

Needless to say, the technology needed to provide a solution to this demanding mission did not exist and needed to be developed, hence the introduction of the ALISTS program (Yu et al., 2013).

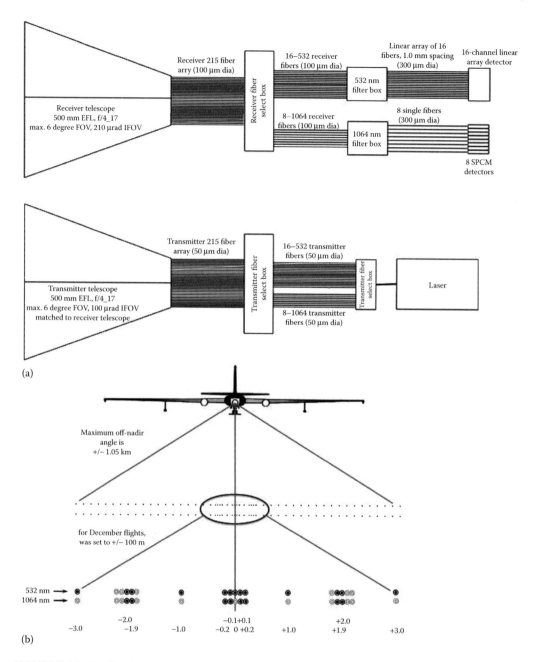

FIGURE 3.36 (a) Showing the optical instrument layout of the MABEL scanner with its twin (transmit and receive) telescopes, each with its fiber bundles and select box. (b) The MABEL viewing geometry is defined by the arrangement of the pulses emitted by its fiber arrays. If the outermost fibers are illuminated, then the swath width would be 1.05 km. When only the tightly arranged inner fibers were used in test flights, then the ground swath would only be 100 m. (Courtesy of NASA, Washington, DC. With Permission.)

The LiDAR instrument has utilized a single Yb:YAG microchip laser transmitter built by Raytheon. This operates at $\lambda = 1,030$ nm wavelength, has a short pulse width of < 1 ns, and a PRF of 10 kHz. The output pulse is then split into 16 separate output beams (subpulses) having a square 4×4 configuration (Figure 3.38). By rotating this particular configuration by $14.5°$ with respect to the line of flight of the carrying aircraft flying at an altitude of 10 km, this resulted in the ground being

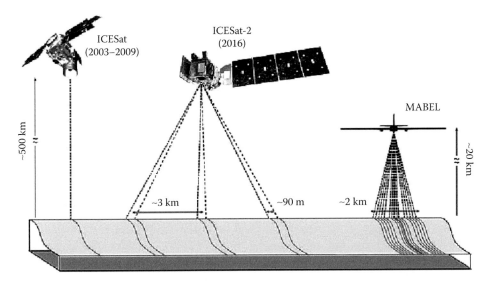

FIGURE 3.37 A diagram providing a comparison of the beam geometry and the ground tracks of the MABEL airborne LiDAR; and the ICESat and ICESst-2 spaceborne profilers. (Courtesy of NASA, Washington, DC. With Permission.)

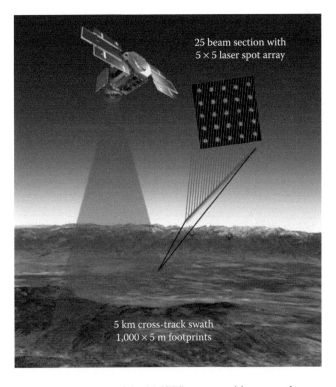

FIGURE 3.38 Showing the basic concept of the ALISTS systems with a square laser spot array being rotated relative to the line of flight to simulate a pushbroom configuration of elevation measurements over a swath of the terrain that would be carried out from a suitably equipped satellite. Note that, although the diagram shows a 5 × 5 array of measurements, the actual arrangement that has been used in the ALISTS system is a 4 × 4 array. (Courtesy of NASA, Washington, DC. With Permission.)

covered by 16 adjacent spots, each 5 m in size and with a 20 m spacing between each pair of spots, so providing an 80 m swath width. On the receiver side, an Intevac InGaAsP 4×4 intensified photodiode forms the basis of a 16-channel single-photon-sensitive detector array, with a 16-channel waveform digitizer recording the measured data. An Applanix GNSS/IMU subsystem has been used for positioning, navigation, and orientation purposes. The ALISTS system has been flown successfully on different aircraft types (e.g., Learjet and Lockheed P-3B) and from different flying heights over a test area in Virginia.

3.6   AIRBORNE LASER SCANNERS ON UAVs

As a consequence of the wholly different operating environment within which UAVs and unmanned aerial systems are operated, the technology of the laser scanners that are mounted on these UAV aircraft is wholly different from that described in the previous sections of this account in which manned aircraft (including fixed-wing aircraft and helicopters) have been used as the operational platforms. In discussing this topic, it is convenient to split UAVs into three distinct groups. These are based on the regulatory frameworks that govern the construction and operation of commercial UAV aircraft in more advanced countries. In turn, this has governed the types of laser scanners that can be operated within each of these groups, which are as follows:

1. Comparatively large and heavyweight commercial or military UAVs that are mostly powered by gasoline or diesel motors
2. Medium-weight UAV aircraft that weigh less than 20 kg that, almost invariably, are powered by electric motors
3. Lightweight and low-powered UAVs that weigh less than 7 kg and mostly utilize battery-driven electric motors

The United Kingdom may be taken as an example of the typical operational environment that exists within European countries. Since 2009, the detailed regulation of flights that are being undertaken for commercial aerial photography and laser scanning by these medium- and light-weight UAVs in the United Kingdom, including the actual permission to fly, has been carried out by the country's Civil Aviation Authority. The regulations set out by this Authority specify in detail the specific conditions under which these flights can be made (Petrie 2013a). In broad terms, permission for flights is granted subject to the UAV *not* being flown

1. At altitudes greater than 400 ft. (120 m) AGL.
2. Beyond a maximum range of 500 m or out of visual range.
3. Either over or within 150 m of an organized open-air assembly of more than 1,000 people.
4. Besides which, the UAV must not be flown over or within 50m of any person not having knowledge of or not having been warned of the UAV flight. This is reduced to 30m for the take-off and landing of the aircraft.
5. Furthermore, a trained and licensed pilot must be able to take manual control of the aircraft and fly it out of danger if needed.
6. Needless to say, no operator is allowed to fly a UAV in various restricted areas without having first obtained permission from the authority.
7. Furthermore, UAV flights' over congested (e.g., urban) areas are very highly restricted, and they also require prior permission to be obtained from the authority.

The quite restricted weights that can be lifted by these medium- and light-weight and comparatively low-powered UAV aircraft also need to be considered (Petrie 2013b). The weights of the typical full-blown airborne topographic laser scanner systems mounted on manned aircraft that have been discussed previously in the first part of this chapter—including their control electronics, obligatory

DGPS/IMU subsystem and almost obligatory camera, and the actual laser scanner—can easily reach several tens of kilograms. Furthermore, the cost of such a complete airborne laser scanning system can lie in the range $500,000 to $1.3 million—which is hugely beyond the cost of a very lightweight UAV, which may only cost a few thousand dollars. Given the fact that accidents with medium- and light-weight UAVs are not uncommon, it is unlikely that operators would be willing to risk damage to such an expensive laser scanner system, given this situation. The severe restrictions in the weight that can be lifted by lightweight UAVs affect not only the laser scanners that can be mounted on these aircraft but also their subsystems such as the GPS/IMU, which is obligatory if the system is to be employed in a meaningful way (Jozkow and Toth, 2014).

It will also be apparent that the mainstream airborne laser scanners that are operated on manned aircraft are used to generate elevation data for topographic mapping purposes over very large areas of terrain. Often these devices are being operated at flying heights up to 6 km (or more) AGL to cover very large areas of terrain in a single flight. The need to measure ranges over such comparatively long distances is, of course, one of the main reasons for conventional airborne laser scanners being comparatively large with regard to their size and weight and having greater power requirements than can be provided or coped with on a lightweight UAV. By contrast, under the current Civil Aviation Authority regulations in the United Kingdom and the similar regulations that are in force elsewhere in Europe that have been outlined earlier, medium-weight (<20 kg) and light-weight (<7 kg) UAV aircraft are being operated within a very different environment. As noted earlier, they are not permitted to be flown at altitudes greater than 400 ft (120 m) AGL, nor can they be operated out of line-of-sight. This means that the range measurements only have to be made over very much shorter distances and over much smaller areas as compared with those encountered with manned aircraft, with consequently lower power requirements. In turn, these restrictions in the flying height and in distances that have to be measured also mean that greater attention needs to be paid to eye safety. Taking into account all of these particular conditions, currently strong efforts are being made to develop suitable laser scanners that are optimized for use on medium- and light-weight UAVs. It is interesting to note that, at the time of writing this account, the basic technologies that are being employed in the laser scanners that are being mounted on lightweight UAVs are, to a large extent, those that have been developed for and utilized in one or other of the following fields: (1) Mobile mapping systems and vehicle navigation and collision avoidance; (2) terrestrial laser scanning; and (3) certain types of robotics applications, rather than those that have been developed specifically for airborne topographic mapping purposes.

3.6.1   Airborne Laser Scanners on Heavyweight UAVs

A number of the larger type of UAV aircraft using gasoline (petrol) engines have been adapted to operate with laser scanners. An example is the Schiebel CAMCOPTER® S-100 UAV helicopter that is built in Austria. This aircraft is more than 3 m in length, it has 110 empty weight or maximum 200 kg take-off weight, and is powered by a 50 hp rotary engine. It is mainly operated by military and naval organizations in a number of countries, in which case, often it is not subject to the regulations and restrictions that apply to civilian users. Schiebel has cooperated with *RIEGL* to produce a version of the aircraft fitted with the *RIEGL* VQ-820-G bathymetric laser scanner that appears to be aimed principally at its naval customers (Figure 3.39a). Another company that has constructed rotary-wing UAV aircraft fitted with laser scanners is Aeroscout from Switzerland. Its Scout B1-100 UAV is 3.3 m in length, weighs 60 kg, and is powered by an 18 hp two-stroke petrol engine. As with the Schiebel company, Aeroscout has formed a cooperative partnership with *RIEGL*. This relationship has included *RIEGL* taking a shareholding in the Aeroscout company. Initially, the Scout UAV was fitted with *RIEGL*'s LMS-Q160 scanner. More recent examples of the UAV have utilized the *RIEGL* VUX-1 scanner coupled with a GNSS/IMU subsystem that features twin GPS antennas (Figure 3.39b). Yet another company that has fitted the *RIEGL* VUX-1 scanner to a similar UAV aircraft is Swiss Drones. Its TC-1235 UAV is 1.7 m in length, weighs 35 kg, and is powered by a small turbine jet engine.

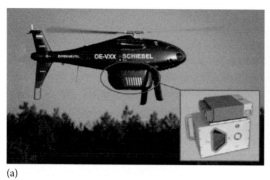 

(a)   (b)

FIGURE 3.39   (a) A Schiebel CAMCOPTER® UAV fitted with a belly pod containing a *RIEGL* CP-820-GU bathymetric laser scanning system. (Courtesy of Schiebel, Vienna, Austria. With Permission.) (b) An Aeroscout B1-100 UAV fitted with a *RIEGL* VUX-1UAV laser scanning system. (Courtesy of *RIEGL*, Horn, Austria. With Permission.)

However, as discussed previously, currently there are severe regulatory restrictions in many countries regarding the commercial operation of these larger UAV aircraft. Taking the United Kingdom as an example, UAV aircraft that weigh more than 20 kg are currently banned from flying in the country's civilian airspace other than in a limited area in west Wales and a still smaller one centered on the military airfield at Boscombe Down in Southern England, where test flying may be carried out. Needless to say, the result is that those UAVs that fall into this category cannot effectively be flown in the United Kingdom by civilian users.

3.6.2   AIRBORNE LASER SCANNERS ON MEDIUM-WEIGHT UAVS

As already mentioned in the previous section, *RIEGL* constructs a specially designed series of scanners for this specific type of application, under its VUX-1 label (Figure 3.40a). This comprises three different models: (1) the VUX-1HA that, in fact, is designed for mounting on mobile land platforms; (2) the VUX-1UAV model that is intended for use on UAVs; and (3) the VUX-1LR that is designed to be mounted on manned or unmanned and ultralight aircraft or helicopters (as mentioned earlier in the context of the Scout and Swiss Drone heavyweight UAVs). The VUX-1UAV model is equipped with a Class 1 laser powering its rangefinder and emitting its pulses at $\lambda = 1,550$ nm for eye safety reasons, having regard to the typical UAV operation from low flying heights. The maximum value of the PRF of the VUX-1UAV's rangefinder is 550 kHz, whereas the speed of the rotating mirror can be set to produce scan values between 10 and 200 Hz (scans per second), even though the field of view (330°) is exceptionally wide. The weight of the VUX-1UAV is also quite low at 3.75 kg; obviously it will be greater if and when a GNSS/IMU subsystem is added to the system. In this context, *RIEGL* also offers its fully integrated VUX-SYS system for sale. This comprises the VUX-1 laser scanner; a GNSS/IMU subsystem; and a purpose-built control unit, the latter two items adding a further 1.6 kg to the weight of the overall VUX-SYS. Going a further step onwards, a complete VUX-SYS system can be installed into a specially-built lightweight VP pod that is designed to fit into the standard mounts that are available on manned helicopters (Figure 3.40b). In which case, the total weight of a VP-1 pod containing a complete VUX-SYS system will rise to 10 kg. The VUX-SYS system is also marketed by *RIEGL* to be fitted to its own in-house supplied RiCOPTER UAV (Figure 3.40c).

In October 2016, *RIEGL* introduced its mini-VUX-1UAV scanner (Figure 3.41), which is a miniaturized version of the VUX-1UAV model. (N.B. Thus, strictly speaking, this new mini-VUX-1UAV scanner should be placed within Section 3.6.3 of this chapter that covers scanners mounted on lightweight UAVs. However, it has been placed here for convenience because of its close relationship to the VUX-1UAV model.) At 1.55 kg, the miniVUX-1UAV weighs less than half

(a)   (b)

(c)

FIGURE 3.40   (a) A *RIEGL* VUX-1 airborne laser scanner. (b) The *RIEGL* VP-1 HelicopterPod containing a VUX-1LR laser scanner. (c) A drawing showing the various components of a VUX-SYS mounted on a RiCOPTER UAV that is built in-house by *RIEGL*. (Courtesy of *RIEGL*, Horn, Austria. With Permission.)

FIGURE 3.41   A handheld *RIEGL* miniVUX-1UAV lightweight laser scanner. (Courtesy of *RIEGL*, Horn, Austria.)

as much as the VUX-1UAV model, which weighs 3.75 kg. Furthermore, its overall dimensions are 242 × 99 × 85 mm. This small size and low weight makes it easier to mount the scanner on lightweight UAVs with their small dimensions and strictly limited payload. Alternatively, the use of the mini-VUX-1UAV could extend battery (and flight) duration on larger UAV systems. With regard to its operational performance, the scanner has a 360° field of view and a maximum operational flying height of 100 m AGL, while its laser rangefinder has a maximum PRF of 100 kHz. The scanner's receiver can detect up to five returning echoes per pulse, and it can also carry out the digitization and online waveform processing of the reflected signals. The basic mini-VUX-1UAV scanner does not feature an integrated IMU, but it does offer a suitable mechanical and electrical interface that will allow one to be attached to the scanner. Obviously, this will add to the resulting system weight, though not too greatly if a suitable micro-electromechanical systems (MEMS)-based IMU is chosen to provide an acceptable performance.

3.6.3 Airborne Laser Scanners on Lightweight UAVs

In attempting to outline and discuss the different approaches and the different types of laser scanners that are being developed and operated on very lightweight UAVs, it is convenient to divide these instruments or systems into four main classes or categories, each employing a different technology, as follows:

1. The first group is that utilizes a simple scanner comprising a laser rangefinder and a rotating mirror that operates within a single 2D scanning plane. These usually have a restricted range from 30 to 50 m, though some examples can measure longer ranges—up to 120 m. Examples of these devices are the simple laser scanners that are produced by Hokuyo (Japan) and SICK (Germany), which are especially popular within the robotics community.
2. A second group comprises multi-layer scanners that measure both distances and angles in four or more horizontal or near-horizontal scanning planes simultaneously. Typical of these are the Ibeo and SICK multilayer scanners that have been developed originally to mount on the front and rear of road vehicles for collision avoidance purposes.
3. A third group of devices comprise the well-known multiple spinning laser scanners (with 16, 32, or 64 individual lasers) that were developed originally by Velodyne for vehicle navigation purposes and which came to prominence after their widespread use on the unmanned (driverless) road vehicles that took part in the DARPA Urban Challenge for autonomous vehicles in 2007. A similar product is the 8-beam M8-1 sensor from Quanergy. Since then, these scanners have been adopted quite widely to form the basis of mobile mapping systems—for example, those built by Topcon and by Navteq. Recently, they have been adapted for use in several scanner systems that are being operated on lightweight UAV aircraft.
4. The fourth group of scanners is the short-range terrestrial 3D laser scanners that are in widespread use by land surveyors. Certain models of these laser scanners have also been used as the ranging and scanning components in a number of mobile mapping systems that are being employed on the kinematic or dynamic surveying of road and railway networks. They are now being used in a similar role mounted on lightweight UAVs.

3.6.3.1 Simple Laser Scanners

As mentioned earlier, the robotics community that deals with the design, construction, operation, and application of unmanned vehicles makes extensive use of laser scanners to measure the proximity of and distance to obstacles, other vehicles, and others. The simple 2D laser scanners from Hokuyo and SICK employ a comparatively low-powered pulse-based (TOF) laser rangefinder utilizing a semiconductor diode laser that emits its radiation in the NIR part of the spectrum. The rangefinder also has a photo-diode receiver that can receive up to three returning echoes (reflections) from a single transmitted pulse. The angular scanning motion is implemented using a continuous rotating mirror equipped with a rotary (angular) encoder. (Figure 3.42a). The specific models from Hokuyo (Figure 3.42b) and SICK that typically are used with lightweight UAVs are comparatively low-cost devices costing circa $6,000.

An early example from 2009 that used this simple 2D laser scanner technology was the Sensei UAV system that has been developed by the Finnish Geodetic Institute. This was based originally on an Align single-rotor UAV equipped with a SICK LMS151 laser scanner, a NovAtel SPAN DGPS/

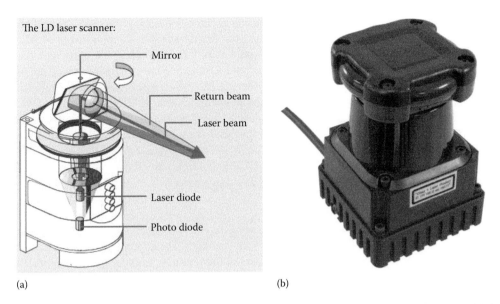

(a)   (b)

FIGURE 3.42 (a) Diagram showing the basic design of the SICK LD 2D laser scanner. (Courtesy of SICK. With Permission.) (b) A Hokuyo UTM-30LX-EW simple 2D laser scanner. (Courtesy of Hokuyo, Osaka, Japan. With Permission.)

IMU subsystem and an Allied Vision Technologies (AVT) Pike CCD camera. This system was applied to forest mapping in Finland, in particular, to acquire data for multitemporal studies of small areas. Since then, the system has been developed further, mounted on a Microdrones MD4 UAV.

Another example of a system that uses this type of simple scanner technology on a lightweight UAV is that developed by the Warwick Mobile Robotics Group at the University of Warwick in the United Kingdom. This group operates a modified Hexakopter UAV built by the German Hi Systems (MikroKopter) company (Figure 3.43a). The laser scanner that is being used on this aircraft is the Hokuyo UTM-30-LX model (Figure 3.42b) emitting its radiation at $\lambda = 905$ nm. It has a maximum

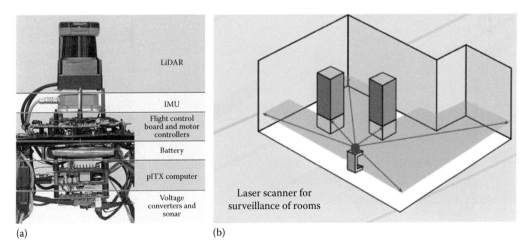

(a)   (b)

FIGURE 3.43 (a) Showing the power, computing, and measuring elements of the UAV system employed by the Warwick Mobile Robotics Group. (Courtesy of Warwick Mobile Robotics, Coventry, UK. With Permission.) (b) Diagram showing the 2D plane that is being traced out by the laser scanner when mounted with its rotational axis vertical, so measuring a profile along the walls and obstacles within a room or building. (Courtesy of SICK AG, Waldkirch, Germany. With Permission.)

distance measuring range of 30 m, with a measuring accuracy of ±3 to 5 cm, and has a pulse repetition rate of 43 kHz. The scanning mechanism provides an angular scanning range of 270°, an angular resolution of 0.25° (360°/1,440 steps), and a scan rate of 40 Hz. This particular scanner weighs a mere 210 g without cable. The Warwick system is intended primarily for indoor use within large buildings (Figure 3.43b)—in which case, the use of GPS receivers for positioning purposes would be excluded. Thus, the system is equipped solely with a Xsens MTi (MEMS-based) IMU and employs the Simultaneous Location and Mapping (SLAM) technique for position determination and for obstacle detection and avoidance within the building. A GoPro Hero2 small-format CMOS camera has also been mounted on the UAV for imaging purposes. The LiDAR-equipped UAV system then maps the interior of the building in 3D by building up a series of near-horizontal 2D profiles stacked one above the other as the UAV ascends vertically.

3.6.3.2 Multilayer Laser Scanners

Multilayer laser scanner units have been developed by the SICK company and its subsidiary, Ibeo— for example, the SICK LD-MRS and the Ibeo Lux scanners. These devices simultaneously scan and measure distances with their laser rangefinders in four parallel layers or planes (Figure 3.44a), instead of the single plane that is being utilized by the simpler scanners discussed in the previous section. Again, tiny diode lasers emitting their radiation at $\lambda = 905$ nm wavelength are used in the unit's rangefinders. This multilayer technology was developed originally to compensate for the pitching motion of road vehicles that were being equipped with scanners for collision avoidance and to detect slopes. Effectively, this resulted in the device having an increased vertical angular coverage, besides the horizontal angular coverage that is being provided by the scanner's main scanning action. The original models of the Ibeo Lux laser scanner unit featured four parallel layers or planes (Figure 3.44a); later models have eight such planes, providing a 6.4° vertical field-of-view. (Figure 3.44b). Different models have maximum ranges of 90 or 200 m and a measuring precision of ±10 cm. The horizontal angular field of the scanner can be as great as 110°. However, 60° is more typical. The weight of an individual unit is approximately 1 kg, but 2 kg is more typical when the GPS/IMU subsystem, computer card, and batteries that are required for UAV operation are added. According to discussions on the Web, the cost of a single Ibeo Lux unit was quoted as circa €13,900 in 2011. In 2013, the cost was quoted as circa $9,000. So far, the scanner has not been used in production road vehicles. However, the cost is expected to reduce greatly if indeed this type of unit is adopted widely by vehicle manufacturers and enters volume production.

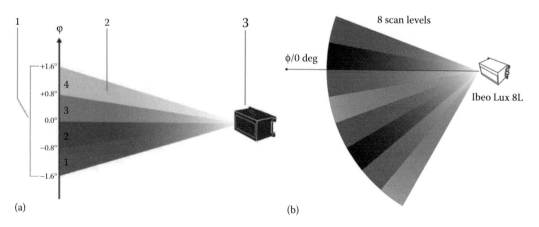

(a)   (b)

FIGURE 3.44 Diagrams showing (a) the angular values of the four layers of the Ibeo Lux laser scanner unit; and (b) the arrangement of the eight layer version of the Lux scanner. (Courtesy of Ibeo Automotive Systems GmbH, Hamburg, Germany. With Permission.)

FIGURE 3.45 The system based on an Ibeo Lux laser scanner that has been developed by the Terra Luma project and operated on one of its rotary-wing UAV aircraft. (Courtesy of University of Tasmania, Hobart, Australia. With Permission.)

Examples of these multilayer laser scanners have been adopted by various agencies for use on lightweight UAVs. An early example was the system based on an Ibeo Lux multilayer scanner that has been developed by the Finnish Geodetic Institute. Initially, this was mounted and operated on the Institute's Sensei UAV platform. The system has since been mounted on an Oktokopter UAV. Another research organization that has developed a similar system is the Terra Luma project at the University of Tasmania in Australia (Wallace et al., 2012). This has utilized a single, four-layer Ibeo Lux scanner pointing downward in the vertical direction. It has been used in combination with a MEMS-based Microstrain 3DM-GX3-35 IMU, a NovAtel DGPS receiver and a video camera to develop a system that was mounted on a Droidworx Oktokopter UAV (Figure 3.45). Since the total weight of the scanner system is 2.4 kg, this limited the flying time to 3 or 4 min. However, more recently, this time has been extended considerably using a larger FoxTech Devourer X8 UAV. The system has been used for various projects concerned with environmental monitoring, including precision agriculture, vegetation mapping, and the monitoring of change in forest plantations utilizing data acquired from multiple flights. The French L'Avion Jaune aerial survey company, which is based near Montpellier in Southern France, has also developed its YellowScan Mapper system based on the Ibeo Lux scanner unit combined with a Septentrio GNSS/IMU positioning and orientation subsystem. This system has been operated on various types of fixed-wing and multirotor UAVs by the L'Avion Jaune company and has been sold to various other companies in Italy and Canada.

3.6.3.3 Multiple Spinning Laser Scanners

As mentioned in the chapter on Terrestrial Laser Scanners, originally the technology of multiple spinning lasers was developed for use on road vehicles by the American company, Velodyne LiDAR Inc. that is based in California. The cost of the original HDL-64E unit has been quoted as US$75,000. In 2010, Velodyne introduced a lower-cost model, the HDL-32E. This operates on the same basic principle as the HDL-64E but has a bank of 32 individual laser ranging units instead of 64 and is correspondingly smaller and lighter (weighing just under 2 kg, compared with the 15 kg for the earlier HDL-64E model). According to various sources on the Web, the cost of a single

---

YellowScan is now an independent system supplier, separated from L'Avion Jaune.

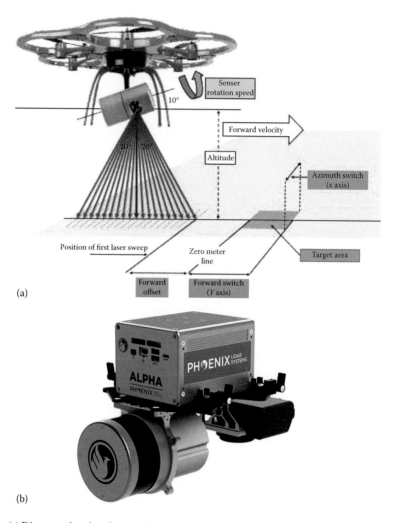

FIGURE 3.46 (a) Diagram showing the scanning pattern resulting from the use of a Veleodyne spinning laser scanner on a UAV with its rotational axis horizontal. (Courtesy of Velodyne, San Jose, CA. With Permission.) (b) The Phoenix LiDAR Systems Alpha AL3-16 laser system based on the Velodyne VLP-16 spinning laser scanner. (Courtesy of Phoenix LiDAR Systems, Los Angeles, CA. With Permission.)

HDL-32E unit was circa $30,000 in its original form. In 2014, Velodyne introduced a further model in the series, called the VLP-16 or *Puck* (Figure 3.46a), using only 16 laser ranging units. Its performance characteristics include a maximum range of 100 m, a PRF of 300,000 points per second, and a vertical field-of-view of +15° to −15°. With a very low weight of only 830 g and a price of $8,000, this model has been of immediate interest to the operators of lightweight UAVs. A still lighter weight version of this device, called the *Puck LITE*, has just been introduced in February 2016. This has the same performance characteristics as the Puck model, but weighs only 590 g.

Needless to say, the advent of these very lightweight and much less expensive models from Velodyne has made them of great interest both to system suppliers and to users wishing to operate laser scanners from UAVs. The Phoenix LiDAR Systems company based in San Clemente, California, has mounted these lighter weight models on a variety of lightweight UAV aircraft. First of all, an HDL-32E unit was mounted on an Oktokopter UAV with the scanner's spinning axis vertical, resulting in the scans being generated in a near-horizontal plane. The resulting system—called the Phoenix AL2—initially featured a NovAtel SPAN-IGM-A1 subsystem to provide the required

DGPS/IMU capability. This unit was then replaced by a more accurate *Spatial FOG* subsystem from the Australian supplier Advanced Navigation, which, as its title suggests, uses FOG technology. Since then, the company has developed its AL3 series comprising the AL3-32 model (using the HDL-32E scanner); the AL3-16 model (using the VLP-16) (Figure 3.46b); and the Aerial Scout (again based on the VLP-16) model, weighing 3.2, 2.5, and 1.85 kg, respectively. Each has the spinning axis of the Velodyne scanner mounted in a horizontal position to allow downward pointing of the scanner from a UAV with a maximum range of 120 m. These systems are available to customers with a wide choice of GPS/IMU subsystems equipped with MEMS or FOG based inertial measurement units.

In the United Kingdom, Mapix Technologies, based in Edinburgh, Scotland, has devised its Routescene LiDAR Pod that is based on the Velodyne HDL-32E laser scanner mounted with its spinning axis in a horizontal position and its laser pulses firing downwards (Figure 3.47a). The Routescene system comes complete with a RealTime Kinematic GNSS/IMU subsystem and ground station. The pod design allows the system to be attached to a wide range of UAV aircraft. Another supplier of a Velodyne-based system is the German UAV constructor, Aibotix, which offers its X6 Hexacopter equipped with a Velodyne HDL-32E laser scanner, again mounted with its spinning axis vertical, as used in the Phoenix AL2 system. This particular combination was in fact shown first at the Intergeo 2013 trade fair held in Essen, Germany. Shortly afterwards, in February 2014, the Aibotix company was bought by Hexagon, and it now operates under the Leica Geosystems brand. The main application of this system would appear to be in the mining industry where the system has been used for the volumetric measurements of stockpiles on a daily basis and for surveys of open-cast mines. Yet another supplier of systems based on the Velodyne VLP-16 scanner set with its spinning axis in a horizontal position is the French company, YellowScan, which offers its YellowScan Surveyor system for sale (Figure 3.47b). This particular model utilizes the Applanix APX-15 single-board GNSS/IMU subsystem employing MEMS inertial technology and weighing only 50 g. Indeed, currently, the YellowScan Surveyor system is being offered for sale by Applanix.

In August 2016, the Ford Motor Company and the Chinese Baidu Inc. Web services company announced that jointly they were investing $150 million in the Velodyne company with a view to it designing and producing (1) on the one hand, a more affordable laser that can be mass-produced for use in automobile applications such as vehicle navigation and collision avoidance, and (2) on the other, to develop more sophisticated laser sensors. If these efforts are indeed successful, then undoubtedly they will have an effect on the use of laser scanner systems on UAV aircraft.

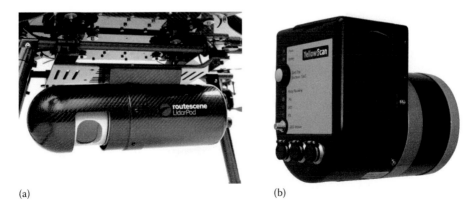

(a) (b)

FIGURE 3.47 (a) The Mapix Routescene LiDAR Pod attached to the underside of a UAV. (Courtesy of Mapix Technologies, Edinburgh, Scotland. With Permission.) (b) The YellowScan Surveyor laser scanner system. (Courtesy of YellowScan. With Permission.)

3.6.3.4   Terrestrial 2D Laser Scanners

As discussed previously, terrestrial (ground-based) laser scanners that measure over short distances—typically with maximum ranges from 50 to 100 m—are in widespread use by the land surveying profession. In terms of their angular coverage, all of the scanner instruments that fall into this particular group belong to the category of panoramic terrestrial laser scanners in terms of their angular coverage. Within the mobile mapping field, a small number of these terrestrial 3D laser scanners have been mounted and operated on mobile mapping vehicles. However, in this particular application, their horizontal (azimuth) angular movements are normally disabled, so the mirrors are only spinning around the instrument's trunnion axis and are only measuring distances and angles in the 2D vertical plane of the instrument's spinning mirror—which effectively makes them into 2D laser scanners (profilers). This same restricted configuration has also been adopted on those terrestrial laser scanners that are being mounted on UAV aircraft.

The FARO Focus 3D (and its Trimble TX5 derivative) is a small and compact terrestrial 3D laser scanner, costing circa €33,000. The instrument can measure over a range of distances from 0.6 to 120 m over a 300° vertical field-of-view, with a quoted accuracy of ±2 mm when used in its static mode. The instrument weighs 5 kg—which is extremely light by the standards of terrestrial 3D laser scanners, most of which weigh between 10 and 15 kg. However, 5 kg is a very considerable payload by the standards of lightweight UAVs. Nevertheless, the FARO Focus 3D scanner instrument has been mounted on certain powerful lightweight UAVs with a suitable payload carrying capacity and has been used to carry out airborne laser scanning operations. In cooperation with FARO Technologies, two Austrian companies—von-oben, which is active in UAV aerial photography, and 4D-IT, which specializes in the digital documentation and visualization of cultural heritage objects—have developed and deployed their Scan-Copter system utilizing a multicopter UAV equipped with a FARO Focus 3D laser scanner (Figure 3.48a). Applications have included the monitoring of landslides as well as the measured surveys and documentation of historic buildings. Recently, the partnership has produced an upgraded version of the system, named Scan-Copter 2.0, which features a very heavy lift UAV providing a much longer time of operation of the FARO Focus 3D scanner—30 min versus the 10 min of the original Scan-Copter. The Scan-Copter 2.0 also features the Applanix APX-15 GNSS/IMU subsystem for positioning and orientation purposes.

In the United Kingdom, a somewhat similar development has been undertaken by Sabre Sea & Land Ltd., which is based near Aberdeen, Scotland. The Skypod unit that has been developed by Sabre—which includes a DGPS/IMU subsystem with twin antennas—can utilize either a FARO Focus S120 or X330 3D scanner. Again, as in several of the above-mentioned projects, there is a strong connection

(a)   (b)

FIGURE 3.48   (a) This FARO Focus 3D terrestrial laser scanner is shown attached to an Oktokopter UAV. (Courtesy of 4D-IT GmbH, Pfaffstätten, Austria. With Permission.) (b) The Sabre Sky-3D system is also based on the use of a Faro Focus 3D terrestrial laser scanner. (Courtesy of Sabre Sea & Land Ltd, Banchory, UK. With Permission.)

with mobile mapping, since the pod unit is designed to be mounted either on a wheeled mobile mapping vehicle or on a lightweight UAV. The latest development from the company is the Sabre Sky 3D model that is designed specifically for operation on a heavy-lift UAV (Figure 3.48b). It can be supplied with a Hokuyo simple 2D laser scanner or an Ibeo multilayer laser scanner or a Velodyne HDL-32E unit as an alternative to the FARO Focus 3D scanner for certain specific applications.

3.7 AIRBORNE BATHYMETRIC LASER SCANNERS

As the title of this section suggests, airborne laser scanners have been devised to carry out mapping of the seabed and the adjacent land in coastal areas. The early systems that have been used for this type of bathymetric mapping were all laser profilers. Their pulse-type laser rangefinders measured the vertical distances from the aircraft to a series of successive points both on the sea surface and on the seabed that was located directly below the aircraft using the reflected signals from these surfaces. These early and largely experimental profiling systems were built in the United States, Sweden, Canada, and Australia; a very detailed account of the development of these systems is given in the paper by LaRoque and West (1990). During the 1980s, several of these early profiling systems were modified, and new systems were built that had a scanning capability added to them. They included the AOL, developed jointly by NASA, NOAA and the U.S. Navy, and the WRELADS-2 built by the Weapons Research Establishment of the Royal Australian Navy (RAN). Optech (later to become Teledyne Optech) was involved in the development of these early scanner systems, building the LARSEN-500 system under the sponsorship of the Canadian Hydrographic Service and the Canadian Centre for Remote Sensing. The Optech company was also involved in the development of the FLASH-1 system for the Swedish Defence Research Establishment (FOA).

In the early 1990s, all of these early experimental efforts came to fruition with the development and the delivery of the first operational systems. These included the Scanning Hydrographic Operational Airborne LiDAR Survey (SHOALS) scanner system built by Optech for the U.S. Army Corps of Engineers (USACE) and its entry into full operational service. Optech was also involved, together with Saab, in the development of the Hawk Eye airborne bathymetric laser scanner systems that were built for the Swedish Department of Defence and entered operational service with Swedish maritime agencies. At the same time, the Laser Airborne Depth Sounder (LADS) system, built by an Australian partnership, entered service with the RAN. The same four countries—the United States, Canada, Sweden, and Australia—that were involved in these early developments are still those among Western countries that are in the forefront of developing airborne bathymetric laser scanners today (Pope et al., 2001). Outside these four countries, experimental systems have also been built in Russia and China (LaRoque and West, 1990) and production systems in Austria (by *RIEGL*). Although all of these above-mentioned systems were built originally to satisfy the requirements of government defense agencies and hydrographic surveying and charting services, there are now a number of commercial operators of airborne bathymetric laser scanning systems offering their services to government agencies and commercial companies on a worldwide basis. A further recent development has been the introduction of bathymetric laser scanners having a restricted water penetration that are designed specifically for use over inland waters with shallow depths and over coastal areas having similar characteristics (Petrie, 2011b; Quadros, 2013).

3.7.1 LASER BATHYMETRIC MEASUREMENTS

The basic principle of measuring the depth of the seabed (or lakebed) below the sea (or lake) surface with an airborne laser scanner usually involves the use of two laser rangefinders emitting pulses simultaneously at different wavelengths—in the NIR and green parts of the electromagnetic spectrum (Guenther et al., 2000). The NIR radiation is reflected from the water surface, whereas the pulse of green radiation passes into and through the water column and is reflected by the seabed back toward the rangefinder (Figure 3.49a). The actual depth that can be measured is limited to 25–70 m, depending

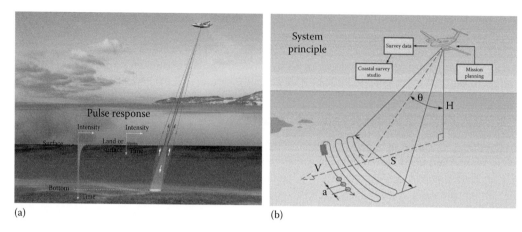

(a)   (b)

FIGURE 3.49 (a) The overall concept and design of an airborne bathymetric laser scanner showing how the pulses from a red/NIR laser are reflected from the sea surface or the land—whereas the pulses from a green laser penetrate the water and are reflected back from the sea floor. (b) The operating principle of the Hawk Eye Mk. II airborne bathymetric and topographic laser scanning system showing the scanning pattern that is used to survey the land, the sea surface, and the sea floor of coastal areas. (Courtesy of Leica Geosystems/Airborne Hydrography AB. With Permission.)

on the specific system that is being used and the clarity or turbidity of the water column through which the radiation is passing. The reflected radiation at both wavelengths is then gathered by the appropriate detectors located within the rangefinder, and the elapsed time between the emitted and received signals is measured for both pulses. From these measurements of the elapsed time, the corresponding ranges can then be derived, knowing the speed of the radiation passing through air and water, respectively. Although this information provides depth values and creates a simple depth profile along the flight line, the addition of a cross-track scanning mechanism provides coverage of a swath or area of the sea surface and the corresponding area of the seabed located below it (see Figure 3.49b). As with an airborne topographic laser scanner, the position and attitude of the airborne platform and the bathymetric laser scanner mounted on it are provided by an integrated GPS/IMU or GNSS subsystem.

3.7.2 System Suppliers—Deeper Water

As noted earlier, currently there are only a few constructors and suppliers of the airborne bathymetric laser scanners that can carry out surveys of deeper water, principally, though not exclusively, for the compilation of maritime navigation charts.

3.7.2.1 Teledyne Optech

In the introduction given earlier, it was noted that Optech (now Teledyne Optech) had played a leading part in the initial development of airborne bathymetric laser scanners, having either constructed or been the supplier of major components of several early systems built for government agencies in the United States, Canada, and Sweden. Several of the complete bathymetric laser scanner systems that have been constructed by Optech since 1990 bear the name SHOALS (Irish and Lillycrop, 1999). There has been a steady development of these systems with successive models being labeled SHOALS-200, SHOALS-400, SHOALS-1000, and SHOALS-3000, respectively.

The original SHOALS system that was supplied to the USACE in 1993 used two pulsed lasers operating at infrared ($\lambda = 1,064$ nm) and green ($\lambda = 532$ nm) wavelengths, respectively, the frequencies having been chosen to optimize the detection of the air/water interface and water penetration, respectively (Irish and Lillycrop, 1999). This system had a laser pulse repetition rate of 200 Hz—hence its later designation as the SHOALS-200. It was flown using a Bell helicopter as the airborne platform. In 1998, the system was modified, including the fitting of a new laser that doubled the pulse rate to 400 Hz—hence the name SHOALS-400. This system was mounted in a Twin Otter fixed-wing aircraft.

In 2003, Optech introduced its SHOALS-1000 model, the first example of which was delivered to the U.S. Navy. Although, in principle, the basic system and the measuring principle remained the same as in the two previous SHOALS models, this new model featured lasers operating at the much higher pulse repetition rate of 1,000 Hz (infrared) and 400 Hz (green). The new system was also much reduced in terms of its size, its weight—205 kg versus 405 kg—and its power requirements—60 A at 28 V instead of the 150 A and 120 A at the same voltage—as compared with the previous -200 and -400 models. The SHOALS-1000 formed a major element of the CHARTS (Compact Hydrographic Airborne Rapid Total Survey) program that was being implemented by the Joint Airborne LiDAR Bathymetry Technical Center of Expertize (JABLTCX), a partnership comprising the USACE, NOAA, and the U.S. Navy's NAVOCEANO organization, which is based at the Stennis Center in Bay St. Louis in Mississippi (Heslin et al., 2003). Further examples of the SHOALS-1000 were supplied to the Japanese Coast Guard's hydrographic and oceanographic survey organization and to the Fugro Pelagos company. The latter company had previously operated a commercial airborne laser bathymetric service using a SHOALS-1000 in partnership with Optech. Fugro had also operated the SHOALS-200 and -400 models on behalf of the JABLTCX through its John E. Chance subsidiary company.

In 2006, a further development took place in the form of the SHOALS-3000 that was delivered to the U.S. Navy (Heslin et al., 2003). Essentially, this was an upgrade to the SHOALS-1000 system producing an improved performance with the measuring rate of the infrared (topographic) laser being increased to 3,000 Hz. This has allowed higher data collection rates and a larger swath width (300 vs. 215 m) for a laser spot area of 4 × 4 m from an altitude of 400 m. This new version of the SHOALS system again formed a major component of the overall CHARTS system. This not only included the improved SHOALS system but also incorporated a CASI-700 pushbroom imaging line scanner supplied by another Canadian company, ITRES Research. The resulting hyperspectral linescan imagery allows the construction of classified (thematic) maps and charts showing the areas of sand, mud, rock, coral, sea grass, and others that are present on the sea floor.

The most recent airborne bathymetric laser scanning system for operation in deeper waters that has been developed by Teledyne Optech is the CZMIL (Figure 3.50). Once again, the initial system has been constructed for the JABLTCX organization for use in the USACE Coastal Mapping Program. It represents a very substantial development and constitutes a substantial upgrade of the Center's existing CHARTS system incorporating a number of innovations. In particular, the CZMIL features a single laser rangefinder capable, through its use of the frequency doubling technique, of generating both the required green and NIR pulses simultaneously. The green laser generates a narrow pulse having a high energy, a short pulse duration, and the much higher PRF of 10 kHz. The scanning mechanism features

FIGURE 3.50 The Optech CZMIL Nova bathymetric scanning system for use in coastal mapping and charting operations. (Courtesy of Teledyne Optech, Vaughan, ON. With Permission.)

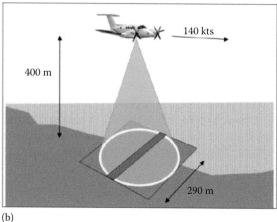

(a)   (b)

FIGURE 3.51   (a) The Fresnel prismatic optical element that is used to produce the circular scanning pattern of the CZMIL system. (Courtesy of Teledyne Optech, Vaughan, ON. With Permission.) (b) Using the Fresnel optical element, the CZMIL laser scanner generates a unidirectional circular pattern of laser pulses at a uniform angular interval to provide the area coverage and spatial density that is required for bathymetric mapping purposes. (Courtesy of JABLTCX. With Permission.)

a novel rotating Fresnel prism (Figure 3.51a) that generates a circular scan that provides two look angles towards those points that are lying in the surf zone and that are often difficult to resolve (Figure 3.51b). Furthermore, the system employs a segmented detector and high-bandwidth electronics that can operate in a scanned flash approach. The result is that the overall system provides a higher density of measured points; a better signal-to-noise ratio; and much better shallow water discrimination, as compared with the particular characteristics of the data that were obtained with the previous CHARTS system. As with the previous CHARTS system, the CZMIL incorporates an additional CASI-1500H pushbroom line scanner supplied by ITRES Research that can carry out hyperspectral imaging. Its imaging capabilities also include those provided by Teledyne Optech's 16 Megapixel, 80 Megapixel, and thermal frame cameras, while the positioning and orientation subsystem is the Applanix POS-AV. The overall system is now being offered for sale to other customers as the CZMIL Nova model (Figure 3.50).

3.7.2.2 Leica-Airborne Hydrography AB

The AHAB company maintained the long-held Swedish interest in airborne laser bathymetry. As mentioned in the introduction to this Section 3.7, this Swedish interest resulted initially in the Flash-1 system of the 1980s. At the beginning of the 1990s, the Saab group received the contract to develop a replacement for the Flash system and, with the help of Optech as its major subcontractor, two Hawk Eye systems were delivered to the Swedish Navy and the Swedish Maritime Administration in 1994 and 1995, respectively. The first of these two systems was later sold to the Indonesian Navy, on whose behalf it was operated by the Blom mapping organization to carry out extensive surveys and chart production of Indonesian coastal waters. The second system was in use for surveys around the Swedish coast that were carried out by the Hydrographics Unit of the Swedish Maritime Administration until 2003. Given the substantial part played by Optech in their development, these two Hawk Eye systems were, in many ways, quite similar to the original SHOALS system.

In 2002, Saab sold the product rights to the Hawk Eye system to three former Saab employees who had been involved in the development of the Hawk Eye laser scanner. They formed the AHAB company and started the development of a new Hawk Eye II system (Figure 3.52a) as well as undertaking part of the upgrading of the Blom TopEye II airborne topographic laser scanner systems mentioned previously. The first Hawk Eye II system was delivered in late 2005 to Admiralty Coastal Surveys AB, a company that was owned jointly by the United Kingdom. Hydrographic Office (49%), Blom ASA (21%), and AHAB (30%). The company offered the system for bathymetric surveys on a worldwide

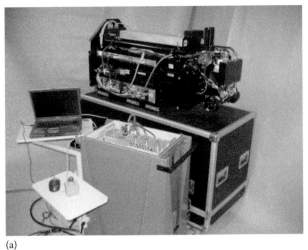 

(a)   (b)

FIGURE 3.52 (a) A Hawk Eye Mk. II airborne bathymetric laser scanner sitting on its transport box, together with its control electronics unit and a laptop computer located in front of it. (Courtesy of Airborne Hydrography AB, Jonkoping, Sweden. With Permission.) (b) The HawkEye Mk. III with its topographic laser scanner and twin bathymetric scanners, the latter designed for the penetration of shallow and deep water, respectively. (Courtesy of Leica Geosystems, St. Gallen, Switzerland. With Permission.)

basis and carried out a number of survey projects successfully. However, the Blom company then bought out its two partners, and, for a period, it operated the system on its own behalf. In 2009, Blom sold its bathymetric survey operation to Pelydryn, a company that was based in Newport, South Wales in the United Kingdom. The Pelydryn company owned three of the Hawk Eye II scanners, which again were operated on a commercial basis world-wide until 2015 when the company went out of business.

The infrared (λ = 1,064 nm) and green (at λ = 1,047 nm) pulsed lasers that have been used in the rangefinders of the Hawk Eye II were bought from a German company. They were then customized and integrated with the optics, receivers, and electronics manufactured by various suppliers to AHAB's own specification. The laser pulse rates in the Hawk Eye II systems that have been operated by Pelydryn are 8 kHz for the infrared (topographic) laser and 1 kHz for the green (bathymetric) laser. The rangefinder is tilted 15°–20° from the nadir pointing forwards in the direction of the line of flight. The position and orientation subsystem that has been used in the Hawk Eye II is an Applanix POS/AV 410 GPS/ IMU subsystem. An IDS uEye 2250 small-format (2 Megapixel) digital frame camera has also been used in conjunction with the laser scanning operations. The total weight of the system is 180 kg. The Hawk Eye II has normally been operated from a height of 200–300 m for International Hydrographic Organisation (IHO) Class 1 surveys (Axelsson, 2010). The measured data could then be displayed in real time on the system monitor and could, if required, be transferred in real time via a wireless communication link to a shore station or a survey ship. The subsequent data processing has been carried out using AHAB's own Coastal Survey Studio software, after which, the data were passed to industry standard hydrographic chart production software such as CARIS for the generation of the final chart.

At the end of 2013, AHAB was acquired by Leica Geosystems. Since then, the Hawk Eye III model has been introduced (Figure 3.52b). This system includes three separate laser systems, each of which operates and collects its own unique datasets simultaneously. They comprise the following:

1. A high-performance topographic scanner system that is operating with PRF values up to 400 kHz
2. A shallow-water bathymetric scanner system, operating at 35 kHz, that is capable of capturing bathymetric data down to about 1.5 times the Secchi depth
3. A deep-water penetrating scanner system operating at 10 kHz that is capable of capturing bathymetric data down to about three times the Secchi depth

As with other similar high-performance bathymetric scanner systems, several receiver channels are used both to enhance the data due to the losses of signal strength that occur in the water volume and to secure an accurate measurement of the water surface. All the channels in the Hawk Eye III system collect the full returning waveform data. These are analyzed in real time before the individual measured points are derived, and the point cloud data are stored to disk. All the bathymetric waveforms are also stored, and, optionally, the topographic waveforms can be stored employing a user-controlled setting. Another noticeable change as compared with the previous models in the Hawk Eye series has been the adoption of the Palmer scanning system resulting in an elliptical scan pattern over the water.

3.7.2.3 Fugro-Laser Airborne Depth Sounder

Following on from the initial WRELADS systems of the 1980s, in 1989, the RAN awarded a contract to a partnership of two Australian companies—Broken Hill Propriety (BHP) Engineering and Vision Systems—for the construction of its LADS system that was completed and brought into operational use by the RAN in 1993. The LADS system was mounted on a Fokker Friendship F27–500 aircraft equipped with twin turbo-prop engines, which has been based at Cairns in Northern Queensland. It has been used continuously for bathymetric surveys, principally of the northern and eastern coasts of Australia, ever since its introduction into service in 1993. In 1998, a second system—LADS-II—entered service mounted on a De Haviland Dash-8 twin turbo-prop aircraft. This second system was operated commercially by a division of the Tenix Corporation, which is a large Australian defense contractor that had bought out the original partnership that had constructed the LADS systems. Besides which, the Tenix LADS company also provided the support for the original RAN LADS system, which was operated jointly by a team of Tenix and RAN personnel. The Tenix LADS-II system was also operated worldwide, including the execution of extensive bathymetric surveys of Alaskan coastal waters for NOAA; surveys of coastal areas for official agencies in the United Kingdom, Ireland, and Norway; and the mapping and charting of the sea around Qatar and Dubai, as well as surveys in Australian waters, supplementing those being carried out by the RAN's LADS system. At the same time as the LADS-II was introduced, the original RAN LADS system was upgraded to a higher specification including a more capable laser system with greater positional capability, increased depth range and more detailed seabed coverage according to the RAN Website.

Fugro's Laser Airborne Depth Sounder (LADS) technology utilizes a Diode Pumped Nd:Yag green laser (532nm). The green (bathymetric) laser rangefinder scans the sea surface and seabed in a rectilinear raster pattern system (rather than the curved arc pattern that is utilized by the SHOALS and Hawk Eye scanners). The laser pulse rate that is employed in the rangefinder of the LADS HD system is 3,000 Hz. As usual, the LADS systems incorporate a GPS/IMU sub-system for the measurement and generation of the required position and attitude data. The LADS HD sensor can be operated from variable heights between 1,200 to 3,000 ft and is capable of varying the swath width independent of operating height to maintain sounding resolution.

In 2009, the Tenix Corporation sold the Tenix-LADS operation to Fugro, which re-named the operational unit as the Fugro-LADS Corporation. Since 2010, Fugro has continued to develop and upgrade the LADS technology and has in this time upgraded the LADS technology from the LADS Mk II, to the LADS Mk 3 (introduced in 2011) and most recently to the LADS HD (introduced in 2016) (Figure 3.53). The LADS HD sensor still retains its main focus on hydrographic surveys for nautical charting but can also be used for coastal zone management projects. The main design objectives for this new model were, on the one hand, to reduce its mass, dimensions and power consumption as compared with the Mk II model. By doing so, this allowed much smaller and more economic twin turbo-prop aircraft such as the Cessna Conquest 441, Cessna C208B and Beechcraft King Air A-90 to be used instead of the much larger Fokker F-27 and DHC Dash-8 twin turbo-prop passenger aircraft that had been employed as the platforms for the Mk I & II LADS systems. The LADS HD model features a higher-powered (7 mJ) laser allied to a higher pulse rate (3,000 Hz) and improved

FIGURE 3.53   The Fugro LADS HD system. (From Fugro LADS Corporation.)

optical components. With the greater power, higher operational altitudes can be employed, result-ing in wider swaths of the sea being covered from each individual flight line. The LADS HD also features a minor adjustment to the rectilinear scan by tilting the scan forward of the nadir direction by up to 5 degrees to improve the water penetration over areas of glassy sea and a stabilized scanner that can compensate for aircraft roll and cross-track, enabling more efficient line planning. The new system has also been interfaced to an Applanix POS/AV 610 GNSS/IMU sub-system for positioning and orientation purposes and includes a Phase One 100Mp camera to provide orthorectified imag-ery captured simultaneously. Most recently, the LADS HD deep-water system has been operated by Fugro in parallel with a *RIEGL* VQ-820-G shallow-water system, each with its own individual IMU, but under the control of a single laptop computer that is also processed simultaneously within Fugro's LADS Data processing software. The combination of these two sensors operated simulta-neously complement each other and delivers seamless high-resolution data on the land and in the shallow water from the *RIEGL* VQ-820-G, whilst collecting deep water ALB depth measurements to depths of up to 80 metres (subject to water clarity), nominally 3 times the Secchi disc measure-ment from the LADS HD sensor.

3.7.3   System Suppliers—Shallow Water

As will be apparent from the above-mentioned discussion, airborne bathymetric laser scanners that are capable of measuring maximum depths and therefore covering the greatest possible area of coastal waters for the production of navigation charts have been available and operational for the past 20 years. However, in 2011, three system suppliers—Teledyne Optech and AHAB and a new entrant, *RIEGL*—all decided quite independently to enter this field of airborne bathymetry with systems that were aimed principally at surveys of inland waters and inshore coastal waters with shallow depths (Petrie, 2011b). In general terms, these shallow water laser scanner systems mostly feature a single green laser operating from a low altitude at $\lambda = 532$ nm with a short pulse duration, resulting in markedly reduced power requirements. In turn, this permits more frequent measure-ments of depth, producing a very high data density and a high spatial accuracy, but having a much more restricted water penetration. Another important consideration in utilizing a laser rangefinder with reduced power is that it allows the scanner to meet the eye safety standards when it is being operated from low altitudes over a populated inland or coastal area.

3.7.3.1   NASA

One can take the view that the forerunner to this particular development was NASA's EAARL (Experimental Advanced Airborne Research LiDAR) system that was designed and built by the Agency's Wallops Flight Facility. The EAARL system was operational for a decade after first entering service in 2001. It was constructed primarily for the purposes of marine science research in relatively shallow water rather than the systematic collection of water depth data for the compilation of nautical charts for navigation purposes that the previous deeper water systems described earlier were designed and used for (Figure 3.54). As a pioneering system with many unusual features, it is appropriate to discuss it in some detail. It differed from the previously described deep-water systems in which it utilized a single (green) laser rather than the twin (NIR + green) lasers of these previous systems. The single Nd:YAG laser, which emitted its pulses at $\lambda = 532$ nm, was comparatively low-powered (70 µJ); had quite a high pulse rate (3,000 Hz); and generated a well collimated pulse, giving a 15–20 cm diameter illumination spot at the water surface. However, the lower power also meant that the maximum depth that could be measured was 20–25 m in clear conditions. The laser rangefinder was pointed vertically downwards, the scan mechanism with its oscillating mirror producing a pattern of depth measurements that was similar to those of the Teledyne Optech ALTM scanners described earlier. When operated from the relatively low flying height of 300 m (1,000 ft.), the EAARL system covered a 240 m wide swath, with each scan providing 120 measured depths, having a 2 m spacing between these points in the cross-track direction. The scan rate was 20 Hz. The total weight of the system was comparatively low at 114 kg (250 lb.), as was its power requirements –400 W at 28V DC.

Besides the actual laser ranger, the EAARL system also included two digital frame cameras, both of which acquired their images at a 1 Hz rate. The first of these was an RGB color camera generating images with a ground sampled distance of 70–90 cm. The second camera generated false-color (green + red + NIR) images with a ground sampled distance of 20 cm. The system also deployed a position and orientation subsystem comprising two dual-frequency, carrier phase GPS receivers, and an integrated miniature digital IMU, which together ensured submeter georeferencing of each laser measurement. The 1 ns temporal resolution of the waveform digitizer equated to

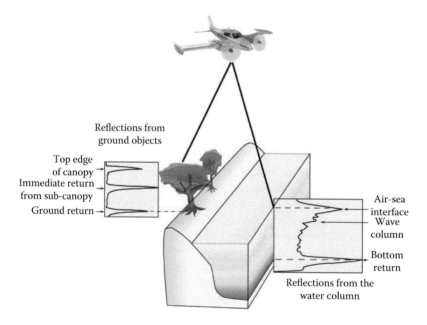

FIGURE 3.54   The overall concept of the EAARL system. (Courtesy of NASA, Washington, DC; Redrawn by M. Shand. With Permission.)

13.9 cm in the air and 11.3 cm in water. The actual ranging accuracy of the system has been quoted as being ±3–5 cm, whereas the horizontal positioning accuracy was quoted as being ≤1 m.

During the period 2011 to 2013, the EAARL system was redeveloped with a view to it providing more detailed and enhanced mapping of shallow and turbid marine environments. To achieve this objective, the laser sampling rate has been increased from 3 to 15 kHz, while the pulse power has been increased from 70 μJ to 0.4 mJ, thus providing a 5x increase in the laser pulse energy. Besides which, the outgoing beam has been optically segmented into three small footprint *beamlets*, the result being a closer spacing of the measured points and a 6x increase in the point density. It is claimed that the system provides high spatial resolution whenever the surface refraction and water column scattering are minimal. The modified system is designated as the EAARL-B system; the original system is now known as the EAARL-A system.

These two EAARL systems have been used extensively for research mapping purposes by U.S. government agencies such as the USGS and NOAA in many coastal areas of South Florida, Puerto Rico, and various American islands in the Caribbean Sea. Numerous other surveys have been undertaken in the immediate aftermath of the many devastating hurricanes that have hit the coastlines of the southern and eastern coasts of the United States in recent years. These have provided many spectacular images that have been displayed on a number of Web sites showing the resulting breaches in the barrier islands and coastal defenses, and the damage that has been wrought in these coastal areas.

3.7.3.2 Teledyne Optech

In 2011, Teledyne Optech introduced its Aquarius shallow-water system (Petrie, 2011b). Previously, in the summer of 2010, the U.S. National Science Foundation had provided the funding for Teledyne Optech to build an additional laser rangefinder for the ALTM Gemini system that was being operated by the National Center for Airborne Laser Mapping (NCALM) based in Houston, Texas. This Center is a joint collaboration between the University of Houston and the University of California, Berkeley, which provides research quality airborne laser mapping and imaging data for use by the U.S. scientific community. It is supported in this mission by the National Science Foundation. The basic idea was that this new rangefinder would operate in the green part of the spectrum and could be swapped with the standard Gemini topographic laser rangefinder for use in shallow water bathymetric mapping (<10 m) and could still utilize the existing Gemini control module and electronics. The system offers a PRF over the range 33 to 70 kHz and a scan rate that can be set at values up to 70 Hz. After its completion, Teledyne Optech then decided to offer the new system—called the ALTM Aquarius—commercially to other users, either as a complete stand-alone system or as an additional swappable module for use with an existing Gemini ALTM system (Figure 3.55). At the time of its introduction, the then new Aquarius system was viewed by Teledyne Optech as being complementary to its existing SHOALS systems, being much more compact and far more affordable for those applications that only required shallow (<10 m) depth measurements in clear water.

3.7.3.3 Leica-Airborne Hydrography AB

The AHAB company introduced its Chiroptera bathymetric laser scanning system at the end of 2011 (Petrie, 2012). The initial system was designed in close collaboration with the Coastal Studies Group of the Bureau of Economic Geology at the University of Texas in Austin. This Bureau forms part of the University's Jackson School of Geosciences and also functions as the Texas State geological survey. The Coastal Studies Group had already gained a great deal of relevant experience, having carried out airborne laser scan surveys of the whole of the coastline of Texas over the previous several years. The initial funding for the development and purchase of the new scanner by the Bureau came from the General Land Office of the State of Texas.

The Chiroptera bathymetric scanner utilizes twin laser rangefinders as the basis of its measuring system. (1) The first of these is the system's bathymetric laser rangefinder with a PRF of up to 18 kHz, which generates its pulses in the green part of the electromagnetic spectrum at the well-established wavelength (λ) of 532 nm for bathymetric depth measurements. (2) The second

FIGURE 3.55 The Optech ALTM Aquarius shallow-water bathymetric laser scanning system with the laser rangefinder and scanner located at right rear and the control electronics cabinet at left rear. In front is the laptop computer that is used to control the overall system, together with a separate small pilot display monitor. (Courtesy of Teledyne Optech, Vaughan, ON. With Permission.)

is a topographic laser rangefinder, similar to that employed in the Leica-AHAB Dragon Eye topographic laser scanner, which uses multiple pulse technology. This latter device can generate its pulses at PRF values of up to 400 kHz in the NIR part of the spectrum, operating at the standard wavelength (λ) of 1,064 nm. Both scanners use a Palmer scan mechanism that generates an elliptical scan pattern of discrete measured points over the sea (or lake) surface, the seabed (or lakebed), and the land surface that are being measured within a scan angle of 20° on either side of the flight line. The scan frequency (or scan rate) is programmable at speeds up to 70 Hz. The German IGI company supplied an AEROcontrol GPS/IMU subsystem and AEROoffice postprocessing software together with its CCNS4 flight management system—and its IGIplan mission planning software, all of which were integrated to form essential components of the overall Chiroptera system.

Since being taken over by Leica Geosystems in October 2013, AHAB has introduced a new Chiroptera II system (Richardson, 2015) that is enclosed in a freshly designed casing similar to that used in its Dragon Eye II topographic laser scanner (Figure 3.56). Instead of the IGI subsystems, this new Chiroptera II model integrates many existing Leica support products such as the NovAtel SPAN GNSS/IMU subsystem, the Leica RCD30 digital camera, and the Leica Mission Pro and Flight Pro flight management software packages. Besides these wholesale changes to the supporting units or subsystems, the PRF values of the topographic and bathymetric scanners have been increased to 500 and 35 kHz, respectively, in the new system. In its latest form, the Chiroptera II system can also utilize the Leica PAV100 stabilized mount.

3.7.3.4   *RIEGL*

The *RIEGL* company entered the field of airborne bathymetric laser scanning with its VQ-820-G model in 2011 (Pfennigbauer et al., 2011). The initial development of the system was carried out with the cooperation of the Hydraulic Engineering Unit of the University of Innsbruck, Austria (Steinbacher et al., 2010). The scanner (Figure 3.57a) is equipped with a single powerful Class 3B green laser operating at the standard wavelength (λ) of 532 nm that emits a very narrow beam (pulse) with a divergence of 1 mrad. This results in a very small diameter pulse that is 1 cm in width when leaving the instrument and 50 cm in diameter on the water surface when the scanner is being

FIGURE 3.56 The Leica-AHAB Chiroptera II airborne bathymetric laser scanner, showing the (larger) twin optical ports installed for the bathymetric and topographic laser scanners respectively and the third (smaller) port for the accompanying digital camera, all located on the underside of the system case. (Courtesy of Leica Geosystems, St. Gallen, Switzerland. With Permission.)

(a)

(b)

FIGURE 3.57 (a) The *RIEGL* VQ-820-G laser scanner is designed for shallow-water bathymetric measurements. The main scanner box is lying on its side; the dark box at top is the laser source for the rangefinder which is fixed to the side of the main scanner box. (b) The compact and highly integrated *RIEGL* VQ-880-G laser scanner showing the optical windows for (i) the green laser rangefinder; (ii) the optional near infrared rangefinder; and (iii) the digital frame camera. (Courtesy of *RIEGL*, Horn, Austria. With Permission.)

operated from a flying height of 500 m. The instrument has a lightweight box-shaped scan head that accommodates (1) the actual scanning mechanism with the rotating optical polygon prism that is typical of *RIEGL* scanners, (2) the laser transmitter optics, and (3) the receiver optics and processing electronics. The actual green laser source is attached to the side of the scan head and is connected to the scan head via an armored fiber-optic cable and the appropriate electrical cabling. The optical axis of the laser rangefinder can be tilted forward or backward at an angle of 20° in the flight direction— as used in the earlier types of Teledyne Optech and AHAB bathymetric scanners. Thus, the scan pattern over the water surface is an arc forming part of an ellipse. The high PRF value of 250 kHz results in a maximum rate of 110,000 measurements per second. These rates have been doubled in the most recent examples of the system. The scan rate can be varied between 100 and 200 lines per second. On the receiver side, full waveform recording of the radiation that has been reflected from the water and lakebed surfaces is performed. The scan head also has the GNSS/IMU subsystem attached to it; this is needed to generate the positional and orientation data that are required by the overall system. Since its initial introduction, the Model VQ-820-G has been made available with an additional topographic laser scanner operating in the NIR part of the spectrum (at $\lambda = 1,064$ nm) at PRF values up to 550 kHz as an optional item. Furthermore, this model forms the basis for the CP-820-GU system that is offered in *RIEGL*'s CP series, together with an integrated GNSS/IMU subsystem.

Another recent development in this particular area from *RIEGL* is the VQ-880-G model (Figure 3.57b). This system comprises a rangefinder employing a powerful Class 4 green laser operating at the standard wavelength (λ) of 532 nm in a circular scan pattern for bathymetric depth measurements in combination with a fully integrated medium-format digital camera and a GNSS/IMU subsystem. As with the VQ-820-G scanner, the VQ-880-G system is available with an optional laser rangefinder operating at $\lambda = 1,064$ nm in the NIR part of the electromagnetic spectrum to provide additional topographic data over the land areas bordering the areas of the shallow lakes or sea coast, which are being surveyed or mapped.

Still another recent development from *RIEGL* in the area of bathymetric surveying is its BathyCopter product (Figure 3.58a) that was introduced in the autumn of 2015. As noted previously, it utilizes a UAV aircraft to carry a single downward-pointing green rangefinder using a Class 2M

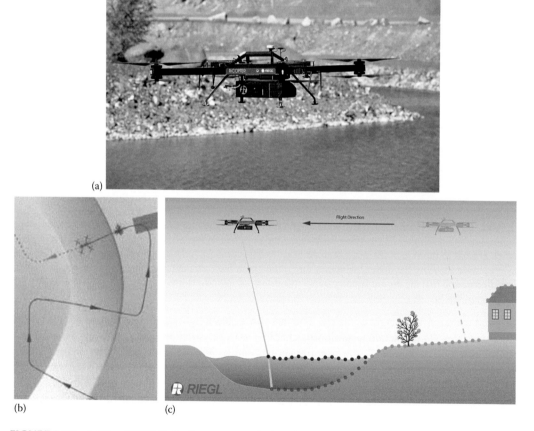

FIGURE 3.58   (a) The *RIEGL* BathyCopter system designed to measure profiles of water bodies such as rivers and canals from a UAV flying at a very low altitude. (b) Showing a typical flight path. (c) A profile being measured across a water body from a BathyCopter UAV aircraft. (Courtesy of *RIEGL*, Horn, Austria. With Permission.)

laser that carries out its measurements at a 4 kHz rate. The system is operated from very low altitudes (typically, 10 to 30 m above the water surface) with the rangefinder being pointed at 8° from the nadir to optimize its water penetration. The system has no scanning capability; thus, essentially it is an airborne profiling system measuring profiles of the surface and the bed of the shallow water body such as a river or canal that is being surveyed (Figure 3.58b and c). The system is offered complete with an integrated IMU unit and a Sony Alpha 6000 digital camera.

3.8 SUMMARY AND CONCLUSION

Over the eight years since the first edition of this book was published in 2008, the technology of airborne laser scanners has developed enormously with the use of rangefinders equipped with microchip and fiber lasers and the employment of detectors of ever increasing sensitivity, down to the level of individual photons. These devices and other related technological developments have resulted in ever faster measuring rates and even higher densities of measured elevation data, while the operation of scanners from still greater flying heights has resulted in more extensive coverage of the terrain topography from an individual flight. A consequence of all these developments is that very many hundreds of these airborne laser scanners are in operation for topographic mapping purposes in all but the poorest countries world-wide. At the other end of the topographic scale, airborne laser scanners have been built specifically for operation from low altitudes to carry out corridor mapping, especially of transport infrastructure and water courses. In this particular area, yet a further development (that did not even exist when the first edition was published) has been the development of laser scanners that can be fitted to lightweight UAVs flying at very low altitudes with particular application to the survey and mapping of small areas.

Besides the extensive topographic mapping that is being undertaken using airborne laser scanners, a corresponding impact has been felt in the area of bathymetric mapping and charting of coastal and lacustrine areas. At least in coastal waters, airborne laser scanning can collect measured depth data with a speed and with a density that far exceeds that which is possible with hydrographic survey vessels, though the latter remain unchallenged in deeper water where the laser pulses cannot reach the sea bed. The last few years have also seen the development of airborne laser scanning technology for specific use over shallow waters, especially in inland lakes and reservoirs.

Turning finally to the use of laser ranging technology to measure elevations and depths from space, at the time of writing, this use is quite limited having regard to the limitations of current technology in the context of the ultra-high speeds and altitudes at which spacecraft are operated. Thus, the methodology is currently confined to profiling with limitations both in coverage and accuracy.

REFERENCES AND FURTHER READING

Abshire, J.B., Sun, X., Riris, H., Sirota, J.M., McGarry, J.F., Palm, S., Yi, D., and Liiva, P., 2005. Geoscience laser altimeter system (GLAS) on the ICESat mission: On-orbit measurement performance. *Geophysical Research Letters*, 32(L21S02): 4.

Al-Bayari, O.A., Al-Hanbali, N.N., Barbarella, M., and Nashwan, A., 2002. Quality assessment of DTM and orthophoto generated by airborne laser scanning system using automated digital photogrammetry. *Proceedings ISPRS Commission III Symposium, Photogrammetric Computer Vision*, Graz, Austria, September 9–13, 2002, 5 pp.

Axelsson, A., 2010. Rapid topographic and bathymetric reconnaissance using airborne lidar. *Proceedings of SPIE*, Vol. 7835, Paper 783503.

Blair, J.B., Rabine, D.L., and Hofton, M.A., 1999. The laser vegetation imaging sensor: A medium-altitude, digitisation-only, airborne laser altimeter for mapping vegetation and topography. *ISPRS Journal of Photogrammetry and Remote Sensing*, 54(2–3): 115–122.

Carabajal, C.C., Harding, D.J., Luthcke, S.B., Fong, W., Rowton, S.C., and Frawley, J.J., 1999. Processing of Shuttle Laser Altimeter range and return pulse energy data in support of SLA-02, Proceedings ISPRS Workshop Mapping Surface Structure and Topography by Airborne and Spaceborne Lasers, La Jolla, CA. *International Archives of Photogrammetry and Remote Sensing*, 32(Part 3-W14): 65–72.

Clifton, W.E., Steele, B., Nelson, G., Truscott, A., Itzler, M., Entwistle, M., 2015. Medium altitude airborne Geiger-mode mapping LIDAR system. *Laser Radar Technology and Applications XX; and Atmospheric Propagation XII, Proceedings of SPIE*, 9465: 6-1–8.

Degnan, J., 2016. Scanning, multibeam, single photon lidars for rapid, large scale, high resolution, topographic and bathymetric mapping. *Remote Sensing*, 8(11): 958. doi:10.3390/rs8110958.

Degnan, J.J., Field, C., Machan, R., Leventhal, E., Lawrence, D., Zheng, Y., Upton, R., Tillard, J., Disque, S., and Howell, S., 2013. Recent advances in photon-counting, 3D imaging LIDARS. *Proceedings 18th International Workshop on Laser Ranging*, Fujiyoshida, Japan, November 11–15, 2013, 6 pp.

Flood, M., 2001. Lidar activities and research priorities in the commercial sector. *International Archives of Photogrammetry and Remote Sensing*, 34–3/W4: 3–7.

Geiger, A., Kahle, H.-G., and Limpach, P., 2007. Airborne laser profiling. ETH Research Database, ETH, Zurich, 15 pp.

Guenther, G.C., Cunningham, A.G., LaRoque, P.E., and Reid, D.J., 2000. Meeting the accuracy challenge in airborne Lidar bathymetry. *Proceedings, 20th EARSeL Symposium—Workshop on LiDAR Remote Sensing of Land and Sea*, Dresden, Germany, June 16–17, 2000, 28 pp.

Harding, D.J. and Carabajal, C.C., 2005. ICESat waveform measurements of within-footprint topographic relief and vegetation vertical structure. *Geophysical Research Letters*, 32(L21S010): 4 pp.

Harding, D.J., Dabney, P., Abshire, J., Huss, T., Jodor, G., Machan, R., Marzouk, J. et al., 2010. The slope imaging multi-polarization photon-counting lidar. *NASA Earth Science Technology Forum*, June 22–24, 2010, Arlington, VA. 7 pp.

Harding, D.J., Gesch, D.B., Carabajal, C.C., and Luthcke, S.B., 1999. Application of the Shuttle Laser Altimeter in an accuracy assessment of GTOPO30, a Global 1 km Digital Elevation Model, Proceedings ISPRS Workshop Mapping Surface Structure and Topography by Airborne and Spaceborne Lasers, La Jolla, CA. *International Archives of Photogrammetry and Remote Sensing*, 32(Part 3-W14): 81–85.

Heslin, J.B., Lillycrop, J., and Pope, R.W., 2003. CHARTS: An evolution in airborne lidar hydrography. U.S. Hydro 2003 Conference, Biloxi, MS, 4 pp.

Irish, J.L. and Lillycrop, J., 1999. Scanning laser mapping of the coastal zone: the SHOALS System. *ISPRS Journal of Photogrammetry and Remote Sensing*, 54(2–3): 123–129.

Jenkins, L.G., 2006. Key drivers in determining LiDAR sensor selection. *5th FIG Regional Conference—Promoting Land Administration and Good Governance*, Accra, Ghana, March 8–11, 2006.

Jensen, H. and Ruddock, K.A., 1965. Applications of a laser profiler to photogrammetric problems. Brochure published by Aero Service Corporation, Philadelphia, PA, 22 pp.

Jozkow, G. and Toth, C., 2014. Georeferencing experiments with UAS imagery. *ISPRS Annals of the Photogrammetry, Remote Sensing and Spatial Information Sciences*, 2(1): 25–29.

LaRoque, P.E. and West, G.R., 1990. Airborne lidar hydrography: An introduction. *Proceedings ROMPE/PERSGA/IHB Workshop on Hydrographic Activities in the ROPME Sea Area and Red Sea*, Kuwait, October 24–27, 1990.

Knowlton, R., 2011. Airborne Ladar Imaging Research Testbed (ALIRT) Technical Notes, MIT Lincoln Laboratory. 2 pp.

Li, Q., Degnan, J., Barrett, T., and Shan, J., 2016. First evaluation on photo-sensitive Lidar data. *Photogrammetric Engineering and Remote Sensing*, 82(7): 496–503.

McDonough, C., Dryden, G., Sofia, T., Wisotsky, S., and Howes, P., 1979. Pulsed laser mapping system for light aircraft. *Proceedings ASPRS Annual Conference*, 10 pp.

McGill, M., Markus, T., Scott, V.S., and Neumann, T., 2013. The Multiple Altimeter Beam Experimental Lidar (MABEL), an airborne simulator for the ICESat-2 mission. Published on-line by NASA in February 2013; Submitted to the Journal of Atmospheric and Oceanic Technology for publication, 21 pp.

Miller, B., 1965. Laser altimeter may aid photo mapping. *Aviation Week and Space Technology*, March 29, 1965.

Nelson, R., Parker, G., and Hom, M., 2003. A portable airborne laser system for forest inventory. *Photogrammetric Engineering and Remote Sensing*, 69(3): 267–283.

Petrie, G., 2006. Airborne laser scanning: New systems and services shown at INTERGEO 2006. *GeoInformatics*, 9(8): 16–23.

Petrie, G., 2011a. Airborne topographic laser scanners: Current developments in the technology. *GeoInformatics*, 14(1): 34–44.

Petrie, G., 2011b. Airborne bathymetric laser scanners: A new generation is being introduced. *GeoInformatics*, 14(8): 18–24.

Petrie, G., 2012. AHAB's Chiroptera: Another new airborne bathymetric scanner! *GeoInformatics*, 15(2): 24–25.

Petrie, G., 2013a. Commercial operation of lightweight UAVs for aerial imaging & mapping: With particular reference to the U.K. *GeoInformatics*, 16(1): 6–17.

Petrie, G., 2013b. Current developments in airborne laser scanners suitable for use on lightweight UAVs: Progress is being made! *GeoInformatics*, 16(8): 16–22.

Pfeiffer, N. and Briese, C., 2007. Geometrical aspects of airborne laser scanning and terrestrial laser scanning. *International Archives of Photogrammetry and Remote Sensing*, 36(Part 3-W52): 311–319.

Pfennigbauer, M., Ullrich, A., Steinbacher, F., and Aufleger, M, 2011. High-resolution hydrographic airborne laser scanner for surveying inland waters and shallow coastal zones. *Proceedings of SPIE*, 8037: 11 pp.

Pope, R.W., Johnson, P., Lejdebrink, U., and Lillycrop, W.J., 2001. Airborne lidar hydrography: Vision for tomorrow. *U.S. Hydro Conference*, Norfolk, VA, May 22–24, 2001.

Quadros, N.D., 2013. Unlocking the characteristics of bathymetric lidar sensors. *Lidar Magazine*, 3(6): 62–67.

Richardson, W., 2015. Initial Flights: Leica/AHAB Chiroptera II. *Lidar News Magazine*, 5(1): 54–57.

Roth, R., 2005. Trends in sensor and data fusion. In D. Fritsch (Ed.) *Photogrammetric Week 05*, Wichmann Verlag, Heidelberg, pp. 253–261.

Roth, R., 2011. Leica ALS70—Point density multiplication for high density surface acquisition. In D. Fritsch (Ed.) *Photogrammetric Week 11*, Wichmann Verlag, Heidelberg, pp. 249–255.

Roth, R. and Thompson, J., 2008. Practical application of multiple pulse in air (MPIA) lidar in large-area surveys. *International Archives of the Photogrammetry, Remote Sensing and Spatial Information Sciences*, 37(Part B1): 183–188.

Shepherd, E.C., 1965. Laser to watch height. *New Scientist*, April 1, 1965, p. 33.

Sinclair, M., Parker, H., Penley, M., and Seaton, P., 2011. Fugro commence new airborne lidar bathymetry trials. FIG Working Week 2011, Bridging the Gap Between Cultures, Marrakech, Morocco, May 18–22, 2011, 8 pp.

Sirota, J.M., Degnan, J., Field, C., Leventhal, E., and Disque, S., 2016. Single photon lidar. An efficient technique for high speed topographic and bathymetric mapping. *Presented Paper, ISPRS Commission I, WG I/2*, Prague, Czech Republic, 3 pp.

Sitar, M., 2015. Beyond 3D—Multispectral Optech Titan opens new applications for lidar. *Lidar News Magazine*, 5(1): 11–15.

Skaloud, J., Vellet, J., Keller, K., and Veyssiere, G., 2005. Rapid large scale mapping using hand held lidar/CCD/GPS/INS sensors on helicopters. *ION GNSS 2005*, Long Beach, CA, September 13–16, 7 pp.

Smith, B. and Sitar, M., 2010. Barriers broken. *Professional Surveyor Magazine*, 30(5): 5.

Steinbacher, F., Pfennigbauer, M., and Aufleger, M., 2010. Airborne hydromapping area-wide surveying of shallow water areas. In A. Dittrich, Ka. Koll, J. Aberle, and P. Geisenhainer (Eds.) *River Flow 2010*, Vienna, Austria: Bundesanstalt für Wasserbau. pp. 1709–1713.

Swatentram, A., Tang, H., Barrett, T., DeCola, P., and Dubayah, R., 2016. Rapid, high-resolution forest structure and terrain mapping over large areas using single photon lidar. *Scientific Reports*, 6: 28277.

Toth, C.K., 2004. Future trends in LiDAR. *ASPRS Annual Conference*, Denver, CO, May 23–28, 2004. Paper No. 232, 9 pp.

Ullrich, A. and Pfennigbauer, M., 2011. Echo digitization and waveform analysis. In D. Fritsch (Ed.) *Photogrammetric Week 11*, Wichmann Verlag, Heidelberg. pp. 217–228.

Vallet, J. and Skaloud, J., 2004. Development and experiences with a fully digital handheld mapping system operated from a helicopter. *International Archives of Photogrammetry and Remote Sensing*, 35(Part B, Commission V): 6.

Vallet, J. and Skaloud, J., 2005. Helimap: Digital imaging/lidar handheld airborne mapping system for natural hazard monitoring. *Sixth Geomatic Week Conference*, Barcelona, Spain, 10 pp.

Wallace, L.O., Lucieer, A., and Watson, C.S., 2012. Assessing the feasibility of UAV-based lidar for high resolution forest change detection. *International Archives of the Photogrammetry, Remote Sensing and Spatial Information Sciences*, 37(Part B7): 499–504.

Young, J.W., 2015. Review of the Optech Galaxy, Titan and CZMIL lidar sensors. *Lidar Magazine*, 5(6): 30–33.

Yu, A.W., Krainak, M., Harding, D., Abshire, J.B., Sun, X., Cavanaugh, J., Valett, S., Ramos-Izquierdo, L., Dogoda, P., and Kamamia, B., 2013. A 16-beam nonscanning swath mapping laser altimeter instrument, *Proceedings of SPIE*, 8599, Solid State Lasers XXII: Technology and Devices, San Francisco, CA, 85990 p.

4 LiDAR Systems and Calibration

Aloysius Wehr

CONTENTS

4.1 LASER SCANNING SYSTEMS

Today several different laser scanners are available that can be distinguished, on the basis of their intended application, as close range and airborne surveying systems. Close range means that laser scanning is carried out over ranges between 0 and 200 m and is typically conducted from fixed locations on the ground. These systems are sometimes referred to as terrestrial laser scanners. Commercial airborne systems typically operate over ranges of several hundred meters to several kilometers from helicopters or fixed-wing airplanes. Laser scanners belong to the family of active sensors, such as the well-known radar, because the target is illuminated by the sensor itself. Therefore, the measurement is independent of external illumination. As similar ranging techniques are applied as for the first microwave radars, laser scanners are also called LiDARs that stands for light detection and ranging. For understanding the functioning of the different LiDARs, for evaluating the performance of the instrument components and for selecting the right components appropriate for specific surveying tasks, the principles of LiDAR systems are described here by subdividing the whole system into functional subunits. By explaining these units, the reader will learn what the measurement potential of such systems is and will make him better able to select the optimum instrument implementation for a given surveying task. In the following, we will focus the discussion on airborne LiDAR systems.

4.1.1 GENERAL SETUP OF AN AIRBORNE TOPOGRAPHIC LiDAR SYSTEM

In the following I discuss a LiDAR system, because a valid surveying result is impossible if a LiDAR is used as an isolated device. In this context, a surveying result means geocoded laser measurements. Figure 4.1 shows that LiDAR systems consist of an airborne and ground segment. The airborne segment includes the

- Airborne platform
- LiDAR
- Position and orientation system (POS)

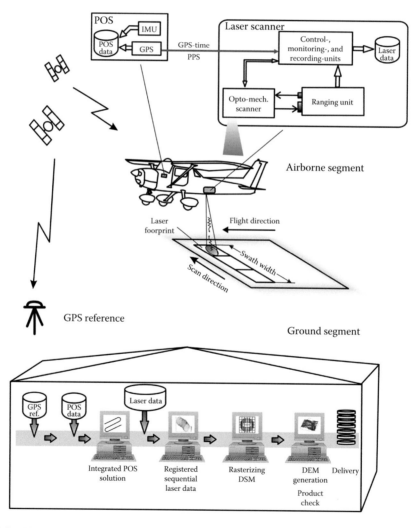

FIGURE 4.1   LiDAR system (airborne and ground segment).

The ground segment comprises

- Global positioning system (GPS) reference stations
- Processing hardware and software for synchronization and registration that is carried out offline

During flight, a LiDAR samples line-of-sight slant ranges referenced to the LiDAR coordinate system, and a POS stores GPS data including carrier phase information and orientation data of an inertial measurement unit (IMU). LiDAR and POS sample data independently. At the same time, on-ground GPS stations gather GPS data and GPS carrier phase data at known earth fixed positions for later offline computing of differential GPS (DGPS) positions of the airborne platform. By using DGPS and inertial data, the position of the laser scanner can be computed with centimeter to decimeter accuracy and its orientation is determined to better than one-hundredth of a degree. This position and orientation data are stored as a function of the GPS time. As the laser scanner data are also stored with timestamps generated from the received GPS signal, the scanner and POS datasets can be synchronized. After synchronization, the laser vector for each sampled ground point can be directly transformed into an earth fixed coordinate system, e.g., World Geodetic System 84 (WGS84). By solving the vector geometry shown in Figure 4.2, geocoded laser data are obtained. $\vec{r}_L$ is the vector from Earth center to the origin of the laser beam given by the position data, and $\vec{s}$ symbolizes the laser beam. Its direction is given by the orientation of the laser scanner unit determined by POS in 3D space and the direction of the laser beam defined by the instantaneous angular position of the laser beam deflection device defined in the scanner coordinate system. If vector $\vec{G}$ is calculated in WGS84, it represents the geocoded laser measurements. The process of calculating $\vec{G}$ is called registration. Today registered laser scanner data with accuracy better than 10 cm in 3D space are possible and state of the art. The accuracy is primarily determined by the accuracy of POS. In the following we will look at the airborne segment first. Figure 4.1 depicts that scanning LiDARs are comprised of the following key components excluding POS:

- Laser ranging unit (LRU)
- Optomechanical scanning device
- Controlling and data sampling unit

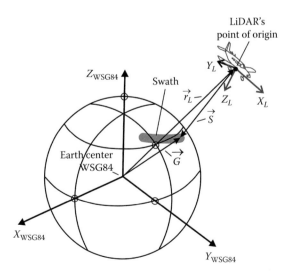

FIGURE 4.2 Vector setup for geocoded LiDAR data.

LRU measures the slant range between the sensor and the illuminated spot on ground. It comprises an emitting laser, an electro-optical receiver and a ranging electronics box. The ranging electronics box drives the laser either with pulsed or continuously modulated signals and computes the slant range from the traveling time of the laser signal from LRU to ground point and back. The transmitting and receiving apertures are mounted so that the transmitting and receiving paths share the same optical path. This assures that object surface points illuminated by the laser are always in the instantaneous field of view (IFOV) of the optical receiver. The angular divergence w of the laser, typically 0.3–2.5 mrad, is smaller than the receiver IFOV so that all of the laser spot is fully contained within the receiver field of view (FOV). To obtain surface coverage on the earth, the laser spot and co-aligned receiver IFOV have to be moved across the flight path during flight. This is realized by the optomechanical scanning device for which a moving mirror is commonly used.

4.2 LASER RANGING UNIT

The ranging unit of a LiDAR is also known as an optoelectronical range finder in geodesy. In the current chapter, the applied ranging principles and the setup are explained and discussed, and the achievable ranging accuracy is estimated by link budget calculations. A detailed performance analysis was published by Abshire et al. (2000) and Gardner (1992).

4.2.1 RANGING PRINCIPLES

Figure 4.3 shows a typical laser ranging setup. Transmitter and receiver are at the same location. The laser transmitter aperture normally is smaller than the receiver aperture. In airborne laser scanning, radar measurement principles are applied, because ranging has to handle high varying ranges. The most direct ranging measurement is determining the time-of-flight of a light pulse, that is, by measuring the traveling time between the emitted and received pulse (Figure 4.4).

According to Figure 4.4 the traveling time t_R of a light pulse is

$$t_R = 2 \cdot \frac{R}{c} \tag{4.1}$$

where:

R is the distance between the ranging unit and the object surface
c is the speed of light

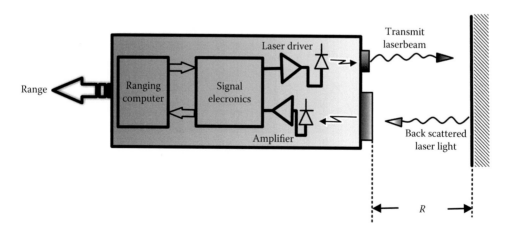

FIGURE 4.3 Two-way ranging setup.

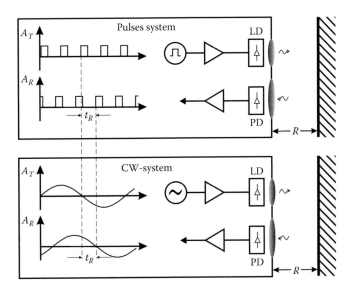

FIGURE 4.4   Traveling time ranging (upper image pulse ranging, lower image phase difference ranging).

As the signal travels the distance two times (Figures 4.3 and 4.4), one talks of two-way ranging. If the traveling time t_R is measured, the range can be directly computed by

$$R = \frac{c}{2} \cdot t_R \qquad (4.2)$$

The range resolution ΔR is determined by the obtainable time resolution Δt_R of the traveling time measuring instrument and is given by

$$\Delta R = \frac{c}{2} \cdot \Delta t_R \qquad (4.3)$$

It is also possible to determine the range, if the laser emits light continuously. Such a signal is called continuous wave (cw) signal. However, ranging can only be carried out, if the light has a deterministic intensity structure. This means that the intensity of laser light is modulated. Assuming the laser light is intensity modulated with a sinusoidal signal (Figure 4.4) that has the period time T, the ratio traveling time t_R to period time T equals the ratio phase difference between transmitted and received signal to 2π:

$$\frac{t_R}{T} = \frac{\phi}{2\pi} \qquad (4.4)$$

Putting Equation 4.4 into Equation 4.2 the range for two-way ranging is given by

$$R = \frac{c}{2} \cdot \frac{T}{2\pi} \cdot \phi \qquad (4.5)$$

As the period time is the reciprocal of the intensity modulation frequency f, R can be calculated by

$$R = \frac{1}{4\pi} \cdot \frac{c}{f} \cdot \phi \qquad (4.6)$$

The ratio c/f is equivalent to the wavelength λ of the ranging signal, so that

$$R = \frac{\lambda}{4\pi} \cdot \phi \qquad (4.7)$$

By measuring the range by the phase difference, the achievable range resolution ΔR is not only dependent on the maximum phase resolution $\Delta\phi$ but also on wavelength λ:

$$\Delta R = \frac{\lambda}{4\pi} \cdot \Delta\phi \tag{4.8}$$

In comparison with pulsed systems (Equation 4.3), Equation 4.8 makes clear that by using cw signals one has an additional physical parameter that can be used to design a system with a desired resolution. This means, even if the phase resolution is kept constant, the range resolution can be improved by applying ranging signals with shorter wavelength. This is not possible with pulsed systems, because in Equation 4.3 the constant of proportionality is the speed of light that cannot be varied. Therefore cw signals are applied if very high range resolutions are required. For example, using a 1 GHz ranging signal, which corresponds to a wavelength λ of 30 cm, and assuming a phase resolution of 0.4°, the range resolution is calculated as 0.2 mm. Achieving this resolution with a pulsed system requires a time resolution of 1 ps that is very demanding and needs very sophisticated time interval counting electronics. This simple calculation makes clear, why cw-systems are used for high precision ranging applications. However, they are usually only applied in close range applications, due to the high power required to continuously emit cw laser light and the limitations imposed by range ambiguity. The unambiguous range R_{unamb}, the maximum distance between targets that can be uniquely differentiated in the cw ranging signal, is given by

$$R_{\mathrm{unamb}} = \frac{\lambda}{2} \tag{4.9}$$

If longer ranges are to be surveyed additional cw signals are used with lower frequencies than the frequency that determines the system resolution. In this case, the highest frequency determines the resolution and the accuracy of the ranging system and the lowest frequency the maximum unambiguous range. To assure a well-functioning cw LiDAR, at least three frequencies are required for airborne applications. Due to this fact and as a ranging accuracy of better than 5 cm is usually not required, cw-systems are rarely used for airborne surveying. Hence, only pulsed laser systems are on the market for airborne surveying, and the remainder of this chapter deals only with pulsed laser systems.

Concerning pulsed laser ranging systems, the number of system parameters has to be extended to fully characterize their performance. Up to now, only the range resolution ΔR has characterized the accuracy potential of a pulsed ranging system. Looking at Figure 4.5, we recognize that several

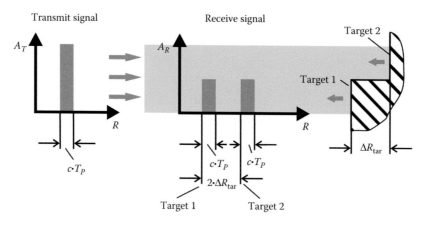

FIGURE 4.5 Target resolution.

targets at different ranges can be illuminated by a single laser pulse. Their different ranges can be resolved, if the laser pulse length is less than half the distance between the two targets. If T_P is the pulse length, the minimum resolvable distance ΔR_{tar} between targets is computed by

$$\Delta R_{\text{tar}} = \frac{c}{2} \cdot T_P \tag{4.10}$$

Let be $T_P = 10$ ns, then $\Delta R_{\text{tar}} = 1.5$ m. This means targets must be more than 1.5 m apart in slant range to be identified as separate targets. Today also shorter laser pulses are possible. 10 ps are state-of-the-art, which theoretically allows a target resolution of 1.5 mm. Here the limiting factor is determined by the processing speed of the ranging electronics, because the range measurement for one target must be completed during this short pulse or parallel gates are necessary. Today's laser scanners commonly generate pulses with pulse lengths of about several nanoseconds.

4.2.2 LASER TRANSMITTER

In laser scanning, the device that offers the electromagnetic signal in the optical spectrum is the laser. Normally semiconductor lasers or solid-state lasers that are pumped by semiconductor lasers are applied in airborne laser scanning. These lasers are optimized, so that very short pulses with high peak power levels and high pulse repetition rates are possible. Peak powers realized in commercially available laser scanners are up to 15 kW with pulse widths of 10 ns. The laser beam is collimated, so that narrow beam divergence and resulting small footprints are possible. Typical diameters for footprints are 30 cm–1 m at flying altitudes of about 1000 m. The footprint diameter d is dependent on the ranging distance and laser beam divergence that is itself a function of the size of the transmitting aperture and the transmitted wavelength λ. A good estimate for the minimal possible footprint is obtained by calculating a diffraction-limited beam. For spatially coherent light at the wavelength λ, the beam divergence is

$$w = 2.44 \frac{\lambda}{D} \tag{4.11}$$

where D is the aperture diameter (Young, 1986). Therefore, the diameter d of the illumination spot on ground is

$$d = D + w \cdot R \tag{4.12}$$

where R is the slant range. As D is small compared with the second term, a good approximation for the foot print d is

$$d = w \cdot R \tag{4.13}$$

At nadir it corresponds with the flying altitude, because R equals the flying height H. Typical laser beam divergences vary between 0.3 and 2.7 mrad. This means that spot diameters on ground between 15 cm and 1.35 m are realized for a flying altitude of 500 m.

All airborne laser scanners available on the market emit light in the near infrared. Typical wavelengths are 900, 1064, and 1550 nm. The wavelengths 900 and 1550 nm are emitted from semiconductor lasers, whereas the 1064 nm signal is generated by diode pumped solid-state neodymium-doped yttrium aluminium garnet (Nd-YAG) lasers. For shallow-water bathymetry, a laser with visible green light (532 nm) and a near infrared laser emitting at 1064 nm are used simultaneously.

The maximum possible emitted laser power is limited by eye safety regulations. Generally speaking, the longer the wavelength and the shorter the pulse width, the more a system is eye safe.

Therefore, 1550 nm laser systems can use higher power levels than a comparable system working at 532, 900 and 1064 nm respectively having the same divergence and pulse width.

4.2.3 RECEIVER

The optical receiver is composed of a receiving optic, an optical detector and a transimpedance amplifier (Keiser, 1983). The optical detector is realized with a semiconductor photodiode that converts the optical signal into an electrical current. The photodiode performance depends on the laser wavelength used. For wavelengths up to 1100 nm, photodiodes made of silicon (Si) can be used and for wavelengths in the 1000 to 1650 nm range germanium (Ge) are commonly used. Instead of Ge, semiconductor alloys such as indium-gallium-arsenide-phosphide (InGaAsP), gallium-aluminium-antimonide (GaAlSb), indium-gallium-arsenide (InGaAs), gallium-antimonide (GaSb), and gallium-arsenide-antimonide (GaAsSb) can also be utilized. A primary criterion in the selection of an appropriate photodiode is its responsivity that should be as high as possible. The unity gain responsivity is given by

$$\Re_0 = \frac{\eta \cdot q}{h \cdot \nu} \tag{4.14}$$

where:
  η is the quantum efficiency
  q is the electron charge
  h is Planck´s constant ($6.625 \cdot 10^{-34}$ J s)
  ν is the frequency of a photon

The responsivity specifies the photocurrent generated per unit optical power incident on the photo-detector. The photocurrent I_{PH} can be computed by

$$I_{PH} = \Re_0 \cdot P_{opt} \tag{4.15}$$

where P_{opt} is the received optical power. The quantum efficiency η is given by the selected photo-todiode, the material it is made of, and the wavelength of the detected light. In Figure 4.6, typical quantum efficiencies for Si, Ge and InGaAs photodiodes are plotted.

Often in LiDAR systems, avalanche photodiodes (APDs) detectors are used that feature higher responsivities (Sun et al., 1992), because they offer an internal amplification, referred to as the avalanche gain M. For APDs the responsivity $\Re$ becomes

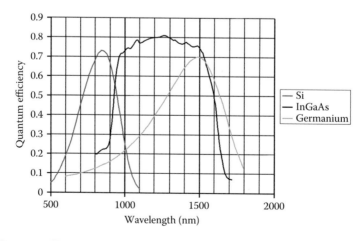

FIGURE 4.6 Quantum efficiencies for different photodiodes.

$$\Re = \frac{\eta \cdot q}{h \cdot \nu} \cdot M \qquad (4.16)$$

For Si photodiodes M is as large as 300. However, for Ge diodes the avalanche gain is limited to 40. The avalanche gain is achieved by building up a region with a high electric field. As soon as the photogenerated electrons or holes traverse this region, they gain so much energy that they ionize bound electrons in the valence band by collision. Moreover, these additional carriers experience accelerations by the high electric field and set free more ionized bound electrons. This physical process is called avalanche effect. As APDs need a high electric field, high bias voltages of several hundreds of volts have to be applied between the anode and cathode of the photodiode. If the bias voltage is lower than the breakdown voltage, a finite total number of carriers are created. APDs used in LiDARs normally work in this mode. Here, the avalanche gain M is a function of the bias voltage. Working above the breakdown voltage causes infinite number of carriers that may destroy the photodiode, because it is not a regular operation mode. However, there are special APDs available that can be used above the breakdown voltage. In this case they are functioning as photon counters. Such photodiodes arranged in an array configuration are used in photon-counting airborne LiDARs, which is a very recent development.

For further signal processing, the photocurrent must be converted into a voltage. This is accomplished by a transimpedance amplifier. Figure 4.7 shows a typical photodetector amplifier configuration. The transimpedance amplifier should exhibit a high conversion gain and a low noise figure. The combination of photodiode and transimpedance amplifier determines the sensitivity of the receiver. The higher the signal-to-noise ratio (SNR) at the output of the transimpedance amplifier, the more sensitive is the receiver and the lower is the standard deviation for ranging, and the more precise is the ranging result.

4.2.4 OPTICS

Today most LiDARs are realized with conventional optics. This means in the transmitting and receiving path optical lenses, mirrors, beam splitters, and interference filters are applied. Figure 4.8 shows a standard setup for pulsed ranging systems. The laser radiation is collimated by a lens controlling the beam divergence. The backscattered signal from the target passes through an interference filter first and is then focused on the detector chip by a focusing lens. The interference filter is required to suppress background light seen within the receiver IFOV that causes additional noise in the receiver and degrades ranging precision. Most background light is sunlight backscattered

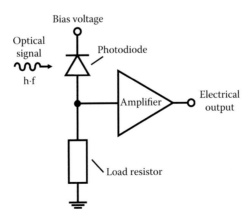

FIGURE 4.7   Photodiode with amplifier.

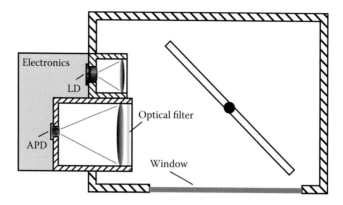

FIGURE 4.8 Optical setup with conventional optics.

from Earth's surface or from objects, such as clouds, within the IFOV. The typical bandwidths for interference filters are about 5–10 nm. Interference filters are also used in cw LiDARs. Here, the purpose is first to circumvent saturation of the detector by background light and second to reduce the received background power down to power levels that make an optimum avalanche amplification possible. This results in reduced noise within the photo current. The still remaining background light, which can well be seen in the photocurrent directly at the photodiode (Wehr, 2007), is filtered out finally by further processing the electrical signal. Here synchronous demodulation or correlation receivers are applied.

As an alternative to a conventional optical design, fibers can be used to relay the optical signal. A configuration using fibers is shown in Figure 4.9. The laser light is directly coupled into a transmit fiber and the received light is focused onto a fiber that is coupled to the detector. By using such a setup, the electronic parts are separated from the optical parts. This makes the adjustment of the optics more simple and robust. Figure 4.10 shows examples of a fiber-coupled laser and APD receiver.

Typically in the transmitter and receiver, the focusing objectives are realized with nonimaging biconvex lenses, because monochromatic light is transmitted and received and the goal is to gather the optical energy as efficiently as possible. A distortion free image is not required.

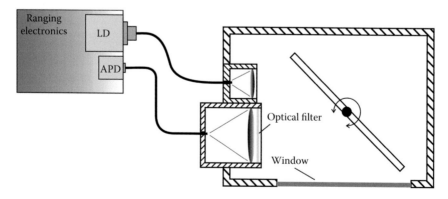

FIGURE 4.9 Principle setup with optical fibers.

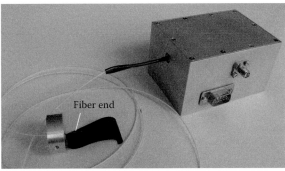

| Laser diode with fiber | APD with pigtail |

FIGURE 4.10 Laser and receiver with fibers.

4.2.5 LINK BUDGET

In Section 4.2.1, the principles of laser ranging were explained. However, degradation of the signal along the round-trip travel path from the laser to the photodetector was not considered. Calculations concerning the signal quality at the output of the receiver are known as a link budget. The signal quality is defined as the SNR at the receiver's output. This ratio determines the achievable standard deviation of the slant range measurement, which is inversely proportional to the square root of SNR:

$$\sigma_R \sim \frac{1}{\sqrt{\text{SNR}}} \tag{4.17}$$

SNR is mostly determined by the optical detector and the signal power incident on the detector element. According to Keiser (1983), the SNR for a photodiode is given by

$$\text{SNR} = \frac{\frac{1}{2}\left(\Re_0 \cdot m \cdot P_r\right)^2}{2q\left(\Re_0 \cdot \left(P_r + P_B\right) + I_D\right) \cdot F(M) \cdot B + \dfrac{4k_B \cdot T \cdot B}{R_{\text{eq}} \cdot M^2} \cdot F_{\text{AMPL}}} \tag{4.18}$$

where:

P_r is the received optical power

P_B is the received optical power of background signals

m is the modulation index

M is the avalanche gain

I_D is the primary bulk dark current

$F(M)$ is the excess photodiode noise factor $\cong M^x$ with $0 < x \leq 1$

B is the effective noise bandwidth

k_B is Boltzmann's constant

T is the absolute temperature

R_{eq} is the equivalent resistance of photodetector and amplifier load

F_{AMPL} is the noise figure of the amplifier

This formula looks quite complicated, and I will not go into much detail. But I think it is necessary to present the total formula for better understanding of the following assumptions and resulting reductions. Regarding commercial pulsed laser systems, the received power levels can be regarded as high and are well above the power of the background signal. In this case, the dark current I_D

and the thermal noise term that is reduced by the high avalanche gain M can be neglected so that Equation 4.18 can be simplified to

$$\text{SNR} = \frac{\frac{1}{2} \cdot \Re_0 \cdot m^2 \cdot P_r}{2q \cdot F(M) \cdot B} \tag{4.19}$$

Furthermore $m = 1$ for pulsed signals, because the laser is switched on and off. This means SNR is directly proportional to the received optical power P_r and the responsivity $\Re_0$ and inversely proportional to the bandwidth B. Bandwidth B is a function of the sample rate and pulse duration T_P, respectively; the shorter the pulse the higher is the required bandwidth. However, a larger bandwidth means lower SNR if P_r is kept constant. This formula makes clear that here we face a typical engineering problem, that needs an engineering tradeoff for each system. By introducing the discussed parameters of Equation 4.19 into proportionality (Equation 4.17), the resulting relation for the ranging standard deviation

$$\sigma_R \sim 2 \cdot \sqrt{\frac{q \cdot B}{\Re_0 \cdot P_r}} \tag{4.20}$$

underlines this fact. Looking at Equation 4.20, one may conclude that the bandwidth should be as small as possible. But in this proportionality the signal form (the pulse shape in a plot of pulse energy versus time) is not regarded. Jelalian (1992) considered the signal form and obtained the following equation for leading edge detection:

$$\sigma_R = \frac{c}{2} \cdot \sqrt{\frac{T_P}{2 \cdot B \cdot \dfrac{E_r}{N_0}}} \tag{4.21}$$

where:
  T_P is the pulse length
  B is the effective bandwidth
  E_r is the received optical energy
  N_0 is the noise power per cycle

Leading edge detection means that the rising edge of the pulse waveform is used as a reference for time delay measurements. As the receiver has a limited bandwidth B, the slope of the received signal has finite gradient (Figure 4.11a). The exact moment at which the received pulse is acknowledged as arrived is dependent on the detection method. If for example, threshold detection is applied, the receive time varies with the threshold level (Figure 4.11a). Even if the threshold level is fixed, the measured receive time of the pulse varies in dependence of the gradient of the leading edge (Figure 4.11b). The slope of the leading edge may vary by signal noise, amplitude variations of the received signal, or spreading of the pulse due to elevation differences within the illumination spot of the laser pulse (Figure 4.11a and b). The first one introduces stochastic errors, whereas the other two cause systematic errors. The amplitude dependent systematic errors can be compensated by applying a constant fraction discriminator. Here, the received signal is split into two channels. In one channel, the signal is delayed by a certain delay time t_d and in the other one, its amplitude is reduced by a factor m. Then, the delayed signal is subtracted from the reduced signal. The time at which the resulting signal crosses zero is independent of the amplitude, and this moment is used for ranging. Such signal processing carries out ranging not on a fixed threshold but on a constant fraction of the total peak power of the received pulse.

The most precise pulse ranging measurements can be expected from full waveform systems, which today are the most advanced commercially available systems. They sample the transient of the return pulse with sampling rates of about 1 GHz, so that no analog information gets lost (Figure 4.11c).

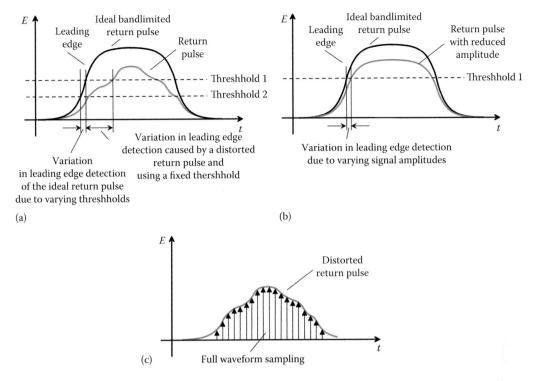

FIGURE 4.11 Pulse time delay measurement. (a) Timing variations caused by distorted signals and different threshholds, (b) Timing variations caused by varying amplitudes and resulting different leading edge slopes, and (c) Full waveform sampling.

The stored data can be processed by different sophisticated algorithms (Jutzi and Stilla, 2005). So it is possible to compensate systematic errors by, for example, calculating the centroid of the return signal and even different elevations can be resolved within the transient of one return pulse. This means that different vegetation elevation levels can be determined in dense forest areas. More details concerning full waveform systems and their data processing are presented in Chapters 3, 5, and 6.

In the following derivations and calculations, an ideal pulse detection is assumed, meaning there are no systematic errors in the measurement caused, for example, by amplitude variations in the return pulses. Assuming now perfect rectangular pulses of width T_P and a matched filter which means that

$$B = \frac{1}{T_P} \tag{4.22}$$

σ_R becomes

$$\sigma_R = \frac{c}{2\sqrt{2}} \cdot \frac{1}{\sqrt{SNR}} \cdot T_P \tag{4.23}$$

This corresponds to proportionality (Equation 4.17), because T_P is normally constant. Introducing Equation 4.19 into Equation 4.23 σ_R amounts to

$$\sigma_R = c \cdot \frac{1}{\sqrt{2}} \sqrt{\frac{q \cdot F(M)}{\Re_0}} \cdot \sqrt{\frac{T_P}{P_r}} \tag{4.24}$$

Equation 4.24 depicts that σ_R is proportional to the square root of T_P and inversely proportional to the square root of the received optical power P_r.

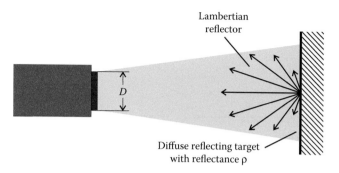

FIGURE 4.12 Ranging setup for diffusing a reflecting target.

For small optical signals onto the detector, this is especially the case in cw-systems, the thermal noise term in the numerator of Equation 4.18 dominates so that

$$\sigma_R = c \cdot \sqrt{\frac{k \cdot T \cdot F_{\text{Ampl}}}{\Re_0 \cdot R_{\text{eq}} \cdot M^2}} \cdot \frac{\sqrt{T_P}}{P_r} \tag{4.25}$$

This means, also here the ranging standard deviation is proportional to the square root of pulse width T_P. However, it is inversely proportional to the received optical signal power P_r.

Equation 4.24 and 4.25 make clear that P_r is the primary parameter that determines the accuracy during flight. Therefore, the determination of P_r is worked out in the following. Figure 4.12 shows the two-way ranging setup.

Assuming a diffuse reflecting target that is equal to or larger than the laser footprint, then the received optical power P_r is computed by

$$P_r = \frac{1}{4} \cdot P_T \cdot \tau_{\text{total}} \cdot \rho \cdot \left(\frac{D}{R}\right)^2 \tag{4.26}$$

where:
P_T is the transmitted optical power
ρ is the reflectivity
D is the diameter of receiving aperture
R is the distance from LiDAR to target
τ_{total} is the total transmission the beam experiences along the two-way path from the transmitting laser aperture to the detector element.

In Equation 4.26, diffuse reflection is approximated by considering Lambertian reflectors (Figure 4.12). This means the laser light is backscattered into the half sphere, and its radiance is independent of the viewing angle. For a pulsed system, the peak power P_{peak} is used as P_T. The total transmission τ_{total} comprises the transmission of the receiver objective, the optical interference filter, the atmosphere, and the scanning device. It should be noted that the atmosphere path is traveled twice and also that the scanning device has a transmit and receive path. P_T and D are constant determined by the instrument design. τ_{total} can be separated into a constant part τ_{system} defined by the LiDAR design and the transmission of the atmosphere, which is a function of the slant range R:

$$\tau_{\text{total}} = \tau_{\text{system}} \cdot 2 \cdot \tau_{\text{atmos}}(R) \tag{4.27}$$

Therefore, P_r can be described by a constant parameter C_{design} defined by the LiDAR design, the target dependent reflectivity ρ, the range dependent parameter τ_{atmos}, and the slant range itself:

$$P_r = C_{design} \cdot 2 \cdot \tau_{atmos} \cdot \frac{\rho}{R^2} \tag{4.28}$$

By neglecting the range dependent influence of the atmosphere, P_r is proportional to the reflectivity of the target and inversely proportional to the square of slant range R. Applying Equation 4.28 in Equation 4.24, one clearly sees that for pulse systems ranging standard deviation σ_R becomes

$$\sigma_R \sim \frac{R}{\sqrt{\rho \cdot \tau_{atmos}}} \tag{4.29}$$

whereas for cw-systems with low receiving power levels, the following proportionality is observed:

$$\sigma_R \sim \frac{R^2}{\rho \cdot \tau_{atmos}} \tag{4.30}$$

However, real LiDAR measurements show that the functional relationship between σ_R and slant range R is located somewhat between proportionality (Equation 4.29 and 4.30). Furthermore, it should be mentioned that there are special detection cases, in which this very general approach does not apply, for example, detection of power lines. Here the fundamental form of the radar equation should be used, which contains the radar cross section σ_{cross} as a parameter. σ_{cross} describes the property of a scattering object. σ_{cross} is a function of the wavelength, the directivity of the backscattered signal, the reflectivity and the size of the target. This parameter is used to model specular reflection and retro reflecting targets. Regarding σ_{cross}, Equation 4.26 becomes accordingly to Wolfe and Zissis (1978)

$$P_r = P_T \cdot \tau_{total} \cdot \frac{1}{w} \cdot \frac{D^2}{R^4} \cdot \sigma_{cross} \tag{4.31}$$

where w is the divergence of the laser beam measured in radians. As w is an instrument constant, it is clear that P_r is inversely proportional to the fourth power of the slant range. This means, the maximum possible accuracy in slant range degrades with $\sigma_R \sim R^2$ as far as pulsed systems are concerned. A more detailed analysis was carried out by Litchi and Harvey (2002) and Jutzi et al. (2002).

The LiDAR link budget calculation presented in this chapter was reduced to the simpler forms to highlight the key parameters that govern instrument performance. These parameters are those with greater relevance for people who will be configuring, using, or procuring a LiDAR system. Calculating the precise ranging accuracy of a system is often difficult because the transmitted power levels are usually not mentioned. Therefore, the equations were reduced to proportionalities so that the influences of the different parameters are seen clearly. Performance of a particular system can be extrapolated to different flying altitudes based on the ranging accuracy achieved at one altitude. Profound calculations are published by Abshire et al. (2000), Bachman (1979), Skolnik (2001), and Sun et al. (1993).

4.3 SCANNING DEVICES

Figure 4.1 depicts that the transmit beam and the coaligned receiver IFOV must be deflected across the flight line to obtain surface coverage. In all commercial LiDARs, optomechanical scanners are implemented to accomplish this. Using conventional optics, two physical principles can be exploited: refraction and reflection. Refraction scanners are realized with two rotating prisms (Figure 4.13).

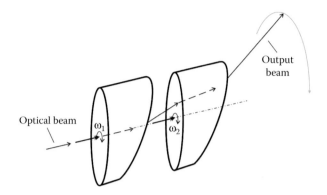

FIGURE 4.13   Principles of rotating prisms.

Such scanners have been used in military LiDARs for power line detection in helicopters. Very fast scans can be realized, because the continuous, unidirectly rotation of each prisms can be performed at high speeds. Figure 4.14 depicts a typical setup with two prisms without drives. With such a setup, all kinds of Lissajous figures are possible by changing the phase between the two prisms. A spiral scanning pattern (Figure 4.15), for example, has been used for power line detection. However, they are not routinely applied in airborne surveying. Mirrors are typically used in commercial airborne LiDARs to carry out the scanning.

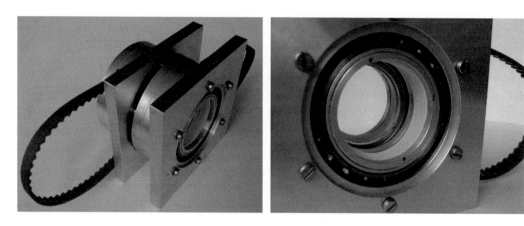

FIGURE 4.14   Rotating prisms without drives.

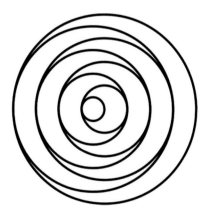

FIGURE 4.15   Spiral scan.

4.3.1 Reflecting Scanning Devices

The current section presents and discusses the different scanner setups exploiting the physical effect of reflection. Typical mirror scanners with the associated scanning patterns are compiled in Table 4.1. For all airborne LiDAR scanners, the primary design goal is to scan sufficiently fast to compensate for the forward velocity of the platform while achieving the desired sampling density in the flight path direction. Oscillating mirrors rotate back and forth to achieve a cross-track scan pattern. For oscillating mirrors, the achievable scanning speed is dependent on the inertia of the moving parts and the available power of the scanner drives, because the mirrors experience accelerated movements. The inertia is a function of the mirror's size and shape. The required size of mirrors is determined by the size of the transmitting and receiving aperture. Here, we face again an engineering problem that means optimizing the parameters scanning speed, accuracy, and laser power, because accuracy can only be improved for a given laser power by enlarging the receiving aperture.

TABLE 4.1
Overview of Mirror Scanners

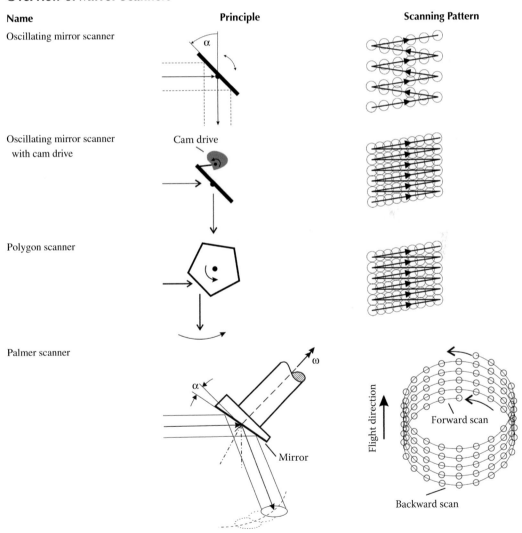

FIGURE 4.16   Facet mirror scanners.

Larger apertures mean larger mirrors and higher inertias. Higher inertias of moving parts require more powerful drives. If the drive power cannot be improved, the mirror must move more slowly.

Faster scans are possible with the polygon scanner that employs a multifacet mirror and the Palmer scanner, because these mirrors do not experience any acceleration once the working rotation speed is reached.

Table 4.1 makes evident that the laser beam deflected by facet mirrors always moves in one direction, either from left to right or vice versa. Figure 4.16 presents typical facet mirror scanners. A rotating multifacet mirror, composed of four facets, is used in the scanners manufactured by *RIEGL*. For oscillating and faceted scanners, the obliquity of the scanning lines with respect to the flight direction and the line spacing in the flight direction are a function of the ratio of the scanning speed to the flying speed. The spacing of the laser spots along the scanning track is dependent on the scanning speed and laser pulse rate.

The Palmer scanning mechanism is the most simple one as far as the optomechanical-, bearing-, and drive-setup are concerned, if large apertures are required. The ideal Palmer scanning pattern is generated on Earth's surface by the translation of a pure circular scan (Table 4.1). A circular scan is realized by mounting a mirror at the phase side of a rotating shaft. If the normal of the mirror has an angle α with the rotation axis and assuming a narrow laser beam radiated in the direction of the rotation axis, an ideal circle is projected back into the direction the laser light is originating from. The circle size increases by angle α. Normally, the shaft is mounted so that the shaft's rotation axis exhibit 45° with regard to incident laser beam. By using such a configuration, the scanning pattern is deflected by 90° (Table 4.1). A detailed ray tracing analysis shows that the deflection of 90° causes a deformation of the ideal circular scanning pattern to an egg-shape one. However, for flight planning, the scanning pattern is approximated with an ellipse. During flight a translated ellipse is observed on Earth's surface. Such scanning devices were first applied in airborne and spaceborne remote sensing missions in the early seventies. For example, the multispectral scanner flown on Skylab uses this scanning concept. During the nineties, this scanning principle was used for laser scanning (Hug, 1994; Krabill et al., 1995, 2002).

A typical Palmer scanner assembly, built at the Institute of Navigation, University of Stuttgart and used in the Scanning Laser Altitude and Reflectance Sensor (ScaLARS), is shown in Figure 4.17. The TopEye Mk II® (Blom Swe AB & TopEye AB, Stockholm, Sweden) also uses the Palmer scanning principle. The helical pattern produced by a Palmer scanner is a redundant one, because the same area on ground is sampled twice. Thiel and Wehr (1999) show that the sampling crossovers formed by the intersecting forward and backward scans can be exploited for intrinsic calibration of the external orientation parameters of POS and LiDAR.

FIGURE 4.17   Palmer scanner.

All discussed scanners are designed so that loss of transmit and received laser energy upon reflection is minimized and the direction of the deflected beam is known with sufficient accuracy to achieve the required position accuracy of the laser spot on the Earth's surface. Regarding, for example, the oscillating mirror scanner, the angular accuracy of rotation about the main axis should typically be better than a hundredth of a degree. Flying at an altitude of 1000 m and assuming a scan angle of 20° off-nadir, an angular accuracy of a hundredth of a degree yields a position accuracy of 20 cm.

4.3.2   Fiber Scanning Technology

The company TopoSys® (Topographische Systemdaten GmbH, Biberach, Germany) was the only one worldwide that manufactures and offers airborne laser scanners using a so called fiber scanner. They were sold under the product name Falcon® (Topographische Systemdaten GmbH, Biberach, Germany). Their latest product in this category was Falcon III®. Fiber scanners were first designed and applied for reconnaissance tasks in military jet planes. Due to the high flying speed, very fast scanning was required that could not be realized with conventional techniques. Therefore, a solution with small moving parts was developed. Figure 4.18 shows that also here mirror scanners are applied. However, due to the small aperture of the fibers, small mirrors can be used. As these mirrors rotate constantly in one direction, high scanning speeds up to 415 Hz were achieved.

Besides the high scanning speed, the main advantage of this configuration is that transmit and receive optics are designed and manufactured identically. They do not have any moving parts. Therefore, they are very robust and stable during surveying missions. Incredibly remarkable is the design idea to transform a circular scan pattern into line scan by using fibers. Identical fiber line arrays are mounted in the focal plane of the receiving and transmitting lenses. By means of two rotating mirrors moved by a common drive, each fiber in the transmitting and receiving path is scanned sequentially and synchronously. In the transmitting path, the laser light is relayed from the central fiber to a fiber of the fiber array mounted in a circle around the central fiber. The optics between mirror scanner and fibers is adjusted in a way that the aperture of the laser fiber is imaged onto the aperture of the instantaneously illuminated fiber of the array. As the received part exhibits the identical optical path, the transmitted laser pulse backscattered from the target is seen by the corresponding fiber in the receiving fiber array. The receiver sees in the IFOV background light besides the laser signal, which must be filtered to improve SNR (Section 4.2.5) and protect the detector against saturation. This is carried out by an optical filter in the central

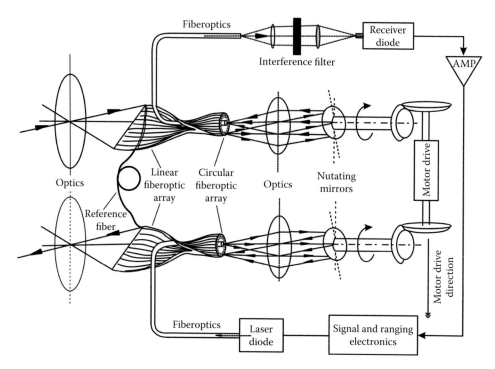

FIGURE 4.18   Principle of TopoSys® fiber scanner.

fiber linked to the photodiode. A total of 127 fibers and a fiber for the reference signal were realized in Falcon III®. The maximum across-track spacing Δx_{across} is dependent on the scan angle θ across the flight direction, which is also called FOV, the flying altitude H above the ground, and the number of fibers N. It is given by

$$\Delta x_{\text{across}} = \frac{\theta}{(N-1)\cdot\cos^2\left(\frac{\theta}{2}\right)}\cdot H \tag{4.32}$$

In the flight direction, the spacing Δx_{along} is determined by the flying speed above ground v and the number of revolutions per unit time f_ω:

$$\Delta x_{\text{along}} = \frac{v}{f_\omega} \tag{4.33}$$

Falcon III® had 127 fibers and an FOV of 28°. Assuming a flying speed of 75 m/s and an altitude of 1000 m, the across-track spacing Δx_{across} equals 4 m and spacing along track is 18 cm with $f_\omega = 415$ Hz. This example demonstrates that the spacing along-track is much smaller than across-track. This is depicted in Figure 4.19. Ideal scan pattern designs try to achieve equally distributed laser points on ground. In the right drawing of Figure 4.19 the scanning pattern is dithered to fill the gaps between fibers by using a swing mode which is realized by controlled synchronous movements of the receive and transmit fiber arrays. Regarding in addition the stochastic movements of the airborne platform, laser spots appear more equally distributed on the ground. This technique improves, for example, the detection of power lines and building edges and was implemented in Falcon III®. The production of the Falcon III has been closed. According to Seidel et al. (2008), this scanner technique that requires very sophisticated know-how is implemented in the commercial and military Helicopter Laser Radar (HELLAS) Obstacle Warning System developed by European

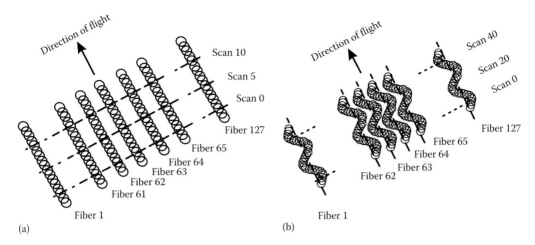

FIGURE 4.19 Scan pattern of TopoSys airborne laser scanner system: (a) without swing mode and (b) with swing mode.

Aeronautic Defence and Space Company (EADS), now. Today it is sold by Fairchild Controls, an EADS North American company under the product name HELLAS AWARENESS Terrain/Obstacle Collision Avoidance System.

4.4 CONTROLLING AND DATA SAMPLING UNIT

Figure 4.1 shows that the control, monitoring, and recording unit is a key device of a LiDAR system. It synchronizes the ranging unit with the scanner, triggering the pulsed laser synchronously with the incremental scanner steps. In addition, this unit stores to hard disk the ranging dataset, including the slant ranges of the return pulses, the return intensity, if available, the instantaneous scanning angles and high precision time stamps. The time stamps, required for later synchronization with the POS data, are derived from the GPS 1 pulse per second (pps) signal. Figure 4.20 presents graphically the information flow. The unit is realized by a powerful data handling device, because data have to be transferred to the disk at high data rates that can easily exceed 1 MB/s. Range calculations and decisions concerning the reliability of ranging are also carried out in the ranging unit.

Data handling for fullwave recording LiDARs is even more demanding, because the full receive pulse shape is digitized rather than simply ranging to its leading edge. First, a sophisticated A/D-board (analog-to-digital conversion) is required that samples the transient of the return pulse at high

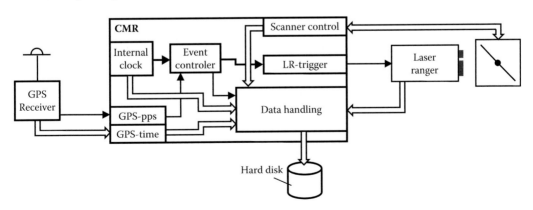

FIGURE 4.20 Information flow of control, monitoring, and recording unit.

rate, for example, at 1 G samples per second. This value is equivalent to a range resolution of only 15 cm. Assuming an amplitude resolution of 16 bits and continuous data logging, one has to deal with a data stream of 2 GB/s. As today's commercial data recorders typically handle data rates of only several hundred MB/s, the waveform data have to be reduced by special algorithms. For example, rather than continuously recording the A/D output from the time of transmission to reception of the return, transmit and receive waveforms can be recorded only when signal is detected above a threshold. In this case, the time interval between the transmit and received pulsed must also be measured and recorded. Applying such preprocessing filters, several hours of surveying are possible using solid-state disks.

4.5 POSITION AND ORIENTATION SYSTEM

Previously it was explained that the position and orientation of the LiDAR must be known very precisely for every instant. A so-called POS is used for this purpose. It comprises a DGPS and an IMU. DGPS requires reference stations on ground. To guarantee reliable phase solutions with centimeter level accuracy, using current GPS equipment and typical commercial processing methods, the reference stations must not be further away than 25 km from the airborne platform. DGPS data processing for LiDAR surveys is normally carried out offline after the flight. Therefore, a telemetry link from the reference stations to the airborne LiDAR is not required. During surveying, the airborne POS data including GPS position and carrier phase information and the IMU data are stored together with GPS time on memory cards or sticks in the airplane. Figure 4.21 shows the components of a typical POS.

Typically, the IMU is positioned as close to the scanner as possible to record orientation and aircraft vibrations at the location of the LiDAR. In Figure 4.22, the IMU is directly mounted on top of the scanner. With IMU systems recording at a bandwidth of 100 Hz, the vibration of the scanner and the carrying structure of the airplane can well be seen in the IMU raw data.

After the mission, position and orientation data together with the instantaneous time have to be computed from GPS-data of the POS receiver and the reference stations and the IMU data. In the beginning of airborne LiDAR, programs were applied, which processed the GPS-data independently of the IMU data. In this case, first GPS-data collected by the POS receiver and the GPS-data of the reference stations are processed by special DGPS-software, which computes

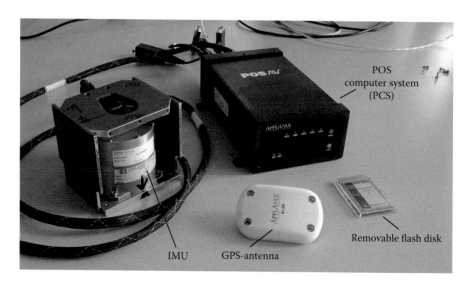

FIGURE 4.21   Applanix POS-AV®. (Courtesy of Applanix, Richmond Hill, ON, Canada.)

FIGURE 4.22 Mounting of IMU.

the LiDAR positions with cm- to dm-accuracy, typically in 0.5 s steps. In a second step, an integrated position and orientation solution is calculated with the DGPS-position data and the IMU-data by another software module, yielding positions determined with cm- to dm-accuracy in 3D and roll, pitch, and yaw angles of the LiDAR determined to better than one hundredth of a degree with a time resolution of a about 5 ms. These POS data together with the instantaneous time at which the position and orientation were sampled are stored in a file. Today, software is applied, which uses together the GPS- and IMU-data for an integrated solution right from the beginning. This improves the solutions for the GPS phase differences. Furthermore, the user has only to work with one program.

4.6 SYNCHRONIZATION

In the preceding sections, the sampling and the recording of the data were explained. Today commercial airborne laser scanning systems are sold with an integrated POS to better guaranty the achievable accuracy. In former systems, the LiDAR and the POS were independent units produced by different companies and for research purposes often several POS were flown together for comparative studies. Here, synchronization of LiDAR and POS datasets is an issue. As little is published on the methods by which commercial surveying companies synchronize the LiDAR and POS datasets, I describe here an approach developed at the Institute of Navigation of the University Stuttgart that is independent of the LiDAR and POS units employed.

Figure 4.23 makes clear that the laser scanner measurements are controlled and stored by the LiDAR control unit (LCU), and the GPS- and IMU-data by the POS Computer System (PCS). LCU and PCS both work in their own time system. PCS is related to the GPS time, whereas the LCU time is defined by its internal computer clock. Assuming that the POS delivers synchronized GPS- and IMU-data, synchronization can be reduced to synchronizing the GPS time with the LCU time. According to Figure 4.23, a software module in the LCU stores the laser scanner data with additional time data controlled by interrupt request 2 (IRQ2) during flight. At the start of each scan or scan line, the local LCU time is linked into the data stream. In parallel to this recording, the time tag of the pps-signal of the POS GPS-unit and the local LCU time at the pps instant are stored in a separate

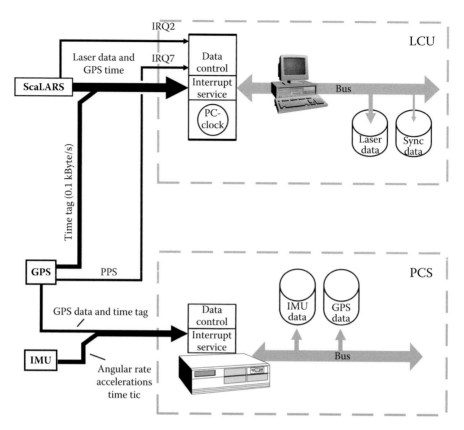

FIGURE 4.23   Synchronization.

protocol file. As the pps-signal of the onboard GPS receiver triggers the hardware interrupt IRQ7, the GPS time and local LCU time are stored at the same instant. LCU time differences from the pps interval can be directly identified as timing errors between the highly accurate and stable GPS-time and the LCU time. They are applied as correction for synchronization errors. This method achieves an operational synchronization of better than 10 μs. As the protocol file comprises the actual synchronization information, an offline synchronization of the POS and LiDAR data is possible. Before the synchronization, the user has three files:

1. LiDAR raw data file containing LCU-time for each scan line start followed by a number of datasets with instantaneous scanning angles, slant range, and intensity.
2. Protocol file.
3. POS file containing the instantaneous 3D position and the orientation angles with regard to GPS time.

Normally, the LiDAR file comprises more data lines per time interval than the POS file, because the sampling rate, equivalent to the laser pulse rate, is much higher. This requires interpolation of the POS positions and orientation angles. In the case of ScaLARS a linear interpolation is applied. After synchronization, data are available in the following possible arrangement:

   GPS time |x, y, z in (WGS84)| heading, pitch, role|slant range|intensity|scanning angle—these data are the input for calculating geocoded data in a process known as registration (Vaughn et al., 1996).

4.7   REGISTRATION AND CALIBRATION

The accuracy of the registered data is very dependent on calibration data of the LiDAR and POS sensors. Incorrect calibration not only introduces systematic errors but can also introduce random noise recognized by higher elevation standard deviations. The importance of calibration is emphasized by the improvement in LiDAR survey accuracy that has been achieved in recent years. This has been achieved primarily, without changing the LiDAR surveying hardware, but rather by improvements in calibration procedures and associated improvements in navigation data accuracy and data synchronization (Burman, 2002; Filin, 2003a, b; Filin and Vosselman, 2004; Huising and Gomes Pereira, 1998).

For better understanding the process of registration, it is assumed first that the inner orientation of the LiDAR is known. In this case, calibration means determination of the exterior orientation described by bore sight misalignment angles in roll $\delta\omega$, pitch $\delta\varphi$, and yaw $\delta\kappa$.

In Section 4.7.1, registration of laser data is explained provided that the system is calibrated. Section 4.7.2 deals in detail with the problem of LiDAR calibration.

4.7.1   REGISTRATION

The registration process is best mathematically described by the simple vector approach already shown in Figure 4.2:

$$\vec{G} = \vec{r}_L + \vec{s} \tag{4.34}$$

where
  $\vec{G}$ is the vector from the earth center to the ground point
  $\vec{r}_L$ is the vector from the earth center to the LiDAR's point of origin
  $\vec{s}$ is the slant ranging vector

The LiDAR's point of origin is a fixed point in 3D space from which the laser beam originates at the instantaneous scanning angle. The actual position of this point is very dependent on the scanning mechanism used, for example, in the case of ScaLARS it is the turning point of the mirror. To make a straight forward registration calculation with a precision as high as possible by applying Equation 4.34, the IMU is normally mounted on the support of the LiDAR instrument or as shown in Figure 4.22 on the suspension of the scanner. The POS data are transformed to LiDAR's point of origin. For this transformation, the user has to enter parameters for the so-called lever arms (Figure 4.24) into the POS processing software, which are two 3D vectors. One vector is from the origin of the LiDAR to the center of the IMU, and the other is from the origin of the LiDAR to the phase center of the GPS-antenna.

In the following, the origin of the LiDAR defines the origin of the LiDAR coordinate system L depicted in Figure 4.2. The x_L-axis points into the flight direction, the y_L-axis points to the right of the airplane, and the z_L-axis points downwards perpendicular to the plane defined by the axes x_L and y_L. Vector $\vec{r}_L$ is measured and described in WGS84. The registered laser measurement points $\vec{G}$ should also be described in WGS84. As $\vec{s}$ is measured in the coordinate system L, it has to be transformed by transformation matrices into WGS84. Regarding the coordinate system used the general approach becomes

$$\underline{G}_{WGS84} = \underline{r}_{L\,WGS84} + \left(\underline{\quad}\right)_H^{WGS84} \cdot \left(\underline{\quad}\right)_{IMU}^H \cdot \left(\underline{\quad}\right)_L^{IMU} \cdot \underline{s}_L \tag{4.35}$$

The product of the matrices $\left(\underline{\quad}\right)_L^{IMU}$ and $\left(\underline{\quad}\right)_{IMU}^H$ describes the orientation of the coordinate system L with respect to the horizontal coordinate system H. Assuming that the LiDAR is perfectly aligned with the IMU—this means the coordinate system L and the coordinate system of the IMU have the same orientation—the matrix $\left(\underline{\quad}\right)_L^{IMU}$ becomes unity. In this case, only the instantaneous

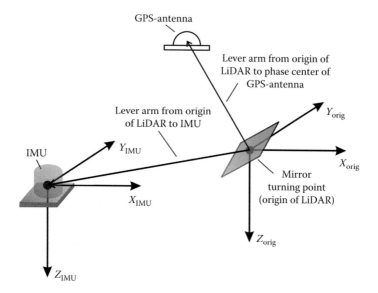

FIGURE 4.24 Lever arms.

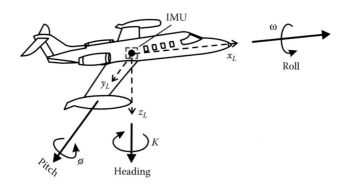

FIGURE 4.25 Orientation angles. (IMU and LiDAR aligned.)

orientation of the airplane or more precisely the orientation of the IMU, determines together with the instantaneous scanning angle the actual direction of the laser beam.

The orientation of the IMU in relation to the horizontal system H (Figure 4.25) is described by the angles roll ω (rotation about the x_L-axis), pitch φ (rotation about y_L-axis), and heading κ (rotation about the z_L-axis). If the rotations are carried out in the following sequence: first rotation about the x_L-axis (roll), second rotation about the y_L-axis (pitch), and finally rotation about the z_L-axis (heading), the matrix $(\_)_{IMU}^{H}$ can be set up as:

$$(\_)_{IMU}^{H} = \begin{pmatrix} a_{11} & a_{21} & a_{31} \\ a_{12} & a_{22} & a_{32} \\ a_{13} & a_{23} & a_{33} \end{pmatrix} \tag{4.36}$$

with

$$\begin{pmatrix} a_{11} \\ a_{12} \\ a_{13} \end{pmatrix} = \begin{pmatrix} \cos(\kappa)\cdot\cos(\varphi) \\ \sin(\kappa)\cdot\cos(\varphi) \\ -\sin(\varphi) \end{pmatrix} \tag{4.37}$$

$$\begin{pmatrix} a_{21} \\ a_{22} \\ a_{23} \end{pmatrix} = \begin{pmatrix} \cos(\kappa) \cdot \sin(\varphi) \cdot \sin(\omega) - \sin(\kappa) \cdot \cos(\omega) \\ \sin(\kappa) \cdot \sin(\varphi) \cdot \sin(\omega) + \cos(\kappa) \cdot \cos(\omega) \\ \cos(\varphi) \cdot \sin(\omega) \end{pmatrix} \tag{4.38}$$

$$\begin{pmatrix} a_{31} \\ a_{32} \\ a_{33} \end{pmatrix} = \begin{pmatrix} \cos(\kappa) \cdot \sin(\varphi) \cdot \cos(\omega) + \sin(\kappa) \cdot \sin(\omega) \\ \sin(\kappa) \cdot \sin(\varphi) \cdot \cos(\omega) - \cos(\kappa) \cdot \sin(\omega) \\ \cos(\varphi) \cdot \cos(\omega) \end{pmatrix} \tag{4.39}$$

The transformation matrix $(\_)_H^{WGS84}$ regards the orientation between the horizon system H and the WGS84. The orientation is defined by the geographical latitude Φ_0 and longitude Λ_0, so that $(\_)_H^{WGS84}$ becomes

$$(\_)_H^{WGS84} = \begin{pmatrix} -\cos\Lambda_0 \cdot \sin\Phi_0 & -\sin\Lambda_0 & -\cos\Lambda_0 \cdot \cos\Phi_0 \\ -\sin\Lambda_0 \cdot \sin\Phi_0 & \cos\Lambda_0 & -\sin\Lambda_0 \cdot \cos\Phi_0 \\ \cos\Phi_0 & 0 & \sin\Phi_0 \end{pmatrix} \tag{4.40}$$

Equation 4.35 and the matrices show that registered LiDAR measurement points can only be computed if the orientation and position of the LiDAR are known. These values are measured by POS. Up to now, it has been assumed that LiDAR and POS have the same orientation. This is not the case in general. Due to tolerances in the mechanical setup, one has to take into account a misalignment between POS and the LiDAR. The misalignment is described by the misalignment angles in roll $\delta\omega$, in pitch $\delta\varphi$, and in heading $\delta\kappa$. As the sequence of rotations is equal to the rotation order of matrix $(\_)_{IMU}^H$, the matrix $(\_)_L^{IMU}$ looks like $(\_)_{IMU}^H$. However, the orientation angles ω, φ, and κ have to be exchange with the misalignment angles $\delta\omega$, $\delta\varphi$, and $\delta\kappa$. Therefore, $(\_)_L^{IMU}$ is given by

$$(\_)_L^{IMU} = \begin{pmatrix} b_{1i} & b_{21} & b_{31} \\ b_{12} & b_{22} & b_{32} \\ b_{13} & b_{23} & b_{33} \end{pmatrix} \tag{4.41}$$

with

$$\begin{pmatrix} b_{11} \\ b_{12} \\ b_{13} \end{pmatrix} = \begin{pmatrix} \cos(\delta\kappa) \cdot \cos(\delta\varphi) \\ \sin(\delta\kappa) \cdot \cos(\delta\varphi) \\ -\sin(\delta\varphi) \end{pmatrix} \tag{4.42}$$

$$\begin{pmatrix} b_{21} \\ b_{22} \\ b_{23} \end{pmatrix} = \begin{pmatrix} \cos(\delta\kappa) \cdot \sin(\delta\varphi) \cdot \sin(\delta\omega) - \sin(\delta\kappa) \cdot \cos(\delta\omega) \\ \sin(\delta\kappa) \cdot \sin(\delta\varphi) \cdot \sin(\delta\omega) + \cos(\delta\kappa) \cdot \cos(\delta\omega) \\ \cos(\delta\varphi) \cdot \sin(\delta\omega) \end{pmatrix} \tag{4.43}$$

$$\begin{pmatrix} b_{31} \\ b_{32} \\ b_{33} \end{pmatrix} = \begin{pmatrix} \cos(\delta\kappa) \cdot \sin(\delta\phi) \cdot \cos(\delta\omega) + \sin(\delta\kappa) \cdot \sin(\delta\omega) \\ \sin(\delta\kappa) \cdot \sin(\delta\phi) \cdot \cos(\delta\omega) - \cos(\delta\kappa) \cdot \sin(\delta\omega) \\ \cos(\delta\phi) \cdot \cos(\delta\omega) \end{pmatrix} \tag{4.44}$$

The misalignment angles are determined by the calibration.

The presented formulas document that the calculation of registered laser measurement points is straight forward, assuming that the system is calibrated and any misalignment between the LiDAR and POS are constant. However, examining airborne LiDAR data georeferenced by the described method, it becomes obvious that there are spatially varying errors present in the form of elevation differences (Bretar et al., 2003; Kager and Kraus, 2001) and tilts between adjacent scanning paths (Rönnholm, 2004). Various solutions for this problem are presented in the literature, for example, Burman (2002), Crombaghs et al. (2000), Schenk (2001a, b), and Vosselmann and Maas (2001) and will also be explained in the next section. Most of the solutions are based on bundle block adjustment. The source of these errors is not fully understood and likely varies between different systems. The cause may be time-varying orientation changes in the LiDAR itself, of the IMU to LiDAR or LiDAR to GPS level arms, or time-varying errors in the POS position and attitude solutions. Moreover, synchronization errors between the orientation angles computed by Kalman filters must be taken into account (Latypov, 2002, 2005; Schenk, 2001a, b).

4.7.2 CALIBRATION

In Section 4.7.1, it was assumed that the bore sight misalignment angles $\delta\omega$, $\delta\varphi$, and $\delta\kappa$ were calibrated. Now, calibration is explained by studying the calibration procedures of ScaLARS (Schiele et al., 2005). ScaLARS uses the Palmer scanning mechanism (Section 4.3.1). Due to its special helical scanning pattern with a constant off-nadir scan angle that can be separated into forward and backward views (Table 4.1), an intrinsic system calibration is possible. Other scanning mechanisms do not offer this feature. However, the basic calibration steps of ScaLARS are also valid for these scanning mechanisms.

For precise geocoding, the set of calibration data is extended by the wobble angle γ_M and an additive parameter s_{Add} of the slant range. The wobble angle corresponds to the maximum scanning angle of a moving or flipping mirror or in other words the FOV or swath width (SW). Therefore, in total five parameters have to be calibrated: the bore sight misalignment angles $\delta\omega$, $\delta\varphi$, and $\delta\kappa$, wobble angle γ_M and the additive parameter s_{Add}. According to Figure 4.26, the instantaneous slant range vector of ScaLARS is calculated in the LiDAR coordinate system L by

$$\underline{s}_L = \left(s + s_{Add}\right) \cdot \begin{pmatrix} \sin(\delta) \cdot \sin(\xi) \\ -\cos(\delta) \\ \sin(\delta) \cdot \cos(\xi) \end{pmatrix} \tag{4.45}$$

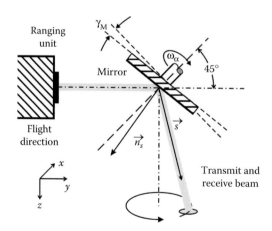

FIGURE 4.26 Palmer scanner of ScaLARS.

with

$$\delta = 2 \cdot \arccos\left(\cos(\alpha_\omega) \cdot \sin(\gamma_M) \cdot \sin(\vartheta) + \cos(\gamma_M) \cdot \cos(\vartheta)\right)$$

$$\xi = -\arctan\left(\frac{\sin(\alpha_\omega) \cdot \sin(\gamma_M)}{\cos(\alpha_\omega) \cdot \sin(\gamma_M) \cdot \cos(\vartheta) + \sin(\gamma_M) \cdot \sin(\vartheta)}\right)$$

where

s is the measured slant range
α_ω is the instantaneous rotation angle of the motor shaft
ϑ is the tilt of the motor shaft that is defined by the scanner design

It is valid 45°. Equation 4.46 makes clear that only two parameters of s_L are of interest for calibration. These are the wobble angle γ_M and the additive parameter for slant ranging s_{Add}. All other calibration parameters are located in the transformation matrix $(\_\_)_L^{IMU}$ that describes the misalignment between IMU and LiDAR. This fact is clearly seen by looking at the equation for registered laser points in WGS84, which is

$$\underline{G}^{WGS84} = \underline{r}_L^{WGS84} + \left(\Lambda_0, \Phi_0\right)_H^{WGS84} \cdot \left(\omega, \kappa, \varphi\right)_{IMU}^{H} \cdot \left(\delta\omega, \delta\kappa, \delta\varphi\right)_L^{IMU} \cdot \underline{s}_L\left(s_{Add}, \gamma_M\right) \qquad (4.46)$$

For better understanding, only the used parameters are written in the symbols for the matrices. On the basis of Equation 4.46, a Gauß-Markoff model is set up to estimate the calibration parameters: bore sight misalignment angles $\delta\omega$, $\delta\varphi$, and $\delta\kappa$, wobble angle γ_M, and the additive parameter s_{Add}. The calibration is carried out by selected flat areas. Calibration tests showed that white marks on runways are optimum targets for this task but other features such as flat roofs are also possible. However, the main objective of these targets is that they exhibit a high contrast to the background and that a representative elevation point can be computed out of the laser data sampled from it. The high contrast is necessary to extract the targets from the surveyed area automatically. These chosen areas for calibration are called *control area* in the following. Figure 4.27 shows, how control areas are used for calibration. Regarding the centroid of the control area, it is obvious that the centroid derived from geocoded LiDAR data is brought to coincidence with the real one if the displacements Δx, Δy, and Δz are zero which means the system is exactly calibrated. Therefore, the objective of the Gauß-Markoff model is to minimize Δx, Δy, and Δz by varying the calibration parameters.

The calibration procedure is going to be explained exemplarily by markers on runways. They can be easily identified out of the LiDAR intensity data (Figure 4.28).

Furthermore, they can be surveyed using GPS surveying instruments (Figure 4.29), and the reference centroid in 3D space can be computed by simple geometry calculations. The centroid determination from LiDAR data require much more effort and sophisticated algorithms, because LiDAR data are not equally distributed over the target's surface (enlarged runway markers in Figure 4.28).

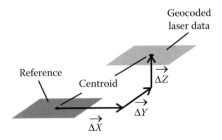

FIGURE 4.27   Control area.

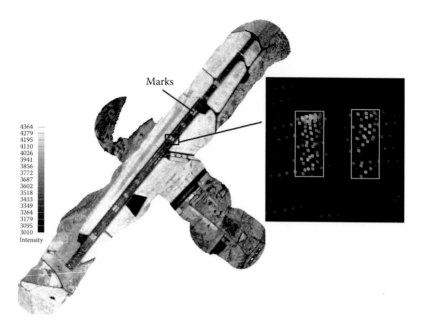

Marks

4364
4279
4195
4110
4026
3941
3856
3772
3687
3602
3518
3433
3349
3264
3179
3095
3010
Intensity

FIGURE 4.28   LiDAR intensity data of runway.

FIGURE 4.29   Surveying a control area with GPS.

After extracting all laser measurements belonging to the control area by special software, the extracted surface is modeled by triangulation. A Delaunay triangulation algorithm was preferred for this task. As the target's surface is approximated by n triangles the coordinates of the centroid can be computed independently on the density of the laser measurements by

$$\vec{x}_{\text{cent}} = \frac{1}{\displaystyle\sum_{i=1}^{n} F_i} \cdot \sum_{i=1}^{n} \left(\vec{x}_i \cdot F_i \right) \tag{4.47}$$

TABLE 4.2
Accuracy of Calibration Parameters Determined for Two Surveying Campaigns

Estimated Parameter Accuracy	γ_M	$\delta\omega$	$\delta\varphi$	$\delta\kappa$	Add
July 2002 ($\gamma_M \approx 10°$)	0.004°	0.006°	0.006°	0.024°	0.03 m
April 2003 ($\gamma_M \approx 7°$)	0.003°	0.005°	0.005°	0.028°	0.03 m

where
F_i is the ith triangle
$\vec{x}_i$ is the corresponding vector to its centroid

The process of using $\vec{x}_{cent}$ and the location of the centroid derived from the GPS measurement for calibration is called calibration with coordinated control points. Applying this method the calibration parameters were determined with an accuracy listed in Table 4.2.

Noncoordinated control points are derived from surveyed features, in which external reference information about shape, size, or position is not available. This means the location of the corresponding centroid is not known a priori and can only be determined by measured LiDAR data. In this case, the parameters wobble angle γ_M and bore sight misalignment angles $\delta\omega$, $\delta\varphi$, and $\delta\kappa$ can be relatively calibrated only. Calibration with noncoordinated control points becomes important if areas are surveyed, whereas in the field measurements are impossible, for example, carrying out LiDAR surveys high in the mountains, in primeval forests, and over the poles. Furthermore, this calibration technique permits to study system drifts along the flying path during the total mission.

4.8 FROM FLIGHT PLANNING TO FINAL PRODUCT

In the preceding chapters, the components of a LiDAR system were presented and their performance attributes were discussed. Here, the different working steps required to obtain a digital elevation model (DEM) as a final product are explained.

Normally the customer defines the basic surveying parameters such as surveying area, the density of laser points on the ground, and the elevation accuracy in the definition phase of a project. In some cases, special surveying objectives are defined for specific target types such as powerlines, building geometries including tilted roofs, or embankments. For these special cases, translation of measurement requirements into the necessary LiDAR performance attributes by the surveyor requires substantial practical experience that is not readily captured by a few system parameters. To get a clear understanding of the workflow comprising the working steps from the flight planning to the final product, the workflow in Figure 4.30 starts with basic surveying parameters as input. Figure 4.30 presents an overview of the processing steps from flight planning to final product. Each step will be discussed in separate sections that follow.

4.8.1 FLIGHT PLANNING

The flight planning stands at the beginning of an airborne surveying project. Several flight planning programs are available on the market. These products are normally sold with flight guidance software, which tells the pilot the flight course and elevation during flight and when he has to start and to stop LiDAR surveying along a flight line. However, the input of this software is not compatible with specifications listed in Section 4.8. Therefore, flight planning for airborne LiDAR surveys needs additional working steps. Furthermore, the GPS reference station and calibration areas must be selected. That is why I prefer the term project or mission planning.

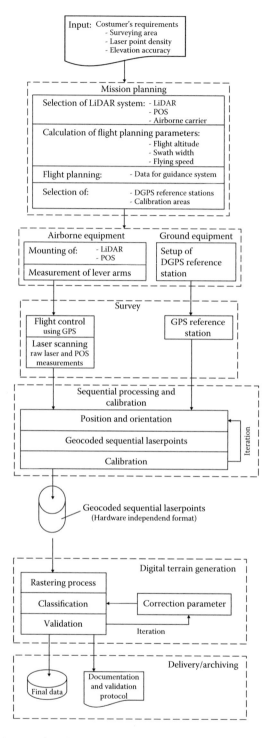

FIGURE 4.30 Flowchart processing steps.

On the basis of the customer's requirements, which are the surveying area size, the density of laser points on ground, and the elevation accuracy, the proper LiDAR system must be chosen first. This means selecting the appropriate LiDAR, POS, and the carrier either an airplane or helicopter platform. Helicopters are used if high laser point densities are necessary, and detailed surveys of small areas are essential. For extended surveys, airplanes are preferred. The performance characteristics of the LiDAR and POS determine the point density across the flight direction, the elevation accuracy, the elevation resolution (Equation 4.10), and the pointing accuracy.

The point density in the flight direction Δx_{along} is given by the speed of the platform v and the scanning speed that is described by the scanning rate f_{sc}:

$$\Delta x_{\text{along}} = \frac{v}{f_{\text{sc}}} \tag{4.48}$$

If the airplane is not known, a speed of 75 m/s is a good choice to start planning. Helicopters can fly more or less with arbitrary speeds.

Assuming a flat terrain and the distances between points along the scan line are equal, the point density across flight direction Δx_{across} can be approximated by

$$\Delta x_{\text{across}} = \frac{\theta}{N} \tag{4.49}$$

where:
  θ is the SW (either expressed in meters or angular degrees)
  N is the number of points per scan line

N is calculated by

$$N = \frac{f_{\text{pulse}}}{f_{\text{sc}}} \tag{4.50}$$

where f_{pulse} is the laser pulse rate. Figure 4.31 depicts a dependence of the spacing between adjacent points on the instantaneous scanning angle α. The surveying parameters have to be dimensioned for the worst case within the scan line. This means for the maximum scanning angle that is half the angular SW θ. Therefore, a more precise formula for point density across flight direction Δx_{across} is derived from Figure 4.31:

$$\Delta x_{\text{across}} = \frac{\theta}{N} \cdot \frac{H}{\cos^2\left(\dfrac{\theta}{2}\right)} \tag{4.51}$$

where H is the flying altitude above ground. In case that terrain features a slope with angle i along the scanning line (Figure 4.32) the maximum spacing across the flight line becomes

$$\Delta x_{\text{across}} = \begin{cases} \dfrac{\theta}{N} \cdot \dfrac{H}{\cos^2\left(\dfrac{\theta}{2}\right) \cdot \cos(i) \cdot \left[1 - \tan\left(\dfrac{\theta}{2}\right) \cdot \tan(i)\right]^2}, & \text{if } i \geq 0 \\[4mm] \dfrac{\theta}{N} \cdot \dfrac{H}{\cos^2\left(\dfrac{\theta}{2}\right) \cdot \cos(i) \cdot \left[\tan\left(\dfrac{\theta}{2}\right) \cdot \tan(i) - 1\right]^2}, & \text{if } i < 0 \end{cases} \tag{4.52}$$

The values Δx_{along} and Δx_{across} are used to determine the minimum point density d_{min} that is given in number of laser spots per square meter.

$$d_{\text{min}} = \frac{1}{\Delta x_{\text{along}} \cdot \Delta x_{\text{across}}} \tag{4.53}$$

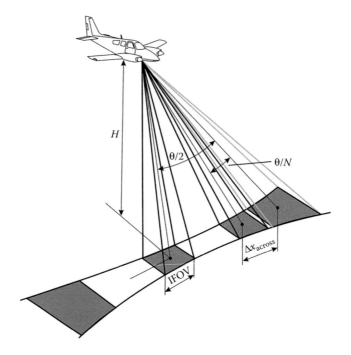

FIGURE 4.31   Geometry for sampling across flight direction.

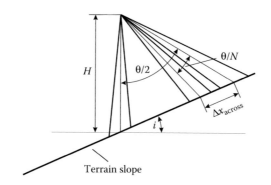

FIGURE 4.32   Terrain with slope across flight line.

The parameters calculated with Equations 4.49 through 4.53 are not only dependent on LiDAR parameters. Rather they are a function of the flying altitude H. The flying height H or the resulting maximum slant range R_{max} that are related by

$$R_{max} = \frac{H}{\cos\left(\dfrac{\theta}{2}\right)} \tag{4.54}$$

determines the achievable accuracy (Equation 4.29 and 4.30). As the accuracy is set by the customer, H is limited. According to Equation 4.52, across track spacing of the laser points limits also H. This means, both limitations must be fulfilled, and the most stringent one determines the SW that is

$$SW = 2 \cdot H \cdot \tan\left(\frac{\theta}{2}\right) \tag{4.55}$$

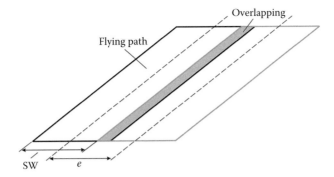

FIGURE 4.33 Overlapping of flight paths.

Now, the key parameters for flight planning are available which are flight altitude H and SW. In addition, the overlapping factor ζ for neighboring paths has to be set. According to Figure 4.33, it is defined by

$$\frac{e}{SW} = 1 - \xi \tag{4.56}$$

where e is the distance between the centerlines of adjacent paths. The overlapping is dependent on the pilot's flying precision. Planning surveys without any overlap is critical, because one has to face the problem that surveyed areas are not completely covered due to rolling of the aircraft and improper straight flight lines. At least a 20% overlap is advisable. For highly resolved surveys an overlapping of at least 50% may also applied if high point densities are essential that cannot not be achieved with the scanning parameters of LiDAR and the forward speed of the surveying carrier.

Table 4.3 compiles the relevant parameters for flight planning.

TABLE 4.3
Compilation of Key Parameters for Flight Planning

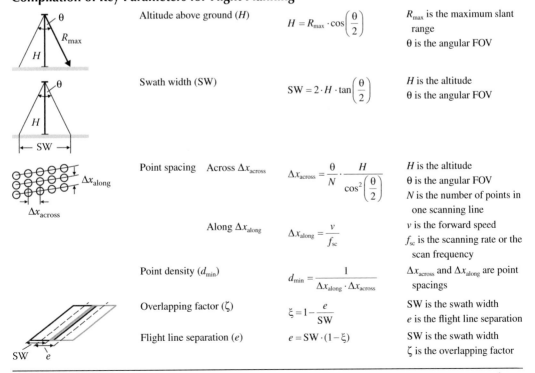

	Altitude above ground (H)	$H = R_{\text{max}} \cdot \cos\left(\dfrac{\theta}{2}\right)$	R_{max} is the maximum slant range θ is the angular FOV
	Swath width (SW)	$SW = 2 \cdot H \cdot \tan\left(\dfrac{\theta}{2}\right)$	H is the altitude θ is the angular FOV
	Point spacing Across Δx_{across}	$\Delta x_{\text{across}} = \dfrac{\theta}{N} \cdot \dfrac{H}{\cos^2\left(\dfrac{\theta}{2}\right)}$	H is the altitude θ is the angular FOV N is the number of points in one scanning line
	Along Δx_{along}	$\Delta x_{\text{along}} = \dfrac{v}{f_{\text{sc}}}$	v is the forward speed f_{sc} is the scanning rate or the scan frequency
	Point density (d_{min})	$d_{\text{min}} = \dfrac{1}{\Delta x_{\text{along}} \cdot \Delta x_{\text{across}}}$	Δx_{across} and Δx_{along} are point spacings
	Overlapping factor (ζ)	$\xi = 1 - \dfrac{e}{SW}$	SW is the swath width e is the flight line separation
	Flight line separation (e)	$e = SW \cdot (1 - \xi)$	SW is the swath width ζ is the overlapping factor

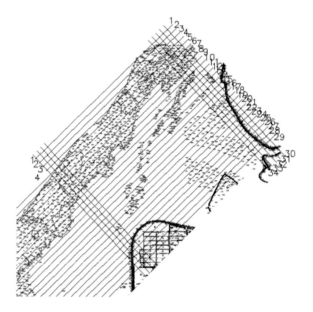

FIGURE 4.34   Flight lines out of a flight planning program.

Figure 4.34 shows a typical graphical output of a flight planning program. The numerical output is transferred to the guidance system of the surveying airplane. In addition, the flight planning result can be used to estimate the expenditure of the project. The length and number of flying lines determines the flight duration. One has to keep in mind that long flight lines are preferable, because after each line the airplane has to fly turns into the next line in which the survey is interrupted. On the other hand, too long lines degrade the accuracy of POS caused by IMU drifts. Flying a straight line for ca. 15 to 20 min., some turns are required.

Based on the flight planning results, the mission planning can be completed by positioning the GPS reference stations that should not be more than 20 km away from each surveying position of the airplane. Therefore, the number of reference stations is determined by the size of surveying area and the possibility to deploy a reference. In inaccessible areas are, for example, primeval forests, summit of glaciers, and so on. Today in many countries, networks of GPS reference stations exist. Paid GPS reference data are available to everybody. In all other cases, the surveyor has to see that enough stations are available.

4.8.2   INSTALLATION OF AIRBORNE EQUIPMENT

In particular if the surveying company has not got an own airplane with fixed LiDAR installations, the LiDAR has to be mounted in the surveying airplane at mission start. The optomechanical assembly is typically mounted in openings that are also used by metric cameras and others. These holes and the mechanical interfaces are more or less standardized. Nevertheless during planning one should check if the LiDAR's FOV is oblique and the available electrical power satisfies the LiDAR's need.

After completing the installation, the so-called lever arms (Section 4.7.1, Figure 4.24) have to be surveyed, which describe the displacements between IMU- and GPS-antenna on top of the airplane. With additional geometrical information concerning the displacement between IMU and surveying origin of LiDAR, all postprocessed POS data are related to this surveying origin.

4.8.3 SURVEY

During survey, the LiDAR and POS data are stored on hard disk or memory cards in the airplane, and the GPS reference stations save their data locally. The pilot is guided by the flight guidance software that uses data out of the flight planning program. He has to pay attention to circumvent roll angles of more than ±20°; otherwise, it could happen that the phase carrier of DGPS cannot be resolved continuously in post processing.

A so-called calibration area (e.g., a runway with markings) should be surveyed at the beginning and the end of a flight. This helps one to identify system drifts. It is a must if the LiDAR and POS system were just new assembled or reassembled and mounted the first time or mounted again in the airplane. The area should exhibit special features that support the calibration precision.

4.8.4 SEQUENTIAL PROCESSING AND CALIBRATION

After the surveying, flight backup copies and working copies of the gathered LiDAR and POS data are stored on different standard hard disk. Normally the LiDAR and POS data are stored in a compressed binary format. This means the data have to be uncompressed and converted to American Standard Code for Information Interchange (ASCII) data format by special program routines. In the first processing step, LiDAR and POS data are treated separately. The POS data processing is discussed first.

The POS data comprises IMU- and GPS-data. Before integrated POS data can be computed, DGPS data using phase carriers have to be calculated by using the GPS-data of the reference stations. The integrated solution with DGPS and IMU data offers the following principle data set:

GPS time | x_{WGS84}, y_{WGS84}, z_{WGS84} | roll, pitch, yaw

The position coordinates are related to the LiDAR origin (Section 4.7.1), and the orientation angles are the rotations about the instantaneous local horizontal system.

In a second step, LiDAR and POS data have to be synchronized by the procedure described in Section 4.6. After synchronization, a file with the following principle content is generated for further processing:

GPS time |x_{WGS84}, y_{WGS84}, z_{WGS84}| roll, pitch, yaw|scan angle, slant range, intensity—this is the output of the processing step *Position and Orientation* presented in Figure 4.30.

Assuming a perfect inner orientation calibration, the LiDAR geocoded laser measurement points on ground can be simply computed by Equation 4.35, if the misalignment angles in roll $\delta\omega$, pitch $\delta\varphi$, and yaw $\delta\kappa$ are known. These angles are caused by misalignment between IMU and LiDAR. Besides other calibration parameters discussed in Section 4.7.2, the misalignment can be determined by the LiDAR data gathered from the calibration areas. Thus, an iteration is foreseen in Figure 4.30. After a new set of calibration parameters is derived, the possible improvement has to be checked and validated.

The outputs of the processing step *Sequential Processing and Calibration* are files containing geocoded—also known as registered—laser measurements in a chronological order:

GPS time x_{Laser_WGS84}, y_{Laser_WGS84}, z_{Laser_WGS84}, intensity

This means that the geometry of the scanning pattern is still present. Figure 4.35 displays sequential laser data surveyed with a Palmer scanning pattern. This pattern is extreme with regard to geometry and time. Due to its egg-shaped pattern, ground points especially along the center line of the flight path are sampled twice, however, at different times. Comparable images are obtained by other scanning mechanisms. In an operational processing environment, the GPS time is not used and therefore normally not stored.

Up to this postprocessing step programs provided by the manufacturer of the LiDAR system are applied, because they must be programmed with a good knowledge of the hardware.

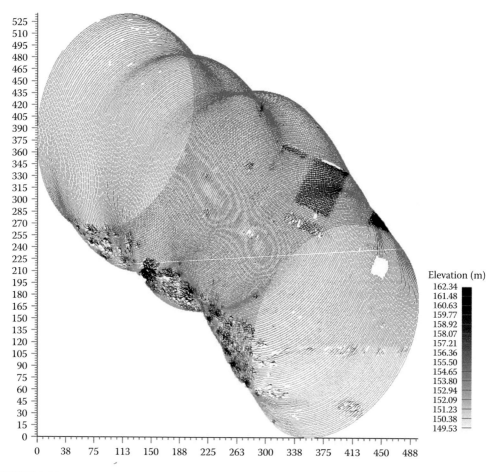

Elevation (m)

162.34
161.48
160.63
159.77
158.92
158.07
157.21
156.36
155.50
154.65
153.80
152.94
152.09
151.23
150.38
149.53

FIGURE 4.35   Sequential laser data.

4.8.5 Digital Terrain Model Generation

For generating final products defined by the customer, the sequential datasets generated in Section 4.8.4 have to be further processed. Final products are, for example, digital surface models, DEMs or extracting other geoinformation, for example, house geometries, shapes of roofs, tree heights, and so on. In other words, this work package must carry out some classification tasks. Today a lot of remote sensing and airborne surveying software is available on the market. But most of them work with rasterized data and cannot handle the so-called point cloud data delivered from LiDARs. Hence in the block *Digital Terrain Model Generation*, the first task is rasterizing sequential data. The rasterizing process is best explained by looking at Figure 4.36. All laser data are sorted into raster cells that are defined by a grid. The sizes of the cells are determined by the resolution projected by the customer. Moreover, the total image size is selected in agreement with the client. In general, a raster size should be larger than or equal to the largest spacing between neighboring laser points to assure that all raster cells contain at least one laser point. The raster cells are also called pixels in the following. Having several elevation measurements in one pixel, sophisticated algorithms are applied to obtain the desired information, because the final output must have only one pixel value.

The generation of classified rasterized data is explained exemplarily for the generation of DEMs. Selecting a pixel size, so that several laser points are sorted in one pixel, one recognizes an

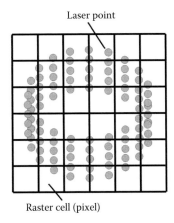

FIGURE 4.36   Rasterizing.

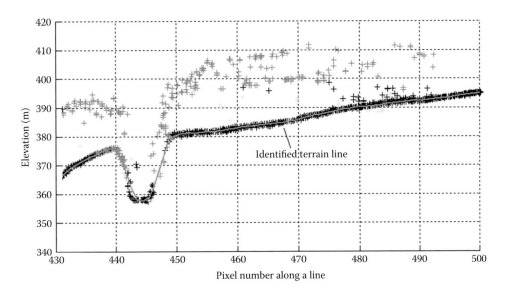

FIGURE 4.37   Pixel line with all laser measurements.

elevation distribution. Even if it is not distinguished between first and last pulse, the user can easily realize from the way how the measurements spread if forest and vegetation or ground is measured (Figure 4.37).

In the case of ground, one line can be clearly identified. The spreading of the elevations is primarily determined by the measurement quality of the LiDAR, whereas in forest areas some measurements represent either backscattered signals from ground and from treetops. Now, applying sophisticated software tools ground points can be classified and the most likelihood value is stored for this pixel. Figure 4.38 shows a rasterized DEM. The scanning pattern is no longer apparent. Such classifications are carried out automatically or semiautomatically with today's software tools. However, a good knowledge of setting filter parameters is necessary to achieve optimum classification results. Therefore, an iteration for optimizing parameters is foreseen in the flowchart (Figure 4.30).

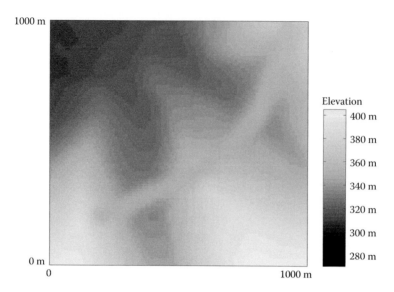

FIGURE 4.38   Rasterized DEM.

The most time consuming part in DEM production is the validation of, for example, the DEMs. First accuracy checks have to be carried out. However, the calibration areas cannot be used for this test, because they fit per definition. Therefore, different, already well-surveyed areas must be selected. By checking sequences of pixel lines, the plausibility of ground lines in forest areas can be examined. Here an absolute check is impossible, because secured measurement data for comparison do not exist in many cases. If validation is completed successfully, the data are stored on appropriate media, for example, CDs or DVDs in agreement with the customer, who also defines the archiving nomenclature.

The different processing steps explained in the preceding chapters were not treated at great length, because the objective was to give an overview about what is needed to achieve a final product from a LiDAR survey. The remarks make clear that successful LiDAR surveying is mainly dependent on the quality and availability of the used software and the implemented algorithms. Looking at the technical data of the different LiDAR systems, the performance of today's LiDARs differ marginally. However, in practice, very different accuracy values concerning geocoded data can be observed. This shows one has to regard the complete LiDAR system as defined in Section 4.1. Today, companies offering LiDARs sell more and more complete systems including software for generating geocoded data. This assures the specified accuracy for the derived geocoded data.

REFERENCES

Abshire, J., Sun, X., Afzal, R., 2000. Mars orbiter laser altimeter: Receiver model and performance analysis. *Applied Optics*, 39(15): 2449–2460.

Bachman, Chr. G., 1979. *Laser Radar System and Techniques*. Norwood, MA: Artech House.

Bretar, F., Pierrot-Oeseilligny, M., Roux, M., 2003. Estimating image accuracy of airborne laser data with local 3D-offsets. *International Archives of Photogrammetry and Remote Sensing*, 34(3-W13): 20–26.

Burman, H., 2002. Laser strip adjustment for data calibration and verification. *International Archives of Photogrammetry and Remote Sensing*, 34(part 3A): 67–72.

Crombaghs, M., Brügelmann, R., de Min, E., 2000. On the adjustment of overlapping strips of laser altimeter height data. *International Archives of Photogrammetry and Remote Sensing*, 33(part 3A): 230–237.

Filin, S., 2003a. Recovery of systematic biases in laser altimetry data using natural surfaces. *Photogrammetric Engineering & Remote Sensing*, 69(11): 1235–1242.

Filin, S., 2003b, Analysis and implementation of a laser strip adjustment model. *International Archives of Photogrammetry and Remote Sensing*, 34(3-W13): 65–70.

Filin, S., Vosselman, G., 2004. Adjustment of airborne laser altimetry strips. *Proceedings ISPRS XXth Congress*, Vol. XXXV Part B, Commission 3, p. 285 ff.

Gardner, C.S., 1992. Ranging performance of satellite laser altimeters. *IEEE Transactions on Geoscience and Remote Sensing*, 30(5): 1061–1072.

Hug, C., 1994. The scanning laser altitude and reflectance sensor—An instrument for efficient 3D terrain survey. *International Archives of Photogrammetry and Remote Sensing*, 30(Part 1): 100–107.

Huising, E.J., Gomes Pereira, L.M., 1998. Errors and accuracy estimates of laser data acquired by various laser scanning systems for topographic applications. *ISPRS Journal of Photogrammetry and Remote Sensing*, 53(5): 245–261.

Jelalian, A.V., 1992. *Laser Radar Systems*. Norwood, MA: Artech House, p. 45.

Jutzi, B., Eberle, B., Stilla, U., 2002. Estimation and measurement of backscattered signals from pulsed laser radar. In S.B. Serpico (Ed.) *Image and Signal Processing for Remote Sensing VIII. SPIE Proceedings*, 4885: 256–267.

Jutzi B., Stilla U., 2005. Measuring and processing the waveform of laser pulses. In: A. Gruen, H. Kahmen (Eds.) *Optical 3-D Measurement Techniques VII*, 1: 194–203.

Kager, H., Kraus, K., 2001. Height discrepancies between overlapping laser scanner strips. *Proceedings of Optical 3D Measurement Techniques V*, October, Vienna, Austria, pp. 103–110.

Keiser, G., 1983. *Optical Fiber Communications*. McGraw-Hill series in electrical engineering, Communications and information theory, McGraw-Hill, Japan.

Krabill, W., Abdalati, W., Frederick, E., Manizade, S., Martin, C., Sonntag, J., Swift, R., Thomas, R., Yungel, J., 2002. Aircraft laser altimetry measurement of elevation changes of the Greenland ice sheet: Technique and accuracy assessment. *Journal of Geodynamics*, 34: 357–376.

Krabill, W.B., Thomas, R.H., Martin, C.F., Swift, R.N., Frederick, E.B., 1995. Accuracy of airborne laser altimetry over the Greenland ice sheet. *International Journal Remote Sensing*, 16(7): 1211–1222.

Latypov, D., 2002. Estimating relative lidar accuracy information from overlapping flight lines. *ISPRS Journal of Photogrammetry & Remote Sensing*, 56(4): 236–245.

Latypov, D., 2005. Effects of laser beam alignment tolerance on lidar accuracy. *ISPRS Journal of Photogrammetry & Remote Sensing*, 59(6): 361–368.

Litchi, D., Harvey, B., 2002. The effects of reflecting surfaces material properties on time-of-flight laser scanner measurements. *Symposium of Geospatial Theory, Processing, and Applications*, Ottawa, ON.

Rönnholm, P., 2004. The evaluation of the internal quality of laser scanning strips using the interactive orientation method and point clouds. *Proceedings ISPRS XXth Congress*, Vol. XXXV Part B, Commission 3, p. 255 ff.

Schenk, T., 2001a. Modeling and analyzing systematic errors of airborne laser scanners. Technical Notes in Photogrammmetry No. 19, Department of Civil and Environmental Engineering and Geodetic Science, The Ohio State University, Columbus, OH, 40 pages.

Schenk, T. 2001b. Modelling and recovering systematic errors in airborne laser scanners. *Proceedings of OEEPE Workshop on Airborne Laserscanning and Interferometric SAR for Detailed Digital Elevation Models*, Stockholm, Sweden.

Schiele, O., Wehr, A., Kleusberg, A., 2005. Operational calibration of airborne laser scanners by Using LASCAL. Papers presented to the *Conference Optical 3-D Measurement Techniques VII* Vol. 1, Eds. Grün, A., Kahmen, H., Vienna, Austria, October 3–5. pp. 81–89.

Seidel, Chr., Schwartz, I., Kielhorn, P., 2008. Helicopter collision avoidance and brown-out recovery with HELLAS. *Electro-Optical Remote Sensing, Photonic Technology, and Applications II, Proceedings of SPIE* Vol. 7114, pp. 71140G-1 to71140G-8.

Skolnik, M.I., 2001. *Radar Systems*. Boston, MA: McGraw-Hill.

Sun, X., Abshire, J., and Davidson, F., 1993. Multishot laser altimeter: Design and performance. *Applied Optics*, 32(24): 4578–4585.

Sun, X., Davidson, F., Boutsikaris, L, and Abshire, J., 1992, Receiver characteristics of laser altimeters with avalanche photodiodes. *IEEE Transactions Aerospace and Electronic Systems*, 28(1): 268–275.

Thiel, K.-H., Wehr, A., 1999. Operational data processing for imaging laser altimeter data. *Proceedings of the Fourth International Airborne Remote Sensing Conference and Exhibition*, ERIM, June 21–24, Ottawa, ON.

Vaughn, C. R., Bufton. J. L., Krabill, W. B., Rabine D. L., 1996. Georeferencing of airborne laser altimeter measurements. *International Journal of Remote Sensing*, 17(11): 2185–2200.

Vosselmann, G., Maas, H.-G., 2001. Adjustment and filtering of raw laser altimetry data. OEEPE Workshop on Airborne Laserscanning and Interferometric SAR for Detailed Digital Elevation Models, Stockholm, Sweden.

Wehr, A., Hemmleb, M., Thomas, M., Maierhofer, C., 2007. Moisture detection on building surfaces by multi-spectral laser scanning. *Proceedings of Optical 3-D Measurement Techniques VIII, Conference at ETH-Zuerich*, July, Switzerland, pp. 9–12, 79–86.

Wolfe, W.L., Zissis, G.J., 1978. *The Infrared Handbook*. Washington, DC: Office of Naval Research, Department oft he Navy, pp. 23-5–23-13.

Young, M., 1986. *Optics and Lasers—Series in Optical Sciences*. Berlin, Germany: Springer, p. 145.

5 Pulsed Laser Altimeter Ranging Techniques and Implications for Terrain Mapping

David Harding

CONTENTS

5.1   INTRODUCTION

The emergence of airborne laser swath mapping (ALSM), initially as a research tool and in the last decade as a commercial capability, has provided a powerful means to characterize the elevation of the Earth's solid surface and its overlying covers of vegetation, water, snow, ice, and structures created by human activity. Other terms used for ALSM include scanning laser altimetry, airborne laser scanning, and Light Detection and Ranging (LiDAR). The rapid collection of georeferenced, highly resolved elevation data with centimeter to decimeter absolute vertical accuracy achieved by ALSM enables unprecedented studies of natural processes and creation of elevation map products for application purposes. Scientific uses of the data span many diverse disciplines, attesting to the value of detailed elevation information. Uses include, but are not limited to, evaluation of natural hazards associated with surface-rupturing faults, volcanic flow pathways, slope instability and flooding, monitoring of coastal change due to storm surges and sediment transport, characterization of ecosystem structure from which estimates of above-ground biomass and assessments of habitat quality are derived, and quantification of ice sheet and glacier elevation change and their contribution to sea level rise. In applied areas, the multilayered map products showing ground topography, even where densely covered by vegetation, as well as canopy density and height, and building footprints and heights make ALSM data uniquely suited for resource management and land-use purposes.

    To most effectively use the products generated from ALSM systems, for both scientific and applied purposes, their principles of operation and the resulting effects on the acquired data need to be well understood. In this chapter, laser ranging methods are examined focusing on the most commonly employed type of laser altimeter technology that uses pulsed laser light. Waveform recording systems and those that generate point clouds are described. For the latter type, discrete return and single-photon detection approaches are addressed. Relationships between ranging methods, instrument parameters, and the character of the resulting elevation data are considered. The emphasis is on vegetated landscapes and, in particular, the measurement of ground topography beneath vegetation cover and characterization of forest canopy structure.

Factors are discussed that affect the amplitude of the received signal, as that bears on the detection of surface returns. These factors also are important considerations in the interpretation of intensity images that are produced along with elevation data by more recent ALSM systems. When properly calibrated, the return intensity can be used to create monochromatic, single-band reflectance images at the laser wavelength. As the laser return is range resolved, the return intensity can be ascribed to different levels within a scene as, for example, from overlying vegetation and the underlying ground surface. This added information has the potential to improve the value of ALSM mapping by providing better differentiation of surface types and features.

The fundamental observation made by pulsed ALSM systems is the range (i.e., distance) from the instrument to a target, determined by timing the round-trip travel time of a pulse of laser light reflected from a surface. Travel time is converted to distance using the speed of light. Combining range with knowledge of the orientation of the transmitted laser pulse, and the position of the instrument in a coordinate frame, yields a vector that defines the location of the reflecting target, usually expressed as a latitude, longitude, and elevation. The laser beam orientation is determined using two devices consisting of an inertial measurement unit (IMU) and an angle encoder. The IMU defines the roll, pitch, and yaw of the instrument aboard an airborne platform, and the angle encoder defines the angular position of the scanner mechanism used to deflect the laser beam across the swath being mapped. The instrument position at the instant the laser pulse is fired is established using a combined solution derived from a differential global positioning system aircraft trajectory and the IMU acceleration data. Here, the focus is on the range measurement and the implications for interpreting information about the elevation structure of a target.

5.2 SIGNAL STRENGTH

Before describing ranging methods, the factors that control the signal strength observed by a laser altimeter need to be addressed. The amplitude of the received signal depends on the wavelength and energy of the transmit pulse, the distance to the target, the target reflectance, the transmission of the atmosphere, the area of the receiver aperture, the throughput efficiency of the receiver, the sensitivity of the detector, and, in the case of analog detection, the amplification gain applied to the detector output.

Of particular relevance here are those factors that are wavelength dependent: detector sensitivity, atmospheric transmission, and target reflectance. Figure 5.1 illustrates the sensitivity of several detector types used, or having potential for use, in laser altimeters. Wavelengths that are commonly used in ALSM systems are indicated: 900 nm, produced by semiconductor lasers; 1064 nm, produced by Nd:YAG lasers; and 532 nm (green visible light), generated by frequency doubling of Nd:YAG output. Other transmitter choices used in ALSM systems, not indicated in Figure 5.1, include 1560 nm semiconductor lasers and 1470 nm Nd:YLF lasers. Atmospheric transmission (Figure 5.2) depends on the density (i.e., optical depth) of aerosols and clouds. Normally, ALSM missions are conducted with a cloud-free path between the instrument and the target, but haze due to water vapor and aerosols can significantly reduce transmission, especially at visible wavelengths. Examples of typical reflectance curves for representative surface materials are illustrated in Figure 5.3, showing strong dependence on wavelength. These spectra show bidirectional reflectance data that are acquired with a phase angle (angular separation of the illumination source and the receiver) of several tens of degrees to be applicable to nadir-viewing, passive imaging systems viewing scenes illuminated at an angle by the sun.

Reflectance data acquired under bidirectional conditions are not directly applicable to assessing laser altimetry signal strength. Laser altimeters acquire data in a geometry unlike traditional passive optical imaging systems reliant on solar illumination. Specifically, the illumination and receiver

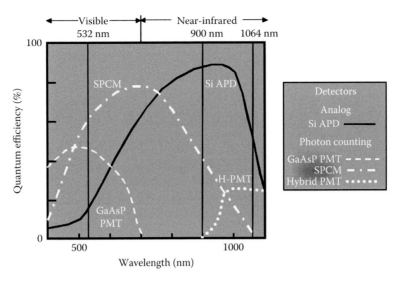

FIGURE 5.1 Representative curves of detector sensitivity as a function of wavelength, expressed as quantum efficiency. The Si APD (silicon avalanche photodiode) example is for a particular detector with sensitivity preferentially shifted toward the near-infrared. Other Si APDs have higher visible sensitivity.

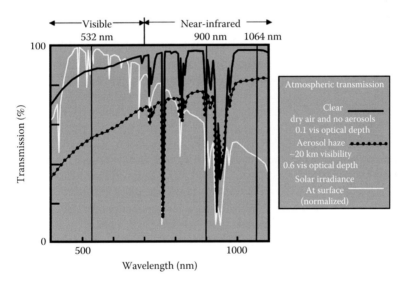

FIGURE 5.2 Wavelength dependence of atmospheric transmission shown for clear and hazy conditions. Also shown is downwelling solar irradiance at the Earth's surface that is a source of noise for laser altimeters. Minima in the transmission curves are due to water vapor absorption. Additional minima in the irradiance spectra are due to absorption occurring in the Sun.

view paths are either exactly parallel, for monostatic laser altimeters that use a common transmit and receiver aperture, or very nearly parallel, for bistatic designs that use adjacent apertures. Observing in this retroreflection *hot-spot* geometry, requires consideration of opposition effects because most natural surfaces, although they are diffuse reflectors that reflect light in all directions, are not Lambertian (equal reflectance in all directions) as is commonly assumed. Enhancement of reflectance due to opposition effects, compared with a nominal bidirectional geometry, is typically a factor of two and can be as large as a factor of three. The increase is a function of surface composition,

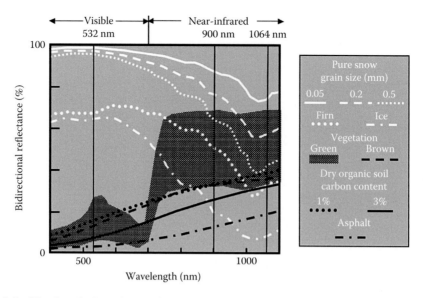

FIGURE 5.3 Wavelength dependence of bidirectional reflectance for representative surface materials. Green vegetation reflectance, a function of leaf structure and chlorophyll content, is represented as a broad band because of the variability observed as a function of species and environmental conditions. Firn is metamorphosing snow transitioning to ice. The spectra are compiled from multiple literature sources.

roughness, and wavelength. For example, Camacho-de Coca et al. (2004) studied the opposition hotspot effect at four wavelengths using the airborne, multiangle, passive imaging POLarization and Directionality of Earth Reflectance instrument flown in the direction of the solar principal plane. Reflectance for corn, barley, and bare soil increases with decreasing phase angle and peaks in the retroreflection orientation at 0° phase angle (Figure 5.4). The solid curves in Figure 5.4 are predicted reflectance based on a simple radiative transfer model that incorporates leaf reflectance, solar zenith angle, and geometric factors that account for scene self-shadowing (casting of shadows by one part of the target onto another part). The model assumes the opposition effect is due to a reduction of observed shadowed surfaces (i.e., increased shadow hiding) as 0° phase angle is approached.

Several studies have documented the increase in laser reflectance from natural terrestrial materials at 0° phase angle (e.g., Kaasalainen and Rautiainen, 2005; Kaasalainen et al., 2006). The increase in reflectance with decreasing phase angle consists of two components: a linear increase over a large angular range, approximated by the dashed lines in Figure 5.5, and a narrow, nonlinear peak in

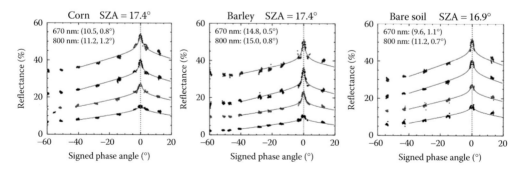

FIGURE 5.4 Phase angle dependence of solar reflectance illustrating the opposition hot-spot effect for corn, barley, and bare soil at 443, 550, 670, and 800 nm (plotted from bottom to top). (From Camacho de Coca, F.C. et al., *Rem. Sens. Environ.*, 90, 63–75, 2004. With Permission.)

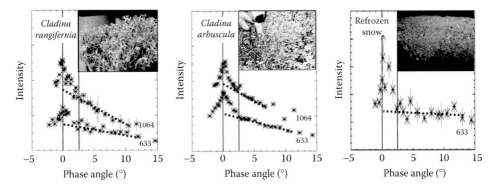

FIGURE 5.5 Phase angle dependence of laser reflectance illustrating the opposition effect at 633 and 1064 nm for lichen, and 633 nm for a snow surface after partial melting and refreezing. (Adapted from Kaasalainen, S. and Rautiainen, M., *J. Geophys. Res. Atmos.*, 110, 2005. With Permission; Kaasalainen, S. et al., *J. Glaciol.*, 52, 574–584, 2006. Reprinted with permission from the International Glaciological Society.)

reflectance beginning at about 2.5°. In the interpretation by Hapke et al. (1996, 1998) and Nelson et al. (1998, 2000), the opposition effect linear increase is due to increased shadow hiding with decreasing phase angle, and the narrow peak is due to the combined effects of shadow hiding and light amplification caused by coherent interference of the transmitted and reflected light. However, uncertainty remains about the physical origin of the laser retroreflection peak, and models of the effect do not fully account for all its attributes (Helfenstein et al., 1997; Shkuratov and Helfenstein, 2001).

Shadow hiding and coherent interference are retroreflection effects present to varying degrees for all surfaces independent of the angle of incidence of the light onto the surface. They are not related to the more commonly known specular reflection phenomenon that controls the brightness of water observed by laser altimeters. A planar interface between transmissive media with differing refractive indices, such as the interface between air and smooth water, reflects and refracts the incident light. The reflection is specular, or mirror-like, and the angles of the incident and reflected rays with respect to the surface normal are equal. The intensity of light specularly reflected from an interface is governed by the Huygens–Fresnel Law that defines reflectivity as a function of incident angle, the refractive index contrast across the interface, and the polarization state of the incident light. Over the typical range of scan angles used by airborne swath mapping laser altimeters, up to 30° of nadir, the reflectance of a water surface is only 2% regardless of polarization state. Although this reflectance is very low, for collimated laser light, the mirror-like reflection preserves a collimated beam, and the light in the reflection direction is much more intense than is the diffuse reflection from nonspecular surfaces.

For the 2% of incident laser energy reflected from a water surface, the intensity of the laser return in the retroreflection direction observed by an altimeter receiver depends on the surface area of facets oriented perpendicular to the laser pulse vector (90° incident angle). For smooth, flat water, not roughened by wind, most surface facets are coplanar and horizontal, resulting in an intense return for laser pulses oriented near nadir and a rapid fall-off of backscattered energy at small off nadir angles, accounting for the loss of signal from inland water bodies observed by airborne scanning laser altimeters at off-nadir angles. Bufton et al. (1983) describe how wind speed, and the resulting angular distribution of surface facets, is the controlling factor on laser backscatter intensity from roughened water surfaces. The remaining 98% of incident light not reflected from a water surface is refracted into the water column. The depth of penetration is defined by the absorption attenuation coefficient that is a function of wavelength. At visible wavelengths, the attenuation coefficient is small, and light penetrates to several tens of meters depth through clear water, accounting for the use of 532 nm lasers for bathymetric mapping. At near-infrared wavelengths the attenuation coefficient of clear water is several orders of magnitude larger due to enhanced molecular absorption, and light penetrates to only several centimeters to decimeters depth, thus acting to warm the uppermost layer of the water column and making 900 and 1064 nm lasers unsuitable for bathymetry purposes.

5.3 ANALOG DETECTION

Figure 5.6 illustrates commonly used laser ranging methods, depicting transmit pulses and received signals from a multistoried forest canopy as recorded by analog detection and photon counting approaches. In analog ranging, a detector converts received optical power into an output voltage yielding signal strength as a function of time. The signal is composed of reflected laser energy and noise sources, consisting of internal detector noise (i.e., dark counts) and background noise from solar illumination (Figure 5.2). In analog systems, high signal-to-noise ratio performance per pulse is achieved (i.e., monopulse detection), using laser pulses with high peak power. The goal is reception of thousands of reflected photons per pulse to exceed the detector noise floor and have sufficient signal to accurately determine the range to the illuminated target.

As the pulse energy must be high, the pulse width (i.e., its duration) is typically relatively broad, commonly being about 7 ns wide (full-width at half the maximum [FWHM] amplitude), equivalent to a FWHM pulse width of ~1 m. Semiconductor lasers or diode-pumped, Q-switched solid-state Nd:YAG laser transmitters operating at rates of hundreds to tens of thousands pulses per second are usually used for this purpose. The detector used for analog ranging is usually a silicon avalanche photodiode, which has high sensitivity across visible to near-infrared wavelengths (Figure 5.1).

5.3.1 WAVEFORM DIGITIZATION

Digitization of the detector output time series using an analog-to-digital converter yields a waveform that fully characterizes the vertical structure of the target (gray shaded signal in Figure 5.6). For a hard target such as bare ground or a building roof, a single return peak is normally observed. The broadening of the received peak, relative to that of the transmit peak, is a measure of the vertical relief of the target within the laser footprint due to surface roughness and/or slope. In some hard target cases, multiple return peaks are observed as for example when a laser footprint intersects a building edge,

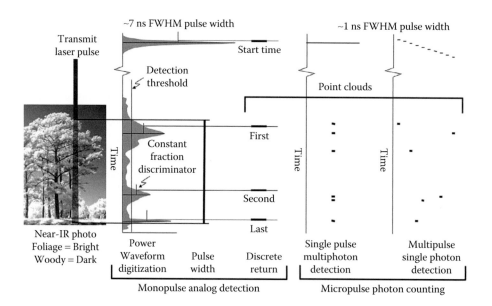

FIGURE 5.6 Illustration of laser ranging methods, depicting transmit pulses and received signals from a multistoried forest canopy as recorded by analog detection and photon counting approaches.

illuminating the top of the building and the adjacent ground. For distributed targets such as vegetation canopies in which multiple vertically distributed surfaces composed of leaves, stems and branches, and underlying ground are illuminated by a single laser pulse, a complexly shaped received signal can result. The signal in that case is a measure of the height distribution of illuminated surfaces weighted by the spatial distribution of laser energy within the footprint and the retroreflectivity of the surfaces at the laser pulse wavelength (Harding et al., 2001). The pulse width of the received signal (Figure 5.6) is the duration from the first to last crossing of the reflected laser energy above a detection threshold, representing the height range of the target from signal start to end. Mapping vegetated surfaces with a scanning waveform-recording system that produces laser footprints contiguous across and along track within a swath produces a three-dimensional, volumetric image of reflected laser energy (Weishampel et al., 2000). As laser energy is intercepted by canopy surfaces, the pulse energy decreases with depth through vegetation. Accounting for this extinction effect, height profiles of normalized canopy plant area (Figure 5.7) can be derived from the waveform data.

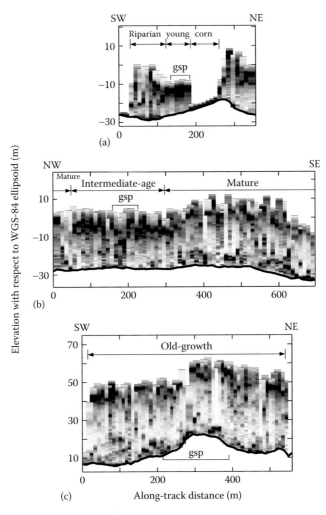

FIGURE 5.7   Examples of waveform-derived canopy height profiles showing normalized plant area from ALSM transects across deciduous forest stands of different ages. (a–c) illustrate differences in heights and vertical structure between stand types that are labelled in each figure. Each vertical bar corresponds to the diameter of one laser pulse (~10 m), and the gray shading corresponds to the relative amount of plant area in 66 cm height increments (darker = denser). (From Harding, D.J. et al. *Rem. Sens. Environ.*, 76, 283–297, 2001. With Permission.)

Normally, in full-waveform systems, both the transmit pulse shape and the received signal shape are digitized and recorded. The range to a specific feature within the received signal, such as the last peak in Figure 5.6 that is due to laser energy reflected from the ground surface, is usually defined as the distance between the center of the transmit pulse signal and the center of the received peak. Different methods for defining the peak center have been applied to waveform data, including use of the centroid (center of mass) or mean of the distribution or the center of a function, such as a Gaussian distribution, fit to the signal (Hofton et al., 2000; Harding and Carabajal, 2005). For waveform-recording systems, a well-defined and consistent transmit pulse shape is an important system attribute as is the signal-handling performance of the receiver, which must record the received pulse shape, free of distorting effects such as detector saturation.

The evolution of laser altimeter systems is shown in time-line form in Figure 5.8, grouped by ranging method and emphasizing research systems developed by NASA, many of which established measurement principles later used in commercial systems. Systems are characterized by operating wavelength (green or near infrared), the platform on which they are flown (airborne or satellite), and the sampling geometry of the laser footprints. Early experiments in waveform sampling of land topography and vegetation structure were performed by the Airborne Oceanographic LiDAR (AOL) (Krabill et al., 1984), acquiring in some cases single profiles and at other times using scanning to map a swath, and by the profiling Airborne Topographic Laser Altimeter System (ATLAS) (Bufton et al., 1991). Later, scanning airborne systems advanced the technology, including the Airborne Topographic Mapper (ATM) (Krabill et al., 2002) that is a follow-on to AOL, the Scanning LiDAR Imager of Canopies by Echo Recovery (SLICER) (Harding et al., 2000), the wide-swath mapping Laser Vegetation Imaging Sensor (LVIS) (Blair et al., 1999), and the Experimental Advanced Airborne Research LiDAR (EAARL) (Brock et al., 2002, 2006; Nayegandhi et al., 2006). The EAARL system, operating at 532 nm, simultaneously acquires bathymetry and topography data using a novel design that splits the received signal into three channels consisting of 90%, 9%, and 0.9% of the return energy. This is done to accommodate large variations in signal strength from diffuse surfaces (land, vegetation, and the bottom of water bodies), and specular water surfaces that normally cause saturation and distortion of the waveforms from smooth water. Recently, several versions of near-infrared, waveform-recording ALSM systems have become available for purchase from commercial vendors.

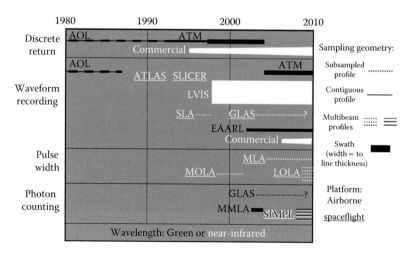

FIGURE 5.8 Timelines of laser altimeter operations for instruments developed by NASA and, commercially, differentiating systems based on the ranging method employed, the transmit pulse wavelength, the sampling pattern, and the type of platform used. For swath mapping systems, their swath width is proportional to the line thickness (Laser Vegetation Imaging Sensor width corresponds to its nominal 4 km swath acquired using 20 m diameter footprints).

Waveform-recording systems intended for characterization of canopy structure (e.g., SLICER, LVIS) normally use large diameter footprints (>10 m) to sample a representative canopy volume in a single footprint, while also illuminating the entire scene with adjacent, contiguous footprints. These are sometimes referred to as large-footprint systems. Other systems geared to high-resolution terrain mapping (e.g., ATM, EAARL, commercial) employ small diameter footprints (<1 m). In forested regions, these small-footprint systems usually yield simpler, but less representative waveforms than do large-footprint systems because the laser footprints intersect only a few distinct surfaces and are noncontiguous.

Satellite laser altimeters to date have not implemented scanning systems, instead acquiring single or a few profiles with footprints that are separated along-track, because of the very high ground speeds of spacecraft and high laser power needed for analog ranging from orbital altitudes (Sun, et al., 2013). Satellite-based waveform-recording systems in Earth orbit include the Shuttle Laser Altimeter (SLA) (Garvin et al., 1998) flown in 1996 and 1997 and the Geoscience Laser Altimeter System (GLAS) (Abshire et al., 2005) aboard NASA's Ice, Cloud and land Elevation Satellite (ICESat) that operated from 2003 to 2009 (Zwally et al., 2002; Schutz et al., 2005). SLA demonstrated techniques for space-based waveform measurements of canopy height and continental-scale topographic profiles that were subsequently used by ICESat to sample the biomass stored in forests and monitor elevation changes of the Earth's ice sheets and glaciers as they respond to global warming. The Mars Orbiter Laser Altimeter (MOLA) (Abshire et al., 2000; Neumann et al., 2003) is an analog detection system that recorded the received pulse width, rather than the complete waveform, to reduce the data volume needing transmission. MOLA profiles acquired continuously over a period of several years from 1997 to 2001 were used to create a topographic map of Mars that has revolutionized understanding of that planet's evolution (Smith et al., 1999, 2001; Zuber et al., 2000). The MESSENGER Mercury Laser Altimeter (MLA) (Ramos-Izquierdo et al., 2005) provided the first topographic mapping of that planet using pulse-width recording from 2011 to 2015. The Lunar Orbiter Laser Altimeter (LOLA) (Riris et al., 2007; Chin et al., 2007), uses a multibeam, pulse-width recording system with five parallel profiles to map the Moon in unprecedented detail from 2009 to the present.

5.3.2 Discrete Returns

Discrete return ranging identifies distinct peaks in the analog detector output time series that exceed a detection threshold. This is usually accomplished by means of a constant fraction discriminator that times the leading edge of the peak at an energy level that is some specified fraction of the peak amplitude. Fifty percent constant fraction discrimination is illustrated in Figure 5.6. This approach substantially reduces the data volume that must be recorded, as compared with full waveforms while identifying the height of prominent surfaces illuminated by the laser pulse. A key attribute of systems that utilize discrete return detection is the number of returns that can be recorded per laser pulse. Earlier versions of such systems recorded only one return, with the timing electronics configured to record either the highest detected distinct peak or the lowest peak. Subsequent systems were configured to record two returns, the highest and the lowest, or multiple returns, up to a maximum of five. Figure 5.6 illustrates a case in which three discrete returns were identified. Most recently, systems have added detection and recording of the return amplitude associated with each discrete return, in some cases reporting the maximum amplitude of the peak and in others reporting an integrated signal strength (i.e., area of the peak).

For discrete return systems, the range is defined as the distance from the transmit pulse leading edge (start time in Figure 5.6) to the leading edge of the distinct peaks. Therefore, in these systems, a well-defined leading-edge on the transmit pulse is most important, and the trailing shape of the pulse is of lesser consequence for ranging accuracy. Each return, when geolocated, yields a discrete location defined by the latitude, longitude, and elevation of a point. Combining the discrete returns from many laser pulses yields a three-dimensional distribution of geolocated points, commonly referred to as a point cloud. Small-diameter footprints (<1 m) are usually used in discrete-return

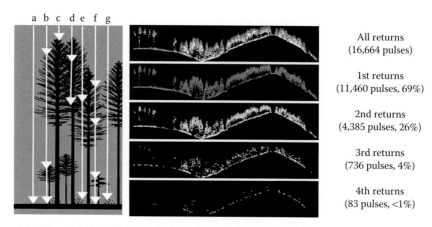

All returns
(16,664 pulses)

1st returns
(11,460 pulses, 69%)

2nd returns
(4,385 pulses, 26%)

3rd returns
(736 pulses, 4%)

4th returns
(83 pulses, <1%)

a: 1st = Ground b: 1st = Highest canopy, 2nd = Mid-story, Last = Ground c: 1st = Highest canopy
d: 1st = Within crown, Last = Within crown e: 1st = Within crown, Last = Under-story
f: 1st = Highest canopy, 2nd = Within crown, 3rd = Mid-story, Last = Ground g: 1st = Under-story

FIGURE 5.9 Schematically illustrating the detection of multiple discrete returns for seven laser pulses transmitted through a forest canopy (left) and a transect through discrete return point cloud data acquired across the Capitol Forest, WA showing data separated by return number. (Courtesy of Bob McGaughey, USDA Forest Service PNW Research Station, Portland, OR. With Permission.)

systems because of the need, when a pulse intercepts multiple surfaces, to have distinct return peaks that are separated in range. In addition, the footprint size is a limiting factor on spatial resolution; small footprints enable high-resolution mapping of topography and canopy structure.

Figure 5.9 is an illustration of point cloud data acquired by a multiple return, discrete return, and small-footprint laser ranging system, depicting acquisition of up to four returns per laser pulse reflected from a forest canopy. The number and height of returns detected for a pulse depend on the distribution of surfaces, and their reflectance at the laser wavelength, encountered along the path of the laser beam. Discrete returns are detected when the area and reflectance of an illuminated surface are sufficient to yield received energy that exceeds the detection threshold. First returns correspond to the leading edge of the first detected signal above the threshold level, and may be from the canopy top, from a layer within the canopy or from the ground. Last returns correspond to the leading edge of the latest detected peak, and may be from the ground, or from a layer within the vegetation canopy. The detection of ground returns in areas of vegetation cover depends on the spatial and angular distribution of open space (gaps) through the canopy, the scan angle of the laser beam, the divergence of the laser beam and the range to the target (defining the diameter of the laser footprint), the footprint sampling density, and the reflectivity of the ground at the laser wavelength. Unlike waveform recording systems that fully characterize vegetation structure, discrete-return representations of canopies are highly instrument-dependent. Measurements of canopy attributes such as height, crown depth, or the distribution of understory layers and gaps obtained using different discrete-return instruments are therefore generally not equivalent, and comparisons for change detection purposes, for example, must be made with care. Lefsky et al. (2002) review the use of waveform-recording and discrete-return systems for estimation of vegetation structural attributes.

The ability to detect separate returns from closely spaced surfaces is dependent on instrument parameters including the laser pulse width (shorter being better), detector sensitivity and response time (i.e., bandwidth), the system signal-to-noise performance, the detection threshold, and the ranging electronics implementation. This is especially relevant for detection of ground returns beneath short-stature vegetation. The elevation recorded by discrete return ranging is dependent on the shape of the received signal's leading edge, illustrated in Figure 5.10 in which waveforms represent four surface types: (a) highly reflective flat ground, (b) lower reflectivity flat ground, (c) ground

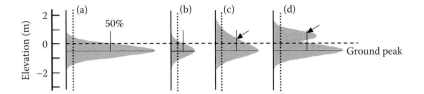

FIGURE 5.10 Schematic waveforms depicting returns from (a) highly reflective flat ground, (b) less reflective flat ground, (c) flat ground with low vegetation cover such as tall grass, and (d) flat ground with a low but distinct vegetation layer such as a forest understory. The detection threshold is dotted, leading edge timing for the ground peak using 50% constant fraction discrimination is dashed, and arrows mark the height of upward biased returns for the vegetated cases.

covered by tall grass causing an upward skew in the received signal, and (d) ground overlain by a distinct layer of low vegetation. Flat surfaces reflect a signal having the same shape as the transmit pulse, in these examples a Gaussian distribution with a 7 ns (1 m) FWMH pulse width. For returns with energy exceeding a detection threshold (dotted lines), 50% constant fraction discrimination yields a range that is insensitive to peak amplitude, so, bright and dark surfaces should have unbiased elevations (a and b, dashed line). However, a laser pulse reflected from low vegetation cover and the underlying ground can yield a composite signal from which only one peak is identified above the threshold, yielding a single discrete return. The vegetation shields the leading edge of the ground return, with the upward-displaced leading edge of the composite signal yielding a return that is close to, but above, the ground (c and d, arrows). This shielding is of course enhanced when leaves are present, not only because this increases plant cover but also because of the brighter reflectance of green vegetation as compared with soil or leaf litter, especially in the near infrared (Figure 5.3). Due to this shielding, discrete return analog systems can yield elevations that are biased high with respect to the ground, in areas with short-stature vegetation cover.

Derivation of gridded digital elevation models from ALSM point cloud data is done to define surfaces using those points associated with specific landscape features. For example, the highest reflective surface, or digital surface model (DSM), is produced from the first returns for each laser pulse (Figure 5.11). In vegetated areas, this surface is a representation of the canopy top. The bald Earth surface, or digital terrain model (DTM), is produced from those returns inferred to be from the ground based on spatial filtering that identifies lowest returns defining a continuous surface. Interpolation is applied to produce a DTM, because returns from the ground can be irregularly distributed and widely spaced especially where vegetation cover is dense. In one interpolation approach, the terrain elevation is first represented as a triangulated irregular network created from the returns classified as ground. The elevation of the triangulated irregular network is then sampled at regularly spaced grid points to derive the DTM. The georeferenced DSM and DTM products provide a simplified representation of the point cloud data in a form amenable for use in raster-based image processing software and Geographic Information Systems. Note, however, as illustrated in Figures 5.9 and 5.11, there is additional information in the point cloud dataset, regarding the structure and density of vegetation within the canopy that is not described by the DSM and DTM surfaces. Vector-based analysis tools are needed to fully assess information concerning canopy structure contained within the irregularly distributed point cloud data.

Examples of the highly detailed images of ground topography and forest canopy structure that can be produced using discrete-return ALSM data are illustrated in Figure 5.12, depicting commercial data acquired for the Puget Sound LiDAR Consortium (Haugerud et al., 2003). The data were acquired with submeter diameter laser footprints having a nominal density of two footprints per square meter using a system that recorded up to four returns per pulse. A heavily forested, topographically rugged area in the Cascade Range, Washington State, is portrayed as a color hillshade image of the ground topography DTM, a gray-scale hill-shade image of the canopy top DSM,

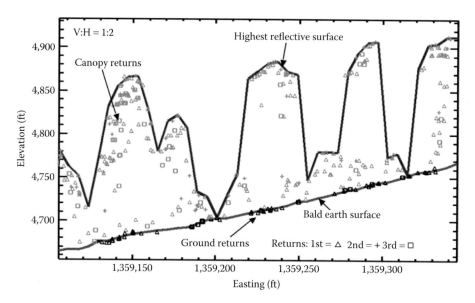

FIGURE 5.11 Derivation of digital elevation model (DEM) products from a multihit, discrete return point cloud, showing the highest reflective DSM produced from first returns and the *bald Earth* topography DTM produced from those returns inferred to be from the ground (black) based on spatial filtering. In this example, depicting returns in a 6 ft wide corridor from individual tree crowns in a forest stand above sloping ground, the surfaces are generated with 6 ft postings using data acquired for the Puget Sound Lidar Consortium. (From Haugerud, R. et al., *GSA Today*, 13, 4–10, 2003. With Permission.) Triangle, plus, and square symbols indicate first, second, and third returns, respectively.

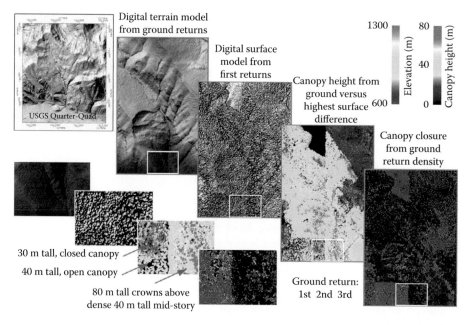

FIGURE 5.12 Discrete return point clouds and the DEMs generated from them provide highly detailed images of ground topography and forest canopy structure. Here commercial ALSM data acquired for the Puget Sound Lidar Consortium (From Haugerud, R. et al., GSA Today, 13, 4–10, 2003. With Permission.) and gridded at six foot postings is used to depict several representations of topography and vegetation cover. The area depicted is in the lower left part of the USGS Quarter Quadrangle shown in the upper left as a DTM. The area indicated by the white rectangles is enlarged in the insets to show image details.

a color image of canopy height (DSM–DTM), and a measure of canopy closure represented by the distribution of ground returns (black areas lack ground returns due to dense canopy cover). The inset enlargements illustrate tree crowns in three stands with different canopy structures produced by varying forest-management practices. The area on the right corresponds to a mature, closed-canopy forest with a low density of ground returns and isolated 80 m tall tree crowns above a 40 m tall midstory canopy layer. The areas on the left are managed stands at different stages of regrowth following clear-cut logging. The upper area is a young, dense stand, whereas the lower area is an older, taller stand that has undergone selective thinning.

Implications for the vertical accuracy of DTMs due to the upward bias of discrete-return, leading-edge ranging are apparent from histograms of elevation differences between ground control points and an ALSM ground topography DTM (Figure 5.13). The ground control was collected using differential global position system surveying at randomly selected locations uniformly distributed across the DTM. For nonvegetated areas, random ranging and geolocation errors yield a distribution of elevation differences that is symmetric and has a mean difference very close to zero. In these bare ground areas, the DTM is unbiased and has a vertical accuracy, expressed as a root-mean-square-error (RMSE) of 19 cm. For areas with tall grass cover, the difference histogram is skewed, with the derived DTM surface sometimes occurring several decimeters above the ground control points yielding a surface that is, on average, too high by 7 cm. For forested areas, the skew can be more pronounced and the distribution can be bimodal. The forest example in Figure 5.13 has a peak at zero for true ground returns, and at −50 cm, due to returns from dense understory vegetation being misclassified as ground. As a consequence, the mean upward bias of the DTM is significant and the RMSE vertical accuracy is substantially degraded. Instrumentation-dependent biases related to vegetation shielding of the ground can be a significant source of error quantifying topographic change by differencing discrete return DTMs acquired separated in time by different systems. Improvements in instrument parameters, such as a narrowed pulse-width to better detect closely space returns, decreased footprint size, increased footprint sampling density, or improved filtering to minimize misclassification of low vegetation as ground are means to lessen this source of error in discrete-return ALSM data.

The AOL system (Krabill et al., 1984) pioneered airborne topographic profiling and mapping using discrete-return analog ranging (Figure 5.8) and directly led to the development of the scanning ATM system (Krabill et al., 1995, 2002). ATM transitioned to a full waveform-recording system in 2004. ATM has conducted comprehensive surveying of Greenland and parts of Antarctic for more than a decade to monitor elevation change associated with changes in ice sheet mass balance (Krabill et al., 2000; Abdalati et al., 2002; Thomas et al., 2004, 2006). ATM has also conducted repeated surveys of U.S. coastlines to document beach elevation change related to storm surges, flooding, and sediment transport (Brock et al., 2002; Sallenger et al., 2003). Given the need

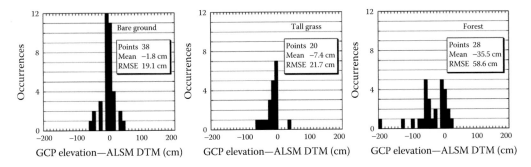

FIGURE 5.13 Cover-type specific histograms of elevation differences between a DTM, produced from ALSM discrete returns inferred to be from the ground, and ground control points measured by differential GPS surveying. As the difference is computed as GCP–DTM, locations where the DTM is biased high yield negative values. The GPS and ALSM data were acquired for the Puget Sound Lidar Consortium. (From Haugerud, R. et al., *GSA Today*, 13, 2003. With Permission.)

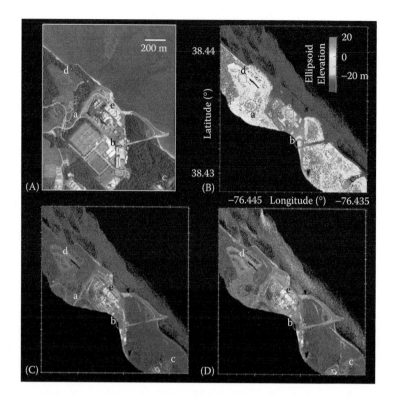

FIGURE 5.14 ATM mapping of the Chesapeake Bay western shoreline at Calvert Cliffs Nuclear Power Plant, Maryland, acquired on September 17, 1997. Individual laser footprints from multiple flight passes are plotted showing elevation (B), and uncalibrated received 532 nm laser energy (whiter being brighter) (D). Also shown is the ATM measurement of uncalibrated reflected 532 nm solar energy acquired between laser pulses (whiter being brighter) (C) and a grey scale aerial photograph from Google Earth (A). Examples of intensity differences include areas where laser backscatter is lower, with respect to the surroundings, as compared with the reflected sunlight (sites: (a) parking lot, (b) building roof, (c) clay tennis court) and where the laser backscatter is higher (sites: (d) grass, (e) cooling tower roof). Moreover, intense specular reflections from offshore waves are observed in the laser backscatter image but not in the reflected sunlight image. (Courtesy of Bill Krabill and Serdar Manizade, NASA Goddard Space Flight Center, Greenbelt, MD. With Permission.)

for repeatable surveying with high spatial resolution and absolute accuracy necessary for change detection, the development and evolution of the ATM instrumentation and associated geolocation processing (Krabill and Martin, 1987; Vaughn et al., 1996) established many of the principles and practices now used in the large number of commercial ALSM instruments operating worldwide. ATM includes a laser return intensity mapping capability that is unique in which the amplitude of the solar background reflectance, observed between laser pulse fires, is also recorded. Thus, solar bidirectional reflectance and laser retroreflectance at 532 nm can be directly compared to gain a better understanding of the hot-spot opposition effect (Figure 5.14).

5.4 PHOTON COUNTING

Unlike analog detectors, which convert received power into an output voltage, photon counting detectors record the arrival of single photons. Their output can be thought of as digital, in which there is either an absence of signal or discrete events upon detection of a photon (Figure 5.6). The principles and methods for photon counting laser ranging were first established in ground-based systems used to range to Earth-orbiting satellites and are now being extended to ALSM systems (Degnan, 2001). For laser ranging implementations, photon counting detectors are used that have very low dark counts, so that the rate of detector noise is very small as compared with the rate of received signal photons. Combining a photon counting detector with timing electronics, the time-of-flight between laser fire and reception of a single photon is recorded. The single photon is a sample from the full distribution of surfaces illuminated by the laser pulse, and accumulation of many single photon ranges can recreate the height structure of a target. During daytime operations, solar illumination reflected from the Earth's surface or from clouds is a significant source of noise for photon counting systems (Figure 5.2), more so than for analog systems in which the detector is the dominant noise source. The rate of solar background noise detection must therefore be controlled. This is usually accomplished by a combination of narrow bandpass filtering, to block detection of energy at all wavelengths other than that of the laser, and use of a small receiver field-of-view, restricting collection of light to the location illuminated by the laser.

Unlike analog systems, in which high-energy laser pulses are used to achieve an adequate signal-to-noise ratio, photon counting systems employ much lower pulse energies with the goal of detecting only a small number of photons per pulse (i.e., micropulse detection). At these low pulse energies, short transmit pulse widths of ≤ 1 ns FWHM (≤ 15 cm) can be achieved.

This narrow pulse width, along with low-jitter detectors and high-resolution timing electronics, makes possible centimeter to decimeter range precision for single detected photons (Priedhorsky et al., 1996; Ho et al., 1999). Photon counting systems, therefore, have the potential to more efficiently acquire ALSM data than can analog systems, using less power, smaller receiver apertures, and/or higher pulse repetition rates (Degnan, 2002). As very short pulses and single photons are the basis for ranging, biases due to the inability of distinguishing closely spaced surfaces, that affects discrete-return leading-edge ranging, can be avoided. Microchip lasers, operating at thousands to tens of thousands of pulses per second, or fiber lasers, operating at hundreds of thousands of pulses per second, are well suited as short-pulse transmitter sources for photon counting systems. Analog discrete returns and photon counting both yield point clouds composed of many individual returns, each positioned at a specific location in three-dimensional space. On account of that commonality, the extensive body of software that has been developed for the processing, analysis and visualization of discrete return point clouds, and derivation of map products is equally applicable to photon counting ALSM data.

5.4.1 SINGLE-PULSE MULTIPHOTON DETECTION

The specific manner in which a photon counting ALSM system is implemented depends on the performance characteristics of the detector used. Specifically, detector dead-time is a controlling factor. Dead-time is the time needed to recover after detection of one photon, returning to a state in which a second photon can be detected with equal sensitivity. Photomultiplier tubes (PMTs) commonly have very short dead times so that successive photons reflected from closely spaced surfaces, such as are encountered in vegetation canopies, can be differentiated. By using PMTs, single-pulse multiphoton detection can be employed in which multiple photons are detected per pulse (Figure 5.6). To date, commercially available PMTs suitable for photon counting ranging are limited to operating at visible wavelengths, restricting their use to ALSM systems employing Nd:YAG lasers frequency doubled to 532 nm (Figure 5.1). An additional limitation of PMTs is limited lifetime due to reduction in detection sensitivity as the total number of detected photons increases through time. An emerging detector technology needing further development, the hybrid PMT (Figure 5.1), offers the potential to conduct low dead-time, photon counting ranging at near-infrared wavelengths without the lifetime limitations of traditional PMTs (Sun et al., 2007).

The first ALSM system to implement PMT-based photon counting ranging (Figure 5.8) was the experimental Multikilohertz Microlaser Altimeter (MMLA) (Degnan et al., 2001) developed with funding from NASA's Instrument Incubator Program. MMLA employed a four-element PMT detector array, with each channel of the array capable of detecting up to four individual photon returns per laser fire. Thus, up to 16 single photon range measurements could be obtained per pulse. A rotating transmissive wedge produced a helical scan pattern thus mapping a narrow swath. Figure 5.15

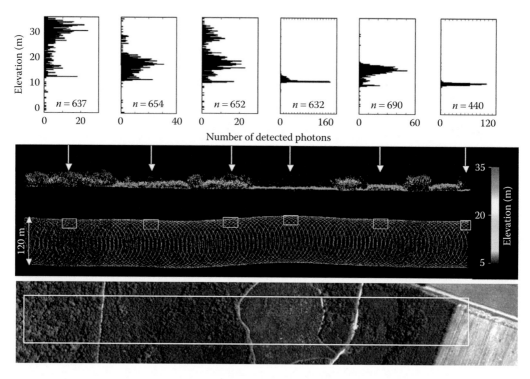

FIGURE 5.15 MMLA photon-counting, helical-scanning airborne data acquired in a 120 m wide swath across the Wicomico Demonstration Forest, MD Eastern Shore on September 12, 2002. Middle: single-photon returns in map and cross-section views; the cross-section portrays laser pulse returns and solar background noise (no noise filtering has been applied) from a ~20 m wide corridor along the top edge of the swath in which returns are densest (~1 return per square meter). Top: height distributions (0.25 m bins) of single-photon returns accumulated over ~20 × 30 m areas (white boxes on the map view) showing canopy vertical structure, distinct ground peaks, and low levels of solar background noise. Bottom: grey scale aerial photograph showing approximate location of the MMLA swath (white rectangle). (Courtesy of Jan McGarry, NASA Goddard Space Flight Center, Greenbelt, MD. With Permission.)

illustrates an MMLA swath acquired during midday (e.g., high solar background) across the Wicomico Demonstration Forest, Maryland Eastern Shore. Map and cross-section representations of the elevation of individual photon returns depict variations in canopy height with distinct changes across stand edges. Accumulating single photon returns from areas approximately 20 × 30 m in size to form elevation histograms yields representations of canopy vertical structure analogous to the waveforms of an analog detection system. The histograms, produced with 25 cm height bins, reveal differences in canopy height and layering related to stand maturity and forest-management practices.

Even with the full leaf-on canopy cover conditions present at the time of data collection, a sufficient number of photons were returned from the ground to form a well-defined last peak beneath the canopy. The accumulation of ground returns in just one, or a few, 25 cm height bins is indicative of the range precision achieved using short-pulse photon counting. Solar background noise is present but at very low levels, as seen by the scarcity of detected photons below the ground surface and above the canopy top. In addition to mapping topography and vegetation structure, MMLA can simultaneously acquire topography and bathymetry data because it operates at 532 nm. MMLA has demonstrated single photon ranges that detect the water surface elevation, the vertical distribution of scattering within the water column, and the depth of the water bottom. The technologies developed in MMLA served as the foundation for follow-on 532 nm PMT-based laser altimeters including airborne instruments (Degnan, 2017) and the Advanced Topographic Laser Altimeter System (ATLAS) that will fly on ICESat-2, scheduled for launch in 2018 (Markus et al., 2017).

5.4.2 Multipulse Single-Photon Detection

In contrast to short dead-time PMT detectors, single-photon counting modules (SPCM) based on a silicon avalanche photodiode detector (Sun et al., 2004) have long dead-times requiring a different approach to photon counting ranging. To avoid a range bias due to photons being received but undetected during the dead-time, a very low-energy per laser pulse is used so that the probability of detecting more than one photon from a target for each pulse is low. A detection probability of less than one is therefore desired, meaning, for some pulses, no photon returns will be detected. By transmitting pulses at very high rates (e.g., up to several hundred kilohertz), individual returns can be rapidly accumulated, yielding return densities comparable with multiphoton detection conducted at lower pulse repetition rates (Figure 5.6). Although the requirement for a low probability of detection per pulse is a significant constraint, the sensitivity of SPCM detectors, unlike PMTs, is not degraded as the total dose of detected photons increases. Similar to PMTs, SPCM detectors are most sensitive at visible wavelengths, achieving even higher sensitivity than PMTs at 532 nm, but their sensitivity falls off less rapidly with increasing wavelength (Figure 5.1), providing some capability for operation in the near-infrared. Single-photon avalanche diode detectors that can compensate for range bias effects due to reception of multiple photons closely spaced in time have been demonstrated (Kirchner and Koidl, 1999), providing a potential alternative to the SPCM detectors.

Although primarily used to profile atmospheric clouds and aerosols, the GLAS atmospheric LiDAR channel operating at 532 nm (Spinhirne et al., 2005) has demonstrated SPCM photon-counting ranging to the Earth's surface from space (Figure 5.8). An airborne instrument, the Slope Imaging Multi-polarization Photon-counting Lidar (SIMPL), based on multipulse single-photon ranging was developed using NASA IIP funding (Dabney et al., 2010). SIMPL is a four beam instrument, operating simultaneously at 532 nm and 1064 nm wavelengths, using a 10 kHz microchip laser, a beam splitter and SPCM detectors (Harding et al., 2011). Received photons are detected in channels with filters oriented parallel and perpendicular to the polarization plane of the transmit pulses in order to measure the depolarization of the reflected laser energy (Yu et al., 2016). Depolarization is a function of the amount of single versus multiple scattering of the photons that occurs during reflection of plane-polarized laser light. Acquiring depolarization data at visible and near-infrared wavelengths directly associated with the ranging data contributes to classification and mapping of surface types based on their height and scattering properties. The non-ranging, analog-detection Airborne Laser Polarization Sensor (ALPS) developed by NASA demonstrated differentiation of needle-leaf and broad-leaf tree species based on 532 nm and 1064 nm depolarization (Kalshoven and Dabney, 1993). With the addition of waveform-recording ranging, ALPS was renamed as the Multiwavelength Airborne Polarimetric Lidar (MAPL) and is being used to test new vegetation remote sensing applications (Tan and Narayanan, 2004; Tan et al., 2005).

5.5 SUMMARY

ALSM is a uniquely capable method for collection of high resolution elevation data. In particular, in vegetated areas, it provides measurements of ground topography and canopy structure unmatched by any other remote sensing technique. The resulting data products are used for scientific investigations and applied mapping purposes. The method used to acquire the fundamental ALSM measurement, the range to the surface, defines the characteristics of the elevation data. Scanning analog-detection systems that record full-waveforms provide three-dimensional representations of backscattered laser pulse energy at visible or near-infrared wavelengths. Analog discrete return systems detect distinct surfaces within that three-dimensional volume that reflect energy exceeding a threshold, yielding point cloud representations of surface elevations. Detection of a discrete return is dependent on instrument parameters and geometric and reflectance attributes of the target. As laser altimeters acquire data at 0° phase angle, with parallel illumination and view angles, the received energy is a function of the retroreflectance of the target. Retroreflectance is enhanced, compared with bidirectional reflectance, by shadow hiding and coherent backscatter opposition effects. Specular reflection from water surfaces yields a high amplitude return for data acquired near-nadir and weak or no signal strength from off-nadir angles. Photon-counting systems, using detectors that record the arrival of single photons and low-energy, short-pulse, high-repetition rate laser transmitters, have the potential to acquire data more efficiently than analog systems. Like discrete-return systems, they record individual returns that, when combined, form point clouds. Unlike discrete-return data that can be biased upward due to the use of leading-edge ranging, photon counting when properly implemented yields unbiased range data that samples the height distribution of surfaces illuminated by the laser pulse. Research systems, many of which have been developed by NASA for airborne and spaceflight use, have established measurement principles and practices now used in commercially available systems that form the basis of a robust, worldwide mapping industry.

ACKNOWLEDGMENTS

This chapter has benefited greatly from discussions over the past two decades with the many experts in laser altimetry at Goddard Space Flight Center. In particular, I would like to thank James Abshire, Bryan Blair, Jack Bufton, Phil Dabney, John Degnan, Jim Garvin, Bill Krabill, and Xiaoli Sun.

REFERENCES

Abdalati, W., Krabill, W., Frederick, E., Manizade, S., Martin, C., Sonntag, J., Swift, R., Thomas, R., Wright, W., and Yungel, J., 2002. Airborne laser altimetry mapping of the Greenland ice sheet: Application to mass balance assessment. *J. Geodyn.*, 34, 391–403.

Abshire, J.B., Sun, X.L., and Afzal, R.S., 2000. Mars orbiter laser altimeter: Receiver model and performance analysis. *Appl. Optic.*, 39, 2449–2460.

Abshire, J.B., Sun, X.L., Riris, H., Sirota, J.M., McGarry, J.F., Palm, S., Yi, D.H., and Liiva, P., 2005. Geoscience laser altimeter system (GLAS) on the ICESat mission: On-orbit measurement performance. *Geophys. Res. Lett.*, 32, L21S02. doi:10.1029/2005GL024028.

Blair, J.B., Rabine, D.L., and Hofton, M.A., 1999. The laser vegetation imaging sensor: A medium-altitude, digitisation-only, airborne laser altimeter for mapping vegetation and topography. *ISPRS J. Photogramm. Rem. Sens.*, 54, 115–122.

Brock, J.C., Wright, C.W., Kuffner, I.B., Hernandez, R., and Thompson, P., 2006. Airborne lidar sensing of massive stony coral colonies on patch reefs in the northern Florida reef tract. *Rem. Sens. Environ.*, 104, 31–42.

Brock, J.C., Wright, C.W., Sallenger, A.H., Krabill, W.B., and Swift, R.N., 2002. Basis and methods of NASA airborne topographic mapper lidar surveys for coastal studies. *J. Coast. Res.*, 18, 1–13.

Bufton, J.L., Hoge, F.E., and Swift, R.N., 1983. Airborne measurements of laser backscatter from the ocean surface. *Appl. Optic.*, 22, 2603–2618.

Bufton, J.L., Garvin, J.B., Cavanaugh, J.F., Ramos-Izquierdo, L., Clem, T.D., and Krabill, W.B., 1991. Airborne lidar for profiling of surface topography. *Opt. Eng.*, 30, 72–78.

Camacho-de Coca, F.C., Breon, F.M., Leroy, M., and Garcia-Haro, F.J., 2004. Airborne measurement of hot spot reflectance signatures. *Rem. Sens. Environ.*, 90, 63–75.

Carter, W., Shrestha, R., and Slatton, K.C., 2004. Photon-counting airborne laser swath mapping (PCALSM). *Proc. SPIE, 4th Int. Asia-Pacific Environ. Rem. Sens. Symp.*, 5661, 78–85.

Chin, G. et al., 2007. Lunar reconnaissance orbiter overview: The instrument suite and mission. *Space Sci. Rev.*, 129, 391–419.

Dabney, P. et al., 2010. The slope imaging multi-polarization photon-counting lidar: Development and performance results. *Geoscience and Remote Sensing Symposium (IGARSS), 2010 IEEE International*, Honolulu, HI, pp. 653–656. doi:10.1109/IGARSS.2010.5650862.

Degnan, J.J., 2001. Unified approach to photon-counting microlaser rangers, transponders, and altimeters. *Surv. Geophys.*, 22, 431–447.

Degnan, J.J., 2002. Photon-counting multikilohertz microlaser altimeters for airborne and space-borne topographic measurements. *J. Geodyn.*, 34, 503–549.

Degnan, J.J., 2016. Scanning, multibeam, single photon lidars for rapid, large scale, high resolution, topographic and bathymetric mapping. *Rem. Sens.*, 8(98), 958. doi:10.3390/rs8110958.

Degnan, J.J., McGarry, J., Zagwodzki, T., Dabney, P., Geiger, J., Chabot, R., Steggerda, C., Marzouk, J., and Chu, A., 2001. Design and performance of an airborne multikilohertz photon-counting, microlaser altimeter. *Proc. Land Surface Mapping and Characterization Using Laser Altimetry, Int. Arch. Photogramm. Rem. Sens.*, XXXIV3-W4, Annapolis, MD, pp. 9–16.

Garvin, J., Bufton, J., Blair, J., Harding, D., Luthcke, S., Frawley, J., and Rowlands, D., 1998. Observations of the Earth's topography from the shuttle laser altimeter (SLA): Laser-pulse echo-recovery measurements of terrestrial surfaces. *Phys. Chem. Earth Solid Earth Geodes.*, 23, 1053–1068.

Hapke, B., DiMucci, D., Nelson, R., and Smythe, W., 1996. The cause of the hot spot in vegetation canopies and soils: Shadow-hiding versus coherent backscatter. *Rem. Sens. Environ.*, 58, 63–68.

Hapke, B., Nelson, R., and Smythe, W., 1998. The opposition effect of the moon: Coherent backscatter and shadow hiding. *Icarus*, 133, 89–97.

Harding, D.J., Blair, J.B., Rabine, D.L., and Still, K.L., 2000. SLICER airborne laser altimeter characterization of canopy structure and sub-canopy topography for the BOREAS Northern and Southern Study Regions: Instrument and Data Product Description. In F.G. Hall and J. Nickeson (Eds.) Technical Report Series on the Boreal Ecosystem-Atmosphere Study (BOREAS), NASA/TM-2000-209891, 93, 45 pp.

Harding, D.J. and Carabajal, C.C., 2005. ICESat waveform measurements of within-footprint topographic relief and vegetation vertical structure. *Geophys. Res. Lett.*, 32, L21S10. doi:10.1029/2005GL023471.

Harding, D.J., Dabney, P.W., and Valett, S., 2011. Polarimetric, two-color, photon-counting laser altimeter measurements of forest canopy structure. *International Symposium on Lidar and Radar Mapping 2011: Technologies and Applications, Proceedings of SPIE 8286*, Nanjing, China, p. 10. doi:10.1117/12.913960.

Harding, D.J., Lefsky, M.A., Parker, G.G., and Blair, J.B., 2001. Laser altimeter canopy height profiles—Methods and validation for closed-canopy, broadleaf forests. *Rem. Sens. Environ.*, 76, 283–297.

Haugerud, R., Harding, D.J., Johnson, S.Y., Harless, J.L., Weaver, C.S., and Sherrod, B.L., 2003. High-resolution topography of the Puget Lowland, Washington—A bonanza for earth science. *GSA Today*, 13, 4–10.

Helfenstein, P., Veverka, J., and Hillier, J., 1997. The lunar opposition effect: A test of alternative models. *Icarus*, 128, 2–14.

Ho, C. et al., 1999. Demonstration of literal three-dimensional imaging, *Appl. Optic.*, 38, 1833–1840.

Hofton, M.A., Minster, J.B., and Blair, J.B., 2000. Decomposition of laser altimeter waveforms. *IEEE Trans. Geosci. Rem. Sens.*, 38, 1989–1996.

Kaasalainen, S., Kaasalainen, M., Mielonen, T., Suomalainen, J., Peltoniemi, J.I., and Naranen, J., 2006. Optical properties of snow in backscatter. *J. Glaciol.*, 52, 574–584.

Kaasalainen, S. and Rautiainen, M., 2005. Hot spot reflectance signatures of common boreal lichens. *J. Geophys. Res. Atmos.*, 110, D20102. doi:10.1029/2005JD005834.

Kalshoven, J.E. and Dabney, P.W., 1993. Remote-sensing of the Earth's surface with an airborne polarized laser. *IEEE Trans. Geosci. Rem. Sens.*, 31, 438–446.

Kirchner, G. and Koidl, F., 1999. Compensation of SPAD time-walk effects. *J. Optic. Pure Appl. Optic.*, 1, 163–167.

Krabill, W.B. and Martin, C.F., 1987. Aircraft positioning using global positioning system carrier phase data. *Navigation*, 34, 1–21.

Krabill, W.B., Abdalati, W., Frederick, E.B., Manizade, S.S., Martin, C.F., Sonntag, J.G., Swift, R.N., Thomas, R.H., Wright, W., and Yungel, J.G., 2000. Greenland ice sheet: High-elevation balance and peripheral thinning. *Science*, 289, 428–430.

Krabill, W.B., Abdalati, W., Frederick, E.B., Manizade, S.S., Martin, C.F., Sonntag, J.G., Swift, R.N., Thomas, R.H., and Yungel, J.G., 2002. Aircraft laser altimetry measurement of elevation changes of the Greenland ice sheet: Technique and accuracy assessment. *J. Geodyn.*, 34, 357–376.

Krabill, W.B., Collins, J.G., Link, L.E., Swift, R.N., and Butler, M.L., 1984. Airborne laser topographic mapping results. *Photogramm. Eng. Rem. Sens.*, 50, 685–694.

Krabill, W.B., Thomas, R.H., Martin, C.F., Swift, R.N., and Frederick, E.B., 1995. Accuracy of airborne laser altimetry over the Greenland ice-sheet. *Int. J. Rem. Sens.*, 16, 1211–1222.

Lefsky, M.A., Cohen, W.B., Parker, G.G., and Harding, D.J., 2002. Lidar remote sensing for ecosystem studies. *Bioscience*, 52, 19–30.

Markus, T. et al., 2017. The ice, cloud, and land elevation satellite-2 (ICESat-2): Science requirements, concept, and implementation. *Rem. Sens. Environ.*, 190, 260–273. doi:10.1016/j.rse.2016.12.029.

Nayegandhi, A., Brock, J.C., Wright, C.W., and O'Connell, M.J., 2006. Evaluating a small footprint, waveform-resolving lidar over coastal vegetation communities. *Photogramm. Eng. Rem. Sens.*, 72, 1407–1417.

Nelson, R.M., Hapke, B.W., Smythe, W.D., and Horn, L.J., 1998. Phase curves of selected particulate materials: The contribution of coherent backscattering to the opposition surge. *Icarus*, 131, 223–230.

Nelson, R.M., Hapke, B.W., Smythe, W.D., and Spilker, L.J., 2000. The opposition effect in simulated plan-etary regoliths. Reflectance and circular polarization ratio change at small phase angle. *Icarus*, 147, 545–558.

Neumann, G.A., Abshire, J.B., Aharonson, O., Garvin, J.B., Sun, X., and Zuber, M.T., 2003. Mars orbiter laser altimeter pulse width measurements and footprint-scale roughness. *Geophys. Res. Lett.*, 30(11), 1561. doi: 10.1029/2003GL017048.

Priedhorsky, W.C., Smith, R.C., and Ho, C., 1996. Laser ranging and mapping with a photon-counting detec-tor. *Appl. Optic.*, 35, 441–452.

Ramos-Izquierdo, L., Scott III, S., Schmidt, S., Britt, J., Mamakos, W., Trunzo, R., Cavanaugh, J., and Miller, R., 2005. Optical system design and integration of the mercury laser altimeter. *Appl. Optic.*, 44, 1748–1760.

Riris, H., Sun, X., Cavanaugh, J.F., Jackson, G.B., Ramos-Izquierdo, L., Smith, D.E., and Zuber, M., 2007. The lunar orbiter laser altimeter (LOLA) on NASA's lunar reconnaissance orbiter (LRO) mission. In R.T. Howard and D. Richards (Eds.) *Sensors and Systems for Space Applications, Proc. SPIE*, 6555, 1–8.

Sallenger, A.H. et al., 2003. Evaluation of airborne topographic lidar for quantifying beach changes. *J. Coast. Res.*, 19, 125–133.

Schutz, B.E., Zwally, H.J., Shuman, C.A., Hancock, D., and DiMarzio, J.P., 2005. Overview of the ICESat mission. *Geophys. Res. Lett.*, 32, L21S01. doi:10.1029/2005GL024009.

Shkuratov, Y.G. and Helfenstein, P., 2001. The opposition effect and the quasi-fractal structure of regolith: I. Theory. *Icarus*, 152, 96–116.

Slatton, K.C., Carter, W.E., and Shrestha, R., 2005. A simulator for airborne laser swath Mapping via pho-ton counting. In R.S. Harmon, J.T. Broach, and J.H. Holloway Jr., (Eds.) *Detection and Remediation Technologies for Mines and Mine-Like Targets, Proc. SPIE*, 5794, 12–20.

Smith, D.E. et al., 1999. The global topography of Mars and implications for surface evolution. *Science*, 284, 1495–1503.

Smith, D.E. et al., 2001. Mars Orbiter Laser Altimeter: Experiment summary after the first year of global mapping of Mars. *J. Geophys. Res. Plan.*, 106, 23689–23722.

Spinhirne, J.D., Palm, S.P., Hart, W.D., Hlavka, D.L., and Welton, E.J., 2005. Cloud and aerosol measurements from GLAS: Overview and initial results. *Geophys. Res. Lett.*, 32, L22S03. doi:10.1029/2005GL023507.

Sun, X., Abshire, J.B., McGarry, J.F., Neumann, G.A., Smith, J.C., Cavanaugh, J.F., Harding, D.J., Zwally, H.J., Smith, D.E., and Zuber, M.T., 2013. Space lidar developed at the NASA Goddard Space Flight Center—The first 20 years. *IEEE J. Sel. Topics Appl. Earth Observ. Rem. Sens.*, 6(3), 1660–1675. doi:10.1109/jstars.2013.2259578.

Sun, X., Krainak, M.A., Hasselbrack, W.B., and La Rue, R.A., 2007. Photon counting performance measure-ments of transfer electron InGaAsP photocathode hybrid photomultiplier tubes at 1064 nm wavelength. In I. Prochazka, A.L. Migdall, A. Pauchard, M. Dusek, M.S. Hillery, and W. Schleich (Eds.) *Photon Counting Applications, Quantum Optics, and Quantum Cryptography, Proc. SPIE*, 6583, 1–14.

Sun, X., Krainak, M.A., Abshire, J.B., Spinhirne, J.D., Trottier, C., Davies, M., Dautet, H., Allan, G.R., Lukemire, A.T., and Vandiver, J.C., 2004. Space-qualified silicon avalanche-photodiode single-photon-counting modules. *J. Mod. Optic.*, 51, 1333–1350.

Tan, S.X. and Narayanan, R.M., 2004. Design and performance of a multiwavelength airborne polarimetric lidar for vegetation remote sensing. *Appl. Optic.*, 43, 2360–2368.

Tan, S.X., Narayanan, R.M., and Shetty, S.K., 2005. Polarized lidar reflectance measurements of vegetation at near-infrared and green wavelengths. *Int. J. Infrared and Millimet. Waves*, 26, 1175–1194.

Thomas, R., Frederick, E., Krabill, W., Manizade, S., and Martin, C., 2006. Progressive increase in ice loss from Greenland. *Geophys. Res. Lett.*, 33, L10503. doi:10.1029/2006GL026075.

Thomas, R. et al., 2004. Accelerated sea-level rise from West Antarctica. *Science*, 306, 255–258.

Vaughn, C.R., Bufton, J.L., Krabill, W.B., and Rabine, D., 1996. Georeferencing of airborne laser altimeter measurements. *Int. J. Rem. Sens.*, 17, 2185–2200.

Weishampel, J.F., Blair, J.B., Knox, R.G., Dubayah, R., and Clark, D.B., 2000. Volumetric lidar return patterns from an old-growth tropical rainforest canopy. *Int. J. Rem. Sens.*, 21, 409–415.

Yu, A.W., Harding, D.J., and Dabney, P., 2016. Laser transmitter design and performance for the slope imaging multi-polarization photon-counting lidar (SIMPL) instrument. *Solid State Lasers XXV: Technology and Devices, Proceedings of SPIE 9726*, San Francisco, CA. doi:10.1117/12.2213005.

Zuber, M.T. et al., 2000. Internal structure and early thermal evolution of Mars from Mars global surveyor topography and gravity. *Science*, 287, 1788–1793.

Zwally, H.J. et al., 2002. ICESat's laser measurements of polar ice, atmosphere, ocean, and land. *J. Geodyn.*, 34, 405–445.

6 Georeferencing Component of LiDAR Systems

Naser El-Sheimy

CONTENTS

Airborne laser scanning (ALS) is becoming a popular choice to determine surface elevation models. ALS systems integrate several surveying technologies. Although there are differences between commercial systems, the basic package remains the same: a global positioning system (GPS) receiver and an inertial navigation system (INS) as the georeferencing component, a laser range finder, and a scanner as the remote-sensing component. The current chapter will mainly discuss the georeferencing component of an ALS.

6.1 INTRODUCTION

Aerial remote sensing, more specifically aerial photogrammetry, in its classical form of film-based optical sensors (analog) has been widely used for high-accuracy mapping applications at all scales. Ground control points (GCPs) were the only required source of information for providing the georeferencing parameters and suppressing undesirable error propagation. In general, the necessity for GCPs was so evident that all operational methods relied on them. Even with the major changes in photogrammetry from analog to analytical and then to the digital mode of operation, it was taken for granted that GCPs were the only source for providing reliable georeferencing information. External georeferencing information was therefore considered as auxiliary data, indicating that it was only useful in minimizing the number of GCPs. The drawback of indirect georeferencing is the cost associated with the establishment of the GCPs. This usually represents a large portion of the overall project budget. In some cases, this cost can be prohibitive; especially when imagery is to be acquired and georeferenced in remote areas such as those found in many developing countries (for more details, see Schwarz et al., 1993).

The use of auxiliary position and navigation sensor data in the georeferencing process has been extensively studied for several decades. The output of these sensors is used to determine the six parameters of exterior orientation, either completely or partially, and thus to eliminate the need for a dense GCP network. These sensors include airborne radar profile recorders, gyros, horizon cameras, statoscope, Hiran, and Shiran. However, the use of these auxiliary data was intended only to support the georeferencing process by reducing the number of GCPs. The accuracy achieved with most of these auxiliary data was limited, so, during the last two decades, the use of such auxiliary data in the georeferencing process has almost disappeared completely from photogrammetry, except for the use of statoscope (for more details, see Ackermann, 1995).

This situation changed fundamentally when GPS locations of the aerial camera at the instant of exposure were included in the block adjustment of aerial triangulation. In principle, the use of GPS data made block triangulation entirely independent of GCPs. For the first time in the history of photogrammetry, the georeferencing process became autonomous, as GCPs were no longer necessarily required (Ackermann, 1995). However, this is only true in the case of the block triangulation scheme with overlapping images. Other sensors cannot be fully georeferenced by GPS alone; examples of these sensors are pushbroom digital scanners, Light Detection and Ranging (LiDAR) systems, and imaging radar systems, which are important in kinematic mapping applications.

Georeferencing of LiDAR data, for example, requires the instantaneous position and attitude information of each range measurement. GCPs alone are not sufficient to resolve all the three position parameters and the three orientations parameters associated with each range measurement, of which there may be thousands in a single image. The data rate of most available GPS receivers is not high enough to support the data rate required for LiDAR systems. Also, for LiDAR systems, the use of GPS in stand-alone mode is not feasible, as no support by block formation is easily realized.

Only recently has direct georeferencing become possible by integrating GPS and INS (termed GPS/INS) such that all the exterior orientation information has become available with sufficient accuracy at any instant of time (Schwarz et al., 1993). The integration of GPS and INS not only puts the georeferencing of photogrammetric data on a new level and frees it from operational restrictions, but also opens the door for new systems such as LiDAR, which would not have been possible without GPS/INS. Together with digital data recording and data processing, it allows the introduction of LiDAR mapping systems.

6.2 KINEMATIC MODELING—THE CORE OF DIRECT GEOREFERENCING

Kinematic modeling is the determination of a rigid body's trajectory from measurements relative to some reference coordinate frame. It combines elements of modeling, estimation, and interpolation. Modeling relates the observable positions to the abstract trajectory. Estimation uses actual observations, that is, it adds an error process to the model and solves the resulting estimation problem in some optimal sense. Interpolation connects the discrete points resulting from the estimation process and generates a trajectory by formulating some appropriate smoothness condition.

A rigid body is a body with finite dimensions, which maintains the property that the relative positions of all its points, defined in a coordinate frame within the body, remain the same under rotation and translation (Goldstein, 1980). The general motion of a rigid body in space can be described by six parameters. They are typically chosen as three position and three orientation parameters. The modeling of rigid-body motion in 3D space can be described by the following equation:

$$r_i^m = r_b^m(t) + R_b^m(t)a^b \tag{6.1}$$

where:
r_i^m are the coordinates of point (i) in the mapping frame (m-frame)
$r_b^m(t)$ are the coordinates of the center of mass (b) of the rigid body in the m-frame at time (t)
$R_b^m(t)$ is the rotation matrix between the body frame (b-frame) and the m-frame at time (t)
a^b is the fixed distance between point (i) and the center of mass of the rigid body

The right-hand side of Equation 6.1 consists of a translation vector $r_b^m(t)$ and a rotational component $(R_b^m(t)a^b)$. The vector a^b can be any vector fixed in the rigid body with its origin at the center of mass of the rigid body. Its rotation is equivalent to rotation about the center of mass of the rigid body.

Figure 6.1 illustrates the basic concept. The coordinate b-frame is fixed to the body and rotates in time with respect to the coordinate m-frame in which the translation vector r_i^m is expressed. The m-frame

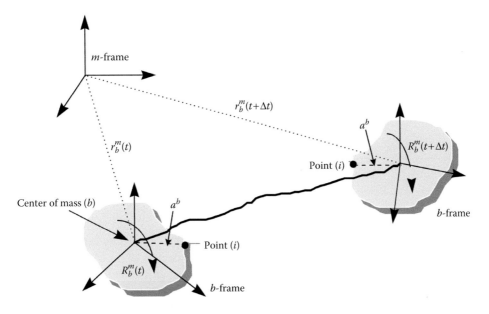

FIGURE 6.1 Modeling rigid-body motion in space.

is, in principle, arbitrary and can be chosen to simplify the problem formulation. The m-frame can be a system of curvilinear geodetic coordinates (latitude, longitude, and height), universal transverse mercator or 3° transverse mercator coordinate systems, or any other earth-fixed coordinate system.

Determining the position and orientation of rigid bodies in 3D space is, in principle, a problem of trajectory determination, which requires measuring systems with the capability to sense six independent quantities from which these parameters can be derived. Most notable among them are strapdown inertial measuring units (IMUs) in an INS and differential GPS (DGPS) positioning and orientation systems.

In general, the INS consists of three gyroscopes and three accelerometers. Gyroscopes are used to sense angular velocity ω_{ib}^b (which can be written in its skew symmetric matrix form as Ω_{ib}^b), which describes the rotation of the b-frame with respect to the inertial frame (i-frame), coordinated in the b-frame. The i-frame is a properly defined inertial reference frame in the Newtonian sense and, thus, can be considered as being nonaccelerating and nonrotating. Accelerometers are used to sense specific force f^b in the b-frame. The first set of measurements, the angular velocities ω_{ib}^b, is integrated with respect to time to provide orientation changes of the body relative to its initial orientation. The second dataset, the specific force measurements f^b, is used to derive body acceleration that, after double integration with respect to time, gives position differences relative to an initial position. Specific force and angular velocity can be used to determine all navigation parameters (r^m, v^m, R_b^m) required for trajectory determination by solving the following system of differential equations (see El-Sheimy, 2000):

$$\begin{pmatrix} \dot{r}^m \\ \dot{v}^m \\ \dot{R}_b^m \end{pmatrix} = \begin{pmatrix} D^{-1}v^m \\ R_b^m f^b - (2\Omega_{ie}^m + \Omega_{em}^m)v^m + g^m \\ R_b^m(\Omega_{ib}^b - \Omega_{im}^b) \end{pmatrix} \tag{6.2}$$

To solve the system, the observables f^b and ω_{ib}^b are needed as well as the scaling matrix D^{-1}, the gravity vector g^m, the Earth rotation rate ω_{ie}^m, and the dimensions of the implied reference ellipsoid. The gravity vector is normally approximated by the normal gravity field, whereas the Earth's rotation is assumed to be known with sufficient accuracy. The scaling matrix D^{-1} is obtained in the integration process using the implied reference ellipsoid.

GPS, on the other hand, is used for trajectory determination. The system outputs in this case are ranges and range rates between the satellites and receiver, derived from carrier phase data. The models that relate the position and velocity with the measurements are well known. In GPS stand-alone mode, a multiantenna system can be used to provide both position and attitude. The feasibility of attitude determination using multiantenna systems has been shown for applications not requiring the highest accuracy (see Cohen and Parkinson, 1992; El-Mowafy and Schwarz, 1994). Similar to the INS model, the GPS trajectory equation can be written in state vector form:

$$\begin{pmatrix} \dot{r}^m \\ \dot{v}^m \\ \dot{R}_b^m \end{pmatrix} = \begin{pmatrix} v^m \\ 0 \\ R_b^m \Omega_{mb}^b \end{pmatrix} \tag{6.3}$$

In Equation 6.3, the angular velocities in the body frame are obtained by differencing between antennas, satellites, and epochs. Note that the translation parameters of the trajectory are obtained by differencing between the master station receiver and the rover receiver, whereas the rotational parameters are obtained by differencing between the rover receivers only.

Both INS and GPS are, in principle, capable of determining position and attitude of the rigid body. In practice, due to the double integration of the INS acceleration data, the time-dependent position errors will quickly exceed the accuracy specifications for many trajectory determination applications. Frequent updating is, therefore, needed to achieve the required accuracies. GPS, on the other hand, can deliver excellent position accuracy, but has the problem of cycle slips, which are, in essence, gross errors leading to a discontinuity in the trajectory. The combination of the two measuring systems, therefore, offers a number of advantages. In the absence of cycle slips, the excellent positioning accuracy of differential GPS can be used to provide frequent updates for the inertial system. The inertial sensors orientation information and the precise short-term position and velocity can be used for cycle slip detection and correction. In general, the fact that nine independent measurements are available for the determination of the six required trajectory parameters greatly enhances the reliability of the system. To optimally combine the redundant information, a Kalman filtering scheme is used, whereby the inertial state vector is regularly updated by GPS measurement.

6.3 DEVELOPMENT OF DIRECT GEOREFERENCING TECHNOLOGY

Direct georeferencing is the determination of time-variable position and orientation parameters for a mobile digital imager. The most common technologies used for this purpose today are satellite positioning by GPS and inertial navigation using an IMU. Although each technology can in principle determine both position and orientation, they are usually integrated in such a way that the GPS receiver is the main position sensor, whereas the IMU is the main orientation sensor. The orientation accuracy of an IMU is largely determined by the gyro drift rates, typically described by a bias (constant drift rate), the short-term bias stability, and the angle random walk. Typically, four classes of gyros are distinguished according to their constant drift rate:

1. Strategic gyros (0.0005–0.0010 deg./h or about 1 deg./month)
2. Navigation-grade gyros (0.002–0.015 deg./h or about 1 deg./week)
3. Tactical gyros (0.1–10 deg./h or about 1 deg./h)
4. Tow-accuracy gyros (100–10,000 deg./h or about 1 deg./s)

Only navigation and tactical grade gyros have been implemented in the georeferencing components of LiDAR systems. Operational testing of direct georeferencing started in the early 1990s (see, for instance, Cannon and Schwarz, 1990; Cosandier et al., 1993; Bossler, 1996; Toth, 1997 for airborne applications, and Bossler et al., 1993; Lapucha et al. 1990 for land-vehicle applications). These early experiments were done by integrating differential GPS with a navigation-grade IMU (accelerometer bias: 2–3 × 10⁻⁴ m/s², gyro bias: 0.003 deg./h) and by including the derived coordinates and attitude (pitch, roll, and azimuth) into a photogrammetric block adjustment. Although GPS was not fully operational at that time, results obtained by using GPS in differential kinematic mode were promising enough to pursue this development. As GPS became fully operational, the INS/DGPS georeferencing system was integrated with a number of different imaging sensors. Among them were the Casi sensor manufactured by Itres Research Ltd. (Cosandier et al., 1993) and a set of digital cameras (see El-Sheimy and Schwarz, 1993). By the end of 1993, experimental systems for mobile mapping existed for both airborne and land vehicles. The evolution of the georeferencing technology during the past decade was driven by ongoing refinement and miniaturization of GPS-receiver hardware and the use of low- and medium-cost IMUs that became available in the mid-1990s. Only the latter development will be discussed here.

The inertial systems used in INS/GPS integration in the early 1990s were predominantly navigation-grade systems, typically strapdown systems of the ring-laser type. When integrated with DGPS, they provided position and attitude accuracies sufficient for all accuracy classes envisaged at that time. These systems came, however, with a considerable price tag (about US$130,000 at that time). With the rapidly falling cost of GPS-receiver technology, the INS became the most expensive component of the georeferencing system. As navigation-grade accuracy was not required for the bulk of the low- and medium-accuracy applications, the emergence of low-cost IMU in the mid-1990s provided a solution to this problem. These systems came as an assembly of solid-state inertial sensors with analog readouts and a postcompensation accuracy of about 10 deg./h for gyro drifts and about 10⁻² m/s² for accelerometer biases. Prices ranged between US$10,000 and 20,000, and the user had to add the A/D portion and the navigation software. Systems of this kind were obviously not suited as stand-alone navigation systems because of their rapid position error accumulation. However, when provided with high-rate position and velocity updates from differential GPS (1 s pseudorange solutions), the error growth could be kept in bounds and the position and attitude results from the integrated solution were suitable for low- and medium-accuracy applications (for details on system design and performance, see Bäumker and Mattissek, 1992; Lipman, 1992; Bader, 1993, among others).

With the rapid improvement of fiber-optic gyro performance, the sensor accuracy of a number of these systems has improved by about an order of magnitude (to 1 deg./h and 10⁻¹ m/s²) in the past 5 years. Typical costs are about US$30,000. Besides the increased accuracy, these systems are more user-friendly and offer a number of interesting options. When integrated with a DGPS phase solution, the resulting position and attitude are close to what is required for the high-accuracy class of applications. When aiming at highest possible accuracy, these systems are usually equipped with a dual-antenna GPS, aligned with the forward direction of the vehicle. This arrangement provides regular azimuth updates to the integrated solution and bounds the azimuth drift. This is of particular importance for flights flown at constant velocity along straight lines, as is the case for photogrammetric blocks. Commercialization of direct georeferencing systems for all application areas has been done by the Applanix Corporation (now a subsidiary of Trimble, see www.applanix.com) and IGI mbH (Ingenieur-Gesellschaft fuer Interfaces (see http://www.igi-systems.com/index.htm). In general, the position and orientation accuracy achieved with these systems is sufficient for all but the most stringent accuracy requirements.

6.4 INS EQUATIONS OF MOTION

Inertial positioning is based on the simple principle that differences in position can be determined by a double integration of acceleration, sensed as a function of time, in a well-defined and stable coordinate frame, that is, by

$$\Delta r(t) = r(t) - r(t_o) = \int_{t_o}^{t}\int a(\tau)\,d\tau\,d\tau \qquad (6.4)$$

where:

$r(t_o)$ is the initial point of the trajectory

$a(\tau)$ is the acceleration along the trajectory obtained from inertial sensor measurements in the coordinate frame prescribed by $r(t)$

Note that the practical implementation of the concept is rather complex. It requires the transformation between a stable Earth-fixed coordinate frame, used for the integration, and the measurement frame defined by the sensitive axes of the IMU. The stable Earth-fixed coordinate frame is often chosen as a local-level frame (ℓ) (ℓ-frame—the z-axis is normal to the reference ellipsoid pointing upwards, the y-axis pointing toward geodetic north). The x-axis completes a right-handed system by pointing east and can be either established mechanically inside the IMU (stable platform concept) or numerically (strapdown concept). In the following, only the strapdown concept will be treated. As inertial sensor measurements are always made with respect to an inertial reference frame, the rotational dynamics between the body frame of the IMU, the inertial reference frame, and different Earth-fixed frames is essential for deriving the acceleration $a(\tau)$ used in Equation 6.4—this is typically achieved through integration of the gyro measurements.

The raw measurements from accelerometers and gyros are specific forces and angular velocities along and about the three axes of the body frame (b-frame), respectively. The navigation frame is that in which the data integration is performed. The local-level frame is often selected as the navigation frame for the following reason (El-Sheimy, 2000):

1. The definition of the local-level frame is based on the normal to the reference ellipsoid; as a result, the geodetic coordinate difference $\{\Delta\varphi, \Delta\lambda, \Delta h\}$ can be applied as the output of the system.
2. The axes of the local-level frame North-East-Down (NED) are aligned with the local north, east, and down directions. Therefore, the attitude angles (roll, pitch, and azimuth) can be obtained directly as an output of the mechanization equations.

According to Newton's second law of motion, the fundamental equation for the motion of a particle in the field of the earth, expressed in an inertial frame, is of the form

$$\ddot{r}^i = f^i + \bar{g}^i \tag{6.5}$$

where:

$\ddot{r}^i$ is the acceleration vector

f^i is the specific force vector

$\bar{g}^i$ is the gravitational vector

Equation 6.5 of motion can be transformed into local-level frame and can be expressed as a set of first-order differential equation (Shin, 2001).

$$\begin{bmatrix} \dot{r}^\ell \\ \dot{v}^\ell \\ \dot{C}_b^\ell \end{bmatrix} = \begin{bmatrix} D^{-1}v^\ell \\ C_b^\ell f^b - (2\omega_{ie}^\ell + \omega_{el}^\ell) \times v^\ell + g^\ell \\ C_b^\ell (\Omega_{ib}^b - \Omega_{i\ell}^b) \end{bmatrix} \tag{6.6}$$

$$D^{-1} = \begin{bmatrix} \dfrac{1}{M+h} & 0 & 0 \\ 0 & \dfrac{1}{(N+h)\cos\varphi} & 0 \\ 0 & 0 & -1 \end{bmatrix} \tag{6.7}$$

The specific force f^b is the raw output measured by the accelerometer and is defined as the difference between the true acceleration in space and the gravitational acceleration. The rotation matrix from b-frame to ℓ-frame, C_b^ℓ, is given as

$$C_\ell^b = R_x(\phi)R_y(\theta)R_z(\psi) \tag{6.8}$$

where ϕ, θ, and ψ are the three components of the Euler rotation angles roll, pitch, and azimuth, respectively, between the ℓ-frame and the b-frame. Similarly, the rotation matrix from the b-frame to the ℓ-frame can be obtained via the orthogonality criteria of direction cosine matrices (DCM):

$$C_b^\ell = (C_\ell^b)^{-1} = (C_\ell^b)^T = R_z(-\psi)R_y(-\theta)R_z(-\phi)$$

$$= \begin{bmatrix} \cos\psi & -\sin\psi & 0 \\ \sin\psi & \cos\psi & 0 \\ 0 & 0 & 1 \end{bmatrix} \begin{bmatrix} \cos\theta & 0 & \sin\theta \\ 0 & 1 & 0 \\ -\sin\theta & 0 & \cos\theta \end{bmatrix} \begin{bmatrix} 1 & 0 & 0 \\ 0 & \cos\phi & -\sin\phi \\ 0 & \sin\phi & \cos\phi \end{bmatrix} \tag{6.9}$$

$$= \begin{bmatrix} \cos\theta\cos\psi & -\cos\phi\sin\psi + \sin\phi\sin\theta\cos\psi & \sin\phi\sin\psi + \cos\phi\sin\theta\cos\psi \\ \cos\theta\sin\psi & \cos\phi\cos\psi + \sin\phi\sin\theta\sin\psi & -\sin\phi\cos\psi + \cos\phi\sin\theta\sin\psi \\ -\sin\theta & \sin\phi\cos\theta & \cos\phi\cos\theta \end{bmatrix}$$

M and N are radii of curvature in the meridian and prime vertical, respectively, and can be expressed as follows:

$$N = \frac{a}{(1-e^2\sin^2\varphi)^{\frac{1}{2}}} \tag{6.10}$$

$$M = \frac{a(1-e^2)}{(1-e^2\sin^2\varphi)^{\frac{3}{2}}} \tag{6.11}$$

where a and e are the semimajor axis and linear eccentricity of the reference ellipsoid, respectively.

The position vector in the ℓ-frame is given by curvilinear coordinates that contain latitude, φ, longitude, λ, and ellipsoidal height, h:

$$r^\ell = \begin{bmatrix} \varphi & \lambda & h \end{bmatrix}^T \tag{6.12}$$

The velocity vector in the ℓ-frame is given as follows:

$$v^\ell = \begin{bmatrix} v_N \\ v_E \\ v_D \end{bmatrix} = \begin{bmatrix} (M+h) & 0 & 0 \\ 0 & (N+h)\cos\varphi & 0 \\ 0 & 0 & -1 \end{bmatrix} \begin{bmatrix} \dot{\varphi} \\ \dot{\lambda} \\ \dot{h} \end{bmatrix} \tag{6.13}$$

where v_N, v_E, and v_D are north, east, and downward velocity components.

The gravity vector in the local-level frame, g^ℓ, is expressed as the normal gravity at the geodetic latitude φ, and ellipsoidal height h (El-Sheimy, 2000)

$$g^\ell = [0 \ 0 \ g]^T, \quad g = a_1(1+a_2\sin^2\varphi + a_3\sin^4\varphi) + (a_4 + a_5\sin^2\varphi)h + a_6h^2 \tag{6.14}$$

where $a_1 - a_6$ are constant values and are listed in Table 6.1.

TABLE 6.1

Constant Coefficients for Normal Gravity

a_1 (m/s²)	9.7803267715	a_4 (m/s²)	−0.0000030876910891
a_2 (m/s²)	0.0052790414	a_5 (m/s²)	0.0000000043977311
a_3 (m/s²)	0.0000232718	a_6 (m/s²)	0.0000000000007211

The rotation rate vector of the *e*-frame with respect to the *i*-frame projected to the *e*-frame is expressed as follows:

$$\omega_{ie}^e = \begin{bmatrix} 0 \\ 0 \\ \omega_e \end{bmatrix} \tag{6.15}$$

Projecting the vector to the ℓ-frame utilizing is given as

$$\omega_{ie}^\ell = C_e^\ell \omega_{ie}^e = \begin{bmatrix} \omega_e \cos\varphi \\ 0 \\ -\omega_e \sin\varphi \end{bmatrix} \tag{6.16}$$

The transport rate represents the turn rate of the ℓ-frame with respect to the *e*-frame and is given using the rate of change of latitude and longitude which are given as follows:

$$\omega_{el}^\ell = \begin{bmatrix} \dot\lambda \cos\varphi \\ -\dot\varphi \\ -\dot\lambda \sin\varphi \end{bmatrix} = \begin{bmatrix} v_E/(N+h) \\ -v_N/(M+h) \\ -v_E \tan\varphi/(N+h) \end{bmatrix} \tag{6.17}$$

Ω_i^ℓ and Ω_{el}^ℓ are skew symmetric matrices corresponding to ω_{ie}^ℓ and $\omega_{e\ell}^\ell$, respectively.

The angular velocity, ω_{ib}^b, is the raw output measured by the gyros and Ω_{ib}^b is its skew-symmetric matrix.

$$\omega_{ib}^b = \begin{bmatrix} \omega_x & \omega_y & \omega_z \end{bmatrix}^T \tag{6.18}$$

The angular velocity $\Omega_{i\ell}^b$ is subtracted from Ω_{ib}^b to remove (1) earth rotation rate and (2) orientation change of the local-level frame. As a result, $\Omega_{i\ell}^b$ is expressed as follows:

$$\Omega_{i\ell}^b = \Omega_{ie}^b + \Omega_{e\ell}^b \tag{6.19}$$

Thus, $\omega_{i\ell}^b$ can be obtained as follows:

$$\begin{aligned} \omega_{i\ell}^b &= C_\ell^b(\omega_{ie}^\ell + \omega_{e\ell}^\ell) = C_\ell^b \omega_{i\ell}^\ell \\ &= C_\ell^b \left[\omega_e \cos\varphi + \frac{v_E}{(N+h)} \quad \frac{-v_N}{(M+h)} \quad -\omega_e \sin\varphi - \frac{v_E \tan\varphi}{(N+h)} \right]^T \end{aligned} \tag{6.20}$$

Consequently, the $\Omega_{i\ell}^b$ can be obtained through $\omega_{i\ell}^b$.

6.5 INS MECHANIZATION EQUATIONS

Solving the vector differential Equation 6.6 will result in a time-variable state vector with kinematic subvectors for position, velocity, and attitude. In the literature, the integration algorithms are often called the mechanization equations. This term dates back to the time when stable platform systems were the norm, and a specific platform orientation was maintained using mechanical systems, as for instance, in the local-level system or the space-stable system. The accelerometers mounted on the platform were isolated from rotational vehicle dynamics, and the computation of the transformation matrix was replaced by a system of gimbals, supplemented by platform commands. Although the algorithms used for strapdown inertial systems can be considered as an analytical form of the platform mechanization, we will use the term integration equations in the following to distinguish between stable platform and strapdown systems.

The integration equations are applied to solve the equations of motion to obtain the necessary position, velocity, and attitude increment. In practice, strapdown IMUs work in discrete form and provide angle and velocity increments, Δv_f and $\Delta\theta_{ib}^b$, respectively, over time interval t_k to t_{k+1} in the body frame. Combining these data with the initial condition of the system, it is possible to provide the navigation information. Figure 6.2 shows the algorithmic flowchart of the integration in the local-level frame.

The integration equations consist of three basic steps:

1. Sensors error compensation: The accelerometer and gyros outputs are corrected for the constant sensor errors utilizing the following equations:

$$\Delta v_f = \begin{bmatrix} \dfrac{1}{(1+S_{gx})} & 0 & 0 \\[2ex] 0 & \dfrac{1}{(1+S_{gy})} & 0 \\[2ex] 0 & 0 & \dfrac{1}{(1+S_{gz})} \end{bmatrix} (\Delta\tilde{v}_f - b_g\Delta t) \tag{6.21}$$

$$\Delta\theta_{ib}^b = \Delta\tilde{\theta}_{ib}^b - b_w\Delta t \tag{6.22}$$

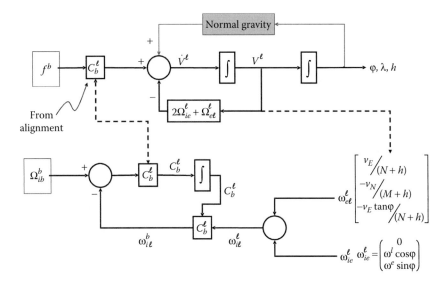

FIGURE 6.2 Flowchart of the integration process in the ℓ-frame.

where:

S_{gx}, S_{gy}, and S_{gz} are the scale factors of the accelerometer

b_g and b_w are the bias of the accelerometer and gyro, respectively

$\Delta \tilde{v}_f$ and $\Delta \tilde{\theta}_{ib}^b$ are the raw output of accelerometers and gyros, respectively

Δv_f and $\Delta \theta_{ib}^b$ are compensated output of accelerometers and gyros, respectively

$\Delta t = t_{k+1} - t_k$ is the time interval between two consecutive computation cycles t_k and t_{k+1}

2. Attitude integration: The transformation matrix C_b^ℓ can be updated by solving the following set of differential equations:

$$\dot{C}_b^\ell = C_b^\ell (\Omega_{ib}^b - \Omega_{i\ell}^e) \tag{6.23}$$

The body-frame angular increment with respect to navigation frame (ℓ-frame) is given

$$\begin{aligned} \Delta \theta_{\ell b}^b &= \begin{bmatrix} \Delta \theta_x & \Delta \theta_y & \Delta \theta_z \end{bmatrix} \\ &= \Delta \theta_{ib}^b - C_\ell^b (\omega_{ie}^\ell - \Omega_{e\ell}^\ell) \Delta t \end{aligned} \tag{6.24}$$

The magnitude of the angular increment is given

$$\Delta \theta = \sqrt{\Delta \theta_x^2 + \Delta \theta_y^2 + \Delta \theta_z^2} \tag{6.25}$$

Due to its computational efficiency, quaternion integration is again chosen for the transformation matrix update. Equations 6.24 and 6.25 are applied to update the quaternion (see El-Sheimy, 2000 for the detailed definition and properties of quaternions).

$$\begin{bmatrix} q_1(t_{k+1}) \\ q_2(t_{k+1}) \\ q_3(t_{k+1}) \\ q_4(t_{k+1}) \end{bmatrix} = \begin{bmatrix} q_1(t_k) \\ q_2(t_k) \\ q_3(t_k) \\ q_4(t_k) \end{bmatrix} + 0.5 \begin{bmatrix} c & s\Delta \theta_z & -s\Delta \theta_y & s\Delta \theta_x \\ -s\Delta \theta_z & c & s\Delta \theta_x & s\Delta \theta_y \\ s\Delta \theta_y & -s\Delta \theta_x & c & s\Delta \theta_z \\ -s\Delta \theta_x & -s\Delta \theta_y & -s\Delta \theta_z & c \end{bmatrix} \begin{bmatrix} q_1(t_k) \\ q_2(t_k) \\ q_3(t_k) \\ q_4(t_k) \end{bmatrix} \tag{6.26}$$

The parameters s and c are given as follows:

$$s = 1 - \frac{\Delta \theta^2}{24} + \frac{\Delta \theta^4}{1920} + \cdots$$

$$c = -\frac{\Delta \theta^2}{4} + \frac{\Delta \theta^4}{192} + \cdots \tag{6.27}$$

The initial value of the quaternion is obtained after determining the initial DCM using Equation 6.28 with the computed initial attitudes during alignment process.

$$\begin{bmatrix} q_1 \\ q_2 \\ q_3 \\ q_4 \end{bmatrix} = \begin{bmatrix} 0.25(C_{32} - C_{23}) / 0.5\sqrt{1 + C_{11} + C_{22} + C_{33}} \\ 0.25(C_{13} - C_{31}) / 0.5\sqrt{1 + C_{11} + C_{22} + C_{33}} \\ 0.25(C_{21} - C_{12}) / 0.5\sqrt{1 + C_{11} + C_{22} + C_{33}} \\ 0.5\sqrt{1 + C_{11} + C_{22} + C_{33}} \end{bmatrix} \tag{6.28}$$

The DCM is updated as follows:

$$C_b^\ell = \begin{bmatrix} (q_1^2 - q_2^2 - q_3^2 + q_4^2) & 2(q_1 q_2 + q_3 q_4) & 2(q_1 q_3 + q_2 q_4) \\ 2(q_1 q_2 + q_3 q_4) & (q_2^2 - q_1^2 - q_3^2 + q_4^2) & 2(q_2 q_3 + q_1 q_4) \\ 2(q_1 q_3 + q_2 q_4) & 2(q_2 q_3 + q_1 q_4) & (q_3^2 - q_1^2 - q_2^2 + q_4^2) \end{bmatrix} \tag{6.29}$$

The Euler angle of the attitudes, roll, pitch, and azimuth are then given as follows:

$$\theta = -\tan^{-1}\left[\frac{C_{31}}{\sqrt{1-C_{31}^2}}\right] \tag{6.30}$$

$$\phi = a\tan 2(C_{32}, C_{33}) \tag{6.31}$$

$$\psi = a\tan 2(C_{21}, C_{11}) \tag{6.32}$$

where $C'_{i,j}s, 1 \le i, j \le 3$ are the (i,j)th element of the DCM matrix and $a\tan 2$ is a four-quadrant inverse tangent function.

3. Velocity and position integration: The body-frame velocity increment due to the specific force is transformed to the navigation frame using Equation 6.33:

$$\Delta v_f^\ell = C_b^\ell \begin{bmatrix} 1 & 0.5\Delta\theta_z & -0.5\Delta\theta_y \\ -0.5\Delta\theta_z & 1 & 0.5\Delta\theta_x \\ 0.5\Delta\theta_y & -0.5\Delta\theta_x & 1 \end{bmatrix} \Delta v_f^b \tag{6.33}$$

The first-order sculling correction is applied utilizing Equation 6.33. The velocity increment is obtained by applying the gravity and the Coriolis correction:

$$\Delta v^\ell = \Delta v_f^\ell = (2\omega_{ie}^\ell + \omega_{el}^\ell) \times v^\ell \Delta t + g^\ell \Delta t \tag{6.34}$$

The velocity integration is then given as

$$v_{k+1}^\ell = v_k^\ell + \Delta v_{k+1}^\ell \tag{6.35}$$

The position integration is obtained using the second-order Runge–Kutta method:

$$r_{k+1}^\ell = r_k^\ell + 0.5 \begin{bmatrix} \dfrac{1}{(M+h)} & 0 & 0 \\ 0 & \dfrac{1}{(N+h)\cos\varphi} & 0 \\ 0 & 0 & -1 \end{bmatrix} (v_k^\ell + v_{k+1}^\ell)\Delta t \tag{6.36}$$

6.6 GPS/INS INTEGRATION

6.6.1 Integration Strategies

The integration of GPS and INS has been investigated for several years in various applications including navigation, mobile mapping, airborne gravimetry, and guidance and control. Both systems are complimentary, and their integration overcomes their individual limitations. In GPS/INS systems, the GPS provides position and velocity, and the INS provides attitude information. In addition, the INS can provide very accurate position and velocity with a high data rate between GPS measurement fixes. Therefore, INS is used to detect and correct GPS cycle slips and also for navigation during GPS signal loss of lock. Finally, the GPS is used for the in-motion calibration of the INS accelerometer and gyro-sensor residual errors. For all INS/GPS applications, navigation information parameters are obtained using kinematic modeling. Thus, the state-space representation is implemented in the mathematical modeling of INS, GPS, and INS/GPS systems. In this context, the Kalman filter (KF) has been commonly used as an optimal estimator and compensator of the INS/GPS system errors.

Two integration strategies can be implemented at the software level using the KF approach. In the first one, a common-state vector is used to model both the INS and the GPS errors. This is often called the centralized filter or tightly coupled approach. It has been applied with good success in Knight (1997), Scherzinger and Woolven (1996), and Moafipoor et al. (2004). If double differenced carrier phases are used for GPS, the state vector is of the form:

$$x = (\varepsilon, \delta r, \delta v, d, b, \delta N)^{\mathrm{T}} \tag{6.37}$$

where:
   ε is the vector of attitude errors
   δr is the vector of position errors
   δv is the vector of velocity errors
   d is the vector gyro drift about gyro axes
   b is the vector of accelerometer bias
   δN the vector of double differenced ambiguities

All vectors, except δN, have three components; δN has $(n-1)$ component, where n is the number of satellites. INS measurements are used to determine the reference trajectory and GPS measurements to update the solution and estimate the state-vector components. Cycle slips can be easily detected in this approach, because they will show up as statistically significant differences between the predicted and the actual GPS measurements. Cycle slip correction is possible in various ways. One can either use a filtering approach as implemented in Wong (1988) or one can directly correct the cycle slip by using INS measurements. In the latter case, the ambiguity parameters do not show up in the state vector but appear as corrections to the phase measurements.

In the second approach, different filters are run simultaneously and interact only occasionally. This is often called the decentralized filter approach. For a discussion of the algorithm, see Hashemipour et al. (1988) and for the implementation, see Scherzinger (2002) and Shin (2005). It has advantages in terms of data integrity and speed, but it is more complex in terms of program management. Again assuming only INS and GPS measurements, two different state vectors are defined—one for INS and one for GPS. They are of the form

$$x_I = (\varepsilon, \delta r, \delta v, \delta d, b)^{\mathrm{T}}$$
$$x_G = (\delta r, \delta v, \delta N)^{\mathrm{T}} \tag{6.38}$$

The interaction between master filter and local filter characterizes the decentralized filtering scheme. The local KF estimates a trajectory from GPS data only. In this case, the derived velocities are used for prediction and the carrier phase measurements for updates. The master filter estimates the trajectory from INS data only. At distinct epochs, the results from the local filter are fused with those of the master filter to obtain an optimal trajectory estimation. The approach has the advantage that each data stream is treated independently, and blunders in one dataset can be detected and corrected by using results from the other. The estimates of each local filter are locally best, whereas the master filter gives a globally best estimate.

6.6.2 Filter Implementation Strategies

Although Kalman introduced his optimal estimation filter (Kalman, 1960), several KF approaches have been implemented for INS/GPS integration. In the early stages of INS and DGPS integration for high-accuracy civil navigation applications (i.e., with benign dynamics), IMUs used were navigation-grade. At that time, the linearized KF (LKF) was commonly applied. With the recent advent in inertial sensor technologies, tactical-grade IMUs and consumer grade IMUs (gyro drift 100–10,000 deg./h) are currently used in several navigation applications. Therefore, for better error control with such systems, the LKF has been replaced by the extended KF (EKF). For low-cost IMUs such as micro-electromechanical systems (MEMS), the sensors have very low accuracy, which causes large navigation errors. This is especially true regarding the initial uncertainties. Hence, the unscented KF (UKF) has been proposed lately for INS/GPS integration as it is able to cope with such large initial errors (Julier and Uhlmann, 2002).

In INS/GPS navigation, both the LKF and EKF algorithms are similar, in that the INS position, velocity, and attitude error models (i.e., the KF navigation error state vector) are obtained by linearizing the INS integration equations, whereas the INS sensor error models (i.e., the KF sensor error state vector) are represented using a stochastic process. For more details about different possible INS sensor stochastic error models, see Nassar and El-Sheimy (2005). In addition, a similar linearization procedure is performed on the observation update model to obtain the measurement update state vector. The error state vector (x) and the corresponding update equations are represented by linearized state-space models of the form:

$$x_k = \Phi_{k,k-1}x_{k-1} + w_{k-1} \tag{6.39}$$

$$z_k = H_k x_k + e_k, \tag{6.40}$$

where:

  k is a time epoch
  Φ is the system state transition matrix
  w is the vector of the system input noise
  z is the update measurements
  H is the measurements design matrix
  e is the update measurements noise

The LKF or EKF approaches can be then summarized as

$$\hat{x}_k^- = \Phi_{k-1}\hat{x}_{k-1}^+ \tag{6.41}$$

$$P_k^- = \Phi_{k-1}P_k^+\Phi_{k-1}^T + \Phi_{k-1} \tag{6.42}$$

$$\hat{x}_k^+ = \hat{x}_k^- + K_k(z_k - H_k\hat{x}_k^-) \tag{6.43}$$

$$P_k^{vv} = (H_k P_k^- H_k^T + R_k) \tag{6.44}$$

$$K_k = P_k^- H_k^T (P_k^{vv})^{-1} \tag{6.45}$$

where:

  P is the covariance matrix of the estimated state vector
  Q is the covariance matrix associated with w
  R is the covariance matrix associated with e
  P^{vv} is called the covariance of the innovation sequence
  K is the Kalman gain matrix

− and + (superscripts) are related to predicted and updated epochs, respectively.

The derivation of such equations can be found in Gelb (1974) and Brown and Hwang (1997). As mentioned earlier, the update measurements (z) are the differences between the GPS position and velocity and the corresponding INS position and velocity.

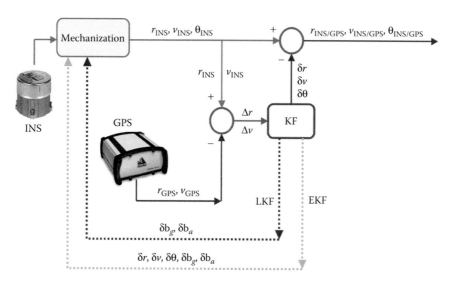

FIGURE 6.3   LKF and EKF in INS/GPS integration.

The LKF differs from the EKF in that the LKF estimated errors are compensated for using a feedforward error control. In the INS/GPS integration case, the LKF estimated navigation errors are fed to the INS navigation output only and not to the INS mechanization. On the other hand, the EKF error compensation is performed using a feedback error control in which the navigation errors are fed to both the INS navigation output and the INS mechanization. Figure 6.3 illustrates the LKF and EKF implementation in INS/GPS integration in which (r, v, θ) represent position, velocity, and attitude, whereas $(\delta b_g, \delta b_a)$ represent the gyros and accelerometers KF estimated errors. However, it should be noted here that the INS sensor error states δb_g and δb_a are fed back to the INS mechanization in both the LKF and EKF.

Due to the two different types of error control processes in of the LKF and EKF, the linearization process is different for each filter. The linearization for LKF is done about the INS mechanization trajectory (i.e., the predicted trajectory), whereas linearization is done about the updated trajectory in the EKF, and in theory, the performance of the EKF should be better than that of LKF. However, this will be only valid for high-quality update measurements otherwise the EKF might diverge. Therefore, the EKF is sometimes riskier than the LKF especially when measurement errors are large (Brown and Hwang, 1997). On the other hand, in the case of using MEMS sensors, the predicted trajectory will be of low accuracy, and thus, the EKF performance will be better than the LKF. In addition, when GPS outages occur in MEMS/GPS navigation application, the LKF is expected to have divergence problems.

6.7   LiDAR GEOREFERENCING—THE MATH MODEL

Georeferencing of LiDAR data can be defined as the problem of transforming the 3D coordinate vector r^S of the laser sensor frame (S-frame) to the 3D coordinate vector r^m of the mapping frame (m-frame) in which the results are required. The m-frame, as mentioned before, can be any Earth-fixed coordinate system such as curvilinear geodetic coordinates (latitude, longitude, and height), universal transverse mercator, or 3° transverse mercator coordinates. The major steps in this transformation are depicted in Figure 6.4 for the airborne case, whereas the carrier could be an airplane or a helicopter.

The S-frame changes position and orientation with respect to the m-frame. Georeferencing is possible if at any instant of time (t), the position of the laser sensor center in the m-frame, that is, $r_S^m (t)$,

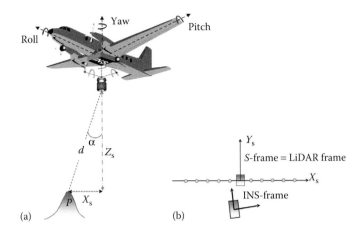

FIGURE 6.4 (a, b) LiDAR scanner angle and distance measurement.

and the rotation matrix between the S-frame and the m-frame $R_S^m(t)$ have been determined. The georeferencing equation can then be written for any object point (i) as

$$r_i^m = r_S^m(t) + R_S^m(t)r_i^S \qquad (6.46)$$

where:

r_i^m is the position vector of an object (i) in the m-frame

(t) is the measurement epoch for measuring the distance d for point (i)

r_i^S is the position vector of an object (i) in the S-frame; which is given as

$$r_i^S = \begin{pmatrix} -d\sin\alpha \\ 0 \\ -d\cos\alpha \end{pmatrix} \qquad (6.47)$$

Equation 6.46 is however, only a first approximation of the actual situation. It implies that the coordinates of the laser sensor center can be directly determined. This is usually not the case because the navigation sensors—GPS antenna/INS body center—cannot share the same location in space with the laser sensor. Thus, small translations and rotations between the different centers have to be considered. The actual situation is shown in Figure 6.4b. It has been assumed that the laser sensor is mounted in the cargo area of the airplane, that the positioning sensor—a GPS antenna is mounted on top of the airplane, and that the attitude sensor an IMU is mounted in the interior of the aircraft, somewhere close to the laser sensor. In this case, aircraft position and attitude are defined by the INS center and the internal axes of the IMU (b-frame).

If the vector between the origin of the INS body frame (b-frame) and the laser sensor center is given in the b-frame as a^b, $r_S^m(t)$ can be written as

$$r_S^m(t) = r_{INS}^m(t) + R_b^m(t)a^b \qquad (6.48)$$

where:

$r_{INS}^m(t)$ is the vector of interpolated coordinates of the INS in the m-frame at time (t) (resulting from the INS/GPS integration)

$R_b^m(t)$ is the DCM, which rotates the b-frame into the m-frame

a^b is the constant vector between the laser sensor center and the center of the INS b-frame, usually determined before the mission by calibration

In addition to transformations between sensors, rotations between different sensor frames have to be taken into account. The INS b-frame cannot be aligned with the S-frame. The constant rotation R_S^b between the two frames is again obtained by calibration. In this case, $R_S^m(t)$ can be written as

$$R_S^m(t) = R_b^m(t)R_S^b \tag{6.49}$$

where R_S^b is the rotation between the S-frame and the INS b-frame as determined from a calibration process.

Applying Equations 6.48 and 6.49 to Equation 6.46, the final georeferencing formula can be written as

$$r_i^m = r_{\text{INS}}^m(t) = R_b^m(t)(R_S^b r_i^S + a^b) \tag{6.50}$$

where R_S^b transforms the vector r_i^S from the S-frame to the b-frame, and $R_S^m(t)$ transforms the vector $(R_S^b r_i^S + a^b)$ from the b-frame to the m-frame.

It should be noted that the vector $r_{\text{INS}}^m(t)$, as well as the rotation matrix $R_S^m(t)$, are time-dependent quantities, whereas the vectors $r_{\text{INS}}^m(t)$, and a^b as well as the matrix R_S^b are not. This implies that the carrier is considered as a rigid body, rotational and translational dynamics of which are adequately described by changes in $r_{\text{INS}}^m(t)$ and $R_S^m(t)$. This means that the translational and rotational dynamics at the three-sensor locations is uniform, in other words, differential rotations and translations between the three locations as functions of time have not been modeled. It also means that the origin and orientation of the three-sensor systems can be considered fixed for the duration of the survey. These are valid assumptions in most cases but may not always be true.

In georeferencing applications, inertial systems function mainly as precise attitude systems and as short-term interpolators for velocity, position, and attitude. Because of the time dependence of all major errors, the long-term velocity and position performance is not sufficient for precise georeferencing. Thus, regular position and/or velocity updates are needed to keep the overall errors within prescribed boundaries. The accuracy of inertial systems depends heavily on the quality of the sensors used which themselves are a function of the system costs. Noise levels in attitude can vary considerably, depending on the type of gyros used. We will therefore distinguish between high-, medium-, and low-accuracy systems. High-accuracy systems typically use navigation grade IMU, whereas medium- and low-accuracy ones typically employ civilian versions of tactical military IMU. Table 6.2 lists the achievable accuracies and the error characteristics of the sensors used in

TABLE 6.2

Characteristics and Achievable Accuracies of Inertial Systems for Direct Georeferencing Applications

Performance	High Accuracy	Medium Accuracy	Low Accuracy
Inertial system	Navigation grade	High-end tactical grade IMU	Low-end tactical grade IMU
Price (US$) of fully integrated GPS/INS systems	100–200 K	75–100 K	50–75 K
Gyro drift rate (deg./h)	$\cong 0.015$	0.1–1	3–10
Accelerometer bias (mg)	50–100	150–200	200–1000
Attitude accuracy	5–20 as	0.1–1 arcmin	2–4 arcmin
Position accuracy (m)	0.05–0.3	0.05–0.3	0.05–0.3
Sensors/applications	Film-based aerial cameras and Land DMM systems	High resolution digital aerial systems and LiDAR	Medium resolution digital aerial systems

these systems. High- and medium-accuracy-grade systems can meet the accuracy requirements for LiDAR systems mounted on fixed-wing aircraft, whereas low-accuracy system can be adopted for low-altitude helicopter-based LiDAR systems.

REFERENCES

Ackermann, F. 1995. Sensor and data integration—The new challenge. In Colomina/Navaro (Eds.) *Integrated Sensor Orientation.* Heidelberg, Germany: Wichmann, pp. 2–10.

Bader, J. 1993. Low-cost GPS/INS. *Proceedings of ION GPS-93.* Salt Lake City, UT, pp. 135–244.

Bäumker, M. and Mattissek, A. 1992. Integration of a fibre optical gyro attitude and heading reference system with differential GPS. *Proceedings ION GPS-92.* Albuquerque, NM, September 16–18, 1992, pp. 1093–1101.

Bossler, J.D. 1996. Airborne integrated mapping system. *International Journal of Geomatics,* 10, 32–35.

Bossler, J.D., Goad, C., and Novak, K. 1993. Mobile mapping systems: New tools for the east collection of GIS information. *GIS'93.* Ottawa, ON, March 23–35, 1993, pp. 306–315.

Brown, R.G. and Hwang, P.Y.C. 1997. *Introduction to Random Signals and Applied Kalman Filtering.* New York: John Wiley & Sons.

Cannon, M.E. and Schwarz, K.P. 1990. A discussion of GPS/INS integration for photogrammetric applications. *Proceedings IAG Symposium # 107: Kinematic Systems in Geodesy, Surveying and Remote Sensing.* Banff, AB, September 10–13, 1990, pp. 443–452.

Cohen, C.E. and Parkinson, B.W. 1992. Aircraft applications of GPS-based attitude determination. *Proceedings of ION GPS-92.* Albuquerque, NM, September 16–18, 1992.

Cosandier, D., Chapman, M.A., and Ivanco, T. 1993. Low cost attitude systems for airborne remote sensing and photogrammetry. *Proceedings of GIS93 Conference.* Ottawa, ON, March 1993, pp. 295–303.

El-Mowafy, A. and Schwarz, K.P. 1994. Epoch by epoch attitude determination using a multi-antenna system in kinematic mode. *Proceedings of KIS-94.* Banff, AB, August 30 to September 2, 1994.

El-Sheimy, N. 2000. *Inertial Navigation and INS/GPS Integration.* Department of Geomatics Engineering, University of Calgary, Calgary, AB.

El-Sheimy, N. and Schwarz, K.P. 1993. Kinematic positioning in three dimensions using CCD technology. *Proceedings IEEE/IEE Vehicle Navigation and Information System Conference (IVHS).* October 12–15, 1993, pp. 472–475.

Gelb, A. 1974. *Applied Optimal Estimation.* Cambridge, MA: MIT Press, Massachusetts Institute of Technology.

Goldstein, H. 1980. *Classical Mechanics.* Boston, MA: Addison Wesley.

Hashemipour, H.R., Roy, S., and Laub, A.J. 1988. Decentralized structures for parallel Kalman filtering. *IEEE Transactions on Automatic Control,* AC-33, 88–94.

Julier, S.J. and Uhlmann, J.K., 2002. The scaled unscented transformation. *Proceedings of IEEE American Control Conference.* Anchorage, AK, pp. 4555–4559.

Kalman, R.E. 1960. A new approach to linear filtering and prediction problems. *ASME—Journal of Basic Engineering,* 82(series D), 35–45.

Knight, D.T. 1997. Rapid development of tightly-coupled GPS/INS systems. *Aerospace and Electronic Systems Magazine.* IEEE, 12(2), 14–18.

Lapucha, D., Schwarz, K.P., Cannon, M.E., and Martell, H. 1990. The use of GPS/INS in a Kinematic Survey System. *Proceedings of IEEE PLANS 1990.* Las Vegas, NV, March 20–23, 1990, pp. 413–420.

Lipman, J.S. 1992. Trade-offs in the implementation of integrated GPS inertial systems. *Proceedings of ION GPS-92.* Albuquerque, NM, September 16–18, 1992, pp. 1125–1133.

Moafipoor, S., Grejner-Brzezinska, D., and Toth C.K. 2004. Tightly coupled GPS/INS integration based on GPS carrier phase velocity update. *ION NTM 2004.* San Diego, CA, January 26–28, 2004.

Nassar, S. and El-Sheimy, N. 2005. Accuracy improvement of stochastic modeling of inertial sensor errors. *Zeitschrift für Geodäsie, Geoinformation und Landmanagement (ZfV) Journal, Wißner, DVW, Germany,* 130(3), 146–155.

Scherzinger, B.M. 2002. Inertially aided RTK performance evaluation. *Proceedings of ION GPS-2002.* Portland, OR, September 24–27, 2002, pp. 1429–1433.

Scherzinger, B.M. and Woolven, S. 1996. POS/MV-handling GPS outages with tightly coupled inertial/GPS integration. *OCEANS '96. MTS/IEEE. 'Prospects for the 21st Century' Conference Proceedings,* 1(23–26), 422–428.

Schwarz, K.P., Chapman, M.A., Cannon, M.E., and Gong, P. 1993. An integrated INS/GPS approach to the georeferencing of remotely sensed data. *PE&RS*, 59(11), 1667–1674.

Shin, E. 2001. Accuracy improvement of low cost INS/GPS for land application. MSc thesis, December 2001, UCGE Report No. 20156.

Shin, E. 2005. Estimation techniques for low-cost inertial navigation. PhD thesis, Department of Geomatics Engineering, University of Calgary, Calgary, AB, UCGE Report No. 20156.

Toth, C.K. 1997. Direct sensor platform orientation: Airborne integrated mapping system (AIMS). *International Archives of Photogrammetry and Remote Sensing, ISPRS Commission III*, XXXII(Part 3–2W3), 148–155.

Wong, R.V.C. 1988. Development of a RLG strapdown inertial survey system. PhD thesis, Department of Geomatics Engineering, University of Calgary, Report No. 20027.

7 Full-Waveform Analysis for Pulsed Laser Systems

Uwe Stilla and Boris Jutzi

CONTENTS

7.1 INTRODUCTION

Aerial photogrammetry and airborne laser scanning (ALS) are the two most widely used methods for generating digital elevation models (DEMs), including digital terrain models (DTMs) that depict ground topography and digital surface models (DSMs) that depict the height of the ground, structures and vegetation cover. In photogrammetry, the distance to a spatial surface is classically derived from a triangulation of corresponding image points from two or more overlapping images of the surface. These points are chosen manually or detected automatically by analyzing image structures. In contrast to photogrammetry, active laser scanner systems allow a direct and illumination-independent measurement of the distance to a surface, otherwise known as the range.

    The interrelationship between aerial photogrammetry and ALS has been intensely discussed within the aerial surveying community in the last decade. Different comparison factors concerning data acquisition (coverage, weather conditions, costs, etc.) and surface reconstruction (accuracy, redundancy, post processing time, etc.) have to be taken into account to choose the optimal method for a certain mapping campaign. An example of a study comparing photogrammetric image matching versus laser scanning for generation of high-quality digital elevation models for glacier monitoring is given in Wuerlaender et al. (2004) and Lenhart et al. (2006). In contrast to a decision to use one or the other technique, in some fields of applications a combined processing of laser data and stereo images is advantageous as shown in the generation of the extraterrestrial digital terrain models of Mars (Albertz et al., 2005; Spiegel et al., 2006) or digital surface models for building characterization.

Conventional pulsed laser scanner systems for topographic mapping are based on time-of-flight ranging techniques to determine the range to the illuminated object. The time-of-flight is measured by the elapsed time between the emitted and backscattered laser pulses. The signal analysis to determine this time typically operates with analog threshold detection. For targets having surfaces at different ranges illuminated by a single laser pulse, more than one backscattered pulse may be detected per emitted pulse. Most ALS systems are able to capture, at a minimum, the range for the first- and last-detected backscattered pulses. Some systems acquire ranges for multiple backscattered pulses, up to as many as five per emitted pulse. First, pulse detection is the optimal choice to measure the hull of partially penetrable objects or so-called volume scattering targets (e.g., canopy of trees). Last pulse detection should be chosen to measure nonpenetrable surfaces (e.g., ground surfaces).

Nowadays, some commercial ALS systems not only capture the range for multiple pulse reflections but also digitize and record the received signal of the reflected laser energy that allows for a so-called full-waveform analysis (Mallet and Bretar, 2009). This offers the possibility to analyze the waveform offline by methods of digital signal processing to extract different surface attributes from the received signal based on the shape of the return pulses.

In the last decade, some waveform analysis investigations were done to explore the structure of vegetation and estimate above ground biomass. For example, NASA developed the waveform-recording Laser Vegetation Imaging Sensor to measure vertical density profiles in forests (Blair et al., 1999). This experimental airborne system operates at altitudes up to 10 km and acquires waveforms for large diameter laser footprints (nominally 20 m) acquired across a wide swath. Another NASA system operating with a large footprint is the spaceborne Geoscience Laser Altimeter System (GLAS) mounted on the Ice, Cloud, and land Elevation Satellite. GLAS measures height distributions of atmospheric clouds and aerosols, and surface elevations of ice sheets, land topography, and vegetation (Brenner et al., 2003; Zwally et al., 2002). It is a profiling system that operates with a footprint diameter of 70 m and measures elevation changes with decimeter accuracy (Abshire et al., 2005; Schutz et al., 2005). In the analysis of data from both systems, surface characteristics are determined by comparing a parametric description of the transmitted and received waveform (Harding and Carabajal, 2005; Hofton et al., 2000). As the laser footprint is large and illuminates multiple surfaces, the resulting return waveform is an integrated, spatially nonexplicit representation of the range to illuminated surfaces separated both vertically and horizontally. The geometric organization of surfaces within a single footprint can therefore not be determined.

In contrast to large footprint systems, small footprint ALS systems illuminate only one or a few surfaces within the footprint, yielding waveforms with distinct return pulses corresponding to specific surfaces. One of the first developed small footprint waveform-recording systems is the Scanning Hydrographic Operational Airborne Lidar Survey instrument used for monitoring nearshore bathymetric environments. Scanning Hydrographic Operational Airborne Lidar Survey has been in full operation since 1994 (Irish and Lillycrop, 1999; Irish et al., 2000). More recently, commercial full-waveform ALS systems for terrestrial mapping have been developed (e.g., Jutzi and Stilla, 2006b; Wagner et al., 2006) that operate with a transmitted pulse width of 4–10 ns and allow digitization and acquisition of the waveform with approximately 0.5–1 GSample/s. Reitberger et al. (2006, 2007) have reported results that show clearly the potential of airborne, small-footprint, full-waveform data for the comprehensive analysis of tree structure and species classification. A set of key attributes have been defined and extracted on the basis of the 3D distribution of the returns in combination with their characteristics in the full-waveform signal, providing information about tree microstructure such as the organization of the trunk and branches.

In this contribution, we focus not on a given application in the context of a dataset from a given sensor but rather than on general principles. Specifically, we describe the different approaches for designing a laser system, modeling the spatial and temporal properties of the emitted lasers pulses, detecting return pulses, and deriving attributes from the waveform. We emphasize aspects of the received waveform that are especially relevant for the newly available small-footprint, full-waveform, commercial systems that yield distinct return pulses when multiple surfaces are illuminated by a laser pulse.

The design of a laser system impacts its measurement capabilities and the way the signal has to be modeled and analyzed. Section 7.2 gives a brief overview of the features that characterize the design of laser ranging systems. Section 7.3 focuses on modeling of the temporal waveform and the spatial beam distribution. Different strategies for pulse detection are explained in Section 7.4. Section 7.5 describes the attributes that can be extracted from a single laser shot and presents an analysis of an entire scene that was recorded with an experimental small-footprint, full-waveform laser ranging system. Section 7.6 addresses some applications and a summary is given in Section 7.7.

7.2 CHARACTERIZATION OF A LASER SYSTEM

Depending on the application, laser systems can be designed in different ways. They may differ in techniques concerning the type of laser used, the modulation, type of measured features, detection method, or design of beam paths (construction). Figure 7.1 sketches a simplified overview of features characterizing a laser system. More detailed descriptions can be found in Kamermann (1993).

7.2.1 LASER TYPE

A laser works as an oscillator and an amplifier for monochromatic radiation (infrared, visible light, or ultraviolet). The operative wavelength of available lasers is located between 0.1 μm und 3 mm. For comparison, it should be mentioned that the visible domain is from 0.37 to 0.75 μm. To achieve a good signal-to-noise ratio (SNR) over long ranges, conventional scanning laser systems emit radiation with high energy. However, this could be a danger for the health of humans due to the focusing of laser radiation on the retina that is most susceptible to damage at visible wavelengths. For this reason most *eye-safe* laser systems used for mapping purposes operate with a wavelength outside the visible spectrum. This allows working with an emitted energy that is many times higher

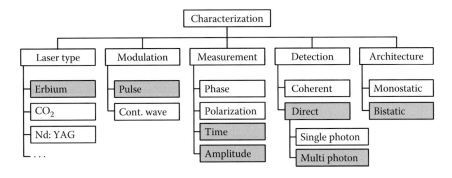

FIGURE 7.1 Features characterizing a laser system, with shaded boxes indicating the characteristics that are the focus of this chapter.

(up to 10^6) compared with the visible domain without the potential for retinal damage. *Eye-safe* lasers of greatest interest for long-range laser scanning are Erbium-fiber lasers, Carbon dioxide lasers, and neodymium:YAG (Nd:YAG) lasers.

Erbium-fiber lasers are optically pumped by a semiconductor diode and the active medium is an Erbium doped fiber. Their construction can be compact while still achieving a high output power. Carbon dioxide lasers use carbon dioxide in gas form as the active medium. Although their construction is simple, their large size and mass is a significant disadvantage. Solid-state Nd:YAG lasers can be pumped by various sources, which defines the characteristics of the emitted laser radiation. In this contribution, we focus on Erbium-fiber lasers (Figure 7.1)

7.2.2 MODULATION TECHNIQUE

Concerning modulation techniques, laser systems can be divided into two groups: continuous wave (cw) and pulsed lasers. A cw laser continuously emits electromagnetic radiation. The temporal energy distribution of the transmitted signal is influenced by amplitude modulation and/or frequency modulation. Depending on the applied modulation technique specific measurement techniques (Section 7.2.3) are required. A pulsed laser emits electromagnetic radiation in pulses of short duration. For laser ranging, it is desirable to emit a pulse as short as possible and with a pulse energy as high as possible to obtain a precise range with a high probability of detection. However, design limitations on maximum peak power introduce a tradeoff that requires a compromise between the length and the energy of the pulse. The length of the pulse (full-width-at-half-maximum, FWHM) is typically between 2 and 10 ns. For applications in remote sensing with long ranges pulsed lasers with higher power density as compared with cw lasers are advantageous. In this contribution, we focus on pulsed laser systems (Figure 7.1).

7.2.3 MEASUREMENT TECHNIQUE

Measurement techniques using laser systems can be distinguished by the exploited signal properties such as phase, amplitude, frequency, polarization, time, or any combination of them. By using an amplitude-modulated cw laser system, the range is measured by exploiting the phase of a sinusoidal modulated signal. A phase difference Φ_d can be determined from a given phase of the transmitted signal and the measured phase of the received signal. With wavelength λ_m of the modulated signal, a corresponding range r can be calculated by $r = \lambda_m \cdot \Phi_d / 4\pi$. If the measurement of the phase difference Φ_d cannot be distinguished from $\Phi_d + n \cdot \pi$, the unambiguity interval of the range measurement will be limited to a maximum range $r_{max} = \lambda_m / 2$. Assuming that the system is able to resolve an angle difference $\Delta\Phi_d$, the range resolution Δr corresponds according to $\Delta r = \lambda_m \cdot \Delta\Phi_d / 4\pi$. To increase the range resolution for a given $\Delta\Phi_d$, the modulation wavelength λ_m has to be decreased. But, this results in a reduction of the unambiguity interval of the range determination.

The problem of ambiguity can be solved by using multiple simultaneous offset sinusoidal modulation frequencies (Multiple-Tone Sinusoidal Modulation). In this case, the maximum modulation wavelength defines the unambiguity and the minimum modulation wavelength defines the range resolution. In addition to this, partially illuminated surfaces with different ranges within the beam corridor result in a superimposed signal depending on the range and the reflectance of the surface. As only a single phase value can be determined at the receiver, the ambiguities caused by the partially illuminated surfaces cannot be resolved (Thiel and Wehr, 2004). An incorrect intermediate value is measured.

The measurement of the amplitude value is feasible for cw lasers as well as for pulsed lasers. The amplitude is influenced by background radiation, the range of the object to the laser system, and the size, reflectance, slope, and roughness of the illuminated surface. In this contribution, we are interested in measuring and analyzing the received pulse waveform, that is, the dependence of the intensity over time (Figure 7.1).

7.2.4 Detection Technique

Detection techniques can be divided into coherent detection and direct detection (Jelalian, 1992). Coherent detection is based on signal amplification due to constructive interference of the wave front of the received signal with that of the reference signal emitted from a cw laser. In direct detection laser systems, the received optical energy is focused onto a photosensitive element that generates an output signal that depends on the received optical power. Two direct detection techniques are appropriate for recording the temporal characteristics of the backscattered signal: multiphoton detection and single-photon detection.

7.2.4.1 Multiphoton Detection

The classical measurement technique for direct detection operates with a photodiode. For optical detectors a PIN (positive–intrinsic–negative diode) or the more sensitive avalanche photodiode are used. The photodiode generates an electrical signal (voltage or current) that is directly proportional to the optical power of incident light composed of multiple photons. Figure 7.2 sketches a pulse resulting from a varying number of photons n over time t. For detailed analysis of the analog signal, a digitizing receiver unit is essential. Analyzing the signal of the emitted short duration laser pulse with only a few nanoseconds pulse width requires a high bandwidth receiver that resolves the signal at GHz rates and a correspondingly high digitizer sampling rate. Increasing bandwidth results in decreasing sensitivity of the photodiode that can be compensated by increasing the power of the emitting laser source. An example of an Nd:YAG laser pulse sampled with 5 GSample/s is given in Figure 7.4a.

7.2.4.2 Single-Photon Detection

The principle of single-photon detection is depicted in Figure 7.3. A short duration pulse is emitted by the laser source. A single photon of the backscattered pulse is detected by the receiver after time interval τ_1. This event blocks the receiver for a certain period of time during which no further photons are able to trigger the receiver. The time-of-flight of this single event is collected into a corresponding time bin of a histogram. After the period of blocking, the receiver is open to detect a new single-photon event. Multiple measurements are repeated and the time-of-flight of each single event (τ_2, τ_3, ...) is registered into the corresponding time bin of the histogram.

Let us assume a stationary scene and a stationary sensor platform. In this case, the statistical properties of the laser radiation do not change with the time and time-average quantities are equal to the ensemble quantities. Under these assumptions the radiation ensembles are stationary and ergodic (Papoulis, 1984; Troup, 1972). The counting of single-photon events with assignment of their time-of-flight into time bins of a histogram is closely related to integration of multiphotons over time (Alexander, 1997; Gagliardi and Karp, 1976; Loudon, 1973). That means the temporal waveform of the pulse can be reconstructed from a histogram of single-photon arrivals over time.

Many transmit pulses are necessary to obtain the waveform with single-photon detection. The quality of the sampled waveform depends on the number of photon counts. Various optical detectors can be used for this purpose, namely photomultiplier tubes, microchannel plate, or avalanche photodiode detectors. Figure 7.4b shows a pulse plotted from a histogram containing the time-of-flight measurements from 16,252 photons distributed in 50 bins, whereas the bin width is 40 ps. Note that

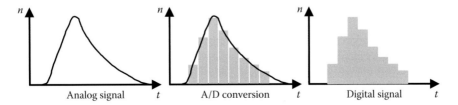

FIGURE 7.2   Digital recording of the signal with multiphoton detection.

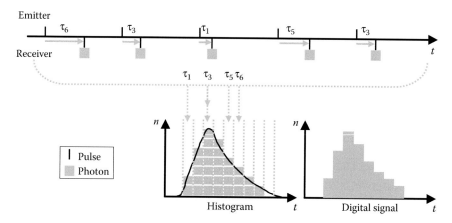

FIGURE 7.3   Digital recording of the signal with single-photon detection.

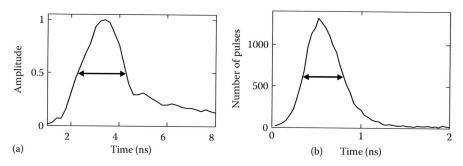

FIGURE 7.4   Examples for pulses backscattered from a diffuse surface. (a) Multiphoton detection (FWHM = 2.1 ns) and (b) single-photon detection (FWHM = 0.4 ns).

the FWHM of the pulse in Figure 7.4a is about five times of the pulse in Figure 7.4b. In this contribution, we focus on multi photon detection (Figure 7.1).

7.2.5   CONSTRUCTION

Depending on the construction of the transmitter and receiver optics, monostatic and bistatic laser systems can be distinguished:

Monostatic laser systems have transmitter and receiver optics collocated on the same optical axis. A disadvantage of this construction is the higher number of components compared with a bistatic laser system, increasing the effort needed to optimally align the components. Advantages of this construction are the isogonal measurement of angles and the exact measurement of ranges, because the illuminated surface area and the observed field-of-view of the receiver are coincident for all ranges.

Bistatic laser systems have a transmitter and receiver optics that are spatially separated and, thus, the illumination and view angles are divergent. In general, both optics are close together and oriented in nearly the same direction. Objects are illuminated via a lens from the transmitter optic and the backscattered radiation is transferred via a separate lens to the receiver optic. An advantage of this measurement system is that the design can be easily constructed. A disadvantage of this design is that depending on the range to the illuminated surface the angle between the transmitter and the receiver optic varies. Furthermore, depending on the range only a partial overlap is obtained between the illuminated surface and the observed field of view.

7.3   MODELING

The received waveform depends on the transmitted waveform of the emitted laser pulse, the spatial energy distribution of the beam, and the geometric and reflectance properties of the surface. To describe the temporal and spatial properties of a pulse, appropriate models that parameterize the pulse attributes have to be introduced.

7.3.1   WAVEFORM OF THE LASER PULSE

Mathematical functions can be used to approximate the shape of laser pulse waveforms. Depending on the system, the shape may be best represented by a rectangular, exponential, or Gaussian distribution. A simple model is given by a by a rectangular distribution $s(t)$ with an amplitude a, the pulse width, and time delay τ

$$s(t) = a \cdot \text{rect}\left\{-\frac{(t-\tau)}{w}\right\} = \begin{cases} a & \text{for} \quad -\tau \leq t \leq w - \tau \\ 0 & \text{else.} \end{cases} \tag{7.1}$$

Especially for short laser pulses, a rectangular model often differs from the measured shape. A waveform with an exponential distribution (e.g., for a Q-switched laser) is applied by Steinvall (2000):

$$s(t) = t^2 \cdot \exp\left\{-\frac{(t-\tau)}{w}\right\} \tag{7.2}$$

A temporally symmetric Gaussian distribution for modeling the waveform of the spaceborne GLAS is proposed by Brenner et al. (2003). The basic waveform $s(t)$ of the used laser system can be described by

$$s(t) = \frac{2a}{w} \cdot \sqrt{\frac{\ln(2)}{\pi}} \cdot \exp\left\{-4 \cdot \ln(2) \cdot \frac{(t-\tau)^2}{2w^2}\right\} \tag{7.3}$$

The width of a pulse w is commonly defined at one-half of the pulse's maximum amplitude, known as FWHM.

7.3.2   SPATIAL ENERGY DISTRIBUTION

The spatial energy distribution of a laser (also known as the beam profile) depends on the pump source, the optical resonator, and the laser medium. In general, beam profiles are modeled by a cylindrical distribution (top-hat form) or by a 2D-symmetric Gaussian distribution (Kamermann, 1993). The measured cylindrical beam distribution of a pulsed Erbium fiber laser that operates at a wavelength of 1550 nm is depicted in Figure 7.5. A Gaussian beam distribution of a Raman-shifted Nd:YAG laser that operates at a wavelength of 1543 nm is depicted in Figure 7.6. Both measurements differ more or less from these idealized distributions.

7.4   ANALYZING THE WAVEFORM

Various detection methods are used to extract attributes of the reflecting surface from the waveform. To obtain the surface attributes, each waveform $s(t)$ is analyzed. The surface within the beam corridor generates a return pulse. To detect and separate this pulse from the noise, a signal dependent threshold is estimated using the signal background noise. For example, in one particular implementation if the intensity of the waveform is above three times the noise standard deviation ($3\sigma_n$) for duration of at least the FWHM of the pulse, a pulse is assumed to be found. The section of the waveform including the detected pulse is passed onto the subsequent processing steps.

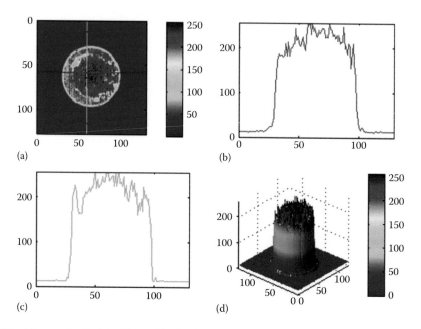

FIGURE 7.5   Measured cylindrical beam distribution (top-hat form). (a) 2D visualization, with the row (red) and column (green) of the maximum intensity indicated; (b) profile of the indicated row; (c) profile of the indicated column; and (d) 3D visualization.

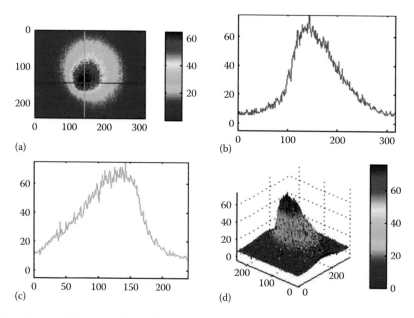

FIGURE 7.6   Measured Gaussian beam distribution. (a) 2D visualization, with the row (red) and column (green) of the maximum intensity indicated; (b) profile of the indicated row; (c) profile of the indicated column; and (d) 3D visualization.

Typical surface attributes to extract from a waveform are *range, elevation variation*, and *reflectance* corresponding to the waveform attributes: *time, width*, and *amplitude*.

A rough surface, that is, a surface of a certain vertical extent, will widen the laser pulse upon reflection. Therefore, the *width* of the pulse is a measure of the *elevation variation* of the surface. Moreover, the widening of the pulse causes the reflected photons to be spread over a greater amount of time, thus reducing the peak amplitude. Therefore, to estimate the *elevation variation* or *reflectance* attributes of a surface, the pulse *width* and *amplitude* have to be known. Estimating just the *amplitude* of a pulse without considering this dependency will lead to inaccurate and noisy *reflectance* values. Determination of the range to a surface can be accomplished with different schemes that include peak detection, leading edge ranging, constant fraction detection, center of gravity detection, Gaussian decomposition, and deconvolution. A discussion of key elements of each approach follows.

7.4.1 PEAK DETECTION

The values of range $r_{P,s}$ and amplitude are determined at the maximum pulse amplitude $a_{P,s}$ (Figure 7.7) and the *width* $w_{P,s}$ is estimated at FWHM of the pulse. Local spikes on the pulse waveform strongly effect the attribute extraction. Therefore, for noisy signals, a smoothing filter is recommended to determine the global maxima.

7.4.2 LEADING EDGE DETECTION

A threshold crossing of the pulse waveform leading edge determines the range value $r_{LE,s}$ (Figure 7.8). The threshold value can be a predefined fixed value, but then the ranging detection strongly depends

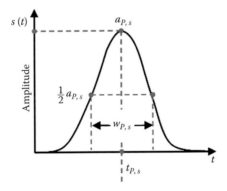

FIGURE 7.7   Attribute extraction with peak detection.

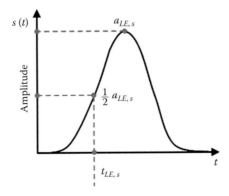

FIGURE 7.8   Attribute extraction with leading edge detection.

on the amplitude and width of the pulse waveform, introducing a ranging bias dependent on pulse shape referred to as range walk. The half of the maximum amplitude $a_{LE,s}$ of the pulse for a threshold is used for range determination.

7.4.3 CONSTANT FRACTION DETECTION

A ranging implementation designed to be insensitive to amplitude-dependent biases applies a constant fraction detection circuit in which the pulse waveform $s(t)$ is inverted and delayed by a fixed time τ and added to the original pulse (Figure 7.9). The combined signal $s_{CFD}(t)$ gives a constant fraction signal with a zero crossing point at t_{CFD} (Figure 7.10):

$$s_{CFD}(t_{CFD}) = 0 \text{ with } s_{CFD}(t) = s(t) - s(t + \tau) \tag{7.4}$$

The determined t_{CFD} is insensitive to the pulse amplitude but depends on the pulse waveform and width (Kamermann, 1993). A suitable value for the delay time τ is the FWHM of the waveform.

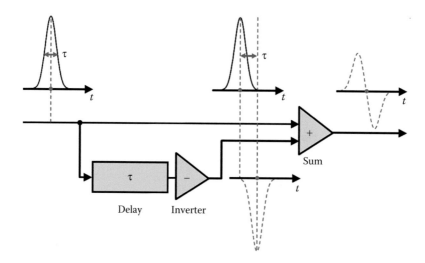

FIGURE 7.9   Simplified schematic visualization of the processing steps for the constant fraction detection.

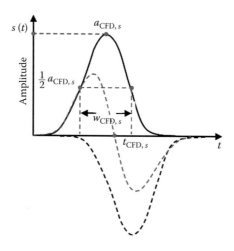

FIGURE 7.10   Attribute extraction with constant fraction detection.

For symmetric waveforms, the traditional constant fraction algorithm delivers unbiased ranging results. However, for an asymmetric noisy waveform, the delayed signal should be reversed in time as well, to avoid ambiguities of the zero crossing point.

7.4.4 CENTER OF GRAVITY DETECTION

The temporal center of gravity of the pulse waveform is determined (Figure 7.11). The time value (*range*) is determined by integrating the pulse waveform $s(t)$:

$$t_{CoG,s} = \int_{t=t_{CoG_1,s}}^{t_{CoG_2,s}} t\, s(t)dt \Bigg/ \int_{t=t_{CoG_1,s}}^{t_{CoG_2,s}} s(t)dt \tag{7.5}$$

It delivers good results for returns with various pulse amplitudes and pulse widths that have low noise. For returns with an asymmetric pulse shape skewed to longer ranges, this method results in a detected range that is slightly longer than the range value obtained with the peak detection.

The following methods to further process the pulse properties are not part of the center of gravity algorithm but are well suited to complement it. Generally, integration over a section of the signal has the advantage of reducing the noise dependence compared with the aforementioned methods relying on single samples. We call the integral of the waveform $s(t)$, shown in the denominator of Equation 7.4, the *pulse strength*. From this, the value a_{CoG} can be calculated assuming a Gaussian and using the Inverse error function (erf^{-1}) and the width w:

$$a_{CoG} = \frac{2\, erf^{-1}(0.5)}{\sqrt{\pi}\, w} \int_{t=t_{CoG_1}}^{t_{CoG_2}} s(t)dt \tag{7.6}$$

Furthermore the *width* $w_{CoG,s}$ is approximated by the width of the central pulse area contributing 0.76 of this *pulse strength*

$$\int_{t=t_{CoG}-\frac{w_0}{2}}^{t_{CoG}+\frac{w_0}{2}} s(t)dt = erf(\sqrt{\ln 2}) \int_{t=t_{CoG_1}}^{t_{CoG_2}} s(t)dt \tag{7.7}$$

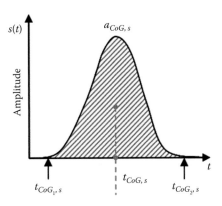

FIGURE 7.11   Attribute extraction with center of gravity detection.

7.4.5 Gaussian Decomposition

Assuming a Gaussian function for the waveform (Equation 7.3) the surface attributes can be extracted by estimating the parameters of the adapted function:

$$s(t) = a_{GD,s} \cdot \exp\{-4 \cdot \ln(2) \cdot \frac{(t - t_{GD,s})^2}{(w_{GD,s})^2}\} \tag{7.8}$$

For a parametric description of the pulse properties a Gaussian decomposition on the waveform can be used. Different methods are known, for example, Expectation Maximation algorithm (Persson et al., 2005), Gauss–Newton or Levenberg–Marquardt algorithm (Hofton et al., 2000; Jutzi and Stilla, 2005; Reitberger et al., 2006). In Figure 7.12, the estimated attributes of the received waveform is depicted.

7.4.6 Deconvolution

Analysis of a received waveform to extract the attributes of the illuminated surface is a difficult task because different processes impact the shape of the waveform. The received waveform $s(t)$ of a laser pulse depends on the transmitted waveform $r(t)$, the impulse response of the receiver, the spatial beam distribution of the laser pulse, and the geometric and reflectance properties of the illuminated surface. The impulse response of the receiver is mainly affected by the photodiode and amplifier, and the spatial beam distribution typically has a Gaussian distribution. Let us assume that a receiver consists of an ideal photodiode and that the amplifier has an infinite bandwidth with a linear response. In that case, the received waveform depends mainly on the transmitted waveform $r(t)$ and the properties of the illuminated surface. The 3D characteristics of the surface can be captured by a time-dependent surface representation, referred to as the surface response $h(t)$. In this case, the received waveform $s(t)$ can be expressed as

$$s(t) = h(t) * r(t) \tag{7.9}$$

where (*) denotes the convolution operation. By transforming $s(t)$ into the Fourier domain and solving the resulting equation for the spectral surface function $\underline{H}(f)$, we obtain

$$\underline{H}(f) = \underline{S}(f)/\underline{R}(f) \tag{7.10}$$

The surface response $h(t)$ is obtained by transforming $\underline{H}(f)$ into the time domain. By applying a Gaussian decomposition method to the surface response, surface attributes can be extracted.

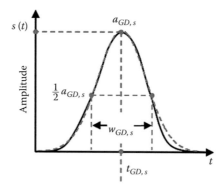

FIGURE 7.12 Attribute extraction with Gaussian decomposition algorithm.

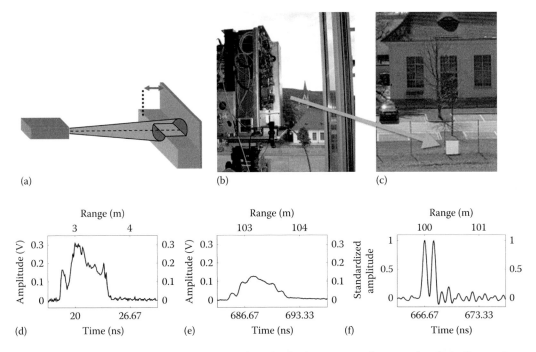

FIGURE 7.13 Example of the measurements and results for the response of a stepped surfaces distance having a step distance of 0.15 m: (a) sketch of the acquisition situation, (b) sensor system, (c) target in a distance of 100 m, (d) transmitted waveform of the emitted pulse, (e) received waveform of the backscattered pulse, and (f) estimated surface response from (d) and (e). Note that the width of the pulses (d, e) is much longer than the distance of the step (f)!

The deconvolution removes the characteristics of the transmitted waveform from the received waveform and enables a description of the observed surface.

For a reliable deconvolution, a high SNR for the received waveform is essential. In addition to this, the waveform has to be captured with a high bandwidth receiver and with an adequate sampling rate of the analog-to-digital converter. Furthermore, it has to be mentioned that large numerical errors may appear depending on the receiver noise. By using a Wiener Filter for deconvolution reduces the noise of the determined surface response (Jutzi and Stilla, 2006a). This method allows discriminating differences in range, for example, given by a stepped surface within the beam, which are smaller than the length of the laser pulse. Experiments have shown that a step smaller than ten times of the pulse length can be distinguished. An example is depicted in Figure 7.13 that shows a transmitted waveform of the emitted pulse, the received waveform of the backscattered pulse and the estimated surface response for stepped surface of 15 cm.

7.5   ATTRIBUTE EXTRACTION

An example of a signal profile applied to multiple pulses is depicted in Figure 7.14a. The waveform parameters for each detected pulse of this signal profile are estimated by a Gaussian decomposition method using the Levenberg–Marquardt algorithm. The extracted attributes are described in the table. The estimated waveform is shown next the original waveform in Figure 7.14.

By comparing the range values in the table, we see, that the distance between the 1st and 2nd pulse is about 10 m and between the 3rd and 4th pulse about 2.5 m. The 3rd pulse shows the highest maximum amplitude and the pulse width of the 1st and 2nd pulse is slightly lower than the 3rd and 4th pulse. It is not possible to classify the type of surfaces illuminated within the single beam corridor using the amplitude, pulse width, and range properties alone. For assigning each return pulse to

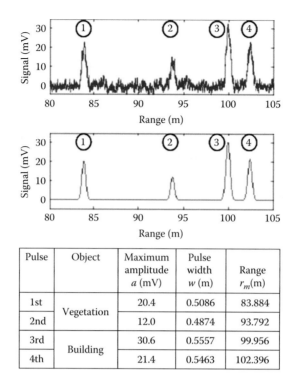

Pulse	Object	Maximum amplitude a (mV)	Pulse width w (m)	Range r_m (m)
1st	Vegetation	20.4	0.5086	83.884
2nd		12.0	0.4874	93.792
3rd	Building	30.6	0.5557	99.956
4th		21.4	0.5463	102.396

FIGURE 7.14 Signal profile of the estimated surface response with multiple reflections.

a specific surface type additional information is required, which can result from the 3D geometrical relationships of the returns within a point cloud.

As an example of information retrieval achieved by combining return pulse properties and the spatial relationship of the returns, a full-waveform dataset of a test scene was captured by scanning along azimuth and elevation and recording the return intensity sampled over time t. Neglecting angular variations of the scan, the measured intensities as a function of time t sampled over azimuth and elevation can be interpreted as a 3D dataset forming a cuboid with Cartesian coordinates x, y, and t. The sampling along the time axis can be recalculated into corresponding range values z. This data cuboid can be analyzed in several different ways.

Figure 7.15 shows a set of image slices (y–t planes). The 2nd slice from the left ($x = 4$) shows vegetation in the center (near range) and building structures on the right side (far range). The gray values correspond to the intensity of the signal. The intensity values along the marked solid line are the intensity values of the waveform shown in Figure 7.14.

Note that although this way of displaying the data suggests that a full 3D representation of the scene has been obtained, this is in fact not the case. Just as with point clouds measured by conventional laser scanners, the data cube represents only 2.5d information. This is because of occlusion effects that are dependent on the target size in relation to the beam footprint size. It is possible that the laser pulse is mostly intercepted by and backscattered from the first illuminated surface along the propagation direction and that following surfaces along the laser vector are hidden, giving weak or no reflections. For instance, a tree with dense foliage may return only a single reflection response per laser pulse even though multiple surfaces are present after the first detected return along the path of the laser vector.

In the following, we use the Wiener–Filter method to extract attributes from received waveforms. The extraction is carried out without considering spatial neighborhood relations. Results of extracted surface attributes from the data cuboid are shown in the xy plane by images of 320 × 600 pixels (Figure 7.16). Figure 7.16a shows the range image, Figure 7.16b shows the width of the pulse, and Figure 7.16c shows the intensity of the pulse. Larger values for an attribute are displayed by brighter

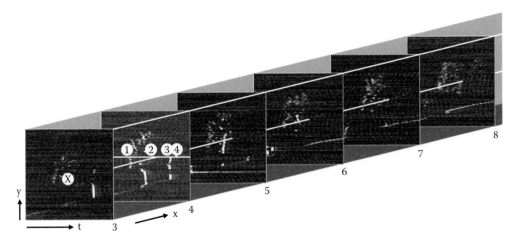

FIGURE 7.15 Vertical image slices with ground, vegetation, and building structures.

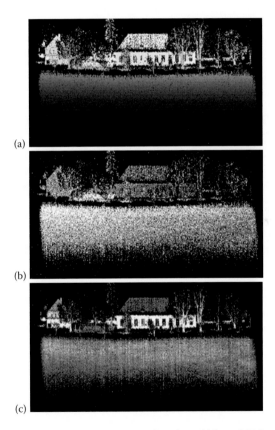

FIGURE 7.16 Extracted attributes of pulses. (a) range, (b) pulse width, and (c) intensity.

pixels. Due to the fact that only a single value can be shown in the 2D images, only the first reflection is considered in cases in which multiple reflection are present for a laser pulse.

The attributes *maximum amplitude* and *pulse width* were extracted using the Wiener Filter and examined for their ability to discriminate different object classes. The entire scene and three objects classes, namely *building*, *trees*, and *meadow*, are shown in Figure 7.17. The left side of the figure depicts range images of the entire scene (Figure 7.17Aa) and of the selected objects classes

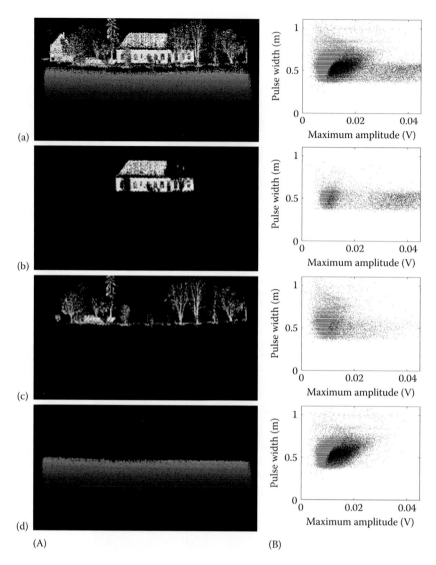

FIGURE 7.17 Comparison of maximum amplitude versus pulse width for selected objects of the measured scene: (a) scene, (b) building, (c) trees, and (d) meadow. Column (A) stands for range image, whereas column (B) for maximum amplitude versus pulse width.

(Figure 7.17Ab–Ad). The right side depicts scatter plots of *maximum amplitude* versus *pulse width* for the entire scene and the object classes (Figure 7.17B).

In Figure 7.17Ba, it can be observed that for the entire scene small values of the *maximum amplitude* have a large spreading of the *pulse width*. By decomposing the scene into the three object classes, it is apparent that the vegetation (trees and meadow) in most cases produces the signal returns with small maximum amplitudes but high values for the pulse width. The building in Figure 7.17Bb shows small values for the pulse width with little variation and a large range in the maximum amplitude values. Furthermore, the *maximum amplitude* of the building shows a cluster of higher values than that from the trees and meadow. These high values may result from high reflectance by the white façade. The trees are depicted in Figure 7.17Bc, in which the pulse width shows high variation with mostly small maximum amplitude values. The meadow (Figure 7.17Bd) induces large variations of the pulse width and small maximum amplitude values. In general, the trees and the meadow produce larger pulse widths than the building.

7.6 APPLICATIONS

7.6.1 Man-Made Objects

In the preceding description laser pulses were analyzed without considering the information of neighboring measurements. For reconstructing man-made objects, the introduction and test of hypotheses about the shape of the surface (e.g., plane, sphere) may efficiently support the analysis of single waveforms (Roncat, 2014). Two different strategies assuming a planar shape in the local neighborhood of the surface and introducing this assumption into the signal analysis should be addressed. Both strategies combine information from top down (surface primitives) and bottom up (signal processing) for an extended analysis of full-waveform laser data.

The first strategy (Kirchhof et al., 2007) uses an iterative processing of waveforms considering a predicted shape of the waveform from the local neighborhood. A presegmentation based on surface attributes is carried out to distinguish between partly penetrable objects (e.g., trees, bushes) and impenetrable surfaces (e.g., roof, wall). Derived range values from presegmentation of the impenetrable surfaces are used to automatically generate surface primitives (e.g., planes).

This allows a refinement of each range value taking the surface geometry in a close neighborhood into account. Furthermore, partly occluded surface areas are extended by prediction of the expected range values. This prediction is further improved by considering the surface slope for the estimated received waveform. Expected pulses are simulated and correlated with the received waveforms. Accepted points that were missed in the first processing step due to weak signal response are associated to the point cloud. The procedure is repeated several times until all appropriate range values are considered to estimate the surface. An example showing intermediate results is given in Figure 7.18. More details and examples can be found in the original literature of Kirchhof et al. (2007).

The second strategy (Stilla et al., 2007) uses a slope compensated stacking of waveforms. Weak pulses with a low SNR are discarded by classic threshold methods and get lost. In signal and image processing, different stacking techniques are used to improve the SNR. For detection of weak laser pulses, hypotheses for planes of different slopes (e.g., angle difference 5°) are generated. According to the slope of the hypothesis, the waveforms in the local neighborhood are shifted in range. From the stack of shifted waveforms, a superimposed signal is calculated. The maxima of superimposed signals from all hypotheses are compared toward verifying a hypothesis. Each signal is assessed by a likelihood value

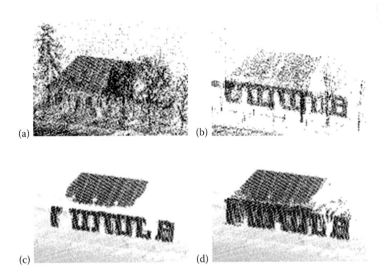

(a) (b)

(c) (d)

FIGURE 7.18 Results of the iterative processing of waveforms: (a) captured point cloud, (b) points belonging to planes, (c) extended point cloud by considering expected waveforms of the planes, and (d) final point cloud after iteration.

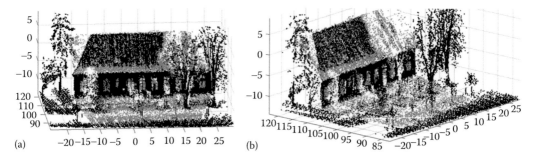

FIGURE 7.19 Result of full waveform stacking (black: original detected 3D points, gray: additional detected 3D points after full waveform stacking): (a) front view (see Figure 7.17a, b) and (b) side view.

with respect to its contribution to the accepted hypothesis. At last, signals will be classified according to likelihood values using two thresholds and visualized by the traffic-light paradigm. Results contain detected pulses reflected from objects, which cannot be predicted by the previously detected point cloud. An example for results processing the same scene as shown in Figure 7.16 is given in Figure 7.18. A more detailed description and investigation can be found in Yao and Stilla (2010).

Both strategies show promising results that encourage to continue the work on analysis of full-waveform laser pulses (Figure 7.19).

7.6.2 Natural Objects

The full waveform technique shows a high potential in applications mapping objects having a volume-scattering characteristic. This is given for natural objects, such as water bodies (bathymetry) or vegetation (forest monitoring). Reitberger et al. (2008a, 2009) have shown that the benefit from full waveform analysis for forest monitoring is twofold. On the one hand, the analysis of the pulse width allows a discrimination of pulses reflected from the stem from other pulses (Figure 7.20a and b). The reconstruction of the stem and its position has a strong impact on the segmentation of single trees in a dense stand (Figure 7.20c) (Reitberger et al., 2007). On the other hand, the higher point density by full waveform analysis in the z-direction (vertical) supports a better segmentation and classification of trees with a smaller stem diameter (Figure 7.20d). These results can be observed for leaf-on

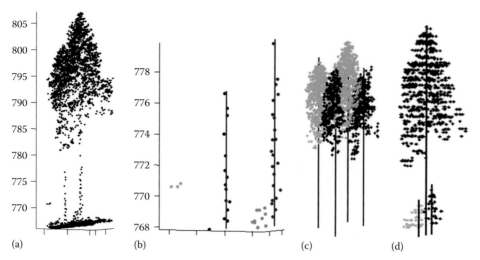

FIGURE 7.20 Single tree detection for forest inventory: (a) point cloud of two trees, (b) classified stem points and estimated stems by RANSAC, (c) segmentation of trees in a cluster, and (d) segmentation of a tree and young generation below.

and leaf-off situation. Furthermore, the experiments have shown that an increasing number of points/ m2 have only a minor effect on the results compared with the decision using full-waveform data or not (Reitberger et al., 2008b).

7.7 SUMMARY

It has been shown that full waveform analysis enables one to extract more information compared with classical analogous pulse detection methods. First, recording of the received waveform offers the possibility for the end user to select different methods to extract range information. The shape of the pulse and the entire signal can be considered for determining the range more accurate. Further improvements on reliability and accuracy can be derived by signal-processing methods based on the transmitted and the received waveform, for example, deconvolution. In addition, attributes of the surface can be derived from a parametric description of the waveform. The attributes maximum amplitude and pulse width may support the discrimination between volume scatterer (vegetation) and hard targets (man-made objects).

REFERENCES

Abshire, J.B., Sun, X.L., Riris, H., Sirota, J.M., McGarry, J.F., Palm, S., Yi, D.H., and Liiva, P. 2005. Geoscience laser altimeter system (GLAS) on the ICESat mission: On-orbit measurement performance. *Geophysical Research Letters*, 32(L21S02). doi:10.1029/2005GL024028.

Albertz, J., Attwenger, M., Barrett, J., Casley, S., Dorninger, P., Dorrer, E., Ebner, H. et al. 2005. HRSC on Mars express–Photogrammetric and cartographic research. *Photogrammetric Engineering & Remote Sensing*, 71(10): 1153–1166.

Alexander, S.B. 1997. Optical communication receiver design. In *SPIE Tutorial Texts in Optical Engineering* (Vol. TT22). Bellingham, WA: SPIE Press.

Blair, J.B., Rabine, D.L., Hofton, M.A. 1999. The laser vegetation imaging sensor (LVIS): A medium-altitude, digitization-only, airborne laser altimeter for mapping vegetation and topography. *ISPRS Journal of Photogrammetry & Remote Sensing*, 54 (2–3): 112–122.

Brenner, A.C., Zwally, H.J., Bentley, C.R., Csatho, B.M., Harding, D.J., Hofton, M.A., Minster, J.B. et al., 2003. Geoscience laser altimeter system (GLAS) - Derivation of range and range distributions from laser pulse waveform analysis for surface elevations, roughness, slope, and vegetation heights. Algorithm theoretical basis document - Version 4.1. http://www.csr.utexas.edu/glas/pdf/Atbd_20031224.pdf.

Gagliardi, R.M. and Karp, S. 1976. Noncoherent (Direct) detection. *Optical Communications*. New York: John Wiley & Sons.

Hecht, J. 1992. *The Laser Guidebook* (2nd ed). Blue Ridge Summit, PA: Tab Books.

Harding, D.J. and Carabajal, C.C. 2005. ICESat waveform measurements of within-footprint topographic relief and vegetation vertical structure. *Geophysical Research Letters*, 32(L21S10). doi:10.1029/2005GL023471.

Hofton, M.A., Minster, J.B., and Blair, J.B. 2000. Decomposition of laser altimeter waveforms. *IEEE Transactions on Geoscience and Remote Sensing*, 38 (4): 1989–1996.

Irish, J.L. and Lillycrop, W.J. 1999. Scanning laser mapping of the coastal zone: The SHOALS system. *ISPRS Journal of Photogrammetry & Remote Sensing*, 54 (2–3): 123–129.

Irish, J.L., McClung, J.K., and Lillycrop, W.J. 2000. Airborne lidar bathymetry: The shoals system. *PIANC Bulletin*, 2000 (103): 43–53.

Jelalian, A.W. 1992. *Laser Radar Systems*. Boston, MA: Artech House.

Jutzi, B. and Stilla, U. 2005. Measuring and processing the waveform of laser pulses. In Gruen, A. and Kahmen, H. (Eds.) *Optical 3-D Measurement Techniques VII*, 1: 194–203.

Jutzi, B. and Stilla, U. 2006a. Range determination with waveform recording laser systems using a Wiener Filter. *ISPRS Journal of Photogrammetry & Remote Sensing*, 61 (2): 95–107. doi:10.1016/j.isprsjprs.2006.09.001.

Jutzi, B. and Stilla U. 2006b. Characteristics of the measurement unit of a full-waveform laser system. Symposium of ISPRS Commission I: From Sensors to Imagery. *International Archives of Photogrammetry, Remote Sensing and Spatial Information Sciences*, 36 (Part 1/A).

Kamermann, G.W. 1993. Laser radar. In: Fox, C.S. (Ed.) *Active Electro-Optical Systems, the Infrared & Electro-Optical Systems Handbook*. Ann Arbor, MI: SPIE Optical Engineering Press.

Kirchhof, M., Jutzi, B., and Stilla, U. 2007. Iterative processing of laser scanning data by full waveform analysis. *ISPRS Journal of Photogrammetry & Remote Sensing*, 63(1): 99–114. doi:10.1016/j.isprsjprs.2007.08.006.

Lenhart, D., Kager, H., Eder, K., Hinz, S., and Stilla, U. 2006. Hochgenaue Generierung des DGM vom vergletscherten Hochgebirge – Potential von Airborne Laserscanning. Arbeitsgruppe Automation in der Kartographie: Tagung 2005, Mitteilungen des Bundesamtes für Kartographie und Geodaesie, Band 36: 65–78 (in German).

Loudon, R. 1973. *The Quantum Theory of Light*. Oxford, UK: Clarendon Press.

Mallet, C. and Bretar, F. 2009. Full-waveform topographic lidar: State-of-the-art. *ISPRS Journal of Photogrammetry and Remote Sensing*, 64(2009): 1–16.

Papoulis, A. 1984. *Probability, Random Variables, and Stochastic Processes*. Tokyo, Japan: McGraw-Hill.

Persson, Å., Söderman, U., Töpel, J., and Ahlberg, S. 2005. Visualization and analysis of full-waveform airborne laser scanner data. In Vosselman, G., Brenner, C. (Eds.) Laserscanning 2005. *International Archives of Photogrammetry, Remote Sensing and Spatial Information Sciences*, 36 (Part 3/W19): 109–114.

Reitberger, J., Heurich. M., Krzystek. P., and Stilla, U. 2007. Stem detection of single trees in forest areas with high-density LiDAR data. In Stilla, U. et al. (Eds.) PIA07 Photogrammetric Image Analysis 2007. *International Archives of Photogrammetry and Remote Sensing*, 36(3/W49B), pp. 139–144.

Reitberger, J., Krzystek, P., and Stilla, U. 2006. Analysis of full waveform LIDAR data for tree species classification. In Förstner, W., Steffen, R. (Eds.) Symposium of ISPRS Commission III: Photogrammetric computer vision PCV06. *International Archives of Photogrammetry, Remote Sensing and Spatial Information Sciences*, 36 (Part 3): 228–233.

Reitberger, J., Krzystek, P., and Stilla, U. 2008a. Analysis of full waveform LIDAR data for the classification of deciduous and coniferous trees. *International Journal of Remote Sensing*, 29(5): 1407–1431. doi:10.1080/01431160701736448.

Reitberger, J., Schnörr, C., Krzystek, P., and Stilla, U. 2008b. Towards 3D mapping of forests: A comparative study with first/last pulse and full waveform LiDAR data. XXI ISPRS Congress, Proceedings. *International Archives of Photogrammetry, Remote Sensing and Spatial Geoinformation Sciences*, 37(B8): 1397–1404.

Reitberger, J., Schnörr, C., Krzystek, P., and Stilla, U. 2009. 3D segmentation of single trees exploiting full waveform LIDAR data. *ISPRS Journal of Photogrammetry and Remote Sensing*, 64(6): 561–574. doi:10.1016/j.isprsjprs.2009.04.002.

Roncat, A. 2014. Backscatter signal analysis of small-footprint full-waveform lidar data. Dissertation. University of Vienna, Department for Geodesy and Geoimnformation.

Schutz, B.E., Zwally, H.J., Shuman, C.A., Hancock, D., and DiMarzio, J.P. 2005. Overview of the ICESat Mission. *Geophysical Research Letters*, 32(L21S01). doi:10.1029/2005GL024009.

Spiegel, M., Stilla, U., and Neukum, G. 2006. Improving the exterior orientation of Mars express regarding different imaging cases. *International Archives of Photogrammetry and Remote Sensing and Spatial Information Sciences*, 36(4): 385–363.

Stilla, U., Yao, W., and Jutzi, B. 2007. Detection of weak laser pulses by full waveform stacking. In: Stilla, U. et al. (Eds) PIA07: Photogrammetric image analysis 2007. *International Archives of Photogrammetry, Remote Sensing and Spatial Information Sciences*, 36(3/W49A): 25–30.

Steinvall, O. 2000. Effects of target shape and reflection on laser radar cross sections. *Applied Optics*, 39 (24): 4381–4391.

Thiel, K.H. and Wehr, A. 2004. Performance capabilities of laser-scanners - An overview and measurement principle analysis. *International Archives of Photogrammetry, Remote Sensing and Spatial Information Sciences* 36 (Part 8/W2): 14–18.

Troup, G.J. 1972. Photon counting and photon statistics. In Sanders, J. H., Stenholm, S. (Eds.) *Progress in Quantum Electronics*, Vol. 2 (Part 1). Oxford, UK: Pergamon.

Wagner, W., Ullrich, A., Ducic, V., Melzer, T., and Studnicka, N. 2006. Gaussian decomposition and calibration of a novel small-footprint full-waveform digitising airborne laser scanner. *ISPRS Journal of Photogrammetry and Remote Sensing*, 60 (2), 100–112.

Wuerlaender, R., Eder, K., and Geist, T. 2004. High quality DEMs for glacier monitoring: Image matching versus laser scanning. In Altan, M.O. (Ed.) *International Archives of Photogrammetry and Remote Sensing*, Vol. 35, Part B7, 753–758.

Yao, W. and Stilla, U. 2010. Mutual enhancement of weak laser pulses for point cloud enrichment based on full-waveform analysis. *IEEE Transactions on Geoscience and Remote Sensing*, 48(9): 3571–3579. doi:10.1109/TGRS.2010.2047109.

Zwally, H.J., Schutz, B., Abdalati, W., Abshire, J., Bentley, C., Brenner, A., Bufton, J. et al. 2002. ICESat's laser measurements of polar ice, atmosphere, ocean, and land. *Journal of Geodynamics*, 34 (3–4): 405–445.

8 Strip Adjustment

Charles K. Toth and Zoltan Koppanyi

CONTENTS

8.1 INTRODUCTION

The primary objective of Light Detection and Ranging (LiDAR) strip adjustment in airborne mapping is to provide quality assurance and quality control for the final geospatial product by reducing, or ultimately eliminating, discrepancies found in strip overlap areas, and thus create a seamless product. In addition, the strip adjustment may allow calibrating the LiDAR system components, as it can be seen as an indirect georeferencing of the mapping system. From this perspective, the adjustment problem has an interpolation-type error characterization, in contrast to the extrapolation-type typical for direct georeferencing, such as used in LiDAR systems.

Ideally, there should be no visible or measurable differences between overlapping LiDAR strips, except for sensor noise, mainly caused by varying performance of the georeferencing component and laser ranging error. However, strip differences frequently occur. The extent is significantly smaller nowadays than it was when the first generation of commercial LiDAR systems was introduced. At that time, only the vertical accuracy of the LiDAR data was specified and no reference was provided for horizontal accuracy; consequently, strip adjustment was aimed at removing only the height differences. Motivated primarily by the generation of digital elevation models (DEMs), the first guidelines to report on LiDAR data quality were only concerned with the vertical accuracy

(ASPRS, 2004), but now, horizontal precision evaluation is part of the process. Strip discrepancies typically show a systematic pattern that provides a basis to model them and subsequently correct them in a strip adjustment process. The differences between various LiDAR strips acquired over the same area are more visible in areas that are rich in objects of simpler geometric shapes, such as man-made objects, for example, buildings, as shown in Figure 8.1; the strip overlap is depicted in Figure 8.2.

LiDAR data are typically collected in strips. In applications, where a rectangular block or corridor is flown in a parallel line pattern, minimal overlap between neighboring strips is typically required to maintain contiguous coverage of the ground. The required margin can vary considerably, although the extent of overlap typically falls in the 10%–30% range. The amount of variation between strips depends on the flight situation, including flight planning and control, weather conditions, and terrain relief. Many strip adjustment techniques are based on sensor parameter calibration and, therefore, to better support these processes, cross-strips are also frequently flown. The reliability of earlier sensor systems necessitated frequent sensor calibration, which required additional dedicated data collection, such as flying cross-strips over an airport or parking areas before and after the survey, resulting in multiple overlap data.

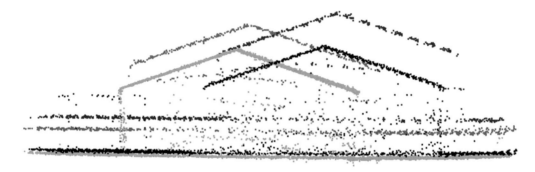

FIGURE 8.1   Strip discrepancies observed from four strips.

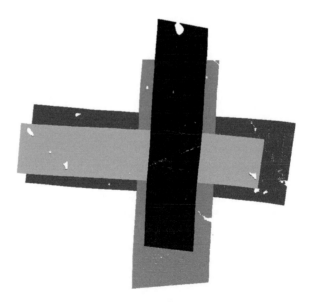

FIGURE 8.2   Multiple strip overlap (cross-strips at different flying height); square marks building area shown in Figure 8.1.

LiDAR systems have undergone remarkable developments since their introduction in the late 1990s and the development of the first strip adjustment techniques. The laser ranging accuracy for hard surfaces has approached the static surveying accuracy, about 1–2 cm (1σ), and the use of multiple returns and the intensity signal have become widespread, and multipulse systems have been introduced recently. More importantly, the pulse rate or pulse repetition frequency (PRF), shown in Figure 8.3, has been increasing (it seems to follow the Moore's law), allowing for better spatial sampling and improved object space reconstruction. Developments in the georeferencing component of LiDAR systems, discussed in detail in Chapter 6, are also significant as the navigation solution is currently the most significant term in the overall LiDAR error budget. Typical topographic LiDAR surveys provide point densities in the 2–10 pts/m² range, although for some applications, such as helicopter-based transmission line surveys, 20+ pts/m² densities are reported. Note that the along and across spatial sampling rates vary depending on the scanning mechanism used in the LiDAR system and the scanning parameter controls, although the parameter settings are generally optimized for approximately even spatial sampling.

If the LiDAR sensor is not modeled properly by calibration parameters, the overlapping strips may have systematic discrepancies. LiDAR users recognized the advantage of strip overlap very early, and methods were developed to assess the discrepancies between strips and then corrections were applied to the data. Initial developments in strip adjustment techniques were further influenced by other factors, such as the LiDAR point density. In the late 1990s, the LiDAR point density was modest compared with current systems. The pulse rate (10 kHz PRF) was by orders of magnitude less than that of the current state-of-the-art systems (1–2 MHz PRF); hence, one approach to increase point density was to fly larger overlaps, for example, 50%, which essentially provided double coverage of the surveyed area (i.e., doubling the effective point density).

The general concept of the LiDAR strip adjustment process is simple: First, differences should be identified and measured between two overlapping strips, and then, using a geometric model, the parameters of a suitable transformation must be determined, which can be subsequently applied to correct or adjust the strips. Unfortunately, the implementation of strip adjustment is not straightforward, as establishing the required correspondence between two strips is rather difficult. This difficulty comes primarily from the irregular distribution of points in the LiDAR point cloud, which means that the same object space is sort of randomly sampled in the spatial domain in every strip and, thus, there are no conjugate points between the two point clouds (even if there were observations of the same object from different strips, they would not be recognized as such,

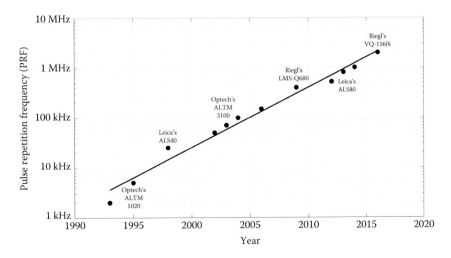

FIGURE 8.3 Increase in LiDAR point density (PRF).

because LiDAR points have no identifiers). Therefore, either interpolation of data is needed (e.g., conversion to a common grid) or shape-based techniques, based on geometrical features extracted from a group of points, should be used instead of conventional point-based methods.

LiDAR strip adjustment methods have evolved over time, and a variety of algorithmic approaches and techniques customized to specific conditions exist. From a conceptual point of view, the strip adjustment methods can be categorized on the basis of several not totally independent characteristics, with the most important ones listed earlier:

- *Coregistration or calibration*: The strip discrepancies can be eliminated or reduced either by applying adequate transformation to the LiDAR strips or by introducing a correction to the sensor modeling, calibration, parameters, and then recreating the LiDAR point cloud from the corrected sensor data. Conceptually, the strip adjustment methods fall into two categories. The techniques in the first group represent a type of rubber-sheeting coregistration solution, which tries to minimize the differences between strips for a given transformation model; this technique is also called data driven, as it is not based on a physical model and is focused only on removing the discrepancies between strips. In contrast, the second approach is concerned with the source of the errors and aims to reduce the strip discrepancies by modifying or adjusting the parameters of the sensor models and sensor orientation, and thus basically implementing an *in situ* calibration of the LiDAR multisensor system.
- *Observing Strip Discrepancies*: Although the visualization of strip discrepancies provides an easy assessment for human evaluation, the quantification of the differences between strips is rather difficult because of the lack of any direct point correspondence. Therefore, many of the LiDAR strip adjustment techniques require the establishment of some correspondence between the two datasets, such as by introducing virtual points in one dataset by interpolation or using features. On the other hand, there are methods that require no correspondence at all and can directly work with irregularly distributed point data.
- *Vertical Only or 3D Correction*: As the LiDAR point accuracy is considerably better in the vertical (sensor) direction, most of the earlier methods were only concerned with the vertical correction, whereas current techniques are usually based on a 3D geometrical model, including 3D shift and rotation.
- *Pairwise or Multiple Surface Matching*: In computer vision, the general problem of matching 3D range data is defined for multiple surfaces with no restriction on the complexity of 3D surfaces. However, this universal model is rarely needed in a typical LiDAR strip adjustment, which is mostly concerned with matching two surfaces, which are less complex and usually represented in a 2.5D irregular grid or raster format (DEM). The sensor parameter-based strip adjustment methods usually impose certain requirements on the data, in terms of overlap configuration.
- *Strip Deformation*: Data driven techniques are generally based on 3D transformations of variable complexity. Many methods assume that the transformation between the LiDAR point clouds (strips) and/or the reference ground surface (DEM) is adequately modeled by a rigid body transformation. More general solutions allow for strip deformation.
- *Use of LiDAR Intensity Data*: Earlier methods could only use range data, as there was originally no intensity data available. As intensity data has become a standard product, it can be used to support the strip adjustment process in various ways, such as providing clues for segmentation or establishing correspondence between strips, typically supporting matching in the horizontal domain.
- *Grid-Based Processing*: Methods vary as to whether they can directly handle irregular data, such as techniques that work on a triangulated irregular network (TIN), or evenly spaced (gridded) raster data, such as a 2.5D DEM may be required, which is obtained by interpolating the LiDAR point cloud, which will potentially introduce additional errors in the data.

- *Point or Feature-Based Techniques*: There are methods that can directly work on the original LiDAR point cloud or on derived raster data, whereas others are based on using features such as planar patches.
- *Object Space Requirements*: Certain techniques impose restrictions on the object space in terms of shape, such as flat areas, buildings, and rolling terrains, where they can be applied.
- *Ground Control Use*: LiDAR strip adjustment by definition is a relative correction, as all the observations are differential, and hence the elimination or reduction of strip discrepancies does not necessarily mean that the absolute accuracy of the LiDAR point clouds will be better after the strip adjustment, regardless of whether data driven or sensor calibration-based methods are used. Therefore, the use of ground control is always necessary to assess the overall point cloud accuracy. Most strip adjustment methods can accommodate the use of ground control. An example for LiDAR specific ground control point is shown in Figure 8.4.
- *Manual or Automated Methods*: Although most of the strip adjustment techniques are automated and require no or little human interaction, there are still a large number of frequently used techniques that are partially based on operator measurements. This is mainly due to the use of ground control, which is less automated in general, although depending on the type of ground control, the need for human interaction can be totally eliminated, such as using LiDAR-specific ground targets or a calibration range.

Despite the large variety of strip adjustment methods, there are several processing steps that are common to most techniques. As a precondition, the primary LiDAR data from an airborne survey are expected to be processed to a point cloud with all available sensor and system calibration data applied before the strip adjustment can begin. In Section 8.2, the typical error sources are discussed; these errors are eliminated during the strip adjustment. Then the main processing steps are:

- The selection of the strip overlap areas is concerned with finding appropriate areas and surface patches in the strip overlap regions. The selection criteria and advices are presented in Section 8.3, also provides a theoretical foundation of the optimal terrain sampling based on information theory.

FIGURE 8.4 LiDAR specific ground control point.

- The strip adjustment techniques can be separated into data driven or sensor model-based strip adjustment. The principles of these two methods are discussed in Section 8.4, and selection on which one to use is based on a combination of performance expectation, application data specifics, and system parameters. For an application with modest accuracy requirements, a simpler data driven technique may be sufficient, whereas more demanding data quality calls for a sensor calibration-based method. For longer, corridor type of surveys, clearly, the second approach is required, as the overlap area is relatively limited for a corridor and only a well-calibrated system can provide better overall data accuracy.
- As part of the strip adjustment, the selected surface patches typically converted into a certain surface representation. Section 8.4.1 presents three main representation types, including point cloud, 2.5D data, and continuous or trend surface. The point cloud representation is the primary point set acquired by LiDAR sensors. 2.5D data are widely used as surface and elevation model, and thus, are typically formatted for strip adjustment; in particular, with reference data. Finally, the continuous or trend surfaces are defined as continuous representation of the selected area used for matching. The last two representations are the basic element of all operator-based strip adjustment processes as well as being used in many automated techniques.
- The discrepancies between the surface patches or strips have to be defined for both adjustment approaches, and generally measured between matching point and/or surfaces. Reliable data matching is an important part of the adjustment. Note that the surface matching algorithm depends on the chosen representation. The concepts of these algorithms are presented in Section 8.4.2.

Finally, Section 8.5 provides a brief overview on the evolution of the strip adjustment techniques, including early, 3D, sensor-based, intensity-based and other strip adjustment methods.

8.2   ERROR SOURCES

The LiDAR systems are multisensory and dynamic systems with many potential error sources, including both systematic and random error components. Due to the complexity and highly interconnected structure of the system, not all the errors can be observed from strip discrepancies. LiDAR systems incorporate at least three main sensors: Global Positioning System (GPS)/Global Navigation Satellite System (GNSS) and inertial navigation system (INS) navigation sensors, and the laser-scanning device. Furthermore, there is a moving component in the laser system (e.g., oscillating or rotating mirrors) with its usual problems of position encoding, wear, and mechanical hysteresis that can further degrade the accuracy of the acquired LiDAR data. In general, the systematic errors in laser scanning data can come from individual sensor calibration (called measurement errors), lack of synchronization, and misalignment between the different sensors. Baltsavias (1999) presents an overview of basic relations and error formulae concerning airborne laser scanning, and Schenk (2001) provides an early error analysis on LiDAR. Even after careful system calibration, some errors could be present in the data, and navigation errors usually dominate. The errors are seen as discrepancies between overlapping strips and ground control surfaces. Most of these systematic errors can be corrected using strip adjustment (with or without ground control) by eliminating the discrepancies between overlapping LiDAR strips using various strip adjustment methods developed over the years. The understanding of the error terms as well as the overall error budget is important and will be discussed next.

LiDAR is based on direct georeferencing and the position of a laser point measured at time t_p is computed by the LiDAR equation:

$$r_M(t_p) = r_{M,\,\mathrm{INS}}(t_p) + R^M_{\mathrm{INS}}(t_p)\left(R^{\mathrm{INS}}_L \begin{bmatrix} 0 \\ \sin(\beta(t_p)) \\ \cos(\beta(t_p)) \end{bmatrix} d_L(t_p) + b_{\mathrm{INS}} \right) \qquad (8.1)$$

where:

$r_M(t_p)$ is the 3D coordinates of the laser point in the mapping frame 3D INS coordinates (origin) in the mapping frame, provided by

$r_{M,\text{INS}}(t_p)$ is the GPS/GNSS/INS (the navigation solution typically refers to the origin of the INS body frame)

$R_{\text{INS}}^M(t_p)$ is the rotation matrix between the INS body and the mapping frame

R_L^{INS} is the boresight matrix between the laser frame and the INS body frame

$d_L(t_p)$ is the range measurement (distance from the laser sensor reference point to the object point)

b_{INS} is the boresight offset vector (vector between the laser sensor reference point and the origin of INS) defined in the INS body frame

$\beta(t_p)$ is the scan angle defined in the laser sensor frame (x_L is flight direction, y_L to the right, and z_L goes down)

The laser point accuracy in a mapping frame can be determined by applying the law of error propagation to Equation 8.1. Earlier discussions on the errors and model parameter recovery can be found in (Baltsavias, 1999; Schenk, 2001; Filin, 2003a, 2003b) and a recent comprehensive discussion on the various terms of the error budget, including analytical and simulation results, can be found in (Csanyi, 2008). The difficulty of applying the theory in practical cases, however, is the time dependency of some of the parameters (note that several parameters are modeled as a function of time). In a simplified interpretation, there is no clear separation between the systematic and stochastic error terms for longer periods of time, as several components could have nonstationary behavior in terms of changing over intervals ranging from 10 to 30 min or even hours. For instance, the navigation part, which typically represents the largest terms in the error budget, may have components changing slowly in time and could be considered as drifts with respect to the other stochastic components with higher dynamics. These could subsequently be modeled as systematic error terms for shorter time periods, such as for the time it takes to fly a strip. In fact, these phenomena form the basis of the need for strip adjustments, as these error terms could change significantly between surveys as well as within surveys, similar to experiences with the GPS/GNSS-assisted aerial triangulation with longer flight lines (Ackerman, 1994). This is also the reason why it is difficult to obtain reliable estimates of the error terms. Table 8.1 lists some of the basic and generally accepted error sources and

TABLE 8.1

Major Error Sources Affecting the Accuracy of LiDAR Point Determination

	Errors	Typical Values
Navigation solution (GPS/GNSS/INS)	Errors in sensor platform position and attitude (shift and attitude errors)	σ_X, σ_Y: 2–5 cm; σ_Z: 4–7 cm; $\sigma_\nu, \sigma_\varphi$: 10–30 as; σ_k: 20–60
Laser sensor calibration	Range measurement error	σ_r: 1–2 cm
	Scan angle error	σ_β: 5 as
	Error in reflectance-based range calibration	[–20–10] cm
Intersensor calibration	Boresight misalignment between the INS body and laser sensor frames (shifts and angular errors)	$\sigma_{Xb}, \sigma_{Yb}, \sigma_{Zb}$: <1 cm; $\sigma_{\nu b}, \sigma_{\varphi b}$: 10 as, σ_{kb}: 20 as
	Error in measured lever arm (vector between GPS/GNSS antenna and INS reference point)	$\sigma_{Xa}, \sigma_{Ya}, \sigma_{Za}$: <1 cm
Miscellaneous errors	Effect of beam divergence (footprint) terrain and object characteristics Time synchronization coordinate system transformations atmospheric refraction sensor mounting rigidity	[0–5] cm

typical values, which can be used for an error propagation analysis for the point cloud accuracy. In addition, the importance of these values is that they provide a lower limit on what can be realistically expected from a strip adjustment-based correction using sensor calibration-based methods.

8.3   SELECTION OF OVERLAPPING AREAS

The determination of strip overlap in a dataset is a simple task, which is generally well supported by most LiDAR data processing software. The selection of the areas in the overlap regions, however, is rather complex as several conditions should be satisfied. First, vegetated areas should be avoided as they provide less reliable surface points and structure. A general rule is that the selected areas should have no multiple returns. Next, there should be a good surface signal in the selected patches, such as moderately sloped terrain or buildings, depending on the type of strip adjustment approach. Preferably, the surface normal vectors over the selected areas should show diversity in terms of significantly deviating from the vertical in several directions. Depending on the characteristics of the surface patch, vertical only, or horizontal, or 3D, discrepancies can be observed; for example, flat areas provide for height difference measurements based on range data, whereas intensity data can help determine horizontal shifts, or a rolling terrain can produce 3D differences. Finally, the selected surface patches should be evenly distributed in space to provide strong geometry for the adjustment. For example, for a cross-strip situation, if possible, the patches are selected in a similar pattern as the tie-points in aerial triangulation. The size of surface patches can vary on a larger scale and is mainly controlled by surface characteristics and the LiDAR point density; typically, it falls into the 10–50 m range. To simultaneously meet all the conditions is clearly not realistic in practice, yet enough attention must be paid to this crucial step, which is usually done in an interactive way. The selection of surface patches significantly reduces the amount of data that will be used in the subsequent processing steps.

Besides the surface characteristics, the LiDAR ground sampling distance, point spacing in both directions, has to be considered to capture sufficient details of the surface. If the ground sampling distance is too large (low point density) and the terrain changes are significant, then these changes might not be observed. Consequently, details are lost that could have been used during the strip adjustment. For this reason, a brief theoretical review is presented to describe the connection between the real and sampled surfaces in terms of representation/reconstruction error.

The surface description obtained from LiDAR data is primarily given as a set of points called bare earth points, which are obtained from the original LiDAR point cloud by filtering. The bare earth points, that is, surface point cloud, are a random or irregularly sampled representation of the actual surface, including natural and man-made objects digital surface model (DSM). Ideally, a true surface elevation data, S_c with respect to a mapping plane can be described as a 2D continuous function:

$$S_c = f(x, y) \tag{8.2}$$

For practical reasons, the discrete representation of the surface should be considered, which is obtained by an evenly spaced 2D sampling of the continuous function and by converting the continuous elevation values to discrete ones:

$$E_{ij}^d = Q_p(S_c) = Q_p(f(x_i, y_j)) \tag{8.3}$$

where:
E_{ij}^d is the discrete surface representation
Q_p is the quantization function (typically a regular step-function), which maps the continuous input parameter space to 2^p discrete levels
x_i, y_j is the 2D sampling locations

The fundamental question is how well the second format describes the first representation. Obviously, for simple surfaces that can be composed from planar and spherical patches, conical shapes, and so on, and thus, can be described by simple mathematical models, the continuous surface representation can be restored from the discrete representation without any error. In reality, however, the Earth's surface and other objects cannot be observed with unlimited precision. Some level of surface detail cannot be discerned and measurement errors exist; hence, the practical question is how to optimize the parameters of a discrete representation, such as sampling distance (sampling spatial frequency) and quantization levels, to achieve given accuracy requirements in terms of acceptable surface deviations. For example, what is the maximum sampling distance to keep the differences between the two representations below a predefined threshold? The answer depends on the application circumstances. For instance, for creating a topological map of a road, the surface roughness is irrelevant and small surface variations need not be considered. However, for road design or maintenance purposes, the details of the elevation changes are equally or even more important than the global nature of the surface, such as the road's absolute location in a mapping frame.

From a strictly theoretical point of view, the problem of how well a discrete representation describes the continuous case is well understood from Shannon's information theory (Shannon, 1948). Probably the most relevant and well-known expression is related to the sampling frequency, which is mostly known as the Nyquist frequency (Shannon, 1949). Rephrasing it for our case in one dimension and ignoring the impact of the signal quantization, it simply states that if a surface has a given maximum detail level (the surface changes are less than a predefined value), then there is an optimal sampling distance, and thus, any discrete representation that has this optimal sampling distance (or shorter) can reconstruct S_c from E_{ij}^d without any degradation. In other words, if f_{max} is the highest spatial spectral frequency for a given surface, then the sampling distance d_s is sufficient for the complete representation of this surface and, consequently, the continuous surface can be restored without any error from the discrete representation in this ideal case. The Nyquist criterion for the 1D case is

$$d_s \leq \frac{1}{2f_{max}} \tag{8.4}$$

Similarly, for 2D representations, the directional spatial frequencies have to be considered and the Nyquist criterion should be satisfied in both directions:

$$d_s^x \leq \frac{1}{2f_{max}^x} \quad \text{and} \quad d_s^y \leq \frac{1}{2f_{max}^y} \tag{8.5}$$

If the Nyquist criterion is satisfied, then the reconstruction of the continuous surface from the discrete values using the required or shorter sampling distances is described by

$$S_c(x,y) = \sum_{i=-\infty}^{\infty}\sum_{j=-\infty}^{\infty} E_{ij} \frac{\sin\left(\pi d_s^x\left(x - \frac{i}{d_s^x}\right)\right)\sin\left(\pi d_s^y\left(x - \frac{j}{d_s^y}\right)\right)}{\pi d_s^x\left(x - \frac{i}{d_s^x}\right)\quad \pi d_s^y\left(y - \frac{j}{d_s^y}\right)} \tag{8.6}$$

In this ideal case, the reconstruction introduces no errors, as the discrete representation provides a complete description of the surface function, see Figure 8.5. In reality, however, it is impossible to achieve this situation for several reasons. These include the characteristics of real surfaces (Toth et al., 2014) and the inherent limitations of the measurement processes in LiDAR systems.

Another less considered yet important aspect of the sampling process is the quantization of the elevation data. What is the smallest elevation difference that can be distinguished? Although the quantization is a nonlinear transformation, but its impact in practice can be safely ignored as

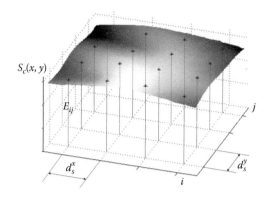

FIGURE 8.5 Surface reconstruction from discrete representation.

quantization step, ranging resolution is usually small. Note that in modern digital systems, the usual numerical representation provides fine representation for a wide signal range, and, consequently, the error introduced by converting the continuous signal into a discrete one is usually negligible.

8.4 FUNDAMENTALS OF THE STRIP ADJUSTMENT

The purpose of this section is to introduce the fundamentals of the basic mathematical models and techniques on a conceptual level. For the simplicity of the discussion, the case for two surfaces is considered, although the generalization to multiple surfaces is possible.

Conceptually, the strip adjustment methods fall into two categories. The strip discrepancies can be eliminated or reduced either by applying adequate transformation to the LiDAR strips or by introducing a correction to the sensor parameters, and then recreating the LiDAR point cloud from the corrected sensor data. The techniques in the first group are a type of rubber-sheeting coregistration solution, which tries to minimize the differences between strips, represented by patches for a given transformation model; this technique is also called data driven, as it is not based on a physical model and is focused only on removing the discrepancies between strips.

The data driven methods are based on establishing a 3D transformation T between P and Q, two corresponding surfaces from two datasets. The data driven approach minimizes the similarity distance (discrepancy) between one dataset and the transformed one:

$$\min_T \langle P, T(Q) \rangle \tag{8.7}$$

where T is the transformation function. The transformation T can be established between two overlapping LiDAR strips or, in general, between a reference surface and a LiDAR strip. Let $\langle .,. \rangle$ define the distance metric that measures the discrepancies between two surfaces. Commonly used distance metric is the Euclidean distance, when the surface representation contains the 3D coordinates of the LiDAR points. In this case, these points can be organized in a $3 \times N$ matrix, where N is the number of points. Then, if $\hat{P}$ and $\hat{Q}$ matrices are constructed from P and Q point clouds, and each $\hat{P}_i, \hat{Q}_i$ row pair in $\hat{P}$ and $\hat{Q}$ are corresponding points, and then Equation 8.7 has the following form

$$\min_{R,t} \sum_{i=1}^{N} \left\| \hat{P}_i - R\hat{Q}_i - t \right\|_2^2 \tag{8.8}$$

where R is the rotation matrix and t is the translation vector. Clearly, Equation 8.7 is the abstract definition of the problem; the exact formula can be defined, once the surface representation detailed in Section 8.4.1 and the surface matching algorithm in Section 8.5.2 are chosen, including the degree of freedom (DoF) of the transformation and the distance metric.

The strip discrepancies are reduced or removed by applying the transformation T to dataset Q, but no improvement in terms of the absolute accuracy of the point cloud can be guaranteed, in general. In contrast, by using a reference surface, the correction applied to the LiDAR strip should and will result in better absolute accuracy, provided the reference surface is of sufficient resolution and accuracy. This typically requires either a 3D test range or an area with precisely surveyed man-made objects, as the accuracy of older existing DEM datasets is not even close to the typical LiDAR vertical accuracy. For multiple strip overlap situations, a quasi-reference surface can be established by averaging the strips and then establishing and applying the transformation parameters for each strip individually. Obviously, this approach only works if the distribution of the strip discrepancies is close to symmetrical.

In contrast, the second approach, called sensor model-based method, is concerned with the source of the errors and aims to reduce the strip discrepancies by modifying or adjusting the parameters of the sensor models and sensor orientation, and thus, basically implementing an *in situ* calibration of the multisensor system:

$$\min_S \langle P(S), Q(S) \rangle \tag{8.9}$$

where $S = \{r_{M,\text{INS}}, R_{\text{INS}}^M, R_L^{\text{INS}}, b_{\text{INS}}\}$ is the parameters of the sensor model from Equation 8.1. Similar to the data driven Equation 8.7, Equation 8.9 is also an abstract formula that takes form once the surface representation and distance metric are defined.

Although the criterion to adjust the parameters is the similar in both models, the sensor model-based computation is adversely affected by a few problems. First, the functional correlation of the parameters should be mentioned, as there is the typical problem of correlation between attitude and position of a sensor platform, such as error in roll and across track offset resulting in the same discrepancy of object point locations. Another problem is whether some parameters can be observed at all, or there exists a dataset in which sufficient observations can be obtained. Another complication is the stability of the sensor parameters, such as the slowly changing offset or drift of the georeferencing solution, expressed as time-dependent terms in Equation 8.1. In summary, there is no general solution for simultaneously calibrating all the sensor parameters; therefore, certain restrictions on sensor parameters and data should be introduced to recover a subset of the calibration parameters.

Most LiDAR systems used in topographic mapping provide only the point cloud as the primary sensor data that serve as the starting point for any further processing, such as bare earth filtration, road or building extraction, and biomass estimation. If the actual raw sensor data are made available to the strip adjustment process, then the sensor parameter recovery could be significantly improved. For example, having the six georeferencing parameters, including position and attitude data, and the scan angle and range measurements for every laser point (all of them are easily available during the LiDAR point cloud creation) can eliminate most of the parameter correlation and observation deficiencies of the sensor calibration-based methods.

The selection whether the data driven transformation or the sensor and system calibration model should be used for the strip adjustment is based on a combination of performance expectation, application data specifics, and system parameters. For example, if the flight data, platform georeferencing solution, is not available, only the data driven approach can be used. Furthermore, for an application with modest accuracy requirements, a simpler data driven technique may be sufficient. But a more demanding data quality calls for a sensor calibration-based method. For longer, corridor type of surveys, clearly the second approach is required, as the overlap area is relatively limited for a corridor and only a well-calibrated system can provide better overall data accuracy. In this case, additional calibration strips are advised, such as a few cross-strips over an area with relatively good surface features.

Regardless of which model is used for strip adjustment, the main difficulty in implementing an adjustment, based on Equation 8.7 or 8.9, is how to establish and measure the discrepancies between the two corresponding 3D datasets. The characteristics of the LiDAR data, including irregular point distribution, point density, error budget, and so on, make the determination of 3D discrepancies between overlapping LiDAR strips or between a LiDAR strip and control information (such as a reference surface) nontrivial. In Section 8.4.1, three common types of surface representations, namely, point clouds, 2.5D and continuous representations are presented. The latter two types of representation require additional algorithms to be derived them from raw LiDAR point clouds, detailed in Section 8.4.1. Once the representations are chosen, the discrepancies can be measured with surface matching algorithms presented in Section 8.4.2.

8.4.1 SURFACE REPRESENTATION

The surface representation in general mapping can be described with the level of details (LOD) (Arefi et al., 2008). DEM, digital terrain model (DTM), or DSM can be defined as the zero level of LOD (LOD0). The next class is the LOD1 that represents the buildings as 3D blocks, and is able to describe vertical surfaces as opposed to the 2.5D surface representation of LOD0. LOD2 provides more details, such as the 3D contours of roofs. LOD3 is a detailed architectural model of the buildings including the representation of façade details. Although LOD1-3 can be extracted from aerial LiDAR datasets and there are several promising automatic and semiautomatic methods, some are presented in Chapters 14 through 19, but they are not used for strip adjustment due to their complexity. However, in the future, their potential in strip adjustment is likely to be exploited; especially in quality assurance and quality control, when the acquired LiDAR data is compared with LOD1-3 models created from other sensors, for instance, terrestrial laser scanner. Currently only LOD0 models, namely surface point clouds, DEMs, DSMs, are typically used in strip adjustment.

The performance of strip discrepancy determination is probably the most critical factor of the whole strip adjustment process. The estimation of strip discrepancies requires defining surface representation, such as point, TIN, grid, analytical surface, and so on (Han et al., 2006). In the next sections, three types of surface representations are presented, namely, the point cloud, 2.5D data structure, and continuous or trend surface.

8.4.1.1 Point Cloud

Several methods can directly estimate the 3D relationship between point clouds. The iterative closest point (ICP) algorithm, discussed in Section 8.4.2.1, is the most widely used approach in practice (reference). If LiDAR intensity data is available, this information can be included in the ICP process to improve the matching performance. The main advantage of the these methods is the direct use of measurement data, as there is no need for interpolation or other data modification or feature extraction, and so on, and thus no information lost or degraded. A minor disadvantage of ICP is the increased requirements for computational power. Tools such as the open source CloudCompare (2017) are widely available.

8.4.1.2 Surface Representation with 2.5D Data Structures

To measure the discrepancies between corresponding surface patches, overlapping areas, 2.5D structures, including TIN and raster surface representation, can be derived from the surface points over these areas. 2.5D data structures are applicable if the surface is defined with Equation 8.2., that is, any 2D surface point has only one elevation value. The main advantage of using 2.5D raster data is that all the image-processing tools can be directly used for processing; the raster elevation data is the equivalent of intensity data of an optical image. In addition, DEM, DSM, and many other geospatial products are typically stored in 2.5D data structure. Various DSM and DEM generation and filtering methods are presented in Chapter 11, and an extensive discussion on digital elevation and surface models can be found in Maune (2007) and El-Sheimy et al. (2005). Here, the most common 2.5D storing and generation techniques are described with a focus on approaches relevant for strip adjustment.

Predominantly, 2.5D data are used for 3 DoF strip adjustment, and the discrepancy between the strips is frequently split into two components: two horizontal displacements and the vertical difference. The importance of these datasets in strip adjustment is that many geospatial products acquired with other technologies, such as stereo photogrammetry, radar, are available in this format, and thus, these datasets can be used for comparisons. Regardless of the source of the elevation data, the raster format is the main distribution structure for national elevation models, such as the National Elevation Dataset in the United States (Gesch et al., 2002), where the DEM is available in the form of GeoTIFF files.

Traditionally, 2.5D data is used in two main structures: TINs and regular grid (raster format). TIN approximates the surface with triangles, where the corners of the triangles are the points of the dataset. The sides of the triangles, typically derived with Delaunay triangulation, do not cross each other. This method finds that optimal triangle arrangement when the sum of all triangles' area and the average of their angles are minimal. TIN is a widely used format, with several advantages for storing and processing surface data. Furthermore, it directly provides for interpolation, and thus, it frequently serves as an intermediate format from which the raster representation is derived.

At the introduction of LiDAR technology, the vast amount of data obtained presented a formidable challenge for the data processing and storage systems, but by now the rapid technological advancements of computer technology make it easy to process and store in any of the two major formats. In addition, the grid or raster format is broadly supported on most computer platforms, offering easy visualization and basic processing tools. To derive gridded surfaces (regular or irregular) from irregularly spaced data over the selected patch or overlapping area requires the estimation of elevation values at the grid points, which is based on the interpolation of the original data. The interpolation methods can be grouped based on how they determine this function. One important aspect is whether the interpolation is local or global. Local techniques use small number of neighboring points to determine the unknown elevation. In contrast, global methods consider the entire surveyed area. In practice, the last approach is not applicable due to the large number of unknown coefficient needed for the global interpolation function, resulting in likely numerical instabilities. Consequently, local methods are preferred.

Local methods separate the area into small regions. Then, interpolation functions are applied within the boundaries of these regions. Local methods can be grouped on the basis of the smoothness of the transitions at the boundaries of the regions. For instance, a simple linear interpolation guarantees that the connecting points between surface patches are the same, but the gradients are not, and thus, the transition may not be smooth. This is called zero-order continuity (C0). In the case of first order continuity (C1), the equality of the first derivatives between the two connecting segments are provided; in the case of second order continuity (C2), the second derivatives are equals too. Applying to surfaces, C0 preserves the elevation, C1 preserves the slope characteristic, and C2 preserves the curvature of the surface. Today, in general, C2 is expected from the interpolation method.

Mostly, because of terrain characteristics and working with undersampled data, no single interpolation technique works best in every situation. This description provides a review of the existing, most commonly used interpolation methods for terrain data.

Inverse distance weighted interpolation determines the unknown values as a weighted average of the surrounding sample points. The weights are a function of the inverse distances between the unknown and the measured points. This method works for both regular and irregular point distribution, and requires a parameter such as search radius to select sample points for the interpolation around an unknown point. Generally, it has drawbacks for irregular point distribution, as shown in Figure 8.6 for LiDAR data by the surface valleys between the LiDAR points.

Natural neighbor interpolation is a more general form of *inverse distance weighted*, as it uses area-based weighting to determine the unknown elevations. Consequently, this widely used

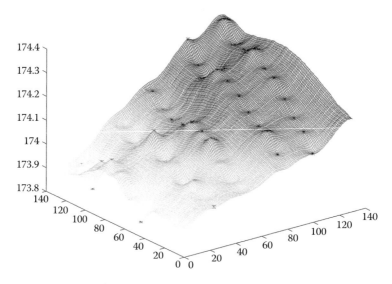

FIGURE 8.6 Dimpled effect around measured points in the case of *inverse distance weighted.*

method works for both irregular and regular point distributions. The algorithm is based on the Delaunay triangulation and Voronoi diagram (Okabe et al., 2000). The Voronoi neighbors of the unknown point are determined in a way that the insertion of the unknown point would result in the Voronoi neighborhood of the measured points. Then these Voronoi neighbors are used to determine the elevation of the unknown point. One advantage of the method is that as the method itself determines the surrounding measured points to be included in the determination of the unknown point, there is no need for additional parameters (search radius) that have drawbacks when the data points have an irregular distribution.

After the points to be included in the calculation are found, the unknown elevation is calculated as a weighted average of these elevations. The weights are based on the common area of the Thiessen polygon of the unknown point (Heywood et al., 1998) and the Thiessen polygons of each selected point before the new point has been inserted to the triangulation. The interpolated surface passes through the sample points and it is constrained by the input data range, and, therefore peaks and valleys appear in the interpolated surface only if they were measured.

Spline interpolation fits a mathematical function, defined by piecewise polynomials of varying orders, to some neighboring measured points to determine the value at the unknown locations, which may be compared with bending a sheet of rubber through the measured points (Loan and Charles, 1997). The lowest order spline is equivalent to linear interpolation, or bilinear transformation for surfaces. Spline interpolation results in a C2 smooth surface that passes through the sample points and, in general, gives a good representation of smoothly changing terrain (no sudden elevation changes, such as buildings and other man-made objects). This is a very useful method if the goal is to derive good quality contours.

Another technique is *kriging*, named after a South African mining engineer D.G. Krige who developed the technique (Oliver and Webster, 1990). Kriging is a geostatistical approach that estimates the unknown values with minimum variances if the sample data fulfills the condition of stationarity, which means that there is no trend in the data, such as main slope. Similar to some other methods, kriging calculates the unknown values as a weighted average of the sample values; however, the weights are based not only on the distance between the sample points and unknown point, but also on the correlation among the sample points. The first step in kriging is the determination of the variogram, which is found by plotting the variances of the elevation values of each sample point with respect to each other sample point versus the distance between points. Once the experimental variogram is computed, the next step is to fit a mathematical function to the variogram that models the trend in it. This fitted polynomial is called the model variogram, and is then used to compute weights for the sample points for the calculation of unknown elevations. This method is called ordinary kriging, and if stationarity of the data is not fulfilled, the more universal kriging can be used. Kriging works with regular and irregular sample point distributions and is an exact interpolation method, as the interpolated surface passes through all the sample points.

Orthogonal transformation-based methods reconstruct a surface function (2.5D) by using a linear combination of a set of orthogonal basis functions. For example, Fourier and wavelet transformations transform the data from time or space domain to frequency domain. By inverse transformation using the coefficients, the original surface can be reconstructed and the surface values can be calculated at the unknown locations too. The numerical methods of forward and backward transformation work only for regular point distribution; therefore, in their original form these series are not directly useable for the interpolation of irregularly distributed sample points. However, the coefficient determination can be formulated as a least squares problem, allowing the extension of the method to irregular point distribution (Toth, 2002). Figure 8.7 shows a LiDAR surface model that was created by a combination of trend surface analysis and Fourier harmonics-based modeling.

The use of interpolation generally introduces error in the data due to the fact that elevation values are calculated at locations where no measurement is available. In rare cases, when the sampling criterion is met, which actually could be easily achieved for flat areas, this error is not present. However, in these cases, the strip adjustment suffers from the lack of the surface signal (terrain undulation), as no 3D discrepancies can be determined between the two surfaces (based on range data). In the general case, the LiDAR data tend to be undersampled and, consequently, if interpolated, errors will be introduced. Therefore, strip adjustment methods that can directly deal with irregular data should be preferred.

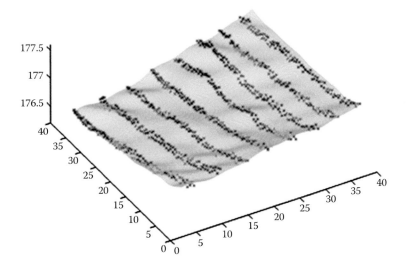

FIGURE 8.7 Fourier series approximation with a third-order polynomial extension.

8.4.1.3 Surface Representation with Continuous or Trend Surfaces

Theoretically, all presented interpolation functions so far are continuous over the data points. In contrast with 2.5D surface interpolation techniques that solve Equation 8.2, in this section, 3D surfaces are presented in the following form:

$$f(x, y, z) = 0 \qquad\qquad (8.10)$$

This implicit function is able to describe more general surfaces as opposed to surfaces given by Equation 8.2. The practical advantage of this surface representation is to allow for 6 DoF matching. Another advantage is that continuous surfaces enable using analytical tools that provide faster and more robust surface matching algorithms. In practice, the estimation of Equation 8.10 is challenging and complex. Consequently, trend surfaces are typically estimated instead of interpolation functions. Trend surfaces approximate the surface by fitting a polynomial to the sample points. This method supports filtering of the data to reduce the error. However, if the trend surface does not correctly model the real surface, then details might be lost. For this reason, regression surface is recommended to be used only for small areas, such as for patches. The proper order of the polynomial can be chosen considering the rule that any cross-section of an n-order surface can have at most $n - 1$ alternating maxima and minima. The coefficients of the fitted polynomial to the sample data can be determined by least squares adjustment, minimizing the square sum of the differences between the z values of measured points and the fitted surface at the sample points. Obviously, increasing the order means a better fit at the sample locations, but between the points it can result in large and sudden changes due to the lack of constraint. Therefore, in common practice, the order of the polynomial normally does not exceed five. Finally, plane fitting is very common for patches.

8.4.2 Surface Matching

Both the data-driven and sensor model-based strip adjustment requires accurate matching of the surfaces to measure the discrepancies that are minimized in Equations 8.7 and 8.9, respectively. Although the visualization of strip discrepancies provides an easy assessment for human evaluation, the quantification of the differences between strips is rather difficult because of the lack of any point correspondence. Therefore, many of surface matching techniques used in LiDAR strip adjustment require the establishment of some correspondence between the two datasets, such as by introducing virtual points in one dataset by interpolation or using features. On the other hand, there are methods that require no correspondence at all and can directly work with irregularly distributed point data, such as ICP.

Surface matching, the automatic coregistration of surface point clouds representing 3D surfaces, is an essential and rather difficult step of any strip adjustment technique. The purpose of this task is to provide the measurements for the identification and determination of strip discrepancies. During surface matching, the surfaces being compared are typically transformed into representations in which the comparison of surfaces is straightforward. The surface matching strategies can be grouped as follows:

- *Vertical-only or 3D matching*: As the LiDAR accuracy is considerably better in the vertical direction, most of the earlier methods were only concerned with the vertical correction, whereas current techniques are usually based on a 3D geometrical model, including 3D shift and rotation.
- *Pairwise or multiple surface matching*: In computer vision, the general problem of matching 3D range data is defined for multiple surfaces with no restriction on the complexity of 3D surfaces. However, this universal model is rarely needed in a typical LiDAR strip

adjustment, which is mostly concerned with matching two surfaces, which are less complex and usually represented in a 2.5D raster format (DEM). The sensor parameter-based strip adjustment methods usually impose certain requirements on the data, in terms of overlap configuration; thus, multiple surface matching techniques can be applied in these cases, such as for overlap areas in multiple cross-strips.

- *Point, surface, or feature-based matching*: There are methods that can directly work on the original LiDAR point cloud or on derived raster data, whereas others are based on using features, such as planar patches. In those cases, when the variation of the surface is moderated, a feature-based methods may promises better results.
- *Ground control use*: LiDAR strip adjustment by definition is a relative correction, as all the observations are differential, and hence the elimination or reduction of strip discrepancies does not mean that the absolute accuracy of the LiDAR point clouds will be better after the strip adjustment, regardless of whether data-driven or sensor calibration-based methods are used. Therefore, the use of ground control is always necessary to assess overall point cloud accuracy. Most strip adjustment methods can accommodate the use of ground control.
- *Manual or automated methods*: Although most of the strip adjustment techniques are self-contained and require no or little human interaction, there are still a large number of frequently used techniques that are partially based on operator measurements. This is mainly due to the use of ground control, which is less automated in general, although depending on the type of ground control, the need for human interaction can be totally eliminated, such as using LiDAR-specific ground targets or a calibration range.

In general, two main parameters have to be defined for LiDAR data matching. The first is how to measure the discrepancies between the surfaces, and the other one is what type of surface representation is used. These two properties are coupled; hence, in the next sections, the methods used for matching are presented based on which surface representation is applied.

8.4.2.1 Point Cloud Matching

The most popular point cloud matching, the ICP algorithm, is commonly used for navigation in computer vision and robotics, and for point cloud coregistration in terrestrial laser scanning applications. The classical ICP (Besl and McKay, 1992; Madhavan et al., 2005) attempts to find the optimal transformation between two point clouds by minimizing the average distance of corresponding point pairs:

$$\min_T \frac{1}{N(Q)} \sum_{i=1}^{N(Q)} \langle match(q_i, P, \langle .,. \rangle), T(q_i) \rangle \tag{8.11}$$

where:

$P = \{(x_i, y_i, z_i)\}, i = 1..N(P), Q = \{(x_j, y_j, z_j)\}, j = 1..N(Q)$, are point clouds
$q_i \in Q, T$ is the transformation function
$N(X)$ is a function that returns the size of point cloud X

In the context of strip adjustment, P, Q are the corresponding patches or overlapping areas. $match(q_i, P\langle .,. \rangle) = p \in P$ is a matching or data association function that finds the closest p point from q_i based on the $\langle .,. \rangle$ distance metric. Typical distance metrics, used in strip adjustment applications are presented in the following. Note that the distance metric applied in the *match* function usually is the same as used for measuring the discrepancy to provide better convergence. P is often referred as the template cloud, and Q is called data cloud that is fitted to the points of P. In the special case, when the discrepancies are measured with Euclidean distance between the corresponding points, Equation 8.11 becomes

$$\min_{R,t} \frac{1}{N(Q)} \sum_{i=1}^{N(Q)} \| p_i - Rq_i - t \|_2^2 \tag{8.12}$$

where:
$p_i = match\left(q_i, P, \|.\|_2\right)$ is the closest point to q_i
R is the rotation matrix
t is the translation vector

Equation 8.12 was the original optimization formula for the ICP algorithm.

Generally, the challenge is to find the appropriate *match* data association function. ICP iteratively tries to find the best correspondences by bringing closer the data cloud to the template cloud. The pseudocode of the ICP algorithm is in Table 8.2.

In the first step, the template and data point clouds are defined as inputs (Step 1), and then initial values for the transformation and other parameters are required (Steps 2 and 3). The next step is to downsample the data by selecting possible candidates (Step 4). Glira et al. (2015a, 2015b) examined four types of selection methods for sensor-based strip adjustment, including random, uniform, normal space, and maximum leverage sampling, and proposed using maximum leverage sampling that chooses those points that has large impact on the parameters. The ICP iteratively solves the problem of Equation 8.10, (Step 5). Next (Step 6), the set of candidate points is initialized (P_t) in each iteration. This set is used for managing the rejection during the data association. Then, q_i goes through the data point cloud (Step 7), and the algorithm assigns p_i to q_i on the basis of the data association function to establish candidate correspondences (Step 8). Note, the $\langle .,. \rangle$ distance metric has to be defined for the data association. There are three main types of commonly used distance metrics. The point-to-point distance metric is the Euclidean distance between the associated points. The point-to-surface distance metric uses the local normal vector during the minimization, which is computed on the basis of the covariance matrix, calculated from the surrounding points located around the examined point within a predefined radius. This metric is widely used and provides fairly robust results. The surface-to-surface metric is based on the local normal vectors computed from the two point clouds. All three approaches may consider additional information, such as intensity data. The *match* data association function typically uses the defined distance metric to find the closest point of p_i for q_i such that $\langle p_i, q_i \rangle = \min$. The points are generally organized in a Kd-tree data

---

TABLE 8.2

Pseudocode of the Iterative Closest Point Algorithm

1   Inputs: P, Q
2   Initial guess: T
3   Initialization: $e \leftarrow Inf$
4    $P_s = select(P); Q_s = select(Q)$
5   While $e < e_t$.
6        $P_t = \{\varnothing\}$
7   For each $q_i \in Q_s$
8            $p_i \leftarrow match(q_i, P_s, \langle .,. \rangle)$ (*data association*)
9   if $\neg is\_rejected(p_i, q_i)$
10            $P_t = P_t \cup p_i$
11        $[e,T] \leftarrow \min_T \frac{1}{N(q)} \sum_i p_i, T(q_i), p_i \in P_t$ (*find optimal transformation*)
12        $Q_s \leftarrow T(Q_s)$

structure to speed up the data association, see Zhang, 1994. As a next step, the associated candidate correspondences can be filtered to reject outlier associations from the further calculation based on a predefined condition (Step 9). The goal of the rejection is to improve the robustness of the system to detect and remove false matchings. The rejection may depend on the difference of the normal vector of the points or the point-to-surface metric. If the possible candidate does not passes the rejection criteria, it is added to P_t, which is the set of the possible candidates (Step 10).

In step (11), the algorithm estimates the optimal parameters for transformation T between the selected and associated points and apply this transformation to all points from Q_s. The type selection of transformation T is optional. Mostly, a rigid body transformation is assumed allowing for a 6DoF transformation; three rotations and three transitions. Alternatively, lower DoF transformations can be used; for instance, 3 transitions when no rotation errors are assumed, or a shift just for vertical correction. The matching function can also reject correspondences based on predefined metrics and thresholds. The minimization problem can include weights to consider the uncertainty of the correspondence matching. The problem can be solved with singular value decomposition if small rotation angles are assumed. In general, for example when the transformation is more complex, other optimization techniques, such as Levenberg–Marquardt method can be used.

The data association is not likely to be correct after the first iteration because the two point clouds are far from each other in the beginning, and thus, the algorithm may not be able to find the best correspondences in the first iteration step. If the initial guess was close to the final result, then the algorithm converges, and the algorithm terminates, if the e error metric is smaller than a predefined e_t threshold (Step 5).

One issue with ICP is that the algorithm, see Equation 8.11, is ill-posed in practical applications, because the data points are not sampled from the same sensor point in the object space across the strips. Thus, the optimization never results in zero on real datasets; in other words, even if the point coordinates are measured perfectly, that is, there is no measurement error. The other disadvantage of the method is that it has large computation needs. In addition, ICP requires good initialization of the transformation parameters due to the multidimensional problem space, which may has several local minima; therefore, the initial values have to be close to the real values. To address this problem, global optimization techniques might be used, however, it requires additional computation. Furthermore, classic ICP expects full overlaps between the areas; note that some newer versions are able to consider smaller or partial overlap. In summary, ICP is an easy to use method for data driven strip adjustment, and also can be used for sensor-based strip adjustment (Glira et al., 2015a, 2015b).

8.4.2.2 Matching with 2.5D Data

Until recently, the most widely used as well as the simplest surface matching technique is based on using raster format surface models, or DEMs (2.5D data), in which case the two datasets are available in the same evenly spaced grid, and, thus, vertical differences between them can be easily computed. The problem with this representation is that for terrains with surface normal vectors considerably deviating from the vertical, the simple vertical difference between the two datasets may not be the right measure for the surface discrepancy, as it does not account for horizontal differences and ignores their impact on the vertical differences. Considering the difference along the surface normal will result in the determination of correct 3D surface discrepancies, in which case, the volume between the two surfaces is minimized during the adjustment (Figure 8.8).

The main advantage of using 2.5D surface data representation is that standard image processing techniques can be directly applied to the data, including area- and feature-based matching methods. Considering height as though it were image intensity is a very useful interpretation from a matching perspective, as in both cases the signal changes and patterns are needed to achieve successful matches. Note that basic image matching solutions usually result only in a 2D offset, and additional data and processing are needed to obtain 3D surface discrepancies. If LiDAR intensity image is used for matching, the vertical discrepancy can be directly obtained from the corresponding range data. A clear disadvantage of the DEM representation with respect to strip adjustment is that it is

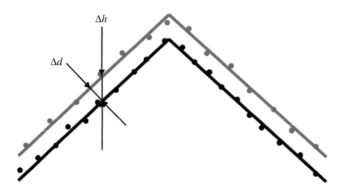

FIGURE 8.8 Distance along surface normal versus height difference.

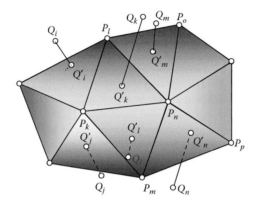

FIGURE 8.9 Distance definition between two point sets.

derived from the primary point cloud data by interpolation and, thus, errors are introduced, which could be further amplified by the fact that the LiDAR point density usually falls below the minimum spatial sampling distance required for complete surface representation.

TIN-based data representation seems to be the natural format for LiDAR, as it can preserve the original 3D information of the point cloud and provide an interpolated value for any location, based on the surface defined by the linked triangles. The determination of surface differences is clearly much more complex than it is for raster data. Surface matching techniques that can handle irregularly distributed points generally considered one dataset in a TIN-based surface representation and then try to minimize the cumulative distance between every point of the other data set and the corresponding triangle along the normal of the triangle (Figure 8.9).

8.4.2.3 Matching Surfaces

In several cases, the surface matching used for strip adjustment is accomplished using the least squares methods to minimize the Euclidean distances between the surfaces and then recover the transformation model parameters. In strip adjustment practice, closeness between the two surface representations is always assumed. Therefore, only a subproblem of the general surface matching, the refinement, or alignment task is addressed. In complex mapping processes, such as the extraction of various natural and man-made object features, the more general problem entails a global search, in which either coarse or no orientation approximations for the two surfaces are available. The general problem of surface matching, also called range data matching, is an important field in computer vision, and a large variety of techniques have been developed over the years; a good review is provided in (Campbell and Flynn, 2001). Here, only the least squares surface matching (Gruen and Akca, 2005) is discussed because of its importance in the geospatial field.

In theory, the functional model of least squares surface matching of two continuous surfaces can be formulated on the basis of using implicit functions as

$$d(x, y, z) = p(x, y, z) - q(x, y, z) \tag{8.13}$$

where:

p and q represent the two surface descriptions of the same object area
d represents the differences between the two samples, introduced by the various sensor errors and environmental conditions

The least squares adjustment minimizes the sum of the squares of the Euclidean distances between all the points of the surface p and the surface q. The computation of these distances requires an interpolation of the second surface, that is, conjugate point determination, usually accomplished by piecewise surface modeling. The conjugate surface, q^0, must be determined in order that the partial derivatives can be estimated to provide the first-order Taylor series coefficients needed for the least squares solution. Thus, Equation 8.13 becomes

$$p(x, y, z) - d(x, y, z) = q^0(x, y, z) + \frac{\partial q^0(x, y, z)}{\partial x} \mathrm{d}x + \frac{\partial q^0(x, y, z)}{\partial y} \mathrm{d}y + \frac{\partial q^0(x, y, z)}{\partial z} \mathrm{d}z \tag{8.14}$$

The transformation T from Equation 8.7 that describes the spatial relationship between the two surfaces is modeled by n_T number of parameters, $m_i = (0, 1,..., n_T - 1)$. If the transformation T is nonlinear, then it should be linearized by Taylor series expansion. Consequently, the differentials are expressed in model parameters as

$$\begin{bmatrix} \mathrm{d}x \\ \mathrm{d}y \\ \mathrm{d}z \end{bmatrix} = \begin{bmatrix} \dfrac{\partial T_x}{\partial m_0} & \dfrac{\partial T_x}{\partial m_1} & \cdots & \dfrac{\partial T_x}{\partial m_{n_T-1}} \\ \dfrac{\partial T_y}{\partial m_0} & \dfrac{\partial T_y}{\partial m_1} & \cdots & \dfrac{\partial T_y}{\partial m_{n_T-1}} \\ \dfrac{\partial T_z}{\partial m_0} & \dfrac{\partial T_z}{\partial m_1} & \cdots & \dfrac{\partial T_z}{\partial m_{n_T-1}} \end{bmatrix} \cdot \begin{bmatrix} \mathrm{d}m_0 \\ \mathrm{d}m_1 \\ \vdots \\ \mathrm{d}m_{n_T-1} \end{bmatrix} \tag{8.15}$$

and should be substituted back to Equation 8.14.

From this point, a conventional or generalized Gauss–Markoff estimation model can be built, in which each point from surface p constitutes one observation equation. This solution is directly applicable to data-driven methods, in which the most typical transformation is the 7-parameter similarity model. When this technique is applied to the sensor calibration-based model, that is, using Equation 8.9, certain restrictions on the model parameters should be introduced to achieve solutions. Finally, the least squares surface matching can be further generalized for multiple surface matching and can be extended with intensity matching and ground control (see Akca and Gruen, 2007).

All the surface matching methods discussed so far impose no restrictions on the surfaces, except to avoid totally flat surfaces (no surface signal, i.e., no terrain relief). If one of the surfaces to be matched is formed from basic shapes, such as planar patches, the matching process can exploit the simple geometry that can be described analytically. Although several shapes, such as linear, spherical, planar, and conical, have simple mathematical descriptions, they are rare in real life and, consequently, planar patches are used in most methods. Obviously, matching planes have a deficiency, as only the distance between the two datasets can be recovered. Therefore, several patches with reasonably varying surface normal vectors are needed to obtain 3D surface discrepancies.

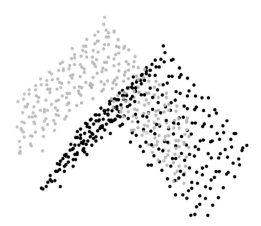

FIGURE 8.10   Matching linear features extracted from the roof intersection of planar roof surfaces.

A good example is saddle roofs, which have quite different surface normal vectors and their flatness is generally good. Similarly, the intersection of planar patches can provide for linear features that can be matched. The advantage of featured-based matching is that there is no imposition on the primary data from where the features were extracted, and thus, it provides a mechanism to match surface data with high-level symbolically described object space data acquired by different sensors; Figure 8.10 shows roof scanned in two different strips.

8.4.2.4   Surface Deformation Modeling

Another key aspect of data matching for strip adjustment is whether strip deformation is allowed or not. Strip deformation occurs if a rigid body model is not adequate to describe the transformation between the ground surface and the LiDAR data. Strips can be deformed for various reasons, but the changing quality of the georeferencing solution is frequently the main cause. Obviously, the concept of strip deformation assumes a spatial or time extent; sensor characteristics and environmental conditions could vary noticeably over longer distances and for extended time periods. For example, the errors of the georeferencing solution can be usually considered static for short time intervals, as discussed earlier. For example, during long and straight strips, such as for strip lengths of 50–100 km and flying time over several hours, the georeferencing solution can show slowly changing errors (drifts), frequently resulting in strip deformation; in particular, this is the case over terrain with significant elevation changes, such as high mountain ranges. Therefore, strips are frequently segmented into smaller sections, which are individually treated using a rigid body model.

Data-driven strip adjustment methods are generally based on a linear 3D transformation, which takes the general form:

$$\begin{bmatrix} x_p \\ y_p \\ z_p \end{bmatrix} = \begin{bmatrix} t_0 & t_1 & t_2 & t_3 \\ t_4 & t_5 & t_6 & t_7 \\ t_8 & t_9 & t_{10} & t_{11} \end{bmatrix} \cdot \begin{bmatrix} x_q \\ y_q \\ z_q \\ 1 \end{bmatrix} \tag{8.16}$$

where the transformation matrix is T in Equation 8.7.

Depending on the complexity of the applied model, which is determined by the data that can be observed and application-specific considerations, the actual number of the independent parameters varies from 1 to 12. The simplest strip correction is based on applying only a vertical shift, which was a typical procedure at the introduction of LiDAR, as the modest point density did not allow for accurate horizontal measurements of strip discrepancies. As technology improved, full 3D offset correction became feasible. The next step was when a similarity transformation was introduced, and since then it has been the most widely used model for data-driven strip corrections. In many cases, the scale parameter is assumed to be unity, and then only the offset and rotation are considered for correction (6-parameter rigid body transformation). The use of the full 12-parameter, 3D affine transformation model is rather rare.

Strip deformation modeling can differ from the sensor calibration-based strip correction methods, in which the deformation phenomenon is modeled through the sensor system parameters. In the general solution, all the terms, including the georeferencing solution and the laser sensor parameters, can be considered as time-dependent parameters, and thus would allow for optimal error modeling and correction. However, as discussed earlier, the implementation of this approach is simply not feasible because of parameter correlation and data dependencies. Therefore, restrictions and simplifications should be introduced to allow for the recovery of the sensor parameters, or more precisely, a subset of these parameters.

In a number of approaches, the georeferencing solution and the laser sensor calibration are separated, and thus the strip adjustment is formulated either for the georeferencing errors or for the sensor parameters. For example, laser sensor calibration parameters may slowly change over longer time periods whereas the georeferencing solution can drift or fluctuate at a much faster rate. Thus, once the laser sensor is calibrated, including range, scan angle, and boresight calibration, the errors in the georeferencing solution can be separately recovered. The georeferencing solution describes the sensor platform trajectory, including three position and three attitude angles as a function of time; for shorter time periods, these parameters could be sufficiently modeled by constant offsets, which leads to the rigid body transformation model discussed earlier. In many cases, this provides an adequate solution, as the georeferencing terms typically account for the largest part of the error budget.

The laser sensor calibration poses more difficulties because of the strong parameter correlation and inseparability of measurements. For example, the ranging error can be only calibrated to the combined accuracy level of the georeferencing solution and the ground control. The usual DEM accuracy is generally below the laser ranging accuracy, which is typically in the 1–3 cm root mean square (RMS) range, whereas the georeferencing positioning accuracy for airborne platforms is in the 5–10 cm range. If a reference DEM (test range) was precisely measured, it will account for at least 1–5 cm error. The removal of systematic scan angle errors is important, as this error results in strip deformation. Using flat surfaces, one can recover the scan angle error, frequently called smiley error, from profile measurements even from a single strip. Both analytical modeling and table lookup methods can be used. The calibration of the laser sensor boresight parameters is based on a rigid body model. To avoid the introduction of any additional error sources, the installation of the laser and georeferencing sensors is expected to provide a stable and rigid relationship between the sensors.

8.5   OVERVIEW OF STRIP ADJUSTMENT TECHNIQUES

A large number of strip adjustment techniques have been developed in the last fifteen years, mostly in the academic environment. Some of them have remained in the research phase, whereas others have made it into mainstream LiDAR production. As the various techniques typically share several similar processing components, there is no simple way to categorize them. The main directions and approaches are presented in the previous section, and the review of actual studies and solutions below follows the historical order of the strip adjustment techniques.

8.5.1   EARLY STRIP ADJUSTMENT

The first widely referenced LiDAR strip adjustment method of transforming overlapping LiDAR strips to make them coincide with each other followed the practice of conventional photogrammetric strip adjustment of airborne imagery (Kilian et al., 1996). Using tie and ground control features (small DEM patches with a size of 10 by 10 m), the process aimed at determining additional

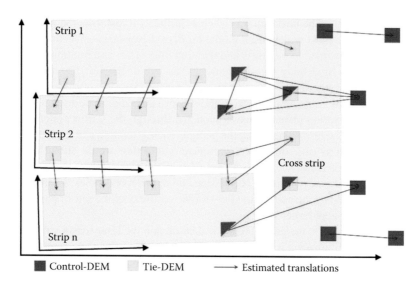

FIGURE 8.11 Tie and control point-based strip adjustment. (From Kilian, J. et al., *Int. Arch. Photogram. Rem. Sens.*, 31, 383, 1996. With Permission.)

transformation parameters for each strip to transform the strips into a homogenous exterior coordinate system, as shown in Figure 8.11. In the first step, the tie and ground control information was extracted, which was then followed by matching the corresponding digital elevation models. The matching results provided three translation parameters for areas with sufficient terrain undulations or object content, or only height difference for flat areas. In the second step, using the tie and ground control information, a standard photogrammetric-style strip adjustment was performed based on a linear drift model that accounted for position and attitude corrections, including three position and attitude offsets with drift parameters, resulting in a 12-parameter model for each strip. The complexity of the model is related to the modest overall data quality of that time, as the LiDAR point spacing was rather sparse and the GPS/GNSS/inertial measurement unit (IMU) solution was rather poor. The technique required substantial user interactions but was able to reduce strip discrepancies from the few meters range to the submeter level.

Several early strip adjustment methods minimize only the vertical discrepancies between overlapping strips or between strips and horizontal control surfaces. To adjust the vertical component was mainly motivated by the poor GPS/GNSS vertical solutions. These strip adjustments can be referred as 1D strip adjustment methods (Crombaghs et al., 2000; Kager and Kraus, 2001; Kornus and Ruiz, 2003). Tie or absolute control features used for this adjustment are flat horizontal surfaces, in which the differences in the vertical direction were estimated by simply averaging about 100+ points (Crombaghs et al., 2000). The problem with this kind of adjustment is that existing planimetric errors are likely to remain in the data.

8.5.2 3D Strip Adjustment

At the introduction of airborne laser scanning, LiDAR data was primarily perceived as a vertical or height measurement, and little attention was paid to the horizontal component. This misconception was mainly related to the low point density, as at a 0.1 pts/m^2 it is almost impossible to notice horizontal discrepancies whereas vertical differences can be easily observed. Therefore, the main concern was to minimize the height differences and simply ignore the often sizeable horizontal offsets. With the advancement of the LiDAR technology, better point density measurements and point accuracy improvements, however, changed the situation and users started to recognize the true 3D nature of the LiDAR data. In particular, the importance of the horizontal component was widely acknowledged. Vosselman and Maas (2001) showed that systematic planimetric errors are often much more significant than vertical errors in LiDAR data and, therefore, a 3D strip adjustment is the desirable solution for minimizing the 3D discrepancies between overlapping strips and at control points. The first 3D strip adjustment methods developed were predominantly based on using raster format LiDAR data and adjusted only for 3D offsets. Also, they were considered data driven techniques, as there was no attempt to rigorously model systematic sensor errors.

The first 3D strip adjustment technique that allowed for the direct use of the irregularly distributed LiDAR data was introduced by Maas (2000, 2002). The method, aimed at providing an efficient technique to determine strip discrepancies, is based on applying least squares matching (LSM) to the LiDAR data structured in a TIN representation. The LSM is performed on appropriately selected overlapping patches taken from different strips. The process tries to iteratively minimize the sum of height differences between the patches by adjusting the shift parameters in all the three coordinate directions. Observation equations are written for every data point of both patches, and the height difference is obtained by projecting a point to the closest triangle in the other patch (Figure 8.9).

The technique requires surface patches with sufficient signal content to allow for the determination of the three shift parameters; basically, at least three noncoplanar surface areas should be in both patches. A rolling terrain with no vegetation is a good candidate for patch selection. Vegetated areas and buildings should be generally avoided because of occlusion. Obviously, the method fails on flat areas without any objects. By using circular patches with about 10 m radius, one can obtain 4–5 cm horizontal and 1 cm vertical precision for the shift parameters at a 0.3 pts/m^2 point density.

An alternative to area-based surface matching is feature-based matching, in which features, typically higher-level data representations, are extracted from the LiDAR point cloud and the matching is performed in the feature domain. Vosselman (2002b) introduced a method to extract linear features, such as roof edge lines, gable roof ridgelines, and ditches, and then used them to determine strip discrepancies. The proper geometrical modeling of the linear features is essential to obtain suitable offset estimates.

8.5.3 STRIP ADJUSTMENTS BASED ON SENSOR CALIBRATION

The parameterization of the strip discrepancies in the physical sensor model and orientation measurement system (the introduction of self-calibration) represented a major milestone in the evolution of LiDAR strip adjustment techniques. Behan et al. (2000) proposed the first 9-parameter model that related to the georeferencing sensor by accounting for three offset, three attitude, and three drift parameters, which provided for GPS/GNSS offset, initial IMU attitude bias, and IMU attitude drift. Burman (2000, 2002) treats the discrepancies between overlapping strips as positioning and orientation errors with special attention given to the alignment error between the INS and laser scanner. The first implementation needed raster data but was later extended to use a TIN structure. Besides providing a solution for stripwise elevation and planimetric differences, the boresight misalignment between the laser scanner and the IMU is determined on the basis of multiple strip discrepancies. The more developed technique that formed the foundation for the TerraMatch product family (Burman, 2002; Soininen and Burman, 2005) can perform matching based on both range and intensity data, and applies corrections directly to the LiDAR point cloud.

Morin and El-Sheimy (2002) introduced a method that is also based on sensor and system error modeling but requires no ground control, as the assumption is that averaging overlapping strips will provide an unbiased estimate of the surface. Toth et al. (2002) presented a method that tries to make overlapping strips coincide, with the primary objective of recovering only the boresight misalignment (three angles) between the IMU and the laser sensor. The technique requires the rasterization of the LiDAR data to perform the strip discrepancy determination. Similar to the previous methods, it only works if a sufficient terrain signal exists.

Filin (2003a, 2003b) presented a comparable method for recovering the systematic errors, which is based on constraining the position of the laser points to planar surfaces (Filin, 2001); linear features can also be used. The model initially considers a large number of error sources (Schenk, 2001), grouped as calibration and system errors. Because of functional correlation and the inseparability of several parameters, the recovery of all the systematic error terms for the total of 14 observations, including three position and three attitude for the GPS/GNSS/IMU system, two for the range and scan angle measurements of the laser sensor (system measurements), and three offset and three misalignment angle errors for the IMU and the laser sensor, however, is not feasible. Therefore, certain assumptions, such as laser sensor scanner controls and object space characteristics, should be made to allow for the determination of a subset of the systematic error parameters. The planar surface patches connecting the LiDAR strips are introduced to the adjustment as ground control objects, with the four surface parameters known, or as tie objects, with surface parameters initially approximated from data and subsequently refined in the adjustment process. An extension of the surface-based strip adjustment method generalizes the control surfaces by allowing for natural and man-made nonplanar surface patches, which are locally approximated by a plane in a piecewise manner (Filin and Vosselman, 2004). The technique requires only the LiDAR point cloud data (no georeferencing data is needed) and is based on minimizing the distance between the laser points and the actual surface. The segmentation is based on the clustering, in which the feature vectors computed for each LiDAR point are classified in an unsupervised way (Filin, 2002). From the initial clusters, a complex validation process will select the tie surface patches based on the fitting accuracy of the analytical surface model to the points in the cluster with adequate upper and lower bound control to avoid under- and oversegmentation. Test results obtained from a block of 20 strips, flown as 10 strip subblocks in perpendicular directions, demonstrated that strip offsets can be reliably recovered and that they are not constant for a block. In addition, the analysis of results revealed that the horizontal offsets in the data are significantly larger than the vertical ones.

Another method introduced by Skaloud and Lichti (2006), which constrains the laser points to planar surface patches, aims at the calibration of the three boresight misalignment angles and the range-finder offset. Similarly to the method introduced by Filin (2003a), the system calibration parameters are modeled within the direct georeferencing equation, and then the laser points are conditioned to lie on a common planar surface patch without the need to know the true surface position and orientation. In contrast to previous techniques, the availability of the system level LiDAR data is assumed, including the sensor trajectory described by three position and three attitude parameters and the laser range and scan angle measurements. The planar surface patch selection and the determination of the surface normal are based on using principal component analysis, which offers computational advantages for larger datasets. Experience showed that the range-offset recovery is somewhat limited, whereas the boresight misalignment angles can be determined to a high accuracy, provided a sufficient number of planar patches with different spatial orientations are available. Numerical results showed remarkably small RMS residuals, such as 2 as for the boresight angles and 1 cm ranging accuracy, whereas height differences fell to the few centimeters range.

8.5.4 Using LiDAR Intensity Data

With the increasing availability of LiDAR intensity data, also frequently called reflectance values, LiDAR strip adjustment methods started to exploit this additional information primarily to support the matching between different strips. As discussed in the previous paragraphs, methods that are exclusively based on the use of the LiDAR point cloud (a mass of points defined by three coordinates) require adequate terrain characteristics, such as planar or smoothly changing surface areas with different orientations, to successfully recover systematic error terms. Large areas with no surface undulations or with limited slope cannot provide for sufficient strip discrepancy determination, in particular in the horizontal direction, and consequently any 3D adjustment will fail in such cases. Intensity data, now a standard output on modern LiDAR systems, complements the blind

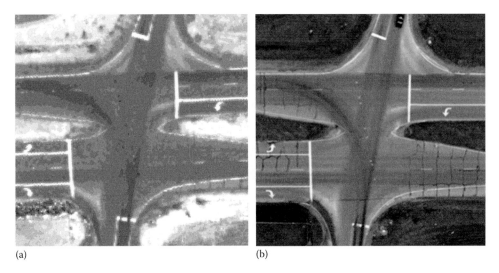

(a)   (b)

FIGURE 8.12   Pavement marking appearance in (a) LiDAR intensity image and (b) reference optical image.

LiDAR point cloud with a conventional image-type of data, which is similar to an image produced by a single spectral band of a hyperspectral camera. As LiDAR intensity generally provides more variation in terms of image texture or contrast, compared with elevation data, therefore, it can support matching in areas where the height differences are limited or nonexistent. A good example, as shown in Figure 8.12, is the transportation road network (which generally represents locally flat areas), in which ubiquitous pavement markings are clearly visible in the intensity image and thus can be routinely matched. Obviously, the image domain matching only provides for the determination of horizontal offsets. As LiDAR intensity and range data are perfectly coregistered, the intensity domain matching results can be directly converted to 3D strip discrepancies.

One of the first methods exploiting the use of intensity data was reported by Maas (2001). In the proposed two-step implementation, the vertical and horizontal offsets are determined by two independent LSM processes of height and intensity data, and then the method described in (Maas, 2000) is used. Although the modest laser point density imposes limits on the matching performance in the image domain, experimental results indicated that a significant improvement can be achieved in the horizontal component, whereas the vertical performance remains unchanged. Vosselman (2002a) provides a comprehensive analysis of the LiDAR intensity signal, including (1) the formation of the signal at the sensor level, such as instantaneous or integrated sampling, (2) the problem of spatial undersampling, in which point spacing is not sufficient to represent the surface, in particular, in urban settings, (3) the impact of multiple reflections at object boundaries, and (4) the impact of the footprint size. The combined effect of these errors results in relatively noisy characteristics of the intensity image and, therefore, feature-based matching, such as matching edges, is suggested for planimetric offset determination. To compensate for the undersampling and finite footprint size, a gray value edge modeling is introduced and the use of longer linear features is advised.

8.5.5   MISCELLANEOUS TECHNIQUES

A modern data driven technique for LiDAR strip adjustment introduced by Bretar et al. (2004) provides a general solution for surface matching and is based on the formation of a homogenous 3D deformation model between two digital surface models (Figure 8.13). The 12-parameter affine transformation is quite different compared with the conventional rigid body transformation, which is based on simple translation and rotation. The determination of the affine model is achieved by local 3D translations using a modified Hough transformation (Hough, 1959). The technique can be

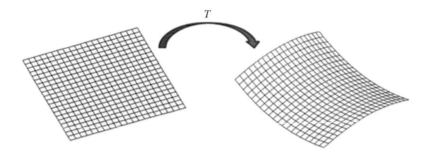

FIGURE 8.13   3D affine model used for surface fitting.

applied to coregister a reference surface to LiDAR data or LiDAR strips to each other or any two digital surface models.

Another data driven method that provides a methodology to avoid directly employing conjugate features in the strip adjustment process was introduced by Han et al. (2006). The technique is based on using the contour tree representation of the surface data and employs a 3D conformal transformation model. In an iterative process, the transformation parameters are sequentially refined until the leaves of the contour tree surface representation reach a minimum. The execution of the process is rather computationally intensive.

8.6   SUMMARY

The success of strip adjustment depends on a lot of factors, including object space specifics, such as surface geometry and material characteristics, and ground control; sensor and system configuration and specifications; airborne survey parameters, including flight parameters and strip patterns; and the method and transformation model used. The combination of all of these will determine the improvement potential of the strip adjustment, that is, to define the performance expectations in terms of error estimates and to assess the quality of the LiDAR product after corrections are applied. In other words, it is imperative to understand the limitations of the strip adjustment process.

First, the surface representation of the surveyed area should be considered, and the LiDAR point cloud should have an average sampling rate close to the sampling distance defined by the Nyquist criterion over the selected surface patches where the strip discrepancies will be determined. It is fair to say that current LiDAR data are typically undersampled and, therefore, care should be exercised when the patches are selected. Next, the object composition of the surveyed area should be considered. Vegetated areas should be avoided and hard surface areas with comparable material signatures should be selected. An additional aspect in selecting the patches is the object shape requirement, if any, which is based on the methodology used for the strip adjustment. For example, some methods require man-made objects with hard surfaces, such as building and roads, whereas others need a sloping terrain with modest or no vegetation. In parallel, the reflectance characteristics of the surface materials should be also considered as they determine whether a good return signal can be obtained. The impact of the footprint size is also very critical; if it is relatively large then the extent to which the object space is restricted can be observed. Finally, the availability of ground control is important; otherwise, the validation of the data in absolute terms is not feasible.

The sensor parameters and configuration jointly define the performance potential of any LiDAR system, and the objective is to approach it as closely as possible under normal operational conditions. The laser sensing unit is generally characterized by its ranging and scanning accuracy. Modern systems support extremely good ranging accuracy, typically 1–2 cm (1σ), up to relatively high flying heights (1000 m AGL). As the moving parts in the laser sensor have shrunk over time, and these smaller components can be better controlled, this has resulted in faster and more even mirror motion, for example. In addition, the encoding performance has improved and, thus, the mirror position can be calibrated to a high accuracy.

In contrast to the laser sensor, the performance of the georeferencing component can vary a great deal and, generally, the georeferencing errors account for the largest contributor of the overall LiDAR error budget. Coincidentally, the georeferencing is a system component that can be significantly influenced by user control and, therefore, is a key part of flight planning. In the simplest approach, the quality of the differential GPS/GNSS solution determines the overall georeferencing performance, provided that, at least, a medium-grade (i.e., tactical grade) IMU is used. The GPS/GNSS data quality of both the airborne platform and ground reference are equally important to the DGPS solution, as well as their separation (base distance or virtual base for network solutions). With proper flight planning and execution, accurate georeferencing solutions can be obtained, resulting in smaller strip adjustment corrections, if any, and, consequently, better overall LiDAR product quality.

The selection of the strip adjustment technique for a particular application primarily depends on the object space characteristics and the expected performance of the strip adjustment, which is mainly determined by the sensor parameters. Additional operating aspects, including flight control, DGPS availability, processing resources and environment, and time requirements may also influence this selection process. From the large variety of the available strip adjustment methods, several perform relatively well in most situations, whereas specific conditions require more application-tailored solutions. In all cases, regardless of the actual strip adjustment method used, the object space and sensor parameters jointly define the quality of the LiDAR-derived surface, which is characterized by the completeness of the surface representation and its accuracy (defined by the point cloud) and, consequently, how well strips can be compared with each other and ultimately corrected for differences.

Strip adjustment techniques continue to evolve primarily driven by technological advances. Most importantly, as multipulse systems develop further, the higher pulse rates will result in better LiDAR point cloud densities. This will substantially improve surface representations, allowing for the observation of smaller object features. In a parallel development, georeferencing technology has made significant progress recently and further advances are expected, resulting in more robust and accurate solutions under almost any conditions.

From the algorithmic point of view, the current trend is to move toward sensor calibration-based strip adjustment techniques. In addition, due to improving feature extraction performance and fusion with optical imagery, strip adjustment methods will probably be combined with higher level feature extraction mapping processes in the future. As sensor performance, including laser and georeferencing sensors, continues to improve, the role of strip adjustment will slowly shift toward QC, becoming a primary tool for assessing relative data quality.

REFERENCES

Arefi, H., Engels, J., Hahn, M., and Mayer, H. 2008. Levels of detail in 3D building reconstruction from LIDAR data. *ISPRS Archives of the Photogrammetry, Remote Sensing and Spatial Information Science*, 37 (Part B3): pp. 485–490.

Ackerman, F. 1994. Practical experience with GPS supported aerial triangulation. *The Photogrammetric Record*, 14(84): 861–874.

Akca, D. and Gruen, A. 2007. Generalized least squares multiple 3D surface matching. *International Society for Photogrammetry and Remote Sensing*, 36(Part 3/W52): 1–7.

ASPRS LiDAR Committee. 2004. ASPRS Guidelines Vertical Accuracy Reporting for LiDAR Data. Retrieved from http://www.asprs.org/society/committees/lidar/Downloads/Vertical_Accuracy_ Reporting_for_ Lidar_Data.pdf.

Baltsavias, E.P. 1999. Airborne laser scanning: Basic relations and formulas. *ISPRS Journal of Photogrammetry and Remote Sensing*, 54: 199–214.

Behan, A., Maas, H.-G., and Vosselman, G. 2000. Steps towards quality improvement of airborne laser scanner data. *Proceedings of the 26th Annual Conference of the Remote Sensing Society,* Leicester, September 12–14, 9 pp, CD-ROM.

Besl, P.J. and McKay, N.D. 1992. A method for registration of 3-D shapes. *IEEE Transactions on Pattern Analysis and Machine Intelligence,* 14(2), 239–256.

Bretar, F., Roux, M., and Pierrot-Deseilligny, M. 2004. Solving the strip adjustment problem of 3D airborne Lidar data. *Proceedings of the IEEE IGARSS'04,* Anchorage, Alaska, September 2004.

Burman, H. 2000. Adjustment of laser scanner data for correction of orientation errors. *International Archives of Photogrammetry and Remote Sensing,* 33 (Part B3): 125–128.

Burman, H. 2002. Laser strip adjustment for data calibration and verification. *International Archives of Photogrammetry and Remote Sensing,* 34 (Part 3A): 67–72.

Campbell, R.J. and Flynn, P.J. 2001. A survey of free-form object representation and recognition techniques. *Computer Vision and Image Understanding,* 81(2): 166–210.

CloudCompare. 2017. http://www.cloudcompare.org/

Crombaghs, M.J.E., Brugelmann, R., and de Min, E.J., 2000. On the adjustment of overlapping strips of laser-altimeter height data. *International Archives of Photogrammetry and Remote Sensing,* 33 (Part B3/1): 224–231.

Csanyi, N. 2008. A rigorous approach to comprehensive performance analysis of state-of-the- art airborne mobile mapping systems, PhD dissertation, The Ohio State University, Columbus, OH.

El-Sheimy, N., Valeo, C., and Habib, A. 2005. *Digital Terrain Modeling: Acquisition, Manipulation, and Applications.* Boston, MA: Artech House.

Filin, S. 2001. Recovery of systematic biases in laser altimeters using natural surfaces. *International Archives of Photogrammetry and Remote Sensing,* 34, (3/W4): 85–91.

Filin, S. 2002. Surface clustering from airborne laser scanning data. *International Archives of Photogrammetry and Remote Sensing,* 34 (3A): 117–124.

Filin, S. 2003a. Recovery of systematic biases in laser altimetry data using natural surfaces. *ISPRS Journal of Photogrammetric Engineering and Remote Sensing,* 69(11): 1235–1242.

Filin, S. 2003b. Analysis and implementation of a laser strip adjustment model. *International Archives of Photogrammetry and Remote Sensing,* 34 (Part 3/W13): 65–70.

Filin, S. and Vosselman, G. 2004. Adjustment of laser altimetry strips. *International Archives of Photogrammetry and Remote Sensing,* 34 (Part 3/W13): 285–289.

Gesch, D., Oimoen, M., Greenlee, S., Nelson, C., Steuck, M., and Tyler, D. 2002. The national elevation data-set. *Photogrammetric Engineering & Remote Sensing,* 68(1): 5–13.

Glira, P., Pfeifer, N., Briese, C., Ressl, C. 2015a. Rigorous strip adjustment of airborne laser scanning data based on the ICP algorithm. *ISPRS Annals of the Photogrammetry, Remote Sensing and Spatial Information Sciences,* Volume II-3/W5, 2015.

Glira, P., Pfeifer, N., Briese, C., and Ressl, C. 2015b. A correspondence framework for ALS strip adjustments based on variants of the ICP algorithm. *Photogrammetrie-Fernerkundung-Geoinformation,* 2015(4): 275–289.

Gruen, A. and Akca, D. 2005. Least squares 3D surface and curve matching. *ISPRS Journal of Photogrammetry and Remote Sensing,* 59(3): 151–174.

Heywood, I., Cornelius, S., and Carver, S. 1998. *An Introduction to Geographical Information Systems.* Upper Saddle River, NJ: Prentice Hall.

Han, D., Lee, J., Kim, Y., and Yu, K. 2006. Adjustment for discrepancies between ALS data strips using contour tree algorithm. *Advances Concepts for Intelligent Vision Systems,* LNCS 4179, pp. 1026–1036.

Hough, P.V.C. 1959. Machine analysis of bubble chamber pictures. *International Conference on High Energy Accelerators and Instrumentation,* CERN, Geneva, Switzerland.

Kager, H. and Kraus, K. 2001. Height discrepancies between overlapping laser scanner strips. *Proceedings of Optical 3D Measurement Techniques* V, October, Vienna, Austria. pp. 103–110.

Kilian, J., Haala, N., and Englich, M. 1996. Capture and evaluation of airborne laser scanner data. *International Archives of Photogrammetry and Remote Sensing,* 31 (Part B3): 383–388.

Kornus, W. and Ruiz, A. 2003. Strip adjustment of LiDAR data. Dresden: *International Archives of Photogrammetry and Remote Sensing,* 34(3/W): 47–50.

Loan, V. and Charles, F. 1997. *Introduction to Scientific Computing.* Upper Saddle River, NJ: Prentice Hall.

Maas, H.-G. 2000. Least squares matching with airborne laserscanning data in a TIN structure. *International Archives of Photogrammetry and Remote Sensing,* 33 (Part B3/1): 548–555.

Maas, H.-G. 2001. On the use of pulse reflectance data for laser scanner strip adjustment. *International Archives of Photogrammetry, Remote Sensing and Spatial Information Sciences,* 33 (Part 3/W4): 53–56.

Maas, H.-G. 2002. Methods for measuring height and planimetry discrepancies in airborne laser-scanner data. *Photogrammetric Engineering & Remote Sensing*, 68(9): 933–940.

Madhavan, R., Hong, T., and Messina, E. 2005. Temporal range registration for unmanned ground and aerial vehicles. *Journal of Intelligent and Robotic Systems*, 44(1): 47–69.

Maune, D. (Ed.) 2007. Digital elevation model technologies and applications. *The DEM Users Manual*, (2nd ed). American Society for Photogrammetry and Remote Sensing, Bethesda, MD.

Morin, K. and El-Sheimy, N. 2002. Post-mission adjustment of airborne laser scanning data. *Proceedings XXII FIG International Congress*, Washington, DC, April 19–26. 12 pp., CD-ROM.

Okabe, A., Boots, B., Sugihara, K., and Chiu, S.N. 2000. *Spatial Tessellations - Concepts and Applications of Voronoi Diagrams*, (2nd edn.). New York: John Wiley.

Oliver, M.A. and Webster, R. 1990. Kriging: A method of interpolation for geographical information system. *International Journal of Geographical Information Systems*, 4(3): 313–332.

Schenk, T. 2001. Modeling and analyzing systematic errors in airborne laser scanners. *Technical Notes in Photogrammetry*, 19: 46.

Shannon, C.E. 1948. A mathematical theory of communication. *Bell System Technical Journal*, 27: 379–423, 623–656.

Shannon, C.E. 1949. Communication in the presence of noise. *Proceedings of Institute of Radio Engineers*, 37(1): 10–21.

Skaloud, J. and Lichti, D. 2006. Rigorous approach to bore-sight self-calibration in airborne laser scanning. *International Journal of Photogrammetry and Remote Sensing*, 61: 47–59.

Soininen, A. and Burman, H. 2005. TerraMatch for MicroStation. Terrasolid, Finland.

Toth, C.K. 2002. Calibrating airborne LiDAR systems. *International Archives of Photogrammetry and Remote Sensing*, 34 (part 2): 475–480.

Toth, C., Csanyi, N., and Grejner-Brzezinska, D. 2002. Automating the calibration of airborne multi-sensor imaging systems. *Proceedings ACSM-ASPRS Annual Conference,* Washington, DC, April 19–26, CD-ROM.

Toth, C.K., Koppanyi, Z., Grejner-Brzezinska, D.A. 2014. Spatial spectrum analysis of various digital elevation models. *Proceedings of American Society for Photogrammetry and Remote Sensing Annual Conference* (*ASPRS 2014*), pp. 327–324.

Vosselman, G. and Maas, H.-G. 2001. Adjustment and filtering of raw laser altimetry data. *Proceedings OEEPE Workshop on Airborne Laserscanning and Interferometric SAR for Detailed Elevation Models.* OEEPE Publications No. 40, pp. 62–72.

Vosselman, G. 2002a. On the estimation of planimetric offsets in laser altimetry data. *International Archives of Photogrammetry and Remote Sensing*, 34 (Part 3A): 375–380.

Vosselman, G. 2002b. Strip offset estimation using linear features. 3rd International LiDAR Workshop, October 7–9, Columbus. Retrieved from http://www.itc.nl/personal/vosselman/papers/vosselman2002.columbus.pdf.

Zhang, Z. 1994. Iterative point matching for registration of free-form curves and surfaces. *International Journal of Computer Vision*, 13(12): 119–152.

9 Accuracy, Quality Assurance, and Quality Control of Light Detection and Ranging Mapping

Ayman Habib

CONTENTS

9.1   INTRODUCTION

The improving capabilities of direct georeferencing technology—that is, integrated Global Navigation Satellite System/Inertial Navigation System (GNSS/INS)—are having a positive impact on the widespread adoption of Light Detection and Ranging (LiDAR) systems for the acquisition of dense and accurate surface models over extended areas. Unlike photogrammetric techniques, LiDAR calibration is not a transparent process and remains restricted to the system's manufacturer. Moreover, derived footprints from a LiDAR system are not based on redundant measurements, which are manipulated in an adjustment procedure. Consequently, one does not have the associated measures (e.g., variance component of unit weight and variance–covariance matrices of the derived parameters) that can be used to evaluate the quality of the final product. In this regard, LiDAR systems are usually viewed as black boxes that lack a well-defined set of quality assurance and quality control (QA/QC) procedures. This chapter introduces the concepts of QA of LiDAR systems and QC of derived data. The most important activity in QA is the system calibration. Therefore, we will present a conceptual approach for LiDAR calibration and the necessary prerequisites for such a calibration. On the other hand, the main focus of this chapter is the introduction of QC procedures to verify the geometric accuracy of a LiDAR point cloud. The main premise of the proposed QC procedures is that overlapping LiDAR strips will be compatible only if there are no biases in the derived surfaces. Therefore, we will use the quality of coincidence of conjugate surface elements in overlapping strips as the basis for deriving the QC measures. The current chapter will start with a brief discussion of LiDAR principles, which will be followed by general remarks regarding QA/QC procedures. Then, an analysis of error sources in a LiDAR system and their impact on the resulting surface will be presented. This analysis will be followed by a discussion of several procedures for QA/QC of LiDAR mapping.

9.2 LiDAR PRINCIPLES

A typical LiDAR system consists of three main components: a GNSS system to provide position information, an INS unit for position and attitude determination, and a laser unit to provide range (distance) information from the laser beam firing point to its footprint onto the ground. In addition to range data, LiDAR systems produce intensity images over the mapped area. Figure 9.1 shows a schematic diagram of a LiDAR system, together with the coordinate systems involved. Equation 9.1 is the basic model that incorporates the LiDAR measurements for deriving positional information for the laser beam footprint (El-Sheimy et al., 2005). This equation relates four coordinate systems: the ground coordinate system, the inertial measurement unit (IMU) body frame, the laser unit coordinate system, and the laser beam coordinate system. This equation is simply the result of a three-vector summation: $r_b^m(t)$ is the vector from the origin of the ground coordinate system (mapping reference frame) to the IMU body frame, r_{lu}^b is the offset between the inertial measurement unit (IMU) body frame and the laser unit—lever arm, and $r_l^{lb}(t)$ is the vector between the laser beam firing point and its footprint onto the ground. The summation of these three vectors, after applying the appropriate rotations ($R_b^m(t), R_{lu}^b, R_{lb}^{lu}(t)$), yields the vector r_l^m, which represents the ground coordinates of the object point under consideration, I. One should note that $r_b^m(t)$

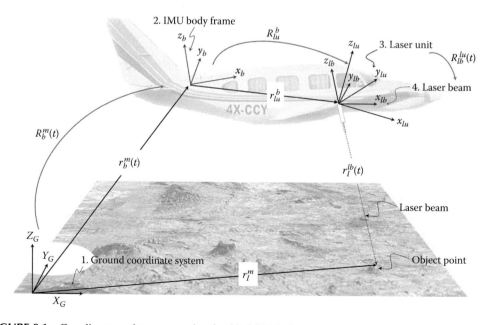

FIGURE 9.1   Coordinates and parameters involved in LiDAR data acquisition.

and $R_b^m(t)$ represent the integrated GNSS/INS position and orientation of the IMU body frame at the time of the pulse (t). On the other hand, the rotation matrix $R_{lb}^{lu}(t)$ is defined by the mirror angles relating the laser beam and laser unit coordinate systems at the time of the pulse. The quality of the derived surface depends on the accuracy of the involved subsystems (i.e., laser, GNSS, and IMU) and the calibration parameters relating these components (i.e., lever arm—r_{lu}^b—and boresight matrix—R_{lu}^b). In general, the system manufacturer provides a range of the expected accuracy of the derived point cloud depending on the specifications of the used subsystems and the precision of the system calibration parameters.

$$r_I^m = r_b^m(t) + R_b^m(t)r_{lu}^b + R_b^m(t)R_{lu}^b R_{lb}^{lu}(t)R_I^{lb}(t) \tag{9.1}$$

Other than the expected accuracy, the quality of the LiDAR-based surfaces can be ensured and checked by QA/QC procedures, which are the main emphasis of this chapter. QA encompasses management activities that are carried out prior to data collection to ensure that the raw and derived data are of the quality required by the user. These management controls cover the calibration, planning, implementation, and review of data collection activities. QC, on the other hand, takes place after data collection to determine whether the desired quality has been achieved. It involves routines and consistent checks to ensure the integrity, correctness, and completeness of the raw and derived data. To illustrate the nature of the QA/QC procedures, one can refer to the respective activities in photogrammetric mapping (Brown, 1966; Fraser, 1997; Habib and Schenk, 2001). The QA procedures for photogrammetric mapping include camera calibration, total system calibration, and flight configuration (e.g., flying height, overlap percentage, and side lap percentage). QC measures for photogrammetric mapping, on the other hand, include internal measures such as the a-posteriori variance component and variance–covariance matrix resulting from the bundle adjustment procedure. External measures include the root-mean-square-error (RMSE) derived from a check point analysis procedure using independently measured targets. After a brief discussion of the error budget, the remaining sections of this chapter will focus on various methods for the QA/QC of LiDAR mapping.

9.3 LiDAR ERROR BUDGET

Errors in the LiDAR-derived coordinates are directly related to the errors associated with the different terms in the LiDAR equation. These terms, or input parameters, can either be measured or estimated from a system calibration procedure. In this section, we are interested in analyzing the effects of random noise and systematic biases in the measurements from the various LiDAR components on the final product. The purpose of such an analysis is to enable the estimation of the quality of the final product from the quality of the system's measurements. Moreover, knowing the expected accuracy of the final product, one will be able to evaluate the outcome of the QC procedure as either acceptable or as an indication of the presence of systematic biases in the data acquisition system. Finally, by analyzing the effects of systematic biases, one might be able to offer some diagnostic judgment regarding the origin of the discrepancies identified from the proposed QC procedures.

For any point measured by a LiDAR system, error propagation can be used to determine the variance–covariance matrix of the LiDAR-derived coordinates given the respective matrices for the different terms in the LiDAR equation. Another issue related to LiDAR error analysis is the nature of the errors resulting from random errors in the input system measurements. Usually, it is expected that random noise will lead to random errors in the derived point cloud. Moreover, it is commonly believed that random noise will not affect the relative accuracy. However, this is not the case for LiDAR systems. In other words, some of the random errors might affect the relative accuracy among the constituents of a derived point cloud. Depending on the parameter being considered, the relative effect of the corresponding noise level in the output might not be the same. As an illustration, Figure 9.2 shows that a given attitude noise in the GNSS/INS-derived orientation (i.e., errors in the derived attitude from the GNSS/INS integration process) will affect the nadir region of the flight trajectory less significantly than the off-nadir regions. Thus, the GNSS/INS attitude error will

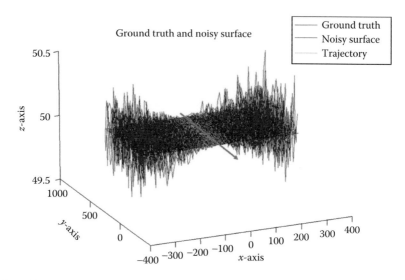

FIGURE 9.2   Effect of attitude errors on a simulated horizontal surface.

affect the relative accuracy of the LiDAR-derived point cloud. The following list gives some diagnostic hints regarding the impacts of noise in the system measurements on the derived point cloud.

- *Position noise*: It will lead to a similar noise level in the derived point cloud. Moreover, the effect is independent of the system parameters (flying height and look angle).
- *Angular noise* (attitude or mirror angles): With this type of noise, the horizontal coordinates are affected more than the vertical coordinates. In addition, the effect is dependent on the system parameters (flying height and look angle).
- *Range noise*: It mainly affects the vertical component of the derived coordinates. The effect is independent of the system's flying height. However, the impact is dependent on the system's look angle.

Systematic biases in the system measurements (e.g., GNSS/INS-derived position and attitude information, mirror angle measurements, and measured ranges) and calibration parameters (e.g., lever arm and boresight parameters relating the system components) lead to systematic errors in the derived point cloud. Table 9.1 provides a summary of the various systematic biases and their impacts on the derived LiDAR coordinates.

TABLE 9.1

Systematic Biases and Their Impacts on the Derived Surface

	Flying Height	Flying Direction	Look Angle
Boresighting offset bias	Impact is independent of the flying height	Impact is dependent on the flying direction (except elevation)	Impact is independent of the look angle
Boresighting angular bias	Impact increases with the flying height	Impact changes with the flying direction	Impact changes with the look angle (except in the across-flight direction)
Laser beam range bias	Impact is independent of the flying height	Impact is independent of the flying direction	Impact depends on the look angle (except in the along-flight direction)
Laser beam angular bias	Impact increases with the flying height	Impact changes with the flying direction (except in the along-flight direction)	Impact changes with the look angle (except in the across-flight direction)

Note: The table assumes a linear scanner flying over a flat horizontal terrain along a straight-line trajectory with constant attitude.

9.4 QA OF LiDAR SYSTEMS

LiDAR-specific QA procedures are established prior to the mapping mission and include flight planning, smart selection of GNSS base stations, ensuring accessibility to all available GNSS signals, and total system calibration. Among these procedures, total system calibration is critical for ensuring the utmost quality of the derived LiDAR coordinates. Calibration procedures can be divided into two main categories: laboratory calibration and *in situ* calibration procedures. Laboratory calibration is usually carried out by the system manufacturer. On the other hand, *in situ* calibration can be carried out by the system's user. Due to the nontransparent nature of LiDAR systems, *in situ* calibration might be challenging to the end users. The calibration methodology developed by Morin (2002) uses the LiDAR equation to solve for the boresight angles and the scan angle correction. These parameters are either estimated using ground control points or by observing discrepancies between tie points in overlapping strips. However, the identification of distinct control and tie points in the LiDAR data is a difficult task due to the irregular nature of the collected point cloud. To alleviate this difficulty, Skaloud and Lichti (2006) presented a calibration technique using tie planar patches in the overlapping strips. The underlying assumption of this procedure is that systematic errors in the LiDAR system will cause incompatibility among conjugate planar regions as well as bending effects in these patches. The calibration process uses the LiDAR equation to simultaneously solve for the plane parameters as well as the boresight angles. However, this approach requires having large planar patches, which might not always be available. In addition, systematic biases, which would not affect the coplanarity of conjugate planar patches, could still remain. To overcome such a problem, control patches can be used for *in situ* calibration of LiDAR systems, as shown in Figure 9.3. By using such control patches, the target function for system calibration should minimize the normal distance between the laser footprint (as derived from the LiDAR equation) and such patches, as illustrated in Figure 9.3. Therefore, one only needs to determine the correspondence between the LiDAR footprint and the control surfaces. The LiDAR equation (Equation 9.1) can be used to estimate the systematic errors (e.g., boresight angles) that minimize this target function. Flight and control surface configurations (e.g., different flight heights in opposite and cross directions together with sloping control surfaces with varying aspects) must be carefully established to enable accurate estimation of these parameters (Habib et al., 2007).

The use of the above-mentioned calibration methodologies, however, is only possible if we are dealing with a transparent system, that is, one in which the raw measurements (i.e., navigation data as well as time-tagged mirror angles and ranges) are available to the user. On the contrary, not all current LiDAR systems provide raw measurements; therefore, as far as the user is concerned, the

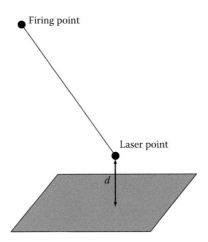

Firing point

Laser point

d

FIGURE 9.3 Target function for LiDAR calibration.

data acquisition system is a black box. For this reason, explicit QA procedures such as this cannot be used to ensure that the derived LiDAR data meet the user's accuracy requirements. The quality of the LiDAR data can, however, be assured through careful flight planning and use of sufficient control. One must then turn to QC procedures to assess the performance of the system and quality of its output after the data have been collected.

9.5 QUALITY CONTROL OF LiDAR DATA

The following sections will provide some tools for the derivation of internal and external QC measures. The internal measures are used to check the relative consistency of the LiDAR data. This is usually conducted by checking the compatibility of the LiDAR footprints within overlapping strips. On the other hand, external QC measures verify the absolute quality of LiDAR data by checking its compatibility with an independently collected and more accurate surface model.

9.5.1 INTERNAL QUALITY CONTROL OF LiDAR DATA

As can be seen in Equation 9.1, there is no redundancy in the LiDAR measurements, leading to the derivation of the coordinates of the laser footprint. Therefore, unlike photogrammetric procedures, one cannot use explicit measures (e.g., a-posteriori variance component and variance–covariance matrix of the derived ground coordinates) to assess the quality of the LiDAR-derived positional information. Hence, alternative QC methods are necessary. The next paragraphs introduce some measures, which could be used for the internal QC of LiDAR data.

With the exception of narrow corridor mapping, LiDAR data are usually acquired from overlapping strips captured along different flight lines, such as those shown in Figure 9.4. A common QC procedure is to assess the coincidence of conjugate features within overlapping strips. Such a procedure ensures the internal quality of the available LiDAR data. There are two main approaches to derive such QC measures: (1) comparing interpolated range or intensity images from the overlapping strips and (2) comparing conjugate features extracted from different strips. The degree of coincidence of such features can be used as a measure to evaluate the data quality and detect the presence of systematic biases. In other words, conjugate features in overlapping strips will coincide if and only if the LiDAR data are quite accurate. Therefore, the separation among conjugate features can be used as a QC measure.

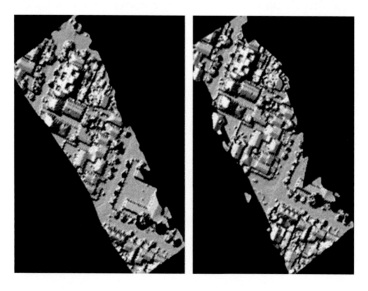

FIGURE 9.4   A pair of overlapping LiDAR strips.

9.5.1.1 Quality Control Using Interpolated Range and Intensity Images

This approach can be applied for either range or intensity data. When using range data, the point clouds from two overlapping strips are interpolated to a regular grid to create two range images. Image differencing is then performed, and the resulting difference image shows the deviations between the two range images. These deviations are then used as a measure of QC; the smaller the deviations, the higher the quality of the datasets being considered. Figure 9.5 shows two interpolated range images and their difference image.

Similarly, intensity data can also be interpolated into a grid to derive two overlapping intensity images. The conjugate features in these images can then be identified, and their 3D coordinates are compared. In addition, horizontal coordinate differences can be derived from the intensity images, whereas the vertical coordinate differences can be determined from the corresponding range images. Differences in the derived coordinates for conjugate features indicate the presence of biases in the data acquisition system. Figure 9.6 illustrates the identification and comparison of conjugate features in interpolated intensity images.

A disadvantage of this QC approach is that the interpolation of irregular LiDAR data leads to artifacts in the derived range or intensity images, especially at the vicinity of discontinuities in the range or intensity data. As a result of such artifacts, incorrect conclusions might be made regarding the quality of the LiDAR point cloud, especially in urban areas where discontinuities in the data are quite common (refer to the significant differences in Figure 9.5 at the building boundaries). Therefore, an alternative QC approach that does not involve the interpolation of LiDAR data should be used.

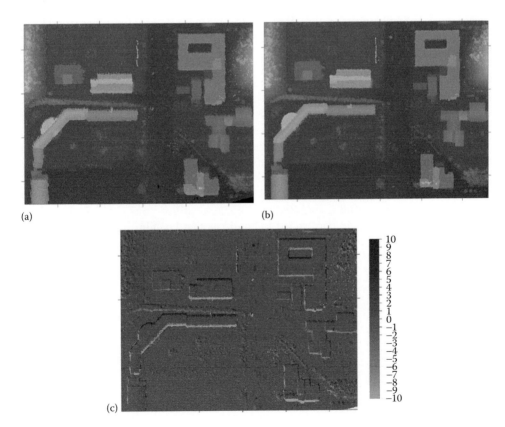

FIGURE 9.5 Image differencing of interpolated range images generated from overlapping strips: (a) interpolated range image from first strip, (b) interpolated range image from second strip, and (c) difference image.

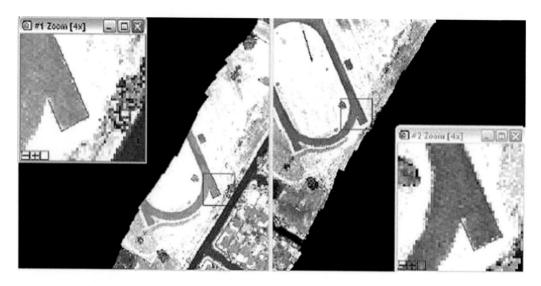

FIGURE 9.6   Comparison of conjugate features in interpolated intensity images.

9.5.1.2   Quality Control by Checking the Coincidence of Conjugate Straight Lines in Overlapping Strips

In the current section, we introduce a QC approach that is based on linear features represented by their end points. It should be noted that this approach does not require lines that have conjugate end points in both LiDAR strips and, depending on the nature of the overlap, conjugate lines may arise from quite different parts of the same object. The quality of the coincidence of conjugate linear features can be used to evaluate the internal quality of LiDAR data. More specifically, the linear features extracted from the irregular LiDAR footprints in the overlapping strips can be used to compute estimates of the conformal transformation parameters that are needed to co-align these features, as shown in Figure 9.7. Deviations from the optimum values (zero shifts, zero rotations, and unit scale factor) can be used as indications of systematic biases in the LiDAR system parameters. Before taking into consideration, the mathematical details of using linear features, which are represented by nonconjugate end points in overlapping strips, for parameter estimation, we will begin with a semiautomated approach for their extraction. By using intensity images, one can develop an interface that extracts the LiDAR point cloud in the vicinity of selected points (i.e., extract the LiDAR points within a given radius from an identified point in the intensity image). The extracted point cloud might correspond to buildings that contain linear features. By using an automated segmentation procedure, planar patches are identified in the extracted point cloud. Then, neighboring patches are intersected to produce infinite line segments. Finally, using the

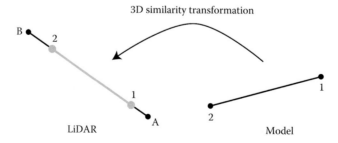

FIGURE 9.7   Conceptual basis of the use of linear features in a line-based approach for the determination of the conformal transformation parameters between two 3D datasets.

segmented patches and a given cylindrical buffer, the axis of which coincides with the infinite line, one can define the end points for the line of intersection. More specifically, LiDAR points within the segmented patches that lie within the given buffer are projected to the line of intersection, and extreme points along the line of intersection are used to define the end points of that line. This procedure should be applied to both of the overlapping strips. After the extraction of the linear features, an automated procedure can be applied to identify conjugate linear features. The identification of conjugate features should consider the normal distance between the lines, parallelism of their direction vectors, and their overlap percentage.

The objective of the proposed approach for the estimation of parameters is to introduce the necessary constraints to describe the fact that a line segment from the first strip (e.g., 12) coincides with the conjugate segment from the overlapping strip (e.g., AB) after applying the transformation parameters (e.g., rotation, shift, and scale), as illustrated in Figure 9.7. For these points, the constraint equations can be written as in Equations 9.2 and 9.3 (i.e., a line point in the first strip should be collinear with the conjugate line in the second strip).

$$\begin{bmatrix} X_T \\ Y_T \\ Z_T \end{bmatrix} + S\,R_{(\Omega,\Phi,K)} \begin{bmatrix} X_1 \\ Y_1 \\ Z_1 \end{bmatrix} = \begin{bmatrix} X_A \\ Y_A \\ Z_A \end{bmatrix} + \lambda_1 \begin{bmatrix} X_B - X_A \\ Y_B - Y_A \\ Z_B - Z_A \end{bmatrix} \tag{9.2}$$

$$\begin{bmatrix} X_T \\ Y_T \\ Z_T \end{bmatrix} + S\,R_{(\Omega,\Phi,K)} \begin{bmatrix} X_2 \\ Y_2 \\ Z_2 \end{bmatrix} = \begin{bmatrix} X_A \\ Y_A \\ Z_A \end{bmatrix} + \lambda_2 \begin{bmatrix} X_B - X_A \\ Y_B - Y_A \\ Z_B - Z_A \end{bmatrix} \tag{9.3}$$

where:
$(X_T Y_T Z_T)^T$ is the translation vector between the strips
$R_{(\Omega,\Phi,K)}$ is the required rotation matrix for the co-alignment of the strips
S, λ_1, and λ_2 are scale factors

Through the subtraction of Equation 9.2 from Equation 9.3, and the elimination of the scale factors, Equations 9.4 and 9.5 can be written to relate the coordinates of the points defining the line segments to the rotation elements of the transformation parameters.

$$\frac{(X_B - X_A)}{(Z_B - Z_A)} = \frac{R_{11}(X_2 - X_1) + R_{12}(Y_2 - Y_1) + R_{13}(Z_2 - Z_1)}{R_{31}(X_2 - X_1) + R_{32}(Y_2 - Y_1) + R_{33}(Z_2 - Z_1)} \tag{9.4}$$

$$\frac{(Y_B - Y_A)}{(Z_B - Z_A)} = \frac{R_{21}(X_2 - X_1) + R_{22}(Y_2 - Y_1) + R_{23}(Z_2 - Z_1)}{R_{31}(X_2 - X_1) + R_{32}(Y_2 - Y_1) + R_{33}(Z_2 - Z_1)} \tag{9.5}$$

These equations can be written for each pair of conjugate line segments, giving two equations, which contribute towards the estimation of two rotation angles (i.e., the azimuth and the pitch angle of the line in question). The roll angle across the line, on the other hand, cannot be estimated. Hence, a minimum of two nonparallel lines is needed to recover the three elements of the rotation matrix. To enable the estimation of the translation parameters and the scale factor, Equation 9.6 can be derived by rearranging the terms in Equations 9.2 and 9.3 and eliminating the scale factors λ_1 and λ_2. A minimum of two noncoplanar lines is required to recover the scale and translation parameters. To recover all seven parameters of the transformation function, a minimum of two noncoplanar line segments is required. For more details regarding this approach, interested readers can refer to Habib et al. (2004). An alternative procedure for the estimation of the rigid body transformation relating two point clouds using corresponding linear features that are represented by nonconjugate points is provided in Renaudin et al. (2011) through a weight modification process.

$$\frac{(X_T + S\,x_1 - X_A)}{(Z_T + S\,z_1 - Z_A)} = \frac{(X_T + S\,x_2 - X_A)}{(Z_T + S\,z_2 - Z_A)}$$

$$\frac{(Y_T + S\,y_1 - Y_A)}{(Z_T + S\,z_1 - Z_A)} = \frac{(Y_T + S\,y_2 - Y_A)}{(Z_T + S\,z_2 - Z_A)}$$

(9.6)

where

$$\begin{bmatrix} x_1 \\ y_1 \\ z_1 \end{bmatrix} = R_{(\Omega,\Phi,K)} \begin{bmatrix} X_1 \\ Y_1 \\ Z_1 \end{bmatrix} \quad \text{and} \quad \begin{bmatrix} x_2 \\ y_2 \\ z_2 \end{bmatrix} = R_{(\Omega,\Phi,K)} \begin{bmatrix} X_2 \\ Y_2 \\ Z_2 \end{bmatrix}$$

9.5.1.3 Quality Control by Checking the Coplanarity of Conjugate Planar Patches in Overlapping Strips

Instead of using linear features, one can incorporate conjugate planar patches in the QC procedure. In this approach, planar patches are represented by equivalent sets of nonconjugate points (i.e., two sets with the same number of points), which can be derived using a planar patch segmentation procedure in the overlapping strips and arbitrarily truncating the segmented points in one of the strips to be of equivalent size to that derived from the other strip. The requirement for having an equivalent number of nonconjugate points along corresponding planar patches is dictated by the utilization of a point-based approach to estimate the conformal transformation parameters for the optimal co-alignment of these patches. In other words, nonconjugate points along corresponding patches will be used in a point-based approach that assumes that these are conjugate points—hence, they will be denoted by pseudo conjugate points. To compensate for the fact that the utilized points along the planar patches are not conjugate, one can expand the variance–covariance matrices for these points in the plane direction, as illustrated in Figure 9.8. The expansion can be carried out using Equation 9.7, in which R is the rotation matrix relating the XYZ and the UVW-coordinate systems. For the latter, the U and V axes are aligned along the plane (i.e., the W-axis is aligned along the normal to the plane). It should be noted that N and M in Equation 9.7 refer to arbitrarily chosen large numbers. Finally, the expanded variance–covariance matrix in the XYZ-coordinate system can be derived according to Equation 9.8. The defined points along conjugate planar patches, together with their expanded variance–covariance matrices, can be used in a point-based conformal transformation to estimate the necessary shifts, rotations, and scale to co-align these patches in the overlapping strips. Deviations from the optimum values (zero shifts, rotations, and unit scale factor) indicate the

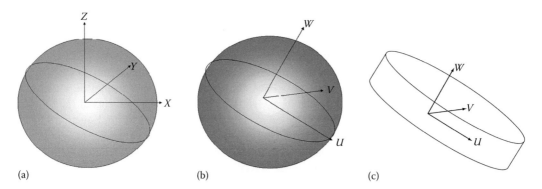

(a)   (b)   (c)

FIGURE 9.8 Conceptual basis for the use of planar patches in a point-based approach for the determination of the conformal transformation parameters between two 3D datasets: (a) the XYZ-coordinate system, (b) the UVW-coordinate system, and (c) the expansion of the error ellipsoid.

presence of biases in the LiDAR unit. It should be noted that a sufficient number of patches with varying slope and aspect values should be available for reliable estimation of the transformation parameters.

$$\sum{}'_{UVW} = R \begin{bmatrix} \sigma_X^2 & \sigma_{XY} & \sigma_{XZ} \\ \sigma_{YX} & \sigma_Y^2 & \sigma_{YZ} \\ \sigma_{ZX} & \sigma_{ZY} & \sigma_Z^2 \end{bmatrix} R^T + \begin{bmatrix} N & 0 & 0 \\ 0 & M & 0 \\ 0 & 0 & 0 \end{bmatrix} \qquad (9.7)$$

$$\sum{}'_{XYZ} = R^T \sum{}'_{UVW} R \qquad (9.8)$$

9.5.1.4 Quality Control by Automated Matching of the Original LiDAR Points in Overlapping Strips

The above-mentioned QC measures require preprocessing of the LiDAR point cloud (e.g., interpolation, planar patch segmentation, and intersection of neighboring planar patches). In addition, feature-based QC procedures (i.e., using corresponding planar and linear features) restrict the process to urban or semiurban areas where these features could be present. Therefore, it is preferred to develop alternative measures, which can deal with the original LiDAR points. The following paragraphs present an alternative approach for evaluating the degree of coincidence of overlapping LiDAR strips, which are represented by irregularly distributed sets of points. The approach presented does not assume one-to-one correspondence between the involved points. More specifically, the approach evaluates the transformation parameters relating the strip pair in question, determines the correspondence between conjugate surface elements, and derives an estimate of the degree of coincidence of these datasets.

To illustrate the conceptual basis of the suggested approach, let us assume that we are given two sets of irregularly distributed points that describe the same surface, as shown in Figure 9.9. Let $S_1 = \{p_1, p_2, ..., p_l\}$ be the first set and $S_2 = \{q_1, q_2, ..., q_m\}$ be the second set, where $l \neq m$ (i.e., there is no assumption of point-to-point correspondence or equivalent number of points in the available sets). These points are randomly distributed and the correspondences between them are not known. Also, the two point sets might be given relative to two different reference frames. The transformation between these reference frames is modeled by a seven-parameter transformation involving three shifts, one scale, and three rotations (X_T, Y_T, Z_T, S, Ω, Φ, and K, respectively). Hence, the problem at hand is to determine the degree of similarity between the two point sets, establish the correspondences between conjugate surface elements, and estimate the transformation parameters relating the respective reference frames. The proposed approach creates a TIN model using the

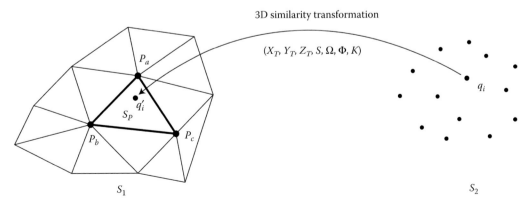

FIGURE 9.9 Comparing two datasets with irregularly distributed sample points.

points in S_1 to form a group of nonoverlapping triangles, as shown in Figure 9.9. The individual triangles in the derived TIN are assumed to represent planar patches.

Now, let us consider the surface patch S_p in S_1, which is defined by three points p_a, p_b, and p_c. The fact that a point q_i in the second set S_2 belongs to the surface patch S_p in the first set can be mathematically described by the determinant constraint in Equation 9.9, as illustrated in Figure 9.9.

$$\begin{vmatrix} x_{q_i} & y_{q_i} & z_{q_i} & 1 \\ x_{p_a} & y_{p_a} & z_{p_a} & 1 \\ x_{p_b} & y_{p_b} & z_{p_b} & 1 \\ x_{p_c} & y_{p_c} & z_{p_c} & 1 \end{vmatrix} = 0 \tag{9.9}$$

The vector $(x_{q_i}, y_{q_i}, z_{q_i})$ in Equation 9.9 represents the point coordinates in the second dataset transformed to the reference frame associated with the first one, using Equation 9.10. Equation 9.9 simply states that the volume defined by the points, $q'_i, p_a, p_b,$ and p_c is zero. In other words, these points are coplanar (i.e., the normal distance between q'_i and the surface patch S_p is zero). After establishing the correspondences between the points in S_2 and the patches in S_1, one can estimate the transformation parameters (implicitly considered in the first row of the determinant in Equation 9.9 through the substitution of Equation 9.10) using a least-squares adjustment procedure. The estimated normal distances between point-patch pairs from the adjustment procedure represent the goodness of fit (degree of similarity) between the two point sets after the co-alignment of their respective reference frames. Moreover, the deviation of the estimated transformation parameters from the optimal ones can be used as an indication of biases in the parameters of the data acquisition system.

$$\begin{bmatrix} x_{q_i} \\ y_{q_i} \\ z_{q_i} \end{bmatrix} = \begin{bmatrix} X_T \\ Y_T \\ Z_T \end{bmatrix} + S R_{(\Omega,\Phi,K)} \begin{bmatrix} x_{qi} \\ y_{qi} \\ z_{qi} \end{bmatrix} \tag{9.10}$$

Thus far, we have established the mathematical model that can be used to derive the transformation parameters relating the reference frames associated with the two point sets. However, the derivation of these estimates requires knowledge of the correspondence between conjugate surface elements in the available datasets (i.e., which points in S_2 belong to which patches in S_1). Figuring out such a correspondence requires careful consideration as the two sets might be given relative to different reference frames. Therefore, the remaining problem is to establish a reliable procedure for the identification of such a correspondence. The proposed solution to this problem is based on a voting scheme, which simultaneously establishes the correspondence between conjugate surface elements and the transformation parameters relating the two datasets.

As mentioned earlier, the parameters of the 3D similarity transformation in Equation 9.10 can be estimated once the correspondences between seven points in S_2 and seven patches in S_2 are known, generating seven constraints of the form given in Equation 9.9. The suggested procedure can start with choosing any seven points in S_2 and matching them with all possible surface patches in S_1, generating several matching hypotheses. For each matching hypothesis, a set of seven equations can be written and used to solve for the transformation parameters. One should repeat this procedure until all possible matches between the points in S_2 and the patches in S_1 are considered. Throughout these combinations, correct matching hypotheses will lead to the same parameter solution. Therefore, the most frequent solution resulting from the matching hypotheses will be the correct set of transformation parameters relating the two datasets in question. Also, the matching hypotheses that led to this solution constitute the correspondences between the points in S_2 and the patches in S_1. A seven-dimensional accumulator array, which is a discrete tessellation of the expected solution space, can be used to keep track of the matching hypotheses and the associated solutions. The correct solution will manifest itself as a peak in this accumulator array. However, exploring all possible matching hypotheses will lead to a combinatorial explosion. For example, if we want to consider all independent combinations between S_1 and S_2, the total number of solutions is $S = (nm)!/7!/(nm-7)!$ where n is the number of patches in the S_1 set, and m is the number of points in S_2. Moreover, the memory requirement for a 7D accumulator array is impractical. This problem is caused by our attempt to simultaneously determine the seven transformation parameters.

To avoid the above-mentioned problems, the parameters can be sequentially and individually solved for in an iterative manner, starting with some approximate values. Consequently, the accumulator array becomes one dimensional, and the memory requirement problem disappears. The total number of point-to-surface patch combinations is reduced to $m \times n$. The suggested procedure can be summarized as follows:

1. Establish some approximate values for the unknown parameters.
2. Select one of the parameters (e.g., X_T). Initialize the corresponding 1D accumulator array to zero votes. The values of the other parameters (Y_T, Z_T, S, Ω, Φ, and K) are considered to be constants.
3. Assume a matching hypothesis between a point q_i in S_2 and a surface patch S_p in S_1. Then, compute a numerical value for X_T by solving the constraint in Equation 9.9.
4. Update the corresponding accumulator array by incrementing the number of votes associated with the parameter estimated in the previous step.
5. Repeat Steps 3 and 4 until all plausible point-to-surface patch correspondences have been explored.
6. Select the distinct peak of the accumulator array, as shown in Figure 9.10. Update the parameter X_T to the peak value.
7. Repeat Steps 2–6 until all parameters have been updated.
8. Repeat the above-mentioned procedure until convergence is achieved (i.e., the parameters have not changed beyond a predefined threshold between successive iterations), as shown in Figure 9.11.

By using the point-to-patch associations that contributed to the peak in the final iteration, one can perform a simultaneous least-squares adjustment to come up with estimates of the transformation parameters relating the two datasets. The above-mentioned procedure can be implemented in a coarse-to-fine strategy that controls the precision of the solution and the permissible range for the

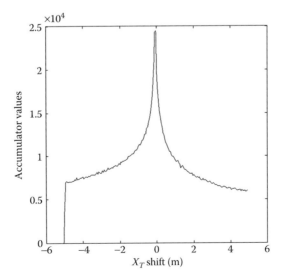

FIGURE 9.10 Accumulator array for the X_T shift of the 3D similarity transformation.

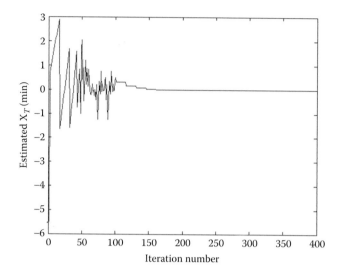

FIGURE 9.11 X_T convergence through the iterative procedure.

parameter under consideration. In the iterative procedure, the accumulator array for each parameter is digitized into several cells that span a defined range. In the earlier iterations, the discrete solution steps (cell sizes) were digitized coarsely to compensate for the poor quality of the approximate values of the unknown parameters. However, in later iterations, the solution range, as well as the digitization steps (cell sizes), can be reduced to reflect the improved quality of the updated transformation parameters. At the end of each iteration, the approximations are updated, and the cell size of the accumulator array is reduced. In this manner, the parameters are estimated with high accuracy. For a more detailed discussion of this procedure, interested readers can refer to Habib et al. (2001, 2006).

9.5.2 EXTERNAL QUALITY CONTROL OF LiDAR DATA

The above-mentioned procedures provide internal and relative QC measures for multistrip LiDAR coverage. However, they do not provide an external and independent measure of the LiDAR data quality. Another common approach to external QC involves check point analysis using specially designed LiDAR targets. Figure 9.12 shows an example of such a target (a white circle inside a black ring). The targets are laid out so that they protrude from troughs in the ground, as shown in Figure 9.12b, and their coordinates are independently surveyed. Unlike the previous internal QC

(a)   (b)

FIGURE 9.12 LiDAR control point targets: (a) target design and (b) target layout in captured LiDAR data. (From Csanyi, N. and Toth, C., *Photogramm. Eng. Rem. Sens.*, 73, 385, April 2007. With Permission.)

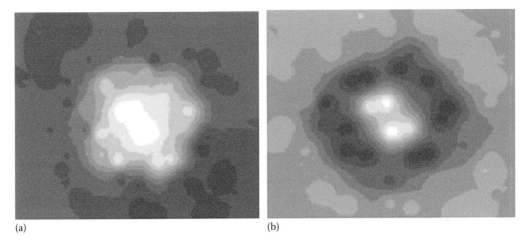

(a) (b)

FIGURE 9.13 Control target in LiDAR imagery: (a) range image and (b) intensity image. (From Csanyi, N. and Toth, C., *Photogramm. Eng. Rem. Sens.*, 73, 385, April 2007. With Permission.)

measures, this procedure can be used for projects involving single LiDAR strips. In other words, it can be used for projects with multiple or single strips.

The targets are then extracted from the range and intensity LiDAR imagery using a segmentation procedure. Figure 9.13 shows how a target of the form shown in Figure 9.12a would look in each type of image; in the range image, it is a light (protruded) circle on a dark (sunken) background in the intensity image, one can see the target's black-and- white pattern. The coordinates of the extracted targets are then compared with the surveyed coordinates using an RMSE analysis. The resulting RMSE value is a measure of the external (absolute) quality of the LiDAR-derived surface. For more details regarding this approach, interested readers can refer to Csanyi and Toth (2007).

Establishing and surveying LiDAR targets is, however, an expensive and time-consuming procedure, and its implementation depends on the accessibility of the site to be mapped. On the positive side, this methodology would enable the estimation of an absolute measure of the quality of the LiDAR data. It should be noted that the internal QC measures can also be used for external QC. In such a case, instead of comparing overlapping strips, one could compare the LiDAR surface with an independently collected surface over the same area. For reliable estimation of the external (absolute) quality, the control surface and targets should be at least three times more accurate than the LiDAR surface (ASPRS, 2004).

9.6 CASE STUDY

To validate the feasibility and applicability of the proposed QC methodologies, a multistrip LiDAR coverage was collected using the OPTECH ALTM 2070 system with an average point density of 2.67 point/m^2 (0.70 m average interpoint spacing) from an altitude of 975 m. The acquired data included three strips that were flown in opposite directions (Figure 9.14). According to the system's specifications, the horizontal and vertical accuracy of the derived point cloud are expected to be in the range of 50 and 15 cm, respectively. Four experiments are presented here for the comparative analysis of the presented methods. First, interpolated range and intensity images are used to determine the quality of coincidence of conjugate surface elements in the overlapping strips. Second, QC is performed by checking the coincidence of conjugate straight lines in the overlapping strips. Third, the analysis is performed by checking the coplanarity of conjugate planar patches in the overlapping strips. Finally, QC is carried out through automated matching of the original LiDAR footprints in the overlapping strips.

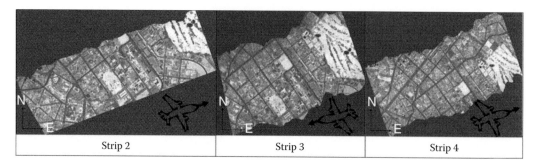

| Strip 2 | Strip 3 | Strip 4 |

FIGURE 9.14 Three overlapping strips used to test the proposed QC procedures.

9.6.1 CHECKING THE COINCIDENCE OF CONJUGATE POINTS IN THE INTENSITY AND RANGE IMAGES

To test this approach, we manually extracted 100 points in the overlapping LiDAR strips. Figure 9.15 shows a sample of the measured point. The differences in the X, Y, and Z coordinates for this point are 0.79, −0.22, and −0.05 m, respectively. The average and standard deviation of these differences for all the points are shown in Table 9.2. The standard deviation of the estimated shifts is significantly large (especially in the Z-direction), which can be explained by the fact that identifying the conjugate points in the overlapping intensity images is sometimes difficult (especially when considering the impact of the surface roughness and slope, together with the LiDAR look angle). Moreover, the interpolation process is another factor that is expected to introduce artifacts leading to inconsistencies in the derived discrepancies. For further illustration, Figure 9.16 shows another sample point, for which the estimated differences in the X, Y, and Z directions are 1.65, 1.11, and −0.63 m, respectively. Therefore, one can conclude that the performance of a QC procedure, which is based on the manipulation of intensity and range images, is not reliable and is subjective to introduced artifacts by the intensity image interpolation.

9.6.2 CHECKING THE COINCIDENCE OF CONJUGATE STRAIGHT LINES IN OVERLAPPING STRIPS

To test this procedure, we manually collected conjugate lines in the three overlapping strips (Figure 9.14). Figures 9.17 through 9.20 illustrate the interface developed for the extraction of these features in two overlapping strips. Table 9.3 summarizes the conformal transformation parameters that were estimated, together with the average normal distance between conjugate linear features before and after applying the transformation. Figure 9.21 shows a sample pair of conjugate linear

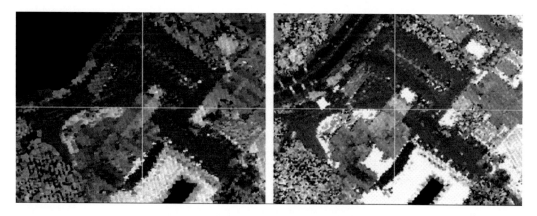

FIGURE 9.15 A sample pair of conjugate points identified in overlapping strips.

TABLE 9.2
Averages and Standard Deviations of the Estimated Discrepancies between Overlapping Strips, Using 100 Points

	Average (m)	Standard Deviation (m)
X	0.45	0.36
Y	0.50	0.37
Z	0.22	0.28

FIGURE 9.16   Another sample pair of conjugate points identified in overlapping strips.

FIGURE 9.17   An example of a building that has been used for the generation of linear features.

FIGURE 9.18 Selected areas in the intensity images that correspond to the building in Figure 9.17.

FIGURE 9.19 Segmented patches in the selected areas in the intensity images that correspond to the building in Figure 9.17.

FIGURE 9.20 Lines extracted from the segmented patches in the selected areas in the intensity images that correspond to the building in Figure 9.17.

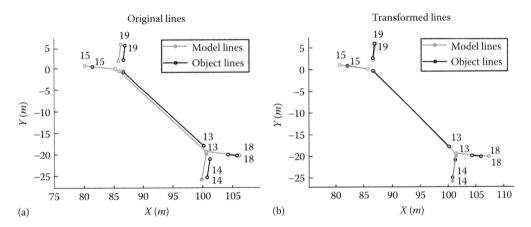

FIGURE 9.21   Sample pair of conjugate linear features in overlapping strips before (a) and after (b) applying the estimated conformal transformation parameters provided in Table 9.3.

TABLE 9.3

Transformation Parameters Estimated Using Conjugate Linear Features in Overlapping Strips, Together with the Normal Distances between the Linear Features before and after Applying the Transformation

	Strips 2 and 3	Strips 3 and 4	Strips 2 and 4
Transformation parameter/number of lines	24	36	24
Scale factor	1.0002	1.0006	1.0013
X_T (m)	−0.56	0.75	0.10
Y_T (m)	0.04	−0.17	−0.16
Z_T (m)	0.03	0.05	0.13
Ω (°)	0.0205	−0.0386	−0.0147
Φ (°)	0.0062	−0.0125	−0.0073
K (°)	0.0261	−0.0145	−0.0113
Normal distance (m) (before)	0.38 ± 0.22	0.49 ± 0.24	0.26 ± 0.14
Normal distance (m) (after)	0.18 ± 0.19	0.18 ± 0.18	0.16 ± 0.11

features in the overlapping strips before and after the application of the conformal transformation parameters. It is quite evident that the degree of coincidence among the conjugate features has significantly improved after applying the estimated transformation parameters (refer to the last two rows of Table 9.3 and Figure 9.21b). A closer look at the numbers reported in Table 9.3 reveals that the estimated discrepancies between the strips, which are mainly in the planimetric coordinates, depend on the flying direction.

9.6.3 Checking the Coplanarity of Conjugate Planar Patches in Overlapping Strips

As mentioned earlier, instead of using the conjugate linear features, one can determine the quality of LiDAR data by checking the coplanarity of conjugate planar patches in the overlapping strips. The coplanarity of these patches can be checked by estimating the conformal transformation parameters (shifts, rotations, and scale) that need to be applied to one strip to ensure the co-alignment of the two strips. To test this procedure, we manually collected conjugate patches in the three overlapping strips (refer to Figure 9.22 for examples of these patches). These patches correspond to the roof of a building that appears in overlapping LiDAR strips. Table 9.4 summarizes the conformal transformation

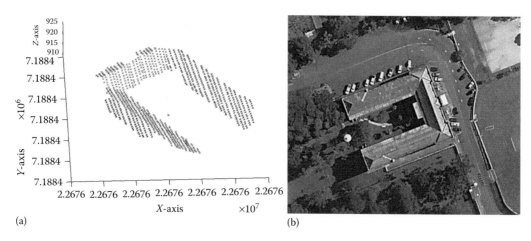

FIGURE 9.22 Planar patches extracted from a LiDAR strip (a) that correspond to the building in the optical image (b).

TABLE 9.4

Transformation Parameters Estimated Using Conjugate Planar Patches in Overlapping Strips

	Strips 2 and 3	Strips 3 and 4	Strips 2 and 4
Transformation parameter/number of patches	21	22	22
Scale factor	1.0000	0.9996	0.9995
X_T (m)	−0.52	0.72	0.08
Y_T (m)	−0.13	−0.17	−0.21
Z_T (m)	0.05	0.09	0.14
Ω (°)	0.0289	−0.0561	−0.0802
Φ (°)	0.0111	−0.0139	−0.0342
K (°)	0.0364	0.0288	0.0784
Average normal distance (m) (after)	0.04	0.03	0.04

parameters estimated using the extracted planar patches. A visual comparison of the results reported in Tables 9.3 and 9.4 shows that the parameters estimated from the line and patch procedures are quite compatible. However, the utilization of the planar patches eliminates the need for the intersection of neighboring planes to derive the linear features, thus saving some processing time.

9.6.4 Automated Matching of the Original LiDAR Footprints in Overlapping Strips

To test this approach, we extracted several areas in overlapping LiDAR strips. Figure 9.23 shows the selected areas in strips 3 and 4. The estimated parameters together with the average normal distances between the conjugate surface elements are provided in Table 9.5.

Based on a comparison of Tables 9.3 through 9.5, it is clear that the proposed approaches yield comparable results. Moreover, these results suggest the presence of systematic biases among the strips. The magnitude of these biases depends on the flying direction (i.e., large biases are detected for strips flown in opposite directions—compare the estimated biases for strips 2 and 3, 3 and 4, and 2 and 4). Also, in spite of the fact that these biases are in the range of the expected accuracy from the involved LiDAR system, this does not mean that they are acceptable. As these measures

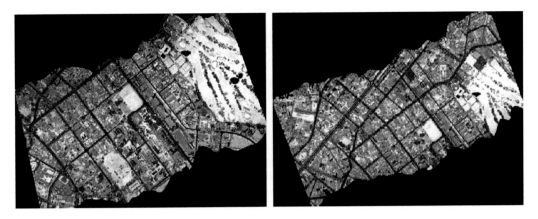

FIGURE 9.23 Locations of areas selected for the alignment and automated matching of original LiDAR footprints in Strips 3 and 4.

TABLE 9.5
Transformation Parameters Estimated through Automated Matching of Original LiDAR Footprints

	Strips 2 and 3	Strips 3 and 4	Strips 2 and 4
Scale factor	0.9996	0.9998	0.9993
X_T (m)	−0.55	0.75	0.19
Y_T (m)	−0.06	−0.13	−0.18
Z_T (m)	0.03	0.12	0.16
Ω (°)	0.0080	−0.0267	−0.0213
Φ (°)	0.0059	−0.0088	−0.0053
K (°)	−0.0009	−0.0003	0.0012
Average normal distance (m) (after)	0.09	0.09	0.10

are estimated from a large sample, the estimated biases should average to zero. In other words, the derived biases are much larger than their respective standard deviations.

9.7   CONCLUDING REMARKS

The current chapter presented alternative procedures for the QA/QC of LiDAR systems and derived data. In general, the QA of LiDAR systems is restricted to flight planning, as the calibration procedures require the raw measurements (e.g., ranges, mirror angles, and navigation data) that are not usually available to the end user. Therefore, QC is an essential procedure to ensure that the data derived from a given system meets the users' requirements. The proposed procedures are based on evaluating the conformal transformation parameters needed to co-align conjugate surface elements in overlapping LiDAR strips. Deviations in the estimated transformation parameters from the theoretical ones (zero rotations and translations and unit scale factor) are used as QC measures to detect the presence of biases in the data acquisition system. When dealing with overlapping LiDAR strips, the deviations are considered as internal (relative) QC measures. On the other hand, external (absolute) QC measures can be derived by comparing the LiDAR surface with another version of the surface that has been independently and accurately acquired.

More specifically, we introduced one approach that utilizes linear features with nonconjugate end points. In addition , the current chapter introduced an approach that utilizes conjugate planar patches, which are represented by nonconjugate points. To compensate for the noncorrespondence of the selected points along the planar patches in the two strips, we artificially expanded their variance–covariance matrix in the plane direction. Finally, we introduced an automated approach that is based on the identification of conjugate surface elements while estimating the transformation parameters. Experimental results with real data have demonstrated the feasibility of the proposed algorithms for detecting biases in the horizontal directions between overlapping LiDAR strips. These errors are speculated to be due to biases in the lever and/or boresight parameters. More specifically, by estimating the discrepancies between overlapping strips, which are flown in opposite directions, one can exclude some of the biases as possible sources of such discrepancies. Moreover, the different approaches have been shown to produce comparable results. When compared with traditional QC techniques (e.g., the utilization of intensity and range images), the proposed approaches delivered more reliable estimates. Moreover, the last approach can be applied to any coverage area, with no requirement for LiDAR targets or structures with linear features (e.g., urban areas).

ACKNOWLEDGMENTS

This research work has been conducted under the auspices and financial support of the British Columbia Base Mapping and Geomatic Services (BMGS) and the GEOIDE Research Network. The author is grateful for the Technology Institute for Development–LACTEC–UFPR, Brazil for supplying the LiDAR data.

REFERENCES

American Society of Photogrammetry and Remote Sensing LIDAR Committee. May 2004. In M. Flood (Ed.) ASPRS Guidelines—Vertical Accuracy Reporting for LIDAR Data and LAS Specifications. Retrieved April 19, 2007, from http://www.asprs.org/society/divisions/ppd/standards/Lidar%20guidelines.pdf.

Brown, D. 1966. Decentric distortion of lenses. *Photogrammetric Engineering and Remote Sensing*, 32(3): 444–462.

Csanyi, N. and Toth, C. 2007. Improvement of LiDAR data accuracy using LiDAR-specific ground targets. *Photogrammetric Engineering and Remote Sensing*, 73(4): 385–396.

El-Sheimy, N., Valeo, C., and Habib, A. 2005. Digital terrain modeling: Acquisition, manipulation and applications. *Artech House Remote Sensing Library*, 200 pp.

Fraser, C. 1997. Digital camera self-calibration. *ISPRS Journal of Photogrammetry and Remote Sensing*, 52(4): 149–159.

Habib, A. and Schenk, T. 2001. Accuracy analysis of reconstructed points in object space from direct and indirect exterior orientation methods. *OEEPE Workshop on Integrated Sensor Orientation*. Institute for Photogrammetry and Engineering Surveying, University of Hanover, Germany, September 17–18, 2001.

Habib, A., Ghanma, M., and Tait, M. 2004. Integration of LiDAR and photogrammetry for close range applications, *Proceedings of the ISPRS XXth Conference*, Istanbul, Turkey, 35B (7), 6 pp.

Habib, A., Lee, Y., and Morgan, M. 2001. Surface matching and change detection using a modified Hough transformation for robust parameter estimation. *Photogrammetric Record*, 17(98): 303–315.

Habib, A., Bang, K., Shin, S., and Mitishita, E. 2007. LIDAR system self-calibration using planar patches from photogrammetric data. *Fifth International Symposium on Mobile Mapping Technology (MMT'07)*, Padua, Italy, May 28–31, 2007.

Habib, A., Cheng, R., Kim, E., Mitishita, E., Frayne, R., and Ronsky, J. 2006. Automatic surface matching for the registration of lidar data and MR imagery authors. *ETRI (Electronics and Telecommunication Research Institute) Journal*, 28(2): 162–174.

Morin, K., 2002. Calibration of airborne laser scanners. M.Sc. thesis, Department of Geomatics Engineering, University of Calgary, November 2003, UCGE Report No. 20179, 125 pp.

OPTECH. ALTM 3100: The Next Level of Performance. Retrieved September 29, 2007, from http:// www. optech.ca/pdf/Specs/specs_altm_3100.pdf.

Renaudin, E., Habib, A., & Kersting, A. P. (2011). Feature-based registration of terrestrial laser scans with minimum overlap using photogrammetric data. *ETRI Journal*, 33(4): 517–527.

Skaloud, J. and Lichti, D. 2006. Rigorous approach to bore-sight self-calibration in airborne laser scanning. *ISPRS Journal of Photogrammetry and Remote Sensing*, 61: 47–59.

10 Data Management of Light Detection and Ranging

Lewis Graham

CONTENTS

The current chapter provides an overview of Light Detection and Ranging (LiDAR) data and data management methodologies. It begins with an overview of raster storage formats to establish a common understanding of uniform data organizations. It then moves into point cloud formats, the standard storage organization for LiDAR data. A brief overview of the LAS data standard is presented, followed by recommendations for managing data in production scenarios. This chapter concludes with an overview of emerging trends in LiDAR data as well as a detailed discussion of the most recent (as of this writing) edition of the LAS 1.4 Standard.

The major industry changes since the first edition of this book include the following:

- The ubiquitous LiDAR point cloud standard, LAS, is now at version 1.4 (a detailed overview of the latest standard is provided in this current chapter).
- Cloud hosting services such as Microsoft Azure and Amazon Web Services are at a much higher level of maturity and significantly more accessible.
- Solid-State Disk (SSD) storage has become very inexpensive, whereas sizes have dramatically increased.
- The average price per square kilometer of aerial LiDAR data has significantly declined.

The impact of these evolutions is discussed in this chapter.

10.1 DATA STRUCTURES AND FORMATS

10.1.1 Uniform Gridded Data

Conventional two-dimensional image data or gridded elevation data are organized on uniformly spaced grids. This allows one to specify an origin and spacing (possibly different for x and y) and then simply list the attribute (such as height) values. For example, an elevation grid could be encoded as

x origin $= 0.0$
y origin $= 0.0$
x spacing $= 0.5$
y spacing $= 1.0$
x rows $= 1,000$
y columns $= 1,000$
$z_1 = 182.3$
$z_2 = 181.7$
etc.

An example of a uniform grid is illustrated in Figure 10.1.

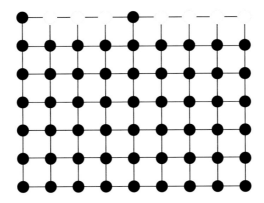

FIGURE 10.1 A uniform grid.

Generally, the header portion of the encoding contains information about the data themselves and is thus termed *metadata*. In the above-mentioned example, we have specified all of the parameters to locate, in a relative way, all of the elevation values (often termed *posts*) in the file. Note that it is not necessary to specify the individual values of each x, y location; they are inferred from the point's location in the data file. We can quantify this by a simple equation (this example is for a row major file). If N is the point count in the file then N = rows × columns. If we consider the first point index to be 0, then the (x, y) coordinates of the ith point in the file are given by

$$x = (\text{i MOD Columns}) * x \text{ spacing} + x \text{ origin}$$

$$y = (\text{i DIV Columns}) * y \text{ spacing} + y \text{ origin}$$

where MOD is the modulo operator, and DIV is the integer value of the division operator. Of course, z is just the value read directly from the data file at position i. External file descriptions must specify if the data are row major or column major. A row major file organizes data by rows meaning that if a row is K elements long, the first K elements of the file will represent the first row, the second K elements the second row, and so forth. On the other hand, a column major file containing J elements per column will be organized such that the first J elements in the file represent column 1, the next J elements column 2, and so forth.

Additional attributes can be added by either considering each record in the file a data structure (e.g., z value, intensity) or by storing sequential records in the file in block fashion. The first strategy is generally called *data interleaving* and is illustrated in Table 10.1. The second approach is termed *Sequential Data Blocks* and is illustrated in Table 10.2.

The gridding of data can result as a natural consequence of the organization of the data formation system or as a result of a *gridding* procedure. A charged-coupled device (CCD) is a sensor that is organized as a grid and hence data collected from such a device naturally lend themselves to a

TABLE 10.1
A Data Interleaved File Format

Post n	Height
	Intensity
	Classification
Post $n + 1$	Height
	Intensity
	Classification
Post $n + 2$	...

TABLE 10.2
Block Organized File

Intensity block	Post *n*
	Post *n* + 1
	....
Height block	Post *n*
	Post *n* + 1
	.....
Classification block	Post *n*
	Post *n* + 1
	...

gridded organization. Another common example from remote sensing is uniform scanning of a photograph using a photogrammetric scanner.

Gridded data have a number of advantages over other organizational structures. Among these are

- Random access to a point at a specific object space location is direct as the location of the point in a storage file can be immediately computed from the desired post coordinates. In other words, this is a very rapid access format.
- Digital signal-processing techniques lend themselves to gridded data. Thus, operations such as low pass spatial filtering are well defined.
- The storage format is compact as two of the coordinates (in elevation grids, X and Y) are implied rather than stored in the file. This compactness of storage is particularly advantageous when the data are in the random access memory of the processing computer.

Alas, for reasons described in the next section, LiDAR data do not lend themselves to gridded storage, forcing us to use different storage, transport, and processing approaches.

10.1.2 Point Cloud Organization

LiDAR data are, in general, not organized on a uniform grid due to the nature of the geometry of the LiDAR data capture device. A topographic laser scanner not only captures data in quasi-random X, Y locations in object space (typically, the view looking down from the sensor) but also can capture multiple elevation values at the same X, Y coordinates (consider scanning up the side of a vertical building wall). A typical section of a LiDAR point cloud is shown as a 3D rendering in Figure 10.2.

Nonuniform data can be *resampled* to a grid or kept in a storage format that supports nonuniform X, Y organization. The process of resampling data is in the domain of *digital signal processing*. Stringent processing procedures such as the Nyquist sampling criteria (Mitra, 2001) and frequency prefiltering must be followed to avoid adding false information (called *aliasing*) to the gridded data. It is generally not considered acceptable to convert point cloud LiDAR data to a gridded format except as a final delivery product. Not only does the resampling introduce errors into the data, but also the data are *flattened* in the vertical dimension. This flattening factor results in the need to produce multiple-gridded representations of data such as first surface (imagine the canopy of trees), ground elevation models, and so forth.

Three-dimensional quasi-random data that are not organized on a grid are often termed *point cloud* as they have ill-defined boundaries similar to a cloud. A comparison of a gridded organization to a cross section of a point cloud is depicted, in top-down (*plan*) view in Figure 10.3.

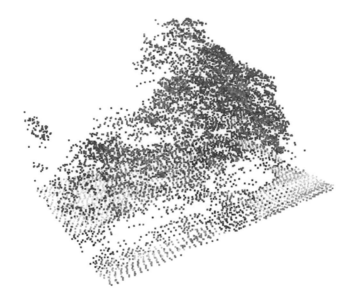

FIGURE 10.2 A 3D section of a typical LiDAR point cloud.

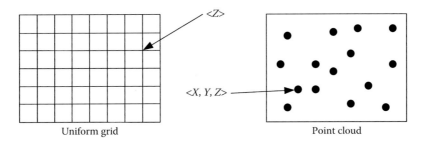

FIGURE 10.3 Comparing a grid to a point cloud.

As the *X*, *Y* locations in the point cloud are not known *a priori* as they are in a grid, all three coordinate values must be encoded for each point. Point cloud organizations are inherently more difficult to work with when performing operations such as searching for a particular point or interpolating a location that lies between points. The application of the Nyquist sampling criteria is not straightforward (Song et al., 2012) and is, in fact, often ignored in practice. However, point clouds have the advantage of being able to precisely encode data from a sensor that collects data in a non-uniform pattern such as most LiDAR systems.

10.1.3 Triangulated Irregular Networks

Point cloud organization is quite common in geographic information systems (GIS) when the individual points are connected via *edges* to their neighbors in a triangular organization. Such a format is termed a *Triangulated Irregular Network* or TIN. The meaning of the term is as follows:

- *Triangular*: The points are all connected such that triangles with three-dimensional vertices are formed
- *Irregular*: As the points are irregularly spaced, the triangles are not uniform
- *Network*: No point is left out of the construction, and thus, any point in the mesh can be reached by traversing the edges

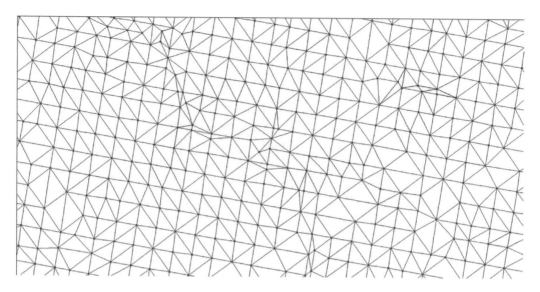

FIGURE 10.4 A triangulated irregular network.

An example of a TIN is depicted in Figure 10.4. The points from which the TIN is formed (the original laser points) are the *vertices*. The connecting lines are called *edges*.

A TIN has the distinct advantage of being able to accurately represent nonregular geomorphology (ridges, drainage, and so forth). For example, a solid model of points in the thalweg of a stream can be constructed using a TIN. This same model cannot be constructed using gridded data without grossly oversampling the grid.

An additional major advantage of a TIN is the ability to introduce model *constraints*. These constraints are points, polylines, and polygons that force the TIN geometry to incorporate the constraints. An example might be a polygon surrounding a static water body. This polygon would have the same elevation at each vertex (as static water must be *flat*). These constraints are referred to as *breaklines*, a bit of a misnomer as the constraints can include other geometry types. Many LiDAR data deliveries require the inclusion of supplementary breaklines for purposes such as valid hydrological flows.

Unfortunately, at the time of the writing of this second edition, there is no standard binary format specification for data in TIN format. Thus, TINs are typically constructed on-the-fly from LiDAR data in LAS format, possibly augmented by model constraints (breaklines).

10.1.4 Point Encoding Approaches

Point cloud elevation data are commonly encoded as either *binary* or American Standard Code for Information Interchange (ASCII). Binary encoding is compact, using the minimum file space possible for each data item. For example, if the height values (z) are stored as 32-bit integers, then this is all the space that is occupied per height value (4 bytes per record). Generally, binary files cannot be read without agreement on both the writer's and reader's part of the specific binary format (i.e., 32 bit integer, signed, little-endian format, and so on).

To overcome this difficulty, elevation data are often delivered in ASCII encodings (or *string*). In this format, the data are written out as text (either 8 bit or 16 bit per character). ASCII format elevation data can be read and edited in a simple word processor such as *Notepad*. ASCII format data are very inefficient for two reasons. The first is space. The elevation value "21,432,124" requires 4 bytes when encoded as a 4-byte integer. However, it requires at least 9 bytes when encoded as ASCII. Eight bytes are required for the actual digits of the value, and one additional character is required

TABLE 10.3
A Comparison of Encodings for a 1 Million Post File

Format	Encoding Example	Size, in bytes
Binary (Int32)	Iiii	4,000,000
Binary (float)	Ffff	4,000,000
Binary (double)	Ddddddddd	8,000,000
ASCII (cm)	487600Ø	7,000,000
Fixed ASCII	0487600	7,000,000
XML	<ht>487600</ht>	15,000,000+

as a *terminator*. A terminator (usually, a byte whose value is zero, called a *null*) allows a computer algorithm to distinguish between the end of one value and the start of the next. Occasionally, one encounters other software (such as programs that use Hollerith encoding) that require fixed format ASCII values. These programs specify, for example, that each value must be 10 characters long with leading zeros. Of course, this is even more inefficient than null terminated ASCII in terms of storage requirements.

The second major inefficiency of ASCII as a storage format is that computers do not do numerical computations in ASCII. This means that every ASCII elevation point (and other attribute data, if present) must be read in and parsed to find the end of the string and then converted to an internal numeric format such as 32 bit integers.

An alternative text-based encoding is extensible markup language (XML). XML has the advantage of ensuring proper reading of data fields (assuming it is a properly formed file) but has the same disadvantages of bloated size and text-to-binary conversion. In fact, XML is even more verbose than ASCII, making it generally unsuitable for high efficiency transport of dense data. Thus, XML data encodings should be limited to metadata.

A comparison of the various encoding formats for a 1 million point elevation file that requires 32 bit integers for storage is provided as Table 10.3. In this example, the heights are encoded in centimeters.

10.1.5 Exploitation versus Transport Formats

When selecting data formats it is necessary to consider the application goals for the data. Data can be classified as belonging to one or more states with the primary state often dictating the optimal format. *Exploitation* formats are used when data are actively being accessed by application software. *Transportation* formats are appropriate when moving data from one application domain to another. Perhaps, the most well-known transportation format is XML. XML is a *tagged* data format that fully specifies each data element in a system neutral format. It is actually a rich, human, and machine readable tagged ASCII format. The advantage of XML is that it can transmit information between disparate systems using an agreed upon tagging scheme. Many different *schemas* have been developed for different disciplines using XML. Geography Markup Language was an emerging standard for GIS use although its presence has waned since the first edition of this book.

In the last few years, JavaScript Object Notation has largely replaced XML as the preferred format for communicating to web application programmer interfaces. There has been some extension into other areas usually reserved for XML but, again, as essentially an organized ASCII representation, it is not suited to representing voluminous datasets.

Where formats such as XML are highly suited for transmitting data between systems, they are ill suited for voluminous data such as LiDAR or image data and for *exploitation* of data. Data exploitation involves the process of actually using the data in a presentation or algorithmic manner. An example might be perusing millions of LiDAR data points, locating all points that were acquired on a specific flight line. Such operations are repeated over and over during LiDAR data processing and therefore require a data organizational format that is optimized for accessing and processing the data in near real time.

The LiDAR community has settled on a compromise format that is very efficient for data transmission of very large datasets and moderately efficient for exploitation. The format is a binary point cloud organization called LAS (a shortening of LASer) that is maintained by the American Society for Photogrammetry and Remote Sensing (www.ASPRS.org).

10.1.6   Important Characteristics of LiDAR Data

Unlike most conventional image datasets, LiDAR data generally provide a significant amount of sensor characteristics to the exploitation software. This is primarily due to the emergent nature of data processing algorithms. Thus, standard LiDAR sensor geocoding software emits LiDAR point data with a rich set of associated attributes (metadata). Examples of LiDAR data attributes are listed in Table 10.4. This is not a comprehensive list but simply an example.

A quick perusal of Table 10.4 reveals how much data are lost to processing analysis when LiDAR data are delivered in so-called *X*, *Y*, *Z* ASCII format. All the information that are critical to *classification* of the data are lost. Thus, it is important that LiDAR specifications always require the delivery of data with fully attributed points.

10.1.7   A Brief Introduction of the LAS LiDAR Data Standard

This is a brief overview of the LAS data specification. A more detailed look at LAS 1.4 is provided later in this chapter. In 2002, representatives from Z/I Imaging Corporation (now a part of Hexagon AB), EnerQuest (now defunct), Optech Corporation (now a subsidiary of Teledyne), Leica Geosystems (also part of Hexagon AB now) and the US Army Corps of Engineers Topographic Engineering Center* (TEC) formed an informal committee to develop a standard interchange format for LiDAR data. LiDAR data were being produced by sensor vendors (primarily Optech and Leica Geosystems at that time) in proprietary binary formats. Software vendors such as Terrasolid OY (Helsinki, Finland) had written special data readers to ingest these sensor-specific formats.

TABLE 10.4

Example LiDAR Per-Point Data Attributes

Attribute	Description
X, Y	The planimetric ground location of the point
Z	The elevation of the point
Intensity	The laser pulse return intensity at the sensor
GPS time	The time (in GPS clock time) of the receipt of the return pulse
Number of returns	Number of returns detected for a given transmitted pulse
Return number, number of returns	The return number of this pulse within an ensemble of returns from a single outbound pulse (e.g., return two of three returns)
Mirror angle	Angle of the scanner mirror at the time of this pulse (only applies to scanning sensors)
Classification	Surface (or other) attribute assigned to this point such as ground, vegetation and so forth
Point source ID	A unique identifier to reference this point back to a collection source

---

* Now Army Geospatial Center (AGC).

EnerQuest, working with TEC, had also developed custom processing software that had to deal with the vendor-specific formats. It was becoming increasingly clear that high performance processing systems would need to integrate hardware and software from a variety of vendors. This set of common needs led to the formation of this original ad hoc laser scanning data format group.

EnerQuest had developed an internal format that they used to exchange data between a number of different application programs that they had developed under a TEC Small Business Innovative Research (SBIR) grant. This format, called LAS (for LASer), was a binary format that abstracted any particular hardware vendor's implementation. EnerQuest and TEC agreed to donate this format to the public domain. The ad hoc committee used this original data specification as the genesis of what would become an American Society for Photogrammetry and Remote Sensing (ASPRS) standard for LiDAR data exchange. The LAS name was retained for this format (this has turned out to be a bit unfortunate as, unknown to our group at the time, LAS is the acronym for the oil industry standard file format *Log ASCII Standard*).

LAS was internally used by the members of the forming committee while simultaneously being offered as a standard to ASPRS. After slight modifications by the ASPRS LiDAR subcommittee, LAS version 1.0 was approved as an ASPRS data standard on May 9, 2003. This standard was officially updated with minor data field changes to LAS 1.1 (March 7, 2005). At the time of this writing, LAS 1.4 is the current version. LAS has been one of the more successful standardization efforts for creating a commercially viable data exchange. LAS is currently supported by all commercial LiDAR data-processing software vendors, allowing seamless interchange of data.

The LAS data format specification can be downloaded in pdf format from www.asprs.org. This specification defines a binary file format for storage of LiDAR point data as a set of record blocks. The header section of the file is a fixed length record that contains information about the *point records* that are stored in variable length record blocks.

Similar to any data standard, LAS was a compromise between storage efficiency and richness of the data contained in the file. LiDAR processing algorithms are still at the point that any metadata regarding the state of the sensor during acquisition can be useful in the *classification* process. For example, if the scan angle of the sensor for a particular point is zero (meaning the point was collected at nadir) then the point probably did not reflect from a vertical surface of a building. LAS is very rich in the collection of per-point metadata. However, the LAS format does not specify an order to the points in the data file and thus random access is not possible in the native LAS format. This organization was by design as it was meant to allow very rapid (real time) storage of data during collection (a so-called streaming format).

It is very important not to make *any* assumptions about the order of points in a LAS file. If a single file represents a single LiDAR flight line, the data will typically be ordered by Global Satellite Navigation System (GNSS) time. However, there is no guarantee that this will be the case. The LAS format does require that multiple returns be sequentially encoded in the file. For example, a pulse that had three returns will be in sequential order of pulse 1 of three followed by pulse 2 of three and finally pulse 3 of three. It is very important when writing software that creates LAS format data to ensure that this rule is followed as LAS does not provide an *associativity* mechanism that allows points to be related.

Software applications are, of course, free to reorganize LAS data during processing. For example, it is quite common to read a LAS data file and then reorganize the data into a spatial arrangement to allow rapid indexing such as a Quad Tree or Octree representation. Some vendors spatially organize LiDAR data in the LAS file structure itself and use one or more user-defined records (Variable Length Records, VLR) to store the spatial index. This allows software that understands the spatial index to rapidly access blocks of LiDAR points while not affecting software that does not use the index. The only drawback to this scheme is that software that relies on the indexing scheme must be able to detect if the data have been randomized by software not using the index scheme. If randomization is detected, the indexing software must reorder the file.

A detailed overview of the current release of LAS (as of the second edition of this book) is contained at the end of this chapter.

10.2 DATA MANAGEMENT AND DATABASE CONSIDERATIONS

10.2.1 PROCESSING CONVENTIONS

LiDAR data attributes should only be modified at points in the processing chain that specifically and reasonably deal with that attribute. For example, *geocoding* software is responsible for encoding the X, Y, and Z values of the data based on GNSS/IMU* input and LiDAR sensor modeling. Once the X, Y, and Z values have been encoded by this modeling (the software that performs these functions is usually supplied by the sensor manufacturer), the coordinates should not be modified without a rigorous, well-understood model. An example of a well-understood model would be changing WGS-84 ellipsoid heights to a gravity model such as North American Vertical Datum, 1988 (NAVD-88) using a transformation model such as one supplied by the United States National Geodetic Survey (NGS). An example of modifying data outside the constraints of a well-understood model would be to attempt to determine the average elevation of vegetation that is not being penetrated by the laser pulses and then moving the LiDAR data points in these areas down by this average height.

Sometimes heuristic processing can be rationalized on the basis of accuracy envelopes. A common example of this is data smoothing. Consider a situation in which a LiDAR dataset has an absolute vertical accuracy requirement of 20 cm with some specified sample standard deviation. Analysis of the data indicates a vertical accuracy of 8 cm with similar sample standard deviation. However, the *noise* in the vertical is causing contours to be created with objectionable aesthetics (due to the in-specification vertical noise). Smoothing the vertical component of the LiDAR data within the vertical accuracy envelope would be deemed an acceptable practice (of course, the stated product sample standard deviation would have to be increased as a result of the smoothing).

Points should never be physically deleted from a LiDAR dataset. Instead, points that are to be excluded from consideration are flagged as *withheld*. LAS 1.0 required changing the classification field to withheld. At version 1.1, LAS added a special attribute bit that is used to indicate the withheld state and the special classification was depreciated. This addition in LAS 1.1 allowed the original class of a withheld point to be maintained (so, e.g., a point in the Ground class can be flagged as withheld without losing the Ground class tag). Points are typically withheld because they are *outliers*, are in overlap regions in which the desire is to maintain uniform data density, are on the edge of flight lines or other similar reasons. It is generally a good idea to do an analysis of outliers to determine their cause. Reflecting from a bird, cloud or aircraft is normal whereas many outliers with unknown cause can be indicative of an impending equipment failure. Similar bits for other flagging purposes have been added with each update of the LAS specification. As of LAS 1.4, bits are defined for *withheld*, *synthetic*, *overlap*, and *Model Key-Point*. A detailed discussion of the LAS 1.4 format is contained in a later section of this chapter.

10.2.2 MANAGING DATA

Designing a data management scheme for LiDAR depends on the stage of the data and the application area. An overview of management schemes is provided as Table 10.5.

The choice of data format during LiDAR processing is dictated by the tools that will be employed. Nearly all software for processing LiDAR and elevation data will read and write LAS data. Although some lesser known binary formats are used in static laser scanning, all topographic processing will read and write LAS format (some software will also support direct input/output of compressed LAS data such as the common compressed LAZ format).

The primary consideration with respect to LiDAR data formats is to ensure that the production chain remains modular and open. Thus, the best approach is to view production as a series of

---

* Inertial Measurement Unit—used to compute the pointing direction of the laser beam at the time of acquisition.

TABLE 10.5
Processing Stages and Management Schemes

Stage	Scheme	Notes
Collection	Typically a vendor proprietary format.	At this stage, data are not geocoded. They are typically in a raw form of time-encoded *range* data packets
Geocoding	1 LAS file per flight line (although it is best practice to keep file sizes below 2 GB)	Data are typically fully geocoded and classified to *never classified* (class 0)
Geometric analysis/correction	1 LAS file per test patch	Test patches are extracted from the flight lines and used for geometric analysis
Editing	Tiled LAS	Tiling scheme is generally determined by the maximum number of data points that the editing software can accommodate while remaining performant. This typically ranges from 1 to 5 million points per tile
Quality check	Tiled LAS	Tiling schemes are occasionally different than editing due to different software or delivery requirements
Delivery	Tiled LAS	Delivery tile schemes are usually dictated by ground gridding schemes (e.g., 1 × 1 km tiles). Often the primary delivery is gridded LAS data with a raster elevation model provided as a *derivative* product.

processing blocks with a neutral data format connection between the blocks. This approach to managing data will allow the processing chain to be upgraded in a module form as various tools improve.

All commercially available LiDAR processing tools in the market today can read and write at least to version 1.2 of LAS. All software used in producing final deliveries to the United States Geological Survey (USGS) 3D Elevation Program (3DEP) must write in complaint LAS 1.4. A number of tools read LAS format and then convert the data into a proprietary format for processing. This sort of scheme has no limitations as long as the data can be converted back to a neutral, open format, preferably during the write back to store operation from the processing tool.

10.2.3   LiDAR and Databases

At the writing of the first edition of this book, spatial databases such as Oracle Spatial and ESRI ArcSDE were beginning to add native support for LiDAR point data. With the advent of these storage technologies, the question arose as to when database storage is preferable over file storage (often called *flat file* storage). A general comparison of file versus database storage is provided in Table 10.6.

TABLE 10.6
Database versus File Storage

Consideration	File	Database
Rapid data input/output	+	−
Transportable at the tile level	Yes	No
Directly compatible with exploitation and viewing tools	Yes	No
Simple data backup schemes	+	+
Random access based on geospatial criteria	No	Yes

At the time of writing the second edition, the use of and even experimentation with database storage for LiDAR data has significantly declined. The primary reason is that there is an insatiable desire by the user community for ever higher data densities and the need during analysis to often drive to particular points for rapid access to information. Tiling data in flat LAS files in organizations such as quad trees and octrees has proven to be a much more performant solution than storage of individual points or point cluster as binary large objects (blobs) in databases.

The general guideline should be to plan to use file-based storage for high-throughput production operations (even as database storage matures) due to the high input/output requirements as well as the fact that data are quite transient during production. There are really no advantages to having LiDAR data resident in a database during production operations and many disadvantages. Even for applications in which point cloud data will be relatively static over long time periods (months, years), database storage is probably a poor choice. Since the time of the first edition of this book, we have seen a migration away from database storage by prior leading advocates such as ESRI, opting instead for simple index files supervising collections of LAS files (e.g., the ESRI LAS Dataset).

10.3   DATA IN PRODUCTION AND WORKFLOW

10.3.1   ORGANIZING DATA FOR PRODUCTION

The most effective current technique for organizing LiDAR data for high-throughput production (where production includes primarily the process of extracting features such as bare earth, buildings and other attributes from the data) remains as a grid of tiles of a size optimized for the processing tool. A variety of commercial off the shelf (COTS) software applications are available for reordering flight line-based LiDAR data into tiles. Examples of software applications include LP360 and the LiDAR 1 CuePac from GeoCue Group, TerraScan from Terrasolid and LAStools from rapidlasso though there are numerous others. The LAS format maintains a field per point record that tracks the source of the points. This Source ID attribute allows the data to be *filtered* from a tile-centric organization back into original source organization such as flight lines if this should prove necessary.

Although many LiDAR-processing companies organize data using a number scheme and file directory discipline, a graphical index will significantly aid maintaining the data inventory. In its simplest form, this index is a GIS layer with an attribute containing the storage location (e.g., full file path) of the associated LiDAR data. More sophisticated management tools can arbitrate multiuser access to data tiles as well as retrieve copies of the data from the storage repository (e.g., GeoCue LiDAR 1 CuePac).

It is good practice to add tools to the processing array that can compute and display the *convex hull* of the LiDAR tile. The difference between a bounding rectangle display of LiDAR tiles and the convex hull display is depicted in Figure 10.5.

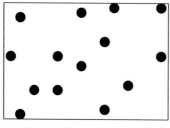 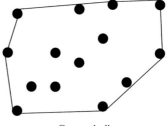

Minimum bounding rectangle   Convex hull

FIGURE 10.5   Minimum bounding rectangle compared with the convex hull.

FIGURE 10.6   LiDAR intensity image depicting voids (in red).

Analyzing LiDAR tiles based on a Minimum Bounding Rectangle rather than the convex hull can give a misleading picture of how well the tiles are edge matched in terms of LiDAR data density. Obviously, the data tile of Figure 10.5 has a void area in the lower right hand corner of the tile that will cause a data gap problem when merged with data from adjacent tiles. This data gap would not be evident by simply viewing the Minimum Bounding Rectangle.

LiDAR data return intensity can be used to create a coverage map of the point data. Used in conjunction with the tile index graphic, the LiDAR *ortho* gives a rapid assessment of the quality of data coverage. An example is depicted in Figure 10.6. The background was set to red to emphasize the *holes* in the LiDAR data. In this case, the coverage is adequate. The holes are water bodies in which the laser packets are being absorbed or scattered away from the sensor, leaving *holiday* areas. With the higher data densities now being routinely collected (e.g., the base data for the 3DEP program are 2 points/m^2), the intensity data become ever more valuable. For this reason, this author advocates for keeping the "I" as a capital in LiDAR data as in Laser Imaging, Detection, And Ranging)! (I bet the printer changes everything back to a lower case "i" as in LiDAR).

Although LiDAR data are considered large as compared with vector-based GIS data, it is actually usually an order of magnitude smaller than associated image data. For example, a countywide LiDAR project of 1000 km^2 with an average point spacing of 1 m will comprise about 26 GB of data (using 28 byte/point record type 6 LAS data storage). This same area covered by 15 cm Ground Sample Distance (GSD) 8 bit per channel color imagery will comprise about 125 GB. Thus, the challenges of managing LiDAR data are somewhat less that the image data typically associated with the project.

10.3.2   Workflow Considerations

It is a good idea to perform any necessary coordinate transformation at one point in the production workflow. LiDAR data are typically collected in the current epoch of WGS-84 with ellipsoid heights. Customers usually request the delivery of processed data in the local coordinate system with geoid heights. It is strongly recommended that these transformations be performed prior to geometric analysis of the data as this approach will consolidate all geometric testing in one part of the workflow.

LiDAR data should be copied at several steps in the processing chain to allow recovery in the event of an error or data disaster. The original flight lines in the collection coordinate system, prior to any geometric adjustments, should always be maintained. This allows a restart in the event of a nonrecoverable, erroneous geometric adjustment of the data. The second phase of data backup should occur immediately following geometric correction of the data. The third phase of data duplication should occur at the point following classification of LiDAR tiles but prior to cutting out the customer delivery. These snapshots of the processed data are in addition to normal system management disciplines.

It is usually not practical to back up LiDAR data to a tape archive (unless the archive system is an extreme performance device capable of 50 MB/s or greater transfer rates) due to the very dynamic nature of the data during processing. The best approach is to duplicate the data across two RAID systems at level 5 in which the systems are in two separate physical locations. At the conclusion of the project, the original LiDAR flight line data (following the initial geocoding process) should be archived as well as the final fully classified and transformed data at the point that customer deliverables were produced. If the final product is a gridded dataset that is to be derived from the point cloud data, the archive should include both the final classified point cloud data (in LAS format) and the gridded customer delivery.

10.4 DATA STORAGE AND RETENTION

10.4.1 Recent Advanced in Storage Technology

In the last few years, offsite rented storage and processing (*cloud* systems) have become an approach worthy of serious consideration for data management. As of mid-2016, Amazon Web Services (AWS) Glacier storage cost $0.007 per GB per month. This translates to $86 per year per TB! This is durable storage with all maintenance (backup, technology move forward, and so forth) taken care of by Amazon. Obviously, this is much less expensive than could be accomplished by a local storage approach.

The problem, of course, is the *last mile*. That is, the bandwidth to push the data to AWS and to retrieve it back for processing. Not only is this the bottleneck in data transfer, it is also the most expensive aspect of using cloud-hosted services. For example, retrieving up to 10 TB per month from AWS costs a breathtaking $0.09 per GB. This means retrieving that 1 TB of data in a single month span would cost $92.16. Clearly this approach will not scale from a financial point of view!

For LiDAR data processing, in which very high and continuous I/O is the nature of the processing flow, local storage is all that makes sense with delivery of final data to customers via a transportable media such as a Universal Serial Bus (USB) connected disk drive. Cloud storage is an excellent option for the permanent archive of processed data as this seldom is retrieved. As of this writing, pushing data up to AWS Glacier was free. Due to the cost and latency of I/O, considering processing in AWS may not make much sense (it does, however, for the *exploitation* of LiDAR data).

Perhaps the technology advancement with the most impact on remote-sensing processing approaches is the rapid advance of solid-state disk (SSD) technology. SSD seems to be following a Moore's Law increase in capacity and performance while following a similar decline in price. In mid-2016, a Peripheral Component Interconnect Express (PCIe) 1 TB Samsung SSD (SM961) could be purchased for less than $500. This drive provides 3.8 GB per second of read performance and 1.8 GB per second write. With LiDAR data processing, I/O speed is often the bottleneck in production. With the advent of inexpensive, high capacity SSD, the most efficient architecture is SSD storage local to the processing workstation/server.

10.4.2 Data Retention Policies

It is very easy to accumulate LiDAR (and imagery, for that matter) data in an open-ended fashion; that is, without a clearly defined date when you will dispose of the data. Therefore, it is very important to establish overall data *policies* in terms of data storage and retention (the same applies, of course, to all types of data that you process and/or deliver).

The recommendation is that you clearly state in your delivery contracts the data retention policy that you propose. For example, if you are processing data for a prime contractor, you might specify

TABLE 10.7
Storage Modes

Mode	Description
On Line	The data are immediately accessible from compute nodes via a connection (network, dedicated storage connection, etc.)
Near online	The data can be brought to an on-line state via an automated mechanism under manual or software control. The most common near online storage has been robotic tape devices though this is now transitioning to high capacity, *archive* performance spinning disk
Offline	The data are not accessible via automatic means or the recovery period of data are measured in multiple hours. This mode typically means data that reside on a physical media that is stored on the shelf such as optical disk (although as of 2016, optical disk is no longer a viable storage media), portable disk drives, non-volatile random access memory devices (*USB Memory Sticks*), tapes and so forth. They must be physically inserted into a reader for use. Off-line storage is rapidly transitioning to cloud hosted solutions such as AWS Glacier or Microsoft Azure

that you will maintain a copy of the final, classified LAS data for 1 year from the final acceptance date. Of course, the prime contractor/client will often specify a data retention clause and this will establish the policy. It is very important to adhere to this policy or you will soon find yourself buried in data of unknown recoverability! A good approach is to specify in your contract a data recovery fee and then store these data in low cost, durable storage that is rented (e.g., AWS Glacier).

The terms online, near online, and offline are used to characterize the levels of data storage typically used in production. These terms are defined in Table 10.7.

Retaining data such that it can be recovered for secondary or replacement use is rather expensive and thus the goal is to keep data retention to a minimal level. The storage media itself are relatively inexpensive but the labor required to ensure viability can be quite high. The two biggest problems with data retention are as follows:

• Stability of the storage media
• Recovery of stored data from the storage media

Stability of the storage media concerns the issue of how long the data can be expected to remain in an uncorrupted state on the media without a refresh. *Refresh* means the data are read from the media, the media are checked and revitalized (if possible), and the data are written back. Some media, such as rewritable optical disk, cannot really be revitalized and must simply be replaced by new media on data refresh.* Hard disks are revitalized by copying the data to a temporary media (always make two separate copies!), reformatting the media, performing an error check, and then copying the data back. Stability is the issue of the reliability of media in terms of aging and exposure to environmental conditions. For example, if you chose writable digital video disk as long-term storage media (a poor choice, by the way, due to its very low data density), you would want to give careful consideration to the expected data recovery life of the media. When selecting a storage media, obtain realistic information on the reliable data shelf life. For example, we often see rewritable digital video disk specified as having a reliable data retention time of 25 or more years; however, the experiential life is more on the order of 5 years. An overview of the expected life of various digital media can be found at the National Archives website and the Optical Storage Technology Association. Surprisingly, little correlation is found between types/cost of spinning hard disks and their reliability (Schroeder and Gibson, 2007).

---

* Although optical disk would never be considered as a storage device in this current age (2016), there are still very large archives of data resident on this type of media. At some point, these data must be copied to modern archival systems.

Another major consideration in establishing a storage policy is the risk of obsolescence of the storage access hardware. For example, you would be hard pressed to find up-to-date hardware and software to recover an archive of 9-track computer tape yet this was the standard data backup system of the 1980s. Therefore, in long-term storage scenarios (e.g., data that will be good source material for change analysis over long time periods), a plan must be in place to rotate the data to new types of media and possibly new data organizational formats (e.g., transforming 1980s proprietary image formats to TIFF).

As you can see, any long-term offline storage (longer than a few years) will require an active storage program. An active storage program requires periodic testing of the storage media and rotation of data to new media. If you are doing a simple media refresh, the data are written back to the same type of media (it may be replacement media such as writing data on magnetic tape to new physical tapes of the same type). If you are performing a technology upgrade, you will be transferring the data to a new type of media. This latter case is the more complex as media density is doubling every 18 months or so, and thus, a repackaging of data might be involved to take advantage of the new, higher density. As noted previously, the cost and efficiency of cloud-hosted bulk storage has significantly improved since the first edition of this book. It is now virtually impossible to favor on-premises long-term storage to cloud-based storage from a Return on Investment perspective.

In general, online and near-online data storage arrangements are much less labor intensive to maintain than offline systems. Of course, the tradeoff is that the hardware required to realize these mechanisms are more expensive than offline storage. It is very important to realize that data redundancy within a single unit such as archiving to an online RAID 5 system is not sufficient protection. If the disk hardware suffers a catastrophic failure such as a lightning strike or physical damage (fall, flood, fire, etc.), then all is lost. Thus, you must maintain a duplicate system at a separate physical location. Large sample size analysis from the Carnegie Mellon Parallel Data Laboratory (Schroeder and Gibson, 2007) seems to indicate that the practical life of hard disk drives is about 5 years. As this represents nearly three generations of hard disk technology, it might be a reasonable plan to completely replace online hard disk technology on a five-year cycle.

So what exactly should be archived and when should these archived data be deleted? If not specified otherwise in the delivery contract, then the suggested base plan of Table 10.8 can be modified to suit your particular needs.

TABLE 10.8
Template Archival Periods

Data	Archive Period	Notes
Raw LiDAR sensor data	One-year postdelivery	
RAW positioning and orientation system (POS) data	One-year postdelivery	
Survey control	Five-year postdelivery	
Geocoded LiDAR data	Until geometric correction is verified	Can be reconstructed from raw data, if need be
Geometrically corrected, tiled LiDAR data	One-year postdelivery	
Geometrically corrected, tiled, classified LiDAR data	Five-year postdelivery	If you are building a speculative LDIAR data archive, these are the data you will retain. Consider AWS Glacier for this repository
Thinned (model keypoint) LiDAR data	Six-month postdelivery acceptance	These data are easily reconstructed from the classified data
Derived products	Five-year postdelivery acceptance	Consider AWS Glacier for this repository. Again, these data are easily reproduced from the corrected, classified LiDAR point cloud data

Most companies tend to save too much data for too long. In addition, these are usually of unknown quality (i.e., only portions of the datasets can be recovered) and hence are of little practical use. It takes a highly disciplined approach to properly maintain data archives, but the time invested is invaluable when that rare occurrence pops up of needing to restore a historic project.

10.4.3 DATA CHECKPOINTING

Data checkpoints are versions of data retained during processing to enable rapid recovery in the event of a processing error. Checkpointing begins with acquisition and occurs at checkpoints throughout production.

Most commercial kinematic LiDAR systems record raw data to a single data drive set during data acquisition (although metadata such as GNSS/IMU data are often recorded to multiple devices). Thus, the first step after a data acquisition mission should be validating and copying the acquired data. You should design a copy system that is deployed with your LiDAR system for this purpose. It is strongly recommended that duplication of data occur prior to any field quality checks (in case a catastrophic failure occurs during the QC process).

Best practices dictate that you do a field check of data prior to commencing processing in your data production facility. A recommended basic approach is depicted in Figure 10.7.

Obviously, the typical processing flow is more complex than is depicted in Figure 10.7. For example, if field personnel could not obtain a satisfactory first look solution for the Positioning and Orientation System (POS), they would probably pass the data (via File Transfer Protocol, FTP) to the processing center to see if experts could obtain a satisfactory solution. The general idea is to always perform Quality Checks as close to the potential error source as possible. It is much more expensive to find an unrecoverable POS error after the collection platform has left the collection

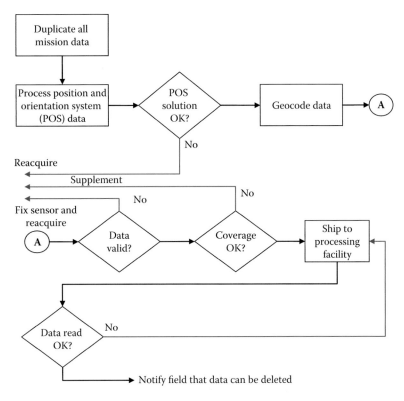

FIGURE 10.7   Field data flow.

area than before! The preservation of two copies of data in the field forms the first data checkpoint. Note that this set of data should be deleted once confirmation has been received from the processing office that all data have been received and are intact. It is strongly recommended that one set of field data be immediately deleted but the second set be retained until the media are recovered for a subsequent mission. This allows a bit of a safety factor in the event that the processing office has made an error in archiving the data.

It is also strongly recommended that two full copies of raw data be maintained at the processing facility for the duration of production of the *entire* project. These data generally comprise the following:

- POS raw data
- LiDAR Range file (LiDAR raw data)
- LiDAR platform calibration data
- Image data (if applicable)
- Control/Checkpoint data (although this is technically not *raw* data, it is convenient to keep it with the raw data as it would be needed for reprocessing)

This recommendation is made on the basis of the fact that the mission can be fully recovered from this dataset should disaster strike during the production phase of the project (e.g., an inadvertent deletion of a top level folder containing all of the project data and so forth). Whether these data are retained after project delivery depends on the contract and your established data retention policies. Be extremely careful in deleting these data as one of the most common reasons for project rejection by discerning clients is the quality of geometric correction. If you retain data at this point, you will be fully able to repeat the calibration/project geometric correction steps.

Establish a temporary mission checkpoint at the conclusion of calibration. This checkpoint allows you to do a full recovery, post-calibration, should you get hopelessly lost in mission geometric correction. These checkpointed data can be deleted as soon as you have confirmed the validity of project geometric calibration.

A checkpoint should be established when geometric correction is complete. Again, it is recommended that you keep two copies of the data as the process of geometric correction can be quite time consuming, and the checkpoint will allow you to recover to this step. After confirming the validity of the checkpointed data, you can delete the checkpoint.

Classification is generally an iterative process of applying automated processing routines (e.g., *macros* in TerraScan or *point cloud tasks* in LP360) followed by interactive classification. Checkpointing during this phase of processing is done on a tile by tile basis and is more appropriately called *versioning*. The number of versions to maintain will depend on your overall scheme for processing as well as the complexity of classification. If you employ a production model of complete automated processing followed by interactive classification (without a return to the project-wide automated processing), then create tile-based file version checkpoints at the end of automated processing. This will allow you to recover a tile to the point of basic classification should an editor make an error in which recovery is quicker from the automated classification (a classic example would be to inadvertently classify outside a polygon when the intent was to classify inside a polygon). Some advanced processing framework software such as GeoCue can automatically create tile checkpoint versions during processing. Automated checkpointing not only saves considerable time but also alleviates the problem of technicians neglecting to create the checkpoints.

The final step of classification is a rigorous tile by tile quality check. This check examines the quality of both the classification and the density of certain classes. For example, if a project requirement is to produce a ground point density of 2 points/m^2 in open regions, this test occurs as part of the final QC during classification (a much better place for this step than final QC prior to delivery). Following successful completion of this QC step, the data are checkpointed. In many situations, this is the most valuable checkpoint within production and may be the data that you store as the long-term archive LAS for the project.

Vector data associated with the project are also carefully checkpointed. These data often are not on the same tiling scheme as the LiDAR data as they tend to be geometric features (e.g., roads) that span tiles. The scheme that you employ will depend on the nature of the repository used for the vector data. If you are collecting to a geodatabase using tools such as ArcGIS, you will have versioning capability included in the collection software. If you are using a file-based application, you will need to save versions of the files at critical points in collection.

10.5 LAS DATA FORMAT

10.5.1 DETAILS OF THE LAS 1.4 DATA FORMAT

This current section provides an overview of the latest version of the specification, LAS 1.4, as well as a history of the LAS format. It is based on an article published in Photogrammetric Engineering & Remote Sensing of the same topic (Graham, 2012). Generally, I will use the pronoun "I" when something is primarily my personal opinion and "we" when the work was consensus work of the LAS Working Group (LWG). The LWG is a working group within the LiDAR Division of the ASPRS. It comprises a group of representatives of LiDAR hardware and software manufacturers as well as several representatives of end-use agencies. The LWG periodically updates the LAS specification as either hardware and/or processing standards change.

The introduction of a specification file format (LAS 1.0) for kinematic LiDAR data in 2002 has been very successful as evidenced by the fact that all major software vendors have adopted the specification, and customers are routinely requiring that LiDAR data be delivered according to this specification. As with any specification, usage has made it apparent that changes need to be made both due to omissions in the original design as well as continuing maturity of the LiDAR industry. These updates have been occurring on an approximate two year cycle.

10.5.2 HIGHLIGHTS OF LAS 1.4

As of mid-2017, version 1.4 is the most recent release of the LAS specification. The update to LAS that resulted in the creation of version 1.4 was primarily motivated by the need to extend the number of possible classes from the 32 available in LAS 1.3 to 256 (ironically, LAS 1.0 did support 256 classes, but this was changed in the update to LAS 1.1).

Of course, any time one starts updating data formats, other features have to be added! The summary of major additions to LAS 1.4 is as follows:

- Extension of offsets and field sizes to support full 64 bit addressing
- Support for up to 15 returns per outgoing pulse
- Extension of the Point Class field to support 256 classes
- Definition of several new ASPRS defined classes
- Extension of the Scan Angle field to 2 bytes to support finer angle resolution
- Addition of a Sensor Channel bit field to support mobile mapping systems
- Addition of Well Known Text (WKT) definitions for Coordinate Reference Systems (CRS)
- Addition of an Overlap bit to allow indicating pulses in the overlap region while maintaining the class definition
- Addition of an (optional) Extra Byte Variable Length Record to describe *extra bytes* stored with each point

10.5.3 SPECIFYING LAS 1.4

Following this current section, I will go in to some detail concerning the major points of the LAS 1.4 specification. This section provides information for those folks who just want to know what they should be specifying when contracting for LiDAR data.

TABLE 10.9
The New Point Data Record Formats (PDRF) of LAS 1.4

PDRF (LAS 1.4 Only)	Major Features
6	Base type
7	Red, green, blue (3 channel) colorization support
8	Red, green, blue, NIR[a] (4 channel) colorization support
9	Waveform packet support
10	Red, green, blue, NIR (4 channel) colorization, and Waveform packet support

[a] NIR—Near infrared.

The most critical features of LAS 1.4 with which most users should be concerned are support for 256 classes, an Overlap region indicator bit, and specification of CRS by the Open Geospatial Consortium method of WKT. As *class* and *Overlap* are per-point attributes, it is necessary to use a new Point Data Record Format (PDRF) to specify these new features. LAS 1.4 adds five new formats (PDRF 6 through PDRF 10) to the specification that all include the new CRS encoding, extended classes, and designation bits. The major differences between these five new formats are delineated in Table 10.9.

As is evident from Table 10.9, most new deliveries will be in PDRF 6.

LAS 1.4 *requires* that the CRS encoding for PDRF 6–10 be WKT as opposed to the old technique of GeoTIFF packets. Thus, you are assured by specifying record types 6–10 that your data will have proper CRS encoding. I feel that this is critical for archived data since GeoTIFF (the old encoding standard) has limited support for describing the vertical components of CRS and limited support for newly emerging CRS definitions.

As of this writing, the USGS is requiring PDRFs that support the designation of overlap via the new Overlap bit supported in PDRFs 6-10. In fact, it is this USGS requirement that has accelerated the move from prior versions of LAS to LAS 1.4.

10.5.4 EMULATING PRIOR LAS VERSIONS

During the specification process for LAS 1.4, there was a strong desire from several developers of open source software to create fields in the LAS 1.4 specification that might make it possible for a prior version of a LAS reader to ingest a LAS 1.4 file, if that file met certain conditions. These conditions are as follows:

- The offset to data pointers must remain at or below 32-bit limitations
- PDRF must be one of the *legacy* formats (0–5)
- The CRS encoding must be GeoTIFF (as opposed to the new method of WKT)
- The CRS must be encoded in a Variable Length Record as opposed to an Extended Variable Length Record
- The legacy reader software must ignore the LAS version number

I was personally opposed to this emulation mode because it introduces (if used) some unreliability to the LAS file and relies on a software defect (namely, ignoring version number) to work at all. Unfortunately, the details (outside of the LAS Working Group) became obscure with some LAS consumers thinking that if this emulation mode were supported, their existing LAS software would be enabled for the new LAS 1.4 features. This, of course, is not possible as the major new features are encoded in the new point types (6–10) that simply are not present prior to LAS 1.4.

I think this emulation feature should be used only for experimental purposes in which the exact implementation is known, *a priori*. My recommendation is that you *never* specify a data delivery of LAS 1.0–LAS 1.3 data within the context of a LAS 1.4 file. If you need LAS 1.0 to 1.3, simply require your vendor to deliver in the proper format.

General rule: Always specify Point Data Record Types of 6 and above and legacy fields set to zero.

10.5.5 THE ORIGINAL SPECIFICATION

The original LiDAR binary file specification grew out of work done by EnerQuest Systems (software development work of which is now owned by Autodesk) in the late 1990s. EnerQuest had developed a proprietary point cloud binary format for laser-scanned data (LASer) to make dealing with the generally unorganized data associated with LiDAR more efficient. Don Wicks, then president of EnerQuest, generously agreed to contribute the EnerQuest LAS work to the public domain. Leica Geosystems, Optech and Z/I Imaging (I was then with Z/I) joined the effort, and the LAS 1.0 specification was born (the bulk of the implementation work was done by Phil Thayer and John Nipper, software developers then at EnerQuest). About a year after this industry group defined the initial specification, the ASPRS agreed to take over ownership.

The LAS specification is a relatively compact binary encoding of point location and point *attribute* data. Rather than store attributes in *referenced* records, the light-weight attribute data of LAS is stored in the same record as the point data. At the time of the definition of LAS 1.0, we did not try to anticipate how users would want to augment the format.

10.6 LAS IN DATA-PROCESSING AND EDITING

10.6.1 THE PHILOSOPHY OF LiDAR PROCESSING

In order to fully appreciate the design of the LAS format, it is necessary to understand the thought process that the original designers of LAS 1.0 had with respect to LiDAR data processing.

A time-of-flight LiDAR system emits a laser pulse and zero or more pulse returns* (*echoes*[†]) are detected in the receiver. These return pulses may be reflected from nearly anything at all (e.g., birds, planes, tree leaves, buildings, ground, etc.). It is the task of the LiDAR data editor to *classify* these data with respect to their impinged surface. Thus, the original LAS specification supported the idea of assigning a *classification code* to an echo. This classification attribute could be set to the surface type such as Building or Ground. Indeed, much of the software challenge in LiDAR processing has been a drive toward developing various algorithms that can automate this processing (so-called LiDAR *filtering* algorithms, an unfortunate misnomer).

The original philosophy behind LiDAR processing was that points can never be added to or deleted from a LiDAR dataset after acquisition. In addition, points should never be *moved* except under rigorously modeled adjustments such as *bore sight* calibration and project geometric correction (for errors introduced by inaccuracies in the positioning system). We (the original LAS design group) felt very strongly that simply *deleting* or moving data that do not fit a desired model is tantamount to fabricating data. To support the editing concept, certain classes were reserved to indicate data adjustments beyond reflectance classification. Thus if, for example, a high point

---

* And, effective with LAS 1.3, possibly wave packets.
[†] I use echo and return interchangeably.

(perhaps reflected from a bird) were to be omitted from the dataset, it remained in the set but its classification code was set to WITHHELD.

It was soon evident that we needed to retain the original classification of points that were being withheld from the dataset (e.g., we needed to designate a point as both "Ground" and "Withheld"), and there was a need to be able to *synthetically* insert points into an existing LiDAR dataset (e.g., digitizing a synthetic LiDAR point to represent the attachment point of a transmission wire to an insulator). Thus, starting with LAS 1.1, the idea of modifier bits was added to the class designation.

Today, the philosophy remains that once LiDAR data have been geocoded in the proprietary software supplied by the hardware vendor, they should only be geometrically moved in a system calibration and/or geometric correction process and that points should never be deleted.

10.6.2 EXTENDING THE LiDAR EDIT PHILOSOPHY

The LAS 1.4 specification incorporates features that allow more robust point marking, adhering to the philosophy that points should never be physically deleted or moved during editing. For example, if points are found to be reflecting from saw grass, they should not be collected up as a group in a point editor and literally *moved* to the ground. We argue that what has really happened is that a collection of points have been marked as *withheld* and a new collection of points *added* based (one would hope!) on an accurate editing surface such as ancillary stereo imagery. This type of point marking has been available in LAS since version 1.1 by supplying appropriate attribute fields in the point records. LAS 1.4 supports four attributes related to point editing (described later).

The LAS specification is being used quite frequently as a general format for point cloud data, regardless of its source. For example, I know of at least one company that generates LAS format data from the output of a stereo image correlator and then applies LiDAR processing tools to the task of editing these data. In the four years (as of 2016) since the introduction of LAS 1.4, all software products that produce dense image matching (DIM) data via Structure from Motion (SfM) algorithms encode the resultant point cloud as LAS (unfortunately, at LAS 1.2!). We very strongly support this use of the LAS specification and have, in fact, added some fields in the 1.4 release to expand support for facilitating these nontraditional uses (primarily the addition of a near-infrared channel for supporting 5 channel 3D image data derived from line imagers).

10.6.3 THE GENERAL LAS LAYOUT

The format contains binary data consisting of a public header block, any number of (optional) VLRs, the Point Data Records, and any number of (optional) Extended Variable Length Records (EVLRs). This general layout is depicted in Table 10.10.

The public header block contains generic data such as point counts and point data bounds. Significantly, it does not contain the definition of the CRS.

The VLRs contain variable types of data, including projection information, metadata, waveform packet information, and user application data. They are limited to a maximum data payload size of 65,535 bytes. The continued existence of VLRs in LAS 1.4 is one of the compromise results of

---

TABLE 10.10

LAS 1.4 Format Definition

PUBLIC HEADER BLOCK

VARIABLE LENGTH RECORDS (VLR)

POINT DATA RECORDS

EXTENDED VARIABLE LENGTH RECORDS (EVLR)

---

supporting emulation mode (discussed earlier). Software writers of LAS 1.4 files in which emulation is not needed (the recommended approach) can encode all variable information into EVLRs.

The EVRLs allow a larger payload than do the VLRs and have the advantage that they can be appended to the end of a LAS file. This allows adding, for example, projection information to a LAS file without having to rewrite the entire file.

A LAS file that contains point record types 4, 5, 9, or 10 could potentially contain one block of waveform data packets that is stored as the payload of any EVLR. Unlike other EVLRs, the Waveform Data Packets (if stored internally to the file) have the offset to the storage header contained within the Public Header Block ("Start of Waveform Data Packet Record").

10.7   LiDAR DATA IN PROJECTS

10.7.1   PROJECTS

In LAS 1.1, we defined the idea of a *Project*. A Project can be a LiDAR job as defined by a number of flights/drives over a project area or it could simply be a number of elevation files being used in a processing job not necessarily even related to LiDAR. A project is uniquely identified by a Globally Unique Identifier (GUID). A GUID (sometimes alternatively referred to as a Universally Unique Identifier, UUID, or as a Uniform Resource Indicator, URI) is generated by a "GUID Generator" algorithm available within the Microsoft development systems or from open sources. A GUID is a 16-byte identifier that is unique across the domain of all possible GUIDs. The Project GUID existed in LAS 1.0, but we clarified its intended use in LAS 1.1.

It is a good idea to assign a Project ID at the time of creation of the project and to ensure that this same project ID is used for all files that are associated with the project.

10.7.2   FILE SOURCE ID

The File Source ID field is used to uniquely identify *this* file within a Project. If the source of the file is a LiDAR flight/take* line, then this field is used as the flight line number. Thus, inclusion of the File Source ID solves a major problem that existed in LAS 1.0; the restriction of flight line information to 256 lines. The File Source ID field supports 65,536 individual sources within a project.

It is important to note that the LAS format can embody any sort of point cloud data in addition to LiDAR data. This means that a project could consist of a heterogeneous collection of a variety of elevation files from any number of sources. For example, we might construct an elevation project with some files originating from LiDAR flights, others from USGS elevation files, still others from photogrammetrically correlated data, and so forth. Thus, the concept of *Project* becomes generalized.

10.7.3   SYSTEM ID

In keeping with the notion of generalizing the LAS specification to apply to any sort of point cloud data, System ID is used to provide information about the manner in which the LAS file was generated. For a LiDAR system, this field should be encoded with a system identifier such as "Optech Orion C." Similarly, this field is used to indicate the method by which a file was generated when the source is not a LiDAR sensor. The specification provides several predefined values such as "MERGE" and "EXTRACTION" and allows the user to provide any definition (within the limits of a 32 byte description). A common example within LiDAR processing would be the generation of a processing tile by merging the content of several flight lines. In this case, the Project ID is set to the overall Project ID assigned to the project, a new and unique File Source ID is created, and the System ID field is set to MERGE.

---

* In Mobile Mapping, a single span of data collection is often referred to as a data "take."

10.7.4 Point Data Records

LAS files are homogeneous with respect to point records. The record *type* is indicated by the PDRF in the Public Header Block. All points within a single file must be the same with the format being specified by the PDRF.

It is important to note that the LAS specification does not impose a point ordering scheme on a LAS file. Thus, the points can (and very often do) appear in any order. For example, if point n is marked as "return 2 of 4" and point $n + 1$ is marked as "return 3 of 4," there is no guarantee that these two points are associated with the same outgoing point (you would have to use GPS time to sort the returns, as noted later). I have seen a lot of software that erroneously assumes that LAS points will be sorted by spatial proximity, by return number or other attributes. This results in, for example, poor display performance when the data are, in fact, not in the assumed order.

LAS 1.4 added five new PDRFs (6–10) to the specification. The salient new features of these types are discussed in a later section. Although LAS 1.4 continues to support PDRFs 0–5, I strongly recommend that you always specify the new point types 6–10. The primary reason for this recommendation is that merging point types 0–5 into files using point types 6–10 will require a type 6–10 PDRF (due to expanded fields). A second reason for this recommendation is that LAS *requires* WKT encoding for the CRS of PDRF 6–10. WKT is a much more reliable CRS representation than GeoTIFF (in fact, there are a number of CRSs that simply cannot be represented in GeoTIFF). Remember, if you have a LAS file and you cannot determine its CRS, the data are useless for absolutely referenced data exploitation.

It is critically important for programers to understand that a point data record in LAS can contain fields beyond those defined by the PDRF. For example, PDRF 0 defines 20 bytes of specified information. If desired, a software implementer can add additional bytes to the end of this point type (e.g., perhaps to encode an accuracy estimator for Z). Thus, one must always refer to the Point Data Record Length field in the Public Header Block to determine the correct number of bytes for each point record.

LAS 1.4 (thanks to a contribution from Dr. Martin Isenburg) adds a way to formally define extensions of the standard PDRFs via a mechanism called *Extra Bytes*. Although nothing in the LAS specification prevents a developer from adding their own PDRF (say PDRF = 21), no other software would be able to read the point data. By using the Extra Bytes facility, software that does not implement Extra Bytes will still continue to correctly read the *base data* of the point record. Those who participated in the review of the canceled LAS 2.0 effort will recognize that this was the total representation method of points in that format.

10.7.5 GPS Time

Most PDRFs support a field for the time of the pulse. This time is encoded as either GPS Week-Second time or Standard GPS time as indicated by the Global Encoding field in the Public Header Block. We recommend that all procurers of LiDAR data *require* that the GPS time be encoded as Standard time. This is because GPS Week Time does not allow one to determine the absolute time that a pulse was transmitted.

Although not clearly described in the LAS, the GPS time is the time when the pulse was *emitted* by the LiDAR sensor. Thus, all returns (see the next section) of a multiple return pulse will have the same GPS time.

When we first defined the original LAS specification, we felt that GPS time, on a per pulse basis, would be used only for certain *close to the sensor* operations. However, many algorithms for data exploitation, particularly in the area of mobile mapping, heavily rely on GPS time. In addition, the GPS time is currently the only mechanism for correctly correlating a set of multiple returns. For this reason, GPS time was included as a field in all of the new PDRFs of LAS 1.4.

10.7.6 RETURNS

LAS is designed for discrete point data (that can optionally support wave form packets). Although it can be used for any 2D or 3D point (or N dimensional with the Extra Bytes facility), the primary emphasis is on kinematic, discrete return, time-of-flight systems LiDAR. All modern LiDAR systems of this type are capable of multiple returns per outgoing pulse. A diagram of a potential multiple return scenario is illustrated for a five return scenario in Figure 10.8.

LAS 1.4 supports up to 15 discrete returns per outgoing pulse. The count of the number of returns per pulse is contained in the field Number of Points by Return in the Public Header Block (not the Legacy Number of Points by Return—this is used for encoding a LAS 1.1–1.3 payload within a LAS 1.4 file body—a legacy implementation that can support only 5 returns). Note that this field is an array that is a histogram of the number of returns for this file. It does not represent the physical capabilities of the system(s) that was used to acquire the data.

The return position of a point within a multiple return source pulse is such an important factor in data analysis that this position is encoded for all Point Data Record Types. PDRF 0–5 (the legacy types from prior LAS versions) support up to 5 returns whereas the new types (6–10) support up to 15.

All returns from a multiple return outgoing pulse will have the same GPS time stamp. This fact is used for correlating the returns of a multiple return pulse. It is entirely possible that returns from the same outgoing pulse will span more than one LAS file. This generally occurs when data are segregated into tiling schemes and different returns of the same outgoing pulse span a tile boundary.

10.7.7 CLASSIFICATION

The LAS 1.0 specification devoted an 8-bit field to CLASS. The purpose of the field was to allow a point to be tagged as to its reflectance surface (e.g., Ground, Low Vegetation, and etc.). A second use for CLASS was to mark points that were special in some sense such as WITHHELD or MODEL KEYPOINT. Other than reserve space in each point record for the field, the value stored in this field was entirely up to the user.

When we initially devised the LAS 1.0 specification, we did not really view the specification as a format that would be conveyed to the ultimate end-users of LiDAR data. It was intended to

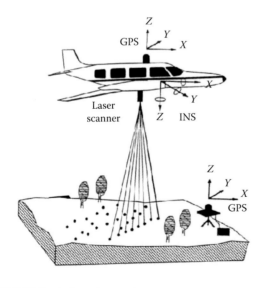
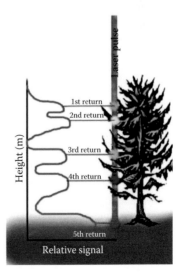

FIGURE 10.8 A five return system.

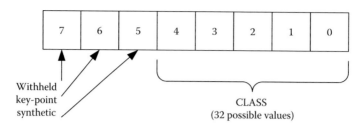

FIGURE 10.9 The classification field of LAS 1.1–1.3.

be a transport protocol between makers of LiDAR sensors and developers of LiDAR processing tools. However (and this is a very good thing), LAS rapidly became the de facto specification for all phases of LiDAR data processing and delivery. This holistic use necessitates specification of the values for the CLASS field. In other words, we all have to agree on which numeric value will signify the various classes.

A second shortcoming of the original CLASS field in the LAS 1.0 specification was that one could not simultaneously maintain a class attribute and a treatment attribute. For example, the field would not support the idea of a point being both Ground and a Model Key Point. Encoding a point as a Key Point destroyed its reflectance categorization.

To solve these two problems, the 8-bit CLASS field of LAS 1.0 was split into two parts; a 5-bit CLASS section and 3 bits used as Flags (Figure 10.9). The classes ranged from zero to 31.

When we made this transition to LAS 1.1, there were typically only a handful of classes defined (usually less than 10). We therefore thought that 32 classes (0–31) would provide plenty of room for any classification scheme. Unfortunately, with the recent widespread use of LiDAR for mobile mapping applications and transmission line coding, the need for additional room for class definitions has become critical.

For this reason, LAS 1.4, for the new PDRFs 6–10, separates CLASS back into a dedicated 8-byte field. Class IDs 0 through 63 are reserved for ASPRS use leaving classes 64 through 255 available for definition by users.

Using ASPRS-defined classes allows any software application to correctly display the meaning of the class without the software vendors needing to communicate this information. Now one could argue that this could be more elegantly communicated by simply encoding the class definition table in a Variable Length Record (in fact, this is what we did in the original version of LAS), but this creates a mapping problem when merging files.

The class definitions for PDRFs 6-10 are listed in Table 10.11.

Two very important changes have occurred in this classification table; the removal of Model Key Point (class 8) and Overlap Point (class 12) as classifications. To avoid confusion with the classification of the old PDRFs 0–5, we have set these two classes to Reserved. Indicating Key Points and Overlap with PDRF 6–10 is performed by attribute bits (see next section).

Note that we have added a few new ASPRS standard class codes such as Rail and Wire types. We anticipate that the full 0–63 class slots reserved for ASPRS will be defined over time as user needs mature.

Obviously another good reason to only accept LiDAR data in formats 6–10 is to ensure that the classification encoding is homogeneous across all of your data.

10.7.8 POINT ATTRIBUTES

Point attribute bits were first expanded to include classification flags in LAS version 1.1. In LAS 1.4 (for PDRF 6–10) an additional flag was added to indicate points in the overlap region. The Point Attributes of PDRF 6–10 are listed in Table 10.12. The first 4 bits are devoted to classification flags.

TABLE 10.11
ASPRS Standard LiDAR Point Classes (PDRFs 6–10)

Classification Value	Meaning
0	Created, never classified
1	Unclassified[a]
2	Ground
3	Low vegetation
4	Medium vegetation
5	High vegetation
6	Building
7	Low point (noise)
8	Reserved
9	Water
10	Rail
11	Road surface
12	Reserved
13	Wire—guard (shield)
14	Wire—conductor (phase)
15	Transmission tower
16	Wire structure connector (e.g., insulator)
17	Bridge deck
18	High noise
19–63	Reserved
64–255	User definable

[a] We are using both 0 and 1 as *Unclassified* to maintain compatibility with current popular classification software such as TerraScan. We extend the idea of classification value 1 to include cases in which data have been subjected to a classification algorithm but emerged in an undefined state. For example, data with class 0 are sent through an algorithm to detect man-made structures—points that emerge without having been assigned as belonging to structures could be remapped from class 0 to class 1.

TABLE 10.12
Point Attributes for PDRF 6–10

Classification Flags	4 bits (bits 0–3)	4 bits
Scanner channel	2 bits (bits 4–5)	2 bits
Scan direction flag	1 bit (bit 6)	1 bit
Edge of flight line	1 bit (bit 7)	1 bit

The first flag bit is the *Synthetic* flag. If this flag is set (value 1), it indicates that the associated point was not collected by means of the method indicated in the System ID field for the file but was added via some other mechanism. The most common use of this field is moving a LiDAR point. We advocate a philosophy that once *bore sighting* and project geometric correction (the equivalent of photogrammetric block bundle adjustment) have been performed, individual LiDAR points cannot have their X, Y, or Z values adjusted. Thus, if one were to move a collection of LiDAR points observed in an editing session to be reflecting from within low vegetation to the ground, the operation would actually proceed as a deletion of the point within the low vegetation (the deletion is actually setting the WITHHELD flag bit, not removing the point from the file) and the addition of new ground points.

These newly added ground points would have their CLASS value set to GROUND and their SYNTHETIC attribute flag bit set to True (a value of 1). The Synthetic flag should be used for any *additive* operation such as digitizing points in LiDARgrammetry, adding a point to indicate a power line connection and so forth.

The second flag is KEY POINT. This allows the marking of a point as one that is to be kept for elevation file formation (perhaps as a contributor to a contour generation routine) while maintaining its class attribute. Thus, the point could maintain a class of Water and also be flagged as a Key Point. Key point algorithms thin data to the minimum points necessary to maintain a specified accuracy criterion. A typical approach is to create a TIN using all of the points (of the desired class such as ground) and then iteratively remove points that do not cause the regenerated TIN to violate a specified error tolerance.

The third flag is the WITHHELD flag. Points should never be deleted from an LAS file but rather flagged as deleted. Thus, for example, in the removal of water points, a point could be classified as Water (CLASS = 9) and then have the WITHHELD bit set to 1 to indicate that the point should be omitted from further modeling.

New to LAS 1.4 is the fourth flag, OVERLAP. It is frequently desired to thin data in the overlap region of LiDAR flight lines (or, in the case of mobile, *takes*). This operation is used to either improve the statistical accuracy of the LiDAR data or to simply thin the multiple density data in these regions. Using a Class to indicate overlap (the technique prior to LAS 1.4) caused problems in that one loses the actually classification of the point (e.g., Ground). This leads to problems in analyzing accuracy in the overlap region. What exactly comprises points that should be designated as Overlap will be determined (as it is now) by the procurer of the LiDAR data. For example, the USGS has an extensive description of how points should be designated as Overlap.

Also new to LAS 1.4 is the ability to designate a scanner channel. The intended use is for mobile mapping systems with multiple laser scanners. Two bits are used for this indicator, and thus, one of four scanners can be designated. For a single-scanner system, the channel must be zero.

The final two bits of the attribute field are Scan Direction and Edge of Flight Line. Unfortunately, even though these are mandatory fields in the LAS point data, I often see them omitted. Scan Direction is defined for rotating mirror scanners and is defined to be positive for counterclockwise rotation if the flight direction were coming out of your eye (looking at the back of the aircraft as it moves along its track). I personally do not know how this parameter might be used in data processing. The last flag, edge of flight line, is far more important for non-360° scanners. It can be used to quickly compute a graphic for the edge of a LiDAR flight line.

10.8   OTHER DATA

10.8.1   USER DATA

Each PDRF supports 1 byte of User Data. This is a data field that can be used for any purpose by processing software. There is no guarantee that data stored in this field will be preserved from one application software package to another. In fact, many software packages use this field for a scratch pad area during processing. Clients ordering LiDAR data may specify that this field be set to zero prior to data delivery.

10.8.2   SCAN ANGLE

The Scan Angle, for PDRF 6–10, has been expanded to two bytes for better scan angle resolution. It is a signed short that represents the rotational position of the emitted laser pulse with respect to the vertical of the coordinate system of the data. Down in the data coordinate system is the 0.0 position. Each increment represents 0.006°. Counterclockwise rotation, as viewed from the rear of the sensor, facing in the along-track (positive trajectory) direction, is positive. The maximum value in the positive sense is 30,000 (180° that is up in the coordinate system of the data). The maximum value in the negative direction is −30,000 which is also directly up. Note that scan angle can now, under this new definition, be used for 360° rotational sensors such as those used on mobile mapping systems.

10.8.3 Intensity, RGB, NIR Channels

The intensity value is the integer representation of the pulse return magnitude. This value is optional and system specific. However, it should always be included if available.

Red, Green, Blue (RGB), and near infrared (NIR) can also be stored with certain point formats. These channels are intended for storing values from colorization of the points by one or more auxiliary optical sensors (cameras). For example, colorizing LiDAR data with RGB reflectance data from a 360° camera is a common operation in mobile mapping.

These data, when included, are always normalized to a 16 bit, unsigned value by multiplying the value by 65,536(intensity dynamic range of the sensor). For example, if the dynamic range of the sensor is 10 bits, the scaling value would be (65,536/1,024). This normalization is required to ensure that data from different sensors can be correctly merged. If intensity and/or color data are not included, these values must be set to zero

LAS is more and more frequently being used to represent point clouds from nonlaser derived data such as 3D points from image correlators. By using PDRF 8, up to five channels of information (by using the available I, R, G, B, NIR fields) can be stored.

10.8.4 Point Source ID

The Point Source ID field (2 bytes) is used to identify the source file in which a point was originally created within a Project dataset. For example, if a project contained a source file of LiDAR points that originated from a flight line and that flight line had been assigned number 77, all points within that original file would keep their Point Source ID of 77 throughout all processing regardless of how many files through which that point might have propagated. This feature allows mapping back to the original source of a point even when an LAS file has resulted from composite operations such as merge and extract. The most common example is that of a processing tile formed by merging numerous flight lines into the tile. By using the Point Source ID, each point can be uniquely identified as to its original source. Thus, this is the field that is used in software when performing operations such as *color by flight line*.

In general, once a Point Source ID has been assigned to a point, it should not change. It should also be noted that Point Source ID is not meaningful outside the context of a single project (e.g., if points from two or more projects are blended, they cannot be segregated using Point Source ID). However, if you have control over all projects that will be merged, you can preplan flight lines such that backtracking, and unique flight lines are maintained.

You will note that for a pure LiDAR project (all data sources originated from LiDAR flight lines), the Point Source IDs must all map back to original flight line numbers except for SYNTHETIC points. This is illustrated in Figure 10.10. In this simple example, we have two flight lines (*strips*) that have been labeled with File Source IDs of 1 and 2. The points from these two flight lines have been MERGED into a file that has been assigned a File Source ID of 500. Within this merged file, the left-most point has been *moved* down. This is accomplished by setting the WITHHELD bit of the original point to 1 (TRUE) and then creating a new point in the desired location. As the new point was created within Source File ID 500, it is assigned a Point Source ID of 500. In addition, this new point has its SYNTHETIC flag set to 1 (TRUE). Notice that in the merged file, each individual point can be traced back to its origin.

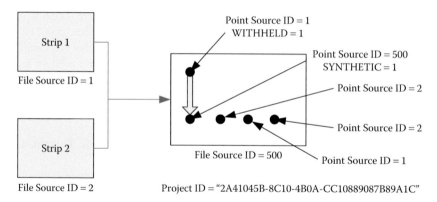

FIGURE 10.10   A composite file with a *moved* point.

10.9   WAVEFORM DATA

LAS 1.3 introduced the storage of *waveform* data to the specification. This capability continues, unchanged in LAS 1.4. A waveform is essentially a digitized representation of the return intensity of a laser pulse with respect to time. The waveform nature of the return was indicated in Figure 10.8.

Primarily for convenience and to reduce data storage requirements, earlier LiDAR systems analyzed waveform return data in front-end electronics and declared the potential location of an impact with an object space object. As digitization hardware improved and software algorithms became more sophisticated, there was a drive from both hardware vendors and data processors (particularly, the academic community) to make waveform data available to processing software. Hence, this capability was introduced in LAS 1.3.

Rather than storing the entire waveform, waveform packets spanning *interesting* events are stored. An interesting event could be anything the hardware vendor decides but is typically amplitude above the noise level within the region of interest range (e.g., if the nominal flying height above ground is 2,000 m, events 200 m from the aircraft would not normally be collected).

To allow dual use of the data (e.g., point mode and waveform mode), each waveform packet is indexed by at least one point (Figure 10.11). This allows software based on discrete returns to continue to function. There is a misconception that this encoding scheme does not allow *returnless* waveform packets to be encoded. However, this is easily achieved by the hardware vendor (in post-processing software) tagging desired *returnless* waveform packets with a synthetic return.

An inspection of the encoding of the waveform packet scheme reveals that we constructed a hybrid approach to storing waveform information. A return pulse can

- Not point to a waveform packet.
- Be the sole pointer into a location within a waveform.
- Point into a waveform packet pointed to by one or more other returns (first example of Figure 10.11).

A single LAS file can encode up to 255 waveform packets schemes. These schemes are encoded within Wave Packet Descriptors (stored in VLRs). A waveform packet descriptor contains the fields listed in Table 10.13. As of version 1.4, compression of the waveform data is not yet supported. The temporal sample spacing is in picoseconds, providing a maximum resolution of 0.03 cm.

Each point indexing a waveform packet includes first-order coefficients for *walking the wave*. This allows you to shift the location of a declared point using a first-order approximation. The common

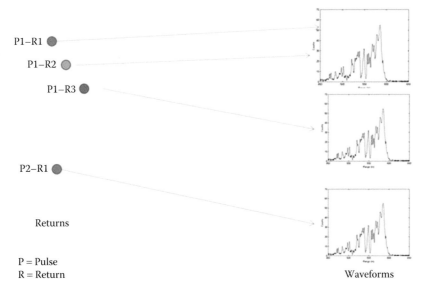

P1–R1

P1–R2

P1–R3

P2–R1

Returns

P = Pulse
R = Return

Waveforms

FIGURE 10.11 Waveform encoding.

TABLE 10.13
Waveform Packet Descriptor User-Defined Record

Item	Format	Size (byte)
Bits per sample	Unsigned char	1
Waveform compression type	Unsigned char	1
Number of samples	Unsigned long	4
Temporal sample spacing	Unsigned long	4
Digitizer gain	Double	8
Digitizer offset	Double	8

operation here would be to analyze a wave packet for the location where you would like to declare a return by shifting the default return of a sensor biased toward high returns (e.g., in high grass) to the first rise in amplitude. You would compute the X, Y, and Z values for your new return by starting at the closest sensor declared return on the wave packet and by using the first-order $X(t)$, $Y(t)$, $Z(t)$ coefficients to position the new pulse. Finally, this new pulse would be marked as *synthetic*.

The exploitation of waveform data in a meaningful way is just starting to appear in algorithms of commercial software (e.g., TerraScan from Terrasolid). As time goes on, we expect this to become more prevalent.

There has been some criticism of the waveform storage of LAS 1.3/LAS 1.4. These criticisms range from too much data being stored to too little data being represented. However, this is a bit of cart before the horse. As commercial software vendors add meaningful processing algorithms, we will get a better handle on what actually needs to be exchanged. In addition, a full understanding of the nature of the problem is necessary prior to making changes. For example, we store wave walking coefficients today on a per return basis. Would a single set of coefficients suffice? A detailed analysis of the curvature of the pulse (caused primarily by atmospheric refraction) would be required prior to making this change.

10.10 NEW DEVELOPMENTS IN LAS FORMAT

10.10.1 (Extended) Variable Length Records

The LAS format has always provided a facility for users to store metadata that are *private* in nature (meaning that other software will simply ignore these records). With the exception of waveform data, prior to LAS 1.4, these VVLRs have occurred (in the file) prior to the Point Data Records. A VLR can hold a payload of up to 65,535 bytes (because the data packet size is limited to 16 bit offsets). To legitimately use a VLR, a user must request a VLR *Key* from the LAS working group (www.lasformat.org). This Key, along with a Record ID number, allows one to uniquely identify a VLR. Note that a registration key is only needed when one intends to add a custom metadata packet to a LAS file (which, for most users, would be quite rare).

For a number of uses, the length of the VLRs proved to be a serious limitation (e.g., when using this facility for storing spatial index information). Thus, in LAS 1.4, we made the Extended Variable Length Record available for general use (prior to LAS 1.4, an EVLR was used only for Waveform Packet information). EVLRs are intended to be placed after the point data records. An EVLR payload size is represented by a *long* (64 bits) so this should definitely solve size limitations!

Unfortunately, VLRs/EVLRs are designed to be sequentially placed in a LAS file (rather than daisy chained). Thus, all VLRs should be sequentially placed ahead of the point data records, and all EVLRs should be sequentially placed after the point data records.

There are several VLR/EVLRs reserved for LAS Specification use. An example is the encoding of CRS information. For the legacy PDRFs 0–5, this CRS encoding can be either GeoTIFF or (preferred) WKT. For PDRF 6–10, the CRS encoding must be WKT. An important note to software developers is that the LAS defined records can be either stored as VLRs or as EVLRs. Thus when searching a file for the CRS records, it is necessary to search the EVLRs if a CRS record was not found in the VLRs.

10.10.2 Extra Bytes

A new LAS VLR* has been defined called *Extra Bytes*. This VLR allows one to parse a Point Data Record that has been augmented with extra fields. As noted earlier in this article, a standard LAS Point Data Record can have data tacked on to the end of each point data record. Prior to LAS 1.4, a writing software would have to communicate to a reading software the content of this extra data using some previously agreed private scheme. The Extra Bytes metadata provide the ability to specify, for each field,

- Field Name
- Description
- Data Type (all standard type as well a several array types)
- A No Data signature
- Minimum permissible value
- Maximum permissible value
- Scale
- Offset

Flags are included to indicate if various attributes apply to a specific field (such as minimum and maximum values).

This facility will prove quite useful for those who wish to make use of *off the shelf* LiDAR software that ingests LAS format data but need to augment that data for other uses. For example, Extra Bytes will provide a canonical method of added *Sigma Z* to a point data record.

---

* Note that I refer to these LAS Spec records as VLRs for convenience. They can actually be encoded as either VLRs or EVLRs.

It is the responsibility of LAS software writers/readers to properly index point data records in LAS files. A software writer must *always* use the Point Data Record Length field in the Public Header Block to determine the size of the point records in a given LAS file. Simply counting bytes for a particular PDRF will result in an indexing error if extra bytes have been appended to a record. Software that is rewriting a LAS file containing Extra Bytes should simply copy these bytes to preserve the records.

It should be noted that although Extra Bytes is a tremendous improvement over no representation of augmented point data records, we do not anticipate that commercial software will attempt to build the semantic engines necessary to handle mixed mode Extra Bytes (i.e., multiple files each with possibly different Extra Bytes metadata). Of course, this issue is certainly not unique to LAS!

10.10.3 A WELL FORMED LAS 1.4 FILE

The current section is primarily my opinion so it will all be written as first person. There is a bit of a difference of opinion among designers of LAS software as to how ambiguous conditions should be handled. Commercial software purveyors tend to focus on error handling whereas the open source camp is more focused on attempting to continue to read the files. My observation, based on working the LAS format since inception and working with dozens of LiDAR processing companies, is that ambiguous conditions must be treated as fatal errors. This is primarily based on the observation that production managers usually do not want production technicians making decisions regarding potential error resolution. Nothing is quite as embarrassing as delivering a final LiDAR project in the wrong CRS! Thus. the following guidelines are recommended for software authors:

1. Although LAS 1.4 will support the encoding of PDRFs 0–5 (the legacy formats) such that, under certain conditions (described in an earlier section), software that has not been updated to LAS 1.4 could read the file, I strongly recommend that this facility never be used for production-type scenarios. There is no need at all for using this facility as one can simply write out a file in one of the legacy formats, if compatibility with these prior formats is needed.
2. In light of 1, the following fields should be always be set and maintained as zero:
    a. Legacy Number of Point Records.
    b. Legacy Number of Points by Return.
3. Always use WKT encoding for the CRS. GeoTIFF has not been maintained in terms of new CRS and vertical representation is often not supported. Note that you will need to set the WKT bit of the Global Encoding field in the Public Header Block.
4. Add an Extra Bytes VLR when augmenting Point Data Records. This eliminates the problem of trying to figure out why your PDRF 0 is longer than 20 bytes!
5. Never store both a GeoTIFF and WKT VLR in the same file. There is very often not a one-to-one mapping from GeoTIFF to WKT, and thus, an intrinsic error exists.
6. Try to avoid using anything other than powers of 10 for the scale values in the Public Header Block. Floating point for X, Y, Z has been purposefully avoided in the LAS definition. Nonpower of ten scaling makes this a pain to avoid when writing software.

10.10.4 ERROR CONDITIONS

Although we attempted to make the LAS 1.4 format unambiguous, we were not completely successful. Again, I will resort to first person in this section as what I say should be an error, other might say to take a different path. Remember, my focus is on rigorous error detection in LiDAR production environments. In any of the following error conditions, the software should, of course, provide an exact message as to the reason an error has been declared.

1. While I strongly recommend against using the *legacy* payload capability of LAS 1.4, one should treat a difference between a nonzero legacy field and its equivalent standard field as an error. For example, if Legacy Number of point records is not zero and is not the same value as Number of point records, an error should be declared.

2. Global Encoding bit WKT is set (nonzero) but a valid WKT VLR does not exist (even if a GeoTIFF packet does exist).

3. PDRF is 6 or greater, and the WKT bit is not set (this is explicitly pointed out as an invalid LAS file in the LAS 1.4 specification). This error must be declared even if a WKT VLR is found in the file.

4. Number of point records does not equal the sum over Number of points by return.

5. Return Number (in a Point Data Record) is zero.

In my opinion, any data inconsistencies within a LAS file should be treated as a *hard* error, meaning that the reader will exit with an error rather than continue to process the file with a warning or, even worse, no message at all. After declaration of a hard error, the file should be fixed *off-line* and reintroduced to the production workflow only after correction. This opinion is based on my experience of seeing many embarrassing data deliveries caused by improperly formed LAS files (particularly in the area of CRS issues).

10.11   FUTURE TRENDS IN LiDAR DATA STORAGE

The most notable advances in data storage will be for analysis and exploitation. In these scenarios, in which the dataset is populated and then remains essentially static, storage in a spatially indexed format offers significant advantages as compared with simple flat file storage. At the time of writing the first edition of this book, it looked as if spatially enabled databases would become the preferred solution for storing data aimed at exploitation. However, data densities have advanced more rapidly than have point oriented spatial databases, and now the preferred format seems to be indexed, tiled flat files. Data are moved from tile storage (most schemes use subtiles within the tiled files) to a rapid spatial searching data structure such as a KD tree to allow fast approximate nearest neighbor searches (ANN).

These systems will allow smooth integration between disparate datasets, enabling a rich array of heterogeneous exploitation tools. An example might be pulling building footprints from a two dimensional vector file and attributing the height by pulling the associated LiDAR points classified as belonging to the Building class from the LiDAR storage scheme.

LiDAR data will continue to grow in dataset size due to two drivers. The first is the desire to collect data (whether it is LiDAR or imagery) at higher and higher density. The second is the trend of adding additional information to the LiDAR return data.

As of 2016, the interest in full waveform LiDAR data seems to have waned. This is perhaps due to a dearth of algorithms that would produce results notably better than those that result from processing waveforms into points at the time of sensor data processing. Outside efforts to create formats more suited to waveform than LAS (such as PulseWaves) seem to have stalled.

REFERENCES

Graham, L. 2005. The LAS 1.1 Standard. *Photogrammetric Engineering and Remote Sensing*, 71: 777–781.
Graham, L. 2012. The LAS 1.4 Specification. *Photogrammetric Engineering and Remote Sensing*, 78: 93–102.
Heller. M. 1990. Triangulation algorithms for adaptive terrain modeling. In *Proceedings of the 4th International Symposium on Spatial Data Handling*, pp. 163–174.
LAS Version 1.1 Specification – see www.asprs.org/las
Mitra, S. 2001. *Digital Signal Processing*. New York: McGraw-Hill, pp. 300–306.

Oracle Spatial 11g. 2007. Advanced spatial data management for enterprise applications, An Oracle White Paper. Retrieved from http://www.oracle.com/technology/products/spatial/pdf/11g_collateral/spatial11g_advdatamgmt_twp.pdf.

Song, Z. et al. 2012. An improved Nyquist-Shannon irregular sampling theorem for local averages. *IEEE Transactions on Information Theory*, 58(9): 6093–6100.

Schroeder, B. and Gibson, G. 2007. Disk failures in the real world: What does an MTTF of 1,000,000 hours mean to you?. In *FAST'07: 5th USENIX Conference on File and Storage Technologies*, San Jose, CA, February 14–16, 2007.

van Kreveld. M. 1997. Digital elevation models and TIN algorithms. In M. van Kreveld, J. Nievergelt, T. Roos, and P. Widmayer, (Eds.), *Algorithmic Foundations of GIS*, Lecture Notes in Comp. Science. Springer-Verlag, 1997.

11 LiDAR Data Filtering and Digital Terrain Model Generation

Norbert Pfeifer and Gottfried Mandlburger

CONTENTS

The current chapter concentrates on airborne Light Detection and Ranging (LiDAR) as source for digital terrain models (DTMs), also known as digital elevation models (DEMs). After defining these terms, we will first consider the specific advantages LiDAR offers for capturing the terrain surface, also in comparison to image-based photogrammetric techniques. The main topic in this chapter will be, however, the extraction of the terrain surface from the LiDAR measurements, either the original point cloud or already preprocessed LiDAR data with terrain elevations sorted into a raster. This process is a classification task, dividing the points or pixels into either ground or off-terrain, where some approaches derive more classes. The usage of full-waveform echo recording for terrain reconstruction will be discussed. Computing a DTM from the classification results is the next step, but the different methods, for example, strict interpolation versus qualified approximation techniques, but will not be discussed at length. However, to make a DTM or DEM usable for many disciplines, for example, hydraulics and geology, the so-called structure or feature lines should also be integrated. Likewise, the amount of data is typically very high and not adapted to the variations in terrain roughness. We will therefore continue with a section on the determination and exploitation of structure lines from the LiDAR data as well as with methods for reducing the data amount, that is, compressing the elevation data, considering error bounds. The quality of DTMs from airborne LiDAR data will be treated in a separate section. Real-world examples, embedded in the text, will demonstrate the earlier. Although DTMs can also be reconstructed from terrestrial laser scanning data, offering more detailed and accurate description, this is only possible for small areas in the order of 1 km^2 and below. Larger areas at very high resolution can be generated from mobile laser scanning data (mounted on cars, integrated into the traffic, or on quads for rough terrain) as well as from Unmanned Aerial Vehicles (UAV). However, covering large areas is to-date feasible only from airborne manned platforms, for example, to collect topographic information over more than 10,000 km^2 in dedicated projects. Therefore, concentration is put on airborne laser scanning, also termed airborne scanning LiDAR.

11.1   DTM AND LASER SCANNING

This section introduces DTMs (Ackermann and Kraus, 2004; El-Sheimy et al., 2005; Maune, 2001) and relates them to laser scanning. It also compares the data acquisition methods laser scanning and photogrammetry (Kraus, 2007; McGlone et al., 2004) with respect to terrain derivation.

11.1.1   DTM Definition

From the many definitions of a DTM (El-Sheimy et al., 2005), we choose one that does not prescribe the data structure but concentrates on the geometrical aspects. A DTM is a continuous function that maps from 2D planimetric position to terrain elevation $z = f(x, y)$. This function is stored digitally, together with a method on how to evaluate it from the stored geometrical and topological entities. Sometimes DTM and DEM are distinguished by recognizing the difference that the first one includes break lines (and other topographic features), whereas the second one does not (El-Sheimy et al., 2005). A break line is a linear feature on the terrain surface, along which the tangent plane of the surface abruptly changes its orientation when moving from one to the other side of this line. Here, the terms DTM and DEM will be used interchangeably. Including or disregarding break lines is rather a question on the level of precision, detail, and morphological quality. All these definitions were coined when ground point datasets had a substantially lower density than those available from airborne laser scanning today. This makes a more detailed consideration necessary.

For using the above-mentioned definition, it is necessary to clarify the nature of the terrain (elevation). It is defined here as the boundary surface between the solid ground and the air. This definition (top soil, pavement, etc.) is also applied by Sithole and Vosselman (2004) for their comparison between different airborne laser scanning (ALS) filtering algorithms. This is typically also the surface of superficial water runoff. This definition is sufficient for well-defined surfaces, but a number of problems are encountered when looking at the details observed and the precision achieved by airborne laser scanning.

  1. The concept refers to the so-called 2.5D approach, in which for one ground position (x, y) only one height (z) may be expressed. For overhangs, which are occasionally recorded, the 2.5D assumption does not hold. Most algorithms for terrain extraction will therefore fail in this special circumstance (Pfeifer, 2005). However, it also plays a role when considering the digital surface model (DSM, defined next). From an application point of view,

for example, orthophoto production, a bridge may also be seen as integral DTM part, again violating the 2.5D assumption. In Sithole and Vosselman (2004), bridges are defined as objects, whereas ramps leading to bridges are considered as part of the terrain.

2. For hard surfaces (rock, concrete, etc.), the boundary surface between ground and air is well defined. However, for many natural surfaces, possibly featuring a dynamic aspect, it is less obvious. From lower flying altitudes, using high pulse repetition rate and small footprint scanners, the reconstruction of plough furrows becomes possible, although not interesting for a general purpose DTM. In addition, elevation of surfaces collected by total station fieldwork over herbaceous vegetation typically refers rather to the soil elevation, whereas laser scanning records within the footprint a mean height between soil and herbaceous layer canopy. The effect of this method depends on the vegetation density. Naesset (2016) reported errors of approximately 1 dm (positive and negative), depending on ground vegetation and terrain form. These quantities are thus relevant given state-of-the-art systems. Before the advent of high-precision terrain elevation capturing by airborne laser scanning, this question did not arise.

3. Large objects built on the terrain, for example, buildings, do not simply shadow the terrain surface but cease its existence. To avoid holes, interpolation of the terrain surface under buildings becomes necessary. This situation is also visualized in Figure 11.7.

Eventually, the application determines, which aspects are to be included in a DTM.

From a geomorphological point of view, the terrain surface is structured by lines. At a so-called break line, the first derivative of the surface is discontinuous, whereas it is smooth elsewhere. In addition, form lines are sometimes defined as soft break lines. In a contour line plot, the form line is the connection of points with maximum curvature. In the *data rich* situation of airborne laser scanning, this is typically no longer considered, as the terrain shape is very well described by the dense LiDAR point cloud.

11.1.2 DIGITAL CANOPY MODELS AND DIGITAL SURFACE MODEL

The DTM is one special surface that can be reconstructed from laser scanning. It is, however, not the only one. DSM and digital canopy models (DCM) represent other surfaces, not necessarily thoroughly physically defined (Schenk and Cshato, 2007). The DCM represents the top surface that is visible from above. It is equivalent to the DTM in open areas and runs on top of the vegetation canopy in forested areas and over house roofs in the built-up environment. For these objects, the DCM is—more or less—well defined, but other objects such as power lines or highly dynamic objects (cars, cattle, etc.) are not treated. A DCM is often computed by interpolating the points corresponding to the first echo of each emitted shot. In the overlapping areas of adjacent strips, the dynamic objects (e.g., moving cars) typically show up twice, both times intermixed with points of the surface below them. Features, such as power lines, light poles, and so on, appear as unfamiliar artifacts in a visualization and therefore hamper proper scene interpretation.

A more coherent DSM can be computed by generating a raster using the maximum height in each cell. This is, however, affected by the measurement noise. Khosravipour et al. (2016) have suggested methods to generate *pit free* (or *spike free*) DSMs, with the aim to provide a closed canopy surface over vegetation, also in cases in which first echoes are reflected from lower branches. Especially with lower flying heights and therefore smaller footprints and a correspondingly high point density, this situation can occur. Hollaus et al. (2010) have developed the *land cover-dependent* DSM, which uses a combination of interpolation in smooth areas as well as in regions with low point density but uses the maximum heights in a grid over rough (e.g., vegetation canopy) surfaces.

The term DSM is used synonymously with DCM, although the *surface* in its acronym does not specify the surface to be modeled. The difference model of a DSM and a DTM is called a

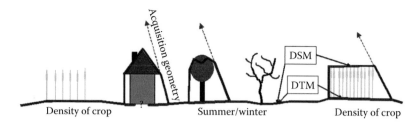

FIGURE 11.1 The DSM runs over buildings and tree crowns, as they are contained in the data acquired by airborne laser scanning. Below houses, the DTM is not properly defined. In open areas, the DTM and the DSM refer to the same surface.

normalized DSM (nDSM); thus, nDSM(x, y) = DSM(x, y) − DTM(x, y). Especially in forestry also the term *nCM*, *normalized Canopy Model*, is used. The differences between DTM and DSM and also some of the factors influencing the DSM as an interpolated surface of all points are shown in Figure 11.1.

11.1.3 Data Structures

The terrain model function $z = f(x, y)$ is computed from the source data, that is, 3D points $p_i = (x_i, y_i, z_i)$, $i = 1,..., n$, with n the number of points. This function needs to be made persistent as the interpolation from the source data is unfeasible for each new evaluation of the terrain. The very large amount of data (billions of points) makes the transformation into another type indispensable. Typical densities of points in open areas, that is, not considering the case of multiple echoes, are 1–20 points/m². Popular data types are the raster, where an elevation is assigned to a small area, that is, a rectangular pixel. The grid is the dual structure, heights are stored at discrete, regularly aligned points, and an interpolation method, typically bilinear or bicubic interpolation, is used to interpolate heights within a grid mesh. The pixel approach does, strictly speaking, not provide a continuous surface. Therefore, grid evaluation is performed, even if the data are stored in a raster image format. A hybrid grid structure is obtained by intermeshing additional lines and special points into the grid structure (Ackermann and Kraus, 2004; Köstli and Sigle, 1986). Examples are given in Figures 11.10 and 11.11. The grid heights are obtained by interpolation methods as inverse distance weighting, moving least squares, and kriging. An alternative method is the triangulation, generating the so-called triangulated irregular network (TIN) data structure. In this case, the original points are used for reconstructing the surface, including also their random measurement errors. For large point sets, deriving the triangulation is time consuming, but the popularity of triangular meshes in computer graphics drives research also to overcome this problem (Figure 11.11). Isenburg et al. (2006) have developed the streaming Delaunay triangulation, which makes processing of large TINs of LiDAR point clouds feasible.

11.1.4 Laser Scanning versus Photogrammetry for Terrain Capture

First, commercial projects to acquire terrain data by laser scanning from airborne platforms were performed in Europe by the company TopScan in the mid-1990s (Wever and Lindenberger, 1999), employing an Optech laser scanner, based on earlier experiences (Lindenberger, 1989). Alternative methods for terrain data capture are tachymetry, as ground technique only suited for very small areas, and aerial photogrammetry. From digital images (airborne and from satellite), the method of automatic image correlation can automatically produce dense point clouds. With digitally acquired images, the overlap between consecutive images can become very high, allowing more robust multiimage matching (Grün and Baltsavias, 1988; Haala and Rothermel, 2012; Hirschmüller, 2008) and generating very dense point clouds. It therefore stands to reason to compare these two data acquisition methods in more detail with respect to terrain reconstruction (Kraus and Pfeifer, 1998; Ressl et al., 2016).

1. Photogrammetry computes ground coordinates by forward intersection. Therefore, first the (automatic) finding of homologous points is necessary. Areas visible from one exposure position only cannot be reconstructed. This applies typically to the ground below forest under leaf-on conditions, where ground points cannot be identified in adjacent images, as different gaps in the foliage are seen from different positions and much of the ground is shielded by the foliage (see also Gil et al., 2013). Furthermore, it is necessary to see patches including texture to generate correspondences, thus also requiring ambient or direct illumination. In laser scanning, one view to the ground is sufficient to record a point. Thus, only in the case of very sparse vegetation, reconstruction of the ground is possible from images (Ressl et al., 2016). Seen from the platform position, buildings cause geometric shadows, which prevent the reconstruction of the ground surface or objects on it. This applies to both, reconstruction from images and laser scanning. However, the polar measurement method alleviates this drawback of airborne data acquisition, as one ray is sufficient.

2. In contrast to photogrammetry, laser scanning is an active measurement technique, and data acquisition is therefore independent of the sun position. Flying can be performed at night time, which is an important aspect next to airports or in high-latitude regions. In addition, laser scanning does not suffer a decrease in accuracy with respect to sun shadows, which are encountered on forest grounds as well as in city areas. Shadow, that is, dark areas, typically lead to increased point noise in image matching.

3. Both laser scanning and photogrammetry provide increased accuracy for decreasing flying heights. In laser scanning, however, the height component is more precise, whereas in photogrammetry planimetry can be measured more accurately. As camera opening angles are typically larger compared with laser scanner opening angles, a smaller number of flying strips is necessary in photogrammetry for covering the same area. However, strip distance and flying height are governed also by other factors such as minimum ground mapping units, point density, and overlap of image strips for increasing automation and reducing shadowed areas.

4. Photogrammetry is a method based on texture, which is high at object discontinuities, that is, edges. Photogrammetrically acquired point clouds are therefore concentrated along distinct points and edges, whereas point clouds acquired by laser scanning are more reliable within surfaces. Using smoothness constraints dense, pixel-wise point clouds can be obtained from image matching (Hirschmüller, 2008), also in weakly textured areas. Over extended areas without texture, for example, fresh snow, laser scanning is the only suitable method for surface reconstruction.

5. Laser scanning is as a method for generating 3D points without overdetermination/ redundancy. A gross error in the laser range measurement therefore can easily lead to a gross error in the DTM. In photogrammetry each point has to be measured at least twice, but a wrong correspondence will also lead to a point off the terrain surface. If the perspective differences between images of the same object are too big, matching fails and no 3D points are generated. Typical examples are peak-like objects, which do rather occur as artificial objects but less in natural terrain. As no object surface point is measured twice by laser scanning, no tie points exist as in photogrammetry. However, surfaces are sampled densely and small areas can be modeled as smooth or planar patches, allowing both correspondences to be established across strips, as well as overdetermination, and therefore reliability, for estimating terrain elevation. This overlap is also the basis for quality control.

DTMs and point clouds from airborne laser scanning are becoming standard products available from national geoportals (e.g., Krishnan et al., 2011) or they are acquired and distributed on the level of federal countries or states. Although the Netherlands was among the first countries with national coverage, completed in 2003 (Brügelmann and Bollweg, 2004), in 2012 the entire country was captured again (Vosselman et al., 2015), and now a regular update cycle is in place. Other countries acquire airborne LiDAR along floodplains or below certain elevation (Artuso et al., 2003).

11.2 GROUND POINT EXTRACTION FROM LASER SCANNING POINT CLOUDS

In the current section, the classification of the acquired point cloud into ground points and off-terrain points, possibly split up in more classes, is discussed. The computation of DTMs, which is an integral part of some of these algorithms, will be treated in the next section. Before presenting an overview on filter algorithms, we will state the problem explicitly. Subsequently, approaches for filtering the airborne laser scanning data are presented. The filters are organized into four groups depending on the concept of the terrain surface used. Two filters will be described in more detail. They have been selected because they have reached certain popularity, they are very different and, thus, highlight the range of approaches and concepts applied.

11.2.1 PROBLEM DEFINITION

Formally, the problem can be stated as follows. Given is the entire set of points $\mathbf{P}$, each point in 3D space: $p_i = (x_i, y_i, z_i) \in \mathbf{R}^3$, $i = 1,..., n$. There can be some additional attribute information a_i such as echo number (first, intermediate, and last echo), intensity (return amplitude), color obtained from a digital image, echo width extracted from full-waveform analysis, or height precision (Otepka et al., 2013). Those points p_i that lie on the terrain surface shall be identified considering the measurement precision. $\mathbf{P_T}$ is the set of terrain points ($\mathbf{P_T} \subseteq \mathbf{P}$), see also Figure 11.2.

The above-mentioned formulation is the classification point of view, which can be applied equally to points and rasters. In the latter case, the pixel position takes the role of x and y and the pixel value the role of z. Alternatively, the problem can be seen as directly obtaining the DTM from the set $\mathbf{P}$, without an interest in the classification itself. In either case, this task is not trivial, because

1. Objects slightly above the ground may have the same geometrical appearance as smaller terrain features. Thus, a criterion only based on points neighboring p_i may fail in such cases.
2. The attribute information a_i typically cannot be used directly to classify the points into ground and off-terrain. Especially with a_i representing the echo number (first, last, etc.), the last of multiple echoes does not always refer to the ground. It may originate from lower vegetation below a forest canopy. The last echo may also feature a multipath effect, effectively leading to a point below the ground surface. However, using only last echoes as start point for ground extraction is a suitable and frequently used approach.
3. Human interpretation uses a significant amount of context knowledge when looking at point clouds or a DSM in 3D perspective views, which is hard to incorporate in algorithms.

11.2.2 RASTERIZATION: PROS AND CONS

Most algorithms for ground extraction employ some geometrical reasoning to decide if a point lies on the terrain surface. Rasterizing the data first, these algorithms can be executed in the domain of digital image processing, in which neighborhood operations run much faster. However, this results in a loss in precision (Axelsson, 1999). In addition, depending on the algorithm, the interpolation of raster heights in areas without points may be necessary. There are a number of options for assigning a height to a cell, which will not be discussed in detail here. These possibilities

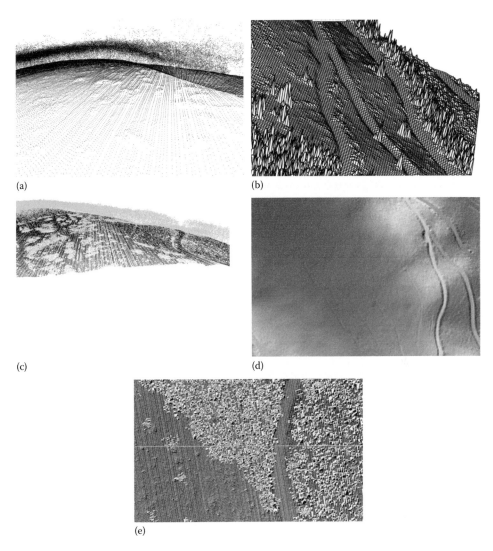

FIGURE 11.2 (a) Original point cloud in perspective view from the side, showing ground points, points on the tree crowns, at intermediate height, and points on features near the ground. (b) DSM of the same dataset, showing streets, bare ground, and wooded areas. (c) Classified point set: ground points (black) and off-terrain points (gray). Note that the features near the ground from (a) have been classified as off-terrain here. (d) Z-coding superimposed to shading of the DTM, plan view. Next to the streets, paths inside the forest also become visible. In (e) a shaded view of a DSM detail together with the original points and a profile of 85 m length are shown.

include taking the height of the lowest/mean/closest to the center/median in height point, and others. Raster algorithms, also for DTM generation, are often used to generate intermediate products in a coarser resolution.

11.2.3 SIMPLE FILTERS

Different concepts for filtering, with different complexity and performance characteristics, have been proposed in literature so far. The simplest filters operate by always taking the minimum elevation within a certain area. If this area is a raster cell, the lowest point falling within this cell defines the cell elevation, leading directly to a DTM in a regular grid structure. These block minimum filters have a number of properties that introduce systematic errors. First, there is an inherent assumption of a flat terrain. For inclined terrain, even without vegetation cover or buildings on top, the height will also be too low. This systematic error depends on both the terrain inclination and the neighborhood size. Whitman et al. (2003) embedded this filter in a multiresolution approach.

11.2.4 MORPHOLOGICAL FILTER

The name for this group of filters is derived from mathematical morphology (Haralick and Shapiro, 1992), which is also explicitly stated in the *Morphological Filter* of Vosselman (2000). A structure element $\Delta h_{max}(d)$ describing admissible height differences as a function of the horizontal distance d is used in the erosion operation (Figure 11.3a). The distance d is the planimetric distance; thus, $d(p_i, p_k) = \sqrt{(x_i - x_k)^2 + (y_i - y_k)^2}$. The smaller the distances between a ground point and its neighboring points, the lesser the height differences accepted between them. The structure element is positioned at each candidate point and this point is identified as off-terrain point if one or more height differences to its neighbors are above the admissible height difference. This structure element has an effect up to the maximum distance d_{max}, thereby defining the neighborhood contributing to the point test. In the publications of Vosselman, d_{max} is typically in the order of 10 m. The structure element encodes information not only on the terrain but also on the points measured by laser scanning. It can be determined from assumptions on the maximum terrain slope found in the area and from the height precision of laser scanning points. With a maximum slope of $\tan(\gamma)$ and a vertical measurement precision of σ_z, it becomes (Vosselman, 2000):

$$\Delta h_{max}(d) = \sqrt{2}\sigma_z + d\tan(\gamma), d \leq d_{max}$$

The term $\sqrt{2}\sigma_z$ is the precision of the height difference of a point pair. The set of ground points is then defined as (Vosselman, 2000):

$$\mathbf{P}_T = \{p_i \in \mathbf{P} \mid \forall p_k \in \mathbf{P}: z_i - z_k \leq \Delta h_{max}(d_{ik})\}, \ d_{ik} = \sqrt{(x_i - x_k)^2 + (y_i - y_k)^2}$$

The structure element can also be determined from terrain training data consisting of correctly classified ground points and vegetation points. The actual height differences between points of the same and of different classes can be used to derive a structure element that avoids omission errors (no real ground point is rejected as new off-terrain point) or commission errors (no real off-terrain point

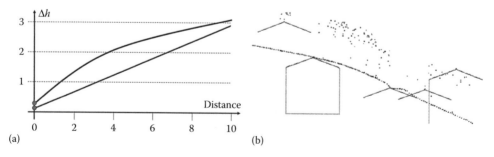

FIGURE 11.3 (a) Two different structure elements showing admissible height differences Δh_{max} versus distance, both axes are shown in meters. (b) Classification of a number of points in a profile by a structure element for an offset of 15 cm and a terrain slope of 20° is shown (straight structure element in left image). For the (two) ground points, no points are below the structure element, whereas for the (three) vegetation points, a number of points are below the structure element. The original data is shown in Figure 11.2e in top view.

is classified as ground point). By averaging those two structure elements, the result is an optimal structure element to minimize commission and omission errors (Vosselman, 2000). In Figure 11.3b, the classification process, using the morphological filter, performed on a profile through the dataset of Figure 11.2 is shown.

11.2.5 EXTENSIONS AND VARIANTS OF THE MORPHOLOGICAL FILTERS

A morphological filter variant is described by Sithole (2001), in which the rotationally symmetric structure element depends on the terrain inclination estimated from averaged height data. For steeper areas, larger admissible height differences are prescribed. Another extension is mentioned in the airborne laser scanning filter overview included in Kobler et al. (2007), in which the structure element is inclined as a whole to follow the terrain and the rotational symmetry is lost. The reason is that height differences downwards have different characteristics compared with the height differences upwards, which is not considered in the other morphological filters.

Kilian et al. (1996) use multiple structure elements with different horizontal and vertical extents in the morphological opening operation. The structure element is, however, only horizontal. At different scales, spikes are thus removed, because for different structure element (i.e., window), sizes different objects (e.g., buildings and cars) are found. Weights are assigned to the points depending on the window size, which are finally used for the classification. A similar method is the progressive morphological filter described by Zhang et al. (2003) that extends the basic concept by a method for choosing the window parameters depending on terrain slope. Chen et al. (2007) use different window size in their morphological filter, and note that small window sizes typically work well for forest areas, but not for large buildings. Large window sizes may also cut off terrain. Based on the existence of step edges in such areas, it is determined if points were removed correctly, or if terrain was wrongly cut off.

Other morphological filters were suggested by Lohmann et al. (2000), Pingel et al. (2013), Li et al. (2014), and Mongus et al. (2014).

11.2.6 PROGRESSIVE DENSIFICATION

The filters in this group work progressively, rebuilding the ground by classifying more and more points as belonging to it. Axelsson (2000) uses the lowest points in large grid cells as the seeds of his approach. This initial set of ground points is triangulated to form the first reference surface.

For each triangle, one additional ground point is determined by investigating the unclassified points within the vertical prism built by that triangle. The offsets of these candidate points to the triangle surface are analyzed. These offsets are the angles α_1, α_2, and α_3 between the triangle face and the edges from the triangle vertices to the new point (Figure 11.4a). If a point is found with offsets below the threshold values, it is classified as a ground point and the algorithm proceeds with the next triangle. These new points are inserted into the triangulation, and the algorithm proceeds with a next round of adding points.

The algorithm stops, if there are no more points in a triangle, if a certain density of ground points is achieved, or if all acceptable points are closer to the surface than a threshold value.

Von Hansen and Vögtle (1999) describe a similar method with two differences. First, the starting points are the lower part of the convex hull of the point sets (Figure 11.4b) and second, the offset is measured as the vertical distance Δh of a candidate point to the reference triangle (Figure 11.4a). Sohn and Dowman (2002) apply the progressive densification first with a downward step, in which points below the current triangulation are added, followed by the upward step, in which one or more points above each triangle are added. The initial triangulation is obtained from the corner points of the entire area. An implementation, *lasground* was made available by Isenburg (Rapidlasso, 2016). All of these filters use a triangulation for accessing the data and the final TIN represents the DTM.

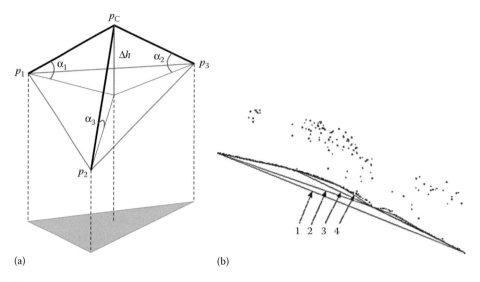

(a)   (b)

FIGURE 11.4   (a) Measures (α_i, Δh) to determine if a candidate point p_C is a ground point, as are the points p_1, p_2, and p_3 forming a triangle. (b) Lower part of the convex hull of a profile through the data of Figure 11.2 and a number of steps in the progressive densification.

11.2.7 SURFACE-BASED FILTER

While the progressive densifications rebuild the terrain surface by adding points, the group of filters presented now start by initially assuming that all points belong to the ground surface and then remove those points that do not fit. Therefore, a general surface model $s_l(x, y)$ is constructed to iteratively approach the terrain surface DTM(x, y), l being the iteration index. The method of Kraus and Pfeifer (1998), known as robust interpolation, will be described in detail now, especially in the context of ground points and points on the vegetation.

In this approach, linear least squares interpolation (also known as linear prediction or simple kriging) is used to describe the surface. The stochastic behavior of the terrain is described by the covariance function $C(p_i, p_k)$ with the variable planimetric distance d as defined earlier:

$$C(p_i, p_k) = C(0)e^{-d(p_i, p_k)^2/c^2}$$

The parameters c and $C(0)$ are computed from the data (Kraus and Pfeifer, 1997; Journel and Huijbregts, 1978), but estimates depending on the terrain type can also be used. The covariance function is used to set up a system of equations for computing the height z of one point $p = (x, y, z)$ from the given points $p_1,..., p_n$. It has to be noted that a trend in the data is subtracted first, for example, by determining an adjusting plane or low-order polynomial approximating all points. This will not be discussed further, and it is assumed that the trend has been removed.

$$z(x, y) = \mathbf{c}^{\mathrm{T}} \mathbf{C}^{-1} \mathbf{z}$$

$$\mathbf{c} = (C(p, p_1), \ ..., \ C(p, p_n))^{\mathrm{T}}, \mathbf{z} = (z_1, \ ..., z_n)^{\mathrm{T}}$$

$$\mathbf{C} = \begin{pmatrix} V_{zzp1} & C(p_1, p_2) & \cdots & C(p_1, p_n) \\ C(p_1, p_2) & V_{zzp2} & & C(p_2, p_n) \\ \vdots & & \ddots & \\ C(p_1, p_n) & C(p_2, p_n) & \cdots & V_{zzpn} \end{pmatrix}$$

In linear least squares interpolation (Kraus and Mikhail, 1972), the surface does not pass exactly through the given points, but random errors can be filtered. The residual at a point is $r_i = z_i - z(x_i, y_i)$. The filtering strongly depends on the variances V_{zzpi} along the diagonal of the **C** matrix. In general, large variances cause large filter values and small variances force the surface to almost strictly interpolate the point. Expressed as weight, the most accurate points shall have the weight one, and it decreases for less accurate points. For the most accurate points, the difference $V_{zzpi} - C(0) = \sigma_z^2$ is the vertical measurement variance (square of precision, which is typically ± 5 to ± 10 cm). Thus, for points with weight w the variance V_{zzpi} becomes

$$V_{zzpi} = \sigma_z^2/w_i + C(0)$$

A method to compute a surface from a point set considering an individual weight per point has been described so far. This method will now be used for classifying ground versus off-terrain points.

The weights w_i are determined iteratively. In the first iteration, all points get the same weight $w_i = 1$, and all points have the same influence on the run of the surface. The values r_i are computed, and they typically show an asymmetric distribution. There is a cluster of negative values, corresponding to points below the surface and originating from the ground points. Because the initial surface is shifted upwards, the ground points form a group of points with negative residuals. The vegetation points do not form a clear cluster, because they are not necessarily on the canopy top surface only, but found at various height layers in the vegetation.

In robust estimation (Koch, 1999), a weight function is used to down-weight observations with larger residuals, which have presumably large errors. This is performed iteratively, so observations that only had large residuals in the first pass but actually fit well to the model (here: the ground surface) can be rehabilitated. However, observations that do not fit to the model obtain lower and lower weights in each iteration. The weight function must fulfill the property of weighing ground points higher (weight 1) and points on the vegetation lower. These properties are achieved with an asymmetric weight function (Figure 11.5a).

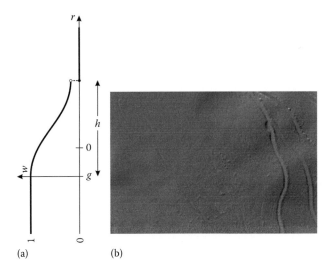

(a)   (b)

FIGURE 11.5   (a) Weight function of the robust interpolation, mapping a residual r to a weight w. (b) Shaded view of the DTM from the dataset of Figure 11.2, filtered by robust estimation. Note that near-ground features have been classified as ground points incorrectly here (opposed to Figure 11.2c and d). There were no points below those features on the terrain surface and also height differences and inclination against the surrounding terrain is small.

$$w(r) = \begin{cases} r < g & : & 1 \\ g \leq r \leq g+h & : & \dfrac{1}{1+(a(r-g))^{b}} \\ r > g+h & : & 0 \end{cases}$$

With the weights $w(r)$, new variances V_{zzpi} are computed for each point. As noted previously, points with large variances, that is, those points with low weights, have less influence on the run of the surface, and the surface is more attracted to the (ground) points with higher weights. In the iterative process, the points on the vegetation get less and less weight. The parameters a and b determine how fast the weight function drops to zero, whereas the parameter g determines where this drop starts, and h is a parameter controlling the exclusion of (definite vegetation) points from the equation system. The results of the robust interpolation applied to the data of Figure 11.2 are shown in Figure 11.5b. For practical usage, a, b, and h can be set to 2 (m^{-1}), 2, and 5 m, respectively. This means that points with a residual of 0.5 m above g get the weight 1/2, and those with a residual of 5 m above g are ignored. The parameter g can be estimated from the residuals themselves. Methods are given in detail in Kraus and Pfeifer (1998). A robust method is to estimate the penetration rate, which does not need to be a very accurate estimation. If the estimation is, for example, 40%, g then becomes the quantile of the residuals found at half of the estimated penetration rate (in the example the 20% quantile of residuals). The entire process is visualized for a profile in Figure 11.6.

As an equation system needs to be solved, the method is applied patch-wise. The functional model of surface interpolation could be replaced by other surface reconstruction techniques, for example, moving least squares in place of kriging, showing better computational performance, but giving less control over filter values.

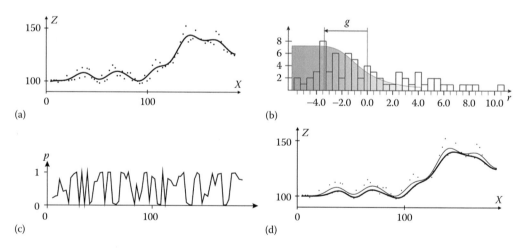

FIGURE 11.6 Profile of ALS data with a hilly terrain and points on the vegetation and the ground. Panel (a) shows the points and the averaging surface computed with equal weights for each point. Panel (b) shows the residuals, overlaid with the weight function giving smaller weights to points that lie higher above the surface. Panel (c) displays the distribution of weights between 0 (no influence on the surface run in the next iteration) and 1 (maximum influence). The surface computed under consideration of the weights is shown in panel (d). The previous (first) iteration is shown in gray.

11.2.8 EXTENSIONS AND VARIANTS OF THE SURFACE-BASED FILTERS

In the work by Pfeifer et al. (2001), the robust interpolation method has been embedded in a hierarchical approach to handle large buildings and reduce computation time. At different levels, thinned out versions of the point cloud are used to compute surface models. The DTM from a coarser level is used to preselect probable ground points for the next layer. The process is illustrated in Figure 11.7.

Elmqvist (2001) uses a snake-approach (Kass et al., 1988), in which the inner forces of the surface determine its stiffness and the external forces are a negative gravity. Iteration starts with a horizontal surface below all points that move, following the negative gravity, upwards to reach the points. Inner stiffness, however, prevents it from reaching up to the points on vegetation or house roofs. Brovelli et al. (2003) have a similar approach to analyzing the residuals obtained from an averaging spline surface interpolation. In a second step, edges are extracted, connected to closed objects, and elevated areas are removed. Zhang et al. (2016) use *cloth simulation* from computer graphics in a similar way, in which the cloth is held together by internal forces

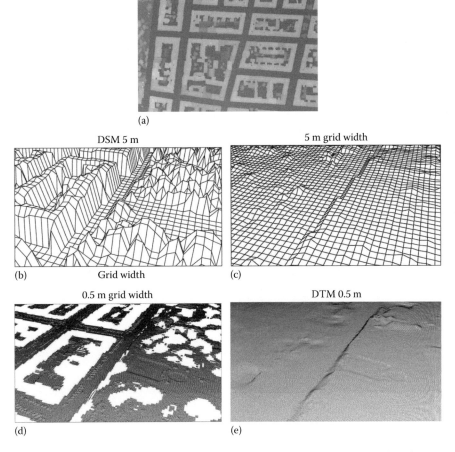

(a)

DSM 5 m   5 m grid width

(b)   Grid width   (c)

0.5 m grid width   DTM 0.5 m

(d)   (e)

FIGURE 11.7 Counter clockwise: (a) Original DSM, (b) coarse DSM, filtered by robust interpolation to (c) coarse DTM, (d) DSM at finer level with points close to coarse level DTM, including cars, and (e) final DTM.

and attracted by the ALS points. Mongus and Zalik (2012) combined the surface-based and the morphological approach. In a hierarchic scheme they determine an approximating surface using thin-plate-splines and apply the morphological operations on the residuals, that is, the height differences between the points and the surface.

All these filters work on point clouds and no explicit neighborhood definition is necessary. The neighborhood is implicitly considered by the interpolation method. A methodological strength of these filters is that surface behavior, for example, smoothness, can be incorporated into the functional model of surface interpolation, whereas the stochastic properties of the laser points—including both random measurement errors and distribution of above ground objects—can be encoded in the weight function.

11.2.9 SEGMENTATION-BASED FILTERS

The fourth group of filters works on segments. As Filin and Pfeifer (2006) note, the processing of laser scanning point clouds can be strengthened by first aggregating information (i.e., building homogeneous segments) and then analyzing segments rather than individual points.

Segmentation can be performed directly in object space, using region-growing techniques. Often the normal vector or its change is used to formulate the homogeneity criterion, resulting in planar surfaces in the first case and smoothly varying surfaces in the second. Alternatively, homogeneity can be formulated in terms of height or height change only. In contrast to the region-growing techniques, segmentation can also be performed in a feature space, which has the advantage that the variability of the features is known before detecting clusters in feature space. In Figure 11.8a, the motivation for segmentation-based filtering is demonstrated. The original point cloud is shown with

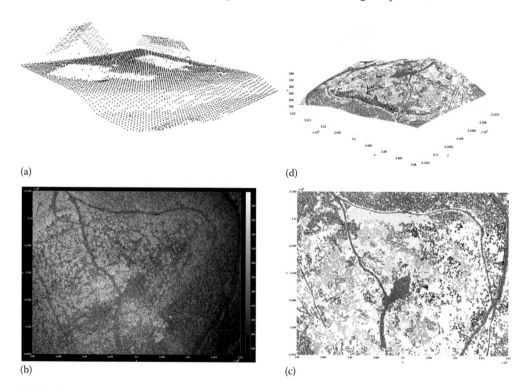

(a)

(d)

(b)

(c)

FIGURE 11.8 (a) Segmented point cloud of a small scene with two houses. Segments are distinguished by color. (b) Z-coded original point cloud of a forest and open area scene Figure 11.2b, white means higher. (c) Segmentation result, with points of the same segment are shown in the same color (only larger segments are displayed). (d) Perspective view of (c).

points from different segments in different colors. Each segment either belongs to the ground or contains only off-terrain points.

Sithole (2005) describes a method building segments by the rational that each point in a segment can be *reached* from each other point in the same segment, in the sense that it would be possible for a person to walk along such a path. The segmentation is performed first along two or more lines in different directions, and the height differences and jumps along this line as well as the horizontal distance are used to classify the segments as raised, lowered, terraced, and into other classes. The common points of line segments in different directions are used to form planimetrically extended segments. The relation to neighboring segments is used to classify the segments, in which *ground* is one class. Shao and Chen (2008) start their segmentation from initial ground points (e.g., locally lowest points) and grow the terrain surface based on slope and slope-increment constraints.

Nardinocci et al. (2003) apply a region-growing technique based on height differences to obtain segments. The geometric and topologic description of the regions can be expressed with graph theory, in which the segments are nodes carrying information on segment size. The edges connect neighboring regions and carry information on height differences. Rules are applied to extract the segments representing the ground, vegetation, buildings, and other objects. Jacobsen and Lohmann (2003) proposed a comparable approach embedded in the eCognition software and, therefore, also on gridded data, and segments are obtained from region-growing. A number of criteria, for example, compactness and height differences to the neighboring segments, are used to detect different types of areas including terrain. Similar segmentation-based filters were described by Schiewe (2001), Ma (2005), and Yan et al. (2012).

Tovari and Pfeifer (2005) use region-growing-based segmentation with the homogeneity formulated in terms of estimated normal vectors and spatial proximity. This segmentation was performed on the data of Figure 11.8b. Adjacent normal vectors were allowed to differ by 5°, and points had to have a maximum distance of 2 m to belong to the same segment. Segments obtained by this method with more than 10 points are shown in Figure 11.8c. As can be seen, segments below forest cover are less dense, but there are many holes due to small ground features and the surface next to the streets is also missing. In the approach of Tovari and Pfeifer, the surface is interpolated, as in the robust interpolation, but a representative (e.g., mean) residual is computed for each segment, and depending on that residual the same weight is given to all points in that segment. Iteration continues and off-terrain segments are assigned lower and lower weights. Another combination of segmentation and an existing filter algorithm is presented by Zhang and Lin (2013). After selecting the seed points in the progressive TIN densification, the seed points are extended by segmentation to seed regions, for which the triangulation is built. Then additional points are inserted on the basis of thresholds with respect to the current triangulation.

Many implementations require rasterized data, and also the segmentation methods are often borrowed from image processing. Generally, these filter methods work on larger entities, that is, not on the single points or pixels, and are therefore less influenced by noise. This is demonstrated in Figure 11.8a, showing two large segments for the ground, one in the foreground and one in the background, a couple of segments for the roofs of the two connected houses, and many single or few point segments for the points on the vegetation.

11.2.10 CLASSIFICATION AND OTHER FILTER APPROACHES

Filtering in the context of point cloud classification is typically based on determining a set of features for each point depending on the distribution of the neighboring points as well as measured features of each point (e.g., echo width). Decision tree or feature distribution-based classifiers are then used to classify the points based on their feature values. Machine learning based on reference data, that is, a point cloud with labels (ground, off-terrain, or more classes), are then used to train the classifier. Wu et al. (2011) used support vector machines. Hu and Yuan (2016) used deep convolutional neural networks, which require a large set of training points, to classify the points.

Jahromi et al. (2011) use artificial neural networks. In a first step, they use simple thresholds on roughness and slope distribution to identify definite ground and object points, which are then used to train the classifier.

Guo et al. (2015) suggested a concept, which has a first step classification (boosting) using features of the points, and in a second step applies segmentation. Labels of the initial classification are used as seed points. Niemeyer et al. (2012) use conditional random fields to classify LiDAR point clouds and consider classification and neighborhood relations, thus, in one step. They also differentiate between different classes of ground (*natural* and *asphalt*).

There are a number of filters that cannot be assigned to one of the above-mentioned filter approaches. The *Repetitive Interpolation* filter of Kobler et al. (2007) works on a prefiltered dataset. It may contain off-terrain points but points on the vegetation canopy are expected to be removed beforehand. It proceeds by randomly picking points and computing a grid-based DTM by sampling the elevations from a triangulation. This process is repeated a number of times, each time choosing randomly different points. The final DTM is then computed from the individual DTMs in an averaging procedure.

Shan and Aparajithan (2005) developed a method specifically for urban areas working on the LiDAR scan line. It is based on the slopes between points. It was extended by Meng et al. (2009) to consider additional directions. Instead of slope, Evans and Hudak (2007) used curvature and removed nonground iteratively, based on curvature thresholds in different levels of the scale space. Bartels and Wei (2010) suggested to remove point by point above the terrain by analyzing the skewness of the elevation distribution in their neighborhood.

It is also noteworthy, that specialized filter algorithms, applicable for certain land use were developed, for example, city areas (Shan and Aparajithan, 2005), areas with low vegetation (Wang et al., 2009), or dense mountain forests (Guan et al., 2014).

11.2.11 FULL-WAVEFORM EXPLOITATION FOR GROUND RECONSTRUCTION

Full-waveform-recording laser scanners allow retrieving more information from an object surface than only its range to the sensor. The width of the echo holds information on the spatial (i.e., vertical) spread of the object surface. For small footprint scanners, this is not influenced strongly by terrain slope because the height differences due to slope are typically much smaller than the length of the pulse. However, low vegetation and other nonground objects lead to a pronounced height variation within the footprint. Echo width can therefore be used as a preclassifier to exclude echoes that cannot originate from the ground (Doneus and Briese, 2006). Examples are shown in Figure 11.9. The ability to discriminate low vegetation from ground in ALS point clouds was also shown by Lin and Mills (2010) and Mücke et al. (2010) suggested it for preassigning probabilities for ground point filtering.

11.2.12 COMPARISON OF FILTER ALGORITHMS

An experimental comparison of filter algorithms was published by Sithole and Vosselman (2004). The conclusions are that filters incorporating a concept of surface perform better, for example, in comparison with the morphological filters. Problems were identified for all filters typically along steep terrain edges und underpasses. In the review of Meng et al. (2010) the list of problems is extended to low vegetation, which was not in the focus of the earlier study. This section is devoted to a comparison of different methodologies for solving the filtering problem.

Comparing the algorithms, it can be stated that the morphological filters mostly investigate height differences only, whereas surface-based filters consider the surface trend also. Therefore, they produce more reliable results. For the morphological filter, a problem faced is that a tradeoff between erosion of ground on steep slopes on the one hand and the inclusion of off-terrain points in flat areas on the other hand has to be made.

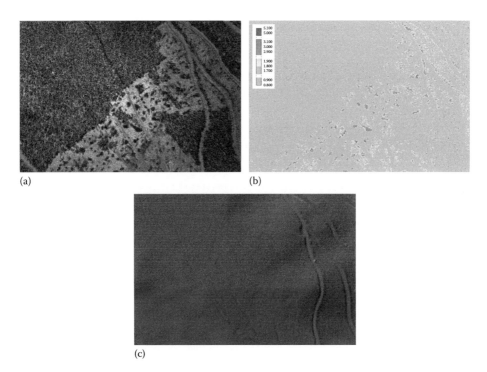

FIGURE 11.9 (a) Amplitude (the so-called intensity) of echoes extracted from the waveform, (b) echo width, and (c) DTM obtained by robust interpolation excluding points with wide echoes (compare Figures 11.2 and 11.5b).

The progressive densification and the surface-based filters use a surface in the generation of the classification. Progressive densification gradually adds points, building a DTM (or the classification) from a few points. The surface-based filters progress by removing points from the entire point cloud. A general advantage of these algorithms is that the geometrical terrain surface properties and the LiDAR data characteristics are, at least, to some extent separated. The structure element of the morphological filters on the other hand includes information on the terrain and on laser scanning characteristics as precision and vertical distribution of off-terrain objects. Each point is classified independently of its neighboring points. The progressive densification algorithms build a terrain model during execution, but the thresholds (angles or vertical distances) also encode terrain information and laser point vertical distribution aspects. In surface-based filters, especially in the robust interpolation, there is a separation between those two. The covariance function of kriging encodes the terrain characteristics, whereas the vertical distribution of ground and off-terrain points is reflected in the weight function. Moreover, the iterative nature of the surface-based approaches, where a hard classification is only derived after some iteration steps, allows a better adaptation to the terrain features. As the surface-based approaches are built on a thorough terrain interpolation method, additional parameters, for example, the range accuracy determined from full-waveform analysis, or knowledge about definite ground points can be considered.

Using a rich set of features and machine learning, local terrain shape is obviously also considered in the point-based classification methods. However, their performance under varying conditions still needs to be assessed.

The segmentation-based algorithms have advantages in areas strongly influenced by human building activities (houses, street dams, embankments, etc.). In these regions, distinct segments, often planes, are found. As these segments belong either entirely to the ground or to buildings, the classification is not affected by edge effects, which can be found in surface-based filters. In wooded areas, this does not hold and segment-based filters lose their advantage. A tight homogeneity criterion produces too many segments (approaching the case of point-based filtering) and a loose homogeneity criterion includes too many off-terrain features, which cannot be split in the segment classification step anymore (under-segmentation).

One general problem remaining for all filter algorithms is that the ground cannot be characterized by geometric properties only. Looking at a local neighborhood, similar characteristics may always be encountered when comparing small terrain features and small objects. To increase automation, information from other sensors or other information from the same sensor will have to be used. Spectral information may be one valuable source, especially if the data is acquired synchronously. The consideration of echo width for improving terrain filtering quality was shown to be successful. Still, this is not widely applied yet.

11.3 DTM DERIVATION AND PROCESSING FROM GROUND POINTS

11.3.1 DIGITAL TERRAIN MODEL INTERPOLATION

After the point cloud classification into terrain points and other points, the task of interpolation of the DTM remains. An overview of methods can be found in Maune (2007), El-Sheimy et al. (2005), and Kraus (2000). As the number of points and their density are typically very high (density often 10 pt./m^2), it is not always necessary to apply the best predictors (e.g., kriging) but rather resort to methods exhibiting better computational performance. However, methods reducing the random measurement errors should be applied, making use of the redundancy in the measurements and thereby improving precision. The interpolated DTM can then be more accurate than a single point measurement.

11.3.2 STRUCTURE LINE DERIVATION

For the representation of models generated from ALS data, mostly rasters, grid models, or TINs are in use, which are determined from the irregularly distributed ALS point cloud. Due to the lack of structure information, the representation of break lines, which are crucial for a high quality delineation of a surface discontinuity, depends, next to the original sampling interval, also on the size of the stored raster, grid, or triangle surface cells, respectively. In contrast to these models without an explicit feature line description, it is essential for high quality terrain models to store these lines explicitly in the DTM (e.g., for hydraulic applications). For this aim, a 3D vector-based description of the lines is necessary. In addition, the line information can be helpful for the task of data reduction, because break lines allow describing surface discontinuities even in models with big raster or triangle cells (see the next section).

The research in the area of break line extraction is focused on methods for the fully automated 2D detection in the so-called range images determined from the ALS point cloud, allowing the application of image processing techniques. Brügelmann (2000) presents one representative method, which introduces a full processing chain starting from an ALS elevation image leading to smooth vector break lines. The break lines are positioned at the maxima of the second derivative. The break line heights are retrieved by interpolation in the elevation image, which is a low pass filter. Borokowski (2004) presents two different break line modeling concepts. One makes use of snakes, whereas the second method utilizes a numeric solution of a differential equation describing the run of the break line. Brzank et al. (2008) present an approach for the extraction and modeling of pair-wise structure lines in coastal areas described by a hyperbolic tangent function. Wehr et al. (2009) describe a method for modeling break lines based on a triangulated DTM derived from the LiDAR ground points. In contrast to the approaches for break line modeling mentioned earlier, which rely on a previously filtered DTM, the approach by Briese (2004) uses the original irregularly distributed point cloud data as input. This enables using the break lines as additional information in the filtering of the ALS bare-earth points. The basic concept of the approach is displayed in Figure 11.10a. Based on a local surface description on either side of the break line, the break line itself can be modeled by

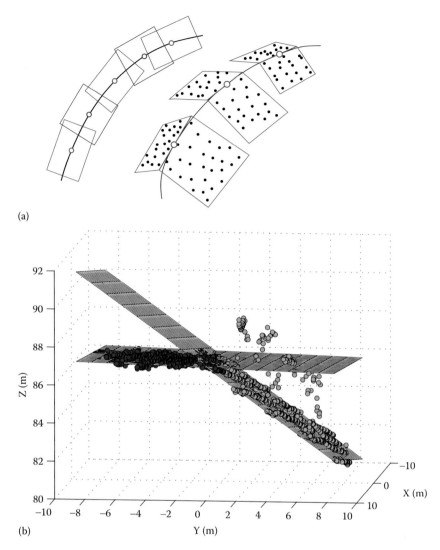

(a)

(b)

FIGURE 11.10 (a) Basic concept for the description of break lines by using overlapping, intersecting patch pairs. (b) Practical example of a robustly estimated plane patch pair (10 by 10 m) based on unclassified ALS data (circles). The elimination of off-terrain points improved the sigma of the adjustment from 0.59 to 0.12 m.

a sequence of locally intersecting surface patch pairs. The individual surface patches are estimated from the ALS points of the respective side by a least squares fit considering a user defined patch length and width. Robust estimation techniques are used to eliminate the influence of off-terrain points on the run of the break line. An example of such a robustly estimated surface patch pair is shown in Figure 11.10b. It can be seen that after robust estimation the off-terrain points do not affect the run of the right (green) patch and that the break line can be accurately approximated by the intersection line of the two planar patch pairs. The described 3D modeling approach requires a 2D approximation of the break line for delineating the patch pairs. To achieve full automation, Briese (2004) introduced a line growing framework. Potential seed segments are first identified via point cloud curvature analysis and the line segments featuring low curvature in and high curvature across line direction are used for line tracking in forward and backward direction. The growing procedure continues until a certain break-off criterion describing the significance of the surface discontinuity (e.g., minimum patch intersection angle) is reached.

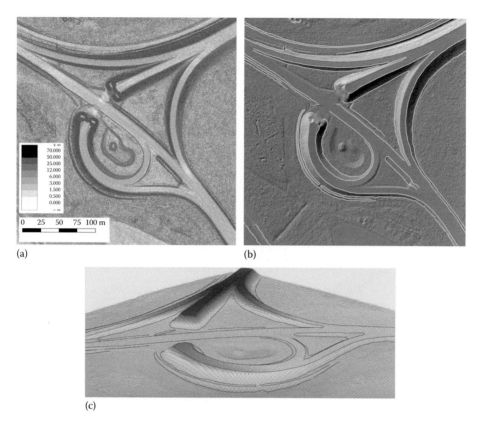

(a)   (b)

(c)

FIGURE 11.11 (a) Color coded DTM slope map serving as basis for edge detection. (b) Automatically detected 2D line approximations (Canny). (c) Modeled 3D break lines meshed into hybrid DTM grid. (From Mandlburger, G. et al., *Publikationen der Deutschen Gesellschaft für Photogrammetrie, Fernerkundung und Geoinformation e. V.* Band 25, 131–142, 2016. With Permission.)

The described approach was further developed by Mandlburger et al. (2016). In addition to planar patch pairs as used by Briese (2004), conical geometries are used to model break lines with high planimetric curvature and polynomial cylinders are employed to cope with nonplanar slopes (across line direction). Furthermore, the line curvature, the line neighborhood topology, and the ALS point density are used to locally adapt the patch length and width. The required 2D break line approximations are obtained by raster-based edge detection (Canny algorithm) using the gradient model of a preliminary DTM as input. Manual editing of the resulting line network is minimized by automatic line merging and gap bridging in a postprocessing step. Exemplary results of the entire procedure are shown in Figure 11.11. Figure 11.11c shows the final result of break line extraction and DTM reconstruction. At pronounced edges the break lines are meshed into the grid of the terrain model, whereas in flat and gently curved areas the terrain surface is represented by the grid only.

11.3.3 DIGITAL TERRAIN MODEL SIMPLIFICATION (THINNING)

Due to advances in Dense Image Matching and ALS sensor technology, the DTM point density has increased dramatically in the recent years. Especially modern ALS sensors allow mapping of topographic details at the price of a highly increased data volume. For that reason, data reduction and surface simplification has become a major issue not only in the field of geodesy but also in computer vision, computer graphics, and applied engineering technology. In any case, the goal is to reduce the amount of data (vertices of a mesh, grid points, break lines, etc.) without losing relevant geometric details.

A survey of different simplification algorithms is published in (Heckbert and Garland, 1997). This paper discusses both curve as well as surface simplification. In the following, we concentrate on the simplification of the DTM as a bivariate functions $z = f(x, y)$. In general, starting with a DTM description using m surface patches, the result of the thinning is a surface described by m_T patches, where $m_T \ll m$. The individual surface patches may either be linear, bilinear, or of higher degree. The surface approximation is achieved using a subset of the original points or any other points of the continuous surface. To describe the approximation error of the simplified surface, the maximum error (L_∞) or the quadratic error (L_2) is in use.

Regular grid methods are the simplest reduction techniques. Hereby, a grid of height values equally spaced in x- and y-direction is derived from the original DTM either by reinterpolation or by resampling (low pass filter). Although these methods work fast, they are not adaptive and therefore produce poor approximation results. So-called feature methods make a pass over the input points and rank them using a certain *importance* measure. Subsequently, the chosen features are triangulated. The quality of a feature-based surface approximation can be increased by adding break lines as constraints of the triangulation. Hierarchical subdivision methods build up a triangulation using a divide-and-conquer strategy to recursively refine the surface. They are adaptive and produce a tree structure, which enables the generation of different Levels of Detail. Better approximation quality can be achieved by applying refinement methods based on general triangulation algorithms such as Delaunay triangulation. Refinement methods represent a coarse-to-fine approach starting with a minimal initial approximation. In each subsequent pass, one or more vertices are added to the triangulation until the desired approximation error is below the intended threshold or the desired number of vertices is used. In contrast, decimation methods (Gotsman et al., 2002; Heckbert and Garland, 1997) work from fine-to-coarse. They start with the entire input model and iteratively simplify it, deleting vertices or faces in each pass. Due to the necessity of building a global triangulation, decimation methods are not suited for processing large DTMs, which arise when using high-density ALS data. The medial axis transformation is used in computer graphics for visualizing ALS point clouds and for thinning of DSMs (Peters and Ledoux, 2016), but this technique is rarely used for DTM data reduction.

The performance of DTM data simplification is highly influenced by the existence of systematic and random measurement errors. For deriving thinned DTM models from raw ALS point clouds, refinement or decimation approaches as described earlier are not the first choice. Systematic errors have to be removed first by thorough sensor calibration and fine adjustment of the ALS-strip data (Glira et al., 2016). The random measurement errors of the ALS points should rather be eliminated by applying a DTM interpolation strategy with measurement noise filtering capabilities. Good results can be achieved using linear prediction (Kraus, 2000) or kriging (Journel and Huijbregts, 1978), respectively, but increased ALS point density also justifies using interpolation techniques featuring better computational performance (e.g., moving least squares). In any case, high reduction ratios can only be achieved for DTMs free of systematic and random errors.

In contrast to the previously mentioned refinement approach, which rely on the original point cloud, the subsequently presented refinement framework uses the filtered hybrid DTM (regular grid and break lines, structure lines, and spot heights) as input (Mandlburger, 2009). The basic parameters for the data reduction are a maximum height tolerance Δz_{max} and a maximum planimetric point distance Δxy_{max}. The latter avoids triangles with overly long edges and narrow angles. The algorithm starts with an initial approximation of the DTM comprising all structure lines and a coarse regular grid (cell size = $\Delta xy_{max} = \Delta_0$), which are triangulated using a Constrained Delaunay Triangulation. Each Δ_0-cell is subsequently refined by iteratively inserting additional points until the height tolerance Δz_{max} is achieved.

The additional vertices can either be inserted hierarchically or irregularly. In case of hierarchical breakdown, each grid cell is divided into four parts in each pass (i.e., quad-tree) if a single grid point within the regarded area exceeds the maximum tolerance. By contrast, only the grid point with the maximum deviation is inserted when using irregular division. Higher compression ratios

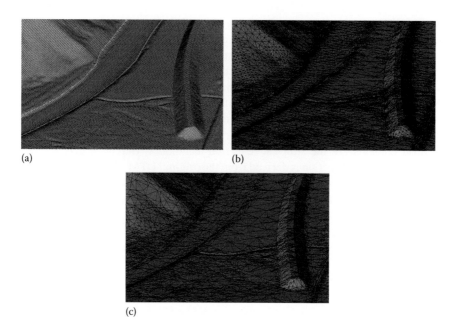

FIGURE 11.12 (a) Original hybrid DEM of Möllbrücke/Drau; base grid width: 2 m. (b) Adaptive TIN using hierarchical division. (c) Adaptive TIN using irregular division. For (b) and (c) $\Delta z_{max} = 0.25$ m and the L_∞-norm was applied as error measure in thinning.

(up to 99% in flat areas) can be achieved with irregular point insertion, whereas the hierarchical mode is characterized by a more homogeneous data distribution. The described framework is flexible concerning the reduction criterion. The decision to insert a point can be based on the analysis of the local DTM slope and curvature or on the vertical distance between the DTM point and the approximating TIN. The results of a practical example are shown in Figure 11.12.

Alternative DTM simplification approaches focusing on geomorphometric analysis use discrete wavelet transformations (Doglioni and Simeone, 2014). Wavelet-based strategies have been applied to regular grid models (Bjorke and Nilsen, 2003) and hybrid DTMs (Beyer, 2000).

11.4 QUALITY

The quality of a DTM derived from laser scanning has three major factors.

1. Quality of the original laser scanning point cloud
2. Quality of the filtering
3. Quality of the DTM interpolation, if applicable, including thinning and break line modeling

The first item is addressed in other chapters on data acquisition and calibration. The starting point is the quality of the trajectory, which is about 5–10 cm in planimetry and 7–15 cm in height, and 0.005° for roll and pitch and 0.008° in yaw (Skaloud, 2007). Rigorous models of strip adjustment with computation of the sensor system parameters as Inertial Measurement Unit misalignment and Global Navigation Satellite System antenna offset, together with the use of control patches, can lead to point clouds with a vertical accuracy of ±5 to 15 cm in height if acquisition is performed from platforms flying below 1000 m a.g.l. and over smooth targets. The accuracy of DTMs (item 3) can be better, as the random measurement error component can be minimized within the interpolation. Furthermore, redundant point information (e.g., from overlapping laser scanner strips) improves the relative accuracy, that is, the precision, of the DTM. Generally speaking, evaluating the accuracy of DTM generation from ALS data is challenging, because obtaining area-wise ground truth that has to be of higher precision than the laser scanning data itself is very expensive. Typically, only a few check points are measured in a few areas, and stationarity of the errors has to be assumed. Concentrating on certain land cover, slope, etc., error measures relevant for the respective application can be derived. An approach for deriving local DTM accuracy measures in a general framework of error propagation is presented by Kraus et al. (2006). The accuracy of every DTM grid point is estimated from the data itself, by considering the original point density and distribution, terrain curvature, and the height accuracy of the original points.

11.4.1 FILTERING QUALITY

In built-up regions, the quality of filtering can be very high, because the distinction between ground points and object points is typically quite clear. At jump edges as found between street level and building roofs there is a clear divide. It is rather the question of the type of surface that the DTM shall represent (e.g., at under passes), which accounts for filtering problems. At break lines (i.e., discontinuous slopes) filter algorithms still have problems (Meng et al., 2010), as the continuity and smoothness are often considered as properties of the terrain in filtering algorithms.

In areas covered by herbaceous vegetation, the quality of filtering cannot be separated from the quality of the original measurements. Laser ranging typically produces only one echo over low vegetation, that is, for heights up to 50 cm, even if the canopy and the ground are separated surfaces. For grass and reed type of vegetation, the scattering volume may even be extended over larger vertical ranges, and one echo is recorded that represents ground and vegetation. In these cases, filtering cannot separate ground and off-terrain points, as the bulk of measurement results show no distinct ground-only echoes. Without additional assumptions, for example, on vegetation height, the quality of DTMs depends on vegetation height and vegetation density within the footprint (Hopkinson et al., 2004; Pfeifer et al., 2004). With smaller footprints this problem is reduced, because less mixture of ground and vegetation occurs within the illuminated area. However, the vertical overlap is not reduced, and only shorter pulse lengths could reduce mixing of ground and vegetation echo. As shown in Doneus and Briese (2006) and Lin and Mills (2010), full-waveform echo width helps identifying echoes from low vegetation. The problem of low vegetation was confirmed in the comparative literature study of Meng et al. (2010).

For areas under forest cover, filtering quality was investigated extensively by Hyyppä et al. (2001), Reutebuch et al. (2003), and Kobler et al. (2007), values are given in the following. One obvious problem is obtaining reference values and the general lack of alternative methods for obtaining area-wise ground height under forest cover. Again, the definition of the terrain surfaces becomes problematic, as tree roots, small gullies, and many others, may be considered as terrain-forming or not.

Sithole and Vosselman (2004) have addressed the question for filtering quality by manually filtering a point cloud using the context knowledge and aerial imagery. The data was made available by EuroSDR (www.eurosdr.net). Eight test sites were chosen to represent urban and rural scenes, "steep slopes, dense vegetation, densely packed buildings with vegetation in between, large buildings (a railway station), multilevel buildings with courtyards, ramps, underpasses, tunnel entrances, a quarry (with breaklines), and data gaps" (Sithole and Vosselman, 2004). Test participants provided results of *their* filter, and the classified ground points were compared with the manual reference by Sithole and Vosselman. Assuming that the manual ground point identification process is error-free, this measures exactly the quality of filtering, without influences of georeferencing or DTM interpolation. However, as some filters produce a DTM (i.e., a surface and not a classification of points), they worked with a 20 cm tolerance band and compared the DTMs with the manually filtered point cloud. First, a qualitative assessment with respect to low and high outliers, very large and small objects (e.g., industrial complexes and cars), buildings on slopes, ramps, bridges, behavior at jump edges found in urban scenes, break lines, and high vegetation was performed. Most algorithms produce poor results at ramps, jump edges, and break lines. Then, the classification result was evaluated quantitatively. Considering all different scenes and all filter results, type I errors ranged from 0% to 60% (false rejection of ground points) and type II errors ranged from 0% to 20% (wrong acceptance of object points as ground). The conclusion of the study is that all filters work well in gently sloped terrain with small buildings and sparse vegetation. The most difficult scenes were complex cityscapes, with multitier buildings, plazas, and so on, and discontinuities in the bare-earth (Sithole and Vosselman, 2004). They further concluded that surface-based filters tend to provide better results.

After Sithole and Vosselman, other groups published comparisons of filtering quality. These comparisons are also evaluated against manually selected ground points, but the team of study authors executed the filter algorithms by themselves. A comparison by Zhang and Whitman (2005) concluded that low relief urban areas are easiest to filter for different algorithms, whereas differences can be found at specific geomorphic features, for example, sand dunes and cliffs. The study of Seo and O'Hara (2008) also included comparative analysis of surface profiles to highlight the properties of different filter algorithms. The comparison of Korzeniowska et al. (2014) went beyond the traditional comparison of counting type I and type II errors, that is, actual ground point wrongly classified as off terrain and vice versa, but also included qualitative assessment—in a quantitative manner—dependent on land cover and concentrating on the spatial distribution of errors. They found that no single algorithm can give the best results over different test sites. Montealegre et al. (2015), testing seven publicly available algorithms, came to a similar conclusion, over Mediterranean forest types.

11.4.2 DIGITAL TERRAIN MODEL QUALITY

In the following, studies representing the final quality of the DTM will be presented. The measure of accuracy is therefore including all aspects discussed so far.

A TopoSys Falcon I laser scanner was used to acquire data over an area in the Vienna city center covering 2.5 km² from an average flying height of 500 m a.g.l. using 19 strips, resulting in 10 points per m², although linearly distributed within each square meter. The DTM was determined by robust interpolation and checked manually for obvious filter errors. With over 800 terrestrially measured points, providing subcentimeter accuracy, the precision of data acquisition, filtering, and DTM generation was checked (Pfeifer et al., 2001). On average, this resulted in an root mean square error (r.m.s.e.) of ±11 cm, but showed a strong dependency on surface type (Figure 11.13). Densely forested park areas (area 5 in Figure 11.13) showed the highest r.m.s.e. with ±15 cm, for sparse trees it was reduced to ±11 cm (1), and reached ±9 cm in open park areas (2 and 3) as well as below parked cars (4). For open asphalt areas (6), a value of ±3 cm was obtained, which is far more precise than the navigation solution of the trajectory and can be explained by georeferencing control in the vicinity of these points. However, it shows the potential of laser scanning after elimination of all systematic errors. One should also keep in mind that random measurement errors in the range measurement are reduced because of qualified surface interpolation.

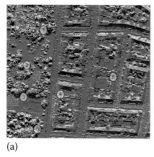

(a)   (b)

FIGURE 11.13   In (a), check points in a city area at different land cover classes are shown. In (b), a perspective view of the scene is shown. In Figure 11.7, the same area is shown.

Specifically, for forest conditions, the outcome of different filter algorithms was compared with ground truth for the DTM within the HIGH-SCAN project (Hyyppä et al., 2001). They reported biases up to 40 cm and standard deviations of the heights below ±40 cm. Ahokas and Hyyppä (2003) used TopoSys and TopEye laser scanning data from flying heights between 100 and 800 m a.g.l. with a point density varying from 1 to 8 points per m². Filtering was performed with a progressive densification method and the DTM, evaluated only at the check points, was calculated from the classified points by local cubic interpolation. For this comparison about 3500 check points were measured. Below forest cover, the standard deviation was between ±12 and ±18 cm (therefore excluding a systematic vertical bias). For grass the standard deviation was between ±6 and ±12 cm, and for asphalt between ±4 and ±10 cm. The systematic shifts encountered in the test areas (which are included in the r.m.s.e. reported for the previous study) reached up to 30 cm of absolute value. Reutebuch et al. (2003) reported similar values for terrain elevation under forest (conifer) cover, namely bias from 20 to 30 cm and similar standard deviation. Tinkham et al. (2011) compared the performance of two freely available filter algorithms in a semi-arid landscape and found, with little differences, a bias of 10 cm and a precision of 25 cm. The study of Naesset (2016) reported even improved values of approximately 10 cm bias and 10 cm precision. For an alluvial forest under leaf-off conditions, Mandlburger et al. (2015) reported a bias of 7 cm and a standard deviation of 6 cm.

In brief, it can be said that under ideal circumstances with (1) proper georeferencing using additional ground control, over (2) hard flat targets, with (3) no need to remove vegetation points, and (4) qualified noise reducing DTM interpolation techniques, an accuracy below ±1 dm can be achieved. For rough terrain, the point density is also strongly influencing accuracy. For areas with low vegetation, a systematic upwards shift of the final DTM can occur, as well as and an increase in the random error. While strongly depending on the vegetation type, both of these values easily reach 1 dm. Under dense forest cover the DTM precision can reach 2 dm. In case of rough ground, this requires a high point density of, for example, 5 pt/m².

11.5 CONCLUSIONS

Extracting the ground from airborne laser scanning data is operational. However, the problems addressed by Sithole and Vosselman (2004), Meng et al. (2010), and Korzeniowska et al. (2014) are not entirely solved, and new algorithms are still being presented to overcome those problems. Considering the proposed filter algorithms, directions for improvement can be determined.

1. Compared with earlier investigations, the accuracy of DTMs from ALS improved. Reasons are not only to be found in the improvement of filter algorithms, but (i) because of improved georeferencing and laser scanner measurement precision and (ii) higher point density.
2. The usage of additional information from full-waveform echo attributes (Doneus et al., 2008) has shown a way to improve the filtering by including additional information and incorporating it into known filter approaches rather than inventing new filter algorithms.
3. As noted, no single algorithm could so far provide the best results over varying land cover classes. On the other hand, filter algorithms for special surface types were developed. A two stage approach, which first identifies which approach to use where, before applying it in the second stage, may lead towards filters requiring less user interaction.
4. Understanding the task of filtering as a point cloud classification gave encouraging results with machine-learning approaches. A challenge for these approaches is the many influencing factors on the point distribution: flight mission parameters, phenology, land cover, and more detailed influence of vegetation on 3D point cloud distribution. Algorithms combining a classification based on the features of each point and neighborhood information, for example, by segmentation, were proposed. This brings surface-based and classification based algorithms together.

In commercial use, a high filtering quality is only achieved by human inspection, typically using shaded relief views of the original and the filtered data, and contour line plots.

For terrain capturing, especially over forested areas and mixed areas, airborne laser scanning is currently the method of choice. The active, polar measurement method, together with the high precision, strongly supports its dominance. Under optimal conditions and for well-defined surfaces, a precision of ± 5 cm is achievable, also over larger areas. In built-up areas featuring many edges and distinct texture, airborne laser scanning is *challenged* by digital aerial imagery and dense image matching.

However, algorithms for data exploitation and filtering are lagging behind the technological development, and the rich information in laser scanning point clouds, augmented by sensor developments described earlier, is still not fully exploited.

REFERENCES

Ackermann, F. and Kraus, K. 2004. Reader commentary: Grid based digital terrain models. *Geoinformatics* 7(6), 28–31.

Ahokas, K. and Hyyppä, J. 2003. A quality assessment of airborne laser scanner data. *IAPRS* 34, 3/W13, Dresden, Germany.

Artuso, R., Bovet, S., and Streilein, A. 2003. Practical methods for the verification of countrywide produced terrain and surface models. *IAPRS* 34, 3/W13, Dresden, Germany.

Axelsson, P. 1999. Processing of laser scanner data—algorithms and applications. *ISPRS Journal of Photogrammetry and Remote Sensing* 54(2–3), 138–147.

Axelsson, P. 2000. DEM generation from laser scanner data using adaptive TIN models. *IAPRS* 33, B4/1, Istanbul, Turkey.

Bartels, M and Wei, H. 2010. Threshold-free object and ground point separation in LIDAR data. *Pattern Recognition Letters* 31(10), 1089–1099.

Beyer, G. (2000). Wavelet-transformation hybrider gelandemodelle mit rasterbasierter kanteninformation. *Photogrammetrie, Fernerkundung, Geoinformation* 126, 23–33.

Bjorke, J. and Nilsen, S. 2003. Wavelets applied to simplification of digital terrain models. *International Journal of Geographical Information Science* 17(7), 601–621.

Borokowski, A. 2004. Modellierung von Oberflächen mit Diskontinuitäten, Habilitation, Veröffentlichung der Deutschen Geodätischen Kommission (DGK), B and C 575.

Briese, C. 2004. Three-dimensional modeling of breaklines from airborne laser scanner data. *IAPRS* 35, B/3, Istanbul, Turkey.

Brovelli, M., Cannata, M., and Longoni, U. 2003. LiDAR data filtering and DTM interpolation within GRASS. *Transactions in GIS* 8(2), 155–174.

Brügelmann, R. 2000. Automatic break line detection from airborne laser scanner data. *IAPRS* 33, B3, Amsterdam, the Netherlands.

Brügelmann, R. and Bollweg, A. 2004. Laser altimetry for river management. *IAPRS* 35, B2, Istanbul, Turkey.

Brzank, A., Heipke, C., Goepfert, J., and Soergel, U. 2008. Aspects of generating precise digital terrain models in the Wadden Sea from LiDAR–water classification and structure line extraction. *ISPRS Journal of Photogrammetry and Remote Sensing* 63(5), 510–528.

Chen, Q., Gong, P., Baldocchi, D., and Xie, G. 2007. Filtering airborne laser scanning data with morphological methods. *Photogrammetric Engineering and Remote Sensing* 73(2), 175–185.

Doglioni, A. and Simeone, V. 2014. Geomorphometric analysis based on discrete wavelet transform. *Environmental Earth Sciences* 71(7), 3095–3108. doi:10.1007/s12665-013-2686-3.

Doneus, M. and Briese, C. 2006. Full-waveform airborne laser scanning as a tool for archeological reconnaissance. *International Conference on Remote Sensing in Archeology*, Rome, Italy, December 4–7, 2006.

Doneus, M., Briese, C., Fera, M., and Janner, M. 2008. Archaeological prospection of forested areas using full-waveform airborne laser scanning. *Journal of Archaeological Science* 35(4), 882–893.

Elmqvist, M. 2001. Ground estimation of lasar radar data using active shape models. *Proceedings of OEEPE Workshop on Airborne Laser Scanning and Interferometric SAR for Detailed Digital Elevation Models*, Stockholm, Sweden, March 1–3, 2001.

El-Sheimy, N., Valeo, C., and Habib, A. 2005. *Digital Terrain Modeling*. Artech House: Boston, MA.

Evans, J. and Hudak, A. 2007. A multiscale curvature algorithm for classifying discrete return LiDAR in forested environments. *IEEE Transactions on Geoscience and Remote Sensing* 45(4), 1029–2007.

Filin, S. and Pfeifer, N. 2006. Segmentation of airborne laser scanning data using a slope adaptive neighborhood. *ISPRS Journal of Photogrammetry and Remote Sensing* 60, 71–80.

Gil, A. L., Núnez-Casillas L., Isenburg, M., Benito A. A., Bello J. J. R., and Arbelo M. 2013. A comparison between LiDAR and photogrammetry digital terrain models in a forest area on Tenerife Island. *Canadian Journal of Remote Sensing* 39(5), 396–409.

Glira, P., Pfeifer, N., and Mandlburger, G. 2016. Rigorous strip adjustment of UAV-based laser scanning data including time-dependent correction of trajectory errors. *Photogrammetric Engineering & Remote Sensing* 82(12), 945–954.

Gotsman, C., Gumhold, S., and Kobbelt, L. 2002. Simplification and compression of 3D meshes. In *Tutorials on Multiresolution in Geometric Modelling*, Iske, A., Quak, E., and Floater, M.S. (Eds.), Springer-Verlag, Heidelberg, Germany, pp. 319–361.

Grün, A. and Baltsavias, E. 1988. Geometrically constrained multiphoto matching. *Photogrammetric Engineering and Remote Sensing* 54(5), 633–641.

Guan, H., Li, J., Yu, Y., Zhong, L., and Ji, Z. 2014. DEM generation from LiDAR data in wooded mountain areas by cross-section-plane analysis. *International Journal of Remote Sensing* 35(3), 927–948.

Guo, B., Huang, X., Zhang, F., and Sohn, G. 2015. Classification of airborne laser scanning data using JointBoost. *ISPRS Journal of Photogrammetry and Remote Sensing* 100, 71–83.

Haala, N. and Rothermel, M. 2012. Dense multistereo matching for high quality digital elevation models. *Journal of Photogrammetry, Remote Sensing and Geoinformation Processing* 4, 331–343.

Haralick, R. and Shapiro, L. 1992. *Computer and Robot Vision*. Addison Wesley: Reading, MA.

Heckbert, P. and Garland, M. 1997. Survey of polygonal surface simplification algorithms. Research report, School of computer science, Carnigie Mellon University, Pittsburgh, PA.

Hirschmüller, H. 2008. Stereo processing by semiglobal matching and mutual information. *IEEE Transactions on Pattern Analysis and Machine Intelligence* 30(2), 328–341.

Hollaus, M., Mandlburger, G., Pfeifer, N., and Mücke, W. 2010. Land cover dependent derivation of digital surface models from airborne laser scanning data. *IAPRS* 38, 3A, Paris, France.

Hopkinson, C., Lim, K., Chasmer, L., Treitz, P., Creed, I., and Gynan, C. 2004. Wetland grass to plantation forest—Estimating vegetation height from the standard deviation of LiDAR frequency distributions. *IAPRS* 36, 8/W2, Freiburg, Germany.

Hu, X. and Yuan, Y. 2016. Deep-learning-based classification for DTM extraction from ALS point cloud. *Remote Sensing* 8(9), 730.

Hyyppä, J. et al. 2001. HIGH-SCAN: The first European-wide attempt to derive single-tree information from laser scanner data. *Photogrammetric Journal of Finland* 17, 58–68.

Isenburg, M., Liu, Y., Shewchuk, J., and Snoeyink, J. 2006. Streaming computation of delaunay triangulations. *ACM Transactions on Graphics* 25(3), 1049–1056.

Jacobsen, K. and Lohmann, P. 2003. Segmented filtering of laser scanner DSMs. *IAPRS*, 34, 3/W13, Dresden, Germany.

Jahromi, A. B., Zoej, M. J. V., Mohammadzadeh, A., and Sadeghian, S. 2011. A Novel filtering algorithm for bare-earth extraction from airborne laser scanning data using an artificial neural network. *IEEE Journal of Selected Topics in Applied Earth Observations and Remote Sensing* 4(4), 836–843.

Journel, A. and Huijbregts, C. 1978. *Mining Geostatistics*. Academic Press: London, UK.

Kass, M., Witkin, A., and Terzopoulus, D. 1988. Snakes: Active contour models. *International Journal of Computer Vision* 1, 321–331.

Khosravipour, A., Skidmore, A. K., and Isenburg, M. 2016. Generating spike-free digital surface models using LiDAR raw point clouds: A new approach for forestry applications. *International Journal of Applied Earth Observation and Geoinformation* 52, 104–114.

Kilian, J., Haala, N., and Englich, M. 1996. Capture and evaluation of airborne laser scanning data. *IAPRS* 31, 3, Vienna, Austria.

Kobler, A., Pfeifer, N., Ogrinc, P., Todorovski, L., Ostir, K., and Dzeroski, S. 2007. Repetitive interpolation: A robust algorithm for DTM generation from aerial laser scanner data in forested terrain. *Remote Sensing of Environment* 108, 9–23.

Koch, K. 1999. *Parameter Estimation and Hypothesis Testing in Linear Models,* 2nd ed. Springer: Berlin, Germany.

Korzeniowska, K., Pfeifer, N., Mandlburger, G., and Lugmayr, A. 2014. Experimental evaluation of ALS point cloud ground extraction tools over different terrain slope and land-cover types. *International Journal of Remote Sensing* 35(13), 4673–4697.

Köstli, A. and Sigle, M. 1986. The random access data structure of the DTM program SCOP. *International Archives of Photogrammetry and Remote Sensing* 26(4), 42–45.

Kraus, K. 2000. *Photogrammetrie, Band 3, Topographische Informationssysteme*. Dümmler: Köln, Germany.

Kraus, K. 2007. *Photogrammetry,* 2nd ed. Walter de Gruyter: Berlin, Germany.

Kraus, K. and Mikhail, E. 1972. Linear least squares interpolation. *Photogrammetric Engineering* 38, 1016–1029.

Kraus, K. and Pfeifer, N. 1997. A new method for surface reconstruction from laser scanner data. *IAPRS* 32, 3/2W3, Haifa, Israel.

Kraus, K. and Pfeifer, N. 1998. Derivation of digital terrain models in wooded areas. *ISPRS Journal of Photogrammetry and Remote Sensing* 53(4), 193–203.

Kraus, K., Karel, W., Briese, C., and Mandlburger, G. 2006. Local accuracy measures for digital terrain models. *The Photogrammetric Record* 21(116), 342–354.

Li, Y., Yong, B., Wu, H., An, R., and Xu, H. 2014. An improved top-hat filter with sloped brim for extracting ground points from airborne LiDAR point clouds. *Remote Sensing* 6(12), 12885–12908.

Lin, Y.-C. and Mills, J. 2010. Factors influencing pulse width of small footprint, full waveform airborne laser scanning data. *Photogrammetric Engineering and Remote Sensing* 76(1), 49–58.

Krishnan, S., Crosby, C., Nandigam, V., Phan, M., Cowart, C., Baru, C., and Arrowsmith, R. 2011. OpenTopography: A services oriented architecture for community access to LiDAR topography. *Proceedings of the 2nd International Conference on Computing for Geospatial Research & Applications*, Washington, DC, May 23–25, 2011.

Lindenberger, J. 1989. Test results of laser profiling for topographic terrain survey. *Photogrammetric Week'89*. Special Publication of the Institute of Photogrammetry, University of Stuttgart, Vol. 13, 25–39.

Lohmann, P., Koch, A., and Schaeffer, M. 2000. Approaches to the filtering of laser scanner data. *International Archives of Photogrammetry and Remote Sensing* 33(B3/1), 534–541.

Ma, R. 2005. DEM generation and building detection from LiDAR data. *Photogrammetric Engineering and Remote Sensing* 71(7), 847–854.

Mandlburger, G., Hauer, C., Höfle, B., Habersack, H., and Pfeifer, N. 2009. Optimisation of LiDAR derived terrain models for river flow modelling. *Hydrology and Earth System Sciences* 13, 1453–1466. doi:10.5194/hess-13-1453-2009.

Mandlburger, G., Hauer, C., Wieser, M., and Pfeifer, N. 2015. Topo-bathymetric LiDAR for monitoring river morphodynamics and instream habitats—A case study at the Pielach River. *Remote Sensing* 7(5), 6160–6195.

Mandlburger, G., Otepka, J., Briese, C., Mücke, W., Summer, G., Pfeifer, N., Baltrusch, S., Dorn, C., and Brockmann, H. 2016. Automatische Ableitung von Strukturlinien aus 3D-Punktwolken. *Publikationen der Deutschen Gesellschaft für Photogrammetrie, Fernerkundung und Geoinformation e. V.* Band 25, 131–142.

Maune, D. 2007. *Digital Elevation Model Technologies and Applications: The DEM Users Manual,* 2nd ed. American Society of Photogrammetry and Remote Sensing: Bethesda, ML.

McGlone, J., Mikhail, E., and Bethel, J. 2004. *Manual of Photogrammetry,* 5th ed. American Society of Photogrammetry and Remote Sensing: Bethesda, MD.

Meng, X., Currit, N., and Zhao, K. 2010. Ground filtering algorithms for airborne LiDAR data: A review of critical issues. *Remote Sensing* 2(3), 833–860.

Meng, X., Wang, L., Silván-Cárdenas, J. L., and Currit, N. 2009. A multi-directional ground filtering algorithm for airborne LIDAR. *ISPRS Journal of Photogrammetry and Remote Sensing* 64, 117–124.

Mongus, D., Lukac, N., and Zalik, B. 2014. Ground and building extraction from LiDAR data based on differential morphological profiles and locally fitted surfaces. *ISPRS Journal of Photogrammetry and Remote Sensing* 93, 145–156.

Mongus, D. and Zalik, B. 2012. Parameter-free ground filtering of LiDAR data for automatic DTM generation. *ISPRS Journal of Photogrammetry and Remote Sensing* 67, 1–12.

Montealegre, A., Lamelas, M., and de la Riva, J. 2015. A comparison of open-source LiDAR filtering algorithms in a Mediterranean forest environment. *IEEE Journal of Selected Topics in Applied Earth Observations and Remote Sensing* 8(8), 4072–4085.

Mücke, W., Briese, C., and Hollaus, M. 2010. Terrain echo probability assignment based on full-waveform airborne laser scanning data. *IAPRS 38, 7A*, Institute of Photogrammetry and Remote Sensing, Vienna University of Technology, Vienna, Austria, pp. 157–162.

Naesset, E. 2016. Vertical height errors in digital terrain models derived from airborne laser scanner data in a boreal-alpine ecotone in Norway. *Remote Sensing* 7, 4702–4725.

Nandigam, V., Baru, C., and Crosby, C. 2010. Database design for high-resolution LiDAR topography data. *Scientific and Statistical Database Management Lecture Notes in Computer Science* 6187, 151–159.

Nardinocci, C., Forlani, G., and Zingaretti, P. 2003. Classification and filtering of laser data. *IAPRS* 34, 3/W13, Dresden, Germany.

Niemeyer, J., Rottensteiner, F., and Sörgel, U. 2012. Conditional random fields for LiDAR point cloud classification in complex urban areas. *ISPRS Annals* I(3), 263–268.

Otepka, J., Ghuffar, S., Waldhauser, C., Hochreiter, R., and Pfeifer, N. 2013. Georeferenced point clouds: A survey of features and point cloud management. *IPSRS International Journal of Geo-Information* 2(4), 1038–1065.

Peters, R. and Ledoux, H. 2016. Robust approximation of the medial axis transform of LiDAR point clouds as a tool for visualisation. *Computers and Geosciences* 90(A), 123–133.

Pfeifer, N. 2005. A subdivision algorithm for smooth 3D terrain models. *ISPRS Journal of Photogrammetry and Remote Sensing* 59(3), 115–127.

Pfeifer, N., Gorte, G., and Oude Elberink, S. 2004. Influences of vegetation on laser altimetry—analysis and correction approaches. *IAPRS* 36, 8/W2, Freiburg, Germany.

Pfeifer, N., Stadler, P., and Briese, C. 2001. Derivation of digital terrain models in the SCOP++ environment. *OEEPE Workshop on Airborne Laserscanning and Interferometric SAR for Detailed Digital Elevation Models*, Stockholm, Sweden.

Pingel, T., Clarke, K., and McBride, W. 2013. An improved simple morphological filter for the terrain classification of airborne LiDAR data. *ISPRS Journal of Photogrammetry and Remote Sensing* 77, 21–30.

Rapidlasso, 2016. https://rapidlasso.com/lastools/. Accessed December 2016.

Ressl, C., Brockmann, H., Mandlburger, G., and Pfeifer, N. 2016. Dense image matching vs. airborne laser scanning–comparison of two methods for deriving terrain models. *Journal for Photogrammetry, Remote Sensing and Geoinformation Science* 2016(2), 57–73.

Reutebuch, S., McGaughey, R., Andersen, H.-E., and Carson, W. 2003. Accuracy of a high-resolution LiDAR terrain model under a conifer forest canopy. *Canadian Journal of Remote Sensing* 29(5), 527–535.

Schenk, T. and Cshato, B. 2007. Fusing imagery and 3D point clouds for reconstructing visible surfaces of urban scenes. *Proceedings of Urban Remote Sensing*, Paris, France, April 11–13, 2007.

Schiewe, J. 2001. Ein regionen-basiertes Verfahren zur Extraktion der Geländeoberfläche aus Digitalen Oberflächen-Modellen. *Photogrammetrie, Fernerkundung, Geoinformation* 2, 81–90.

Seo, S. and O'Hara, C. G. 2008. Parametric investigation of the performance of LiDAR filters using different surface contexts. *Photogrammetric Engineering and Remote Sensing* 74(3), 343–362.

Shao, Y.-C. and Chen, L.-C. 2008. Automated searching of ground points from airborne LiDAR data using a climbing and sliding method. *Photogrammetric Engineering and Remote Sensing* 74(5), 625–635.

Sithole, G. 2001. Filtering of laser altimetry data using a slope adaptive filter. *IAPRS* 34, 3/W4, Annapolis, MD.

Sithole, G. 2005. Segmentation and classification of airborne laser scanner data, Dissertation, TU Delft, ISBN 90 6132 292 8, Publications on Geodesy of the Netherlands Commission of Geodesy, Vol. 59.

Sithole, G. and Vosselman, G. 2004. Experimental comparison of filter algorithms for bare—Earth extraction from airborne laser scanning point clouds. *ISPRS Journal of Photogrammetry and Remote Sensing* 59, 85–101.

Shan, J. and Aparajithan, S. 2005. Urban DEM generation from raw LiDAR data. *Photogrammetric Engineering and Remote Sensing* 71, 217–226.

Skaloud, J. 2007. Reliability of direct georeferencing—beyond the achilles' heel of modern airborne mapping. *Photogrammetric Week'07*. Wichmann, Heidelberg, Germany.

Sohn, G. and Dowman, I. 2002. Terrain surface reconstruction by the use of tetrahedron model with the MDL criterion. IAPRS 34, 3A, 336–344, Graz, Austria.

Tinkham, W. T., Huang, H., Smith, A. M. S., Shrestha, R., Falkowski, M. J., Hudak, A. T., Link, T. E., Glenn, N. F., and Marks, D. G. 2011. A comparison of two open source LiDAR surface classification algorithms. *Remote Sensing* 3, 638–649.

Tovari, D. and Pfeifer, N. 2005. Segmentation based robust interpolation—a new approach to laser data filtering. *IAPRS* 36, 3/W19, Enschede, the Netherlands.

von Hansen, N. and Vögtle, T. 1999. Extraktion der Geländeoberfläche aus flugzeuggetragenen Laserscanner-Aufnahmen. *Photogrammetrie, Fernerkundung, Geoinformation* 4, 229–236.

Vosselman, G. 2000. Slope based filtering of laser altimetry data. *IAPRS* 33, B3/2, Amsterdam, the Netherlands.

Vosselman, G., Oude Elberink, S., Post, M., Stoter, J., and Xiong, B. 2015. From nationwide point clouds to nationwide 3D landscape models. *Photogrammetric Week'15*, Wichmann, Stuttgart, Germany, pp. 247–256.

Wang, C., Menenti, M., Stoll, M.-P., Feola, A., Belluco, E., and Marani, M. 2009. Separation of ground and low vegetation signatures in LiDAR measurements of salt-marsh environments. *IEEE Transactions on Geoscience and Remote Sensing* 47(7), 2014–2023.

Wehr, A., Petrescu, E., Duzelovic, H., and Punz, C. 2009. Automatic break line detection out of high resolution airborne laser scanner data. *Conference on Optical 3-D Measurement Techniques IX*, Vienna, Austria, Vol. 2, pp. 72–78.

Wever, C and Lindenberger, J. 1999. Experiences of 10 years laser scanning. *Photogrammetric Week 1999*, Stuttgart, Germany.

Whitman, D., Zhang, K., Leatherman, S. P., and Robertson, W. 2003. Airborne laser topographic mapping: Application to hurricane storm surge hazards. In *Earth Sciences and the City*, Heiken, G., Fakundiny, R., Sutter, J. (Eds.), American Geophysical Union, Washington, DC, pp. 363–376.

Wu, J, Liu, L., and Liu, R. 2011. Automatic DEM generation from aerial LiDAR data using multiscale support vector machines. *Proceedings of SPIE 8006, MIPPR 2011: Remote Sensing Image Processing, Geographic Information Systems, and Other Applications*, Guilin, China, November 11, 2011.

Yan, M., Blaschke, T., Liu, Y., and Wu, L. 2012. An object-based analysis filtering algorithm for airborne laser scanning. *International Journal of Remote Sensing* 33(9), 7099–7116.

Zhang, J. and Lin, X. 2013. Filtering airborne LiDAR data by embedding smoothness-constrained segmentation in progressive TIN densification. *ISPRS Journal of Photogrammetry and Remote Sensing* 81, 44–59.

Zhang, K., Chen, S., Whitman, D., Shyu, M., Yan, J., and Zhang, C. 2003. A progressive morphological filter for removing nonground measurements from airborne LiDAR data. *IEEE Transactions on Geoscience and Remote Sensing* 41(4), 872–882.

Zhang, W., Qi, J., Wan, P., Wang, H., Xie, D., Wang, X., and Yan, G. 2016. An easy-to-use airborne LiDAR data filtering method based on cloth simulation. *Remote Sensing* 8(6), 501.

Zhang, K. and Whitman, D. 2005. Comparison of three algorithms for filtering airborne LiDAR data. *Photogrammetric Engineering and Remote Sensing* 71(3), 313–324.

12 Forest Inventory Using Laser Scanning

*Juha Hyyppä, Xiaowei Yu, Harri Kaartinen, Antero Kukko,
Anttoni Jaakkola, Xinlian Liang, Yunsheng Wang, Markus
Holopainen, Mikko Vastaranta, and Hannu Hyyppä*

CONTENTS

12.1 INTRODUCTION

12.1.1 History of Profiling Measurements over Forests

The concept of producing forest stand profiles (i.e., height profiles) with high-precision instruments was demonstrated as early as 1939 (Hugershoff, 1939), and the concept was implemented with laser profilers around 1980 (Solodukhin et al., 1977; Nelson et al., 1984; Aldred and Bonnor, 1985; Schreier et al., 1985; Maclean and Krabill, 1986; Currie et al., 1989). In North America, the development of laser profilers for forestry stemmed from bathymetric Light Detection and Ranging (LiDAR) studies. Since then, studies have used laser measurements, for example, for estimating the tree height, stem volume, and biomass (Nelson et al., 1984; Aldred and Bonnor, 1985; Schreier et al., 1985; Maclean and Krabill, 1986; Nelson et al., 1988; Currie et al., 1989; Nilsson, 1990). Nelson et al. (1984) proposed the use of the laser-derived profiles for the retrieval of stand characteristics. They also showed that the elements of the stand profile are linearly related to crown closure and may be used to assess tree height. Schreier (1985) concluded that the near-infrared laser can produce terrain and vegetation canopy profiles, and examined laser intensity for vegetation discrimination. Nelson et al. (1988) demonstrated that the tree height, stem volume, and biomass can be predicted with reasonable accuracy by using reference plots and averaging. They proposed the use of both density and height related LiDAR metrics for stand attribute estimation that later on lead to the so-called area-based method in airborne laser scanning (ALS) of forests (Næsset et al. 1997b). Aldred and Bonnor (1985) presented laser-derived tree height estimates within 4.1 m of the field-measured stand heights at a 95% level of confidence. Currie et al. (1989) estimated the height of the flat-topped crowns with an accuracy of about 1 m. Nilsson (1990) proposed that the data of a laser mounted on a boomtruck correlates with volume changes, such as thinnings. In Hyyppä (1993) and Hyyppä and Hallikainen (1993), a radar-based profiling (i.e., waveform) system and feasibility for forest measurements (tree height, basal area, and volume) were depicted. To obtain more information on the history of prior laser ranging measurements over forest, the reader is referred to, for example, Nelson et al. (1997), Nilsson (1996), Lim et al. (2003), Holmgren (2003), and Nelson (2013).

12.1.2 Background of Airborne Laser Scanning in Forestry

An airborne LiDAR system for estimating tree heights and stand volume was investigated by Mats Nilsson (1996). A helicopter-borne laser operated in a scanning mode and different laser footprints, and sampling densities were tested at three independent registration times. As a conclusion, the mean tree height was underestimated by 2.1–3.7 m, and optimal laser footprint size was found to change across data acquisition times.

Global Navigation Satellite System (GNSS) and inertial measurement unit (IMU) development lead to direct georefencing allowing the positioning of scanning laser beams. That was especially supported by military development.

After that small-footprint ALS for forests included the determination of terrain elevations (Kraus and Pfeifer, 1998; Vosselman, 2000), standwise mean height and volume estimation (Næsset, 1997a, 1997b), individual-tree-based height determination and volume estimation (Brandtberg, 1999; Hyyppä and Inkinen, 1999; Ziegler et al., 2000, Hyyppä et al. 2001a), tree-species classification (Brandtberg et al., 2003; Holmgren and Persson, 2004), and measurement of forest growth and detection of harvested trees (Hyyppä et al., 2003b; Yu et al., 2003, 2004a). Early laser-scanning experiences in Canadian, Finnish, Norwegian, and Swedish forestry can also be found in Wulder (2003), Hyyppä et al. (2003a), Næsset (2003), and Nilsson et al. (2003). Early Scandinavian summary of laser scanning in forestry can be read in Næsset et al. (2004) and summary of the methods used for forest inventory can be also read in Hyyppä et al. (2008), and in Koch (2010). Focus on full waveform topographic LiDAR is depicted in Mallet and Bretar (2009).

FIGURE 12.1   Laser point cloud from sparse forest. (Courtesy of Hannu Hyyppä. With Permission.)

A whole book concentrating on forestry applications of ALS has been edited by Maltamo et al. (2014). Since 2002, there have been almost annual meetings of LiDAR-based forest mensuration society (Silvilaser series, annual until 2013, and then in two years interval).

In addition to small-footprint (0.2–2 m) ALS, several large footprint systems have been developed. Today, mainly the small-footprint systems are commercially attractive, we will focus our discussion on small-footprint ALS. (Figure 12.1).

The book chapter is divided into eight chapters intending to give a very broad understanding of laser-based forest inventory. A summary of user requirements of forestry is given in Chapter 2. Chapter 3 is a brief introduction to currently applied laser scanning platforms and system, including airborne, mobile, terrestrial, and Unmanned Aircraft Vehicle (UAV) platforms, and their implication to forestry. Chapter 4 introduces the airborne LiDAR scattering process in forests. Chapter 5 depicts canopy height retrieval and corresponding accuracy obtained in height measurements—all ALS-based forest inventories are based on accurate measurements of canopy height. Chapter 6 focuses on retrieval of tree or stand parameters from laser scanning data including airborne, mobile, and terrestrial laser scanning (TLS). Chapter 6 also includes the use of multispectral ALS for tree species classification. Chapter 7 gives examples of change detection possibilities by using multitemporal aerial laser surveys. Chapter 8 gives concluding discussion and future possibilities. The book chapter is adapted using previous CRC Press, Taylor & Francis Group articles Hyyppä et al. (2009, 2015).

12.2   USERS' REQUIREMENTS FROM FORESTS

About 100 years ago, forest inventory was considered to be the determination of the volume of logs, trees, and stands, and a calculation of the increment and yield. More recently, forest inventory has expanded to cover assessment of various issues including wildlife, recreation, watershed management, and other aspects of multiple-use forestry. However, a major emphasis of forest inventory still lies in obtaining information on the volume and growth of trees, forest plots, stands, and large areas. In the following we will depict in more detail the major attributes of forests relevant from the point of view of laser scanning.

12.2.1 INDIVIDUAL TREE ATTRIBUTES

Forest can be characterized by its attributes (parameters, features, and variables). Basic attributes for a tree are, for example, height, diameter at breast height (DBH), upper diameter (diameter of the tree at the height of, for example, 6 m), height of crown base (height from the ground to the lowest green branch or to the lowest complete living whorl of branches), species, age, location, basal area (cross-sectional area defined by the DBH), volume, biomass, growth, and leaf area index. Some of them can be directly measured or calculated from these direct measurements, whereas others need to be estimated (predicted) through statistical or physical modeling.

Traditionally, individual tree attributes, such as, height, diameters at different height along the stem, crown diameter, are measured in the field, Figures 12.2 and 12.3, and Table 12.1. The conventional strategy for collecting such data is a plot-wise field inventory, especially by measuring the diameters because diameter is convenient to measure and one of the directly measurable dimensions from which tree cross-sectional area, surface area, and volume can be computed. Various instruments and methods have been developed for measuring the tree dimension in the field (Husch et al., 1982; Päivinen et al., 1992; Gill et al., 2000; Korhonen et al., 2006), such as caliper, diameter tape, and optical devices for diameter measurements; level rod, pole, hypsometers for tree height measurements; increment borer for diameter growth measurements. The method used in obtaining the measurements is largely determined by the accuracy required. Sometimes it is necessary to fell the tree to obtain more accurate measurements, such as the only way to actually measure the stem volume is through destructive sampling of a tree. As a result, direct and indirect methods have been developed for the estimation of such variables. As examples of the volume estimation methods include graphical method in which cross-section area at different heights along trunk have been plotted over height on paper and the area under the curve is equivalent to the volume, and the use of volume equations which estimate volume through a relationship with measurable parameters such as height and/or DBH. In practice, tree volume is estimated from DBH and possibly together with height and upper diameter for each tree species. The models for volume, especially based on diameter information, for individual trees exist in literature in each country. For example, in Finland, there are volume models v for main tree species based on different inputs (diameter d, height h, diameter at the height of 6 m, d6)

- $v = f(d)$
- $v = f(d, h)$
- $v = f(d, h, d6)$

(a)   (b)

FIGURE 12.2 Collecting individual tree information using hypsometer for height (a), and using caliper (b) for breast height diameter. (Courtesy of Hannu Hyyppä. With Permission.)

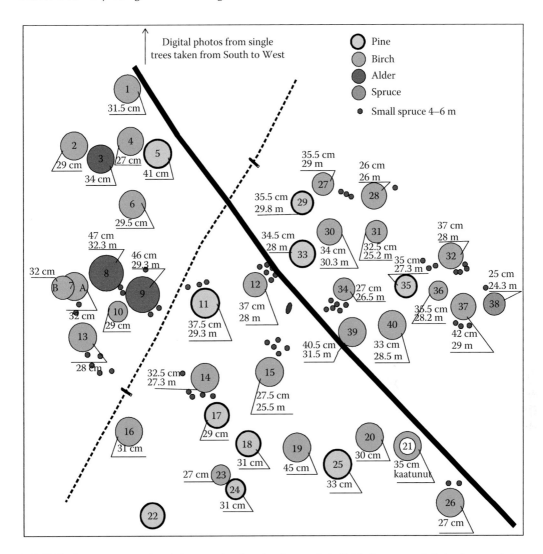

FIGURE 12.3 An example of the sample plot showing each individual trees and their characteristics. (Courtesy of Hannu Hyyppä. With Permission.)

Better measurements of a tree are required when the interest is in growth over time rather than size at a particular time. Estimates of height increment are usually satisfactory if height is measured by height sticks, but may be unsatisfactory if measured by hypsometer (Husch et al., 1982). Estimates of diameter increment are much more reliable, particularly, if the point of measurement on stems is marked permanently. Past growth of diameter can be obtained from increment borings or cross-section cuts (Poage and Tappeiner, 2002). Past height increment may be determined by stem analysis (Uzoh and Olive, 2006). For species for which the internodal lengths on the stem indicate a year's growth, past height growth may be determined by measuring internodal lengths (Husch et al., 1982). Determination of growth is most commonly obtained by repeated measurements at the beginning and at the end of a specified period and by taking the difference (Husch et al., 1982; Uzoh and Olive, 2006). In principle, only past growth can be measured, but usually, future growth is of interest and it has to be predicted by using growth model (Hynynen, 1995; Hökkä and Groot, 1999; Hall and Bailey, 2001; Matala, 2005).

The accuracy of field measurements is usually high at a point level. Päivinen et al. (1992) reported that DBH could be measured with a standard deviation ranging from 2.3 to 4.6 mm and height with

TABLE 12.1

Attributes of Trees to Be Measured

Attribute	Unit	Typically Expected Accuracy for Measurement[a]
Height	M	0.5–2 m
Diameter at breast height (DBH)	Mm	5–10 mm
Upper diameter (e.g., at height 6 m)	Mm	5–10 mm
Height of crown base	M	0.2–0.4 m
Species		Species-specific volume estimates
Age	Years	5 years
Location	m	0.5–2 m
Basal area	m^2	see diameter accuracy
Volume	m^3	10%–20%
Biomass	kg/m^3	10%–20%
Growth	For example, mm (increment borer)	1 mm

Source: Hyyppä, J. et al., *Remote Sensing of Forests from LiDAR and Radar, in Remote Sensing Handbook*, Boca Raton, FL, CRC Press, pp. 397–427, 2015.

[a] Depends strongly on the use of the data.

a standard deviation of 67 cm. The 5 years height increment of Scot pine and Norway spruce was measured with a standard deviation of 27 cm and 20.5% in the estimate of volume increment for a 65-year-old Scot pine stand.

12.2.2 STAND ATTRIBUTES

The common attributes used to describe stand and plots of even-aged forests of a single species include age, number of trees per hectare, mean diameter, basal area per hectare (sum of the cross-sectional areas per hectare), mean diameter, mean height (arithmetic mean height), dominant height (referring e.g., to the mean height of the 100 trees per hectare with the largest DBH, dominant trees), Lorey's mean (each tree is weighted by its cross-sectional area) or weighted height, volume per hectare, mean form factor (co-efficient to relate volume of trees using a product of basal area and Lorey's mean height), current annual increment per hectare, mean annual increment per hectare, and growth.

A more accurate way to estimate the volume of the plot is to sum up the volumes of individual trees using individual tree models for each tree species and strata separately. In forest inventory, a practical way to measure the basal area per hectare relative to the plot is to use a relascope, which is an instrument used in the forest inventory to discriminate between trees on the basis of whether or not the tree subtends an angle equal to or greater than that of the relascope when viewed from the sampling point (Philip, 1983). Volumes per hectare of even-aged forests of a single species can be predicted using stand volume table. The commonest stand volume table is derived from a simple linear regression of volume per hectare on both the basal area per hectare and height representative of the forest (mean height or dominant height). To get an estimate for a stand, several plots need to be measured.

12.2.3 OPERATIVE COMPARTMENTWISE INVENTORY

There exist several types of operational forest inventory methods ranging from national/continent-wise forest inventory to compartmentwise forest inventory. In this presentation we concentrate on compartmentwise inventory due to its high commercial impact. Compartmentwise forest inventory is a widely used method in Finland, both in public and privately owned forests. The basic unit of forest inventories is a forest stand, which is used as the management-planning unit. The size of a forest stand is normally 0.5–3 ha. The forest stand is defined as a homogenous area according to relevant stand characteristics, for example, site fertility, composition of tree species, and stand age. Forest inventory data are mostly collected with the aid of field surveys, which are both expensive and time-consuming. The compartments are typically measured separately by analyzing sample plots placed on the stands. From each plot, tree and stand attributes are measured. Finally, the standwise attributes describing the density and tree dimensions are derived from these plot measurements. The method is also sensitive to subjective measurement errors. Remote sensing is normally used for nothing more than delineation of compartment boundaries. The total costs of compartmentwise inventory in Finland were 17.9 €/ha in 2000, of which 7.9 €/ha, that is, 45% of the costs, consisted of field measurements (Uuttera et al., 2002; Holopainen and Talvitie, 2006). For the total stem volume per hectare, basal area per hectare, and mean height, the required accuracy is roughly 15%.

In practice, the accuracy ranges from 10% to 30% depending on the heterogeneity of the forests (mixture of tree species, several strata, dense undervegetation, and varying elevation). The stem volume for each tree species and each strata is obtained with significantly lower accuracy.

The forest inventory has been performed to most of the stands several times. Thus, there exist information from previous inventories, and the minimum data requirement for standwise forest inventory is presently the total volume, basal area, and mean height for each tree species from the dominant tree storey. Other needed attributes can then be derived from these data. In near future, more accurate quantity and quality information of individual trees in the stand can be used as a base for felling and in transportation round timber from forest straight to the right manufacturing factories according to the demand of raw material. One important benefit from improved accuracy of forest data is the ability to better plan the forest operations as well as the supply chain. As these activities constitute a significant part of the cost for raw material for the industry, it is of vital importance to control these cost effectively.

12.3 LASER SCANNING PLATFORM IN BRIEF

The focus in this chapter is on explaining various platforms used today for small-footprint laser scanning covering, Table 12.2:

- ALS
- TLS
- Mobile Laser Scanning (MLS)
- UAV Laser Scanning
- Personal Laser Scanning (PLS)

UAV and PLS are considered to be included inside MLS, although the characteristics of them are different to MLS.

12.3.1 Airborne Laser Scanning, Airborne LiDAR

ALS is a method based on LiDAR measurements from an aircraft, in which the precise position and orientation of the sensor is determined, and therefore the point cloud (x, y, z) of the reflecting objects can be determined. Today, ALS is becoming a standard technique in the mapping and monitoring of forest resources. By using ALS-based inventory, 5%–20% error level in main forest stand attributes at stand level has been obtained. ALS is a promising technique also for efficient and accurate above-ground biomass retrieval because of its capability of direct measurement of vegetation 3D structure. Above-ground biomass correlates strongly with canopy height. Two main processing principles

TABLE 12.2

Small-Footprint Laser Scanning Platforms

Airborne Laser Scanning

Point density	Point density 0.5–40 pts/m²
Elevation accuracy	5–30 cm
Planimetric accuracy	20–80 cm
Operating range	Few hundred m to several km
Feasible	Cost-effective for areas larger than 50 km²
Characteristics	Homogenous point clouds
System providers	Optech, Leica, and Riegl

Mobile Laser Scanning

Point density	Point density in the range of 100 to several thousand pts/m²
Accuracy	Point accuracy of few centimeters (egg) when collected with good GNSS coverage
Operating range	Applicable range of few tens of m
Feasible	Collecting large datasets for road environment
Characteristics	Relatively high variation (density) in the range data
System providers	StreetMapper, Optech, Riegl, Trimble, Nokia Here, Topcon, IGI, and MDL

Terrestrial Laser Scanning

Point density	Point density in the range of 10,000 pts/m² at the 10 m
Accuracy	Distance accuracy of few mm to 1–2 cm
Operating range	Applicable range of few tens of meters
Feasible	Operational scanning range from 1 to several hundreds meters
Characteristics	Feasible for small areas less than few tens of meter distance
System providers	Processing time challenging: image processing techniques applied; small variation in data, for example, distance variation low, thus surface normal can be calculated
	Faro, Leica, Riegl, Topcon, Trimble, Zoller, and Fröhlich

Source: Hyyppä, J. et al., *Remote Sensing of Forests from LiDAR and Radar, in Remote Sensing Handbook*, Boca Raton, FL, CRC Press, pp. 397–427, 2015.

of ALS data for forestry exist: area-based estimation (area-based approaches [ABA]) relying on the point cloud metrics derived from ALS data, and individual tree detection (ITD) (Hyyppä and Inkinen, 1999). The differences of the methods are becoming smaller and smaller. Originally, ABA did not use neighborhood information in the data, but Niemi and Vauhkonen (2016) used canopy-based textures in the area-based prediction.

Today, there are several technological solutions offering ALS. Single photon systems (Degnan, 2016) require only one detected photon compared with hundreds or even thousands photons needed in conventional laser pulse time of flight or waveform LiDARs. Multispectral ALS provides point clouds with multiple intensities, allowing imaging also at night time, Figure 12.4.

12.3.2 Mobile Laser Scanning

MLS is based on LiDAR measurements from a moving platform, in which the precise position and orientation of the sensor is determined using a navigation system, exactly similar to ALS, and therefore, the position of the reflecting objects can be determined either from pulse travel time or phase information. An MLS system consists of one or several laser scanners. Navigation system consists of various sensors for positioning and determining the rotation angles of the system, whereas GNSS and IMU being the most important parts of the system. Additional mapping sensors, such as cameras, thermal imagers, and spectrometers, can be added into the MLS system. The application of MLS in forestry is being recently studied (Lin et al., 2010; Holopainen et al., 2013; Liang et al., 2014c). In the near future, MLS can be seen as a practical means to produce tree maps and field

FIGURE 12.4 RGB image demonstrating the quality of the multispectral ALS. (Courtesy of Leena Matikainen, FGI, Finland. With Permission.)

reference in urban forest environments, and in boreal forests and managed forests. To be able to position itself under the forest canopy, MLS uses robotic Simultaneous Localization and Mapping techniques together with IMU. MLS scientific tests in forestry are currently rapidly developing. Current work has concentrated on reporting MLS positioning accuracies inside forest canopies and accuracy for DBH determination (Figure 12.5).

FIGURE 12.5 Research in the field of MLS requires testing of various sensors together. (Courtesy of Harri Kaartinen. With Permission.)

12.3.3 Terrestrial Laser Scanning

TLS, known also as ground-based LiDAR, has been shown during the last 10 years a promising technique for forest field inventories at tree and plot level. The major advantage of using TLS in forest field inventories lies in its capacity to document the forest in detail with small random variation in the point cloud data. The first commercial TLS system was built by Cyra Technologies (acquired by Leica in 2001) in 1998, and the first papers related to plot-level tree attribute estimation were reported in early 2000s. Currently, TLS has shown to be feasible for collecting basic tree attributes at tree and plot-level, such as DBH and tree position (Maas et al., 2008; Brolly and Kiraly, 2009; Murphy et al., 2010; Lovell et al., 2011; Liang et al., 2012). By reconstructing tree stem, it is possible to derive high-quality stem volume and biomass estimates comparable in accuracy with the best national allometric models (Liang et al., 2014a). TLS data also permit time series analyses by documenting entire plots consecutively over time (Liang et al., 2012b). TLS is starting to be operational in plot-level forest inventories as more and more appropriate software becomes available, best practices become known and general knowledge of these findings is more widely spread. Currently, a EuroSDR (European Spatial Data Research) TLS comparison project is under finishing. It will give accuracy of about 20 different processing solutions for TLS data. TLS is especially suitable for permanent field plots where changes need to be mapped. TLS also allows the development of local tree models based on TLS data. Local tree models can be used to improve stand attribute predictions (Figure 12.6).

12.3.4 UAV-Based Laser Scanning

Mini-UAV-based ALS data collection has been possible since Jaakkola et al. (2010). Jaakkola built the first mini-UAV laser scanner on an radio-controlled (RC)-controlled helicopter already in 2009 and it was demonstrated for European Mapping companies in Terrasolid Users Event in Finnish Lapland in February 2010. The scientific paper of the development can be read in Jaakkola et al. (2010). Mini-UAVs (<20 kg) have been before that used for mapping purposes by using, for example, aerial images. Zhao et al. (2006) depicted a remote controlled helicopter supplied with navigation

FIGURE 12.6   An example of output of the single-scan TLS. (Courtesy of Harri Kaartinen. With Permission.)

sensors, namely GPS, and a laser range finder. In Jaakkola et al. (2010), the first mini-UAV (FGI [Finnish Geodetic Institute] Sensei), including the laser scanner, intensity recording, spectrometer, thermal camera, and conventional digital camera, was depicted. Laser scanners in the Sensei included an Ibeo Lux and a Sick LMS151 profile lasers. Jaakkola et al. (2010) evaluated the accuracy and feasibility of a mini-UAV-based laser scanning system for tree measurements. The standard deviation of individual tree heights was approximately 30 cm. Jaakkola et al. (2010) also demonstrated a method to derive the biomass change of a coniferous tree from a multitemporal laser point cloud with an R^2 value of 0.92. Jaakkola's work was later on commercialized by start-up company Sharper Shape and applied to corridor mapping (Matikainen et al., 2016). Wallace et al. (2012) continued the work of Jaakkola et al. (2010) in Tasmania with a similar system and concluded that the standard deviation of tree height was shown to reduce from 0.26 m, when using data with a density of 8 points per m², to 0.15 m when the higher density data was used. Tree locations up to 0.5 m and crown width to 0.6 m accuracy has been reported (Wallace et al., 2012). In Wallace et al. (2014), use of UAV laser scanner data resulted in 98% of trees being repeatedly and correctly delineated from the point cloud, and trees measured with a location accuracy of 0.48 m and a crown area accuracy of 3.3 m².

UAVs having Laser Scanner(s) are expected to be capable to derive in addition to the tree height and crown dimensions also diameter breast height and stem curve information. They can either fly above the canopy or inside the canopy. The system flying inside the forest should have field of view (FOV) close to 360 and those flying above the canopy should preferably have a FOV close to 30°–80°. 360 FOV allows scanning of the canopy from top to bottom while flying inside the canopy. Example of such systems is Riegl VUX implemented, for example, in Riegl Ricopter. The problems related to the use of such systems related to the safety of data collection. Chisholm et al. (2013) fly along forest road with a UAV laser scanner and measured trees along the road. They could estimate (DBH) of 12 trees that were detected by the LiDAR. The method detected 73% of trees greater than 20 cm DBH within 3 m of the flight path. Smaller and more distant trees could not be detected reliably. The UAV-based DBH estimates of detected trees were measured with an RMS error of 25.1% (about 10 cm).

UAV laser scanning is today dramatically developing. Gordon Petrie made his review (2013) in Geoinformatics about UAV and laser scanning, and he summarized it by saying: "Until now, airborne laser scanners from the mainstream system suppliers have not been operated from lightweight UAVs on a commercial basis, mainly due to the size and weight of the scanner systems and the limited payload of these very lightweight aircraft. However, currently, considerable efforts are being made to develop laser scanners that are suitable for use on these aircraft. These are utilizing the technologies that are already being employed in mobile mapping systems, terrestrial (ground-based) laser scanning and autonomous (driverless) vehicles." Thus, the situation is now rapidly changing and MLS systems are also implemented in UAV. Typical systems implemented in UAV include, for example, Velodyne-16. It should be noticed that autonomous driving sensors are currently driving source for UAV sensor development. Mini-UAV-laser scanning is also feasible for small-area surveys, for example, collecting airborne data from areas of several hectares. In future, corridor-type applications done by UAV laser scanning can provide cost-effective solution (Figure 12.7).

12.3.5 Personal Laser Scanning

Laser scanning sensors are becoming smaller and more ubiquitous. PLS can be, for example, a backpack-type mobile laser scanner or even a smartphone-embedded laser scanner. Last five years have witnessed rapid progresses in sensor miniaturization. The newest systems weight approximately 2–5 kg. The main advantage of PLS lies in its easy movement in various terrain conditions and the fast data collection. By providing professional quality scanning and navigation systems, the collected data can document objects in detail and high precision. PLS has the potential to improve the mapping efficiency compared with conventional field measurement and compensate the limitations of other laser scanning (LS) techniques, for example, the TLS is not easy to transfer from site to site and the vehicle-based MLS has to be utilized in areas where the terrain conditions are

FIGURE 12.7 The quality of UAV laser scanning is increasing. Someday UAV laser scanning can be used to collect field plot reference. (Courtesy of Anttoni Jaakkola. With Permission.)

easy for the vehicle movement. These characters are very attractive for forest mapping and ecosystem services. In Liang et al. (2014b), a professional-quality, PLS for collecting tree attributes was demonstrated for the first time.

Small laser-based devices are also developed for smartphones. In 2013, we demonstrated the feasibility of using the Microsoft Kinect depth sensor for tree stem measurements, and reconstruction for the purpose of forest inventory. We tested about 100 reference stem diameter measurements made with tape, caliper, and Microsoft Kinect. Color (i.e., RGB) and range images acquired by a Kinect system were processed and used to extract tree diameter measurements for the reference tree stems. Kinect-derived tree diameters agreed with the calliper measurements to 2.50 cm root-mean-squared-error (RMSE) and 10% (RMSE %), and with tape measurements to 1.90 cm and 7.3%. The stem curve from the ground to the diameter breast height agreed with a bias of 0.7 cm and random error of 0.8 cm with respect to the reference trunk. As a highly portable and inexpensive system, Kinect provides an easy way to collect tree stem diameter and stem curve information vital to forest inventory.

12.4 AIRBORNE LASER BEAM INTERACTION WITH FOREST CANOPIES

The laser pulse hit on the forest canopy can be simple or complex, Figure 12.8. In the simplest case a laser pulse may be scattered directly from the top of a very dense vegetation canopy resulting in a single return. As forest canopy is not a solid surface and as there exist gaps in the canopy cover, the situation becomes more complex when a laser pulse hits forest canopy and passes through the top of the canopy and intercepts with different parts of the canopy such as trunk, branches, leaves before reaching the ground. This series of events may result in several returns being recorded for a single laser pulse, which is called multiple returns. In most cases first and last returns are recorded. First returns are mainly assumed to come from the top of the canopy and last returns are mainly coming from ground, which is important for extracting the terrain surface. Multiple pulses produce useful forest information of the forest structure.

Trunks, branches, and leaves in dense vegetation tend to cause multiple scattering or absorption of the emitted laser energy so that fewer backscattered returns are reflected directly from ground (Harding et al., 2001; Hofton et al., 2002). This effect increases when canopy closure, canopy depth, and structure complexity increase because laser pulse is greatly obscured by the canopy.

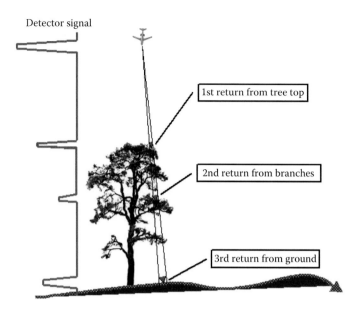

FIGURE 12.8 Interaction of laser pulse with forest canopy resulting multireturns. The amplitudes and echo widths are not to be considered as typical ones. Due to fading, absorption, and scattering, the amplitude and echo width can change dramatically between waveforms. (Courtesy of Xiaowei Yu. With Permission.)

In practice, the laser system specification and configurations can also play an important role in how laser pulse interacts with forest. For example, it has been found that small footprint laser tends to penetrate tree crown before reflecting a signal (Gaveau and Hill, 2003), ground returns were decreased as the scanning angle increased (TopoSys, 1996), penetration rate is affected by laser beam divergence (Aldred and Bonnor, 1985; Næsset, 2004), higher flight altitude alters the distribution of laser returns from the top and within the tree canopies (Næsset, 2004), and the distribution of laser returns through the canopy varies with the change of laser pulse repetition frequency (Chasmer et al., 2006). Goodwin et al. (2006) used three different platform altitudes (1000, 2000, and 3000 m), two scan angles at 1000 m (10° and 15° half maximum angle off nadir), and three footprint sizes (0.2, 0.4, and 0.6 m) in eucalyptus forests at three sites, which varied in vegetation structure and topography. They reported that higher platform altitudes record a lower proportion of first/last return combinations that will further reduce the number of points available for forest structural assessment and development of digital elevation models, and for discrete LiDAR data increasing platform altitudes will record a lower frequency of returns per crown resulting in larger underestimates of the individual tree crown area and volume. Furthermore, the sensitivity of the laser receiver, wavelength, laser power, and total backscattering energy from the tree tops are also the factors that may influence the ability of laser pulses to penetrate and distribution of laser returns from forest canopy (Baltsavias, 1999).

One of the most crucial factors for an exact range measurement is the echo detection algorithm used (Wagner et al., 2004; Wagner, 2005). As the length of the laser pulse is longer than the accuracy needed (a few meters versus a few centimeters), a specific timing in the return pulse needs to be defined. In a nonwaveform ranging system, analog detectors are used to derive discrete, time-stamped trigger pulses from the received signal in real time during the acquisition process. The timing event should not change when the level of signal varies, which is an important requirement in the design of analog detections as discussed by Palojärvi, 2003. On the contrary, in the case of commercial ALS systems detailed information concerning the analog detection method is normally lacking, even though different detection methods may yield quite different range estimates. For full-waveform digitizing ALS systems several algorithms can be used in the postprocessing stage (e.g., leading edge discriminator/threshold, center of gravity, maximum, zero crossing of the second derivative, and constant fraction). The most basic technique for pulse detection is to trigger a pulse whenever the rising edge of the signal exceeds a given threshold (leading edge discriminator).

12.5   EXTRACTION OF CANOPY HEIGHT

The basics for modern ALS-based forest measurements rely on the acquisition of the CHM (Canopy Height Model), DTM (Digital Terrain Model) corresponding to the ground surface, and DSM (Digital Surface Model) corresponding to treetops. The random errors in the DTM will result in errors of tree height and CHM.

The most appropriate technique to obtain a DSM relevant to treetops is to classify the highest reflections (i.e., by taking the highest point within a defined neighborhood) and interpolate missing points, for example, by Delaunay triangulation. Then, the CHM is obtained by subtracting the DTM from the corresponding DSM. CHM is also called a normalized DSM. The DSM is typically calculated by means of the first pulse echo and the DTM with the last pulse echo.

The processing of the point cloud data into canopy heights or normalized heights is an effective way of increasing the usability of the laser data. Laser canopy heights are simply obtained as the difference between elevation values of laser hits and estimated terrain elevation values at the corresponding location.

For the final ground point classification, there are several DTM filtering techniques developed. Mathematic morphology is one applied technology. By using operators, such as erosion, dilation, closing, and opening, can produce DTM and DSM. Vosselman (2000) applied a maximum admissible height difference function for a defined distance. Points within the maximum height difference function were included as ground points. Progressive densification strategy starts with step 2 (initial DTM point cloud selection), and then iteratively increase the amount of accepted terrain points. Axelsson (2000) developed a progressive triangulated irregular network (TIN) densification method that is implemented into Terrascan software. In Terrascan, laser point clouds are first classified to separate ground points from all other points. The program selects local low points on the ground and makes an initial triangulated model. New laser points are then added to the model iteratively and the actual ground surface is then described more and more precisely. Maximum building size, iteration angle, and distance parameters determine which points are accepted. Kraus and Pfeifer (1998) developed a DTM algorithm for which laser points between terrain points and nonterrain points were distinguished using an iterative prediction of the DTM and weights attached to each laser point, depending on the vertical distance between the expected DTM level and the corresponding laser point. The method officially goes to category surface-based filtering, in which the starting point is that all given points belong to the terrain class and then iteratively remove the points that do not fit to surface model. The method is implemented in SCOP++ (Kraus and Otepka, 2005). In addition, the filtering can be on the basis of segments, that is, segment-based filtering. Either object or feature based segmentation can be done and then filtering is performed to each segment separately. In addition, waveform and intensity can be used to assist in the ground filtering.

The detailed comparison of the filtering techniques used for DTM extraction was made within the International Society for Photogrammetry and Remote Sensing (ISPRS) comparison of filters (Sithole and Vosselman, 2004). Kraus and Pfeifer (1998) reported an RMSE of 57 cm for DTM in wooded areas by using Airborne Laser Terrain Mapper (ALTM) 1020 and average point spacing of 3.1 m. Hyyppä et al. (2000) reported a random error of 22 cm for fluctuating forest terrain (variation for couple of tens of meters) using TopoSys-1 and nominal point density of 10 pts/m². Reutebuch et al. (2003) reported random errors of 14 cm for clear-cut, 14 cm for heavily thinned forest, 18 cm for lightly thinned forest, and 29 cm for uncut forest by using TopEye data with 4 pts/m². However, in dense forests, DTM errors of up to 10 to 20 m can occur (Takeda, 2004). Results described in a paper by Hyyppä et al. (2005) can be partly used to optimize the laser flight parameters with respect to the desired quality. The paper analyzed the effects of the date, flight altitude, pulse mode, terrain slope, forest cover, and within-plot variation on the DTM accuracy in the boreal forest zone.

It was already demonstrated in the 1980s by using small-footprint systems that the use of a laser leads to an underestimation of tree height (Nelson et al., 1988). Thus, detection of the uppermost portion of a forest canopy is expected to require a sufficient density of laser pulses to sample the tree tops and a sufficient amount of reflecting material occupying each laser pulse footprint to cause a detectable return signal (Lefsky et al., 2002). Examples of reported tree underestimation values and the accuracy of individual tree assessments are given in the following including assessments in which errors have not been calibrated or compensated for with reference data. For a comparison of mean tree height obtained in a forest inventory, the reader is referred to, for example, Næsset et al. (2004). Hyyppä and Inkinen (1999) reported individual tree height estimation with an RMSE of 0.98 m and a negative bias of 0.14 m (nominal point density about 10 pts/m^2), whereas Persson et al. (2002) reported an RMSE of 0.63 m and a negative bias of 1.13 m. Both forest sites were mainly consisted of Norway spruce and Scots pine. Persson et al. (2002) explained their greater underestimation of the average tree height as resulting from a lower ALS sampling density (about 4 pulses/m^2). Næsset and Økland (2002) concluded that the estimation accuracy was significantly reduced by a lower sampling density. Gaveau and Hill (2003) reported a negative bias of 0.91 m for sample shrub canopies and of 1.27 m for sample deciduous tree canopies. Leckie et al. (2003) presumed that some of the 1.3 m underestimation could be accounted for by the undergrowth. Yu et al. (2004b) reported a systematic underestimation of tree heights of 0.67 m for the laser acquisition carried out in 2000 and 0.54 m for another acquisition in 1998. The underestimation corresponded to 2–3 years' annual growth by those trees. Of that, the elevation model overestimation (due to undervegetation) was assumed to account for about 0.20 m. Maltamo et al. (2004) used 29 pines, the height of which was measured with a tacheometer giving more precise field measurements than conventional methods, and found a 0.65 m underestimation of the height for single trees, including annual growth that was not compensated for in the plot measurements. They also found that the random error of 0.50 m for individual tree height measurements was better than reported earlier (Hyyppä and Inkinen, 1999; Persson et al., 2002). In the studies of Rönnholm et al. (2004) and Brandtberg et al. (2003), it was shown that the tree height can be reliably estimated even under leaf-off conditions for deciduous trees.

Yu et al. (2004b) studied the effect of laser flight altitude on the tree height estimation at individual tree level in a boreal forest area mainly consisting of Norway spruce, Scots pine, and silver and downy birch. The test area (0.5 km by 2 km) was flown over at three altitudes (400, 800, and 1500 m) with a TopoSys II scanner (beam divergence 1 mrad) in spring 2003. A field inventory was performed on 33 sample plots (about 30 m × 30 m) in the test area during summer 2001. Evaluations of estimation errors due to flight altitudes, including beam size and point density, were carried out for different tree species. The results indicate that the accuracy of the tree height estimation accuracy decreases (from 0.76 to 1.16 m) with the increase in flight height (from 400 to 1.5 km). The number of detectable trees also decreases. Point density had more influence on the tree height estimation than footprint size; for more details the reader is referred to Yu et al. (2004b). Birch was less affected than coniferous trees by the change in the flight altitude in this study. Persson et al. (2002) reported that the estimates of tree height were not affected much by different beam diameters ranging from 0.26 to 2.08 m. With a larger beam diameter of 3.68 m acquired at a 76% higher altitude, the underestimations of tree heights were greater than with other beam diameters, which is probably due to the decreased point density. Nilsson (1996) did not find any significant effects of beam size on the height estimates over a pine-dominated test site. Aldred and Bonnor (1985) reported increased height estimates as the beam divergence increased, especially for deciduous trees. In the study conducted by Næsset (2004), it was concluded that first pulse measurements of height are relatively stable regardless of flight altitude/beam size when the beam size varies in the 16–26 cm range. Goodwin et al. (2006) used three different platform altitudes (1000, 2000, and 3000 m), two scan angles at 1000 m (10° and 15° half maximum angle off nadir), and three footprint sizes (0.2, 0.4, and

0.6 m) in eucalyptus forests at three sites, which varied in vegetation structure and topography, and observed no significant difference between the relative distribution of laser point returns, indicating that platform altitude and footprint size have no major influence on canopy height estimation. The results seem to indicate that relatively good canopy height information can be collected with various parameter configurations.

12.6 EXTRACTION OF TREE AND STAND ATTRIBUTES FROM LASER SCANNING

12.6.1 POINT HEIGHT METRICS

The prediction of stand variables is typically based mainly on point height metrics calculated from ALS data, Table 12.3. Nelson et al. (1988) divided features related to the height and density, which is the foundation of the area-based technology. Features such as percentiles calculated from a normalized point height distribution, mean point height, densities of the relative heights or percentiles, standard deviation, and coefficient of variation are generally used (Hyyppä and Hyyppä, 1999; Næsset, 2002). The percentiles are down to the top heights calculated from the vertical distribution of the point heights, that is, the percentile describes the height at which a

TABLE 12.3
Typical Point Height Metrics Used in Forest Attribute Derivation

No.	Feature	Explanation
Point Height Metrics		
1	meanH	Mean canopy height calculated as the arithmetic mean of the heights from the point cloud
2	stdH	Standard deviations of heights from the point cloud
3	P	Penetration calculated as a proportion of ground returns to total returns
4	COV	Coefficient of variation
5	H10	10th percentile of canopy height distribution
6	H20	20th percentile of canopy height distribution
7	H30	30th percentile of canopy height distribution
8	H40	40th percentile of canopy height distribution
9	H50	50th percentile of canopy height distribution
10	H60	60th percentile of canopy height distribution
11	H70	70th percentile of canopy height distribution
12	H80	80th percentile of canopy height distribution
13	H90	90th percentile of canopy height distribution
14	maxH	Maximum height
15	D10	10th canopy cover percentile computed as the proportion of returns below 10% of the total height
16	D20	20th canopy cover percentile computed as the proportion of returns below 20% of the total height
17	D30	30th canopy cover percentile computed as the proportion of returns below 30% of the total height
18	D40	40th canopy cover percentile computed as the proportion of returns below 40% of the total height
19	D50	50th canopy cover percentile computed as the proportion of returns below 50% of the total height
20	D60	60th canopy cover percentile computed as the proportion of returns below 60% of the total height
21	D70	70th canopy cover percentile computed as the proportion of returns below 70% of the total height
22	D80	80th canopy cover percentile computed as the proportion of returns below 80% of the total height
23	D90	90th canopy cover percentile computed as the proportion of returns below 90% of the total height

Source: Hyyppä, J. et al., *Remote Sensing of Forests from LiDAR and Radar, in Remote Sensing Handbook*, Boca Raton, FL, CRC Press, pp. 397–427, 2015.

certain number of cumulative point heights occur. Density related features are calculated from the proportion of vegetation hits compared with all hits. A hit is seen as a vegetation hit from trees or bushes if it has been reflected from over some threshold limit above ground level. All the features are calculated separately for every echo type. The reason for this is that the sampling between echo types is somewhat different (Korpela et al., 2010). Table 12.3 gives a list of typical point height metrics used. The point height metrics can be calculated inside any unit of the data. Typical units are individual trees, group of trees, micro-stand, or stand.

12.6.2 Approaches for Obtaining Forest Data from Point Clouds

Approaches aimed at obtaining forest and forestry data from point cloud data have been divided into two groups: (1) ABA and (2) ITD/single-tree detection approaches.

ABA prediction of forest variables is based on the statistical dependency between the variables measured in the field and ALS point height metrics, such as listed in Table 12.3. ABA is based on the height and density data correlated with the forest variables. Conventionally, ABA did not use neighborhood information in the data, but Niemi and Vauhkonen (2016) used canopy-based textures in the area-based prediction. Without neighborhood information, the high-quality orientation data provided by IMU are mostly neglected. The sample unit in the ABA is most often a grid cell, the size of which depends on the size of the field-measured training plot. Stand-level forest inventory results are obtained by summing and weighting the grid-level predictions inside the stand. Stand boundaries need to be determined by, for example, segmentation.

Originally in the ITD, individual trees were detected and tree-level physical variables, such as height, crown size, and tree species, were directly measured, and other variables, such as stem volume, biomass, and DBH, were predicted. Later on, it was possible to combine the physically-derived parameters with point cloud metrics if plot or tree level data was calibrated by proper field reference. The basic unit is an individual tree. Then the stand-level forest inventory results are aggregated by summing up the treewise data.

Even though early literature emphasizes the difference of the techniques, technically the approaches are quite similar, the major difference is (1) at segmentation level (whether tree or stand), (2) at physical parameters used (whether tree level heights, crown dimensions, and tree species information are applied in predictions), and (3) at field reference data. Most of the early results can be explained by the use of the applied reference data. Some of the first studies assumed that ITD should work without field reference, and the comparison between methods complained the costs of ITD laser scanning data, but not the costs of field reference needed in ABA. Today, it can be seen that the use of ITD techniques improves the estimation accuracy at microstand or stand level if the use of reference data is used properly. Actually, the point height metrics can be calculated inside any unit of the data. Typical units are individual trees, group of trees, microstand, or stand. Thus, depending on the unit size, either individual trees, group of trees, microstands, or stands are segmented to find the wanted processing level. After that, the same ABA estimation techniques (e.g., regression or nonparametric estimation) can be applied to any level from individual trees to stands. The reference data should be at the same level at which is the processing level. It is also possible to carry out calibration both at individual tree level and at plot or stand level. The latter is needed to calibrate out the small trees not visible in the ITD analysis. Currently, the ABA is operationally applied in the Nordic countries when carrying out stand-wise forest inventories. Some 3 Mha of Finnish forests is inventoried every year by applying ABA. White et al. (2013) report on best practices for using the ABA in a forest management context.

One of the first tests with ABA in Finland was performed by Suvanto et al. (2005). Regression models were developed by using laser height metrics for diameter, height, stem number, basal area, and stem volume of 472 reference plots. The predicted accuracies were 9.5%, 5.3%, 18.1%, 8.3%, and 9.8%, respectively, at stand level. Current forest management planning inventories in Scandinavia require species-specific information. Maltamo et al. (2006) added predictor features from aerial photographs and existing stand registers to ALS height metrics resulting in plot-level volume estimation accuracy from 13% to 16% depending on the predictors used. Similarly, Packalén and Maltamo (2007) used the k nearest neighbor and Most Similar Neighbor methods (k-MSN) method to impute species-specific stand variables by using ALS metrics and aerial photographs to the same dataset as in Suvanto et al. (2005), and the species-specific volume estimates at the stand level were 62.3%, 28.1%, and 32.6% for deciduous Scots pine (*Pinus sylvestris*, L.) and Norway spruce (*Picea abies*, L.), respectively. Thus, there are limitations with current technology, and especially species-specific tree size (height and diameter) distribution information are needed. One possible way is to use more detailed data and use ITD type processing and multispectral ALS data.

In addition to nearest neighbor (NN), k-MSN methods, Random Forest (RF) classifier has also been applied in ABA (Yu et al., 2011). RF is a nonparametric regression method based on regression trees, each constructed using a different random sample of the training data, and choosing splits of the trees from among the subsets of the available features, randomly chosen at each node. The samples that are not used in training are called *out-of-bag* observations. They can be used to estimate the feature's importance by randomly permutating out-of-bag data across one feature at a time and then estimating the increment in error due to this permutation. The greater the increment, the more important the feature.

First, ITD and ABA comparison was performed in Hyyppä et al. (2001b). Both methods were computed with similar field data. Point cloud metrics was used in ABA and physically-derived variables in ITD. ITD was found to be more accurate. In Hyyppä et al. (2012), both physically-derived variables and point cloud metrics were applied in plot level estimation. Individual tree-based features improved the ABA's accuracy significantly as they had very high correlation, for example, with the reference stem volume. When calculating the importance of the features, most of the individual tree-based features were among the best features. When estimating plot level mean height, the best laser-derived feature was the mean height derived by using the individual tree technique. When estimating DBH, the best laser-derived features were (1) mean canopy height, (2) penetration to the ground, (3) mean tree height (derived from the extracted individual trees), and (4) mid percentiles. For the estimation of stem volume, the best laser-derived feature was the stem volume derived from extracted individual trees, followed by the basal area derived from extracted individual trees. It is possible to easily derive further laser point height metrics and individual tree-based features, such as those based on textures (Niemi and Vauhkonen, 2016).

12.6.3 Tree Locating with ALS

Tree locations can be obtained by detecting image or point cloud local maxima (Gougeon and Moore, 1989; Dralle and Rudemo, 1996; Hyyppä and Inkinen, 1999; Wulder et al., 2000; Hyyppä et al. 2001a). In laser scanning, the aerial image is for example, replaced by the crown DSM, the CHM, or normalized point cloud. Provided that the filter size and image smoothing parameters are appropriate for the tree size and image resolution, the approach works relatively well with coniferous trees (Gougeon and Leckie, 2003). In Scandinavia, the filtering should be very modest due to narrow tall trees. After local maxima have been found, the edge of the crown can be found using region segmentation, edge detection or local minima detection (Pinz, 1991; Uuttera et al., 1998; Hyyppä and Inkinen, 1999; Hyyppä et al., 2001a; Culvenor, 2002; Persson et al., 2002). Full crown delineation is also possible with techniques such as shade-valley-following (Gougeon, 1995), edge curvature analysis (Brandtberg and Walter, 1999), template matching (Pollock, 1994; Larsen and Rudemo, 1998), region growing (Erikson, 2003), and point cloud based reconstruction (Pyysalo and Hyyppä, 2002) (Figure 12.9).

Hyyppä and Inkinen (1999), Friedlaender and Koch (2000), Ziegler et al. (2000), and Hyyppä et al. (2001a) demonstrated the individual-tree-based forest inventory using laser scanner tree finding with maxima of the CHM and segmentation for edge detection. They also presented the basic LiDAR-based individual tree crown approach in which, from individual trees, location, tree height,

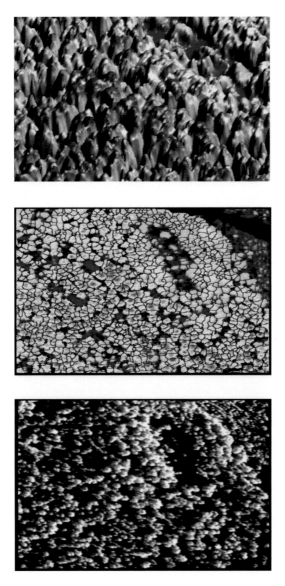

FIGURE 12.9 Canopy height model, segmented canopy height model, and corresponding aerial image is shown. (Courtesy of Antero Kukko. With Permission.)

crown diameter, and species are derived using laser, possibly in combination with aerial image data, especially for tree species classification, and then other important variables, such as stem diameter, basal area, and stem volume, are derived using existing models. According Hyyppä et al. (2001b), first segmentation methods tested in Finnish, Austrian, and German coniferous forests could segment 40%–50% of the trees correctly. Persson et al. (2002) could link 71% of the tree heights with the reference trees. Other attempts to use DSM or CHM image for ITD have been reported by Brandtberg et al. (2003), Leckie et al. (2003), Straub (2003), Popescu et al. (2003), Tiede and Hoffmann (2006), and Falkowski et al. (2006). Andersen et al. (2002) proposed to fit ellipsoid crown models in a Bayesian framework to the point cloud. Morsdorf et al. (2003) presented a two-stage procedure in which tree locations were defined using the DSM and local maxima, and crown delineation was performed using k-means clustering in the 3D point cloud. Pitkänen et al. (2004) proposed three methods for ITD: smoothed CHM with the knowledge of the canopy height, elimination of candidate tree locations based on the predicted crown diameter and distance and valley depth between two locations studied (maximum elimination) and modified scale-space method used for blob detection. The maximum elimination method gave the best results of tree detection, however, with the cost of including several parameters to keep the number of false positives low. Sohlberg et al. (2006) presented methods for controlling the shape of crown segments, and for residual adjustment of the CHM. The method was applied and validated in a Norway spruce dominated forest having heterogeneous structure. The number of trees detected varied with social status of the trees, from 93% of the dominant trees to 19% of the suppressed trees. Since then numerous new methods have been developed. Villikka et al. (2007) proposed a first method combining ABA and ITD approaches. Since then ABA and ITD methods have been even more converging (Hyyppä et al., 2005, 2006; Breidenbach et al., 2010; Hyyppä et al., 2012). The methods are still lacking the proper use of physics. Tree height measured by ITD is a physical measure of the height of the tree. In the same way, we can use knowledge-based approach to link tree height to the potential size of the crown. There are still significant improvements possible.

Most of the current approaches for tree detection are based on finding trees from the CHM, which is calculated as a maximum of canopy height values within each raster cell. Thus, the CHM corresponds to the maximum canopy height of the first pulse data. Recently, other approaches have been proposed (Hyyppä et al., 2012) utilizing the canopy penetration capability of the last pulse returns with overlapping trees and correcting past information is this area (Hyyppä et al., 2008). When trees overlap, the surface model corresponding to the first pulse stays high, whereas with last pulse, even a small gap results in a drop in elevation, that is, the trees can be more readily discriminated. The first pulse works, when the whole laser beam penetrates the gap between the crowns, so that the drop is detectable after filtering. With last pulse, the drop in elevation is substantially larger and the drop can be detected even with overlapping trees as the last pulse is more sensitive to lower canopy levels. With numerical tests it was shown that with the DBH class 5–10 cm, the last pulse resulted in 10% better detection of trees. The results confirm, thus, that there is substantial information for tree detection in last pulse data. Currently, in raster-based processing, other than CHM information has been largely neglected. The advantages of first pulse data obviously include the lower number of commission errors and the high quality of tree separation when the crowns are not overlapping, whereas the advantage of last pulse is in the separation of trees whose crowns overlap. A hybrid model, utilizing the advantages of both pulse types should be developed. For the comparison of various ITD techniques, the reader is referred to Kaartinen and Hyyppä (2008), Kaartinen et al. (2012), Vauhkonen et al. (2011), and Wang et al. (2016).

12.6.4 Individual Tree Height Derivation

To test the individual tree extraction methods using the same remote sensing datasets, the EuroSDR Organization and the ISPRS initiated the Tree Extraction Project in 2005 to evaluate the quality, accuracy, and feasibility of automated tree extraction methods based on high-density laser scanner data and digital aerial images. The project was hosted by the FGI. Twelve partners from the United States of America, Canada, Norway, Sweden, Finland, Germany, Austria, Switzerland, Italy, Poland, and Taiwan participated in the test included in the Tree Extraction Project. The partners were requested to extract trees using the given test datasets having point densities between 2 and 8 pts/m^2. Another objective of the study was to find out how the pulse density impacted on individual tree extraction. The results were published in the project final report (Kaartinen and Hyyppä, 2008) and in Kaartinen et al. (2012).

Tree height quality analysis showed that the variability of point density was negligible when compared with variability between the methods. With the best models, a RMSE of 60 cm to 80 cm was obtained for tree height. The results with the best automated models were significantly better than those attained when using the manual process. Both underestimation of tree height and standard deviation were decreased in general as the point density increased. The overestimation produced by the Model Norway in regard to tree height was due to the correction applied to the tree height in the preprocessing phase. The methods capable of finding more trees in the lower classes are obviously suffering; the uncertainty regarding the heights of the extracted tree in the lower levels is greater.

ALS is the major data source for forest inventories today. On the other hand, TLS and MLS are studied to be able to get plot and tree level references for ALS data. The use of TLS at the plot level cannot be used for field reference without large number of TLS plots because the visibility of treetops with TLS is limited in dense forests. However, in Huang et al. (2011), a −0.26 m bias and a 0.76 m RMSE was reported for one plot (212 stems/ha, sparse stand) by using the multiscan approach. In Hopkinson et al. (2004), an approximate 1.5 m underestimation of tree heights was reported for two medium-density plots (465 and 661 stems/ha) using the multiscan approach. Maas et al. (2008) depicted a −0.64 m bias and a 4.55 m RMSE for nine trees locating on four plots (212–410 stems/ha) using the single- and multiscan approaches. Fleck et al. (2011) concluded a 2.41 m RMSE for 45 selected trees on one plot (392 stems/ha) using multiscan data. The observation of tree tops from the TLS data is possible on sparse sample plots using many scans, as reported in Huang et al. (2011) and Fleck et al. (2011), but not in dense sample plots. In the ISPRS/EuroSDR Tree extraction test (Kaartinen and Hyyppä, 2008), TLS was able to collect tree height information with the level of 10 cm, but very high density scanning was made (both as point cloud density and in the number of scans applied).

One of the advantages of using MLS for plot level inventories lies in the fact that MLS can see many of the invisible tree tops for TLS. As MLS platform moves all the time during the data acquisition, the gaps to tree tops are more likely visible with MLS than with TLS. This preliminary conclusion still requires detailed scientific studies to be confirmed.

12.6.5 DIAMETER DERIVATION

As diameter cannot be directly measured from ALS, TLS and MLS have been studied to provide accurate reference diameters at tree and plot level. The most popular processing method today for determining diameter is to cut a slice of data from the original point cloud and to model tree stems from this layer by point clustering or circle finding (Simonse et al., 2003; Aschoff and Spiecker, 2004; Thies et al., 2004; Watt and Donoghue, 2005; Maas et al., 2008; Brolly and Kiraly, 2009; Tansey et al., 2009; Huang et al., 2011). This approach assumes that all trees present a clear stem at the same height and this assumption is typically not valid in most mixed forests having branches at different heights. A study of a mixed deciduous stand using this approach showed difficulties even in the manual stem detection in a TLS data layer (Hopkinson et al., 2004). Results from studies of different types of forests have still high variation. In Liang et al. (2012b), the DBH estimation results reported at the tree level from five plots had bias of 0.16 cm, and the RMSE of 1.29 cm. In Lindberg (2012), the bias and RMSE of tree-level DBH estimates from six plots were 0.16 cm and 3.8 cm, respectively. In Liang et al. (2014c), a MLS system was tested. The root mean square errors of the DBH estimates were 2.4 cm. These results indicate that the MLS system has the potential to accurately map forest plots as long as the positioning inside the forests are solved.

12.6.6 STEM CURVE DERIVATION

The tree stem curve depicts the tapering of the stem as a function of the height. If the stem curve can be determined, it is also possible to derive biomass and stem volume estimates from the trees without knowledge of conventional allometric models relating, for example, diameter and height information to volume and biomass. Early research work of TLS for stem curve includes nine pine trees studied in Henning and Radtke (2006), a spruce tree in Maas et al. (2008), and two trees, one pine and one spruce, in Liang et al. (2011). The RMSE of the stem curve measurements was 4.7 cm in Maas et al. (2008) utilizing single-scan data. In Liang et al. (2011), the RMSE of the stem curve estimation of the pine tree was 1.3 cm and 1.8 cm with the multi- and single-scan data, respectively; and the RMSE of the curve measurement of the spruce tree was 0.6 cm and 0.6 cm using the multi- and single-scan data, respectively.

The first detailed study on the plot-level automatic measuring of the stem curves of different species and different growth stages using TLS was reported in 2014 (Liang et al., 2014a). Twenty-eight trees, sixteen pines, and twelve spruces, were selected from nine sample plots. The plots were scanned utilizing the multiscan approach and trees were felled and the stem curves were manually measured in the field. The stem curves were automatically measured with a mean bias of 0.15 cm and mean RMSE of 1.13 cm at the tree level. The highest diameters measured were between 50.6% and 74.5% of the total tree height, with a mean of 65.8% for pine trees and 61.0% for spruce trees. The upper part of the reconstructed stems should be estimated.

12.6.7 TREE SPECIES

The integration of laser scanning and aerial imagery can be based on simultaneous or separate data capturing. There is high synergy between the high-resolution optical imagery and laser-scanner data for extracting forest information: laser data provides accurate height information, which is missing in nonstereo optical imagery, and also supports information on the crown shape and size depending on point density, whereas optical images provide more details about spatial geometry and color information usable for classification of tree species and health. In practice, tree species derivation is the major reason for the synergetic use needed between aerial image and laser scanner data.

Holmgren and Persson (2004) tested species classification of Scots pine and Norway spruce using laser data (point spacing 0.4 to 0.5 m) at individual tree level. The proportion of correctly classified trees on all the plots was 95%. Moffiet et al. (2005) suggested that the proportion of laser singular returns is an important predictor for the tree species classification for species such as Poplar Box and Cypress Pine. Although a clear distinction between these two species was not always visually obvious at the individual tree level, due to some other extraneous sources of variation in the dataset, the observation was supported in general at the site level. Sites dominated by Poplar Box generally exhibited a lower proportion of singular returns compared with sites dominated by Cypress Pine. Brandtberg et al. (2003) used laser data under leaf-off conditions for the detection of individual trees and tree species classification results using different indices. The results suggest a moderate to high degree of accuracy in deciduous species classification. Liang et al. (2007) used a simple technique, the difference of the first and last pulse return under leaf-off conditions, to discriminate between deciduous and coniferous trees at individual tree level. Classification accuracy of 90% was obtained. Reitberger et al. (2006) described an approach to tree species classification based on features derived by a waveform decomposition of full waveform LiDAR data. Point distributions were computed in sample tree areas and compared with the numbers that result from a conventional signal detection. Unsupervised tree species classification was performed using special tree saliencies derived from the LiDAR points. The classification into two clusters (deciduous, coniferous) led to an overall accuracy of 80% in a leaf-on situation.

Conventionally, tree species information is extracted from high spatial resolution color-infrared aerial photographs (Brandtberg, 2002; Bohlin et al., 2006). Persson et al. (2006) derived tree species for trees through combining features of high resolution laser data (50 pts/m^2) with high resolution multispectral images (ground resolution 0.1 m). Tree-species classification experiment was conducted in southern Sweden in a forest consisting of Norway spruce (*P. abies*), Scots pine (*P. sylvestris*), and deciduous trees, mainly birch (*Betula* spp). The results implied that by combining a laser-derived geometry and image-derived spectral features, the classification could be improved and accuracies of 95% were achieved. Packalén and Maltamo (2006) used combination of laser scanner data and aerial images to predict species specific plot level tree volumes. A nonparametric k-Most Similar Neighbor application was constructed by using characteristics of canopy height distribution approach of laser data and textural and spectral values of aerial image at plot level.

There are several problems related to merging image and laser data, such registration leading to errors in matching same tree from different datasets, multiple costs in flying when data are taken separately and limited flight hours when data collected simultaneously. In real-life, the obtained tree species accuracy is typically much less than in previous studies. In research papers, isolated and dominant trees are often used. Thus, practical single sensor solutions or forest mapping that is capable of providing high-quality species-specific information are needed. In Yu et al. (2017), multispectral ALS data acquired with Optech Titan scanner over a boreal forest, mainly consisting of Scots pine (*P. Sylvestris*), Norway spruce (*P. abies*), and birch (*Betula* spp.), in southern Finland were collected with 1903 ground measured trees from 22 sample plots. Point cloud and intensity metrics used as predictors for tree species were extracted from segmented tree objects and used in RF classification. The best overall tree species classification accuracy achieved for correctly detected trees was 88.4% using all features. Point cloud features alone achieved the lowest accuracy of level 77.9%. Corresponding value for intensity features of all channels was 86.8%. Examination of classification accuracy showed that isolated and dominant trees can be detected with a detection rate of 92.8% and classified with high overall accuracy of 90.9%. Corresponding detection rate and accuracy were 82.6% and 89.6% for group of trees, 26.8% and 72.0% for trees next to a bigger tree, and 8.3% and 66.7% for trees under a bigger tree, respectively. Channel 1064 nm seems to contain more information for separating pine, spruce, and birch, followed by channel 1550 nm and channel 532 nm with an overall accuracy of classification of 82.0%, 78.0%, and 70.1%, respectively. The results suggest that the use of multispectral ALS data, such as in Titan, has great potential to lead to a single-sensor solution for forest mapping.

12.7 FOREST CHANGE

12.7.1 METHODS AND QUALITY OF FOREST GROWTH

Tree growth consists of the elongation and thickening of roots, stems, and branches (Husch et al., 1982); growth causes trees to change in weight, volume, and shape (form). Usually only the growth of the tree stem is considered by using the growth characteristics of the tree, diameter, height, basal area, and volume. In most cases volume growth is the most interesting characteristic and it has to be derived from the change observed in other characteristics. In practice height and diameter growth of individual trees are determined in the field from repeated measurements of permanent sample plots and from increment core measurements (e.g., boring) (Husch et al., 1982).

Laser data based methods for forest growth are relatively simple in principle. The height growth can be determined by several means: from the difference in the height of individual trees determined from repeated measurements (Yu et al., 2003) (Figure 12.10), from height difference of repeated DSMs (Hirata, 2005), from repeated height histograms (Næsset and Gobakken, 2005), or from difference of the volumes of individual trees. The changes in forests that affect the laser scanning response include the vertical and horizontal growth of crowns, the seasonal change of needle and leaf masses, the state of undergrowth and low vegetation, and the trees moving with the wind

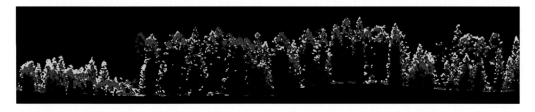

FIGURE 12.10 Profiles of 150 m sections. Yellow: laser scanner of year 2003, Red: laser scanner of year 1998. Cutting tree was showed in 2003 data but not in 1998. Growth of young trees was clearly visible on left part of the profiles. (Courtesy of Xiaowei Yu. With Permission.)

(especially for taller trees). Thus, the monitoring of growth using ALS can be relatively complicated in practice. The technique applied should be able to separate growth from other changes in the forest, especially those due to selective thinning or naturally fallen trees. The difference between DSMs is assumed to work in areas with wide and flat-topped crowns. In coniferous forests with narrow crowns, the planimetric displacement between two acquisitions can be substantial. Height histograms can be applied to point clouds corresponding to individual trees or plots or stands, but the information contents of histograms are corrupted if, for example, thinning has occurred or the parameters of laser surveys are significantly different, thus change detection based on height histograms does not work in practice.

Yu et al. (2003, 2004b) demonstrated the application of laser data to forest growth at plot and stand level using an object-based tree-to-tree matching algorithm and statistical analysis. St-Onge and Vepakomma (2004) concluded that sensor-dependent effects such as echo triggering are probably the most difficult to control in multitemporal laser surveys for growth analysis purposes. Due to rapid technological developments, it is very likely that different sensors will be used, especially over long-term intervals that are needed in forest inventories (e.g., 10 year time interval). Næsset and Gobakken (2005) concluded that over a 2-year period, the prediction accuracy for plotwise and standwise change in mean tree height, basal area, and volume was low when a point density of about 1pt/m^2 and canopy height distribution technique were used. They also reported that certain height measurements, such as maximum height, seemed less suitable than many other height metrics because maximum height tends to be less stable—most probably due to low pulse density, narrow beam size, and relatively short growth period (2 years).

Yu et al. (2005) showed that height growth for individual trees can be measured with an accuracy better than 0.5 m using multitemporal laser surveys conducted in a boreal forest zone for a 4-year time series and higher point density. In Yu et al. (2006) 82 sample trees were used to analyze the potential of measuring individual tree growth of Scots pine in the boreal forest. Point clouds, having 10 pts/m^2 and illuminating 50% of the tree tops (i.e., the beams covering 50% of tree tops) were acquired in September 1998 and May 2003 with TopoSys 83 kHz LiDAR system. Three variables were extracted from the point clouds representing each tree included the difference of the highest z value, difference between the DSMs of tree tops and difference of 85%, 90%, and 95% quantiles in the height histograms corresponding to a crown. An R^2 value of 0.68 and standard deviation of 43 cm were derived with the best model. The results confirmed that it is possible to measure the growth of an individual tree with multitemporal laser surveys. They also demonstrated a better algorithm for tree-to-tree matching that is needed in operational individual-tree-based growth estimation in areas with narrow trees. The method is based on minimizing the sum of distances between tree tops in an N-dimensional data space. The experiments showed that the location of trees (derived from laser data) and height of the trees were together adequate to provide a reliable tree-to-tree matching. In future the crown area should also be included in the matching as the fourth parameter.

In Yu et al. (2008) extended dataset were used to estimate the tree mean height and volume growth at plot level in a boreal forest. Laser datasets were collected with a TopoSys laser scanner in 1998, 2000, and 2003 with a nominal point density of 10 points/m^2. Three techniques were used to predict the growth values based on individual tree top differencing, DSM differencing and canopy height distribution differencing. The regression models were developed for mean height growth and volume growth using single predictor derived from each method and using selected predictors from all methods. The best results were obtained for mean height growth (adjusted R^2 value of 0.86 and standard deviation of residual of 0.15 m) using the individual tree top differencing method. The corresponding values for volume growth were 0.58 and 8.39 m^3 ha^{-1} (35.7%), respectively, using DSM differencing. Combined use of three techniques yielded a better result for volume growth (adjusted $R^2 = 0.77$) but did not improve the estimation for mean height growth. In the tree top differencing methods, the most problematic part is to find pairs of tree tops that represent the same tree (tree-to-tree matching), Figure 12.11.

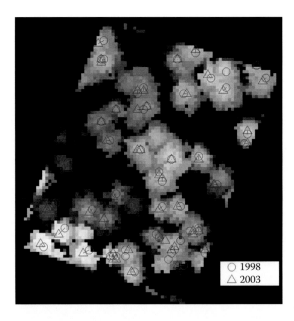

FIGURE 12.11 Result of tree matching using three variables (x, y, and H) for one test plot 3 (match rate is 70.5% for all trees). Background image is 2003 CHM. The location of the trees identified from the 1998 and 2003 data are marked with circles and triangles. Matched trees are linked by line. (Courtesy of Xiaowei Yu. With Permission.)

12.7.2 METHODS AND QUALITY OF HARVESTED TREE DETECTION

Laser data can also be used for change detection. Yu et al. (2003, 2004b) also examined the applicability of airborne laser scanners in monitoring harvested trees (Figures 12.12 and 12.13), using datasets with a point density of about 10 pts/m² over a 2-year period. The developed automatic method used for detecting harvested trees was based on image differencing. First, a difference image was calculated by subtracting the latter CHM (or DSM) from former CHM (or DSM). The resulting difference image represented the pixelwise changes between the two dates. Clustered high positive differences presented harvested trees. Most of the image values were close to zero (ground) or a little below zero due to the tree growth. To identify harvested trees, a threshold was selected and applied to the different image for distinguishing no changes or little changes from big changes.

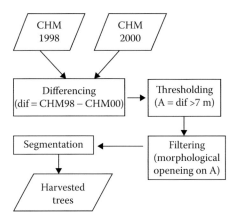

FIGURE 12.12 Flowchart for harvested tree detection. (Courtesy of Xiaowei Yu.)

FIGURE 12.13 Change detection of harvested trees from two laser surveys, (a) DSM from year 2000, (b) DSM from year 2003, (c) difference of DSMs, (d) filtered difference image showing the harvested trees. (Courtesy of Xiaowei Yu. With Permission.)

A morphological opening was then performed to reduce noise-type fluctuation. Location and number of harvested trees was determined on the basis of the segmentation of the resulting image. Out of 83 field-checked harvested trees, 61 were detected automatically and correctly. All the mature harvested trees were detected.

12.8 OUTLOOK

ITD and ABA are converging in processing of ALS data. The point height metrics can be calculated inside any unit of the data. Typical units are individual trees, group of trees, microstand, or stand. Thus, depending on the unit size, either individual trees, group of trees, microstands or stands are segmented to find the wanted processing level. After that the same estimation techniques (e.g., regression or nonparametric estimation) can be applied to any level from individual trees to stands. The reference data should be at the same level similar to the processing level. It is also possible to carry out calibration both at individual tree level and at plot or stand level. The latter is needed to calibrate out the small trees not visible in the ITD analysis. Currently, the ABA is operationally applied in the Nordic countries when carrying out standwise forest inventories. Some 3 Mha of Finnish forests is inventoried every year by applying ABA. White et al. (2013) report on best practices for using the ABA in a forest management context. Tree species problems related to airborne laser data are today solved with multispectral ALS data, and future will show whether such data will lead into operative stage.

More automated and more cost-effective techniques are needed to provide field inventory data. Recently, there are several techniques currently been studied to automize plot level field data collection. Both single-scan and multiscan methods have been studied in connection of TLS measurement. Currently, TLS has shown to be feasible for collecting basic tree and plot attributes, such as tree position, DBH, stem curve, tree species, volume, and biomass. The accuracy is equivalent to those based on key measurements and use of best allometric models. The state-of-the-art in deriving plot level data from TLS can be found from, for example, Liang et al. (2016). Other techniques to automize plot level data collection, that has been recently studied, include, for example, MLS, photogrammetric terrestrial point clouds, and use of crowdsourcing. In addition, there are automation developments of key forest parameters. MLS allows collection of about 50 plots per day with about accuracy of 2–4 cm for DBH. MLS, as TLS, can provide stem curve, but it has not yet been demonstrated. MLS allows problematic areas to be measured with higher detail. Due to GNSS shadows in forests, Simultaneous Localization and Mapping needs to be integrated to MLS measurements. MLS techniques seem to be optimum for forest inventory service companies, which needs up-to-date information of forest plots. Photogrammetric terrestrial point clouds provide also 2–4 cm accuracy, but data collection and processing is more time consuming compared with MLS. Moreover, UAV laser scanning is today studied to meet the requirement of field inventory.

ACKNOWLEDGMENTS

Financial support from the Academy of Finland projects "Interaction of Lidar/Radar Beams with Forests Using Mini-UAV and Mobile Forest Tomography" (No. 259348), "Centre of Excellence in Laser Scanning Research (CoE-LaSR)," laserscanning.fi, (No. 272195) and "Competence-Based Growth Through Integrated Disruptive Technologies of 3D Digitalization, Robotics, Geospatial Information and Image Processing/Computing—Point Cloud Ecosystem," pointcloud.fi @pointcloudfi, (No. 293389) are acknowledged.

REFERENCES

Aldred, A.H. and Bonnor, G.M., 1985, Application of airborne lasers to forest surveys. Information Report PI-X-51, Canadian Forestry Service, Petawawa national Forestry Institute, 62 p.

Andersen, H-E., Reutebuch, S., and Schreuder, G., 2002, Bayesian object recognition for the analysis of complex forest scenes in airborne laser scanner data. In *International Archives of the Photogrammetry, Remote Sensing and Spatial Information Sciences*, 34, Part 3A, ISPRS, Annapolis, pp. 35–41.

Aschoff, T. and Spiecker, H., 2004. Algorithms for the automatic detection of trees in laser scanner data. *International Archives of Photogrammetry, Remote Sensing and Spatial Information Sciences*, 36, Part W2 ISPRS, Freiburg, pp 71–75.

Axelsson, P., 2000, DEM Generation from laser scanner data using adaptive TIN models. In *International Archives of Photogrammetry and Remote Sensing*, 33, Part B4, ISPRS, Amsterdam, pp. 110–117.

Baltsavias, E.P., 1999, Airborne laser scanning: Basic relations and formulas. *ISPRS Journal of Photogrammetry and Remote Sensing*, 54(2–3), 199–214.

Bohlin, J., Olsson, H., Olofsson, K., and Wallerman, J., 2006, Tree species discrimination by aid of template matching applied to digital air photos. EARSeL SIG Forestry. In *International Workshop 3D Remote Sensing in Forestry Proceedings*, February 14–15, 2006, Earsel, Vienna, Austria, pp. 199–203.

Brandtberg, T., 1999, Automatic individual tree-based analysis of high spatial resolution remotely sensed data. PhD Thesis, Acta Universitatis Agriculturae Sueciae, Silvestria 118, Swedish University of Agricultural Sciences, Uppsala, Sweden, 47 p.

Brandtberg, T., 2002, Individual tree-based species classification in high spatial resolution aerial images of forests using fuzzy sets. *Fuzzy Sets and Systems*, 132(3), 371–387.

Brandtberg, T. and Walter, F., 1999, An algorithm for delineation of individual tree crowns in high spatial resolution aerial images using curved edge segments at multiple scales. In *Proceedings on International Forum on Automated Interpretation of High Spatial Resolution Digital Imagery for Forestry*, Hill, D.A. and Leckie, D.G., (Eds.), Natural Resources Canada, Canadian Forest Service, Pacific Forestry Centre, Victoria, BC, Columbia, pp. 41–54.

Brandtberg, T., Warner, T., Landenberger, R., and McGraw, J., 2003, Detection and analysis of individual leaf-off tree crowns in small footprint, high sampling density LiDAR data from the eastern deciduous forest in North America. *Remote Sensing of Environment*, 85, 290–303.

Breidenbach, J., Naesset, E., Lien, V., Gobakken, T., and Solberg, S., 2010, Prediction of species specific forest inventory attributes using a nonparametric semi-individual tree crown approach based on fused airborne laser scanning and multispectral data. *Remote Sensing of Environment*, 114(4), 911–924.

Brolly, G. and Kiraly, G., 2009. Algorithms for stem mapping by means of terrestrial laser scanning. *Acta Silvatica et Lignaria Hungarica*, 5, 119–130.

Chasmer, L., Hopkinson, C., and Treitz, P., 2006, Investigating laser pulse penetration through a conifer canopy by integrating airborne and terrestrial LiDAR. *Canadian Journal of Remote Sensing*, 32(2), 116–125.

Chisholm, R.A, Jinqiang, C., Shawn, K.Y.L., and Ben M.C., 2013, UAV LiDAR for below-canopy forest surveys. *Journal of Unmanned Vehicle Systems*, 1(1), 61–68.

Culvenor, D.S., 2002, TIDA: An algorithm for the delineation of tree crowns in high spatial resolution digital imagery of Australian native forest. PhD Thesis, University of Melbourne, Melbourne, Australia.

Currie, D., Shaw, V., and Bercha, F., 1989, Integration of laser rangefinder and multispectral video data for forest measurements. In *Proceedings of IGARSS'89 Conference*, July 10–14, 1989, Vancouver, Canada, 4, pp. 2382–2384.

Degnan, J., 2016, Scanning, multibeam, single photon LiDARs for rapid, large scale, high resolution, topographic and bathymetric mapping. *Remote Sensing*, 8(11), 958. doi:10.3390/rs8110958

Dralle, K. and Rudemo, M., 1996, Stem number estimation by kernel smoothing of aerial photos. *Canadian Journal of Forest Research*, 26, 1228–1236.

Erikson, M., 2003, Segmentation of individual tree crowns in colour aerial photographs using region growing supported by fuzzy rules. *Canadian Journal of Forest Research*, 33, 1557–1563.

Falkowski, M.J., Smith, A.M.S., Hudak, A.T., Gessler, P.E., Vierling, L.A., and Crookston, N.L., 2006, Automated estimation of individual conifer tree height and crown diameter via two-dimensional spatial wavelet analysis of LiDAR data. *Canadian Journal of Remote Sensing*, 32(2), 153–161.

Fleck, S., Mölder, I., Jacob, M., Gebauer, T., Jungkunst, H.F., and Leuschner, C., 2011, Comparison of conventional eight-point crown projections with LiDAR-based virtual crown projections in a temperate old-growth forest. *Annals of forest science*, 68, 1173–1185.

Friedlaender, H. and Koch, B., 2000, First experience in the application of laser scanner data for the assessment of vertical and horizontal forest structures. In *International Archives of Photogrammetry and Remote Sensing*, July 2000, ISPRS Congress, Amsterdam, the Netherlands, 33, Part B7, pp. 693–700.

Gaveau, D. and Hill, R., 2003, Quantifying canopy height underestimation by laser pulse penetration in small-footprint airborne laser scanning data. *Canadian Journal of Remote Sensing*, 29, 650–657.

Gill, S.J., Biging, G.S., and Murphy, E.C., 2000, Modeling conifer tree crown radius and estimating canopy cover. *Forest Ecology and Management*, 126(3), 405–416.

Goodwin, N.R., Coops, N.C., and Culvenor, D.S., 2006, Assessment of forest structure with airborne LiDAR and the effects of platform altitude. *Remote Sensing of Environment*, 103(2), 140–152.

Gougeon, F.A., 1995, A crown-following approach to the automatic delineation of individual tree crowns in high spatial resolution aerial images. *Canadian Journal of Remote Sensing*, 21(3), 274–284.

Gougeon, F.A. and Leckie, D., 2003, Forest information extraction from high spatial resolution images using an individual tree crown approach. Information report BC-X-396, Natural Resources Canada, Canadian Forest Service, Pacific Forestry Centre, 26 p.

Gougeon, F.A. and Moore, T., 1989, Classification individuelle des arbres à partir d'images à haute résolution spatiale. *Télédétection et gestion des resources*, 6, 185–196.

Hall, D.B. and Bailey, R.L., 2001, Modeling and prediction of forest growth variables based on multilevel nonlinear mixed models. *Forest Science*, 47(3), 311–321.

Harding, D.J., Lefsky, M.A., Parke, G.G., and Blair, J.B., 2001. Laser altimeter canopy height profiles. Methods and validation for closed canopy, broadleaf forests. *Remote Sensing of Environment*, 76(3), 283–297.

Henning, J. G. and Radtke, P.J., 2006, Detailed stem measurements of standing trees from ground-based scanning LiDAR. *Forest Science*, 52(1), 67–80.

Hirata, Y., 2005, Relationship between crown and growth of individual tree derived from multi-temporal airborne laser scanner data. In *Conference on Silvilaser 2005*, September 29–October 1, 2005, Powerpoint Presentation, Virginia Tech, Blacksburg, VA.

Hofton, M. A., Rocchio, L., Blair, J. B., and Dubayah., R., 2002, Validation of vegetation canopy LiDAR subcanopy topography measurements for a dense tropical forest. *Journal of Geodynamics*, 34(3–4), 491–502.

Hökkä, H. and Groot, A., 1999, An individual-tree basal area growth model for black spruce in second-growth peatland stands. *Canadian Journal of Forest Research*, 29, 621–629.

Holmgren, J., 2003, Estimation of forest variables using airborne laser scanning. PhD Thesis. Acta Universitatis Agriculturae Sueciae, Silvestria 278, Swedish University of Agricultural Sciences, Umeå, Sweden.

Holmgren, J. and Persson, Å., 2004, Identifying species of individual trees using airborne laser scanning. *Remote Sensing of EnvironmentREmote Sensing of Environment*, 90, 415–423.

Holopainen, M., Kankare, V., Vastaranta, M., Liang, X., Lin, Y., Vaaja, M., Yu, X. et al., 2013, Tree mapping using airborne, terrestrial and mobile laser scanning–A case study in a heterogeneous urban forest. *Urban forestry & urban greening*, 12(2013), 546–553.

Holopainen, M. and Talvitie, T., 2006, Effects of data acquisition accuracy on timing of stand harvests and expected net present value. *Silva Fennica*, 40(3), 531–543.

Hopkinson, C., Chasmer, L., Young-Pow, C., and Treitz, P., 2004, Assessing forest metrics with a ground-based scanning LiDAR. *Canadian Journal of Forest Research*, 34, 573–583.

Huang, H., Li, Z., Gong, P., Cheng, X., Clinton, N., Cao, C., Ni, W., and Wang, L., 2011, Automated methods for measuring DBH and tree heights with a commercial scanning LiDAR. *Photogrammetric engineering and remote sensing*, 77, 219–227.

Hugershoff, R., 1939, Die Bildmessung unde ihre forstlichen Anwendungen. *Der Deutsche Forstwirt*, 50(21), 612–615.

Husch, B., Miller, C.I., and Beers, T.W., 1982, *Forest Mensuration*, 2nd ed. New York: John Wiley & Sons.

Hynynen, J., 1995, Modelling tree growth for managed stands. The Finnish Forest Research Institute. Research Papers 576. 59 + 76 p.

Hyyppä, J., 1993, Development and feasibility of airborne ranging radar for forest assessment. PhD Thesis, Helsinki University of Technology, Laboratory of Space Technology, Espoo, Finland, 112 p.

Hyyppä, H. and Hyyppä, J., 1999, Comparing the accuracy of laser scanner with other optical remote sensing data sources for stand attribute retrieval. *Photogrammetric Journal of Finland*, 16(2), 5–15.

Hyyppä, J. and Hallikainen, M., 1993, A helicopter-borne eight-channel ranging scatterometer for remote sensing, Part II: Forest inventory. *IEEE Transactions on Geoscience and Remote Sensing*, 31(1), 170–179.

Hyyppä, J., Hyyppä, H., Leckie, D., Gougeon, F., Yu, X., and Maltamo, M., 2008, Review of methods of small-footprint airborne laser scanning for extracting forest inventory data in boreal forests. *International Journal of Remote Sensing*, 29(5), 1339–1366.

Hyyppä, J., Hyyppä, H., Maltamo, M., Yu, X., Ahokas, E., and Pyysalo, U., 2003a, Laser scanning of forest resources - some of the Finnish experience. In *Proceedings of the ScandLaser Scientific Workshop on Airborne Laser Scanning of Forests*, September 3–4, 2003, Umeå, Sweden, pp. 53–59.

Hyyppä, J. and Inkinen, M., 1999, Detecting and estimating attributes for single trees using laser scanner. *The Photogrammetric Journal of Finland*, 16, 27–42.

Hyyppä, J., Karjalainen, M., Liang, X., Jaakkola, A., Karila, K., Kaartinen, H., Kukko, A. et al., 2015, *Remote Sensing of Forests from LiDAR and Radar, in Remote Sensing Handbook*. Boca Raton, FL: CRC Press, pp. 397–427.

Hyyppä, J., Kelle, O., Lehikoinen, M., and Inkinen, M., 2001a, A segmentation-based method to retrieve stem volume estimates from 3-dimensional tree height models produced by laser scanner. *IEEE Transactions of Geoscience and Remote Sensing*, 39, 969–975.

Hyyppä, J., Mielonen, T., Hyyppä, H., Maltamo, M., Yu, X., Honkavaara, E., and Kaartinen, H., 2005, Using individual tree crown approach for forest volume extraction with aerial images and laser point clouds. In *Proceedings of ISPRS Workshop Laser Scanning 2005*, September 12–14, 2005, Netherlands, GITC bv, Enschede, the Netherlands, 36, Part 3/W19, pp. 144–149.

Hyyppä, J., Pyysalo, U., Hyyppä, H., Haggren, H., and Ruppert, G., 2000, Accuracy of laser scanning for DTM generation in forested areas. In *Proceedings of SPIE*, September 5, 2000, Orlando, FL, 4035, pp. 119–130.

Hyyppä, J., Schardt, M., Haggrén, H., Koch, B., Lohr, U., Scherrer, H.U., Paananen, R. et al., 2001b, HIGH-SCAN: The first European-wide attempt to derive single-tree information from laser scanner data. *The Photogrammetric Journal of Finland*, 17, 58–68.

Hyyppä, J., Yu, X., Hyyppä, H., and Maltamo, M., 2006, Methods of airborne laser scanning for forest information extraction. EARSeL SIG Forestry. In *International Workshop 3D Remote Sensing in Forestry Proceedings*, February 14–15, 2006, Earsel, Vienna, pp. 63–78.

Hyyppä, J., Hyyppä, H., Yu, X., Kaartinen, H., Kukko, A., and Holopainen, M., 2009, Forest inventory using small-footprint airborne lidar. In *Topographic Laser Ranging and Scanning: Principles and Processing*, Shan, J. and Toth, C.K. (Eds.), CRC Press, Boca Raton, FL, pp. 335–370, 590.

Hyyppä, J., Yu, X., Hyyppä, H., Vastaranta, M., Holopainen, M., Kukko, A., Kaartinen, H. et al., 2012. Advances in forest inventory using airborne laser scanning. *Remote Sensing*, 4, 1190–1207.

Hyyppä, H., Yu, X., Hyyppä, J., Kaartinen, H., Honkavaara, E., and Rönnholm, P., 2005, Factors affecting the quality of DTM generation in forested areas. In *Proceedings of ISPRS Workshop Laser scanning 2005*, September 12–14, 2005, Netherlands, GITC bv, Enschede, the Netherlands, 36, Part 3/W19, pp. 85–90.

Hyyppä, J., Yu, X., Rönnholm, P., Kaartinen, H., and Hyyppä, H., 2003b, Factors affecting object-oriented forest growth estimates obtained using laser scanning. *The Photogrammetric Journal of Finland*, 18, 16–31.

Jaakkola, A., Hyyppä, J., Kukko, A., Yu, X., Kaartinen, M., Lehtomäki, M., and Lin, Y., 2010, A low-cost multi-sensoral mobile mapping system and its feasibility for tree measurements. *ISPRS Journal of Photogrammetry and Remote Sensing*, 65(6), 514–522.

Kaartinen, H. and Hyyppä, J., 2008, EuroSDR/ISPRS Project, Commission II, "Tree Extraction". Final Report, EuroSDR. European Spatial Data Research, *Official Publication* No 53.

Kaartinen, H., Hyyppä, J., Yu, X., Vastaranta, M., Hyyppä, H., Kukko, A., Holopainen, M. et al., 2012, An international comparison of individual tree detection and extraction using airborne laser scanning. *Remote Sensing*, 4(4), 950–974.

Koch, B., 2010, Status and future of laser scanning, synthetic aperture radar and hyperspectral remote sensing data for forest biomass assessment. *ISPRS Journal of Photogrammetry and Remote Sensing*, 65(6), 581–590.

Korpela, I., Orka, H., Hyyppa, J., Heikkinen, V., and Tokola, T., 2010, Range and AGC normalization in airborne discrete-return LiDAR intensity data for forest canopies. *ISPRS Journal of Photogrammetry and Remote Sensing*, 65(4), 369–379.

Korhonen, L., Korhonen, K.T., Rautiainen, M., and Stenberg, P., 2006, Estimation of forest canopy cover: A comparison of field measurement techniques. *Silva Fennica*, 40(4), 577–588.

Kraus, K. and Otepka, J., 2005, DTM modeling and visualization - The SCOP approach. In *Photogrammetric Week '05*, Fritsch, D. (Ed.), Herbert Wichmann Verlag, Heidelberg, Germany, pp. 241–252.

Kraus, K. and Pfeifer, N., 1998, Determination of terrain models in wooded areas with airborne laser scanner data. *ISPRS Journal of Photogrammetry and Remote Sensing*, 53, 193–203.

Larsen, M. and Rudemo, M., 1998, Optimizing templates for finding trees in aerial photographs. *Pattern Recognition Letters*, 19(12), 1153–1162.

Leckie, D., Gougeon, F., Hill, D.,Quinn, R., Armstrong, L., and Shreenan, R., 2003, Combined high-density LiDAR and multispectral imagery for individual tree crown analysis. *Canadian Journal of Remote Sensing*, 29(5), 633–649.

Lefsky, M., Cohen, W., Parker, G., and Harding, D., 2002, LiDAR remote sensing for ecosystem studies. *Bioscience*, 52, 19–30.

Liang, X., Hyyppä, J., and Matikainen, L., 2007, First-last pulse signatures of airborne laser scanning for tree species classification. ISPRS Workshop on Laser Scanning 2007 and Silvilaser 2007. In *International Archives of Photogrammetry, Remote Sensing and Spatial Information Sciences*, September 12–14, 2007, 36(3/W52), ISPRS, Espoo, pp. 253–257.

Liang, X., Hyyppä, J., Kaartinen, J., Holopainen, M., and Melkas, T., 2012b. Detecting changes in forest structure over time with bi-temporal terrestrial laser scanning data. *ISPRS International Journal of Geo-Information*, 1(3), 242–255.

Liang, X., Kukko, A., Kaartinen, H., Hyyppä, J., Yu, X., Jaakkola, A., and Wang, Y., 2014b. Possibilities of a personal laser scanning system for forest mapping and ecosystem services, *Sensors*, 14(1), 1228–1248.

Liang, X., Hyyppä, J., Kankare, V., and Holopainen, M., 2011, Stem curve measurement using terrestrial laser scanning. In *Proceedings of Silvilaser 2011*, University of Tasmania, Australia, 6 p.

Liang, X., Hyyppä, J., Kukko, A., Kaartinen, H., Jaakkola, A., and Yu, X. 2014c. The use of a mobile laser scanning for mapping large forest plots. *IEEE Geoscience and Remote Sensing Letters*, 11(9), 1504–1508.

Liang, X., Kankare, V., Hyyppä, J., Wang, Y., Kukko, A., Haggrén, H., Yu, X. et al., 2016, Terrestrial laser scanning in forest inventories. *ISPRS Journal of Photogrammetry and Remote Sensing*, 115, 63–77.

Liang, X., Kankare, V., Yu, X., and Hyyppä, J., 2014a, Automated stem curve measurement using terrestrial laser scanning. *IEEE Transactions on Geoscience and Remote Sensing*, 52(3), 1739–1748.

Liang, X., Litkey, P., Hyyppä, J., Kaartinen, H., Vastaranta, M., and Holopainen, M., 2012a, Automatic stem mapping using single-scan terrestrial laser scanning. *IEEE Transactions on Geoscience and Remote Sensing*, 50(2), 661–670.

Lin, Y., Jaakkola, A., Hyyppä, J., and Kaartinen, H., 2010, From TLS to VLS: Biomass estimation at individual tree level, *Remote Sensing*, 2(8), 1864–1879.

Lindberg, E., 2012, Estimation of canopy structure and individual trees from laser scanning data [WWW Document]. URL http://pub.epsilon.slu.se/8888/ (accessed 3.21.13).

Lim, K., Treitz, P., Wulder, M., St. Onge, B., and Flood, M., 2003, LiDAR remote sensing of forest structure. *Progress in Physical Geography*, 27(1), 88–106.

Maclean, G. and Krabill, W., 1986, Gross-merchantable timber volume estimation using an airborne LiDAR system. *Canadian Journal of Remote Sensing*, 12, 7–18.

Maas, H.G., Bienert, A., Scheller, S., and Keane, E., 2008, Automatic forest inventory parameter determination from terrestrial laser scanner data. *International Journal of Remote Sensing*, 29(5), 1579–1593.

Magnussen, S., Eggermont, P., and LaRiccia, V.N., 1999, Recovering tree heights from airborne laser scanner data. *Forest Science*, 45(3), 407–422.

Mallet, C. and Bretar, F., 2009, Full-waveform topographic LiDAR: State-of-the-art. *ISPRS Journal of Photogrammetry and Remote Sensing*, 64(1), 1–16.

Maltamo, M., Malinen, J., Packalen, P., Suvanto, A., and Kangas, J., 2006, Nonparametric estimation of stem volume using airborne laser scanning, aerial photography, and stand-register data. *Canadian Journal of Forest Research*, 36, 426–436.

Maltamo, M., Mustonen, K., Hyyppä, J., Pitkänen, J., and Yu, X., 2004, The accuracy of estimating individual tree variables with airborne laser scanning in boreal nature reserve. *Canadian Journal of Forest Research*, 34, 1791–1801.

Maltamo, M., Naesset, E., and Vauhkonen, J. (Eds.), 2014, *Forestry Applications of Airborne Laser Scanning, Concepts and Case Studies*. Dordrecht, the Netherlands: Springer, 464 p.

Matala, J., 2005, Impacts of climate change on forest growth: A modelling approach with application to management. Ph D dissertation, Faculty of Forestry, University of Joensuu, Finland.

Matikainen, L., Lehtomäki, M., Ahokas, E., Hyyppä, J., Karjalainen, M., Jaakkola, A., Kukko, A., and Heinonen, T., 2016, Remote Sensing methods for power line corridor surveys. *ISPRS Journal of Photogrammetry and Remote Sensing*, 119, 10–31.

Moffiet, T., Mengersen, K., Witte, C., King, R., and Denham, R., 2005, Airborne laser scanning: Exploratory data analysis indicates potential variables for classification of individual trees or forest stands according to species. *ISPRS Journal of Photogrammetry and Remote Sensing*, 59, 289–309.

Morsdorf, F., Meier, E., Allgöwer, B., and Nüesch, D., 2003, Clustering in airborne laser scanning raw data for segmentation of single trees. In *International Archives of the Photogrammetry, Remote Sensing and Spatial Information Sciences*, 34, Part 3/W13, ISPRS, Munich, pp. 27–33.

Murphy, G.E., Acuna, M.A., and Dumbrell, I., 2010, Tree value and log product yield determination in radiata pine (Pinus radiata) plantations in Australia: Comparisons of terrestrial laser scanning with a forest inventory system and manual measurements. *Canadian Journal of Forest Research*, 40, 2223–2233.

Næsset, E., 1997a, Determination of mean tree height of forest stands using airborne laser scanner data. *ISPRS Journal of Photogrammetry and Remote Sensing*, 52, 49–56.

Næsset, E., 1997b, Estimating timber volume of forest stands using airborne laser scanner data. *Remote Sensing of Environment*, 61, 246–253.

Næsset, E., 2002, Predicting forest stand characteristics with airborne scanning laser using a practical two-stage procedure and field data. *Remote Sensing of Environment*, 80, 88–99.

Næsset, E., 2003, Laser scanning of forest resources – The Norwegian experience. In *Proceedings of the ScandLaser Scientific Workshop on Airborne Laser Scanning of Forests*, September 3–4, SLU, Umeå, Sweden, pp. 35–42. http://pub.epsilon.slu.se/9060/

Næsset, E., 2004, Effects of different flying altitudes on biophysical stand properties estimated from canopy height and density measured with a small-footprint airborne scanning laser. *Remote Sensing of Environment*, 91(2), 243–255.

Næsset, E. and Gobakken, T., 2005, Estimating forest growth using canopy metrics derived from airborne laser scanner data. *Remote Sensing of Environment*, 96(3–4), 453–465.

Næsset, E., Gobakken, T., Holmgren, J., Hyyppä, H., Hyyppä, J., Maltamo, M., Nilsson, M. et al., 2004, Laser scanning of forest resources: The Nordic experience. *Scandinavian Journal of Forest Research*, 19(6), 482–499.

Næsset, E. and Økland, T., 2002, Estimating tree height and tree crown properties using airborne scanning laser in a boreal nature reserve. *Remote Sensing of Environment*, 79, 105–115.

Nelson, R., 2013, How did we get there? An early history of forestry LiDAR. *Canadian Journal of Remote Sensing*, 39(1), 6–17.

Nelson, R., Krabill, W., and Maclean, G., 1984, Determining forest canopy characteristics using airborne laser data. *Remote Sensing of Environment*, 15, 201–212.

Nelson, R., Krabill, W., and Tonelli, J., 1988, Estimating forest biomass and volume using airborne laser data. *Remote Sensing of Environment*, 24, 247–267.

Nelson, R., Oderwald, R., and Gregoire, T., 1997, Separating the ground and airborne laser sampling phases to estimate tropical forest basal area, volume, and biomass. *Remote Sensing of Environment*, 60, 311–326.

Niemi, M. and Vauhkonen, J., 2016, Extracting canopy surface texture from airborne laser scanning data for the supervised and unservised prediction of area-based forest characteristics. *Remote Sensing*, 8(7), 582.

Nilsson, M., 1990, Forest inventory using an airborne LiDAR system. In *Proceedings from SNS/IUFRO workshop in Umeå*, Remote Sensing Laboratory, Report 4, February 26–28, 1990, (Umeå: Swedish University of Agricultural Sciences), pp. 133–139.

Nilsson, M., 1996, Estimation of tree heights and stand volume using an airborne LiDAR system. *Remote Sensing of Environment*, 56, 1–7.

Nilsson, M., Brandtberg, T., Hagner, O., Holmgren, J., Persson, Å., Steinvall, O., Sterner, H., Söderman, U., and Olsson, H., 2003, Laser scanning of forest resources – The Swedish experience. In *Proceedings of the ScandLaser Scientific Workshop on Airborne Laser Scanning of Forests*, September 3–4, 2003, SLU, Umeå, Sweden, pp. 43–52. http://pub.epsilon.slu.se/9060/

Packalén, P. and Maltamo, M., 2006, Predicting the plot volume by tree species using airborne laser scanning and aerial photographs. *Forest Science*, 52(6), 611–622.

Packalén, P. and Maltamo, M., 2007, The k-MSN method in the prediction of species specific stand attributes using airborne laser scanning and aerial photographs. *Remote Sensing of Environment*, 109, 328–341.

Päivinen, R., Nousiainen, M., and Korhonen, K., 1992, Puutunnusten mittaamisen luotettavuus. English summary: Accuracy of certain tree measurements. *Folia Forestalia*, 787, 18 p.

Palojärvi, P., 2003, Integrated electronic and optoelectronic circuits and devices for pulsed time-of-flight laser rangefinding. PhD Thesis, Department of Electrical and Information Engineering and Infotech Oulu, University of Oulu, Finland, 54 p.

Persson, Å., Holmgren, J., and Söderman, U., 2002, Detecting and measuring individual trees using an airborne laser scanner. *Photogrammetric Engineering & Remote Sensing*, 68(9), 925–932.

Persson, Å., Holmgren, J., and Söderman, U., 2006, Identification of tree species of individual trees by combining very high resolution laser data with multi-spectral images. EARSeL SIG Forestry. In *International Workshop 3D Remote Sensing in Forestry Proceedings*, February 14–15, 2006, Earsel, Vienna, Austria, pp. 91–96.

Petrie, G., 2013. Current developments of laser scanner suitable for use on lightweight UAVs. *Geoinformatics*, 8, 16–22.

Philip, M., 1983, Measuring trees and forests, the division of forestry, University of Dar Es Salaam, 338 p.

Pinz, A.J., 1991, A computer vision system for the recognition of trees in aerial photographs. Multisource data integration in remote sensing. In *NASA Conference Publication*, 3099. pp. 111–124.

Pitkänen, J., Maltamo, M., Hyyppä, J., and Yu, X., 2004, Adaptive methods for individual tree detection on airborne laser based height model. *International Conference NATSCAN "Laser-Scanners for Forest and Landscape Assessment - Instruments, Processing Methods and Applications,"* October 3–6, 2004, Freiburg, Germany, In *International Archives of Photogrammetry, Remote Sensing and Spatial Information Sciences*, ISPRS, Freiburg, 36(8/W2), pp. 187–191.

Poage, N.J. and Tappeiner, J.C., 2002, Long-term patterns of diameter and basal area growth of old-growth Douglas-fir trees in western Oregon. *Canadian Journal of Forest Research*, 32, 1232–1243.

Pollock, R.J., 1994, A model-based approach to automatically locating individual tree crowns in high-resolution images of forest canopies. In *Proceedings of First International Airborne Remote Sensing Conference and Exhibition*. September 11–15, 1994, Strasbourg, France, 3, pp. 357–369. ERIM International.

Popescu, S., Wynne, R., and Nelson, R., 2003, Measuring individual tree crown diameter with LiDAR and assessing its influence on estimating forest volume and biomass. *Canadian Journal of Remote Sensing*, 29(5), 564–577.

Pyysalo, U. and Hyyppä, H., 2002, Reconstructing tree crowns from laser scanner data for feature extraction. *International Society for Photogrammetry and Remote Sensing - ISPRS Commission III Symposium (PCV'02)*, September 9–13, 2002, ISPRS, Graz, Austria.

Reitberger, J., Krzystek, P., and Heurich, M., 2006, Full-waveform analysis of small footprint airborne laser scanning data in the Bavarian forest national park for tree species classification. In *International Workshop 3D Remote Sensing in Forestry Proceedings*. February 14–15, 2006, Earsel, Vienna, Austria, pp. 218–227.

Reutebuch, S., McGaughey, R., Andersen, H., and Carson, W., 2003, Accuracy of a high-resolution LiDAR terrain model under a conifer forest canopy. *Canadian Journal of Remote Sensing*, 29, 527–535.

Rönnholm, P., Hyyppä, J., Hyyppä, H., Haggrén, H., Yu, X., Pyysalo, U., Pöntinen, P., and Kaartinen, H., 2004, Calibration of laser-derived tree height estimates by means of photogrammetric techniques, *Scandinavian Journal of Forest Research*, 19(6), 524–528.

Schreier, H., Lougheed, J., Tucker, C., and Leckie, D., 1985, Automated measurements of terrain reflection and height variations using airborne infrared laser system. *International Journal of Remote Sensing*, 6(1), 101–113.

Simonse, M., Aschoff, T., Spiecker, H., and Thies, M., 2003, Automatic determination of forest inventory parameters using terrestrial laser scanning. In *Proceedings of the ScandLaser Scientific Workshop on Airborne Laser Scanning of Forests*, SLU, Umeå, pp. 252–258.

Sithole, G. and Vosselman, G., 2004, Experimental comparison of filter algorithms for bare-Earth extraction from airborne laser scanning point clouds. *ISPRS Journal of Photogrammetry and Remote Sensing*, 59, 85–101.

Sohlberg, S., Næsset, E., and Bollandsas, O., 2006, Single tree segmentation using airborne laser scanner data in a structurally heterogeneous spruce forest. *Photogrammetric Engineering and Remote Sensing*, 72(12), 1369–1378.

Solodukhin, V., Zukov, A., and Mazugin, I., 1977, Possibilities of laser aerial photography for forest profiling, *Lesnoe Khozyaisto* (Forest Management), 10, 53–58. (in Russian).

St-Onge, B. and Vepakomma, U., 2004, Assessing forest gap dynamics and growth using multitemporal laser scanner data. *International Conference NATSCAN "Laser-Scanners for Forest and Landscape Assessment - Instruments, Processing Methods and Applications,"* October 3–6, 2004, Freiburg, Germany, In *International Archives of Photogrammetry, Remote Sensing and Spatial Information Sciences*, 36, Part 8/W2, pp. 173–178.

Straub, B., 2003, A top-down operator for the automatic extraction of trees—concept and performance evaluation. In *Proceedings of the ISPRS Working Group III/3 Workshop "3-D Reconstruction from Airborne Laserscanner and InSAR Data,"* October 8–10, 2003, Dresden, Germany, ISPRS, Dresden, pp. 34–39.

Suvanto, A., Maltamo, M., Packalen, P., and Kangas, J., 2005, Kuviokohtaisten puustotunnustenennustaminen laserkeilauksella. *Metsätieteen aikakauskirja*, 4, 2005, 413–428.

Takeda, H., 2004, Ground surface estimation in dense forest. *The International Archives of Photogrammetry, Remote Sensing and Spatial Information Sciences*, 35, Part B3, ISPRS, Istanbul, pp. 1016–1023.

Thies, M., Pfeifer, N., Winterhalder, D., and Gorte, B.G.H., 2004, Three-dimensional reconstruction of stems for assessment of taper, sweep and lean based on laser scanning of standing trees. *Scandinavian Journal of Forest Research*, 19, 571–581.

Tiede, D. and Hoffman, C., 2006, Process oriented object-based algorithms for single tree detection using laser scanning. In *International Workshop 3D Remote Sensing in Forestry Proceedings*, February 14–15, 2006, Vienna, Austria.

TopoSys, 1996, Digital Elevation Models, Services and Products, 10 p.

Uuttera, J., Haara, A., Tokola, T., and Maltamo, M., 1998, Determination of the spatial distribution of trees from digital aerial photographs. *Forest Ecology and Management*, 110, 275–282.

Uuttera, J., Hiltunen, J., Rissanen, P., Anttila, P., and Hyvönen, P., 2002, Uudet kuvioittaisen arvioinnin menetelmät – Arvio soveltuvuudesta yksityismaiden metsäsuunnitteluun. *Metsätieteen aikakauskirja*, 3, 523–531. (In Finnish)

Uzoh, F. C.C. and Oliver, W.W., 2006, Individual tree height increment model for managed even-aged stands of ponderosa pine throughout the western United States using linear mixed effects models. *Forest Ecology and Management*, 221, 147–154.

Vauhkonen, J., Ene, L., Gupta, S., Heinzel, J., Holmgren, J., Pitkänen, J., Solberg, S. et al., 2011, Comparative testing of single-tree detection algorithms under different types of forest. *Forestry*, 85(1), 27–40.

Villikka, M., Maltamo, M., Packalén, P., Vehmas, M., and Hyyppä, J., 2007, Alternatives for predicting tree-level stem volume of Norway spruce using airborne laser scanner data. *Photogrammetric Journal of Finland*, 20(29), 33–42.

Vosselman, G., 2000, Slope based filtering of laser altimetry data. *International Archives of Photogrammetry and Remote Sensing*, 33, B3/2, 935–942.

Wagner, W., 2005, Physical principles of airborne laser scanning. University course: Laser scanning, data acquisition and modeling, October 6–7, 2005.

Wagner, W., Ullrich, A., Melzer, T., Briese, C., and Kraus, K., 2004, From single-pulse to full-waveform airborne laser scanners: Potential and practical challenges. In *International Archives of Photogrammetry, Remote Sensing and Spatial Information Sciences* 2004, 35, Part B3, ISPRS, Istanbul, pp. 201–206.

Wallace, L., Arko, L., Christopher, W., and Darren, T., 2012, Development of a UAV-LiDAR system with application to forest inventory. *Remote Sensing*, 4(6), 1519–1543.

Wallace, L., Robert, M., and Arko, L., 2014, An assessment of the repeatability of automatic forest inventory metrics derived from UAV-borne laser scanning data. *Transactions on Geoscience and Remote Sensing*, 52(11), 7160–7169.

Wang, Y., Hyyppä, J., Liang, X., Kaartinen, H., Yu, X., Lindberg, E., Holmgren, J. et al., 2016, International benchmarking of the individual tree detection methods for modelling 3-D canopy structure for silviculture and forest ecology using airborne laser scanning. *IEEE Transactions on Geoscience and Remote Sensing*, 54(9), 5011–5027.

Watt, P.J. and Donoghue, D.N.M., 2005, Measuring forest structure with terrestrial laser scanning. *International Journal of Remote Sensing*, 26, 1437–1446.

White, J.C., Wulder, M.A., Varhola, A., Vastaranta, M., Coops, N.C., Cook, B.D., Pitt, D, and Woods, M., 2013, A best practices guide for generating forest inventory attributes from airborne laser scanning data using the area-based approach. Information Report FI-X-10. Canadian Forest Service, Canadian Wood Fibre Centre, Pacific Forestry Centre, Victoria, BC. 50 p.

Wulder, M., 2003, The current status of laser scanning of forests in Canada and Australia. In *Proceedings of the ScandLaser Scientific Workshop on Airborne Laser Scanning of Forests*, September 3–4, 2003, SLU, Umeå, Sweden, pp. 21–33.

Wulder, M., Niemann, K.O., and Goodenough, D., 2000, Local maximum filtering for the extraction of tree locations and basal area from high spatial resolution imagery. *Remote Sensing of Environment*, 73, 103–114.

Yu, X., Hyyppä, J., Kaartinen, H., and Maltamo, M., 2004a, Automatic detection of harvested trees and determination of forest growth using airborne laser scanning. *Remote Sensing of Environment*, 90, 451–462.

Yu, X., Hyyppä, J., Hyyppä, H., and Maltamo, M., 2004b, Effects of flight altitude on tree height estimation using airborne laser scanning. *International Conference NATSCAN "Laser-Scanners for Forest and Landscape Assessment—Instruments, Processing Methods and Applications,", 3–6*, October 2004, Freiburg, Germany, In *International Archives of Photogrammetry, Remote Sensing and Spatial Information Sciences*, 36 Part 8/W2, pp. 96–101.

Yu, X., Hyyppä, J., Kaartinen, H., Hyyppä, H., Maltamo, M., and Rönnholm, P., 2005, Measuring the growth of individual trees using multitemporal airborne laser scanning point clouds. In *Proceedings of ISPRS Workshop Laser Scanning 2005*, September 12–14, 2005, Netherlands, GITC bv, Enschede, the Netherlands, 36, Part 3/W19, pp. 204–208.

Yu, X., Hyyppä, J., Kaartinen, H., Maltamo, M., and Hyyppä, H., 2008, Obtaining plotwise mean height and volume growth in boreal forests using multitemporal laser surveys and various change detection techniques. *International Journal of Remote Sensing*, 29(5), 1367–1386.

Yu, X., Hyyppä, J., Kukko, A., Maltamo, M., and Kaartinen, H., 2006, Change detection techniques for canopy height growth measurements using airborne laser scanner data. *Photogrammetric Engineering & Remote Sensing*, 72(12), 1339–1348.

Yu, X., Hyyppä, J., Rönnholm, P., Kaartinen, H., Maltamo, M., and Hyyppä, H., 2003, Detection of harvested trees and estimation of forest growth using laser scanning, In *Proceedings of the Scandlaser Scientific Workshop on Airborne Laser Scanning of Forests*, September 3–4, 2003, Swedish University of Agricultural Sciences, Umeå, Sweden, Final ed. pp. 115–124.

Yu, X., Hyyppä, H., Vastaranta, M., Holopainen, M., and Viitala, R., 2011, Predicting individual tree attributes from airborne laser point clouds based on the random forests technique. *ISPRS Journal of Photogrammetry and Remote Sensing*, 66(1), 28–37.

Yu, X., Hyyppä, J., Litkey, P., Kaartinen, H., Vastaranta, M., and Holopainen, M., 2017, Single-sensor solution to tree species classification using multispectral airborne laser scanning. *Remote Sensing*, 9(2), 108. doi:10.3390/rs9020108.

Zhao, X., Liu, J., and Tan, M., 2006, A remote aerial robot for topographic survey. *Proceedings on International Conference on Intelligent Robots and Systems, IEEE/RJS*, Beijing, China, October, 9–15, pp. 3143–3148.

Ziegler, M., Konrad, H., Hofrichter, J., Wimmer, A., Ruppert, G., Schardt, M., and Hyyppä, J., 2000, Assessment of forest attributes and single-tree segmentation by means of laser scanning. *Laser Radar Technology and Applications V*, 4035, 73–84.

13 Integration of LiDAR and Photogrammetric Data

Triangulation and Orthorectification

Ayman Habib

CONTENTS

13.1 INTRODUCTION

The steady evolution of mapping technology is leading to an increasing availability of multisensor geospatial datasets at a reasonable cost. For decades, analog frame cameras have been the traditional source of mapping data. The development of softcopy photogrammetric workstations, together with the improved performance of charge-coupled device and complementary metal–oxide–semiconductor arrays, is stimulating the direct incorporation of digital imaging systems in mapping activities. However, current single-head digital frame cameras are incapable of providing imagery while simultaneously maintaining the geometric resolution and ground coverage of analog sensors. In spite of this limitation, the low cost of medium-format digital frame imaging systems has led to their frequent adoption by the mapping community, especially when combined with Light Detection and Ranging (LiDAR) systems. To offset the limitations of single-head digital frame cameras, multihead frame cameras and line cameras are being utilized in spaceborne imaging systems and some airborne platforms. On another front, the improved performance of global navigation satellite system (GNSS)/inertial navigation system (INS) technology is having a positive impact in reducing the control requirements for photogrammetric triangulation. In addition to this contribution, the increased accuracy of direct georeferencing systems is leading to the widespread implementation of LiDAR systems for 3D data collection. The complementary nature of the spectral and spatial data acquired by imaging and LiDAR systems is motivating their integration for a better description of the object space. However, such integration is only possible after accurate coregistration of the collected data to a common reference frame. The current chapter introduces algorithms for a multiprimitive and multisensor triangulation environment, which is geared towards taking advantage of the complementary characteristics of spatial data available from the above-mentioned sensors. The triangulation procedure ensures the alignment of the involved data to a common reference frame, as defined by the control utilized. Such alignment leads to the straightforward production of orthophotos. The main advantages of orthophotos include the possibility of using them as an inexpensive alternative to base maps and the capability of easily integrating them with other forms of geospatial data such as surface models and existing geographic information systems databases. In addition, orthophotos can be draped on top of LiDAR surface models to produce realistic visualization of urban environments. Differential rectification, which is the standard procedure for orthophoto generation, leads to serious artifacts (ghost images/double mapping problems) when dealing with large scale imagery over urban area. To eliminate these artifacts, true orthophoto generation techniques should be adopted. In this regard, this chapter outlines several methodologies for true orthophoto generation using LiDAR and photogrammetric data collected over urban areas. The devised methodologies are tested and proven to be efficient through experiments with real multisensor data.

13.2   SYNERGISTIC CHARACTERISTICS OF LiDAR AND PHOTOGRAMMETRIC DATA

A diverse range of spatial data acquisition systems are now available onboard satellite, aerial, and terrestrial mapping platforms. The diversity starts with analog and digital frame cameras and continues to include linear array scanners. In the past few years, imaging sensors witnessed vast development as a result of enormous advancement in digital technology. For example, the increasing sensor size and storage capacity of digital frame cameras led to their application in traditional and new mapping functions. However, due to technical limitations, current single-head digital frame cameras are not capable of simultaneously providing geometric resolution and ground coverage similar to those associated with analog frame cameras. To alleviate this limitation, multihead frame cameras and push-broom scanners (line cameras) have been developed and used onboard satellite and aerial platforms. Rigorous modeling of the imaging process used by push-broom scanners is far more complicated than the modeling of frame sensors (Ebner et al., 1994, 1996; Habib and Beshah, 1998; Lee et al., 2000; Habib et al., 2001; Lee and Habib, 2002; Lee et al., 2002; Poli, 2004; Toutin, 2004a, b). For example, in the absence of georeferencing parameters, the narrow angular field of view associated with current push-broom scanners is causing instability in the triangulation procedure due to high correlations among the exterior orientation parameters (EOP). Therefore, there has been a tremendous body of research dealing with the derivation of alternative models for line cameras, to circumvent such instability (El-Manadili and Novak, 1996; Gupta and Hartley, 1997; Wang, 1999; Dowman and Dolloff, 2000; Ono et al., 2000; Tao and Hu, 2001; Hanley et al., 2002; Fraser et al., 2002; Fraser and Hanley, 2003; Habib et al., 2004; Tao et al., 2004). In addition to line cameras, the low cost of medium-format digital frame cameras is leading to their frequent adoption by the mapping community, especially when combined with LiDAR systems. Apart from passive imaging systems, LiDAR scanning is rapidly taking its place in the mapping industry as a fast and cost-effective 3D data acquisition technology. The increased accuracy and affordability of GNSS/INS systems are the main reasons behind the expanding role of LiDAR sensors in mapping activities.

Considering the characteristics of the acquired spatial and spectral data by passive imaging and LiDAR systems, one can argue that their integration will be beneficial for an accurate and better description of the object space. As an illustration of the complementary characteristics of passive

TABLE 13.1

Photogrammetric Weaknesses, Contrasted with LiDAR Strengths

LiDAR Pros	Photogrammetric Cons
Dense information along homogeneous surfaces	Almost no positional information along homogeneous surfaces
Day or night data collection	Day time data collection only
Direct acquisition of 3D coordinates	Complicated and sometimes unreliable matching procedures
The vertical accuracy is better than the planimetric accuracy	The vertical accuracy is worse than the planimetric accuracy

TABLE 13.2

LiDAR Weaknesses, Contrasted with Photogrammetric Strengths

Photogrammetric Pros	LiDAR Cons
High redundancy	No inherent redundancy
Rich in semantic information	Positional information; difficult to derive semantic information
Dense positional information along object space break lines	Almost no positional information along break lines
The planimetric accuracy is better than the vertical accuracy	The planimetric accuracy is worse than the vertical accuracy

imaging and LiDAR systems, Tables 13.1 and 13.2 list the advantages and disadvantages of each system, contrasted with the corresponding cons and the pros of the other system. As can be seen in these tables, the disadvantages of one system can be compensated for by the advantages of the other system (Baltsavias, 1999; Satale and Kulkarni, 2003). However, the synergic characteristics of both systems can be fully utilized only after ensuring that both datasets are georeferenced relative to the same reference frame (Habib and Schenk, 1999; Chen et al., 2004).

Traditionally, photogrammetric georeferencing has been either indirectly established with the help of Ground Control Points (GCP) or directly defined using GNSS/INS units onboard the imaging platform (Cramer et al., 2000; Wegmann et al., 2004). On the other hand, LiDAR georeferencing is directly established through the GNSS/INS components of the LiDAR system. In this regard, this chapter presents some alternative methodologies for the utilization of LiDAR features as a source of control for photogrammetric georeferencing. These methodologies have two main advantages. First, they ensure the coalignment of the LiDAR and photogrammetric data to a common reference frame as defined by the GNSS/INS unit of the LiDAR system. Moreover, LiDAR features eliminate the need for GCP to establish the georeferencing parameters for the photogrammetric data. The possibility of utilizing LiDAR data as a source of control for photogrammetric georeferencing is furnished by the accuracy of the LiDAR point cloud (e.g., the horizontal accuracy is usually in the range of few decimeters whereas the vertical accuracy is in the subdecimeter range). LiDAR data can be used for image georeferencing if and only if we can identify common features in both datasets. Therefore, the first objective of the developed methodologies is to select appropriate primitives (i.e., point, linear, or areal features). Afterwards, the mathematical models, which can be utilized in the triangulation procedure, to relate the LiDAR and photogrammetric primitives will be introduced. Another objective of the proposed methodologies is that they should be flexible enough to allow for the incorporation of primitives identified in scenes captured by frame and line cameras. In other words, the developed methodologies should handle multisensor data, while using a wide range of primitives (i.e., they are multiprimitive and multisensor triangulation methodologies). The outcome of the multisensor triangulation contributes to a straightforward incorporation of photogrammetric and LiDAR data for orthophoto generation. However, to ensure the generation of accurate true orthophotos when dealing with large scale imagery over urban areas, one must make use of additional criteria to detect portions of the LiDAR surface that are invisible in the involved imagery. Therefore, the triangulation discussion will be followed by an investigation of potential true orthophoto generation methodologies that use LiDAR and imagery acquired by frame and line cameras.

The chapter will start by offering a brief discussion of LiDAR and photogrammetric principles. The discussion of photogrammetric principles will focus on the possibility of incorporating frame and line cameras in a single triangulation mechanism. Then, the selection of primitives for relating the LiDAR and photogrammetric data, as well as the respective mathematical models for their incorporation in the triangulation procedure, will be outlined. Afterwards, alternative methodologies for true orthophoto generation will be introduced. Finally, the feasibility and the performance of the developed multiprimitive and multisensor triangulation methodologies will be outlined in the experimental results section (in which real data is used), together with some concluding remarks.

13.3 LiDAR PRINCIPLES

LiDAR has been conceived to directly and accurately capture digital surfaces. The commercial market for LiDAR has developed significantly only within the last two decades (Faruque, 2003; Lemmens, 2015). The affordability, increased pulse frequency, and versatility of new LiDAR systems are causing an exponential profusion of these systems in the mapping industry. A LiDAR system is composed of two main units; these are the laser scanning and ranging as well as GNSS/INS units, shown in Figure 13.1. Positional information derived from LiDAR systems is based on calculating the range from the laser unit to the object space. As shown in Figure 13.1, the measured range and laser beam pointing direction are coupled with the position and orientation information, as determined by the onboard GNSS/INS unit, to directly determine the position of the laser footprint, through a vector summation process, using Equation 13.1 (El-Sheimy et al., 2005). As can be seen in Equation 13.1, there is no inherent redundancy in the computation of the captured LiDAR surface. Therefore, the overall accuracy depends on the accuracy and calibration quality of the different components comprising the LiDAR system.

$$\vec{X}_G = \vec{X}_o + R_{\text{yaw,pitch,roll}} R_{\Delta\omega,\Delta\phi,\Delta\kappa} \vec{P}_G + R_{\text{yaw,pitch,roll}} R_{\Delta\omega,\Delta\phi,\Delta\kappa} R_{\alpha,\beta} \begin{bmatrix} 0 \\ 0 \\ -\rho \end{bmatrix} \quad (13.1)$$

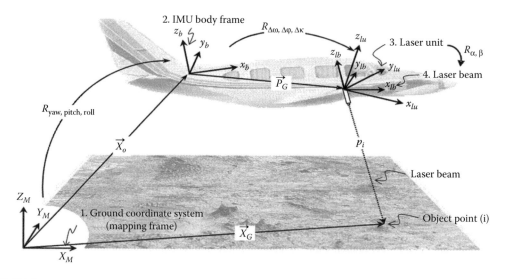

FIGURE 13.1 Coordinate systems and parameters involved in the direct georeferencing of LiDAR systems.

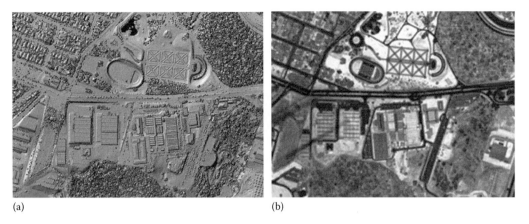

(a)   (b)

FIGURE 13.2   Visualization of LiDAR coverage: Shaded-relief of range data (a) and intensity image (b).

Besides the derived positional information, the intensity of the signal echo is also recorded by most LiDAR scanners, as shown in Figure 13.2. The intensity map can be used for object space segmentation and feature extraction. However, it is still difficult to derive semantic information regarding the captured surfaces (e.g., material and type of observed structures) from the intensity image (Baltsavias, 1999; Schenk, 1999; Wehr and Lohr, 1999). It is expected that the integration of photogrammetric and LiDAR data will improve the semantic content of the LiDAR data and will offer another check for the quality control of LiDAR surfaces.

13.4   PHOTOGRAMMETRIC PRINCIPLES

In this section, a brief overview of photogrammetric principles will be presented. As photogrammetric operations using frame cameras are well established, the focus of this section will be on the utilization of line cameras in photogrammetric triangulation, while investigating the possibility of incorporating frame and line cameras in a single triangulation mechanism. The majority of imaging satellites make use of a line camera that has a single linear array in the focal plane, in contrast to the 2D array of a frame camera, as shown in Figure 13.3. A single exposure of this linear array covers a narrow strip in the object space. Continuous coverage of contiguous areas on the ground is achieved by repetitive exposure of the linear array while observing a proper relationship between the sensor's integration time, ground sampling distance, and the velocity of the platform. In this regard, a distinction will be made between a scene and an image. An image is obtained through a single exposure of the light-sensitive elements in the focal plane. A scene, on the other hand, covers a 2D area in the object space and might be composed of a single image or multiple images, depending on the nature

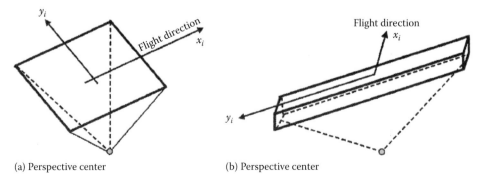

(a) Perspective center   (b) Perspective center

FIGURE 13.3   Imaging sensors for frame (a) and line (b) cameras.

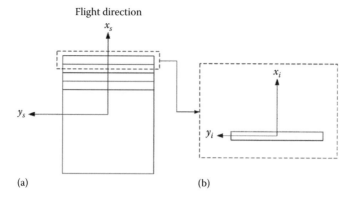

FIGURE 13.4 Scene (a) and image (b) coordinate systems for a line camera.

of the camera utilized. Based on this distinction, a scene captured by a frame camera consists of a single image. A scene captured by a line camera, on the other hand, is composed of multiple images.

For frame and line cameras, the collinearity equations mathematically describe the constraint that the perspective center, the image point, and the corresponding object point are aligned along a straight line. For a line camera, the collinearity model can be described by Equation 13.2. It should be noted that the collinearity equations involve the image coordinates (x_i, y_i), which are equivalent to the scene coordinates (x_s, y_s) when dealing with a scene captured by a frame camera. For line cameras, however, the scene coordinates (x_s, y_s) need to be transformed into image coordinates. In this case, the x_s value is used to indicate the moment of exposure of the corresponding image. The y_s value, on the other hand, is directly related to the y_i image coordinate, as shown in Figure 13.4. It should be noted that the x_i image coordinate in Equation 13.2 is a constant that depends on the alignment of the linear array in the focal plane.

$$x_i = x_p - c\,\frac{r_{11}^t(X_G - X_O^t) + r_{21}^t(Y_G - Y_O^t) + r_{31}^t(Z_G - Z_O^t)}{r_{13}^t(X_G - X_O^t) + r_{23}^t(Y_G - Y_O^t) + r_{33}^t(Z_G - Z_O^t)}$$

$$(13.2)$$

$$y_i = y_p - c\,\frac{r_{12}^t(X_G - X_O^t) + r_{22}^t(Y_G - Y_O^t) + r_{32}^t(Z_G - Z_O^t)}{r_{13}^t(X_G - X_O^t) + r_{23}^t(Y_G - Y_O^t) + r_{33}^t(Z_G - Z_O^t)}$$

where:

(X_G, Y_G, Z_G) are the ground coordinates of the object point
(X_O^t, Y_O^t, Z_O^t) are the ground coordinates of the perspective center at the moment of exposure
r_{11}^t to r_{33}^t are the elements of the rotation matrix at the moment of exposure
(x_i, y_i) are the image coordinates of the point in question
(x_p, y_p, c) are the Interior Orientation Parameters (IOP) of the imaging sensor

Another difference between the collinearity equations for frame and line cameras is the multiple exposures associated with a line camera scene, in contrast to the single exposure for an image captured by a frame camera. Due to the multiple exposures, the EOP associated with a line camera scene are time dependent and vary depending on the image considered within the scene. For practical reasons, the bundle adjustment of scenes captured by line cameras does not consider all the EOP involved as this would lead to an excessive number of parameters to be solved, requiring an extensive amount of control. Moreover, one should expect that the EOPs associated with the different images to be relatively close to each other due to smooth nature of the platform trajectory. To avoid such a problem, bundle adjustments of scenes captured by line cameras make use of one of two parameter reduction approaches (Ebner et al., 1994, 1996). In the first approach, the system's trajectory and attitude are assumed to follow a polynomial trend, with time as the independent variable. Thus, the number of unknown EOPs is reduced to the number of coefficients involved in the assumed polynomials. Available knowledge of the trajectory of the imaging system can be used as prior information regarding the polynomial coefficients, leading to what is known as the physical sensor model (Toutin, 2004a, 2004b). Another approach to reducing the number of EOPs involved is based on the use of orientation images that are equally spaced along the system's trajectory. The EOPs at any given time are modeled as a weighted average of the EOPs associated with the neighboring orientation images. In this way, the number of EOPs for a given scene is reduced to the number of parameters associated with the orientation images involved.

It should be noted that the imaging geometry associated with line cameras (including the reduction methodology used for the involved EOPs) is more general than that of frame cameras. In other words, the imaging geometry of a frame camera can be derived as a special case of that for a line camera. For example, an image captured by a frame camera can be considered to be a special case of a scene captured by a line camera, in which the trajectory and attitude are represented by a zero-order polynomial. Alternatively, when working with orientation images, a frame image can be considered to be a line camera scene with one orientation image. The general nature of the imaging geometry of line cameras lends itself to straightforward development of multisensor triangulation procedures capable of incorporating frame and line cameras (Habib and Beshah, 1998; Lee et al., 2002).

Having discussed LiDAR and photogrammetric principles, the focus will now be shifted towards algorithm development for the direct incorporation of LiDAR data in photogrammetric triangulation. To address this, Section 13.5 will deal with the selection, representation, and extraction of appropriate primitives from LiDAR and photogrammetric data. This discussion will be followed by the derivation of the necessary mathematical models used to incorporate these primitives in the photogrammetric triangulation.

13.5 TRIANGULATION PRIMITIVES

A triangulation process relies on the identification of common primitives to relate the involved datasets to the reference frame defined by the control information. Traditionally, photogrammetric triangulation has been based on point primitives. When considering photogrammetric and LiDAR data, relating a LiDAR footprint to the corresponding point in the imagery is almost impossible, as shown in Figure 13.5. Therefore, point primitives are not appropriate for the task at hand. Linear and areal features are other potential primitives that can be more suitable for relating LiDAR and photogrammetric data. Linear features can be manually identified in the imagery, whereas conjugate LiDAR lines can be extracted through planar patch segmentation and

(a)   (b)

FIGURE 13.5   Imagery (a) and LiDAR (b) coverage of an urban area.

intersection. Alternatively, LiDAR lines can be directly identified in the intensity images produced by most of today's LiDAR systems. It should be noted that linear features extracted through planar patch segmentation and intersection are more accurate than features extracted from intensity images. The lower quality of features extracted from intensity images is caused by the interpolation process utilized to convert the irregular LiDAR footprints to a raster grid (Ghanma, 2006). Instead of using linear features, areal primitives in photogrammetric datasets can be defined by their boundaries, which can be identified in the imagery. Such primitives include, for example, rooftops, lakes, and other homogeneous regions. In LiDAR datasets, areal regions can be derived through planar patch segmentation techniques.

Another issue that is related to the selection of primitives is their representation in both photogrammetric and LiDAR data. Image space lines can be represented by a sequence of image points along the feature, which can be manually identified, as shown in Figure 13.6a. This is an appealing representation as it can handle image space linear features in the presence of distortions, which cause deviations from straightness in the image space. Furthermore, such a representation allows the inclusion of linear features in scenes captured by line cameras, as perturbations in the flight trajectory also lead to deviations from straightness in image space linear features corresponding to object space straight lines (Habib et al., 2002). It should be noted that the intermediate points selected along corresponding line segments in overlapping scenes need not be conjugate. As for the LiDAR data, object space lines can be represented by their end points, as shown in Figure 13.6b. The points defining the LiDAR lines need not be visible in the imagery.

When using areal primitives, photogrammetric planar patches can be represented by three points (e.g., corner points, as in Figure 13.7a) along their boundaries. These points should be identified in all overlapping imagery. Similar to the representation of linear features, this representation is valid for scenes captured by both frame and line cameras. LiDAR patches, on the other hand, can be represented by the laser beam footprints defining the patches, as shown in Figure 13.7b. These points can be derived directly through planar patch segmentation techniques. Having settled the issues of primitive selection and representation, Section 13.6 will focus on explaining the proposed mathematical models to relate the photogrammetric and LiDAR primitives within the triangulation procedure.

(a)   (b)

FIGURE 13.6 Image space linear features defined by sequences of intermediate points (a), whereas corresponding LiDAR lines are defined by their end points (b).

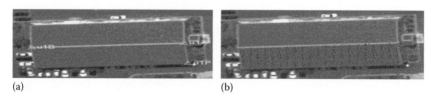

(a)   (b)

FIGURE 13.7 Image space planar features are represented by three points (a), whereas LiDAR patches are defined by the points comprising the patches (b).

13.6 MATHEMATICAL MODELS

13.6.1 UTILIZING STRAIGHT LINEAR PRIMITIVES: COPLANARITY-BASED CONSTRAINT

The current section will focus on deriving the mathematical constraint used to relate LiDAR and photogrammetric lines, which are represented by the end points in the object space and a sequence of intermediate points in the image space, respectively. From this perspective, the photogrammetric datasets will be aligned to the LiDAR reference frame through the direct incorporation of LiDAR lines as the source of control. The photogrammetric and LiDAR measurements along corresponding linear features can be related to one another through the coplanarity constraint of Equation 13.3. This constraint indicates that the vector from the perspective center to any intermediate point along the image line is contained within the plane defined by the perspective center of that image and the two points defining the LiDAR line. In other words, for a given intermediate point, k'' the points $\{(X_1, Y_1, Z_1), (X_2, Y_2, Z_2), (X''_o, Y''_o, Z''_o), \text{ and } (x_{k''}, y_{k''}, 0)\}$ are coplanar, as illustrated in Figure 13.8.

$$(\vec{V}_1 \times \vec{V}_2) \cdot \vec{V}_3 = 0 \tag{13.3}$$

where:

$\vec{V}_1$ is the vector connecting the perspective center to the first end point along the LiDAR line
$\vec{V}_2$ is the vector connecting the perspective center to the second end point along the LiDAR line
$\vec{V}_3$ is the vector connecting the perspective center to an intermediate point along the corresponding image line

It should be noted that the earlier constraint can be introduced for all the intermediate points along the image space linear feature. Moreover, the coplanarity constraint is valid for both frame and line cameras. For scenes captured by line cameras, the EOPs involved correspond to the image associated with the intermediate point under consideration. For frame cameras with known IOPs, a maximum of two independent constraints can be defined for a given image. However, in

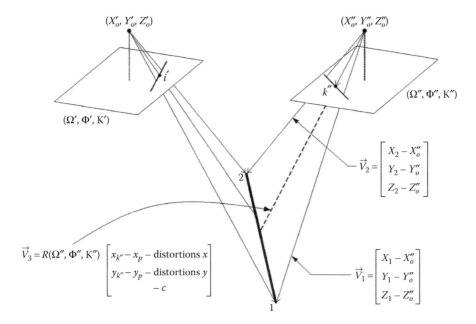

FIGURE 13.8 Perspective transformation between image and LiDAR straight lines, and the coplanarity constraint for intermediate points along the line.

self-calibration procedures, additional constraints aid in the recovery of the IOPs, as the distortion pattern changes from one intermediate point to the next along the image space linear feature. The coplanarity constraint also aids in the recovery of the EOPs associated with line cameras. This contribution is attributed to the fact that the system's trajectory affects the shape of the linear features in the image space.

For an image block, a minimum of two noncoplanar line segments are needed to establish the datum of the reconstructed object space, namely, the scale, rotation, and shift components. This requirement assumes that a model can be derived from the image block, and is explained by the fact that a single line defines two shift components across the line, as well as two rotation angles (as defined by the line heading and pitch). Having another noncoplanar line helps in the estimation of the remaining shift and rotation components, as well as the scale factor.

13.6.2 Utilizing Straight Linear Primitives: Collinearity-Based Constraint

Instead of the above-mentioned approach, existing bundle adjustment procedures, which are based on the collinearity equations, can be used to incorporate linear features after the modification of the variance–covariance matrices associated with the line measurements in either the object space or the image space. Before getting into the implementation details of this approach, we will start by outlining the method of representing the linear features. Similar to the previous approach, derived object space lines from the LiDAR data are represented by two points. Image space linear features, on the other hand, are represented by two points that are not conjugate to those defining the line in the object space, as illustrated in Figure 13.9. However, the image and object space points are assigned the same identification code. The fact that these points are not conjugate is tolerated by manipulating their variance–covariance matrices in either the image space or the object space. In other words, the variance–covariance matrix of a given point is expanded along the linear feature to allow for its movement along the direction of the line. For example, Equations 13.4 through 13.8 are used for the expansion of the error ellipse associated with an end point in the

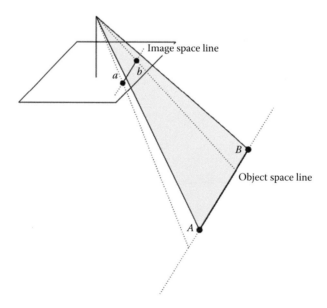

FIGURE 13.9 Representation of image and object lines for the collinearity-based incorporation of linear features in a bundle adjustment.

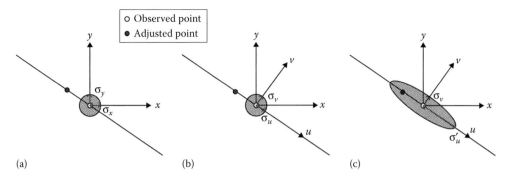

FIGURE 13.10 Steps involved in the expansion of the error ellipse associated with an end point of an image line. (a) Error ellipse in the (x, y) coordinate system, (b) error ellipse in the (u,υ) coordinate system, and (c) error ellipse in the (u,υ) coordinate system after expanding the error along the line orientation.

image space (refer to Figure 13.10 for a conceptual illustration). First, we need to define the xy- and $u\upsilon$-coordinate systems. The xy-coordinate system is the image coordinate system, shown in Figure 13.10a. The $u\upsilon$-coordinate system, on the other hand, is a coordinate system with the u-axis aligned along the image line, as illustrated in Figure 13.10b. The relationship between these coordinate systems is defined by a rotation matrix (Equation 13.4), which is defined by the image line orientation. Using the law of error propagation, the variance–covariance matrix of a line end point in the $u\upsilon$-coordinate system can be derived from its variance–covariance matrix in the xy-coordinate system, Equation 13.5, according to Equation 13.6. To allow for the movement of the point along the line, its variance along the u-axis is replaced by a large value (simply by multiplying the variance by a constant m), as in Equation 13.7 and shown in Figure 13.10c. Finally, the modified variance–covariance matrix in the xy-image coordinate system is calculated through the utilization of the inverse rotation matrix, as in Equation 13.8. The same procedure can be alternatively used to expand the variance–covariance matrices of the end points of the LiDAR lines.

$$\begin{bmatrix} u \\ \upsilon \end{bmatrix} = R \begin{bmatrix} x \\ y \end{bmatrix} \tag{13.4}$$

$$\sum_{xy} = \begin{bmatrix} \sigma_x^2 & \sigma_{xy} \\ \sigma_{xy} & \sigma_y^2 \end{bmatrix} \tag{13.5}$$

$$\sum_{u\upsilon} = R \sum_{xy} R^T = \begin{bmatrix} \sigma_u^2 & \sigma_{u\upsilon} \\ \sigma_{u\upsilon} & \sigma_\upsilon^2 \end{bmatrix} \tag{13.6}$$

$$\sum_{u\upsilon}' \begin{bmatrix} m \times \sigma_u^2 & \sigma_{u\upsilon} \\ \sigma_{u\upsilon} & \sigma_\upsilon^2 \end{bmatrix} \tag{13.7}$$

$$\sum_{xy}' = R^T \sum_{u\upsilon}' R \tag{13.8}$$

The decision as to whether expand the variance–covariance matrix in the image space or object space depends on the procedure under consideration (e.g., single photo resection, bundle

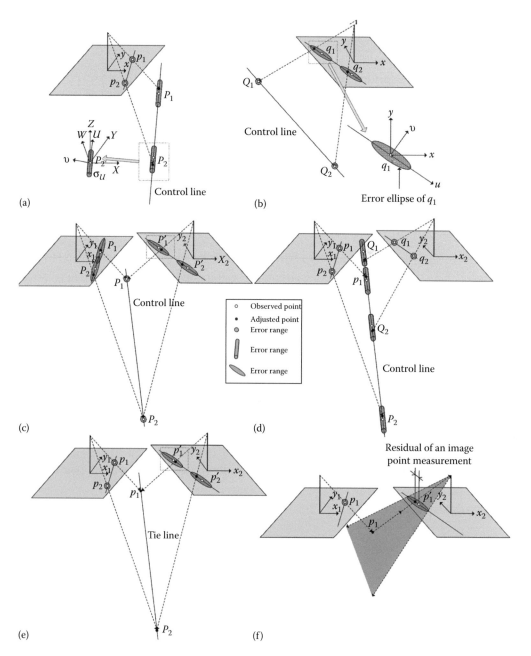

FIGURE 13.11 (a–f) Variance expansion options for line end points, for their incorporation in a collinearity-based bundle adjustment procedure.

adjustment of an image block using control lines, or bundle adjustment of an image block using tie lines), as explained in the following:

1. Single photo resection using control lines (Figure 13.11a): In this approach, the image line is represented by two end points with their variance–covariance matrices defined by the expected image coordinate measurement accuracy. The variance–covariance matrices of the end points of the control line, on the other hand, are expanded to compensate for the

fact that the image and the object end points are not conjugate. It should be noted that this approach can also be used in single photo resection of a scene captured by a line camera.

2. Single-photo resection using control lines (Figure 13.11b): In this approach, the object line is represented by its end points, whose variance–covariance matrices are defined by the expected accuracy of the procedure used to define the points. The variance–covariance matrices of the image line end points, on the other hand, are expanded along the direction of the line. One should note that this approach is not appropriate for scenes captured by line cameras, as the image line orientation cannot be rigorously defined at a given point, due to perturbations in the flight trajectory.

3. Bundle adjustment of an image block using control lines (Figure 13.11c): In this approach, the object line is represented by its end points, with their variance–covariance matrices defined by the procedure used to provide the control lines. The variance–covariance matrices of the image lines, on the other hand, are expanded to compensate for the fact that the end points of the image lines are not conjugate to those defining the object line. It should be noted that this approach is not appropriate for scenes captured by line cameras as the image line orientation cannot be rigorously defined.

4. Bundle adjustment of an image block using control lines (Figure 13.11d): In this case, the image lines in overlapping images are represented by nonconjugate end points whose variance–covariance matrices are defined by the expected image coordinate measurement accuracy. To compensate for the fact that these points are not conjugate, the selected end points are assigned different identification codes. The object line, on the other hand, is defined by a list of points whose variance–covariance matrices are expanded along the line. The number of object points used depends on the number of selected points in the image space. It should be noted that this approach can be used for scenes captured by frame or line cameras as it does not require the expansion of the variance–covariance matrices in the image space.

5. Bundle adjustment of an image block using tie lines (Figure 13.11e): In this approach, all variance–covariance matrices associated with the end points of the image lines are expanded, except for two points, which are used to define the object line. As this approach requires an expansion of the error ellipses in the image space, it is not appropriate for scenes captured by line cameras, as the image line orientation cannot be rigorously defined due to perturbations in the flight trajectory. Figure 13.11f illustrates the conceptual basis of the derivation of the object point through the proposed variance–covariance manipulation. An image point used to define the object line, and its error ellipse, which is established by the image coordinate measurement accuracy, together with the corresponding perspective center, define an infinite light ray. An image point and its expanded error ellipse, on the other hand, together with the corresponding perspective center, define a plane. The intersection of the aforementioned light ray and plane is used to estimate the ground coordinates of the point in question.

13.6.3 Utilizing Planar Patches: Coplanarity-Based Constraint

The current section will focus on deriving the mathematical constraint used to relate LiDAR and photogrammetric patches, which are represented by a set of points in the object space and three points in the image space, respectively. As an example, let us consider a surface patch that is represented by two sets of points, namely the photogrammetric set $S_{PH} = \{A, B, C\}$ and the LiDAR set $S_L = \{(X_P, Y_P, Z_P), P = 1 \text{ to } n\}$, as shown in Figures 13.12 and 13.14a. As the LiDAR points are randomly distributed, no point-to-point correspondence can be assumed between the datasets. For the photogrammetric points, the image and object space coordinates are related to one another through the collinearity equations. LiDAR points belonging to a certain planar patch, on the other hand, should coincide with the photogrammetric patch representing the same object space surface,

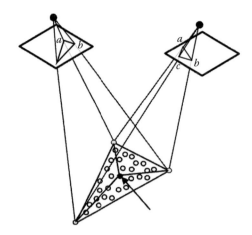

FIGURE 13.12 Coplanarity of photogrammetric and LiDAR patches.

as illustrated in Figure 13.12. The coplanarity of the LiDAR and photogrammetric points can be mathematically expressed through the determinant constraint in Equation 13.9.

$$V = \begin{vmatrix} X_P & Y_P & Z_P & 1 \\ X_A & Y_A & Z_A & 1 \\ X_B & Y_B & Z_B & 1 \\ X_C & Y_C & Z_C & 1 \end{vmatrix} = \begin{vmatrix} X_P - X_A & Y_P - Y_A & Z_P - Z_A \\ X_B - X_A & Y_B - Y_A & Z_B - Z_A \\ X_C - X_A & Y_C - Y_A & Z_C - Z_A \end{vmatrix} \qquad (13.9)$$

The above-mentioned constraint is used as the mathematical model to incorporate LiDAR points into the photogrammetric triangulation. In physical terms, this constraint describes the fact that the normal distance between any LiDAR point and the corresponding photogrammetric surface should be zero, or the volume of the tetrahedron composed of the four points should be zero. This constraint can be applied to all LiDAR points comprising a surface patch.

It should be noted that the above-mentioned constraint is valid for both frame and line cameras. Another advantage of this approach is the possibility of using such a constraint for LiDAR system calibration. If the system model is available with explicit values of the systematic error terms, the original LiDAR points (X_P, Y_P, Z_P) in Equation 13.9 can be replaced with the LiDAR raw measurements, as in Equation 13.1. In this case, the systematic error terms can be solved for in the bundle adjustment procedure, and photogrammetric planar patches serve as the source of control for the LiDAR calibration. Therefore, additional control information is needed to establish the georeferencing parameters for the photogrammetric data, which, in turn, serve as the control for LiDAR calibration.

To be sufficient as the only source of control, LiDAR patches should be able to provide all the datum parameters; three translations (X_T, Y_T, Z_T), three rotations (ω, ϕ, κ), and one scale (S) parameter. By inspecting Figure 13.13, it is evident that a patch normal to one of the axes will provide the shift in the direction of that axis, as well as the rotation angles across the other axes. Therefore, three nonparallel patches are sufficient to determine the position and orientation components of the datum. To enable scale determination, the three planar patches should not intersect at a single point (e.g., they should not be facets of a pyramid). Alternatively, the scale can be determined by incorporating a fourth plane, as shown in Figure 13.13.

13.6.4 Utilizing Planar Patches: Collinearity-Based Constraint

In this approach, existing bundle adjustment procedures, which are based on the collinearity equations, can be used to incorporate areal control features after the manipulation of the

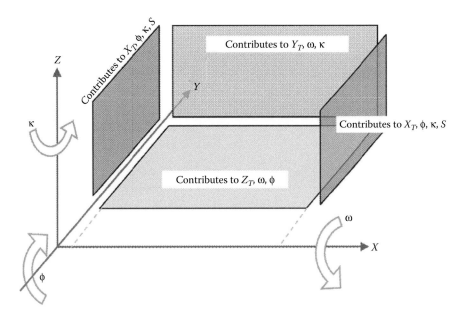

FIGURE 13.13 Optimal configuration for establishing the datum using planar patches as the source of control.

variance–covariance matrices defining the control patches. Similar to the previous approach, the planar patch in the image space is defined by a set of vertices that can be identified in overlapping imagery. The control patch, on the other hand, is defined by an equivalent set of object points (i.e., there are an identical number of vertices defining the image and object patches). However, the object points need not be conjugate to those identified in the image space, as illustrated in Figure 13.14b. To compensate for the fact that these sets of points are not conjugate, the variance–covariance matrices of the object points are expanded along the plane direction (Equations 13.10 through 13.14, Figure 13.15). In these equations, the XYZ-coordinate system defines the ground reference frame, and the UVW-coordinate system is defined by the orientation of the control patch in the object space. Specifically, the UV-axes are defined to be parallel to the plane through the patch (i.e., the W-axis is aligned along the normal to the patch). The expansion of the error ellipsoid along the plane is facilitated by multiplying the variances along the U and V directions by large numbers (e.g., m_U and m_V in Equation 13.13). It should be noted that this approach is suitable for scenes captured by both frame and line cameras.

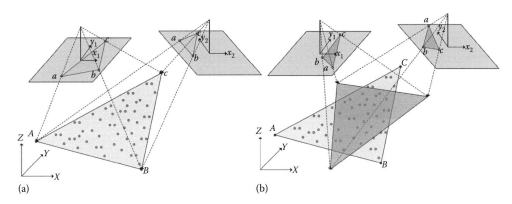

(a)   (b)

FIGURE 13.14 Coplanarity-based (a) and collinearity-based (b) incorporation of LiDAR patches in photogrammetric triangulation.

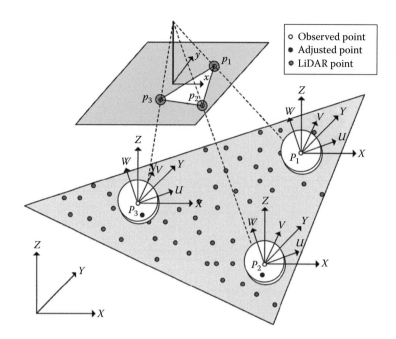

FIGURE 13.15 Conceptual basis of the collinearity-based incorporation of areal control patches in existing bundle adjustment procedures.

$$\begin{bmatrix} U \\ V \\ W \end{bmatrix} = R \begin{bmatrix} X \\ Y \\ Z \end{bmatrix} \tag{13.10}$$

$$\sum_{XYZ} = \begin{bmatrix} \sigma_X^2 & \sigma_{XY} & \sigma_{XZ} \\ \sigma_{YX} & \sigma_Y^2 & \sigma_{YZ} \\ \sigma_{ZX} & \sigma_{ZY} & \sigma_Z^2 \end{bmatrix} \tag{13.11}$$

$$\sum_{UVW} = R \sum_{XYZ} R^T = \begin{bmatrix} \sigma_U^2 & \sigma_{UV} & \sigma_{UW} \\ \sigma_{VU} & \sigma_V^2 & \sigma_{VW} \\ \sigma_{WU} & \sigma_{WV} & \sigma_W^2 \end{bmatrix} \tag{13.12}$$

$$\sum_{UVW}' = \begin{bmatrix} m_U \times \sigma_U^2 & \sigma_{UV} & \sigma_{UW} \\ \sigma_{VU} & m_V \times \sigma_V^2 & \sigma_{VW} \\ \sigma_{WU} & \sigma_{WV} & \sigma_W^2 \end{bmatrix} \tag{13.13}$$

$$\sum_{XYZ}' = R^T \sum_{UVW}' R \tag{13.14}$$

In summary, Sections 13.6.1 through 13.6.4 introduced various approaches for performing multi-sensor and multiprimitive triangulation. These methodologies are intended to incorporate photogrammetric datasets acquired by frame and line cameras, in addition to LiDAR derived features, in a single triangulation mechanism. Having discussed multisensor triangulation for the alignment of the LiDAR and photogrammetric data to a common reference frame, Section 13.7 will address the utilization of this data for orthophoto generation.

13.7   LiDAR AND PHOTOGRAMMETRIC DATA FOR ORTHOPHOTO GENERATION

Orthophoto production focuses on the elimination of the sensor tilt and terrain relief effects from captured perspective imagery. Uniform scale and the absence of relief displacement make orthophotos an important component of geographic information systems databases, in which the user can directly determine geographic locations, measure distances, compute areas, and derive other useful information about the area in question. Recently, with the increasing availability of overlapping LiDAR and photogrammetric data, which is acquired from aerial and satellite imaging platforms, there has been a persistent need for an orthophoto generation methodology that is capable of dealing with imagery acquired from such systems. Moreover, the increasing resolution of available data mandates the development of true orthophotos, in which occluded portions of the LiDAR surface are reliably detected in the involved imagery.

Differential rectification has traditionally been used for orthophoto generation, and is illustrated in Figure 13.16. The differential rectification procedure starts by projecting the digital surface model (DSM) cells (as derived from the LiDAR data) onto the image space (Novak, 1992). This is followed by resampling the gray values at the projected points from neighboring pixels in the image plane. Finally, the resampled gray values are assigned to the corresponding cells in the orthophoto plane. It should be noted that the projection from the DSM cells to the image plane is a straightforward process when working with images acquired by frame cameras (i.e., the projection involves solving two collinearity equations in two unknowns, which are the image coordinates). However, object-to-image projection requires an iterative process when working with scenes captured by line cameras,

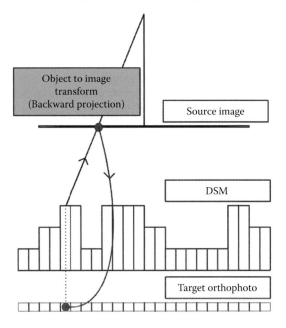

FIGURE 13.16   Orthophoto generation through differential rectification.

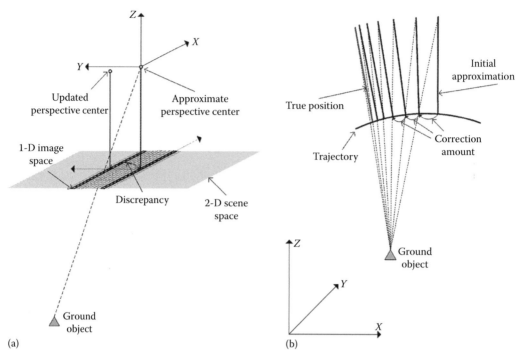

FIGURE 13.17 Iterative procedure for determining the exposure station that captured a given object point (b), starting from an approximate location (a).

as shown in Figure 13.17. The iterative procedure is essential to estimate the location of the exposure station that captured the object point in question (Habib et al., 2006).

The performance of differential rectification procedures has been quite acceptable when dealing with medium resolution imagery over relatively smooth terrain. However, when dealing with high resolution imagery over urban areas, differential rectification produces artifacts in the form of ghost images (double-mapped areas) at the vicinity of abrupt surface changes (e.g., at building boundaries and steep cliffs), as shown in Figure 13.18. Double-mapped areas are caused by the fact that gray values of occluded areas are erroneously imported from the perspective imagery. The effects of these artifacts can be mitigated by true orthophoto generation methodologies. The term *true orthophoto* is generally used for orthophotos in which surface elements that are not included in the digital terrain model are rectified to the orthogonal projection. These elements are usually buildings and bridges (Amhar et al., 1998). Kuzmin et al. (2004) proposed a polygon-based approach for the detection of these obscured areas to generate true orthophotos. In this method, conventional digital differential rectification is first applied, then hidden areas are detected through the use of polygonal surfaces generated from a Digital Building Model. With the exception of the methodology proposed by Kuzmin et al., the majority of existing true orthophoto generation techniques are based on the z-buffer method, which is briefly explained in Section 13.7.1 (Catmull, 1974; Amhar et al., 1998; Rau et al., 2000, 2002; Sheng et al., 2003).

13.7.1   *z*-BUFFER METHOD

The z-buffer algorithm proposed by Amhar et al. (1998) has mainly been used for true orthophoto generation. As can be seen in Figure 13.19a, double-mapped areas arise when two object space points (e.g., A and D, B and E, or C and F) are competing for the same image location (e.g., d, e, or f, respectively). The z-buffer method resolves the ambiguity regarding which object point should be assigned to a certain image location by considering the distances between the perspective center and the object points in question. Among the competing object points, the point closest to the

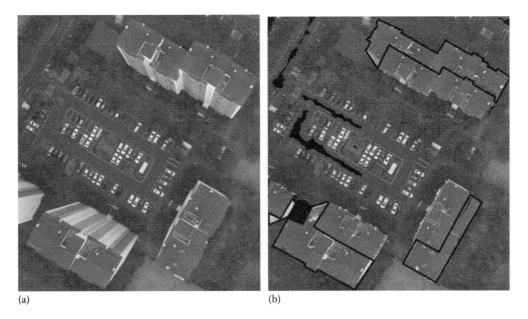

FIGURE 13.18 Double-mapped areas where relief displacement in the perspective image (a) causes duplication of the gray values projected onto the orthophoto plane (b).

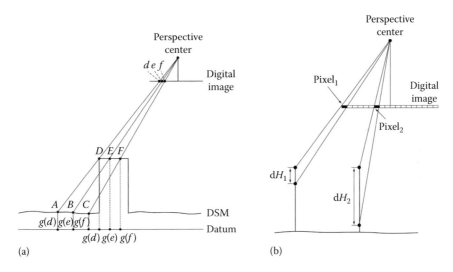

FIGURE 13.19 (a) True orthophoto generation using the z-buffer method, and (b) optimal spatial interval for the pseudoground points.

perspective center of an image is considered to be visible, whereas the other points are considered to be invisible in that image. It should be noted that the z-buffer method can be applied to scenes captured by both frame and line cameras.

The z-buffer has several limitations, which include its sensitivity to the sampling interval of the DSM, as it is related to the ground sampling distance of the imaging sensor. Another significant drawback of the z-buffer methodology is the false visibility associated with narrow vertical structures. This problem is commonly known in the photogrammetric literature as the *M-portion problem*, which can be resolved by introducing additional artificial points (pseudoground points) along building facades (Rau et al., 2000, 2002). Pseudoground points should be carefully introduced along the building facades to avoid false visibility while optimizing the computational speed, as illustrated in Figure 13.19b. If the pseudoground points are too dense, the true orthophoto generation process will be time intensive. On the other hand, pseudoground points that are too sparse can cause false visibilities. As shown in Figure 13.19b, the spacing between two successive pseudoground points should be small enough to handle facades located at image boundaries (i.e., the relative displacement between their projections onto the image space should not exceed one pixel). The separation between successive pseudoground points should increase as one moves towards the nadir point, where no pseudoground points are needed. For more detailed analysis of the limitations of the *z*-buffer method, interested readers can refer to Habib et al. (2007).

13.7.2 ANGLE-BASED METHOD

For an orthogonal projection, the top and bottom of a vertical structure are projected onto the same location. However, in a perspective projection, the top and bottom of that structure are projected as two points. These points are spatially separated by the relief displacement. This displacement takes place along a radial direction from the image space nadir point (Mikhail, 2001); relief displacement is the cause of occlusions in perspective imagery. The presence of occlusions can be discerned by sequentially checking the off-nadir angles to the lines of sight connecting the perspective center to the DSM points, along a radial direction starting with the object space nadir point, as shown in Figure 13.20. More specifically, as one moves away from the nadir point, the angle between the projection ray and the nadir line should increase. Whenever there is a sudden decrease in the off-nadir angle, invisibility is detected. This invisibility persists until the off-nadir angle exceeds that of the last visible point, as illustrated in Figure 13.20. This approach for true orthophoto generation is referred to as the angle-based method. The angle-based methodology for true orthophoto generation can be applied, with minor variations, to scenes captured by both frame and line cameras. For scenes captured by frame cameras, the angle-based method inspects the off-nadir angles starting from the nadir point and moving towards the point in question. For a line camera scene, on the other hand, we have multiple exposure stations, which, in turn, lead to multiple nadir points. Therefore, for scenes captured by line cameras, the angle-based method starts from the object point and moves inward towards the nadir point of the image associated with the object point under consideration. For more details regarding the angle-based method of orthophoto generation, interested readers can refer to Habib et al. (2006, 2007).

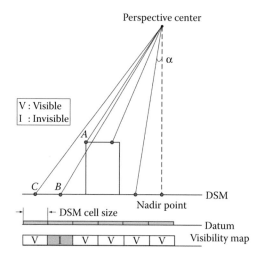

FIGURE 13.20   True orthophoto generation using the angle-based method.

13.8 CASE STUDY

To validate the feasibility and applicability of the earlier methodologies, multisensor datasets will be analyzed. The conducted experiments will involve three types of sensors, namely, a digital frame camera equipped with an integrated GNSS and INS, a satellite-based line camera, and a LiDAR system. These experiments will focus on investigating the following issues:

- Validity of using the line-based georeferencing procedure for scenes captured by frame and line cameras.
- Validity of using the patch-based georeferencing procedure for scenes captured by frame and line cameras.
- Impact of integrating satellite scenes, aerial scenes, LiDAR data, and GNSS exposure positions in a unified bundle adjustment procedure.
- Qualitative investigation of the methodologies presented for true orthophoto generation.

The first dataset is composed of three blocks, each consisting of six digital frame images, captured in April 2005 by the Applanix Digital Sensor System (DSS) over the city of Daejeon in South Korea from an altitude of 1500 m. The DSS camera has 16 megapixels (9 pm pixel size) and a 55 mm focal length. The position and the attitude of the DSS camera were tracked using the onboard GNSS/INS unit. However, only the position information was available for this test. The second dataset consists of an IKONOS stereo-pair, which was captured in November 2001, over the same area. It should be noted that these scenes are raw imagery that did not go through any geometric correction and are provided for research purposes. Finally, a multistrip LiDAR coverage, corresponding to the DSS coverage, was collected using the OPTECH ALTM 2050 system with an average point density of 2.67 point/m^2 from an altitude of 975 m. An example of one of the DSS image blocks and a visualization of the corresponding LiDAR coverage can be seen in Figure 13.21. Figure 13.22 shows the IKONOS coverage and the locations of the DSS image blocks.

To extract the LiDAR control, a total of 139 planar patches and 138 linear features were manually identified through planar patch segmentation and intersection, respectively. Figure 13.21 shows the locations of the features extracted from the middle LiDAR point cloud, within the IKONOS scenes. The corresponding linear and areal features were digitized in the DSS and IKONOS scenes. To evaluate the performance of the different georeferencing techniques, a set of 70 GCP was also acquired; refer to Figure 13.22 for the distribution of these points. The performance of the point-, line-, patch-based, and GPS-assisted georeferencing techniques was evaluated using root-mean-square-error (RMSE) analysis. In the experiments, a portion of the available GCPs was used as control in the bundle

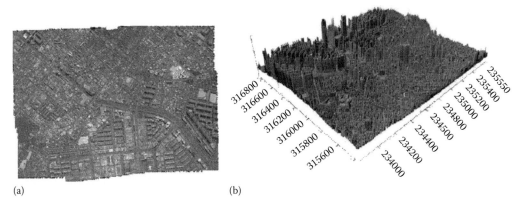

(a)   (b)

FIGURE 13.21 DSS middle image block (a) and the corresponding LiDAR cloud (b); the circles in (a) indicate the locations of linear and areal primitives extracted from the LiDAR data.

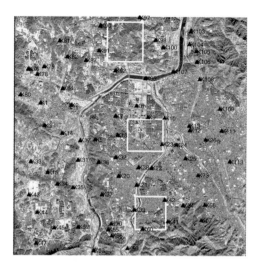

FIGURE 13.22 IKONOS scene coverage, showing the three patches covered by the DSS imagery and LiDAR data, together with the distribution of the GCP.

adjustment, whereas the rest was used as check points. It is worth mentioning that none of the control points are visible in any of the available DSS imagery for this test.

To investigate the performance of the various georeferencing methodologies, we conducted the following experiments (the total RMSE values resulting from these experiments are reported in Table 13.3):

- Photogrammetric triangulation of the IKONOS scenes, while varying the number of GCP utilized (second column in Table 13.3).
- Photogrammetric triangulation of the IKONOS and DSS scenes, while varying the number of GCP utilized (third column in Table 13.3).
- Photogrammetric triangulation of IKONOS and DSS scenes, while considering the GNSS observations associated with the DSS exposures and varying the number of GCP utilized (fourth column in Table 13.3).
- Photogrammetric triangulation of IKONOS and DSS scenes, while varying the number of LiDAR lines (45 and 138 lines) and changing the number of GCP (fifth and sixth columns in Table 13.3).
- Photogrammetric triangulation of IKONOS and DSS scenes, while varying the number of LiDAR patches (45 and 139 patches) and changing the number of GCP (seventh and eighth columns in Table 13.3).

In Table 13.3, N/A means that no solution was attainable (i.e., the control provided was not sufficient to establish the datum for the triangulation procedure). The reported RMSE values in the last four columns of Table 13.3 are based on the coplanarity-based approaches for incorporating linear and areal features. The utilization of the collinearity-based approaches for incorporating linear and areal features resulted in almost identical results. It is worth mentioning that the collinearity-based incorporation of linear and areal features can be used in existing point-based triangulation packages. In such a case, the input files should be modified by expanding the variance–covariance matrices of the utilized points along the linear/areal features. For more clarity, the results in Table 13.3 are visually aggregated in Figure 13.23, in

TABLE 13.3
RMSE (m) for the Checkpoints in Multisensor and Multiprimitive Triangulation

			IKONOS + 18 DSS Frame Images				
			Control Point Plus				
	IKONOS Only			Control Lines		Control Patches	
Number of GCPs	Control Points Only	Control Points Only	DSS GNSS	138	45	139	45
0	N/A	N/A	3.07	3.07	3.09	5.41	5.86
1	N/A	N/A	3.37	3.01	3.08	5.39	6.40
2	N/A	N/A	3.08	3.13	3.20	4.82	5.25
3	N/A	21.32	2.86	2.86	2.85	2.93	3.09
4	N/A	19.96	2.76	2.71	2.75	2.64	3.05
5	N/A	4.34	2.73	2.72	2.74	2.63	2.72
6	3.67	3.35	2.70	2.69	2.74	2.61	2.71
7	3.94	3.04	2.62	2.70	2.73	2.55	2.56
8	3.59	3.42	2.58	2.57	2.53	2.45	2.72
9	4.08	2.54	2.51	2.55	2.52	2.44	2.51
10	3.07	2.50	2.50	2.56	2.53	2.43	2.49
15	3.15	2.41	2.46	2.46	2.43	2.40	2.44
40	2.01	2.09	2.11	2.11	2.10	2.03	2.05

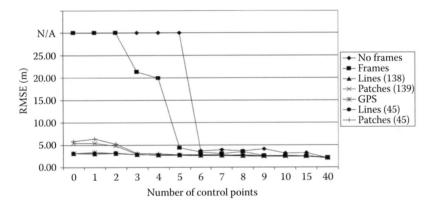

FIGURE 13.23 Check-point analyses from the bundle adjustments involving IKONOS and DSS imagery as well as LiDAR features, for various control configurations.

which the Total RMSE values are plotted against the number of control points, for the different georeferencing techniques. By examining Table 13.3 and Figure 13.23, one can make the following remarks:

- Utilizing points as the only source of control for the triangulation of the stereo IKONOS scenes required a minimum of six GCPs.
- Including DSS imagery in the triangulation of the IKONOS scenes reduced the control requirement for convergence to three GCPs. Moreover, the incorporation of the GNSS observations at the DSS exposure stations enabled convergence without the need for any

GCP. Therefore, it is clear that incorporating satellite scenes with a few frame images enables photogrammetric reconstruction with a reduced GCP requirement.

- LiDAR linear features are sufficient to georeference the IKONOS and DSS scenes without the need for an additional source of control. As can be seen in the fifth and sixth columns of Table 13.3, incorporating additional control points in the triangulation procedure did not significantly improve the reconstruction outcome. Moreover, by comparing the fifth and sixth columns, one can see that increasing the number of linear features from 45 to 138 does not significantly improve the quality of the triangulation outcome.

- LiDAR patches are sufficient to georeference the IKONOS and DSS scenes without the need for an additional source of control. However, as it can be seen in the seventh and eighth columns of Table 13.3, incorporating a few control points had a significant impact in improving the results (through the use of 3 GCPs and 139 control patches, the total RMSE was reduced from 5.41 to 2.93 m). Incorporating additional control points (i.e., beyond three GCPs) did not have a significant impact. The improvement in the reconstruction outcome achieved by using a few GCPs can be attributed to the fact that the majority of the utilized patches were horizontal or with mild slopes, as they represented building roofs. Therefore, the estimation of the model shifts in the X and Y directions was not accurate enough. Incorporating vertical or steep patches could have solved this problem; however, such patches were not available in the dataset provided. Moreover, by comparing the seventh and eighths columns, it can be seen that increasing the number of control patches from 45 to 139 did not significantly improve the quality of the triangulation outcome.

- By comparing the different georeferencing techniques, it can be seen that the patch-based, line-based, and GNSS-assisted georeferencing techniques produced better outcomes than the point-based georeferencing technique. Such an improvement attests to the advantage of adopting multisensor and multiprimitive triangulation procedures.

In an additional experiment, we utilized the EOP derived from the multisensor triangulation of frame and line camera scenes together with the LiDAR surface to generate orthophotos. Figure 13.24 shows sample patches, in which the IKONOS and DSS orthophotos are laid side by side. As can be seen in Figure 13.24a, the generated orthophotos are quite compatible, as indicated by the smooth continuity of the observed features between the DSS and IKONOS orthophotos. Figure 13.24b reveals changes in the object space between the moments of capture of the IKONOS and DSS imagery. Therefore, it is evident that the multisensor triangulation of imagery from frame and line cameras improves the quality of the derived object space, while offering an environment for the accurate georeferencing of temporal imagery. Following the geo-referencing, the involved imagery can be analyzed for change detection applications using the derived and properly georeferenced orthophotos.

Figures 13.18, 13.25, and 13.26 illustrate the comparative performance of the differential rectification, z-buffer, and angle-based methods of orthophoto generation using frame and line cameras, respectively. As can be seen in these figures, differential rectification produces ghost images (refer to the highlighted areas in Figure 13.18b for frame cameras and Figure 13.26b for line cameras). However, ghost images are not present in the true orthophotos generated by the z-buffer and angle-based methods (refer to the highlighted areas in Figure 13.25a and b for frame cameras and Figure 13.26c and d for line cameras). However, visual inspection of Figures 13.25a, b, 13.26c, and d reveals the superior performance of the angle-based method when compared to that of the z-buffer method.

(a)   (b)

FIGURE 13.24 Change detection between DSS (left) and IKONOS (right) orthophotos. A smooth transition between the two orthophotos can be observed in (a), whereas discontinuities are observed in (b) due to changes in the object space.

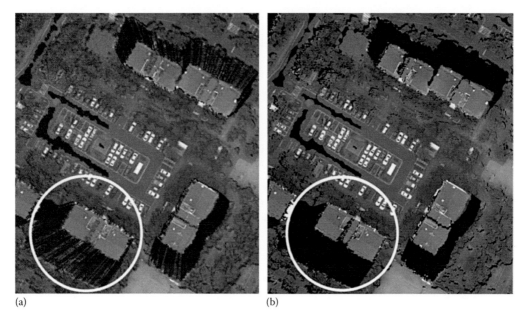

(a)   (b)

FIGURE 13.25 True orthophotos, corresponding to the illustrated image in Figure 13.18a, generated using z-buffer (a), and angle-based (b) techniques.

FIGURE 13.26 A scene captured by a line camera (a), and orthophotos generated using the differential rectification (b), z-buffer (c), and angle-based (d) techniques.

13.9 CONCLUDING REMARKS

The continuous advancement in mapping technology demands the development of commensurate processing methodologies to take advantage of the synergistic characteristics of available geospatial data. In this regard, it is quite evident that integrating LiDAR and photogrammetric data is essential to ensure accurate and better description of the object space. The current chapter presented different methodologies for aligning LiDAR and photogrammetric data relative to a common reference frame using manually extracted linear and areal primitives. The developed methodologies are suited to the characteristics of these datasets. Moreover, the introduced methodologies are general enough to be directly applied to scenes captured by both line and frame cameras. The experimental results have shown that the utilization of LiDAR derived primitives as the source of control for photogrammetric georeferencing yields slightly better results when compared with point-based

FIGURE 13.27 Photo-realistic 3D model produced by rendering a true orthophoto onto LiDAR data.

georeferencing techniques. Moreover, it has been shown that the incorporation of sparse frame imagery, together with satellite scenes, improves the results by taking advantage of the geometric strength of frame cameras to improve the inherently weak geometry of line cameras onboard imaging satellites. Therefore, the combination of aerial and satellite scenes improves the coverage extent as well as the geometric quality of the derived object space. The incorporation of LiDAR data, aerial images, and satellite scenes in a single triangulation procedure also ensures the coregistration of these datasets relative to common reference frame, which is valuable for orthophoto generation and change detection applications. Future work should focus on the automated extraction of the triangulation primitives from the image and LiDAR data.

In addition to the multiprimitive triangulation of multisensor data, this chapter outlined several methodologies for orthophoto generation, to take advantage of the synergistic characteristics of photogrammetric and LiDAR data. With the increasing resolution of geospatial data, true ortho-photo generation techniques should be adopted to ensure the absence of double-mapped areas (ghost images). Such artifacts are expected even when working with satellite scenes. The true ortho-photos generated, after being draped on top of the LiDAR DSM, can be used to produce photo-realistic 3D models, as in Figure 13.27.

ACKNOWLEDGMENTS

This research work has been conducted under the auspices and financial support of the GEOIDE Research Network. The author would like to thank Mr. Paul Mrstik, Terrapoint, Canada Inc. for the valuable technical support and providing the LiDAR and image datasets, which was utilized in the experimental results section.

REFERENCES

Amhar, F. et al., 1998. The generation of true orthophotos using a 3D building model in conjunction with a conventional DTM, *International Archive Photogrammetry and Remote Sensing*, 32: 16–22.
Baltsavias, E., 1999. A comparison between photogrammetry and laser scanning, *ISPRS Journal of Photogrammetry and Remote Sensing*, 54(1): 83–94.

Catmull, E., 1974. A subdivision algorithm for computer display of curved surfaces, PhD dissertation, Department of Computer Science, University of Utah, Salt Lake City, UT.

Chen, L., Teo, T., Shao, Y., Lai, Y., and Rau, J., 2004. Fusion of LiDAR data and optical imagery for building modeling, *International Archives of Photogrammetry and Remote Sensing*, 35(B4): 732–737.

Cramer, M., Stallmann, D., and Haala, N., 2000. Direct georeferencing using GPS/Inertial exterior orientations for photogrammetric applications, *International Archives of Photogrammetry and Remote Sensing*, 33(B3): 198–205.

Dowman, I. and Dolloff, J., 2000. An evaluation of rational functions for photogrammetric restitutions, *International Archives of Photogrammetry and Remote Sensing*, 33(B3): 254–266.

Ebner, H., Kornus, W., and Ohlhof, T., 1994. A simulation study on point determination for the MOMS-02/d2 space project using an extended functional model, *Geo-Information Systems*, 7(1): 11–16.

Ebner, H., Ohlhof, T., and Putz, E., 1996. Orientation of MOMS-02/D2 and MOMS-2P Imagery, *International Archives of Photogrammetry and Remote Sensing*, 31(B3): 158–164.

El-Manadili, Y. and Novak, K., 1996. Precision rectification of SPOT imagery using the direct linear transformation model, *Photogrammetric Engineering and Remote Sensing*, 62(1): 67–72.

El-Sheimy, N., Valeo, C., and Habib, A., 2005. *Digital Terrain Modelling: Acquisition, Manipulation and Applications*, Artech House, Norwood, MA.

Faruque, F., 2003. LiDAR image processing creates useful imagery. *ArcUser Magazine*, January–March 2003, 6(1).

Fraser, C. and Hanley, H., 2003. Bias compensation in rational functions for IKONOS satellite imagery, *Photogrammetric Engineering and Remote Sensing*, 69(1): 53–57.

Fraser, C., Hanley, H., and Yamakawa, T., 2002. High-precision geopositioning from IKONOS satellite imagery, *Proceeding of ACSM-ASPRS 2002*, Washington, DC, unpaginated CD-ROM.

Ghanma, M., 2006. Integration of photogrammetry and LiDAR. PhD Thesis, Department of Geomatics Engineering, University of Calgary, Alberta, Canada, April 2006, UCGE Report 20241, 141 pp.

Gupta, R. and Hartley, R., 1997. Linear push-broom cameras, *IEEE Transactions on Pattern Analysis and Machine Intelligence*, 19(9): 963–975.

Habib, A., Bang, K., Kim, C., and Shin, S., 2006. True orthophoto generation from high resolution satellite imagery. In *Innovations in 3D Geo Information Systems*, Lecture Notes in Geo Information and Cartography, pp. 641–656. Abdul-Rahman, A., Zlatanova, S., and Coors, V., (Eds.). Springer, Heidelberg, Germany.

Habib, A. and Beshah, B., 1998. Multi sensor aerial triangulation, *Proceeding of ISPRS Commission III Symposium*, July 6–10, 1998, Columbus, OH, pp. 37–41.

Habib, A., Kim, E., and Kim, C., 2007. New methodologies for true orthophoto generation, *Photogrammetric Engineering and Remote Sensing*; *2007*, 73(1): 25–36.

Habib, A., Kim, E., Morgan, M., and Couloigner, I., 2004. DEM Generation from high resolution satellite imagery using parallel projection model, *Proceeding of the XXth ISPRS Congress*, Commission 1, TS: HRS DEM Generation from SPOT-5 HRS Data, July 12–23, 2004, Istanbul, Turkey, pp. 393–398.

Habib, A., Lee, Y., and Morgan, M., 2001. Bundle adjustment with self-calibration of line cameras using straight lines, *Joint Workshop of ISPRS WG I/2, I/5 and IV/7: High Resolution Mapping from Space 2001*, September 19–21, 2001, University of Hanover, Hanover, Germany, unpaginated CD-ROM.

Habib, A., Morgan, M., and Lee, Y., 2002. Bundle adjustment with self-calibration using straight lines, *Photogrammetric Record*, 17(100): 635–650.

Habib, A. and Schenk, T., 1999. A new approach for matching surfaces from laser scanners and optical sensors, *International Archives of Photogrammetry and Remote Sensing*, 32(3W14): 55–61.

Hanley, H., Yamakawa, T., and Fraser, C., 2002. Sensor orientation for high resolution satellite imagery, Pecora 15/Land Satellite Information IV/ISPRS Commission I/FIEOS, November 10–15, 2002, Denver, CO, unpaginated CD-ROM.

Kuzmin, P., Korytnik, A., and Long, O., 2004. Polygon-based true orthophoto generation, *XXth ISPRS Congress*, July, 12–23, Istanbul, Turkey, pp. 529–531.

Lee, C., Thesis, H., Bethel, J., and Mikhail, E., 2000. Rigorous mathematical modeling of airborne pushbroom imaging systems, *Photogrammetric Engineering and Remote Sensing*, 66(4): 385–392.

Lee, Y. and Habib, A., 2002. Pose estimation of line cameras using linear features, *Proceedings of ISPRS Commission III Symposium Photogrammetric Computer Vision*, September 9–13, 2002, Graz, Austria.

Lee, Y., Habib, A., and Kim, K., 2002. A study on aerial triangulation from multi-sensor imagery, *Proceedings of the International Symposium on Remote Sensing (ISRS) 2002*, October 30 to November 1, 2002, Sokcho, Korea.

Lemmens, M. J. P. M., 2015. Photon LiDAR, *GIM International*, September 29.

Mikhail, E., Bethel, J., and McGlone, J., 2001. *Introduction to Modern Photogrammetry*, John Wiley & Sons, New York.

Novak, K., 1992. Rectification of digital imagery, *Photogrammetric Engineering and Remote Sensing*, 58(3): 339–344.

Ono, T., Hattori, S., Hasegawa, H., and Akamatsu, S., 2000. Digital mapping using high resolution satellite imagery based on 2-D affine projection model, *International Archives of Photogrammetry and Remote Sensing*, 33(B3): 672–677.

Poli, D., 2004. Orientation of satellite and airborne imagery from multi-line pushbroom sensors with a rigorous sensor model, *International Archives of Photogrammetry and Remote Sensing*, 34(B1): 130–135.

Rau, J., Chen, N., and Chen, L., 2000. Hidden compensation and shadow enhancement for true orthophoto Generation, *Proceedings of Asian Conference on Remote Sensing*, December 4–8, 2000, Taipei, CD-ROM.

Rau, J., Chen, N., and Chen, L., 2002. True orthophoto generation of built-up areas using multi-view images, *Photogrammetric Engineering and Remote Sensing*, 68(6): 581–588.

Satale, D. and Kulkarni, M., 2003. LiDAR in mapping. Map India Conference GISdevelopment.net. Available at http: #www.gisdevelopment.net/technology/gis/mi03129.htm (last accessed March 28, 2006).

Schenk T., 1999. Photogrammetry and laser altimetry. 32, Part 3-W14, *Proceedings of ISPRS Workshop Mapping Surface Structure and Topography by Airborne and Spaceborne Lasers*, November 9–11, 1999, La Jolla, CA, pp. 3–12.

Sheng, Y., Gong, P., and Biging, G., 2003. True orthoimage production for forested areas from large-scale aerial photographs, *Photogrammetric Engineering and Remote Sensing*, 69(3): 259–266.

Tao, V. and Hu, Y., 2001. A comprehensive study of rational function model for photogrammetric processing, *Photogrammetric Engineering and Remote Sensing*, 67(12): 1347–1357.

Tao, V., Hu, Y., and Jiang, W., 2004. Photogrammetric exploitation of IKONOS imagery for mapping applications, *International Journal of Remote Sensing*, 25(14): 2833–2853.

Toutin, T., 2004a. DTM generation from IKONOS in-track stereo images using a 3D physical model, *Photogrammetric Engineering and Remote Sensing*, 70(6): 695–702.

Toutin, T., 2004b. DSM generation and evaluation from quickbird stereo images with 3D physical modeling and elevation accuracy evaluation, *International Journal of Remote Sensing*, 25(22): 5181–5192.

Wang, Y., 1999. Automated triangulation of linear scanner imagery, *Joint Workshop of ISPRS WG I/1, I/3 and IV/4 on Sensors and Mapping from Space*, September 27–30, 1999, Hanover, Germany, unpaginated CD-ROM.

Wegmann, H., Heipke, C., and Jacobsen, K., 2004. Direct sensor orientation based on GPS network solutions, *International Archives of Photogrammetry and Remote Sensing*, 35(B1): 153–158.

Wehr, A. and Lohr, U., 1999. Airborne laser scanning—an introduction and overview, *ISPRS Journal of Photogrammetry and Remote Sensing*, 54(2–3): 68–82.

14 Feature Extraction from Light Detection and Ranging Data in Urban Areas

Frédéric Bretar

CONTENTS

14.1 INTRODUCTION

14.1.1 BACKGROUND

Automatic mapping of urban areas from aerial images is a challenging task for scientists and surveyors because of the complexity of urban scenes. From a photogrammetric point of view, aerial images can be used to produce 3D points provided that their acquisition has been performed in a (multi-) stereoscopic context [1]. Altitudes are processed using automatic correlation algorithms to generate digital surface models (DSMs) [2,3].

For the last decade, Light Detection and Ranging (LiDAR) data have become an alternative data source for generating a tridimensional representation of landscapes. Basically, LiDAR data are acquired by strips and finally gathered as 3D point clouds (Figure 14.1). Raw LiDAR data (3D points) are not expressed in an easy topology as there is no relationship between neighboring points, unlike the intrinsic 4/8-connexity of image-based representations. The access to the data is

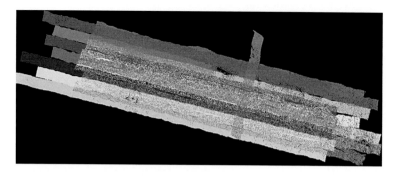

FIGURE 14.1 Example of a global survey over a city area, Amiens, France. Each strip is colored and has a width of ~170 m.

obtained using Kd-tree structures or graphs. Moreover, airborne LiDAR systems do not provide textural information that can be easily exploited.[*]

Although the spatial distribution of a LiDAR point cloud is not uniform, punctual 3D information is an accurate representation of the topography. But a higher description level of landscapes is often wished to fill geographic information system (GIS) databases (vector-based features, higher order semantic). Beyond the production of digital surface or terrain models, the understanding of airborne LiDAR scenes depends on our capability of finding specific shapes to derive a higher order cartography (vector based) than 3D points.

14.1.2 LINE EXTRACTION

In the context of building reconstruction, lines and planar surfaces are particularly of interest. The extraction of lines from a LiDAR point cloud is a difficult task as LiDAR points are randomly distributed over surfaces: Depending on the point density, building edges are approximately delineated. Points considered as edge points may not belong to the true building edge. Nevertheless, the extraction of 2D or 3D lines is considered in different problems such as the reconstruction of building footpoints [4] or the registration of LiDAR data with regard to images. In the case of a joint registration of LiDAR and images, we need to detect invariant features in both datasets before estimating a transform between reference frames [5]. On the one hand, 3D segments can be generated by intersecting preprocessed 3D planar surfaces considered as adjacent ones [6]. On the other hand, 3D segments can be extracted directly from the point cloud [7]. The idea of such approaches (which is very sensitive to parameters) is to classify each LiDAR point depending on its local geometric environment by computing 3D moments. The eigenvalues and the eigenvectors of a covariance matrix deliver geometric characteristics of the distribution of each LiDAR point. In the case of an ideal line, two eigenvalues are null whereas the third one is greater than zero. In the case of a plane, two eigenvalues are similar to a nonzero value, whereas the third one is zero. In the case of an edge line, one eigenvalue is zero, whereas there is a significant difference between the two other ones.

Building footprints are fundamental information for mapping applications. Different steps are involved in the process:

1. Generating a DSM
2. Computing the DSM gradient (Canny–Deriche filter)
3. Processing a hysteresis thresholding to identify high gradient values that are meant to describe roof gutters
4. Chaining segments

---

[*] Intensity values are not considered here since researches are not conclusive regarding their usefulness.

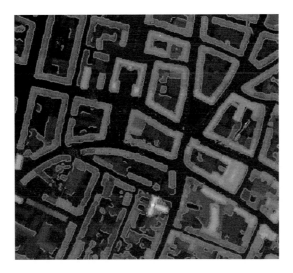

FIGURE 14.2 Extracted contours.

This simple flowchart provides initial solutions for further developments (Figure 14.2). Indeed, the interpolation influence can be significant on the surface quality, particularly when working on urban areas. Moreover, connecting segments are still a challenging problem in the image-processing community.

14.1.3 SURFACE EXTRACTION: HOUGH TRANSFORM

The current chapter focuses on the extraction of 3D planar primitives. It is critical to extract high quality of 3D planar surfaces as they serve as important primitives to reconstruct 3D building models.

A classical methodology for extracting geometrical entities is the Hough transform. A plane P that includes a given LiDAR point (x, y, z) can be written in spherical coordinates as

$$P : [0, \pi[\times [0, \pi[\rightarrow \mathbb{R}$$
$$(\theta, \phi) \rightarrow \rho = \cos\theta\cos\phi x + \sin\theta\cos\phi y + \sin\phi z$$

(14.1)

Therefore, each LiDAR point (x, y, z) is represented as a complex surface in the Hough space. This surface is represented in a discretized volume (accumulator) in the (ρ, θ, ϕ)-directions. Looking for 3D planar surfaces in the object space is equivalent to looking for 3D points in the Hough space (ρ, θ, ϕ). They are the local maxima of the accumulator volume. Depending on the discretization step, calculations can become time consuming. Within the building reconstruction context, the parameter space should be limited by some constraints that are often used, such as

1. Plane normal vector is perpendicular to at least one of the directions of the building facades, if they are available.
2. Roof slopes are beneath 45° for instance.
3. ρ is limited to an embedded volume delimited by the DSM.

In a cadastral map, roofs may have a complex shape. The extraction of relevant planes within the accumulator is not an obvious process. First, a plane may include two disconnected sets of LiDAR points within a region. Vosselman proposes to avoid this problem by comparing the generated triangulated irregular network (TIN) from assigned points to a global TIN: Only the connected piece with minimal surface is considered to be a planar face [8]. Selected points are then removed from the initial point set, and the process is repeated. Another preprocessing step could be to refine the designed cadastral parcel in light of the DSM. Jibrini has proposed to perform a 3D segmentation of the accumulator volume by watershed [9] but prefers searching plane hypothesis satisfying building model hypothesis [10]. With no a priori knowledge on the building shape, this methodology proposes plane hypotheses that are all intersected before searching for the best polyhedra describing the roof surface [11].

The aim of the current chapter is to present two approaches for detecting 3D roof facets; the first one is based on the analysis of the 3D point cloud, whereas the second one integrates aerial images. We would like to show that searching for planar primitives in a LiDAR point cloud has limitations with regard to a joint use of aerial images and LiDAR data. Two approaches will be detailed and discussed: a robust RANSAC* based approach and eventually a hybrid image segmentation approach.

14.2 RANSAC-BASED APPROACH FOR PLANE EXTRACTION

14.2.1 BACKGROUND

The RANSAC algorithm introduced by Fischler and Bolles is a general robust approach to estimate model parameters [12]. Instead of using the largest amount of data to obtain an initial solution and then attempting to eliminate the invalid data points, RANSAC uses the smallest feasible dataset and enlarges this set with consistent data when possible. With applications to roof facet detection, classical RANSAC would be formulated as follows (cf. Algorithm 14.1): (1) randomly select a set of $N \in \mathbb{N}$ planar surfaces P within a LiDAR point cloud S and keep a count of the number of points (also called supports) of Euclidean distances from the associated planes P less than a critical distance d_{cr}. (2) A least-square estimation of the final plane (P_{final}) is performed with the set of supports (M_k) of maximum cardinal. (3) The set M_k is then removed from the initial point cloud S. The algorithm runs until card(S) <3, where card (S) is the cardinal of set S.

Theoretically, all triplets of S have to be tried to ensure that the best plane was drawn at each iteration. N_{th} is therefore defined as

$$N_{th} = \frac{Card(S)!}{3!(card(S)-3)!}$$

(14.2)

Nevertheless, this approach may be extremely time consuming. Most often it is not worthy to try all possible draws [13]. In other words, for a given probability t of drawing a correct plane P (i.e., three points without outlier), we would like to maximize the probability w that any selected point is an inlier (w^3 for three points). We can derive a relationship between t, w, and N:

$$(1-t) = (1-w^3)^N \Leftrightarrow N = \frac{\log(1-t)}{\log(1-w^3)}$$

(14.3)

The number of draws N can therefore be calculated directly from the knowledge of t and w. If t is generally kept constant to 0.99, w has to be estimated with a priori knowledge. The general idea of our approach is to improve the efficiency of a classical RANSAC approach by focusing the drawing of triplets on presegmented areas. In our context, main plane directions correspond to roof facet orientations. As a result, focusing the consensus onto regions sharing the same normal orientation will constrain the probability w to follow specific statistical rules as developed in Section 14.2.3.2.

---

* Random sample consensus.

ALGORITHM 14.1   Classical RANSAC for Detecting Roof Facets

begin
   repeat
      # Selection of the sets of supports
      while $n \leq N$ **do**
         Randomly select a plane P(3 points)
         # Selecting points within a critical distance of the plane
         $M_n = \{m \ E \ S \ / \ \|m - P(m)\|^2 \leq d_{cr}\}$
         $++n$
         # Select of the set of highest cardinal
         $\exists k \leq N / \forall n \leq N, \text{card}(M_k) > \text{card}(M_n)$
         # Estimation of the final plane over all planes
         $P_{final} = \arg \min_{P'} \sum_{m \in M_k} \|m - P'(m)\|^2$
         # Removing previous supports from the main point cloud
         $S \leftarrow S \backslash M_k$
      until *card*(S) < 3
   end

14.2.2 Surface Normal Vector Clustering Based on Gaussian Sphere Analysis

Computing local surface properties in a 3D scene may help the automatic shape recognition step, providing relevant feature descriptions for segmentation purpose. There are several methods for obtaining local surface properties from range data depending on the 3D scene. For terrestrial applications when curved objects are present, quadric or super-quadric fitting becomes appropriate [14].

Here, we propose a planar segmentation of the LiDAR point cloud by analyzing the Gaussian sphere of the scene. Normal vectors are calculated for each LiDAR point by extracting a spherical neighboring. A plane is estimated using a robust regression of M- estimators' family with norm $L_{1,2}$ [15]. Unlike the standard least-square method that tries to minimize $\Sigma_i r_i^2$ where r_i is the ith residual, M-estimators try to reduce the effect of outliers by substituting the squared residuals r_i^2 for another function of the residuals, thus yielding to minimize $\Sigma_i \rho(r_i)$ where ρ is a symmetric, positive function with a unique maximum at zero, and is chosen to be less increasing than square. This optimization is implemented as an iterative reweighted least power algorithm. In a robust cost model, nothing special needs to be done with outliers. They are just normal measurements that happen to be down-weighted owing to their large deviation.

The mass density of the normal vectors on the Gaussian sphere is described by an Extended Gaussian Image (EGI) [16]. First, the Gaussian sphere can be approximated by a tessellation of the sphere based on regular polyhedrons. Such tessellation is computed from a geodesic dome based on the icosahedron divisions (Figure 14.3a). We define Φ as a face of the Gaussian sphere. The EGI can be computed locally by counting the number of surface normals that belong to each cell. The values in the cells can be thought of as a histogram of the orientations. The angular spread (related to the number of faces) depends on the error that we tolerate for the coherence of the normal vectors in a cell. As a result, within a specific cell, normals n_x, n_y, and n_z will be distributed following a certain density of probability $p(n_x, n_y, n_z)$, which will be analyzed in Section 14.2.3.

Algorithm 14.2 summarizes the EGI generation, where N is the set of normal vectors of the scene, Φ is an EGI's face, and $[\Phi] \in \mathbb{R}^3$ is the face's normal vector (Figure 14.4a).

ALGORITHM 14.2   Computation of an Extended Gaussian Image

Data: $N°$
Result: EGI
begin
    # Select each normal vector
    foreach $\vec{n} \in N°$ **do**
        # Attribution of each normal vector to an EGI's face
        $[\Phi] = Argmin_{\Phi \in EGI} \| \vec{n} - \Phi \|^2$
        # Increment of the 3D histogram of orientations
    ++EGI[(Φ]
end

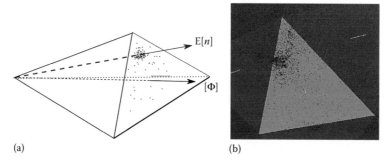

(a)   (b)

FIGURE 14.3   Generation of an EGI. (a) Details of an EGI's face [Φ]. E[n] is the mean of the orientation vectors. (b) Details of a Gaussian sphere (dark points).

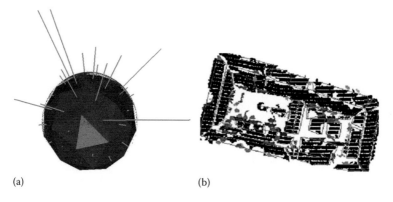

(a)   (b)

FIGURE 14.4   Aspects of the normal-driven RANSAC algorithm. (a) Example of an EGI: orientation histogram collected on a geodesic dome derived from the icosahedron. This is a discrete approximation of the EGI. The length of the vectors attached to the center of a cell is proportional to the number of surface normals that fall within the range of directions spanned by that cell. Blue points are the projection of normals onto the Gaussian sphere. (b) Projection of LiDAR points (strips) and extracted facets (projected polygons) onto focusing masks (coded in gray level depending on their surface).

A Gaussian sphere only describes facet orientations, and not their locations. In order to separate two parallel roof facets, an image-based representation is used. Normal vectors belonging to each EGI's face Φ are resampled in a 3-channel image, each of them representing a component of the normal vector. There are as many images as EGI faces. Once the image background is removed (binary segmentation), a labeling algorithm is applied to extract the contours of coplanar regions. Connected areas are finally extracted, and a set of masks containing single planar regions is generated. Several masks are therefore associated with a single EGI's face Φ. They are denoted by C_Φ.

The set of all C_Φ describes the coarse delineation of roof facets. We prefer to under-sample the EGI so that each focusing mask should contain more LiDAR points than the real facet does. RANSAC will therefore be applied sequentially on these sets of masks to extract 3D planar surfaces. This focusing step is then used to estimate two main parameters of RANSAC: the number of draws N to be done and the critical distance d_{cr}.

14.2.3 PARAMETER ESTIMATION

The methodology developed in the current section aims at automatically estimating the parameters of the RANSAC algorithm, which are the critical distance and the number of draws. The first one depends on the S/N ratio of the data, whereas the second one depends on the statistical distribution of points on each face of the Gaussian sphere.

14.2.3.1 Critical Distance

We noticed in Section 14.2.1 that supports were considered in a set M_n only if their distances to the associated random plane were less than a critical distance d_{cr}. This distance may be seen as the standard deviation of the supports with regard to the 3D plane. d_{cr} is therefore defined for each focusing mask C_Φ, as proportional to the square root of the final residuals of a least-square fitted plane estimated from the entire LiDAR points $m \in C_\Phi$. If $\{m'\}$ is the set of orthogonal projection of LiDAR points $\{m\}$ onto the fitted plane, then

$$d_{cr} = \sqrt{\sum_{m \in C_\Phi} \left\| m - m' \right\|^2} \tag{14.4}$$

14.2.3.2 Number of Draws

Within a focusing mask C_Φ, normal vectors $\vec{n} \in \mathbb{R}^3$, considered as a random variable, have an empirical probability density function $pc_{-\Phi}(\vec{n})$. We define w_{C_Φ} as the probability for a LiDAR point belonging to C_Φ, to be a support of the final plane (inlier). Picking up a point whose orientation is close to the mathematical expectation $\mathbf{E}[\vec{n}]$ is the standard deviation of the distribution; we have

$$w_{C_\Phi} = \int_{\mathbf{E}[\vec{n}] - \vec{\sigma}}^{\mathbf{E}[\vec{n}] + \vec{\sigma}} pc_{-\Phi}(\vec{n}) \, d\vec{n} \tag{14.5}$$

Practically, we consider the component of $\vec{n}$ as three independent random variables and

$$w_{C_\Phi} = \int_{\mathbf{E}(n_x) - \sigma_{n_x}}^{\mathbf{E}(n_x) + \sigma_{n_x}} p_x(n_x) \, d n_x \cdot \int_{\mathbf{E}(n_y) - \sigma_{n_y}}^{\mathbf{E}(n_y) + \sigma_{n_y}} p_y(n_y) \, d n_y \cdot \int_{\mathbf{E}(n_z) - \sigma_{n_z}}^{\mathbf{E}(n_z) + \sigma_{n_z}} p_z(n_z) \, d n_z \tag{14.6}$$

The probability density function of each component is explicitly calculated as the derivative of the empirical cumulative distribution function that is described, for the x component as

$$F_{K_x}(x) = \begin{cases} 0 & \text{if } x < \inf n_i \\ \dfrac{n_i}{K_x} & \text{if } n_i < x \leq n_{i+1} \\ 1 & \text{if } x > \sup n_i \end{cases} \qquad (14.7)$$

where
  n_i is the proportion of values less than x
  K_x the number of realizations of the random variable n_x

Finally, an optimal number of draws are calculated for each focusing mask C_Φ, using Equation 14.3 and the previous calculation of w_{C_Φ}, and keeping $t = 0.99$.

14.2.4 RESULTS

Figure 14.5 shows some results of the extraction method. The normal-driven RANSAC approach enhances the classical RANSAC by parsing the initial 3D scene in pseudo-homogeneous planar regions. This strategy reduces the number of draws in each focusing areas and makes RANSAC valuable for large LiDAR datasets. We can notice that the more accurate the normal vector map (in terms of homogeneity intra regions), the less the number of draws.

The approximation of the Gaussian sphere by a regular polyhedron, that is, its discretization with a fixed step, has advantages with regard to other methods such as the K-means one. The main

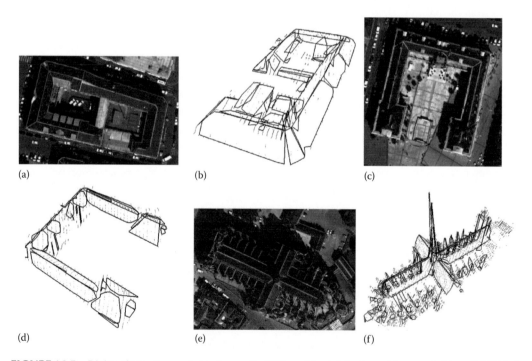

(a)   (b)   (c)

(d)   (e)   (f)

FIGURE 14.5 Right column (a, c, and e) is the result of 3D roof facet detection with normal-driven RANSAC. Left column (b, d, and f) is the corresponding aerial image.

one is no doubt to manage with the extension of each cell. It represents the tolerance granted around one direction for gathering normal vectors. A rough discretization provides large areas of similar normal vectors. Indeed, this clustering stage is only considered as a focusing stage. Therefore, there is no need to have accurate boundaries as the final facet estimation is performed directly onto the point cloud. As a result, a good parameterization includes a not-so-fine Gaussian sphere approximation to provide large focusing areas. By analyzing the distribution of the 3D points within these focused areas will provide the best robust plane.

This methodology provides patches of planar surfaces in which borders are the convex envelop of LiDAR points. These borders have no physical reality, but these primitives bring strong geometric constraints to advanced reconstruction strategies [11].

14.3   JOINT USE OF IMAGE AND LiDAR DATA FOR ROOF FACET EXTRACTION

Three-dimensional primitives are initial features for further reconstruction strategies. By integrating aerial images with LiDAR data in a primitive detection process may highly enhance the resulting facets. First, the image geometry integrates the continuity of described objects: Intensity changes generally correspond to object ruptures. Most of the time, roof edges are visible in aerial images. Then, images describe sets of radiometric elements (pixels) that have a strong correlation between each other when describing the same surface (radiometry similarity). Having investigated the potential of extracting planar entities solely from 3D LiDAR data, we present in the following sections, an attempt to integrate image radiometry, LiDAR geometry, and LiDAR semantic in an image segmentation framework for detecting 3D roof facets.

14.3.1   BACKGROUND

Segmenting an image I consists of determining a partition $\Delta_N(I)$ of N regions $R_{i \in [1,N]}$. A region is a connected set of pixels that satisfies certain predefined homogeneity criteria satisfying

$$\Delta_N(I) = \bigcup_{i \in [1,N]} R_i, R_i \bigcap_{i \neq j} R_j = \phi, \forall i, R_i \text{ is connected}$$

The segmentation problem may be considered under various points of views seeing that a unique and reliable partition does not exist. Beyond classical region-growing algorithms, approaches based on a hierarchical representation of the scene retained our attention. These methodologies open the field of multiscale descriptions of images [17]. Here, we are interested in obtaining an image partition wherein roof facets are clearly delineated and understandable as unique entities.

A data structure for representing an image partition is the Region Adjacency Graph (RAG). The RAG of an N-partition (an N-RAG) is defined as an undirected graph $G = (V, E)$ where V (card $[V] = N$) is the set of nodes related to an image region and E the set of edges related to adjacency relationships between two neighboring regions. Each edge is weighted by a cost function (or energy) that scores the dissimilarity between two adjacent regions. The general idea of a hierarchical ascendant segmentation is to merge sequentially the most similar pair of regions (or the one that minimizes the cost function) until a single region remains. The fusion of two regions (or the contraction of the RAG minimal edge) creates a node in the hierarchy and two father–child relationships in the case of a binary tree. The root of the tree corresponds to image H and the leaves to the image initial segmentation. A hierarchy H can be considered as a union of partition sets. Figure 14.6 sketches the generation process of the hierarchy from an initial partition of the image.

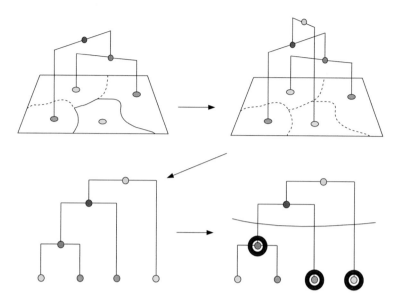

FIGURE 14.6   Construction of a hierarchy based on a RAG (top left).

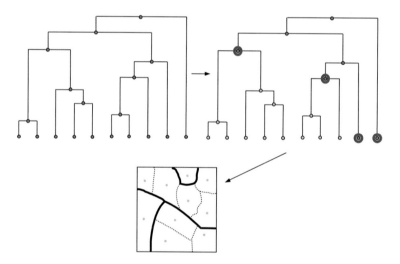

FIGURE 14.7   Sketch of a cut in a hierarchy (dendrogram). Circles are the selected cut nodes and correspond to the presented image partition.

A cut of H is a set of nodes that intersects all branches of H once and only once. A cut is not necessarily horizontal and provides a partition of I (Figure 14.7). Both the hierarchy and the cut depend on the segmentation context, and their definitions will vary depending on the applications.

14.3.2   THEORY

The shape of the hierarchy (therefore, the region-merging order) constrains the existence of an eligible partition. In other words, initial regions that theoretically belong to a roof facet must be mutually merged until a node in the hierarchy appears as a roof facet entity. If it appears that subregions of a facet merge with adjacent regions that do not belong to their supporting facet, the embedded geometry is broken.

14.3.2.1 Cost Function

The shape of the hierarchy, as well as the existence of the adequate eligible *cut*, which is a partition of roof facet regions, depends on the region-merging order, and therefore, on the definition of the cost function ε. We can write ε as a sum of two terms ε_r, ε_l, and ε_s, respectively, related to the image radiometry, to the LiDAR geometry, and to the semantic extracted from LiDAR data.

ε_r is defined to minimize the loss of information when describing the image from n to $n-1$ regions. We retained the cost function given by Haris [18] for merging two neighboring regions R_i and R_j:

$$\varepsilon_r(R_i, R_j) = \frac{\|R_i\|_r \|R_j\|_r}{\|R_i\|_r + \|R_j\|_r} (\mu_i - \mu_j)^2 \qquad (14.8)$$

where

$\|\cdot\|$ is the number of pixels in each region

$\mu = 1/\|R\|_r \sum_k I(k)$ is the average value of the radiometries at image sites k of the region

In our context, ε_l is defined to take advantage of both the accuracy and the regularity of LiDAR measurements onto roof surfaces to make appear in the hierarchy nodes corresponding to roof facet entities. It is therefore expected that all regions within a single roof facet merge into one reliable region before any mis-merging occurs. Higher levels of the hierarchy are not of interest here. The adequation of LiDAR points to lie on a roof facet is measured by estimating a plane onto those included in $R_i \cup R_j$. A nonrobust least-square estimator is applied specifically to neighboring regions such that they do not merge when the estimated plane is corrupted by noncoplanar points. Such is the case when attempting to merge two regions apart from the roof top before other couples of regions belonging to the same roof facet with possible significant radiometric dissimilarities. If $\| N_i \|_l$ (resp. $\| N_j \|_l$) is the number of LiDAR points in region R_i (resp. R_j) and r_p the residuals of a LiDAR point to the estimated plane, ρ_l^2 is the average square distance of LiDAR points to the estimated plane with

$$\rho_l^2 = \frac{1}{\|N_i\|_l + \|N_j\|_l} \sum_{p=1}^{\|N_i\|_l + \|N_j\|_l} r_p^2$$

If we consider a similar weighting factor as for ε_r depending on the number of LiDAR points $\|\cdot\|_l$ in image regions, ε_l is expressed as

$$\varepsilon_l(R_i, R_j) = \frac{\|N_i\|_l \|N_j\|_l}{\|N_i\|_l + \|N_j\|_l} \rho_l^2$$

3D LiDAR points that have been previously processed to extract a binary semantics: ground and off-ground points. Theoretically, the off-ground class includes building and vegetation. However, we will only consider the segmentation algorithm that focuses on buildings. The process is performed with a high level of relevancy over urban areas owing to the sharp slope breaking onto building edges [19]. An image region will be classified as ground if it contains at least one projected LiDAR ground point. Otherwise, the region is considered to be a built-up arca. This binary semantics provides a reliable ground mask that can be integrated into the initial segmentation. Two regions of different classes are kept disjoint until the highest levels of the hierarchy. The methodology is summarized in Figure 14.8. Finally, we can write

$$\varepsilon_s(R_i, R_j) = \begin{cases} \infty & \text{if } R_i \text{ or } R_j \text{ is a } \textbf{ground} \text{ region} \\ 0 & \text{otherwise} \end{cases}$$

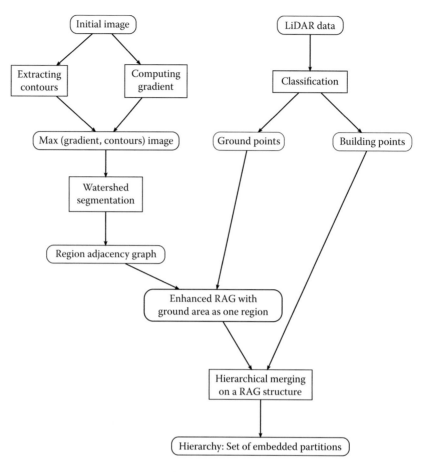

FIGURE 14.8　Flowchart of the joint segmentation methodology.

14.3.2.2　Optimal Eligible Cut

A roof facet is defined as a 3D planar polygon the average square distance to LiDAR support points (ρ_I^2) of which is less than a threshold, s. The final partition is obtained by recursively exploring the binary tree structure from its root comparing ρ_I^2 of each node to s.

14.3.3　RESULTS

Planar 3D primitives extracted by the joint segmentation process LiDAR or image are presented in Figures 14.9 and 14.10. For these applications, an orthorectified image has been used in order to avoid the facade detection problem. Red points are LiDAR points belonging to a single strip. We clearly distinguish the optic fiber LiDAR device used for processing the planar primitives. We used a threshold, s, of 0.5 m in the first figure and 0.7 m in the second case. This reconstruction considers LiDAR points belonging to an image region larger than 30 pixels, and the orientation of which is greater than 30° from the zenith direction. The presented 3D scenes give a realistic representation of the buildings wherein hyperstructures such as dormer windows are particularly visible. In order to enforce the region borders to lie on real discontinuities, we applied a contour detection algorithm (hysteresis thresholding) on the gradient image. The gradient was computed with a Canny–Deriche operator. The watershed algorithm is finally applied on a combination of both images (maximum of gradient and contour images).

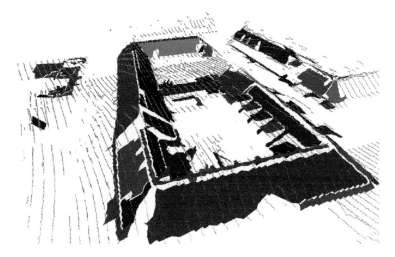

FIGURE 14.9 Extracted 3D facets of a complex building from the joint segmentation process. LiDAR points are visible on the picture.

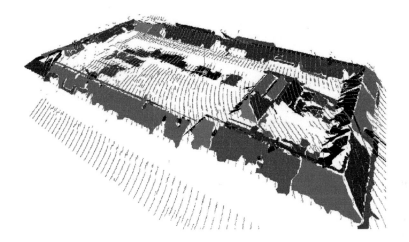

FIGURE 14.10 Extracted 3D facets of a complex building from the joint segmentation process. LiDAR points are visible on the picture.

When comparing Figure 14.9 with Figure 14.5b, we can notice a better detection of planar primitives as well as a fine description of plane orientation. The introduction of a semantic term in the energy produced a sharp separation between points belonging to the ground and the others and can be observed in Figure 14.11b in which regions belong only to built-up areas. The delineation of planar entities is also improved with regard to those extracted directly from the point cloud without image information. Figure 14.11 is a visual comparison between both methodologies developed in this chapter. The patch effect is removed in Figure 14.11d, and region frontiers are coherent with regard to the image-based building edges. One can remark on some remaining small regions near the building edges that can be attributed either to radiometric artifacts introduced during the orthorectification process (resampling) or to building shadows the radiometry of which is similar to the roof's.

The actual implementation of the methodology depends on parameter, s, which is set a priori. The threshold theoretically varies with each building, and even with each roof facet depending on its regularity (dormers, chimneys, antennas, etc.).

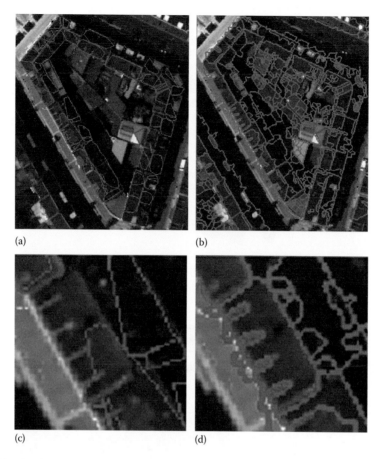

(a)   (b)

(c)   (d)

FIGURE 14.11   Comparison of 3D facets extracted from the normal-driven RANSAC algorithm to the hierarchical segmentation process. (a) 3D facets extracted from normal-driven RANSAC and projected onto an orthoimage. (b) Result of the joint segmentation process. Polygons are eligible roof facets. (c) Zoom in image (a) of a particular facet extracted from the normal-driven RANSAC algorithm. (d) Zoom in image (b) of the same facet calculated from the joint segmentation process.

14.4   CONCLUSION

We have presented in this chapter two methodologies for extracting 3D planar primitives in a building reconstruction context: The first one is based on the well-known RANSAC paradigm, and the other one investigates the potentialities of using LiDAR data and aerial images jointly. The algorithm developed in Section 14.2 shows that the recognition of planar surfaces in a point cloud is a challenging task. First, the point density may be a limiting factor for the building description: the higher the point density, the better the roof facet description. However, the orientation of planar surfaces is of quality due to the robustness of the LiDAR measurements. Second, the topology is not well adapted to generate fine edges as we cannot predict a LiDAR impact to hit a particular area. As a result, the integration of aerial images may overcome this problem. The radiometric correlation between two neighboring pixels is high. It ensures to retrieve the continuity of particular linear structures and tends to describe a single facet with neighboring radiometries. The hierarchical segmentation framework allows one to integrate as a unique energy function, information from LiDAR data and aerial images. Even if this approach need improvements, results are qualitatively better than the RANSAC approach.

Photogrammetry and airborne LiDAR are not antagonist techniques. On the contrary, a relevant combination of images and LiDAR data is the future research direction for improving performances of feature-extraction algorithms. The use of full-waveform LiDAR data may also be relevant in urban areas beyond the densification of the point cloud. Among many dependencies, waveforms vary depending on the roof material and on its geometry. Future researches would consist of establishing neighboring relationships between waveforms by extracting morphological characteristics.

REFERENCES

1. M. Kasser and Y. Egels. *Digital Photogrammetry*. Taylor & Francis, New York, 2002.
2. C. Baillard and O. Dissard. A stereo matching algorithm for urban digital elevation models. *Photogrammetric Engineering and Remote Sensing*, 66(9):1119–1128, 2000.
3. M. Pierrot-Deseilligny and N. Paparoditis. A multiresolution and optimization-based image matching approach: An application to surface reconstruction from spot5-hrs stereo imagery. In *Proceedings of the ISPRS Conference Topographic Mapping from Space* (*with Special Emphasis on Small Satellites*), ISPRS, Ankara, Turkey, February 2006.
4. K. Zhang, J. Yan, and S.C. Chen. Automatic construction of building footprints from airborne lidar data. *IEEE GRS*, 44:2523, 2006.
5. T. Schenk and B. Csatho. Fusion of lidar data and aerial imagery for a more complete surface description. In Kalliany, R. and Leberl, F. (Eds.), *Proceedings of the ISPRS Commission III Symposium on Photogrammetric and Computer Vision*, Volume XXXIV of *The International Archives of the Photogrammetry, Remote Sensing and Spatial Information Sciences*, Institute for Computer Graphics and Vision, Graz University of Technology, Graz, Austria, pp. 310–317, 2002.
6. A. Habib, M. Ghanma, M. Morgan, and R. Al-Ruzouq. Photogrammetric and lidar data registration using linear features. *Photogrammetric Engineering and Remote Sensing*, 71(6):699–707, 2005.
7. H. Gross and U. Thoennessen. Extraction of lines from laser point cloud. In *Proceedings of the ISPRS Commission III Symposium on Photogrammetric and Computer Vision*, Volume XXXVI in *The International Archives of the Photogrammetry, Remote Sensing and Spatial Information Sciences*, ISPRS, Bonn, Germany, pp. 86–91, 2006.
8. G. Vosselman and S. Dijkman. 3D building reconstruction from point cloud and ground plans. In *Proceedings of the ISPRS Workshop on Land Surface Mapping and Characterization Using Laser Altimetry*, Volume XXXIV of *International Archives of Photogrammetry and Remote Sensing*, Annapolis, MD, pp. 37–43, October 2001.
9. H. Jibrini. Reconstruction automatique de batiments en modeles polyhedriques 3-D a partir de donnees cadastrales vectorisees 2-D et un couple d'images aeriennes a haute resolution. PhD Thesis, Ecole Nationale Superieure des Telecommunications de Paris, 2002.
10. H. Jibrini, M. Pierrot-Deseilligny, N. Paparoditis, and H. Maitre. Automatic building reconstruction from very high resolution aerial stereopairs using cadastral ground plans. In *Proceedings of the XIXth ISPRS Congress, The International Archives of the Photogrammetry, Remote Sensing and Spatial Information Sciences*, ISPRS, Amsterdam, the Netherlands, July 2000.
11. F. Taillandier. Automatic building reconstruction from cadastral maps and aerial images. In Stilla, U., Rottensteiner, F., and Hinz, S. (Eds.), *Proceedings of the ISPRS Workshop CMRT 2005: Object Extraction for 3D City Models, Road Databases and Traffic Monitoring—Concepts, Algorithms and Evaluation*, Vienna, Austria, pp. 105–110, August 2005.
12. M.A. Fischler and R.C. Bolles. Random sample consensus: A paradigm for model fitting with applications to image analysis and automated cartography. *Graphics and Image Processing*, 24(6):381–395, 1981.
13. R. Hartley and A. Zisserman. *Multiple View Geometry in Computer Vision*, Cambridge University Press, Cambridge, UK, 2002.
14. A. McIvor and R.J. Valkenburg. A comparison of local surface geometry estimation methods. In *MVA IAPR Conference on Machine Vision Applications*, 10:17–26, 1997.
15. G. Xu and Z. Zhang. *Epipolar Geometry in Stereo, Motion and Object Recognition*, Kluwer Academic, Dordrecht, the Netherlands, 1996.
16. B.K.P. Horn. Extended Gaussian images. *PIEEE*, 72(12):1656–1678, 1984.
17. L. Guigues, J.-P. Cocquerez, and H. Le Men. Scale sets image analysis. *International Journal of Computer Vision*, 68(3):289–317, 2006.

18. K. Haris, S. Efstratiadis, N. Maglaveras, and A.K. Katsaggelos. Hybrid image segmentation using watersheds and fast region merging. *IEEE Transactions on Image Processing*, 7(12):1684–1689, 1998.

19. F. Bretar, M. Chesnier, M. Pierrot-Deseilligny, and M. Roux. Terrain modeling and airborne laser data classification using multiple pass filtering. In *Proceedings of the XXth ISPRS Congress*, Volume XXXV Part B of *The International Archives of the Photogrammetry, Remote Sensing and Spatial Information Sciences*, International Society for Photogrammetry and Remote Sensing (ISPRS), Istanbul, Turkey, pp. 314–319, July 2004.

15 Global Solutions to Building Segmentation and Reconstruction

Jie Shan, Jixing Yan, and Wanshou Jiang

CONTENTS

15.1 INTRODUCTION

Automated reconstruction of building roof models is yet a challenging step towards 3D city modeling. Being able to directly collect dense, accurate 3D point clouds over urban objects, LiDAR (Light Detection and Ranging) technology provides an effective and beneficial data source to this end. In spite of many efforts made in the past two decades (Haala and Kada, 2010; Wang, 2013; Tomljenovic et al., 2015), building roof reconstruction remains to be an open issue, largely due to the insufficiency of the data density and the complexity of the actual building roofs (Wang, 2013; Xiong et al., 2014).

Reported methods for building roof reconstruction mostly fall into two categories: model driven and data driven (Yan et al., 2016). In terms of model-driven methods, a building is assumed to be an assembly of simple roof primitives (e.g., gable roof and hipped roof). The shape and topology of these primitives are predefined in a library of roof or building models (Tarsha-Kurdi et al., 2007b; Huang et al., 2013). To extract roof primitives from LiDAR point clouds, various techniques have been developed, including invariant moments (Maas and Vosselman, 1999), graph matching (Verma et al., 2006; Oude and Vosselman, 2009; Xiong et al., 2014), Support Vector Machine (Satari et al., 2012; Henn et al., 2013), random sample consensus (RANSAC) (Henn et al., 2013; Xu et al., 2015), and Reversible Jump Markov Chain Monte Carlo (Huang et al., 2013). However, these approaches tend to fail when reconstructing complex roof shapes. To alleviate this problem, complex roof shapes are usually decomposed into simple ones that are defined in the model library (Kada and McKinley, 2009). Among the reported studies, 2D-plans (Kada and McKinley, 2009; Henn et al., 2013), graph matching technique (Verma et al., 2006; Oude and Vosselman, 2009; Xiong et al., 2014), and convexity analysis (Lin et al., 2013) are used for this purpose. Unlike the aforementioned approaches (Huang et al., 2013), extract roof primitives one by one from LiDAR points by using the Reversible Jump Markov Chain Monte Carlo technique. Although these model-driven approaches do not require highly dense LiDAR points, they are limited to roof primitives predefined in the model library. If the roof shape is complex or not predefined in the model library, the model reconstruction may fail (Kim and Shan, 2011; Yan et al., 2016).

Among reported studies, the data-driven approaches are more flexible and work well for complex roof models. It is often assumed that the reconstructed roof model is a polyhedron consisting of multiple planes. Geometric features of a roof model, such as edges and vertices as well as their topology can be determined by the intersection of adjacent roof planes (Sampath and Shan, 2010; Perera and Maas, 2014). To reconstruct a building model, it usually starts with roof (plane) segmentation of building LiDAR points, followed by an assembly of segmented individual roof planes. In terms of roof segmentation, the point clouds of a building are usually segmented into disjoint planar regions. Various approaches such as data clustering, region growing, energy minimization, and model fitting can be used for this purpose. A review of these approaches can be found in Awwad et al. (2010), Haala and Kada (2010), Sampath and Shan (2010), Wang (2013), and Yan et al. (2014).

Data clustering is in principle a statistical technique that classifies the point clouds into primitives based on certain precalculated local surface properties or features. Filin and Pfeifer (2006) propose a slope adaptive neighborhood for such calculation. Considering the variations in local point density, Sampath and Shan (2010) use the Voronoi neighborhood to estimate the local surface properties, whereas Lari et al. (2011) use a cylindrical neighborhood for this purpose. As for clustering the feature vectors, mode-seeking (Filin and Pfeifer, 2006), conventional mean-shift (Melzer, 2007), and fuzzy k-means (Sampath and Shan, 2010) are applied. In spite of the popularity and efficiency of this approach, it suffers the difficulty in neighborhood definition and is sensitive to noise and outliers.

Region growing is a region-based segmentation method that partitions point clouds into disjoint homogenous regions. It starts with a selected seed point or region and expands it to neighboring points. Gorte (2002) selects the triangles in triangulated irregular networks as seed regions and extends them to neighboring triangles. Zhang et al. (2006) perform a local plane fitting at points and select the points with good planarity as seed points. To obtain robust seed points, Chauve et al. (2010) develop an iterative Principal Component Analysis to estimate local planarity. You and Lin (2011) present a noniterative approach using tensor voting for this purpose. Unlike the aforementioned approaches, Dorninger and Pfeifer (2008) determine seed clusters (regions) by a hierarchical clustering approach. As for the expanding of seed regions, similarity measures such as distances of points to planes (Zhang et al., 2006; Dorninger and Pfeifer, 2008; Chauve et al., 2010) and angle between two normal vectors (Dorninger and Pfeifer, 2008; Awwad et al., 2010) are used. Nevertheless, region growing is susceptible to the selection of seed regions (Awwad et al., 2010) and difficult to stop when transitions between two regions are smooth (Sampath and Shan, 2010).

The energy minimization approach is a global solution that formulates the segmentation as an optimization problem (Yan et al., 2014). Its objective function may consist of fidelity to data, continuity of feature values, and compactness of segment boundaries (Kim and Shan, 2011; Yan et al., 2014). A widespread application of this approach to image segmentation can be found in Vitti (2012). As for the segmentation of LiDAR data, multiphase level set approach is adopted to segment planar roof primitives under an energy minimization formulation (Kim and Shan, 2011). Compared with the RANSAC (Fischler and Bolles, 1981) based approaches, it is global and can segment multiple roof planes at one time. However, a common shortcoming of this approach is that poor segmentation may occur when the energy function converges to a local minimum.

As the reconstructed models are dependent on a robust estimate on the planar primitives, robust model fitting methods such as RANSAC and Hough transform (Duda and Hart, 1972) are also applied to roof segmentation. LiDAR points that fit a mathematical plane with most inliers are extracted and regarded as a planar segment. This approach is robust to noise and outliers, but it tends to result in spurious planes (Vosselman and Dijkman, 2001; Tarsha-Kurdi et al., 2007a; Sampath and Shan, 2010). With the help of available building ground plans, Vosselman and Dijkman (2001) split the dataset into small parts before applying Hough transform to prevent the detection of spurious planes. Some extended RANSAC methods considering local surface normals (Bretar and Roux, 2005; Schnabel et al., 2007; Awwad et al., 2010; Chen et al., 2012) are also developed for this purpose. Considering spatial connectivity, Zhang et al. (2006) and Chauve et al. (2010) combine model fitting and region growing to suppress spurious planes. Nevertheless, most of the model fitting approaches are order-dependent and based on a single model. Segmentation results are dependent on the order in which the planes are extracted. When multiple planes are present, each plane instance needs to be sequentially extracted. As a result, points at transitions between roof planes will be assigned to the one that is first extracted. In most cases, this approach performs well with some additional constraints. However, for complex roof structures, it tends to result in mistakes, such as spurious planes (segments that do not exist in reality), oversegmented planes (one actual plane is segmented into multiple segments), or undersegmented planes (multiple actual planes are segmented into one segment).

Once roof planes are correctly segmented from LiDAR point clouds, building roof models can be created by stitching neighboring roof planes. The difficulty of such procedure is to determine the topologic relations among the segmented roof planes (Kim and Shan, 2011). To address this issue, roof topology is often described as an adjacency matrix (Sampath and Shan, 2010) or roof topology graph (Verma et al., 2006; Oude and Vosselman, 2009; Perera and Maas, 2014; Xiong et al., 2014), which can be derived from a connectivity analysis of the segmented roof planes. As a result, spatially connected roof planes can be intersected to determine roof edges and vertices. However, the roof topology derived from the aforementioned methods is based on a local decision, that is, the connectivity analysis is independently carried out for every two or several segmented roof planes in a sequential manner. Due to incomplete roof planes arising from occlusion or sparse point clouds, such derived roof topology could be incorrect or inconsistent.

In this chapter, building segmentation and reconstruction are respectively described as a minimization problem of appropriate energy functions. To reduce the number of mistakes resultant from building segmentation, a global plane fitting approach for roof segmentation from LiDAR point clouds is developed. Starting with a conventional plane fitting approach (e.g., plane fitting based on region growing), an initial segmentation is first derived from roof LiDAR points. Such initial segmentation is then optimized by minimizing a global energy function consisting of the distances of LiDAR points to initial planes (labels), spatial smoothness between data points, and the number of planes. As a global solution to building segmentation, the proposed approach can determine multiple roof planes simultaneously (Yan et al., 2014).

Based on the segmented roof planes, a global solution to building topological reconstruction from LiDAR point clouds is then developed (Yan et al., 2016). Starting with segmented roof planes from building LiDAR points, a binary space partitioning algorithm is used to partition the bounding box of the building into volumetric cells, whose geometric features and their topology are simultaneously determined. To determine building surfaces and their topology, a global energy function considering surface visibility and spatial regularization between adjacent cells is constructed and minimized. As a result, the cells are labeled as either inside or outside the building, where the planar surfaces between the inside and outside form the reconstructed building model.

The remainder of the chapter is structured as follows. Section 15.2 introduces the framework and describes the workflow of the global solutions. Section 15.3 starts with an introduction of a global optimization approach to roof segmentation, followed by the assessment and discussion of this approach. Section 15.4 first introduces a space partitioning approach using segmented roof planes and detected facades, in which the bounding box of a building is partitioned into volumetric cells. We then formulate the roof reconstruction task as an inside-outside labeling problem of cells derived from space partitioning. A global solution based on graph cuts is presented, and an evaluation of this approach is conducted. Section 15.5 presents our concluding remarks.

15.2 WORKFLOW OVERVIEW

The current chapter is primarily about the segmentation and reconstruction of building roofs from LiDAR points. It is assumed that urban objects such as buildings and trees have been extracted from LiDAR data. For an individual building, the workflow of the proposed approach can be described as follows. Building roof segmentation task is first formulated as a minimization of a global energy function consisting of the distances of LiDAR points to planes (labels), spatial smoothness between data points, and the number of planes. Once roof planes are segmented from building LiDAR points, the bounding box of the building is recursively partitioned into (volumetric) cells, in which roof planes are intersected and possible roof features (e.g., roof faces and edges) as well as their topology can be determined. To look for the correct roof faces and edges, another global energy function considering surface visibility and spatial regularization is formed and minimized via graph cuts. As a result, the cells are labeled as inside or outside the building, where the planar surfaces (i.e., roof faces) and edges between the inside and outside cells form the reconstructed roof model.

Figure 15.1 illustrates the workflow of the proposed approach. First, by using the segmented roof planes (Figure 15.1a) and the facades derived from building boundary and step edges (Figure 15.1b), the bounding box of a building is partitioned into cells (Figure 15.1c). Then, a binary labeling of these cells is conducted via graph cuts (Figure 15.1d). Building surfaces that correspond to the faces of the reconstructed building model are extracted (Figure 15.1e). As a result, roof faces and edges

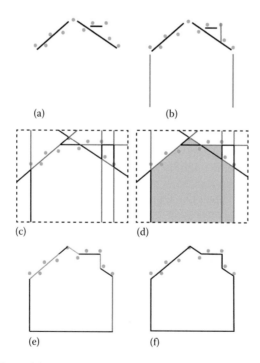

(a)   (b)

(c)   (d)

(e)   (f)

FIGURE 15.1 The workflow of the proposed global approach: (a) roof LiDAR points and segmented roof planes (black lines), (b) detected facades (gray lines), (c) cells derived from space partitioning using segmented roof planes and detected facades (dashed rectangle is the bounding box of the building), (d) binary (inside/outside) labeling of cells, in which the inside ones are shown in polygons filled with gray, (e) extracted building surfaces shown in different shades, and (f) roof faces after merging spatially connected building surfaces.

as well as their topology are all determined at one time. Finally, spatially connected coplanar roof faces are merged and the roof model is reconstructed (Figure 15.1f). Comparing with most existing approaches, the proposed approach is a global solution, that is, roof faces and edges as well as their topology are simultaneously determined.

15.3   BUILDING SEGMENTATION

In this section, the roof segmentation task is formulated as a minimization of a global energy function. Starting with a conventional plane fitting approach (e.g., plane fitting based on region growing), an initial segmentation is first derived from roof LiDAR points. Such initial segmentation is then optimized by minimizing a global energy function consisting of the distances of LiDAR points to initial planes (labels), spatial smoothness between data (LiDAR) points and the number of roof planes. As a global solution, the proposed approach can determine all roof planes simultaneously.

15.3.1   FORMULATION

The segmentation task can be expressed as a labeling problem and formulated in terms of energy minimization. Equation 15.1 provides such a formulation (Delong et al., 2012; Isack and Boykov, 2012)

$$E(L) = \overbrace{\sum_{p\in P} D_p(L_p)}^{\text{data cost}} + \overbrace{\sum_{p,q\in N} w_{pq}\cdot\delta(L_p \neq L_q)}^{\text{smooth cost}} + \overbrace{\sum_{l\in L} h_l\cdot\delta(l\in L')}^{\text{label cost}} \qquad (15.1)$$

where:

L is a given set of labels (planes)

$\delta\,(.)$ is an indicator function

Let P represent a set of data points, the multiple labeling task is to assign a point $p \in P$ a label $L_p \in L$ such that the labeling L minimizes the energy $E(L)$, where L' is the set of labels appearing in L and N is an assumed neighborhood for data points. Three energy terms are considered in the energy formula. The data cost term (the first term in Equation 15.1) measures the discrepancy between data points and labels. It is the sum of the distances of points to their assigned labels. The smooth cost term (the second term in Equation 15.1) measures the label inconsistency between neighboring points. It is the sum of weight w_{pq} of each pair of neighboring points p and q that are assigned to different labels. The label cost term (the third term in Equation 15.1) measures the number of labels appearing in L. It is the sum of the label cost h_l of each label $l \in L'$. Figure 15.2 illustrates a labeling of data points and their fitted lines. Two lines A and B are fitted to data points, and the value of label costs is $H_A + H_B$. As a result, the data cost can be calculated as dist(a, A) + dist(b, A) + dist(c, A) + dist(d, B) + dist(e, B) + dist(f, B) and the value of smooth costs is w_{cd}, where dist$(.)$ is the perpendicular distance of a point to its fitted line.

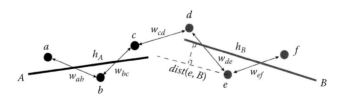

FIGURE 15.2   Labeling of LiDAR points and their fitted lines. The double-arrowed lines link each pair of neighboring points. LiDAR points and their corresponding fitted lines are shown with the same shade.

15.3.2 Cost Terms

As a plane corresponds to a unique label, the multiplane optimization can be stated as a multilabel optimization problem. Accordingly, the three terms in Equation 15.1 can be adapted for our task. This section will discuss how to construct the energy function for multiplane optimization.

The data cost term penalizes the disagreement between a point and its assigned label. In the chapter, the perpendicular distances of LiDAR points to the segmented planes are used to measure the disagreement. For a plane $ax + by + cz + d = 0$ ($a^2 + b^2 + c^2 = 1$), its parameter vector is (a, b, c, and d). Assuming that the plane fitting errors of inliers follow the Gaussian distribution, the data cost between a LiDAR point p (x_p, y_p, z_p) and its assigned label $L_p(a_p, b_p, c_p, d_p)$ is calculated as

$$D_p(L_p) = -\ln\left(\frac{1}{\sqrt{2\pi}\Delta d} \cdot \exp\left[-\frac{\text{dist}(p,L_p)^2}{2\Delta d^2}\right]\right) \tag{15.2}$$

where

$$\text{dist}(p,L_p) = \begin{cases} ax_p + by_p + cz_p + d_p & L_p \neq L_{\text{outlier}} \\ 2\Delta d & L_p = L_{\text{outlier}} \end{cases} \tag{15.3}$$

In the aforementioned equation, L_{outlier} is an extra label for noise or outliers, and Δd is a distance threshold in plane fitting. Notably, the distance between LiDAR points and label L_{outlier} is a constant value, which is set to $2\Delta d$ in this paper. It means that the LiDAR points with distances to their corresponding planes larger than $2\Delta d$ are more likely to be noise or outliers.

The smooth cost term penalizes the label inconsistency between neighboring points. To minimize the energy function, neighboring points are encouraged to have similar labels (Delong et al., 2012). For this purpose, the neighborhood in the derived triangulated irregular networks is used. If a pair of neighboring LiDAR points p and q fits the same plane, the smooth cost between them is 0; otherwise, the smooth cost is 1. The smooth cost term considers the discontinuity at plane transition under a global fitting and produces spatially consistent segments. As adjacent points are more likely to fit the same model, w_{pq} is set inversely proportional to the distance between points (Isack and Boykov, 2012), that is,

$$w_{pq} = \exp(-\|p - q\|) \tag{15.4}$$

The label cost term penalizes the number of labels. Fewer labels are encouraged to be used to represent data compactly (Delong et al., 2012). The term is introduced to reduce the number of redundant planes in the segmentation results. When minimizing the energy function, if the costs (data costs and smooth costs) of eliminating a label l are getting smaller than h_l, then the label l will be eliminated or merged with other labels. A large label cost can help to eliminate redundant planes, but small segmented planes may be missed after the label optimization. To ensure that small roof planes are kept after the label optimization, h_l is written as

$$h_l = \begin{cases} n \times \left(\frac{1}{2} - \ln\frac{1}{\sqrt{2\pi}\Delta d}\right) & L_p \neq L_{\text{outlier}} \\ 0 & L_p = L_{\text{outlier}} \end{cases} \tag{15.5}$$

where n is the minimum number of LiDAR points required for a valid plane. Except for the labels for noise or outliers, every label has the same label cost. However, the label costs derived from

Equation 15.5 may be smaller for some large redundant planes. To alleviate this issue, an adaptive label cost is used. The similarities between segmented planes are checked. If the angle and distance (parameter d in the plane equation) differences between segmented planes are less than given thresholds, n is replaced with the average number of inliers of the similar planes, and their corresponding label costs are updated. Notably, very similar planes can be merged even under a small label cost.

15.3.3 Graph Cuts-Based Minimization

The above-mentioned energy minimization problem can be resolved via graph cuts. The graph cuts technique constructs a graph for the energy function such that the minimum cut on the graph corresponds to the minimum of energy (Boykov et al., 2001; Kolmogorov and Zabin, 2004). We adopt a popular graph cuts-based method, that is, the extended α-expansion algorithm (Delong et al., 2012) to minimize the energy function in Equation 15.1. Given a set of initial labels, the algorithm converts the multiple labeling problems into sequences of independent binary labeling problems and minimizes the energy function using graph cuts. As the input to the energy function in Equation 15.1 is a set of labels (planes), the planes derived from existing model fitting methods can be taken as the initial labels. In our work, the plane fitting approach described in Chauve et al. (2010) is adopted to produce initial planes without constraints on the plane normals. The following algorithm illustrates the process of multiplane optimization. Having initialized the three cost terms (Step 5–7), the graph cuts-based method, that is, the extended α-expansion algorithm is used to minimize the energy function (Step 9) to find the optimal labeling of data points. Similar to the propose expand and reestimate labels (PEARL) algorithm (Isack and Boykov, 2012), the optimization is conducted in an iterative manner to refine the model parameters of the derived labels. Notably, some coplanar segments may be merged to a same label in the optimization. To separate them, a connectivity analysis needs to be performed after label optimization (Step 11).

Algorithm of multilabel optimization

1. Input: a set of LiDAR points P and its corresponding labeling L^0.
2. Input: neighborhood system N.
3. Input: threshold n and Δd.
4. Output: optimal labeling L.
5. Initialize smooth costs (Equation 15.4) and set $t = 0$.
6. Derive the initial planes M_t from L^t.
7. Initialize data costs (Equation 15.2) and label costs (Equation 15.5).
8. Check the similarities between segmented planes and update their corresponding label costs.
9. Set $t = t + 1$ and run the extended α-expansion to compute the optimal labeling L^t.
10. If the energy decreases, go back to Step 2.
11. Compute the connected components of L^t and remove the labels with less than n points.
12. Update L^t and set $L = L^t$.

To illustrate the procedure of multilabel optimization, a sample building (Figure 15.3a) from the International Society for Photogrammetry and Remote Sensing (ISPRS) test dataset (Cramer, 2010) is used. We add a number of random noise points to the raw building LiDAR points to evaluate the robustness of the algorithm. Figure 15.3b presents the initial segmentation result, in which many roof planes, including small ones are correctly separated from the noisy LiDAR points. However, spurious planes (e.g., segment C) are introduced, and one roof plane (e.g., segment A) may be partitioned into several segments. Similarly, segmentation of several transition regions between roof planes results in dangling portions (e.g., segment B). After the first iteration

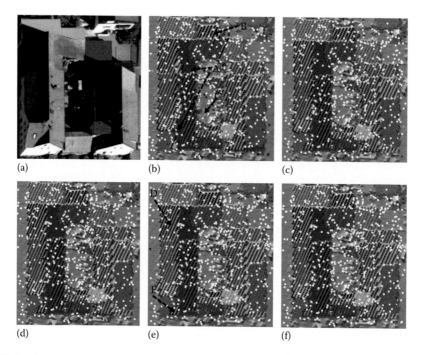

FIGURE 15.3 Segmentation of roof LiDAR points (random noise accounts for 25% of the data): (a) reference image, (b) initial segmentation, (c) iteration 1, (d) iteration 3, (e) iteration 6 (convergence), and (f) final segmentation. Building point clouds are colored by segmented planar segments, with white dots for unsegmented LiDAR points.

(Figure 15.3c), most of the spurious planes are eliminated (e.g., segment C in Figure 15.3b), and the redundant the planar segments belonging to the same roof plane are merged (e.g., segment A in Figure 15.3b). More notably, the introduction of the smooth cost term can remove most of the dangling portions of the planar segments (e.g., segment B in Figure 15.3b). In the subsequent iterations, the remaining artifacts are gradually eliminated, and the iteration converges after six times (Figure 15.3e). The use of the label cost term may produce fewer labels, that is, some coplanar roof planes are merged (e.g., planar segments D and E in Figure 15.3e). Therefore, a connected component analysis of the segmented planes needs to follow to separate the coplanar segments. Given a distance threshold based on the point density, the components far away from other components are separated and assigned a new label. At the end, the labels (planes) consisting of less than a given number of LiDAR points are discarded. Figure 15.3f gives the final segmentation of the complex building. Figure 15.4 shows the changes of the energy and the number of planes during the iteration. It is shown that most of the roof segmentation is completed after the first iteration in which the number of identified planes and the energy drops sharply. In the subsequent iterations, only a few LiDAR points change their labels, leading to a quick convergence.

15.3.4 RESULTS AND DISCUSSIONS

15.3.4.1 Results

Two airborne LiDAR datasets over city Yangjiang (China) and Wuhan (China) are used for evaluation. Both of them were captured with a Trimble Harrier 68i system. The test site in Yangjiang (Figure 15.5a) is located in residential areas with detached houses. Its point density is 2–3 pts/m^2. The test site in Wuhan (Figure 15.5b) is seated to the south of Wuhan University campus and characterized by dense buildings with tree clutter. Its point density is about 5 pts/m^2.

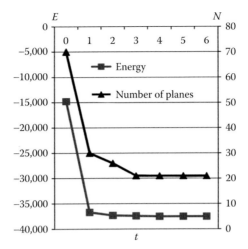

FIGURE 15.4 Energy (*E*) and the number of planes (*N*) versus the number of iterations (*t*) in multilabel optimization.

(a)   (b)

FIGURE 15.5 Ortho images of the two test sites: (a) Yangjiang and (b) Wuhan.

Two necessary thresholds Δd (distance threshold) and n (the minimum number of LiDAR points required for a valid plane) must be selected before segmentation, where n is related to the density of the LiDAR points and the size of considered roof planes. In this study, the distance thresholds in the initial segmentation and label optimization are set to be the same: $n = 4$ and 10 for Yangjiang and Wuhan, respectively, and $\Delta d = 10$ cm for both datasets. Figure 15.6 provides an overview of the segmentation results using the proposed global approach.

Common mistakes in building segmentation are missed planes (one actual plane is not segmented), spurious planes, oversegmented planes, and undersegmented planes. To evaluate the performance of the proposed approach, six metrics are used in this study. Completeness (Comp) is the percentage of reference roof planes that are correctly segmented. It is sensitive to the number

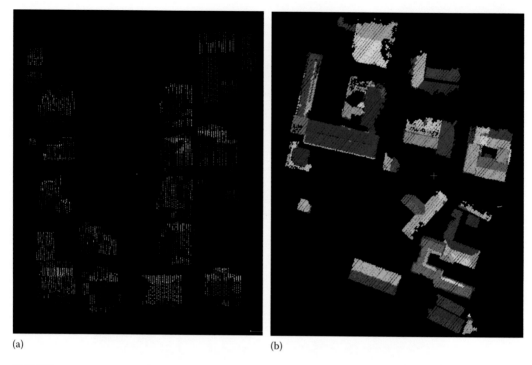

(a)   (b)

FIGURE 15.6 Overview of the segmentation results color coded by planar segments: (a) Yangjiang and (b) Wuhan.

of missed roof planes. Correctness (corr) is the percentage of correctly segmented planes in the segmentation results. It is sensitive to the number of spurious planes. The two metrics are defined as

$$\text{comp} = \frac{TP}{TP + FN}$$

$$\text{corr} = \frac{TP}{TP + FP}$$

(15.6)

where:

TP is the number of true positives, that is, the number of planes found both in reference and segmentation

FN is the number of false negatives, that is, the number of reference planes not found in segmentation

FP is the number of false positives, that is, the number of detected planes not found in the reference

To be counted as a true positive, a minimum overlap of 50% with the reference is required.

To evaluate the effect of the incorrect segmentation, two additional metrics, detection cross-lap (DCL) rate, and reference cross-lap (RCL) rate (Shan and Lee, 2005; Awrangjeb et al., 2010) are adopted. DCL rate is the percentage of detected planes that overlap multiple reference roof planes. RCL rate is the percentage of reference planes that overlap multiple detected planes. They are defined as

$$DCL = \frac{N'_d}{N_d}$$

$$RCL = \frac{N'_r}{N_r}$$

(15.7)

where N_d' is the number of detected planes that overlap more than one reference roof plane, N_r' is the number of reference roof planes that overlap more than one detected plane, N_r and N_d are, respectively, the number of reference planes and detected planes. However, the DCL and RCL metrics are mainly used for an object-based evaluation. As supplementary metrics, boundary precision and boundary recall (BR) are used (Estrada and Jepson, 2009). Boundary precision measures the portion of boundary points in segmentation that correspond to a boundary point in the reference. BR measures the portion of boundary points in the reference that correspond to a boundary point in the segmentation. For LiDAR points, the two metrics can be defined as

$$\text{boundary precision} = \frac{|B_s \cap B_r|}{|B_s|}$$

$$\text{boundary recall} = \frac{|B_s \cap B_r|}{|B_r|}$$

(15.8)

where:

  B_r and B_s are, respectively, the set of boundary points in the reference and segmentation
  $|\ |$ denotes the number of points in a dataset

The reference planes and boundary points are derived by manually labeling (segmenting) the roof points.

To make a comparison, Figure 15.7 presents some representative segmentation results before and after the label optimization. Evaluation metrics of these buildings are listed in Table 15.1.

15.3.4.2 Discussions

We first focus on the evaluation of completeness and correctness that is related to the number of missed and spurious planes. Table 15.1 shows that the optimization improves the completeness from 77.0% to 92.9% and in the same time improves the correctness from 92.6% to 97.7%, which means a considerable decrease in the number of missed and spurious planes.

A number of observations can be noted based on the statistics from the initial segmentation and its optimization (Table 15.1). First, as the outcome of the initial segmentation defines the maximum number of plane segments (i.e., the set of possible labels), the optimization process cannot generate more number of plane segments. Under this restriction, the optimization process essentially may possibly merge initial segments, split a segment, and in the same time combine the results with other existing segments, or include initially unsegmented LiDAR points to an existing segment, all under the restriction that no more number of segments are created. This fact is shown in the two #SP columns of Table 15.1 in which a fewer number of plane segments are present in the optimization outcome. Second, the optimization can segment many initially unsegmented LiDAR points into planar segments (#UC columns). In average, more than half (253 vs. 115) initially unsegmented points (#UC) are involved in the definition of the roof planes, which considerably improves the completeness from 92.6% to 97.7%. This is beneficial as more individual LiDAR points can contribute to the subsequent roof reconstruction. It should be noted that such property is likely due to the use of the smooth cost term (Equation 15.1) that enforces a transition at regions with sparse LiDAR data. Third, many spurious planes (e.g., the middle image in Figure 15.6c) resultant from the initial segmentation are resolved. Table 15.1 shows that the proposed optimization approach keeps the high completeness rate (92.9%) and, at the same time, improves the correctness from 92.6% to 97.7%, which means a very small false alarm rate can be expected.

It should be noted that the optimization cannot contribute to the case that roof planes are not segmented in the initial segmentation. As shown in Figure 15.7b, f, and g (rectangular regions in the reference images), the plane segments not segmented in initial segmentation are also not present in the optimization results. As such, the improvement in the completeness of these buildings is mainly due to the increase in the number of LiDAR points involved in the definition of roof planes.

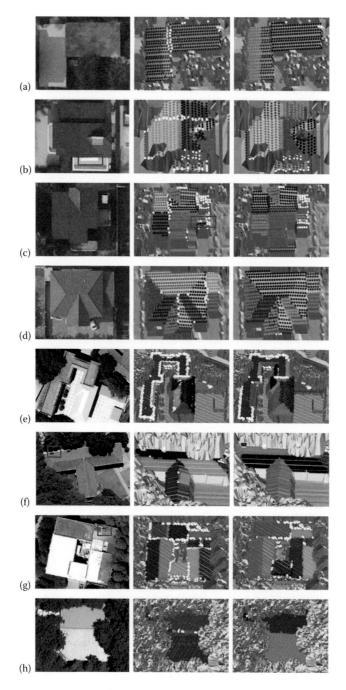

FIGURE 15.7 Segmentation of roof LiDAR points. Building point clouds are colored by planar segments, with white dots representing unsegmented points. From left to right: reference images, initial segmentation, and the optimized results. (a–d) Yangjiang; (e–h) Wuhan.

Next, we discuss the DCL, RCL, BR, and boundary precision, which are related to undersegmented and oversegmented roof planes. Conventional roof segmentation methods such as RANSAC and region growing usually extract planes one after the other from LiDAR points. Although these methods are robust to noise and outliers, they tend to result in mistakes at plane transitions, which in turn causes topological inconsistency among segmented planes. Figure 15.8 illustrates several

TABLE 15.1
Quality of Segmentation Results[*]

Bld	#Pts	#Pls	Methods	#SP	#UC	%Comp	%Corr	%RCL	%DCL	%BP	%BR
a	396	3	Initial	3	26	66.7	100	0	0	50.0	36.7
			Optimized	3	4	100	100	0	0	93.3	93.3
b	424	10	Initial	13	90	90.0	76.9	20	0	71.2	58.4
			Optimized	11	28	90.0	81.8	0	0	82.7	88.8
c	445	11	Initial	11	81	72.7	81.8	0	0	82.0	58.7
			Optimized	9	55	81.8	100	0	0	100	80.4
d	447	10	Initial	11	29	100	90.9	0	0	86.4	74.6
			Optimized	10	5	100	100	0	0	99.6	95.6
e	3930	18	Initial	22	1066	61.1	95.5	22.2	0	72.4	30.9
			Optimized	16	491	88.9	100	0	0	98.3	65.1
f	1494	8	Initial	20	44	62.5	100	50	0	65.2	73.7
			Optimized	7	15	87.5	100	0	0	98.2	89.7
g	3358	19	Initial	22	628	63.2	95.5	15.8	0	76.6	39.2
			Optimized	17	307	94.7	100	0	0	99.8	84.5
h	2273	3	Initial	6	57	100	100	33.3	0	41.3	61.2
			Optimized	3	14	100	100	0	0	93.7	89.7
Average			Initial	–	253	77.0	92.6	17.7	0	68.1	54.2
Average			Optimized	–	115	92.9	97.7	0	0	95.7	89.7

[*] Bld: building ID (Figure 15.7); # Pts: number of lidar points in a building; # Pls: number roof planes in lidar data; Initial: initial segmentation results; Optimized: label optimization results; # SP: the number of segmented planes; #US: the number of unsegmented lidar points; %Comp: completeness; %Corr: correctness; %DCL: detection cross-lap rate; %RCL: reference cross-lap rate; %BP: boundary precision; %BR: boundary recall.

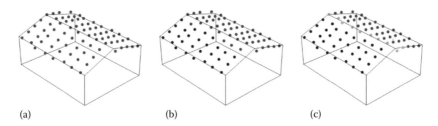

(a) (b) (c)

FIGURE 15.8 Incorrect segmentations at a roof transition. Roof point clouds are colored by segmented planar segments. (a) Two neighboring roof planes are segmented into one. (b) A roof plane and its neighboring points of another plane are segmented into one segment. (c) The transition region is segmented into one segment.

examples of incorrect segmentation. Figure 15.8a shows that two planes are merged into one. The type of incorrect segmentation in Figure 15.8b is common (cf. the elliptical region in Figure 15.7a), in which two planes are separated at a wrongly defined boundary. Figure 15.8c is a typical over-segmentation at plane transitions (cf. the elliptical region in Figure 15.7h). It can be satisfactorily resolved through label optimization. However, the optimization of Figure 15.8a may fail as the label optimization cannot increase the number of segments.

The under and oversegmentation can be measured by the %DCL and %RCL in Table 15.1. A good segmentation should have both as low as possible, so that the detected planes match the actual roof planes well, and the segment boundaries adhere to roof edges. It is seen that the improvement is rather substantial as the RCLs diminish after label optimization, whereas the low DCLs are kept (0%). Accordingly, all the %BR) are increased by 16%–56.6%, whereas the average boundary precision (%) is increased from 68.7% to 95.1%. From the earlier discussion, it can be noted that the mistakes at plane transition can be considerably reduced. Notably, such improvement is mainly due to the use of the smooth cost in optimization. As neighboring points are encouraged to fit to the same plane, the label optimization procedure can produce segments with compact boundaries, which in turn reduces the cross-lap rates among derived patches.

The previous discussion reveals that the global optimization approach can produce spatially coherent segments and reduce the number of mistakes in roof segmentation. Notably, in addition to the aforementioned artifacts, other factors such as roof materials or features may also affect the segmentation quality. For example, some roofs may have a height texture of several centimeters (Oude and Vosselman, 2011), which may influence the selection of distance threshold in model fitting. Moreover, due to the lack of sufficient returns, some small or narrow roof features (e.g., chimneys and parapets) may be classified into noise or outliers.

15.4 BUILDING RECONSTRUCTION

The building reconstruction task can be formulated as an optimal inside/outside labeling of cells derived from space partitioning, in which the surfaces between the inside and outside cells form the reconstructed building model. As shown in Figure 15.1d, each cell is labeled as inside (i.e., occupied cell) or outside (i.e., empty cell). As a result, the facets corresponding to the edges between occupied and empty cells form a surface of the reconstructed building. Such a binary labeling problem can be formulated as a global energy minimization problem (Chauve et al., 2010; Verdie, 2013), which can be resolved via the min-cut theorem in graph cuts (Boykov et al., 2001).

15.4.1 Space Partitioning

Space partitioning is a necessary step for surface reconstruction via graph cuts (Chauve et al., 2010; Oesau et al., 2014). It can be accomplished via various approaches, such as voxel rasterization (Paris et al., 2006; Chehata et al., 2009), 3D Delaunay triangulation (Labatut et al., 2007, 2009), and space subdivision by hyperplanes (Chauve et al., 2010; Oesau et al., 2014). Note that the surfaces of the reconstructed building are the subset of the volumetric surfaces derived from space partitioning. To reconstruct the roof model from the building LiDAR points, we use the segmented roof planes obtained in Section 15.3 to partition the building space.

However, as shown in Figure 15.9, building facades (vertical walls) may not be available in airborne LiDAR points. To resolve this issue, building boundary and step edges are first extracted, from which facades planes are derived (Sampath and Shan, 2010; Perera and Maas, 2014). In this study, building boundary is traced from roof LiDAR points using a concave hull algorithm (Park and Oh, 2013). As for step edges, they are derived from a height discontinuity analysis (Sohn et al., 2008) of neighboring roof LiDAR points. Then, an orientation regularization of the detected boundary and step edges is conducted (Zhang et al., 2012). Figure 15.9 presents an example of this solution. For the building shown in Figure 15.9a, the extracted building boundary and step edges are shown in Figure 15.9b. They are then

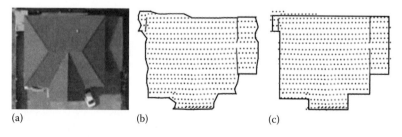

(a)   (b)   (c)

FIGURE 15.9 Detected building boundary (outer line segments) and step edges (inner line segments): (a) building image, (b) building boundary and step edges, and (c) orientation regularization.

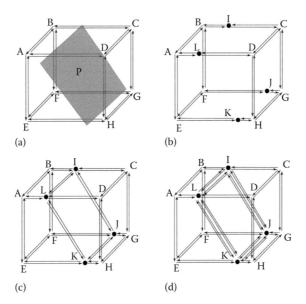

FIGURE 15.10 Workflow of space partitioning with a splitting plane: (a) bounding box ABCDEFGH and given splitting plane P; (b) newly generated vertices I, L, J, and K; (c) newly generated edges IL, IJ, JK, and LK; (d) newly generated face LIJK.

regularized with the help of a dominant direction derived from the longest boundary (cf. Figure 15.9c). Finally, these line features are extruded vertically to create facades as shown in Figure 15.1b.

Having extracted building planes, we first construct a cell complex (Damiand, 2015) for the bounding box of the building (Figure 15.10a). Similar to the half-edge data structure, the cell complex is a B-rep structure to represent topology and geometry. A benefit of such data structure is that it can provide an efficient manipulation of topology in volumes. As shown in Figure 15.1c, the next step is to use a binary space partitioning algorithm (Thibault and Naylor, 1987) to partition the bounding box into spatially connected cells. In this study, we use building planes (segmented roof planes and derived facades) to partition the building space into connected cells. Figure 15.10 illustrates the procedure of space partitioning with a splitting plane (building plane). This procedure is carried out independently for every building plane in a recursive manner. Finally, the bounding box of reconstructed building can be decomposed into connected cells. Benefiting from the cell complex data structure, the topology between derived cells can be preserved (cf. Figure 15.10d). Figure 15.11 presents an example of space partitioning with building planes. Given 21 building planes resultant from facades and roof segmentation (cf. Figure 15.11a), the bounding box of the building is decomposed into 177 volumetric cells (cf. Figure 15.11b).

15.4.2 VISIBILITY ANALYSIS

In this study, the binary labeling problem is noted as a visibility analysis of roof LiDAR points to determine whether a cell is in front of or behind the fitted roof planes. In other words, a cell in front of a roof plane with more supporting LiDAR points is more likely to be visible (i.e., outside cells), whereas a cell behind a roof plane with more LiDAR points is more likely to be invisible (i.e., inside cells).

Figure 15.12 illustrates the visibility analysis of roof LiDAR points, in which 12 cells are derived from the space partitioning step using building planes. Given a LiDAR point t and its fitting roof plane P, a line of sight from an assumed camera center o to the point t is created (i.e., line ot). Along this line of sight, the cells behind roof plane P are invisible (e.g., cells h and g), whereas others are visible (e.g., cell d). As a result, the LiDAR point t is taken as a point supporting visibility of cell d, and a point supporting invisibility of cells h and g. For each LiDAR point, this process is conducted.

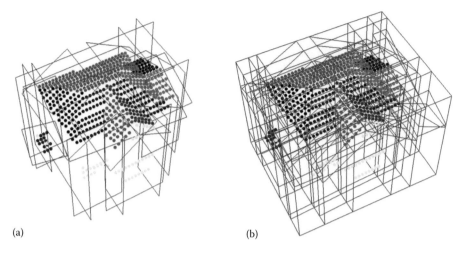

FIGURE 15.11 Space partitioning with building planes: (a) LiDAR points and input building planes (shown by their bounding boxes), (b) volumetric cells.

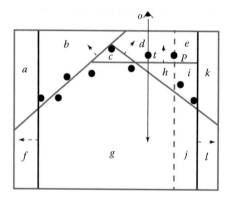

FIGURE 15.12 Visibility analysis of cells: the two dark solid lines are the facade planes derived from the building boundary; the dash line is the facade plane derived from step edge; the short dash lines with arrows are the normal directions of corresponding building planes.

As a result, for each cell p, a visibility vector $\mathbf{X}_p$ (x_1, x_2) can be derived, in which x_1 and x_2 are, respectively, the number of supporting points of the cell's visibility and invisibility.

15.4.3 ENERGY FUNCTION

Based on the visibility analysis of cells, the binary labeling task can be formulated as a minimization of the energy function:

$$E(f) = \sum_{p \in C} V_p(f_p) + \sum_{(p,q) \in N} \left(w_{pq} + \beta \cdot A(p,q) \right) \cdot \delta(f_p \neq f_q) \tag{15.9}$$

where:
   f is a binary labeling of cells
   $\delta(.)$ is an indicator function
   N is the derived neighborhood from cell adjacency graph
   C represents the set of cells

The binary labeling task is to assign a cell $p \in C$ a label f_p (1 for outside/visible cells and 0 for inside/invisible cells) such that the labeling f minimizes the energy $E(f)$. Two energy terms, that is, data term (the first term) and regularization term (the second term) are considered in the energy formula. The data term measures the discrepancy between cell visibility and labels. It is actually the sum of the number of supporting points of cell's visibility and invisibility. The regularization term measures the label inconsistency between neighboring cells. It is the sum of the weight w_{pq} for each pair of neighboring cells p and q that are assigned to different labels and the surface areas of the reconstructed model. w_{pq} is the similarity between neighboring cells p and q, and $A(p, q)$ is the area of facet shared by p and q. Note that β is a parameter favoring small area surfaces (Verdie, 2013).

For each cell, $p \in C$, $x_1(p)$ and $x_2(p)$ are, respectively, the number of supporting points of cell's visibility and invisibility of cell p. The data cost between cell p and its assigned label f_p can be calculated as

$$V_p(f_p) = \begin{cases} x_2(p) + d_{vis} & f_p = 1 \text{ (visible cell)} \\ x_1(p) + d_{vis} & f_p = 0 \text{ (invisible cell)} \end{cases} \tag{15.10}$$

where d_{vis} is a constant value that is set to 1 in the study. In the regularization term, w_{pq} is a measure of similarity between visibility vectors of cells p and q. However, due to the sparse LiDAR points, there may be no points supporting the visibility/invisibility of some cells, that is, cell's visibility vector is $(0, 0)$. To address this issue, w_{pq} is calculated as

$$w_{pq} = \max\left[\min\left(x_1(p), x_1(q)\right), \min\left(x_2(p), x_2(q)\right), d_{vis} \right] \tag{15.11}$$

By using the min-cut theorem, the above-mentioned energy minimization problem in Equation 15.9 can be resolved.

15.4.4 SURFACE EXTRACTION AND MERGING

According to the earlier binary labeling of cells, the surfaces between the inside and outside cells can be extracted and form an estimated model (cf. Figure 15.1e). However, roof faces may be partitioned into several planar surfaces in the process of space partition. Therefore, spatially connected coplanar roof faces should be merged after surface extraction. As discussed in Section 15.4.1, the roof faces and edges as well as their topology have been determined after space partitioning. Based on the spatial connectivity of extracted surfaces, spatially connected coplanar surfaces can be merged (cf. Figure 15.1f) and the roof model is reconstructed.

Given segmented roof LiDAR points, Figure 15.13 illustrates the workflow of roof model reconstruction. Roof and facade planes are first derived. They partition the building bounding box into cells, in which the possible roof faces and edges are determined (Figure 15.13a). Based on the energy function defined in Equation 15.9, a binary labeling of cells is conducted. As a result, cells are labeled as the outside (empty cells in Figure 15.13b) and inside (occupied cells in Figure 15.13b), where the surfaces between inside and outside cells are extracted (Figure 15.13c). Finally, the spatially connected coplanar surfaces are merged, and the estimated roof model is reconstructed as shown in Figure 15.13d.

15.4.5 RESULTS AND DISCUSSION

Based on the segmented roof planes in Section 15.3, Figure 15.14 presents the reconstructed roof models of two test sites. There are 23 buildings in Yangjiang test site and 16 buildings in Wuhan test site. However, due to occlusion from trees, some building boundaries are extremely irregular, which may lead to a wrong regularization of building boundary (cf. the building in the rectangular region in Figure 15.14b).

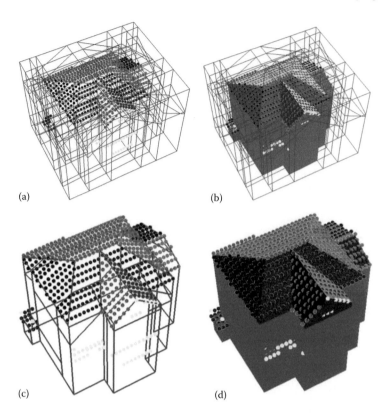

FIGURE 15.13 Segmented roof LiDAR points and reconstructed model: (a) space partitioning, (b) binary labeling of cells, (c) surface extraction, and (d) reconstructed model. Roof point clouds are colored by their planar segments.

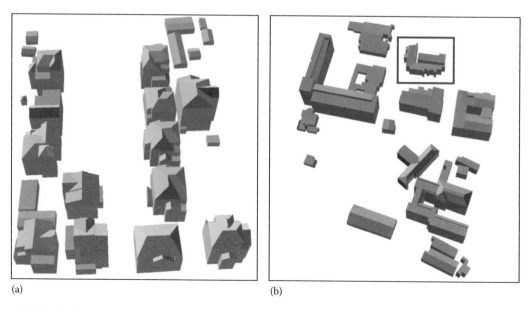

FIGURE 15.14 Reconstructed roof models: (a) Yangjiang and (b) Wuhan.

15.4.5.1 Comparative Analysis

To evaluate the performance of our approach, we compare it with one recently reported model-based graph matching approach (Xiong et al., 2014). Figure 15.15 illustrates the qualitative evaluation on eight buildings of Yangjiang (a–d) and Wuhan (e–h) with varying point densities and roof structures. Both approaches perform well with roof LiDAR points, and most of the roof planes are successfully

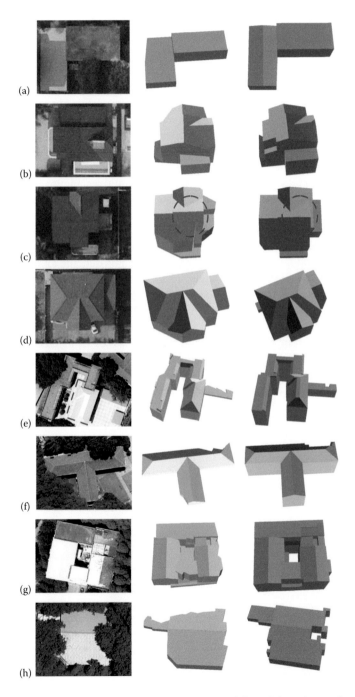

FIGURE 15.15 Comparison of roof model reconstruction. From left to right: reference images, results from the graph matching approach, and results from the global solution.

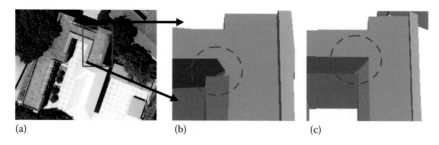

FIGURE 15.16 Topologic reconstruction (Figure 15.15e) between two roof primitives: (a) image, (b) from the graph matching approach, and (c) from the global approach.

reconstructed. However, due to the limitation of roof primitives defined in model library, the graph matching approach may fail when reconstructing complex roof structures. As shown in Figure 15.15e, the flat top of the mansard roof visible in the reference image is not reconstructed by the graph-matching method, whereas the global solution yields the correct result. Besides, as the topology among adjacent roof planes belonging to different roof primitives is not fully considered, the graph matching approach may fail when assembling roof primitives. Figure 15.16 further illustrates the situation of Figure 15.15e. As shown in the rectangular region in Figure 15.9a, there are two pitched roof primitives on the large flat roof plane, and both of them are successfully reconstructed as shown in Figure 15.16b and c. However, as the topology between the two pitched roof primitives is not considered, the graph matching approach fails at the intersection of the two roof primitives (cf. the elliptical region in Figure 15.16b). As for the global approach, the topology between the two pitched roof primitives is well preserved (Figure 15.16c).

However, the proposed approach is dependent on the completeness of input roof planes. Completeness is the percentage correctly segmented roof planes from roof LiDAR points. Table 15.2 lists the completeness of reconstructed roof model using our global approach as shown in Figure 15.15. It is observed in Table 15.2 that the completeness of segmented roof planes is 92.9%. Most of the roof planes are correctly segmented from roof LiDAR points. As for the reconstructed models, the completeness of reconstructed roof planes is 87.5%, a decrease of 5.4% from the segmentation results, and an increase of 6.2% compared with the graph-matching approach. Note that

TABLE 15.2
Completeness of Reconstructed Roof Models*

Bld	#Pls	%CompA	%CompB-1	%CompB-2
a	3	66.7	100	100
b	10	70	90	90
c	11	90.9	81.8	72.7
d	10	90	100	100
e	18	94.4	88.9	88.9
f	8	75	87.5	75
g	19	63.2	94.7	73.7
h	3	100	100	100
Average		81.3	92.9	87.5

* Bld: building ID (cf. Figure 15.15); #Pls: the number of reference roof planes; %CompA: the completeness of reconstructed roof planes using the graph matching approach; %CompB-1: the completeness of segmented roof planes; %CompB-2: the completeness of reconstructed roof planes using the global approach.

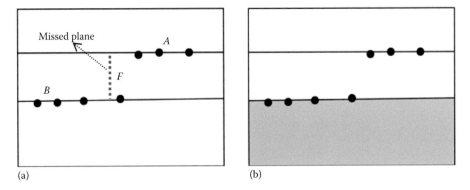

FIGURE 15.17 Profile of a reconstructed flat roof: (a) missed facade F and (b) labeled gray cell.

most of the missed planes in the reconstructed buildings are flat roofs whose neighboring facade planes are missed due to the failure in extracting step edges. As a result, some roof edges and topology cannot be derived from space partitioning. To reconstruct a visibility consistency surface, minimization of the energy function may get rid of roof planes with incomplete edges. As for the model-based graph matching approach, flat roof planes can be correctly reconstructed even when their neighboring facade planes are missed (cf. Figure 15.15c).

Figure 15.17 illustrates the outside/inside labeling of the derived cells when a facade is missed. For example, due to the absence of facade plane F between roof plane A and B, the roof edges determined by the intersection of plane A and F, B and F are missed (Figure 15.17a). As a result, roof plane A is not reconstructed in Figure 15.17b. Besides, roof planes may be missed in the segmentation results. Figure 15.18 illustrates the reconstructed roof models using incomplete roof planes. Although some segmented roof planes are missed, the global approach still outputs a visibility-consistent building surface.

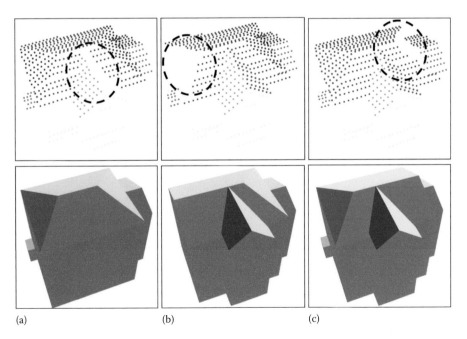

FIGURE 15.18 Reconstructed roof models (building in Figure 15.15d) from incomplete roof planes. (a–c) Roof planes in the elliptical areas (top) are missed.

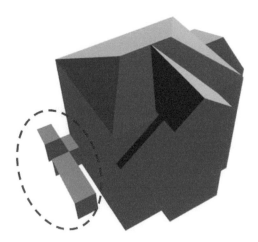

FIGURE 15.19 Reconstructed building in Figure 15.15d without considering the regularization term in Equation 15.9.

15.4.5.2 Parameter Sensitivity

Binary labeling problem can be resolved via various approaches. The advantage of the global approach is that spatial regularization (i.e., the regularization term in Equation 15.1) of surfaces is considered. To demonstrate this, we first illustrate a reconstructed roof model without considering spatial regularization. As shown in Figure 15.19, minimization of the energy function makes a hard assignment of labels to cells, which results in a topologically incorrect roof model in which its surface is not closed. However, this problem can be well fixed using the global approach. As shown in Figure 15.18a–c, the global approach tends to output roof models with closed surfaces even when some roof planes are missed.

The β in Equation 15.9 is an essential parameter in the global optimization approach. A smaller β results in a building model with large surface area, whereas a larger one produces a building model with small surface area. Figure 15.20 provides an example of reconstructed roof models under different β values. As shown in Figure 15.20a–c, the number of planar surfaces derived from the reconstructed model decreases with the increase in parameter β. Finally, only a cubic building model is created (Figure 15.20d), and the surface area of the reconstructed model is minimal. From the earlier discussion, it can be noted that the global approach is able to create the building roof model in different levels of details (Yang et al., 2016). The detail of the roof models is sensitive to the value of β, and may not conform to the specification of levels of details defined in Gröger et al. (2012).

From the previous discussion, the global approach can determine roof edges and topology without a recursive local connectivity analysis of the segmented roof planes. This can avoid

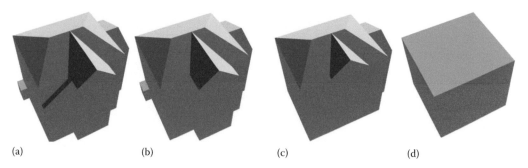

(a)   (b)   (c)   (d)

FIGURE 15.20 Reconstructed roof model under different β values: (a) $\beta = 0$, (b) $\beta = 0.55$, (c) $\beta = 1.0$, and (d) $\beta = 1.5$.

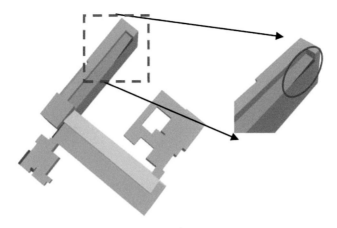

FIGURE 15.21   Building roof model derived from wrong labeling of cells.

incorrect topology in ambiguous cases. However, the global approach is somewhat sensitive to missing building planes, which may in turn miss roof edges in the process of space partitioning. Notably, other factors such as incorrect labeling of cells may also affect the quality of the reconstructed roof model. As shown in Figure 15.21, due to the lack of sufficient LiDAR points supporting the visibility of derived cell, a narrow small part of the building in the elliptical area is wrongly labeled as outside or visible space.

15.5   CONCLUSIONS

The current chapter presents global solutions for roof segmentation and reconstruction from airborne LiDAR point clouds. Roof segmentation is treated as a minimization of energy function consisting of the distances of LiDAR points to planes (labels), spatial smoothness between data points, and the number of planes. Experimental results show that mistakes in roof segmentation such as undersegmented planes, oversegmented planes, spurious planes, and missed planes are considerably reduced. The noticeable incorrect segmentation problems at plane transition regions can be satisfactorily resolved. Compared with the existing model fitting methods for roof segmentation, the label optimization solution is a global one. It simultaneously determines all roof planes, avoiding the order-dependency problem in sequential methods, such as region growing and RANSAC-based ones. The most important benefit of such approach is that multiple roof planes can compete for LiDAR points at their transition regions to reach an optimal segmentation. Another distinctive property of the global approach is that the spatial smoothness between data points is considered, not only reducing the number of unsegmented points but also eliminating the spurious planar segments.

Once roof segmentation is completed, segmented roof planes are used to partition the building space into volumetric cells. Possible geometric features of the reconstructed roof model and their topology are simultaneously determined. To determine the building surfaces and their topology, a global energy function considering surface visibility and spatial regularization between adjacent cells is constructed and minimized via graph cuts. As a result, the cells are labeled as either inside or outside the building, where the planar surfaces between the inside and outside form the reconstructed building model. Compared with the existing reconstruction approaches based on local topology analysis, the presented approach is a global solution. Roof faces and edges as well as their topology are simultaneously determined at one time by minimizing the energy function. This can avoid topologic errors resultant from local connectivity analysis among roof planes. Another property of the global approach is that it is robust to incomplete roof planes due to occlusion leading to visibility-consistent building surfaces even using incomplete roof planes. As a final demonstration

FIGURE 15.22   Two perspective views of the reconstructed building models of the Wuhan dataset.

of the capabilities of the presented global solutions, Figure 15.22 displays two perspective views of the reconstructed buildings models of the Wuhan dataset.

However, the global solution also has limitations. In terms of roof segmentation, it cannot produce more number of planes than what it starts with. This suggests an intentionally slight over-segmentation may be applied to start this method. Like many model fitting methods, a relatively small distance threshold would help for this end. Since the multilabel optimization solution provides a scalable framework for LiDAR point segmentation, additional constrains can be integrated into this formulation. For example, surface normal may be introduced to further consider the transition of planes and even nonpolyhedral roofs. As for building reconstruction, the global solution is helpless when the input segmentation result completely misses a plane. Under such a situation, the reconstruction outcome will be erroneous and a model-based solution, for example, the graph-matching approach, might be considered as a more reliable alternative.

REFERENCES

Awrangjeb, M., Ravanbakhsh, M., Fraser, C.S., 2010. Automatic detection of residential buildings using LiDAR data and multispectral imagery. *ISPRS Journal of Photogrammetry and Remote Sensing*, 65, 457–467.

Awwad, T.M., Zhu, Q., Du, Z., Zhang, Y., 2010. An improved segmentation approach for planar surfaces from unstructured 3D point clouds. *The Photogrammetric Record*, 25, 5–23.

Boykov, Y., Veksler, O., Zabih, R., 2001. Fast approximate energy minimization via graph cuts. *IEEE Transactions on Pattern Analysis and Machine Intelligence*, 23, 1222–1239.

Bretar, F., Roux, M., 2005. Extraction of 3D planar Primitives from Raw Airborne Laser Data: A Normal Driven RANSAC Approach, IAPR Machine Vision and Application, Tsukuba, Japan.

Chauve, A.L., Labatut, P., Pons, J.P., 2010. Robust piecewise-planar 3D reconstruction and completion from large-scale unstructured point data. *2010 IEEE Conference on Computer Vision and Pattern Recognition (CVPR)*. IEEE, pp. 1261–1268.

Chehata, N., Jung, F., Stamon, G., 2009. A graph cut optimization guided by 3D-features for surface height recovery. *ISPRS Journal of Photogrammetry and Remote Sensing*, 64, 193–203.

Chen, D., Zhang, L., Li, J., Liu, R., 2012. Urban building roof segmentation from airborne LiDAR point clouds. *International Journal of Remote Sensing*, 33, 6497–6515.

Cramer, M., 2010. The DGPF-test on digital airborne camera evaluation overview and test design. *Photogrammetrie-Fernerkundung-Geoinformation*, 2010, 73–82.

Damiand, G., 2015. *Linear Cell Complex, CGAL User and Reference Manual*. CGAL Editorial Board.

Delong, A., Osokin, A., Isack, H.N., Boykov, Y., 2012. Fast approximate energy minimization with label costs. *International Journal of Computer Vision*, 96, 1–27.

Dorninger, P., Pfeifer, N., 2008. A comprehensive automated 3D approach for building extraction, reconstruction, and regularization from airborne laser scanning point clouds. *Sensors*, 8, 7323–7343.

Duda, R.O., Hart, P.E., 1972. Use of the Hough transformation to detect lines and curves in pictures. *Communications of the ACM*, 15, 11–15.

Estrada, F.J., Jepson, A.D., 2009. Benchmarking image segmentation algorithms. *International Journal of Computer Vision*, 85, 167–181.

Filin, S., Pfeifer, N., 2006. Segmentation of airborne laser scanning data using a slope adaptive neighborhood. *ISPRS Journal of Photogrammetry and Remote Sensing*, 60, 71–80.

Fischler, M.A., Bolles, R.C., 1981. Random sample consensus: A paradigm for model fitting with applications to image analysis and automated cartography. *Communications of the ACM*, 24, 381–395.

Gorte, B., 2002. Segmentation of TIN-structured surface models. *International Archives of Photogrammetry Remote Sensing and Spatial Information Sciences*, 34, 465–469.

Gröger, G., Kolbe, T., Nagel, C., Häfele, K., 2012. OGC city geography markup language (CityGML) encoding standard V2. 0. OGC Doc.

Haala, N., Kada, M., 2010. An update on automatic 3D building reconstruction. *ISPRS Journal of Photogrammetry and Remote Sensing*, 65, 570–580.

Henn, A., Gröger, G., Stroh, V., Plümer, L., 2013. Model driven reconstruction of roofs from sparse LiDAR point clouds. *ISPRS Journal of Photogrammetry and Remote Sensing*, 76, 17–29.

Huang, H., Brenner, C., Sester, M., 2013. A generative statistical approach to automatic 3D building roof reconstruction from laser scanning data. *ISPRS Journal of Photogrammetry and Remote Sensing*, 79, 29–43.

Isack, H., Boykov, Y., 2012. Energy-based geometric multi-model fitting. *International Journal of Computer Vision*, 97, 123–147.

Kada, M., McKinley, L., 2009. 3D building reconstruction from LiDAR based on a cell decomposition approach. *International Archives of Photogrammetry, Remote Sensing and Spatial Information Sciences*, 38, W4.

Kim, K.H., Shan, J., 2011. Building roof modeling from airborne laser scanning data based on level set approach. *ISPRS Journal of Photogrammetry and Remote Sensing*, 66, 484–497.

Kolmogorov, V., Zabin, R., 2004. What energy functions can be minimized via graph cuts? *IEEE Transactions on Pattern Analysis and Machine Intelligence*, 26, 147–159.

Labatut, P., Pons, J.-P., Keriven, R., 2007. Efficient multi-view reconstruction of large-scale scenes using interest points, Delaunay triangulation and graph cuts. In *IEEE 11th International Conference on Computer Vision*. ICCV 2007, IEEE, pp. 1–8.

Labatut, P., Pons, J.P., Keriven, R., 2009. Robust and efficient surface reconstruction from range data. In *Computer Graphics Forum*. Wiley Online Library, pp. 2275–2290.

Lari, Z., Habib, A., Kwak, E., 2011. An adaptive approach for segmentation of 3D laser point cloud. *ISPRS - International Archives of the Photogrammetry, Remote Sensing and Spatial Information Sciences*, Calgary, Canada, pp. 103–108.

Lin, H., Gao, J., Zhou, Y., Lu, G., Ye, M., Zhang, C., Liu, L., Yang, R., 2013. Semantic decomposition and reconstruction of residential scenes from LiDAR data. *ACM Transaction Graph*, 32, 66.

Maas, H.-G., Vosselman, G., 1999. Two algorithms for extracting building models from raw laser altimetry data. *ISPRS Journal of Photogrammetry and Remote Sensing*, 54, 153–163.

Melzer, T., 2007. Non-parametric segmentation of ALS point clouds using mean shift. *Journal of Applied Geodesy*, 1, 159.

Oesau, S., Lafarge, F., Alliez, P., 2014. Indoor scene reconstruction using feature sensitive primitive extraction and graph-cut. *ISPRS Journal of Photogrammetry and Remote Sensing*, 90, 68–82.

Oude, E.S., Vosselman, G., 2011. Quality analysis on 3D building models reconstructed from airborne laser scanning data. *ISPRS Journal of Photogrammetry and Remote Sensing*, 66, 157–165.

Oude, E.S.J., Vosselman, G., 2009. Building reconstruction by target based graph matching on incomplete laser data: Analysis and limitations. *Sensors*, 9(8), 6101–6118.

Paris, S., Sillion, F.X., Quan, L., 2006. A surface reconstruction method using global graph cut optimization. *International Journal of Computer Vision*, 66, 141–161.

Park, J.-S., Oh, S.-J., 2013. A new concave hull algorithm and concaveness measure for n-dimensional datasets. *Journal of Information Science and Engineering*, 29, 379–392.

Perera, G.S.N., Maas, H.-G., 2014. Cycle graph analysis for 3D roof structure modelling: Concepts and performance. *ISPRS Journal of Photogrammetry and Remote Sensing*, 93, 213–226.

Sampath, A., Shan, J., 2010. Segmentation and reconstruction of polyhedral building roofs from aerial LiDAR point clouds. *IEEE Transactions on Geoscience and Remote Sensing*, 48, 1554–1567.

Satari, M., Samadzadegan, F., Azizi, A., Maas, H.G., 2012. A multi-resolution hybrid approach for building model reconstruction from LiDAR data. *The Photogrammetric Record*, 27, 330–359.

Schnabel, R., Wahl, R., Klein, R., 2007. Efficient RANSAC for point-cloud shape detection. In *Computer Graphics Forum*. Wiley Online Library, pp. 214–226.

Shan, J., Lee. S.D., 2005. Quality of building extraction from IKONOS imagery. *Journal of Surveying Engineering*, 131 (1), 27–32.

Sohn, G., Huang, X., Tao, V., 2008. Using a binary space partitioning tree for reconstructing polyhedral building models from airborne LiDAR data. *Photogrammetric Engineering & Remote Sensing*, 74, 1425–1438.

Tarsha-Kurdi, F., Landes, T., Grussenmeyer, P., 2007a. Hough-transform and extended ransac algorithms for automatic detection of 3D building roof planes from LiDAR data. *International Archives of Photogrammetry, Remote Sensing and Spatial Information Systems*, 36, 407–412.

Tarsha-Kurdi, F., Landes, T., Grussenmeyer, P., Koehl, M., 2007b. Model-driven and data-driven approaches using LiDAR data: Analysis and comparison. *International Archives of Photogrammetry, Remote Sensing and Spatial Information Systems*, 87–92.

Thibault, W.C., Naylor, B.F., 1987. Set operations on polyhedra using binary space partitioning trees. *ACM SIGGRAPH Computer Graphics*, 21(4), 153–162.

Tomljenovic, I., Höfle, B., Tiede, D., Blaschke, T., 2015. Building extraction from airborne laser scanning data: An analysis of the state of the art. *Remote Sensing*, 7(4), 3826–3862.

Verdie, Y., 2013. Urban scene modeling from airborne data. Université Nice Sophia Antipolis.

Verma, V., Kumar, R., Hsu, S., 2006. 3D building detection and modeling from aerial LiDAR data. *IEEE Computer Society Conference on Computer Vision and Pattern Recognition*, New York, pp. 2213–2220.

Vitti, A., 2012. The Mumford–Shah variational model for image segmentation: An overview of the theory, implementation and use. *ISPRS Journal of Photogrammetry and Remote Sensing*, 69, 50–64.

Vosselman, G., Dijkman, S., 2001. 3D building model reconstruction from point clouds and ground plans. *International Archives of Photogrammetry Remote Sensing and Spatial Information Sciences*, 34, 37–44.

Wang, R.S., 2013. 3D building modeling using images and LiDAR: A review. *International Journal of Image and Data Fusion*, 4(4), 273–292.

Xiong, B., Oude, E.S., Vosselman, G., 2014. A graph edit dictionary for correcting errors in roof topology graphs reconstructed from point clouds. *ISPRS Journal of Photogrammetry and Remote Sensing*, 93, 227–242.

Xu, B., Jiang, W., Shan, J., Zhang, J., Li, L., 2015. Investigation on the weighted RANSAC approaches for building roof plane segmentation from LiDAR point clouds. *Remote Sensing*, 8 (1), 5. doi:10.3390/rs8010005.

Yan, J., Jiang, W., Shan, J., 2016. A global solution to topological reconstruction of building roof models from airborne LiDAR point clouds. *ISPRS Annals of the Photogrammetry, Remote Sensing and Spatial Information Science*, 3, 379–386.

Yan, J., Shan, J., Jiang, W., 2014. A global optimization approach to roof segmentation from airborne LiDAR point clouds. *ISPRS Journal of Photogrammetry and Remote Sensing*, 94, 183–193.

Yang, B., Huang, R., Li, J., Tian, M., Dai, W, Zhong, R., 2016. Automated reconstruction of building LoDs from airborne LiDAR point clouds using an improved morphological scale space. *Remote Sensing*, 9(1), 14.

You, R.-J., Lin, B.-C., 2011. Building feature extraction from airborne LiDAR data based on tensor voting algorithm. *Photogrammetric Engineering and Remote Sensing*, 77, 1221–1232.

Zhang, K., Yan, J., Chen, S.-C., 2006. Automatic construction of building footprints from airborne LiDAR data. *IEEE Transactions on Geoscience and Remote Sensing*, 44, 2523–2533.

Zhang, Y., Xiong, X., Shen, X., 2012. Automatic registration of urban aerial imagery with airborne LiDAR data. *Journal of Remote Sensing*, 16, 579–595. (in Chinese).

16 Building and Road Extraction from LiDAR Data

Franz Rottensteiner and Simon Clode

CONTENTS

16.1 INTRODUCTION

Mapping of topographic objects, especially man-made structures such as buildings and roads, has always been one of the most important tasks in photogrammetry and remote sensing. Due to the increasing demand for spatial data as they are provided by geographical information systems (GIS) or 3D City Models, the automation of that task has gained considerable importance in photogrammetric research, and Light Detection and Ranging (LiDAR) provides one of the most important data sources for that purpose. LiDAR sensors deliver a 3D point cloud with the intensity of the returned signal, sometimes providing multiple pulses or even the full waveform of the returned signal (Mallet and Bretar, 2009). The information about objects in a scene is only contained implicitly in these data, and a considerable amount of processing is required to make this semantic information explicit.

Automated object extraction as it is understood in this chapter aims at the automatic detection of topographic objects and their geometric reconstruction by vectors that can easily be used in GIS or computer-aided design systems. It is usually carried out in two steps:

1. Object detection is essentially a classification of the input data using a model of the appearance of the object in the sensor data. It results in detected objects with coarse boundaries in object space.
2. Object reconstruction or object vectorization is carried out in the previously detected object regions to reconstruct the shape of the detected objects by vectors.

The current chapter will focus on the extraction of buildings and roads from airborne LiDAR data. We start with a generic context-based classification technique that assigns a semantic class label to each LiDAR point. The points assigned to the classes *building* and *road* can be used as the basis for the definition of distinct objects of the specified type and, subsequently, for vectorization.

Compared with aerial imagery, LiDAR has both advantages and disadvantages with respect to automatic object extraction (Schenk and Csatho, 2002). First, LiDAR directly provides 3D points, whereas with aerial imagery, matching is required to obtain the third dimension. Being an active sensing technique, LiDAR does not have any problems with cast shadows, and occlusions are less problematic because LiDAR sensors typically have a smaller opening angle than aerial cameras. Due to its explicit 3D nature, LiDAR data gives better access to deriving geometrical properties of surfaces such as surface roughness, which is a distinguishing feature of man-made objects. However, LiDAR only provides limited information about the reflectance properties of the object surfaces, whereas modern aerial cameras deliver multispectral images. The boundaries of objects are usually better defined in aerial imagery than they are in LiDAR data especially if they correspond to height discontinuities, and finally, the resolution of LiDAR data, though astonishingly high, does not yet match the resolution of aerial imagery. Given the pros and cons of LiDAR and aerial imagery, these data sources are frequently combined to improve the robustness of automatic object extraction. For instance, the majority of contributions to the benchmark on urban object detection organized by the International Society for Photogrammetry and Remote Sensing (ISPRS) were based on a combination of image and LiDAR data; however, the 3D reconstruction of buildings in the same benchmark was dominated by methods based on LiDAR (Rottensteiner et al., 2014).

16.2 CONTEXT-BASED CLASSIFICATION OF LiDAR DATA AND THE DETECTION OF BUILDINGS

In this section, object detection from LiDAR is posed as a classification problem. We assume the LiDAR data to consist of N points $\mathbf{P}_i$ with $i \in \{1, ..., N\}$, each being described by its 3D coordinates x_i, y_i, and z_i, and, potentially, the intensity i_i and additional (e.g., full waveform) features directly assigned to that point. The set of all laser points collected is represented by $\mathbf{x} = \{\mathbf{P}_1, \mathbf{P}_2, ..., \mathbf{P}_N\}$. It is the goal of classification to assign each $\mathbf{P}_i$ to a class C_i from a given set of classes $C = \{C^1, C^2, ..., C^K\}$, where K is the total number of classes. In this notation, subscripts refer to a specific point $\mathbf{P}_i$, whereas superscripts represent a specific class label C^k corresponding to a specific object type, for example, *building*. The classification process typically does not only take into account the original data, but feature vectors $\mathbf{f}_i$ that are derived from the original features of point $\mathbf{P}_i$ as well as from other points in a local neighborhood of that point. To make this dependency explicit, we will use the notation $\mathbf{f}_i(\mathbf{x})$. We will start this section with short reviews of LiDAR features and classification techniques. After that, a context-based classification technique for LiDAR points is presented. Finally, we will show how these classification results can be used to detect individual buildings.

16.2.1   LiDAR Features

The most important LiDAR features is the *height above ground* (Chehata et al., 2009), distinguishing objects on the ground from other ones. Computing this feature requires the definition of a terrain height that is subtracted from a LiDAR point's height z_i. The simplest way of estimating the terrain height is to use the height of the lowest point inside a local neighborhood, but this definition is prone to errors in the presence of very large buildings or undulating terrain. It is preferable to generate a digital terrain model (DTM) first, from which the terrain height at the LiDAR point's planimetric position can be interpolated. If a grid-based digital surface model (DSM) is derived from the LiDAR points, morphological opening can be used to obtain an approximation for the DSM (Weidner and Förstner, 1995); to overcome problems of selecting an appropriate filter size for scenes containing undulating terrain and buildings of heterogeneous size, a method using different filter sizes while excluding large buildings based on simple rule-based algorithm for building detection can be applied (Rottensteiner et al., 2005a). Apart from such simple methods, any DTM filtering algorithm can be used in this context, for example, progressive triangulated irregular network (TIN) densification (Axelsson, 2000) or hierarchical robust linear prediction (Rottensteiner and Briese, 2002); cf. (Sithole and Vosselman, 2004; Pfeifer and Mandlburger, 2017) (this volume) for overviews.

Modern LiDAR sensors are capable of delivering multiple echoes for each emitted laser pulse, and a point $\mathbf{P}_i$ as defined in this chapter corresponds to an individual echo. For such an echo, the *echo index* and the *total echo count* are frequently delivered. These features are indicators for object classes that can be partly penetrated by the laser beam, in particular vegetation (Höfle et al., 2012). The *intensity* of the returned laser beam provides information about the surface reflectance properties in the wavelength of the laser. The advantage of LiDAR intensities is that they are hardly affected by natural illumination conditions and cast shadows. However, the intensities are often very noisy (Vosselman, 2002) because the diameter of the laser beam footprint is usually smaller than the sampling distance. For LiDAR systems recording the full waveform of the reflected laser pulse, the waveform has to be decomposed into individual echoes, and additional features such as the amplitude, echo width, or the signal cross-ratio can be determined (Wagner et al., 2006; Mallet and Bretar, 2009), again particularly useful for differentiating vegetation from other object types.

A large group of LiDAR features is determined from all points within a local neighborhood. The first option for defining a neighborhood-based feature is the *density* of points inside a local sphere. Another group of features can be derived from the analysis of the covariance matrix $\mathbf{M}$ of the coordinates of all points $\mathbf{P}_i$ inside a certain neighborhood:

$$\mathbf{M} = \sum_i \left(\mathbf{P}_i - \mathbf{P}_c \right) \cdot \left(\mathbf{P}_i - \mathbf{P}_c \right)^T \tag{16.1}$$

where $\mathbf{P}_c$ is the center of gravity of all $\mathbf{P}_i$ inside that neighborhood. The Eigenvalues $\lambda_1, \lambda_2, \lambda_3$ of $\mathbf{M}$ with $\lambda_1 \leq \lambda_2 \leq \lambda_3$ are indicators to the local configuration of points. If all Eigenvalues take large values, the points inside the neighborhood considered are distributed more or less evenly in 3D. The Eigenvalue λ_3 can be interpreted as the square sum of residuals of a planar fit, and consequently, λ_3 can be interpreted as a feature related to *planarity*. Finally, a situation in which the points are aligned on a linear structure (e.g., in the case of a power line) is indicated by two small Eigenvalues λ_2, λ_3. In addition to using the Eigenvalues themselves, ratios of Eigenvalues, for example, *omnivariance, planarity, anisotropy, sphericity,* and *eigenentropy, scatter* can be used (Pauly et al., 2003; Chehata et al., 2009; Weinmann et al., 2015). The Eigenvector corresponding to the smallest Eigenvalue can be interpreted as the normal vector of the adjusting plane, so that it gives access to the local *surface slope*. The local variation of the surface normal vectors gives access to the local *curvature* or *surface roughness* (Brunn and Weidner, 1997). Again, if a grid-based DSM is available, surface slope, curvature, and roughness features can be computed on the basis of linear filter techniques, taking into account derivative filters or the Laplacian of Gaussian operator (Maas, 1999). In (Rottensteiner et al., 2005a, 2007), a local homogeneity measure R of the first derivatives is used as a feature related to surface roughness:

$$R = \mathbf{L} * \left\{ \frac{1}{\sigma_x^2} \cdot \left[\left(\frac{\partial^2 z}{\partial x^2} \right) + \left(\frac{\partial^2 z}{\partial x \partial y} \right)^2 \right] + \frac{1}{\sigma_y^2} \cdot \left[\left(\frac{\partial^2 z}{\partial y \partial x} \right)^2 + \left(\frac{\partial^2 z}{\partial y^2} \right)^2 \right] \right\} \tag{16.2}$$

In Equation 16.1, the DSM is represented by a height grid $z(x, y)$, whose derivatives are used to determine R, $\mathbf{L}$ is a lowpass filter, for example, a binomial filter kernel of size $n \times n$, and $\bar{\sigma}_x^2$ and $\bar{\sigma}_y^2$ are the variances of the smoothed second derivatives. They can be derived from an estimate of the variance σ_z^2 of the DSM heights by error propagation. R can also be interpreted as a measure of local planarity.

Finally, for classification techniques operating on point cloud segments rather than on individual points, *shape features* of the segments can be considered as well (Brunn and Weidner, 1997; Rottensteiner and Briese, 2002).

For the computation of features based on the local configuration of points, the definition of the neighborhood is of crucial importance. The neighborhood is typically defined on the basis of a distance criterion. Taking into account 3D distances results in *spherical (3D) neighborhoods,* whereas *cylindrical (2D) neighborhoods* are based on planimetric distances only. In the case of airborne LiDAR, both definitions may be useful, because spherical neighborhoods give access to local surface properties, whereas a cylindrical neighborhood may give access to specific properties of objects that can be partially penetrated by the laser beam. Beyond that, one can differentiate neighborhoods based on the k nearest neighbors (*knn*) from fixed neighborhoods characterized by all points inside a sphere or cylinder of radius r. A neighborhood definition based on *knn* is more adaptive to varying point densities, whereas it may deliver unreliable features in the case of isolated points. In any case, the number k of neighbors or a radius r for the definition of a neighborhood can be interpreted as a *scale* parameter. The selection of that scale parameter may be difficult, but it is crucial for an appropriate feature definition. Thus, Weinmann et al. (2015) suggest adapting the scale to the local point density. In many cases, it is useful to consider neighborhood definitions at multiple scales to capture the local structure of the point cloud for neighborhoods of different sizes. This leads to the definition of *multiscale* feature vectors (Niemeyer et al., 2014).

16.2.2 CLASSIFICATION TECHNIQUES FOR LiDAR DATA

Classification techniques can be characterized according to the entities for which a class label is predicted. In the case of LiDAR data, the classification can be applied either to each pixel of the DSM (if the DSM is represented by a grid), to each LiDAR point, or to DSM or point cloud segments. The main advantage of using segments is that points having similar properties are grouped by the segmentation process so that the number of entities to be classified is much smaller and, consequently, the classification is much faster. Furthermore, features computed for segments are usually more accurate than features for individual points. On the other hand, segment-based classification procedures cannot recover from errors in the segmentation process that lead to situations in which object boundaries do not coincide with segment boundaries. The errors introduced may reach 1%–3%, depending on the average size of a segment (Shapovalov et al., 2010). Using DSM pixels or segments rather than LiDAR points or point cloud segments has the advantage that image processing techniques can be applied for feature computation and that the definition of neighborhoods is straightforward. On the other hand, DSM grids only allow for 2.5D processing, which causes problems at object boundaries characterized by height discontinuities and with overhanging object parts. Point-based and segment-based processing can also be combined, either in a sequential process as in (Rottensteiner et al., 2005a, 2007) in which building segments defined on the basis of the results of pixel-based classification are verified in a second classification process, or in a joint hierarchical classification process as described in (Xiong et al., 2011).

Another criterion for characterizing classification techniques is the way in which the knowledge about the objects is introduced into the process. In *model-based* techniques, the knowledge about the objects is modeled explicitly. For instance, rule-based classification techniques define intervals for the features of a certain class based on model knowledge about the appearance of the objects in the data, which usually leads to a series of thresholding operations. Such techniques are easily implemented and still used rather frequently, in particular for binary classification tasks; see (Rottensteiner et al., 2014) for some recent examples. However, selecting the thresholds properly is often critical, which may lead to a poor transferability of such methods. Additionally, many rule-based algorithms follow a hierarchical procedure, selecting a subset of features for an initial classification and using the other features to resolve any ambiguities (Rottensteiner and Briese, 2002); such a procedure makes it impossible to recover from previous errors in the classification process. The latter problem can be overcome by approaches based on fuzzy logic (Vögtle and Steinle, 2003) or by probabilistic methods based, for example, on the theory of Dempster–Shafer (Rottensteiner et al., 2005a, 2007) or Bayesian reasoning (Maas, 1999). These approaches are considered to be model-based if the underlying fuzzy membership or probability or probability mass functions are based on heuristics such as a model for the distribution of building heights or other features.

The problem of appropriate parameter selection is solved by *supervised classification* techniques in a principled way. In such approaches, the knowledge about the objects in a scene is represented implicitly in a set of training samples for each class that has to be given to the classifier. From these training samples, each consisting of a feature vector and a class label, the classifier learns a model of the data. In the testing phase, the trained model is used to predict the class label C_i of a feature vector $\mathbf{f}_i(\mathbf{x})$. In the case of *probabilistic supervised classification*, the class assignment is based on the maximum a posteriori principle, that is, the classification result will be the class for which the posterior probability $P(C_i \mid \mathbf{f}_i(\mathbf{x}))$ of the class label given the feature vector becomes a maximum. *Generative probabilistic classifiers* use the theorem of Bayes for determining the posterior $P(C_i \mid \mathbf{f}_i(\mathbf{x}))$ (e.g., Bishop, 2006):

$$P\left(C_i \mid \mathbf{f}_i(\mathbf{x})\right) = \frac{P\left(\mathbf{f}_i(\mathbf{x}) \mid C_i\right) \cdot P(C_i)}{\sum_j P\left(\mathbf{f}_i(\mathbf{x}) \mid C_j\right) \cdot P(C_j)} \tag{16.3}$$

In Equation 16.4, $P(\mathbf{f}_i(\mathbf{x}) \mid C_i)$ is the *likelihood*, indicating how probable it is to observe the feature vector $\mathbf{f}_i(\mathbf{x})$ under the assumption of a class C_i, and $P(C_i)$ is the *prior*, corresponding to the probability of class C_i if no data are given. The likelihood $P(\mathbf{f}_i(\mathbf{x}) \mid C_i)$ is frequently modeled by multivariate Gaussian distributions or, if a class corresponds to multiple clusters in feature space, by Gaussian mixture models (Bishop, 2006). In the training phase, the parameters of the likelihood function and the prior have to be learned from the data. The product of likelihood and prior in the numerator of Equation 16.3 is the joint probability of class labels and features. Consequently, learning the parameters of the model in Equation 16.3 is equivalent to learning a model of the distribution of the data. From this model, one can draw synthetic datasets, hence the term *generative model*. However, learning a model of the distribution of the data might require a large amount of training samples. This is avoided by *discriminative probabilistic classifiers* such as multiclass logistic regression, which model the posterior $P(C_i \mid \mathbf{f}_i(\mathbf{x}))$ directly (e.g., Bishop, 2006; Niemeyer et al., 2011). *Nonprobabilistic discriminative classifiers* are not based on probabilities at all but learn other functions that discriminate the classes on the basis of the feature vectors. Some of the best classifiers currently known belong to this group, for instance, Support Vector Machines (Cortes and Vapnik, 1995; Mallet, 2010) or Random Forests (RF) (Breiman 2001; Chehata et al., 2009). Most recently, deep learning methods such as convolutional neural networks have been gaining importance (Farabet et al., 2013). Convolutional neural networks can be interpreted as techniques for learning high-level feature representations of images; they have been applied to DSM in combination with image data in (Paisitkriangkrai et al., 2015).

All methods discussed so far classify each point or segment independently without considering its neighborhood beyond integrating features based on the local point configuration. Such a purely local approach will lead to inhomogeneous classification results in areas where the features correspond to the regions in feature space where the distributions of the features for different classes overlap. To overcome such problems, one can apply *context-based classification* techniques. Models for spatial dependencies between the object classes can be trained to improve the results, because some object classes are more likely to occur next to each other than others. A sound statistical model of context leads to undirected *graphical models* such as Markov Random Fields (MRFs) (Geman and Geman, 1984). In a graphical model, the classification problem is represented by a graph in which the class label of each entity (point or segment) corresponds to a node, whereas neighboring entities are connected by edges. These edges represent the statistical dependencies between the nodes connected by them. In an MRF, the class label of an object is statistically dependent on its neighbors, whereas the data of different objects are assumed to be conditionally independent. This leads to a model for the joint posterior $P(\mathbf{C} \mid \mathbf{x})$ of the class labels of all points, collected in a vector $\mathbf{C} = (C_1, \ldots C_N)^T$, given the data $\mathbf{x}$ (Geman and Geman, 1984):

$$P(\mathbf{C} \mid \mathbf{x}) = \frac{1}{Z} \cdot \prod_{i \in n} \varphi(C_i, \mathbf{x}_i) \cdot \prod_{i,j \in e} \psi(C_i, C_j) \qquad (16.4)$$

In Equation 16.4, n and e denote the sets of nodes and edges of the graph, respectively. The terms denoted by $\varphi(C_i, \mathbf{x}_i)$ are called *unary* or *node potentials*. They connect the class label of an entity (node) i with the feature vector $\mathbf{x}_i$ of that entity. We use the symbol $\mathbf{x}_i$ to indicate that the feature vector may also contain functions of the original data assigned to a LiDAR point $\mathbf{P}_i$. In a Bayesian setting, one can use a probabilistic model for the likelihood of a generative local classifier for that purpose. The terms denoted by $\psi(C_i, C_j)$ are called *binary* or *pairwise* or *edge potentials*. Linking the class labels of neighboring entities, usually they are based on a model that favors neighboring entities to belong to the same class such as the Potts model (Bishop, 2006). The term Z in Equation 16.4 is a normalization constant called the partition function. According to the model for the posterior in Equation 16.4, the class labels mutually influence each other, so that an independent classification of individual entities is impossible. Thus, to classify the data, one has to determine the optimum configuration of labels $\mathbf{C}$ by maximizing the posterior in Equation 16.4 over all entities simultaneously. This can only be achieved in an approximate way by methods such as graph cuts (Boykov et al., 2001; Boykov and Jolly, 2001) or Loopy Belief Propagation (LBP) (Frey and McKay, 1998).

MRF tend to oversmooth the classification result because the pairwise terms will favor identical class labels at neighboring entities independently from the data. This problem is overcome by Conditional Random Fields (CRF) (Kumar and Hebert, 2006), in which the joint posterior $P(\mathbf{C} \mid \mathbf{x})$ of all class labels given the data is modeled by Equation 16.5:

$$P(\mathbf{C} \mid \mathbf{x}) = \frac{1}{Z} \cdot \prod_{i \in n} \varphi(C_i, \mathbf{x}) \cdot \prod_{i,j \in e} \psi(C_i, C_j, \mathbf{x}) \qquad (16.5)$$

At a first glance, Equation 16.5 looks very similar to Equation 16.4, but there are two differences. First, in the unary potential for a node, here also called *association potential* φ (C_i, **x**), depends on all data **x** rather than only on the data observed at node i. This is usually taken into account by defining feature vectors $\mathbf{f}_i(\mathbf{x})$ (*node feature vectors*) that may depend on a more or less large neighborhood of the entity i as described in Section 16.2.1. Unlike with MRF, the association potentials are usually based on the probabilistic output of a *discriminative* local classifier, that is, φ (C_i, **x**) $\propto P(C_i \mid \mathbf{f}_i(\mathbf{x}))$; nonprobabilistic classifiers can be used nevertheless if one can derive a probabilistic output in a postprocessing stage, which is often the case (Kumar and Hebert, 2006). The second difference between Equations 16.4 and 16.5 is that the binary potential, here also called *interaction potential* ψ (C_i, C_j, **x**), also depends on the data **x**. Again, this potential can be based on the probabilistic output of a classifier, but in this case, there are two options. Methods based on data-dependent smoothing apply a model for the probability of the two class labels to be equal given the data, that is, ψ (C_i, C_j, **x**) $\propto P(C_i = C_j \mid \boldsymbol{\mu}_{ij}(\mathbf{x}))$, in which we have defined a feature vector $\boldsymbol{\mu}_{ij}(\mathbf{x})$ for the edge connecting the two nodes i and j (*edge feature vector*). Typical choices for $\boldsymbol{\mu}_{ij}(\mathbf{x})$ are the difference of the feature vectors of nodes i and j, that is, $\boldsymbol{\mu}_{ij}(\mathbf{x}) = \mathbf{f}_i(\mathbf{x}) - \mathbf{f}_j(\mathbf{x})$, the Euclidean distance of the node feature vectors, or the concatenation of the two feature vectors, that is, $\boldsymbol{\mu}_{ij}(\mathbf{x}) = (\mathbf{f}_i^T(\mathbf{x}) \, \mathbf{f}_j^T(\mathbf{x}))^T$, but one can also define other features that are independent from the node feature vectors (Kumar and Hebert, 2006). A simple heuristic model of that group is the contrast-sensitive Potts model (Boykov and Jolly, 2001) in which the degree of smoothing depends on the Euclidean distance of the two node feature vectors.

As an alternative to data-dependent smoothing, one can use the probabilistic output of a classifier predicting the joint local configuration of the labels at nodes i and j, that is, ψ (C_i, C_j, **x**) $\propto P(C_i, C_j \mid \boldsymbol{\mu}_{ij}(\mathbf{x}))$. Such a model, although being more general than data-dependent smoothing, requires a sufficient amount of training samples for each pair of classes that might occur at neighboring nodes, which may be prohibitive in some applications. In any case, the use of classifiers for modeling the interaction potentials requires fully labeled patches of training data to have a sufficient amount of representative training samples for class transitions.

It has to be noted that MRF or CRF offer a framework that can lead to quite different classification techniques. Applications of that framework differ by the definition of the graph structure, by the way they model the potentials, and by the definition of the (node and edge) features. In the context of the classification of LiDAR data, the first applications of MRF and CRF were in robotics and mobile terrestrial laser scanning (Anguelov et al., 2005; Lim and Suter, 2007; 2009; Munoz et al., 2008). One of the first applications of CRF to airborne LiDAR data is shown in (Lu et al. 2009), in which the main goal of classification is DTM generation from a grid-based DSM. Shapovalov et al. (2010) presented a CRF-based classification of point cloud segments derived by mean shift segmentation. Lafarge and Mallet (2012) use an MRF for the classification of a point cloud. As their main interest is in building detection, they apply a simple heuristic model for the unary potentials that requires no training. The method for point-based classification described in Section 16.2.3 is based on a CRF (Niemeyer et al., 2014). As an alternative to the approaches mentioned so far, Ortner et al. (2008) apply statistical sampling to obtain the most likely configuration of building objects given a DSM. This allows for the inclusion of a strong geometrical model (all building footprints are modeled to be rectangular) in a statistical framework, which is particularly useful if the quality of the data, for example, in terms of geometrical resolution, is relatively poor.

Point cloud classification is only a first step for the extraction of objects such as buildings. The second step requires the definition of contiguous objects, for example, represented by boundary polygons. This can be achieved by modified convex hull algorithms (Sampath and Shan, 2007) or alpha shapes (Dorninger and Pfeifer, 2008), which have to be applied to segments of points classified as buildings. Alternatively, coarse building outlines can be determined on the basis of 3D line segments extracted in the subset of the point cloud classified as buildings (Lafarge and Mallet, 2012) or on a clustering of point cloud segments followed by a classification of boundary points (Poullis, 2013). In Section 16.2.4, a method for the definition of building objects on the basis of the output of a point-based CRF and a MRF is presented (Niemeyer et al., 2014).

16.2.3   Context-Based Classification of LiDAR Data

To apply the CRF framework to the context-based classification, one first has to define the structure of the underlying graph, that is, one has to decide what constitutes the nodes and the edges of the graph. Furthermore, one has to select models for the potentials in Equation 16.5, which includes the definition of the feature vectors used for classification. Finally, one has to decide about how to train the CRF and how to carry out inference, that is, how to estimate the most probable class configuration $\mathbf{C}$ by maximizing the posterior $P(\mathbf{C} \mid \mathbf{x})$ according to Equation 16.5.

16.2.3.1   Graph Structure

If a point-wise classification is envisaged, the nodes of the graph underlying the CRF correspond to the class labels of individual LiDAR points, that is, for each point $\mathbf{P}_i$ the graph will contain a node for that point's class label C_i. This definition avoids problems of segmentation algorithms that might not deliver segment boundaries corresponding to object boundaries at the cost of a considerably higher computational complexity. The edges of the graph are introduced on the basis of neighborhood relations. As with the definition of the neighborhood for computing features (cf. Section 16.2.1) there are several options. In the case of airborne LiDAR data, a neighborhood definition on the basis of 2D distances is preferred because it will maintain links between points at the roof or tree canopy level and points on the ground, which carry important information (Niemeyer et al., 2011). Thus, an edge is inserted into the graph between each point and its *knn* in 2D, with a fixed value of k. Note that an efficient construction of the graph requires a spatial index such as a *kd-tree* for an efficient nearest neighbor search.

16.2.3.2   Association Potential

In the association potential, the data are represented by node feature vectors $\mathbf{f}_i(\mathbf{x})$. For each node i, such a vector is determined taking into account the points in a certain neighborhood. The particular definition of the node features depends on the dataset to be classified (cf. Section 16.2.3.5). As pointed out in Section 16.2.2, the association potential $\varphi(C_i, \mathbf{x})$ linking the data to the class labels can be modeled on the basis of the probabilistic output of any local classifier (Kumar and Hebert, 2006). One possible choice that has been shown to be well suited for the (noncontextual) classification of LiDAR data (Chehata et al., 2009) is RF (Breiman 2001). An RF is a bootstrap ensemble classifier based on decision-trees. It consists of a number n_T of such trees, each determined from a subset of the training data only. These subsets are called bootstrap datasets, and they are drawn randomly with replacement from the entire pool of training samples. In this context, it is important that the bootstrap datasets used to train the individual trees are drawn independently from each other. Each internal node of any tree contains a test of a subset of features and corresponding thresholds, splitting the data into two parts. In the training phase, a series of potential splits are randomly chosen, and the split criterion optimizing a certain criterion, for example the Gini gain (Breiman, 2001), is selected in each node. The selection of the node tests is based on a subset of the bootstrap dataset only; the remaining training data are presented to the tree, and each leaf node is assigned to the class for which the maximum number of training samples is collected in this leaf after applying all the learned splits. The classification is performed by presenting the feature vector $\mathbf{f}_i(\mathbf{x})$ of an unknown sample to all the trees. Each tree casts a vote for the most likely class. If the number of votes cast for a class C^l is n_l, the association potential can be defined according to

$$\varphi\left(C_i = C^l, \mathbf{x}\right) = e^{\frac{n_l}{n_T}} \tag{16.6}$$

where n_T is the total number of trees. Note that the ratio n_l/n_T can be interpreted as the posterior of the RF classifier for the class label C_i taking the specific value C^l. The exponent is used to avoid potentials of value 0 for unlikely classes. Of course, this is not the only possible model for the association potentials. For a comparison of RF to generalized linearized models, refer to (Niemeyer et al., 2014).

16.2.3.3 Interaction Potential

The interaction potential $\psi\,(C_i,\,C_j,\,\mathbf{x})$ in Equation 16.5 models the dependencies of the class labels of two adjacent nodes i and j considering the observed data $\mathbf{x}$, the latter being represented by interaction feature vectors $\boldsymbol{\mu}_{ij}(\mathbf{x})$. Given the definition of the graph (cf. Section 16.2.3.1), the neighborhoods used for computing the feature vectors $\mathbf{f}_i(\mathbf{x})$ and $\mathbf{f}_j(\mathbf{x})$ at neighboring nodes i and j will overlap considerably, so that using the difference of both node feature vectors to define the interaction feature vectors $\boldsymbol{\mu}_{ij}(\mathbf{x})$ would lead to a vector with most elements close to zero. Alternatively, $\boldsymbol{\mu}_{ij}(\mathbf{x})$ can be determined by concatenating features of both node feature vectors $\mathbf{f}_i(\mathbf{x})$ and $\mathbf{f}_j(\mathbf{x})$ and computing some differences (cf. Section 16.2.3.5). The model for the interaction potential used in this section is also based on a RF classifier. The difference to the RF used for the association potential is that in this case, the RF is trained to deliver the most likely combination of class labels for each pair of neighboring points (i.e., points connected by an edge). If K classes are discerned, the RF has to be trained to separate K^2 such potential configurations. Note that this requires a sufficient amount of representative training data for any possible class transition between neighboring points. To determine the interaction potential for the edge connecting nodes i and j in the graph, the corresponding interaction feature vector $\boldsymbol{\mu}_{ij}(\mathbf{x})$ is presented to the RF classifier, and the number of votes n_{lm} for the occurrence of class labels C^l and C^m at these two nodes is counted. The interaction potential becomes:

$$\psi\left(C_i = C^l, C_j = C^m, \mathbf{x}\right) = e^{\frac{n_{lm}}{n_T}} \tag{16.7}$$

where n_T is the total numbers of trees of the RF. The association and interaction potentials are weighted equally, but a relative weighting factor could be introduced and determined by a procedure such as cross validation (Shotton et al., 2009). Again, this is not the only possible model for the interaction potential; for alternative definitions based on the contrast-sensitive Potts model or generalized linear models, refer to (Niemeyer et al., 2014) and (Weinmann et al., 2015).

16.2.3.4 Training and Inference

In the context of graphical models, inference is the task of determining the optimal label configuration based on maximizing $P(\mathbf{C}\,|\,\mathbf{x})$ according to Equation 16.5 for given parameters of the potentials. In the experiments reported in Section 16.2.3.5, we use the standard message passing algorithm LBP (Frey and MacKay, 1998) as implemented by Schmidt (2012).

In the training process, two independent RFs have to be trained for the association and the interaction potentials. In the experiments reported in Section 16.2.3.5, the RF implementation considers the Gini gain for training the individual trees. The number of features used for a test is set to the square root of the number of input features, following Gislason et al. (2006). To avoid biases for classes having many training samples, the training set is balanced by randomly selecting the same number of samples for each class (Chen et al., 2004). Note that the model for the interaction potentials described in Section 16.2.3.3 requires fully labeled subsets of point clouds, so that a sufficient number of representative training samples for each possible class transition is available.

16.2.3.5 Results and Discussion

The performance of our method is evaluated on the LiDAR dataset of Vaihingen, Germany (Cramer, 2010), in the context of the ISPRS Test Project on Urban Classification and 3D Building Reconstruction (Rottensteiner et al., 2014). It was acquired using a Leica ALS50 system with a mean flying height of 500 m above ground and a 45° field of view. The average strip overlap is 30%, and the point density in the test areas is approximately 8 points/m². Multiple echoes and intensities were recorded. However, only very few points (2.3%) are multiple returns, as the acquisition was in summertime under leaf-on conditions. For the benchmark, three test sites with different scenes are considered. Area 1, situated in the center of Vaihingen, is characterized by dense, complex buildings, and some trees. Area 2 consists of a few high-rising residential buildings surrounded by trees, whereas Area 3 is a purely residential neighborhood with small, detached houses. The point cloud, consisting of 780,879 points, was labeled manually for evaluation purposes. The following classes were discerned: *gable roof (GR), flat roof (FR), façade (F), road (R), grassland (GL), low vegetation (LV),* and *tree (T)*. The class *LV* also contains the cars in the scene. In addition, a contiguous part of the overall point cloud consisting of 263,368 points was labeled manually to be used for training.

The feature vectors $\mathbf{f}_i(\mathbf{x})$ derived for each point contained the LiDAR intensity of the respective point, the ratio of the echo number of that point and the number of echoes in the corresponding waveform, and the height above the terrain. In addition, several features computed from all points in a neighborhood of a fixed radius r near $\mathbf{P}_i$ were determined. First of all, we determined an approximated plane based on a spherical neighborhood and used the sum, mean, and standard deviation of residuals as well as the direction and variance of the normal vector as feature. We also used the variance of point elevations in both, a cylindrical and a spherical neighborhood, the point density in a spherical neighborhood, and the ratio of the point densities in a cylindrical and a spherical neighborhood, respectively. The features derived from the Eigenvalues of the covariance matrix of the points in a spherical neighborhood included the three Eigenvalues ($\lambda_1, \lambda_2, \lambda_3$), omnivariance, planarity, anisotropy, sphericity, eigenentropy, and scatter (λ_1/λ_3) (Chehata et al., 2009). This was complemented by the principal curvatures κ_1 and κ_2 and the mean and Gaussian curvatures in a spherical neighborhood as well as the variation of intensity, omnivariance, planarity, anisotropy, sphericity, point density, number of returns, and the curvature-based features in a spherical neighborhood. The DTM for the feature height above DTM is generated using robust filtering (Kraus and Pfeifer, 1998) as implemented in the commercial software package SCOP++.[*] The features considering the local point distribution are computed for multiple scales with radii $r = 1, 2, 3,$ and 5 m. The number of scales was chosen empirically; using more scales did not improve the classification results. In total, the feature vector $\mathbf{f}_i(\mathbf{x})$ for node i used for the 3D classification consists of 131 elements.

The interaction feature vectors $\boldsymbol{\mu}_{ij}(\mathbf{x})$ contain the height and intensity differences between the neighboring points $\mathbf{P}_i$ and $\mathbf{P}_j$, but we also include original features of both nodes, namely the features defined at the point itself and the features derived for a neighborhood at scale $r = 1$ m. Node features derived from larger neighborhoods are not considered because the overlap between the neighborhoods of the two points would be too large to carry much information for the classification of the class configuration. As a consequence, each interaction feature vector $\boldsymbol{\mu}_{ij}(\mathbf{x})$ consists of 72 elements.

The result of the CRF-based classification is the assignment of an object class label to each LiDAR point. The graph is built by linking each point to its three nearest neighbors in 2D. For both the association and the interaction potentials, we use RF consisting of 300 trees, a number that was found empirically. For the training of the association potentials 3000 training samples were used per class. Another set of 3000 samples per class relation was used to train the interaction potential, having to discriminate 49 different class interactions. The results of the 3D point cloud classification for Area 3 of the Vaihingen data are depicted in Figure 16.1.

For the quantitative evaluation, we compared the class labels predicted by the classification process with the reference labels to determine a confusion matrix of the results. From the confusion matrix, we derived the overall accuracy (OA), that is, the percentage of correct class labels, as well as the *Completeness*, the *Correctness* and the *Quality* of all classes (Heipke et al., 1997):

$$Completeness = \frac{TP}{TP + FN} \tag{16.8}$$

[*] http://www.trimble.com/imaging/inpho/.aspx?tab=Geo-Modeling_Module (accessed 22 June 2016).

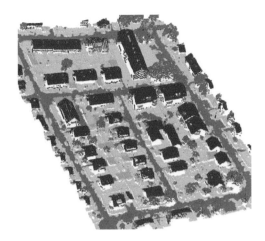

FIGURE 16.1 3D view of the classification results for study area 3 with seven classes: *grassland* (khaki), *road* (gray), *gable roof* (purple), *low vegetation* (light green), *façade* (dark purple), *flat roof* (orange), and *tree* (dark green). (Reprinted from Niemeyer, J. et al., *ISPRS J. Photogra. Remote Sens.*, 87, 152, 2014. With Permission.)

$$Completeness = \frac{TP}{TP + FP} \qquad (16.9)$$

$$Quality = \frac{Completeness \cdot Correctness}{Completeness + Correctness - Completeness \cdot Correctness} \qquad (16.10)$$

where (Equations 16.9 through 16.11) TP denotes the number of true positives, that is, the number of entities of a class that are correctly detected (i.e., TP is the element in the main diagonal of the confusion matrix corresponding to the class being analyzed), FN is the number of false negatives, that is, the number of entities of the class being analyzed in the reference dataset that were not detected automatically, and FP is the number of false positives, that is, the number of entities of the class being analyzed that were detected, but do not correspond to an entity of that class in the reference dataset. The results of the quantitative evaluation based on the reference are shown in Table 16.1; given this definition of the confusion matrix, TP + FN is the sum of the elements of the column of the confusion matrix corresponding to the class in question, whereas TP + FP is the sum of the elements of the corresponding row of the confusion matrix. The *quality* is a parameter indicating the tradeoff between FP and FN.

The results shown in Table 16.1 are quite good considering the rather large number of classes compared with other studies (Chehata et al., 2009), and the fact that the LiDAR data were captured under leaf-on conditions. The OA is 83.4% for the three test areas. The completeness, correctness, and quality numbers show that the classes occurring most frequently in the data are detected rather well, in particular *GR* and *road* (*quality* of 88.1% and 81.7%, respectively). One limiting factor was the fact that only a small amount of training samples was available for some classes or for some class transitions, in particular related to façades. There are also relatively few LiDAR points per façade plane, so that the characteristics of façades (vertical planes) are not captured in the feature vectors well. Consequently, *façade* is the class with the lowest completeness and correctness values. *LV* is another class that cannot be differentiated very well from the other classes, in particular from *grassland* and *tree*. The main difference between *LV* and *tree* is height, but of course the height distributions overlap; this is also true for *LV* and *GL*. Despite being a class of relatively high variability (including small bushes, low trees, etc.), *LV* is another class for which relatively few training samples are available. The error occurring most frequently according to Table 16.1 is a confusion between *grassland* and *road*. These two classes are only differentiated by the intensity, but in the test data, the difference is relatively weak, and the classes are also difficult to discern for humans. Another remaining problem are clusters of tree points near the ridges of GRs, which is partly caused

TABLE 16.1

Confusion Matrix and Derived Quality Metrics of the Results of CRF-Based LiDAR Classification for the Three Vaihingen Test Areas

Class	GR	FR	F	R	GL	LV	T	Corr.
GR	112,116	742	2,567	6	327	318	3,568	93.7% (−0.2%)
FR	3,909	44,913	5,043	107	2,724	1,372	958	76.1% (+1.8%)
F	1,071	147	16,932	116	662	1,497	7,971	59.6% (+10.3%)
R	24	50	332	187,725	13,182	274	182	93.0% (+3.5%)
GL	31	347	1,338	26,434	149,110	5,166	1,827	80.9% (+1.8%)
LV	549	1,267	2,863	1,198	10,329	39,107	21,453	50.9% (+0.7%)
T	1,993	1,231	3,430	4	370	2,607	101,390	91.3% (−0.5%)
Comp.	93.7% (+0.1%)	92.2% (−1.0%)	52.1% (−0.9%)	87.1% (+1.3%)	84.4% (+5.3%)	77.7% (+2.5%)	73.8% (+2.3%)	OA: 83.4% (+2.0%)
Qu.	88.1% (−0.1%)	71.5% (+1.0%)	38.5% (+4.2%)	81.7% (+3.8%)	70.4% (+5.0%)	44.4% (+1.3%)	69.0% (+1.7%)	

Source: Niemeyer, J. et al., *ISPRS J. Photogra. Remote Sens.*, 87, 152, 2014. With permission.

Notes: Classes: *gable roof* (*GR*), *flat roof* (*FR*), *façade* (*F*), *road* (*R*), *grassland* (*GL*), *low vegetation* (*LV*), and *tree* (*T*). Rows: classification results; Columns: reference classes. Numbers in parentheses show the improvement of the respective value caused by including the interaction potentials. Except for Completeness (*Comp.*), Correctness (*Corr.*), Quality (*Qu.*) and Overall Accuracy (OA), the table shows numbers of points.

by the smooth tree canopies. Nevertheless, these classification results provide a good basis for subsequent tasks such as the detection and vectorization of buildings and roads.

To assess the influence of integrating contextual relations into the classification process, the results of the CRF-based classification are compared with the output of a purely local RF classifier, that is, without considering the interaction potential. The differences in the quality indices are shown in parentheses in Table 16.1, where positive values indicate an improvement caused by considering context. The interactions increase the OA by 2.0%. This does not look very much at a first glance, which to a certain degree can be attributed to the facts that RF is a strong classifier per se and that context is implicitly considered in the RF-based classification by using multiscale features. However, at a second glance, the pairwise potentials do have a non-negligible effect, namely in improving the quality of the results for some of the classes. Although in some cases, an improvement of completeness is contrasted by a decrease of correctness and vice versa, the *quality*, indicating the best tradeoff between FP and FN, is improved for all classes except *GR*, for which it decreases by a marginal value. Most notably, the differentiation of *grassland* and *road* is improved considerably (*quality*: +5.0% and +3.8%, respectively), and the problematic classes *façade* and, to a lesser degree, *LV* also benefit from considering context (*quality*: +4.2% and +1.3%, respectively). To sum up, modeling of interactions is useful to obtain a more reliable classification compared with points being labeled individually.

16.2.4 BUILDING DETECTION

The classification of LiDAR points as described in Section 16.2.3 only delivers a class label for each point, but it does not deliver information about building *objects*. Section 16.2.4.1 describes how building objects can be defined based on the output of the classifier. The evaluation of this method is presented in Section 16.2.4.2.

16.2.4.1 Context-Based Definition of Building Objects

It is the goal of the methods described here to derive a 2D representation in the form of a binary building mask from the labeled point cloud. Individual buildings correspond to connected components of building pixels in that mask, and polygons describing the building outlines can be derived from such a representation easily. For that purpose, a 2D grid aligned with the XY plane of the object coordinate system is defined, and all LiDAR points are projected to this grid. However, due to the irregular distribution of the points, some pixels remain empty. To deliver accurate object masks or boundaries, these holes in the image data must be closed. On the other hand, small clusters of FP building points have to be eliminated as well. Morphological operators turned out not to be very accurate, in some cases eliminating correct building areas while at the same time resulting in FPs. A better solution can be achieved by another graphical model, this time based on a grid.

The goal is a classification of each grid point based on the output of the CRF. Each pixel of the grid corresponds to a node, whereas a 4-neighborhood on the grid is used to define the edges of the graph. This second classification step applies a MRF model according to Equation 16.4. The unary potential of the MRF is derived from the probabilistic output of the CRF described in Section 16.2.3. LBP, the method used to determine the optimal label configuration in the CRF, delivers a *belief* for each node, which, after normalization, can be interpreted as the marginal posterior probability of the class labels for that node. The unary potential of a pixel i is the average normalized belief of all points falling into this pixel. For pixels without LiDAR points, we assume all classes to be equally probable. This results in the following mathematical formulation:

$$\log \varphi\left(C_i = C^l, \mathbf{x}_i\right) = \begin{cases} \frac{1}{\|N_p\|} \cdot \sum_{p \in N_p} bel_{CRF}\left(C_p = C^l\right) & \text{if} \quad \|N_p\| > 0 \\ \frac{1}{K} & \text{if} \quad \|N_p\| = 0 \end{cases} \tag{16.11}$$

In Equation 16.11, bel_{CRF} $(C_p = C^l)$ is the *belief* of the original CRF that the class label of point p takes the value C^l given the data. N_p is the set of LiDAR points falling into pixel i, whereas $\| N_p \|$ is the number of points in N_p. K is the number of classes. The pairwise potential is represented by a Potts model favoring neighboring pixels i and j to have the same labels:

$$\log \psi \left(C_i = C^l, C_j = C^m \right) = \begin{cases} \lambda & \text{if} \quad C^l = C^m \\ 0 & \text{if} \quad C^l \neq C^m \end{cases} \qquad (16.12)$$

In Equation 16.11, the parameter λ is the weight of the interaction potential and, thus, influences the degree of smoothing. The smoothing effect of the MRF closes holes of pixels without corresponding LiDAR points. As the unary potentials of these *empty* pixels are initialized on the assumption of an equal distribution of the class labels, the Potts model helps to transfer the information of the neighboring pixels to this pixel without LiDAR points. As the parameter of the Potts model is chosen manually, no further training is required. Again, LBP is used for obtaining the optimal configuration of class labels in the MRF. From the final multilabel image, the binary building mask is derived by considering only the class *building*, defined as the union of *GR* and *FR* of the point-based classification. In a postprocessing step, only building pixels that are classified reliably are maintained to obtain only reliable objects. For that purpose, we compute the difference between the maximum and the second largest belief and only maintain building pixels with a difference in belief larger than a user-defined threshold.

16.2.4.2 Results and Discussion

The results of the 3D classification of the Vaihingen data (Section 16.2.3.5) serve as input for the generation of 2D building masks at a grid width of 0.5 m. As the class *façade* would disappear in a 2.5 analysis, it is not considered in classification and its beliefs are discarded. Furthermore, the class *LV* is aggregated with *tree*, and *GR* with *FR* by adding the corresponding beliefs. Thus, we distinguish between *road*, *grassland*, *vegetation*, and *building*. The weighting factor in Equation 16.12 is set to $\lambda = 1$. A binary building mask is derived from the classification results. In postprocessing, a pixel classified as *building* is discarded in the building mask if the relative difference between the belief of class *building* and the second largest belief is below 50%.

The resultant building masks for the tree test areas were evaluated in the context of the ISPRS benchmark for urban object detection and 3D building reconstruction based on a 2D reference for building outlines. The results of the evaluation are shown in Figure 16.2. Table 16.2 shows the *completeness, correctness,* and *quality* (Equations 16.8 through 16.10) of the results both on a per-area and on a per-object level (Rottensteiner et al., 2014).

The results of the evaluation presented in Figure 16.2 and Table 16.2 indicate that the approach presented here works well for the detection of buildings. Most of the FPs are caused by tree areas wrongly classified as buildings. This is due to the similar features: LiDAR points covering trees are mainly distributed on the canopy and not within the trees, which leads to a nearly horizontal and planar point distribution. The relatively large FN area of the building situated in the north of Area 3 is covered by only very few points due to the properties of the roof material. The area-based *completeness* and *correctness* values for buildings are between 90.8% and 96.7%, the *quality* takes values from 86.3% to 88.9%. The object-based metrics, counting a building as a TP if at least 50% of its area is contained in the reference (Rottensteiner et al., 2014), reveal that the buildings in Areas 1 and 3 were detected reliably with completeness and correctness values between 83.9% and 94.1%. The objects in Area 2 suffer from a relatively poor correctness value of 63.2%, but these numbers have to be interpreted with care because they are based on a very small number of reference buildings (14). If only building objects larger than 50 m^2 are considered, all objects in Area 2 and 3 were detected, and all of them are correct. Only in Area 1, there is one larger FP building (again, two trees labeled as *building*), which results in a quality

(a)   (b)   (c)

FIGURE 16.2 Results of the evaluation of the building detection results. Left to right: Results for (a) Area 1, (b) Area 2, (c) Area 3 respectively. Yellow: TP, red: FP, and blue: FN. (Reprinted from Niemeyer, J. et al., *ISPRS J. Photogra. Remote Sens.*, 87, 152, 2014. With Permission.)

TABLE 16.2

Evaluation of Building Detection Results for the Three Vaihingen Test Areas

Building	Object	Object ≥ 50 m²	Per area
Area	A1/A2/A3	A1/A2/A3	A1/A2/A3
Completeness	86.5%/85.7%/83.9%	100.0%/100.0%/100.0%	90.8%/91.4%/91.6%
Correctness	89.2%/63.2%/94.1%	96.6%/100.0%/100.0%	94.5%/96.4%/96.7%
Quality	78.3%/57.1%/79.7%	96.6%/100.0%/100.0%	86.3%/88.4%/88.9%

Source: Niemeyer, J. et al., *ISPRS J. Photogra. Remote Sens.*, 87, 152, 2014. With permission.
Notes: A1/A2/A3: Test areas 1, 2, and 3, as described in Rottensteiner et al. (2014).

of 96.6% rather than 100%. These results would indicate that the most relevant building per plot can be identified very reliably by the method described in this section, whereas smaller buildings can still largely be detected.

16.3 BUILDING RECONSTRUCTION FROM LiDAR DATA

The goal of automated building reconstruction is the generation of computer-aided design models of the buildings. For visualization purposes, the most common way of modeling buildings is by boundary representation (B-rep) (Mäntylä, 1988), that is, the representation of a building by its bounding faces, edges, vertices, and their mutual topological relations. For reconstruction, it is assumed that the location of buildings is known at least in an approximate way. This can be the result of automatic building detection, as described in Section 16.2. As an alternative, building locations can be provided by an existing 2D GIS database (Vosselman and Dijkman, 2001). This has the advantage of giving relatively precise positions of the building outlines, which are not very well defined in LiDAR data. On the other hand, the up-to-dateness of existing GIS data can become problematic.

Two general strategies can be applied for the geometrical reconstruction of buildings. First, parametric primitives can be instantiated and fit to the data, for example, in rectangular regions derived by splitting a 2D ground plan (Vosselman and Dijkman, 2001; Kada and McKinley, 2009), or based on stochastic sampling (Huang et al., 2013). This corresponds to a top–down or model-based approach. It has the advantage that the resulting building models are visually appealing because the primitives usually are characterized by regular shapes. Regularization of the resulting building models thus is an implicit part of the reconstruction process. On the other hand, this can result in an overregularization: buildings having nonrectangular footprints are usually not accurately reconstructed in this way.

The second strategy applies polyhedral models, which can be generated by extracting planar segments in a DSM and grouping these planes (Dorninger and Pfeifer, 2008; Perera and Maas, 2014); cf. also (Rottensteiner et al., 2014) for a comparison of some recent examples. This corresponds to a data-driven or bottom-up strategy, though some algorithms based on graph matching may introduce model knowledge about predefined object types in this context (Oude and Vosselman, 2009; Xiong et al., 2015). As building outlines are difficult to locate precisely in LiDAR data, ground plans are also often used (Vosselman and Dijkman, 2001). Lafarge and Mallet (2012) combine planes and other 3D surface primitives with a mesh-based representation. The local arrangement of planes near the roof plane boundaries is determined on the basis of a MRF.

Polyhedral models are more generally applicable than parametric primitives, and a higher level of detail can be achieved if they are used. However, polyhedral models are sometimes not as visually appealing as models generated by primitives, because geometric regularities are not an implicit part of the model, so that regularization has to be applied in a separate processing stage (e.g., Rottensteiner, 2006; Xiong et al., 2015).

In this section, the strategy based on generic polyhedral models will be followed. The locations of buildings are assumed to be known with an accuracy of 1–3 m. Further, we assume that the LiDAR data are sampled into a DSM in the form of a regular grid of width Δ, for example, by linear prediction (Kraus and Pfeifer, 1998). The work flow for the geometric reconstruction of buildings as presented in this section consists of three steps:

1. Detection of roof planes based on a segmentation of the DSM to find planes that are expanded by region growing.
2. Grouping of roof planes and roof plane delineation: Coplanar roof segments are merged, and hypotheses for intersection lines and step edges are generated based on an analysis of the neighborhood relations.
3. Consistent estimation of the building parameters using all available sensor data. This includes model regularization by introducing hypotheses about geometric constraints into the estimation process.

In the course of the reconstruction process, many decisions have to be taken, for example, with respect to the actual shape of the roof polygons or to the mutual configuration of roof planes. Traditionally, such decisions are based on comparing geometric entities, for example, distances, with thresholds. To avoid threshold selection as far as possible, decisions can be taken based on hypothesis tests, for example, about the incidence of two geometric entities. Thus, the selection of thresholds is replaced by the selection of a significance level for the hypothesis test. This requires rigorous modeling of the uncertainty of the geometrical entities involved. The concept of uncertain projective geometry (Heuel, 2004) is well suited for that purpose and is applied in all stages of the workflow, as will be presented in the subsequent sections.

16.3.1 Detection of Roof Planes in LiDAR DSMs

The problem of detecting planar segments in DSMs has been tackled in many different ways. For instance, local planes can be estimated for each LiDAR point (or each DSM pixel) from the points in a small neighborhood. Each plane corresponds to a point in a space whose axes correspond to the plane parameters. *Clustering techniques* detect planes by finding clusters of such points in the parameter space (e.g., Dorninger and Pfeifer, 2008). An alternative is to use Random Sample

Consensus (Fischler and Bolles, 1981) in an iterative way (e.g., Tasha-Kurdi et al., 2008). A third group of algorithms defines seed regions for planes based on some measure of local coplanarity. For each seed region, the parameters of a plane passing through all points of the seed region are estimated, and the seed regions are expanded by neighboring points found to be situated in this plane (Rottensteiner, 2003). In this section, a method for roof plane detection in DSM grids based on the third principle is described.

In Section 16.2.1, a measure R for surface roughness was introduced (Equation 16.2). This measure can be used to detect seed regions for planar segment in the DSM grid: pixels having a low value of R are surrounded by coplanar points. A straightforward method for roof plane detection could start by generating a binary image of homogeneous pixels (pixels with $R < R_{min}$) inside the coarse building outlines. Assuming connected components of homogeneous pixels to be seed regions, a plane can be fitted to all DSM pixels belonging to such a seed region, and seed regions that are actual planar (i.e., having a root mean square error [r.m.s.e.] of planar fit smaller than a threshold, e.g., ± 0.15 m) are grown. In each of the iterations of region growing, all pixels being adjacent to any seed region are tested according to whether they are inside the adjusting plane of that seed region, and if they are, they are added to that planar region. This test can be a hypothesis test for the incidence of a point and a plane (Heuel, 2004; Rottensteiner et al., 2005b). The process is terminated when no more pixels can be added to any plane.

This naive approach does not work very well, because it is impossible to find an appropriate value for the threshold R_{min}: If a very low value is chosen, some planes will not be detected, whereas if it is too high, planes that should be separated will be merged. As a consequence, it is advisable to use the method outlined earlier in an iterative way, using different values for R_{min} in this process. In the first iteration, only the $\alpha\%$ most homogeneous pixels are used to determine the seed regions. Using for instance $\alpha = 5\%$ or $\alpha = 10\%$ results in very small seed regions and an oversegmentation of the DSM, whereas smaller roof planes might not yet be detected because these thresholds are very tight. The region-growing procedure is repeated several times, each time using a higher value for α, and not considering pixels that have already been assigned to a roof plane. In each iteration, less and less significantly coplanar seed regions are used. The iteration process is completed when a threshold α_{max} for α is reached. The value for this threshold has to take into account that pixels at the transition between neighboring roof planes and pixels at step edges, especially at the building outlines, will usually not be classified as homogenous. A typical value is $\alpha_{max} = 70\%$.

As stated earlier, this iterative scheme will result in an oversegmentation. Thus, neighboring planar segments have to be checked according to whether they are coplanar or not, and coplanar segments have to be merged. This can be achieved by an F-test comparing the r.m.s.e. of planar fit achieved using two models (identical vs. separate planes) (Rottensteiner et al., 2005b). Figure 16.3 shows two examples for the results of planar segmentation after merging of coplanar segments.

16.3.2 Grouping of Roof Planes and Roof Plane Delineation

16.3.2.1 Initialization of the Roof Boundary Polygons

After the roof planes have been detected, their boundary polygons are determined. First, a Voronoi diagram of the planar segments is determined. The boundaries of the planes in the Voronoi diagram deliver the initialization of the boundary polygons of these planes. The heights of the vertices of the boundary polygons are derived from the roof plane parameters. Each boundary polygon is split into an ordered set of polygon segments so that each segment separates exactly two neighboring planes. Then, each polygon segment is classified according to whether it corresponds to a step edge, to an intersection line, or, for example, in the case of dormers, to both. This classification has to take into account the uncertainty of both the two neighboring planes and the approximate positions of the vertices of the initial polygon segments.

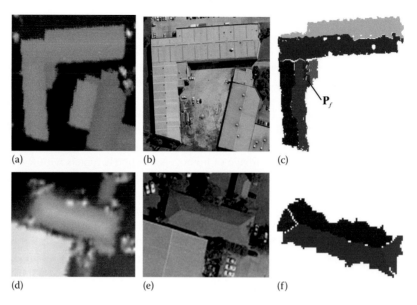

FIGURE 16.3   (a) DSM for building 1 (b) Orthophoto showing building 1 (c) Planar segments of building 1 (d) DSM for building 2 (e) Orthophoto showing building 2 (f) Planar segments of building 2. (From Rottensteiner, F. et al., *Int. Arch. Photogram. Rem. Sens. Spatial Inform. Sci.*, 36, 3/W19, 221, 2005. With Permission.)

All vertices of an initial boundary polygon segment are tested for incidence with the intersection line of the two neighboring planes. If all the vertices of the polygon segment are found to be incident with the intersection line, the segment is classified as an intersection. If the polygon segment is an outer boundary or if no vertex is found to be on the intersection line, the polygon segment is classified as a step edge. If some vertices are determined to be on the intersection and others are not, the polygon segment will be split up into several new segments, each having a different classification. Of these new segments, very short ones and intersection segments the average distance of which from the approximate polygon is larger than the segment length are discarded. For polygon segments classified as intersections, the intersection line might cut off considerable portions of the two roof planes if the planes are nearly horizontal (e.g., the building in upper part of Figure 16.3). Thus, another hypothesis test checks whether the cut-off parts of the roof planes are coplanar with the neighbor. Only in this case, the classification as an intersection will be accepted; otherwise, a step edge will be assumed. Figure 16.4 shows the results of the classification of the polygon segments for the two buildings in Figure 16.3.

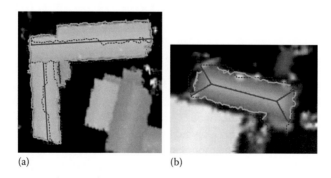

FIGURE 16.4   (a) building 1, (b) building 2. Classification of the roof boundary polygons and approximate step edges. Dashed blue lines: initial boundary polygons. Green solid lines: original step edges. Red solid lines: intersections. (From Rottensteiner, F. et al., *Int. Arch. Photogram. Rem. Sens. Spatial Inform. Sci.*, 36, 3/W19, 221, 2005. With Permission.)

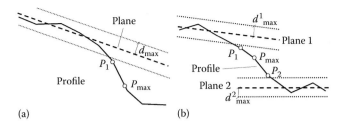

FIGURE 16.5 (a) Step edge detection at building outlines. $\mathbf{P_1}$: the first point on the profile outside the tolerance band of width d_{max}. $\mathbf{P_{max}}$: the step edge point determined as the point of maximum height gradient. (b) Step edge between two planes. In this case, $\mathbf{P_1}$ and $\mathbf{P_2}$ are the first points outside the tolerance bands of the two planes. (From Rottensteiner, F. et al., *Int. Arch. Photogram. Rem. Sens. Spatial Inform. Sci.*, 36, 3/W19, 221, 2005. With Permission.)

16.3.2.2 Refinement of Roof Polygons

If two planes intersect, the part of the roof boundary polygons separating these two planes is replaced by the intersection line. Step edges have to be positioned more precisely in the LiDAR data. The original polygon segments at step edges are sampled at the DSM grid width. For each vertex $\mathbf{P}$ of such a polygon segment, a profile of the DSM passing through $\mathbf{P}$ is analyzed to determine a step edge point. The profile is orthogonal to the approximate step edge, and it is ordered so that it starts from the interior of the roof plane (Figure 16.5).

At outer boundaries, the analysis starts with a search of the first point $\mathbf{P_1}$ in the profile found not to be incident with the roof plane (Rottensteiner et al., 2005b). If no such point is found, the profile is supposed not to contain a step edge. Otherwise, the point $\mathbf{P_{max}}$ of maximum height difference is searched for in the profile, starting from $\mathbf{P_1}$. As the terrain has to be lower than the roof, the search for $\mathbf{P_{max}}$ is stopped if the height difference between neighboring points in the profile becomes positive, which may happen if the roof boundary is occluded by a high tree. To eliminate points on *LV* next to a roof, $\mathbf{P_{max}}$ is discarded if it is above the roof plane. If no valid step edge point $\mathbf{P_{max}}$ is found, the step edge is assumed to be a straight line between the two closest step edge points visible to the sensor.

For step edges separating two roof planes, the first points $\mathbf{P_1}$ and $\mathbf{P_2}$ of the profile that are not incident with the two planes are determined. If the order of $\mathbf{P_1}$ is found to be *behind* $\mathbf{P_2}$ in the profile, no step edge point can be determined; otherwise, the position $\mathbf{P_{max}}$ of the step edge is determined as the position of the maximum height difference between $\mathbf{P_1}$ and $\mathbf{P_2}$.

The detected step edge polygons are quite noisy (green lines in Figure 16.4) and need to be generalized. First, step edge points having a distance larger than two times the original LiDAR point distance from both their predecessor and successor are eliminated as outliers. The remaining step edge polygon is approximated by longer polygon segments using a simple recursive splitting algorithm. Polygon segments containing less than three step edge points are discarded, and in an iterative procedure, neighboring polygon segments are merged if they are found to be incident based on a statistical test (Rottensteiner et al., 2005b). In a second merging step, we search for very short polygon segments. If both neighbors of such a segment are found to be incident, the short segment is eliminated and its neighbors are merged. Finally, the vertices of the generalized step edge polygon are determined by intersecting neighboring polygon segments; only if these segments are nearly parallel, their end points are connected by a new polygon edge.

16.3.2.3 Improving the Planar Segmentation

Replacing the approximate roof boundary polygons by the intersection lines and by the generalized step edges can affect the planar segmentation and also the neighborhood relations. Consequently, after generating the roof polygons, all pixels inside a roof polygon assigned to another roof plane and all pixels assigned to a roof plane that are outside the roof polygon are eliminated from their

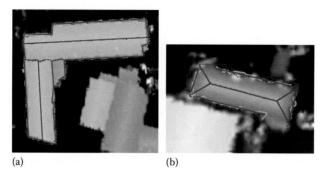

(a)   (b)

FIGURE 16.6 (a) building 1, (b) building 2. Initial roof polygons (dashed blue lines), approximate step edges (green solid lines), and generalized roof polygon segments (red solid lines) after improving the planar segmentation. (From Rottensteiner, F. et al., *Int. Arch. Photogram. Rem. Sens. Spatial Inform. Sci.*, 36, 3/W19, 221, 2005. With Permission.)

respective planes. This is followed by a new iteration of region growing, in which first the regions are only allowed to grow within their polygons. Spurious segments such as the plane $\mathbf{P}_f$ in Figure 16.3 might be eliminated in this process. Having improved the segmentation, the boundary classification and step edge detection are repeated. Figure 16.6 shows the final positions of all roof polygon segments for the buildings in Figure 16.3.

16.3.2.4   Combination of Roof Polygon Segments

Up to now, all segments of all roof boundary polygons were handled individually. Before combining these segments, their consistency is checked. If within a roof boundary there are two consecutive segments l_1 and l_2 classified as intersections, l_1 and l_2 must intersect. If replacing the segment endpoints $\mathbf{P}_1$ and $\mathbf{P}_2$ by the intersection point $\mathbf{I}$ changes the direction of one of the segments, a new step edge has to be inserted between $\mathbf{P}_1$ and $\mathbf{P}_2$ (Figure 16.7). A second consistency test checks whether for each roof polygon segment separating two planes, there is a matching opposite segment of the same type belonging to the neighboring plane. If this is not the case, it has to be inserted (Rottensteiner et al., 2005b).

Now all the polygon segments have to be combined. This involves an adjustment of the vertices at the transitions between consecutive polygon segments. First, all polygon segments intersecting at one planimetric position have to be found. Figure 16.8 shows an example involving three roof planes. There are altogether three polygon segments. One of them is the intersection line between two roof planes, and the other two are step edges. There are two intersection points $\mathbf{P}_1$ and $\mathbf{P}_2$ having the same planimetric position but different heights. For adjustment, we consider all the planes in the vicinity of the vertices $\mathbf{P}_1$ and $\mathbf{P}_2$, that is, the roof planes (ε_1, ε_2, and ε_3) and the walls corresponding to the step edge segments (ε_{13}, ε_{23}). For each plane, we observe the distance between the plane and the point $\mathbf{P_i} = (X, Y, Z_i)^T$ to be 0. The weight of such an observation are determined from the standard deviation σ_i of the distances between the initial position $\mathbf{P}_i^0 = (X^0, Y^0, Z_i^0)^T$ and the respective planes. The observation equations for a roof plane, u, giving support to height Z_i and for a wall, w, look as follows:

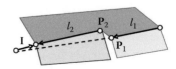

FIGURE 16.7 Consistency check: a new step edge is inserted between l_1 and l_2. (From Rottensteiner, F. et al., *Int. Arch. Photogram. Rem. Sens. Spatial Inform. Sci.*, 36, 3/W19, 221, 2005. With Permission.)

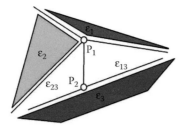

FIGURE 16.8 Vertex adjustment. (From Rottensteiner, F. et al., *Int. Arch. Photogram. Rem. Sens. Spatial Inform. Sci.*, 36, 3/W19, 221, 2005. With Permission.)

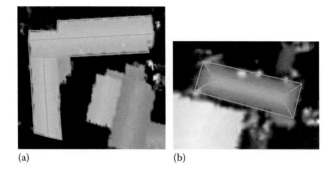

(a)   (b)

FIGURE 16.9 Combined roof polygons after local adjustment of vertices. (From Rottensteiner, F. et al., *Int. Arch. Photogram. Rem. Sens. Spatial Inform. Sci.*, 36, 3/W19, 221, 2005. With Permission.)

$$0 + v_{ui} = A_u(X^0 + \delta X) + B_u(Y^0 + \delta Y) + C_u(Z_i^0 + \delta Z_i) + W_u$$

$$0 + v_w = A_w(X^0 + \delta X) + B_w(Y^0 + \delta Y) + W_w$$

(16.13)

where:

v denotes the correction of the observation

(A_u, B_u, C_u, W_u) and (A_w, B_w, W_w) are the plane parameters of a roof or a wall, respectively

$\delta X, \delta Y, \delta Z_i$ are the corrections to the coordinates of the vertices that are to be determined in the adjustment.

Equation 16.13 is used in an iterative least squares adjustment. This model assumes that all walls intersect in one vertical line. Due to errors in step edge extraction, small step edge segments might have been missed, and the extracted step edge segment might pass by the intersection point at (X, Y). To find such segments, we compute the normalized corrections $v_w^n = (v_w/\sigma_w)$ of the wall observations after each iteration and exclude the wall with a maximum value of v_w^n if $v_w^n > 3.5$. For all excluded walls, a new step edge segment is introduced between the original end point of the step edge and the adjusted position of the vertex. Figure 16.9 shows the resulting roof boundaries for the buildings in Figure 16.3. From these combined roof polygons, a topologically consistent building model in B-rep is generated.

16.3.3 CONSISTENT ESTIMATION OF BUILDING PARAMETERS AND MODEL REGULARIZATION

Besides resulting in a more regular visual appearance, geometric regularities help to improve the geometric accuracy of the models if the sensor geometry is weak. The geometric constraints are not an implicit part of the building model but rather are added as additional information to the estimation of the building parameters and thus only have to be considered in building parts in which enough evidence is found in the data. In parameter estimation, geometric regularities can be considered in the adjustment by constraint equations. This strategy will result in models precisely fulfilling such *hard constraints* (Brenner, 2005). The alternative is to add *soft constraints*, that is, direct observations for entities describing a geometric regularity, to the adjustment of the sensor-based observations. In this case, the constraints will not be fulfilled exactly, but there will be residuals to the observations. Using the second strategy, robust estimation techniques can be applied to the soft constraints to determine whether a hypothesis about a regularity fits the sensor data or not (Rottensteiner, 2006).

The result of the previous building reconstruction steps is a polyhedral building model in B-rep. The faces of the model are labeled as being a roof face, a wall, or the floor. The model parameters are the coordinates of the model vertices and the plane parameters of the model faces. The topology of the model and some meaningful initial values for its parameters are known. The coarse model has to be analyzed for geometric regularities, and the model parameters have to be estimated. In the subsequent sections, the observations and the adjustment model used for parameter estimation are described.

16.3.3.1 Observations Representing Model Topology

Parameter estimation is based on a mapping between the B-rep of the polyhedral model and a system of GESTALT (shape) observations representing the model topology in adjustment. GESTALT observations are observations of a point $\mathbf{P}$ being situated on a polynomial surface (Kager, 2000) that is parameterized in an observation coordinate system (u, v, w) related to the object coordinate system by a shift $\mathbf{P_0}$ and three rotations $\Theta = (\omega, \phi, \kappa)^T$. The observation is $\mathbf{P}$'s distance from the surface, which has to be 0. Using $(u_R, v_R, w_R)^T = \mathbf{R}^T (\Theta) \cdot (\mathbf{P} - \mathbf{P_0})$, with $\mathbf{R}^T (\Theta)$ being a transposed rotational matrix parameterized by Θ, and modeling walls to be strictly vertical, there are three possible formulations of GESTALT observation equations:

$$r_u = \frac{m_u u_R + a_{00} + a_{01} m_v v_R}{\sqrt{1 + a_{01}^2}}$$

$$r_v = \frac{m_v v_R + b_{00} + b_{10} m_u u_R}{\sqrt{1 + b_{10}^2}} \qquad (16.14)$$

$$r_w = \frac{m_w w_R + c_{00} + c_{10}(m_u u_R) + c_{01}(m_v v_R)}{\sqrt{1 + c_{10}^2 + c_{01}^2}}$$

In Equation 16.14, r_i is the corrections of the fictitious observations of coordinate i, and $m_i \in \{-1, 1\}$ are mirror coefficients. With GESTALT observations, one is free to decide which of the parameters in Equation 16.14 are to be determined in the adjustment and how to parameterize a surface. Different GESTALTs can refer to identical transformation or surface parameters, which will be used to handle geometric regularities. Here, the rotations are 0 and constant. $\mathbf{P_0}$ is a point situated inside the building and constant. For each face of the B-rep of the building model, a set of GESTALT observations is defined, taking one of the first two Equations from 16.14 for walls and the third one for roofs. The unknowns to be determined are the coordinates of each vertex $\mathbf{P}$ and the plane parameters $(a_{jk}, b_{ik} c_{ij})$. As each vertex is neighbored by at least three faces, the vertex coordinates are determined by these GESTALT observations and need not be observed directly in the sensor data. These observations link the vertex coordinates to the plane parameters and thus represent the building topology in the adjustment.

16.3.3.2 Observations Representing Geometric Regularities

Geometric regularities are considered by additional GESTALT equations, taking advantage of specific definitions of the observation coordinate system and specific parameterizations of the planes.

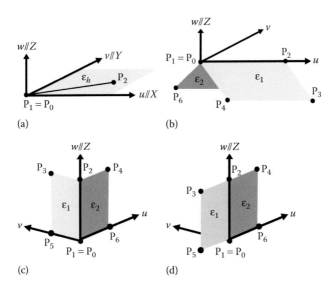

FIGURE 16.10 Geometric regularities: (a) horizontal edge, (b) symmetric roof planes intersecting in a horizontal edge, (c) perpendicular walls, and (d) identical walls. (From Rottensteiner, F. et al., *Int. Arch. Photogram. Rem. Sens. Spatial Inform. Sci*, 36, 3/W19, 221, 2005. With Permission.)

In our current implementation, geometric regularities can occur between two planes of the model that intersect in an edge or between two vertices connected by an edge. The observation coordinate system is centered in one vertex $\mathbf{P_1}$ of that edge and the w-axis is vertical, thus $\omega = \phi = 0 = $ constant. Four types of geometric regularities are considered (Figure 16.10). The first type, a horizontal roof edge, involves the edge's end points: Its two vertices $\mathbf{P_1}$ and $\mathbf{P_2}$ must have identical heights. The two points are declared to be in a horizontal plane ε_h that is identical to the u, v—plane of the observation coordinate system. One observation is inserted for $\mathbf{P_2}$: $r_w = w_R = Z_2 - Z_1$.

The other cases involve the two neighboring planes of an edge. One of the axes of the observation coordinate system is defined to be the intersection of these two planes ε_1 and ε_2. There is one additional unknown rotational angle κ describing the direction of the u-axis. For each vertex $\mathbf{P}_i$ of the planes, GESTALT observations are added for ε_1 or ε_2. For the edge's second vertex $\mathbf{P_2}$, two observations (one per plane) are added. The GESTALT observations for ε_1 and ε_2 are parameterized in a specific way.

In the first case, the edge is the horizontal intersection of two symmetric roof planes ε_1 and ε_2. There is only one tilt parameter c_{01}^1. Symmetry is enforced by selecting $m_v = -1$ for ε_2:

$$\varepsilon_1 : r_w = \frac{w_R + c_{01}^1 \cdot v_R}{\sqrt{1 + (c_{01}^1)^2}}; \quad \varepsilon_2 : r_w = \frac{w_R - c_{01}^1 \cdot v_R}{\sqrt{1 + (c_{01}^1)^2}} \tag{16.15}$$

In the second case, the edge is the intersection of two perpendicular walls —ε_1: $r_u = u_R$, ε_2: $r_v = v_R$. There is no additional surface parameter to be determined. In the third case, two walls are identical, and the edge does not really exist in the object: ε_1: $r_v = v_R$, ε_2: $r_v = v_R$. There is no additional surface parameter. $\mathbf{P_1}$ or $\mathbf{P_2}$ might become undetermined, so that direct observations for the coordinates of these vertices have to be generated.

The GESTALT observations corresponding to the geometrical constraints can be subject to robust estimation for gross error detection. If the sensor observations contradict the constraints, the respective GESTALT observations should receive large residuals, which can be used to modulate the weights in an iterative robust estimation procedure (Kager, 2000). Thus, wrong hypotheses about geometric regularities can be eliminated. Whether or not a hypothesis about a constraint is introduced is decided by analyzing the coarse model, comparing geometric parameters such as height differences or angles with thresholds. For instance, if two neighboring walls differ from 90° by less than a threshold ε_α, a constraint about perpendicular walls can be inserted.

16.3.3.3 Sensor Observations and Observations Linking the Sensor Data to the Model

The observations described so far link the plane parameters to the vertices or to the parameters of other planes. To determine the surface parameters, observations derived from the LiDAR data are necessary. LiDAR points give support to the determination of the roof plane parameters. As a LiDAR point is not a part of the model, its object coordinates have to be determined as additional unknowns. Each LiDAR point gives four observations, namely, its three coordinates and one GESTALT observation for the roof plane the point is assigned to. To determine the parameters of the wall planes, the vertices of the original step edge polygons (cf. Section 16.3.2.2) are used as observations in the same way as the original LiDAR points are used to determine the roof plane parameters.

16.3.3.4 Overall Adjustment

Using all the observations discussed in the previous sections, an overall adjustment process is carried out in which the weights of the observations are determined based on their standard deviations given a priori. Robust estimation is carried out by iteratively reweighting the observations depending on their normalized residuals in the previous adjustment (Kager, 2000). The reweighting scheme is only applied to the LiDAR and step edge point observations and to the observations modeling geometric constraints to eliminate gross observation errors and wrong hypotheses about geometric regularities. The surface parameters and the vertex coordinates determined in the adjustment are used to derive the final building model (Rottensteiner, 2006).

16.3.4 Results and Discussion

The method for building reconstruction described in this section was tested using a subset of the Fairfield dataset, which will be described in greater detail in Section 16.4.2.2. Reference data were generated in a semiautomatic working environment using digital aerial images and LiDAR data. The precision of the building vertices was ±17 cm in X and Y and ±5 cm in Z (Rottensteiner, 2006). From the LiDAR data, a DSM with a grid width of $\Delta = 0.5$ m was generated. Eight buildings of different size and complexity were reconstructed, and the reconstruction results were compared with reference data. These buildings were chosen to highlight the method's potential to handle buildings of both regular and irregular shapes.

The roof polygons before adjustment are shown in Figure 16.11. In general, the models look quite good except for building 8, which is partly occluded by trees. There are some errors in the outlines of buildings 1 and 2. Buildings 4, 6, 7, and 8 and the main part of building 3 should have a rectangular footprint that is not preserved in the initial models. The initial models, the original LiDAR points, and the step edge points provide the input for the overall adjustment. Hypotheses about geometric regularities were introduced just on the basis of a comparison of angles and height differences to thresholds. Robust estimation was applied to the soft constraints and to the LiDAR and step edge points.

Figure 16.12 shows the final results of building reconstruction and a comparison with the reference data. Compared with Figure 16.11, the building models appear to be more regular. For buildings 1–6, the number of extracted roof planes is correct. The intersection lines are very accurate, and step edges are in general also determined quite well. Some small roof structures are generalized, for example, the outline of the smallest roof plane of building 1 or of roof plane a of building 2. The step edge between planes a and b of building 2 was not determined very precisely because that

FIGURE 16.11 Initial roof boundary polygons for the eight buildings superimposed to the DSM. The buildings are shown in different scales, according to the extents shown in the figure. (From Rottensteiner, F., *Int. Arch. Photogram. Rem. Sens. Spatial Inform. Sci.*, 36, 3, 13, 2006. With Permission.)

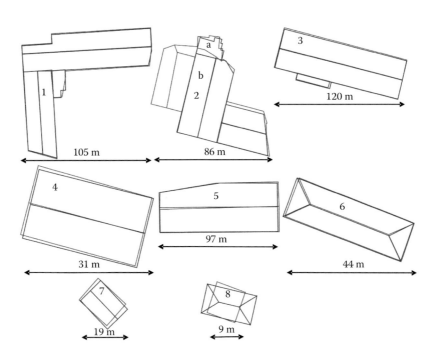

FIGURE 16.12 Final roof boundary polygons (red) and reference data (blue). A part of building 2 is missing in the reference data as it does not exist in the aerial images used to create the reference data. (From Rottensteiner, F., *Int. Arch. Photogram. Rem. Sens. Spatial Inform. Sci.*, 36, 3, 13, 2006. With Permission.)

TABLE 16.3

Evaluation of the Geometric Accuracy of the Roof Vertices after Adjustment with Regularization

B	P	RMS$_{XY}$ (m)	RMS$_Z$ (m)	Δ_{XY} (m)	Δ_Z (m)
1	5	0.76	0.12	0.24	0.01
2	5	2.27	0.20	0.00	−0.02
3	3	0.82	0.10	0.07	0.16
4	2	0.60	0.02	0.13	0.03
5	2	1.31	0.08	−0.08	−0.02
6	4	0.48	0.09	0.36	0.17
7	2	1.43	0.14	0.44	0.03
8	0	2.74	—	−0.02	—

Source: Rottensteiner, F., *Int. Arch. Photogram. Rem. Sens. Spatial Inform. Sci.*, 36, 3, 13, 2006.

Notes: B, building; P, number of planes; RMS$_{XY}$, RMS$_Z$ combined RMS values in planimetry/height; and Δ_{XY}, Δ_Z, improvement of RMS$_{XY}$/RMS$_Z$ caused by regularization.

step edge was poorly defined. Roof plane *a* was horizontal, its western vertex being higher and its eastern vertex lower than roof plane *b*, with a maximum height difference of only 0.3 m. Building 7 was reconstructed as being flat. The intersection of the two roof planes is only 0.15 m lower than the eaves, which is the reason why the two planes were merged. Building 8 was also reconstructed as a flat roof. It was the smallest building in the sample with only a few LiDAR points on the roof planes, and both ends occluded by trees. The outlines at the occluded ends are not very well detected either. Apart from the visual inspection of the building models, a numerical evaluation of these results was carried out. Root mean square (RMS) values of the coordinate differences of corresponding vertices in the reconstruction results and the reference data were computed for each roof plane. For buildings 7 and 8, only the outlines were evaluated.

Table 16.3 gives the RMS values for all the test buildings. The large value for RMS$_{XY}$ for building 2 of ±2.27 m is caused by the erroneous step edge; the combined value without that edge would be ±1.43 m. For most buildings, RMS$_{XY}$ is better than the average point distance across the flight direction. Apart from problems with low step edges, errors occurred at the outlines of some of the larger building due to occlusions: as the test area was at the edge of the swath, the positions of the step edges were very uncertain. The height accuracy is good, with the largest value of ±0.20 m occurring at building 2, again at the problematic step edge. Table 16.3 also gives the impact of the overall adjustment to the RMS values. With building 5, the RMS values get worse by a small value after adjustment, but in most cases, the RMS values are improved by the overall adjustment. The improvement can be up to 45% (building 6).

16.4   ROAD EXTRACTION FROM LiDAR DATA

Compared with other remote sensing data sources, airborne LiDAR data are less frequently applied for road extraction (e.g., Alharthy and Bethel, 2003; Hu et al., 2004; Zhu et al., 2004; Clode et al., 2007; Vosselman and Liang, 2009). A big advantage of LiDAR is that it can partly penetrate vegetation, so that LiDAR data are particularly useful for the road extraction in forests (Azizi et al., 2014). Sometimes, LiDAR data are only used to support road extraction from imaging sensors (Hu et al., 2004; Zhu et al., 2004). In this section, we will describe a method for road extraction from LiDAR data alone that follows the two-step procedure of detection followed by reconstruction.

16.4.1 ROAD DETECTION

For road detection, similar features and classification principles as for building detection can be used (cf. Sections 16.2.1 and 16.2.2). In fact, one could use the results of the point-based LiDAR classification as described in Section 16.2.3 directly as the basis for vectorization. Nevertheless, in this section, a simple rule-based classification algorithm is described. In addition to the LiDAR point cloud, it requires the generation of a DTM, which, in the experiments reported here, is based on hierarchical morphologic filtering described in (Rottensteiner et al., 2005a). The DTM having been generated, the original LiDAR points are classified according to whether or not they are situated on a road. A binary road image is generated from the classification results, and postclassification is carried out to correct some errors in the original classification.

16.4.1.1 Rule-Based Classification of Road Points

A rule-based classification technique is used to classify the LiDAR points into the classes *road* or *nonroad*. As specified in Section 16.2, we denote the set of Laser points by $\mathbf{x} = \{\mathbf{P}_1, \mathbf{P}_2, \ldots, \mathbf{P}_N\}$, in which each point $\mathbf{P}_k$ is described by its 3D coordinates x_k, y_k, and z_k, the intensity of the laser echo, i_k, and the echo index e_k and echo count c_k, thus $\mathbf{P}_k = (x_k, y_k, z_k, i_k, e_k, c_k)^T$. We can safely assume that a road point corresponds to the last echo of an emitted pulse; thus, for road points, we have $e_k = c_k$. Furthermore, road points have to be situated on the DTM, thus, for road points, the height above ground has to be small; this is true except for elevated roads, bridges, and tunnels. Finally, roads have very specific reflectance properties in the wavelength of the LiDAR system. In LiDAR intensity images, roads are clearly visible as dark thick connected lines, and thanks to the typically uniform and consistent nature of road material along a section of a road, the intensities of road points are not as noisy as the intensity of points situated on other objects (Clode et al., 2007). These three features (echo index, height above DTM, and intensity) are used to define a subset $\mathbf{S}_I \subseteq \mathbf{x}$ of the LiDAR points that are potential road points:

$$\mathbf{S}_1 = \{\mathbf{P_k} \in \mathbf{x} : \left(\left| z_k - \mathrm{DTM}(x_k, y_k) \right| < \Delta h_{\max} \right) \wedge (i_{\min} < i_k < i_{\max}) \wedge (e_k = c_k)\} \qquad (16.16)$$

where:

DTM (x_k, y_k) is the height value of the DTM at location (x_k, y_k)
$\Delta h_{\max}$ is the maximum allowable difference between z_k and the DTM
$i_{\min}$ and $i_{\max}$ are the minimum and maximum acceptable LiDAR intensities at any road point.

The thresholds for the intensities have to be adapted to the road material to be detected, which can be achieved by training. The result of Equation 16.16 is a set of LiDAR points ($\mathbf{S}_1$) that were reflected from a road along with some other false positive (nonroad) detections. If more than one type of road material is to be detected in the surveyed region, different subsets $\mathbf{S'}_1$ can be created according to the individual reflectance properties of the different materials. A combined and complete set $\mathbf{S}_1$ can be created by considering the union of the different subsets $\mathbf{S'}_1$.

Roads are depicted as a continuous network of points that form thick lines. Due to this continuous nature of a road network, LiDAR points that have struck the middle of the road are expected to be surrounded by other points that have struck the road. The percentage of road points within a local neighborhood (e.g., defined by a circle of radius d around any point $\mathbf{P}_k$) will be called the *local point density*. For points in $\mathbf{S}_1$ that are situated in the middle of a road, the local point density should be close to 100%. For any point that lies on the edge of a road, it can be considered to be close to 50% and 25% for a LiDAR point in the corner of a sharp 90° bend. By testing all points against a chosen minimum local point density $\rho_{\min}$, a new subset of points, $\mathbf{S}_2$, is described as per Equation 16.17:

$$\mathbf{S}_2 = \left\{ \mathbf{P}_k \in \mathbf{S}_1 : \frac{\left| \left\{ \mathbf{P}_j \in \mathbf{S}_1 : \left\| \mathbf{P}_k - \mathbf{P}_j \right\|_2 < d \right\} \right|}{\left| \left\{ \mathbf{P}_j \in \mathbf{x} : \left\| \mathbf{P}_k - \mathbf{P}_j \right\|_2 < d \right\} \right|} > \rho_{\min} \right. \qquad (16.17)$$

where:

d is the maximum distance from $\mathbf{P}_k$ or the radius of the local neighborhood

$|\{...\}|$ denotes the number of points $\mathbf{P}_j$ in the respective set

$\|\mathbf{P}_k - \mathbf{P}_j\|_2$ is the Euclidean distance from $\mathbf{P}_j$ to $\mathbf{P}_k$

An upper bound should be placed on the possible values of d so that it is any value less than or equal to half of the expected maximum road width (Clode et al., 2007). Applying Equation 16.17 will remove isolated nonroad points from the set of road candidate points.

16.4.1.2 Generation of a Binary Road Image and Postclassification

A binary image, $F(x, y)$, is created from the final subset $\mathbf{S}_2$ with a pixel size Δ loosely corresponding to the original average LiDAR point density. The pixel values $f(x, y)$ of the binary image are determined according to whether a point $\mathbf{P}_k \in \mathbf{S}_2$ exists inside the area represented by the pixel at position (x, y) or not

$$f(x, y) = \begin{cases} 1 \text{ if } \exists\, \mathbf{P}_k \in \mathbf{S}_2 : \left(x - \dfrac{\Delta}{2} < x_k \le x + \dfrac{\Delta}{2} \right) \wedge \left(y - \dfrac{\Delta}{2} < y_k \le y + \dfrac{\Delta}{2} \right) \\ 0 \text{ otherwise} \end{cases} \quad (16.18)$$

The pixels in that binary image characterized by $f(x, y) = 1$ represent roads. Many small gaps exist between these road pixels. This is caused by reflections from other objects such as vehicles and overhanging trees and by the fact that in some image regions, the pixel size will be smaller than the LiDAR point density. These gaps are removed using a two-step approach based on morphologic filtering. First, a morphological closing with a small structural element is initially performed to connect neighboring road pixels. The not-road image (i.e., $1 - f(x, y)$) is then labeled using a connected component analysis, and the values of all pixels corresponding to not-road segments with a small area are switched to 1 in the binary image $f(x, y)$, which results in a road image that contains all public roads, private roads, car parks, and some noise. In a second stage of processing, another label image is created from this binary image to identify individual continuously connected road segments. This time, small road segments are erased in the binary image of road pixels, thus removing most of the noise present. This ensures that our detected roads are continuous in nature.

Car parks are not considered to be roads. However, as car parks and roads have similar surface and reflectance properties, it is difficult to detect and eliminate car parks. By defining a maximum acceptable road width prior to processing, very wide unconnected car parks can be removed from the binary image. As roads form a network of long, thin connected objects, the area ratio of each individual road segment and the corresponding minimum bounding rectangle will decrease as the length of the smallest side in the minimum bounding rectangle increases. Large isolated blobs can be detected in the image using this ratio, thus allowing the removal of any unconnected car parks from the final binary image of road pixels.

16.4.1.3 Results and Discussion

The road detection method described in this section was evaluated using a test dataset captured over Fairfield (Australia) using an Optech ALTM 3025 laser scanner. The dataset covers an area of 2×2 km^2. Both the first and the last echoes of the laser beam were recorded with an average point separation of about 1.2 m. An RGB orthophoto with a resolution of 0.15 m was also available. DSMs with a grid width $\Delta = 1$ m were generated for both the first and the last pulse data, and these grids in addition to a *pseudo-NDVI image* generated from the LiDAR intensities and the red band of the orthophoto were used for obtaining a coarse DSM using the method described in (Rottensteiner et al., 2005a). The orthophoto was used for generating a reference by manual digitization. The guideline used during digitizing was that public roads were to be classified as roads, but car parks and private

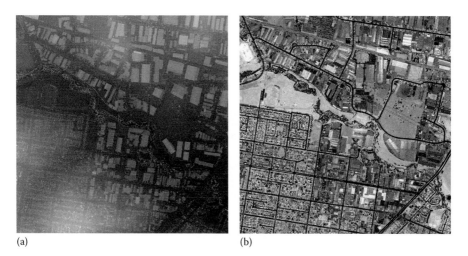

(a)   (b)

FIGURE 16.13   LiDAR DSM (a) and intensity image (b) of the Fairfield dataset. (DSM Reprinted from Rottensteiner, F. et al., *Inform. Fusion*, 6, 4, 283, 2005. With Permission.)

roads (driveways and roads leading to car parks) were not. Figure 16.13 shows the LiDAR DSM and the intensity image of the Fairfield dataset.

Figure 16.14 shows the results of the classification. The classification results from the workflow described in the previous section are displayed as a binary image of road pixels in Figure 16.14a (as a negative). The spatial distribution of the TP, FP, and FN pixels (cf. Section 16.2.3.5, especially Equations 16.8 and 16.9) are displayed in Figure 16.14b in yellow, red, and blue, respectively. A perusal of the spatial distribution reveals that the majority of the FP detections correspond to car parks, whereas the majority of FN detections have occurred at the ends of detected roads or on the edge of the image. These road components are disconnected from the road network and exist in small sections due to overhanging trees and were removed when small, disconnected components were removed.

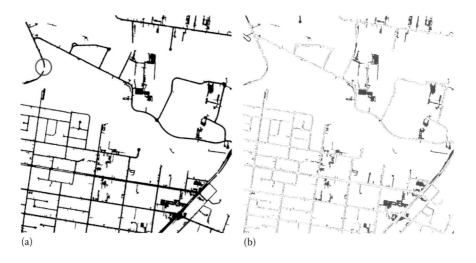

(a)   (b)

FIGURE 16.14   (a) Results of road network classification. The circle highlights a problem that occurs with bridges. (b) Spatial distribution of the classified pixels. TP pixels are shown in yellow, FP appear in blue, and FN appears in red. (From Clode, S. et al., *Photogram. Eng. Rem. Sens.*, 73, 5, 517, 2007. With Permission.)

A per-pixel evaluation of these results resulted in a completeness of 88% and a correctness of 67%, which corresponds to a quality of 61% (cf. Equations 16.8 through 16.10). Several problems were encountered during the classification phase of the algorithm. One of them was the detection of elevated roads and bridges (cf. the circle in Figure 16.14a). Our road network model assumed that roads lie on the DTM, an assumption that is not true for bridges. Another problem is car parks. Due to the industrial nature of sections of the test data, there are quite a few large car parks connected to the road network, and the classification method did not succeed in eliminating most of them; the majority of these car parks still remain. This difference is reflected in the correctness value, and better results are expected in nonindustrial areas.

16.4.2 ROAD VECTORIZATION

In LiDAR data of a resolution better than or equal to 1 m, roads appear as two-dimensional areas rather than one-dimensional lines; they have width as well as length. The centerline can no longer be observed directly, but has to be derived by other methods, for instance from the road boundaries. Note that, although the classification method described in Section 16.4.1 uniquely exploits the properties of LiDAR data, the method of road vectorization described in this section is based on a binary road image and could be applied to such a road image derived by any classification method. Unlike other methods (e.g., Zhu et al., 2004), it is not restricted to roads that form a grid-like pattern. It does overcome the problem of common line detection methods such as the Hough transform that detects the diagonal of a straight road segment rather than the centerline. Other road parameters, such as the width and direction, can also be extracted.

The method is based on the convolution of the binary road image with a phase coded disk (PCD). The PCD is a complex kernel that uses phase to code for the angle of the line. By convolving the original image with a PCD, the centerline, direction, and width can be accurately extracted at any point along the detected centerline. The PCD is defined by Equation 16.19 (Clode et al., 2007):

$$O_{\text{PCD}} = e^{j2\tan^{-1}(b/a)} = e^{j2\vartheta} \tag{16.19}$$

The variables a and b are x and y coordinates relative to the center of the PCD. Further, $a^2 + b^2 \leq r^2$, $\vartheta = \tan^{-1}(b/a)$, $j^2 = -1$, and r is the radius of the disk. The constant 2 in the exponent has been introduced into the definition of the PCD to ensure that pixels that are diametrically opposite in their direction from the center of the kernel during the convolution process indicate the same direction after convolution and do not cancel out. The convolution of the PCD with the binary image takes the form:

$$Q(x, y) = F(x, y) \otimes O_{\text{PCD}} \tag{16.20}$$

where:
  $Q(x, y)$ is the resultant image
  $\otimes$ is the convolution operation
  O_{PCD} is the PCD
  $F(x, y)$ is the binary road image

The result of the convolution defined in Equation 16.20 yields a magnitude image M and a phase image Φ that are defined by Equations 16.19 and 16.20, respectively:

$$M = \left| F(x, y) \otimes O_{\text{PCD}} \right| \tag{16.21}$$

$$\Phi = \frac{1}{2} \arg(F(x, y) \otimes O_{\text{PCD}}) \tag{16.22}$$

The magnitude M and phase Φ images can be used to determine the desired parameters of a road. The result $q(x, y)$ of the convolution at any position (x, y) is given by the complex integral over the entire disk:

$$q(x, y) = \int_\pi^\pi \int_0^r f(x, y) e^{j2\vartheta}\, u\, \mathrm{d}u\, \mathrm{d}\vartheta \qquad (16.23)$$

In Equation 16.23, the variable u is a substitute variable that has been introduced to represent the radius of the PCD as r is in the limits of the integral and the function $f(x, y)$ is understood to be translated to appropriate polar coordinates. Remembering that the road image is binary, the result of Equation 16.23 is identical to the integral over only the road area covered by the disk (i.e., all areas where $f(x, y) = 1$). Thus, for the model of a straight line segment of width w and orientation ϕ passing through the area covered by the PCD, the integral in Equation 16.23 can be solved analytically by changing the limits of the innermost integral so that the area defined is only the road contained within the disk. The integral consists of two parts: some areas can be integrated over the full radius (r) of the PCD, whereas others can only be integrated to a distance of $w/2 \cos(\vartheta)$ (Figure 16.15). The variable R is used to represent the limits of the integral at all different angles of ϑ over the road section contained within the PCD (Clode et al., 2007):

$$R = \min\left(\frac{w}{2\cos(\vartheta)}, r\right) \qquad (16.24)$$

Further, we substitute $\vartheta = \phi + \alpha$, thus relating the directional angle to the road direction ϕ. The integral over the road can be described as the integral over the disk where $f(x, y) = 1$:

$$q(x, y) = \int_\pi^\pi \int_0^R e^{j2(\phi+\alpha)} u\, \mathrm{d}u\, \mathrm{d}\alpha \qquad (16.25)$$

Roads correspond to ridges in the magnitude image M of the convolution, with the relative maximum of M corresponding to the road centerline. The radius r of the PCD must be larger than the maximum road width to be detected. It can be shown (Clode et al., 2007) that the phase of $q(x, y)$ in

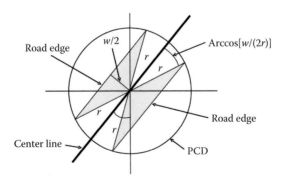

FIGURE 16.15 Extracting line information from the PCD. The thick line represents the road center line, passing through the center of the PCD. The road orientation is ϕ. The integral over the road pixels in Equation 16.23 can be separated into two parts: the two wedges of opening angle $180° - 2 \arccos(w/2r)$, in which the integral is to be taken over the full radius r of the PCD, and the yellow triangles, in which the integral is limited by the road edges. (From Clode, S. et al., *Photogram. Eng. Rem. Sens.*, 73, 5, 517, 2007. With Permission.)

Equation 16.25 only depends on the road orientation ϕ, and thus, ϕ is related to the phase Φ of the complex convolution (Equation 16.22) by

$$\phi = \frac{\Phi}{2} \tag{16.26}$$

Furthermore, evaluating the integral in Equation 16.25 results in an equation relating the magnitude M to the road width w and the PCD radius r (Clode et al., 2007):

$$M = \left| w^2 \cos^{-1}\left(\frac{w}{2r}\right) - 2w\sqrt{r^2 - \frac{w^2}{4}} \right| \tag{16.27}$$

A graph can be generated for a PCD of a fixed radius, thus enabling the width of a road to be determined at any point. Figure 16.16 illustrates the relationship between width and magnitude. If the road width w is to be determined from the magnitude M, ambiguity resolution is required. For instance, a magnitude of 600 could imply a width of 17 or 32. A constraint placed on r will resolve this ambiguity as well as avoid saturation problems that will occur when the road width is greater than the kernel radius.

16.4.2.1 Road Tracking and the Generation of a Road Network

The final road network consists of a set of road segments, each of them represented by three polylines, namely the centerline and two road edges. Vectorization consists of three steps. First, the road centerline of each road segment is extracted. Second, the extracted road centerlines are joined or intersected with other centerlines of the neighboring segments to create a continuous network of road centerlines. An exact intersection is created by splitting the crossing road centerlines at the intersection point. Where a T-junction is encountered, the road segment at the base of the T is extended straight ahead until the centerlines of both roads intersect. A list of all crossing points and T-junctions is kept in conjunction with the connecting road segments to maintain the topology of the road network. Third, the polylines representing the road edges are constructed as being parallel to the centerline at a distance of half the road width. The polygon formed by the two edge polylines defines the road segment. The set of all road segments and their topology describes the continuous road network.

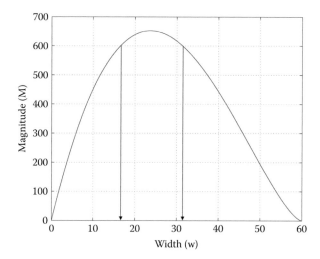

FIGURE 16.16 Graphical representation of the relationship between the magnitude of the convolution and the width of the road for a PCD of radius 30. (From Clode, S. et al., *Photogram. Eng. Rem. Sens.*, 73, 5, 517, 2007. With Permission.)

16.4.2.1.1 *Vectorizing the Road Centerlines*

The ridge of the magnitude image is traced to extract the road centerlines. Tracing is achieved by initially masking the magnitude image with the binary road mask to limit the search space and remove noisy edge areas. The maximum of the magnitude image is found, and the corresponding line direction is read from the phase image. The tracing algorithm moves along the line pixel by pixel ensuring that the point is still a maximum against its neighbors until the line ends at a pixel of zero magnitude. A polygon is created that consists of a series of centerline segments. The magnitudes of pixels to the side of each centerline segment within the calculated road width are set to zero as the centerline is traced, ensuring a similar path is not retraced. Once the road segment is completed, the process is repeated from the original maximum but with the diametrically opposed line direction. Then, the maximum of the remaining untraced magnitude pixels is found and tracing is recommenced from this pixel. The process is repeated until all relevant pixels have been traced. Thus, one centerline segment after the other is extracted until the masked magnitude image is completely blank.

16.4.2.1.2 *Connecting Road Segments and Determining Junctions*

Due to noise in the magnitude image, the tracing of some road centerlines is terminated prematurely by the tracing algorithm. In such cases, the tracing algorithm will extract another centerline that will terminate close to the original prematurely terminated centerline end. Connection of road centerline ends is performed by concatenating road segment chains that have ends that are both close and pointing to each other. The segments are close if they are within one road width of the centerline end. A check is made that the midpoint of the link actually lies on a ridge in the original magnitude image, ensuring that dead end streets do not erroneously get connected. Once two road segments have been connected, the process is repeated until no more concatenation can occur. During concatenation, all road crossings are found by identifying crossing road segments. These segments are split at the intersection, and the road crossing is created, leaving only the determination of *T*-junctions. This requires that the end of each centerline be checked against all other centerlines. In a manner similar to the way individual centerlines were concatenated to form longer centerlines, the closest point on any other centerline is found for each centerline end point. If the direction to the closest other centerline position matches the current centerline direction, the road segment is extended if the midpoint corresponds to a road in the binary image.

16.4.2.1.3 *Vectorizing the Road Edges*

The edges of the road are also represented as polygons that are created by calculating the width of the road and the orientation of the road ϕ. The road width is then smoothed by applying a low pass filter to the widths of each road segment. At each point along the centerline of each road segment, two new road edge points are created based on the road orientation ϕ plus or minus $90°$ and half the smoothed road width. Two edge polylines are defined for each road segment, thus defining the road segment polygon. At the end of each centerline, the intersections of the accompanying road edges are calculated. The edges within the intersection are kept to ensure that the connection between the centerline and the edges within a road segment is kept intact. To complete the visualization at the intersections, blanked road edge ends on the same side of the centerline are joined.

16.4.2.2   Results and Discussion

Again, the Fairfield dataset (cf. Section 16.4.1.3) is used for evaluation. Figure 16.17 shows the result of the convolution of the binary road image for Fairfield (Figure 16.14) with the PCD. Figure 16.17a displays the magnitude image M with the highest values displayed as white and the lowest as black. The centerlines of the road image are represented by white ridges running through the magnitude image. Figure 16.17b displays the resultant phase image Φ, which is related to the road direction as described previously.

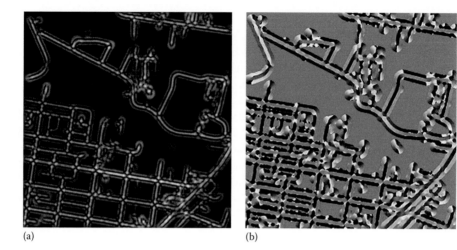

(a)   (b)

FIGURE 16.17   Results of the convolution with the PCD. (a) Magnitude M with the highest values displayed as white and the lowest as black. (b) Phase image Φ. (From Clode, S. et al., *Photogram. Eng. Rem. Sens.*, 73, 5, 517, 2007. With Permission.)

Vectorization of the convolution results was performed by the tracing algorithm as previously described. The results of the vectorization process can be seen in Figure 16.18. The centerline vectors appear visually to be a good approximation of the road network. In areas where there are very few car parks, the vectors are smooth, predictable, and continuous. The results show that roads that have successfully been classified as dual carriageways have also been vectorized as dual carriageways in these areas. The curvilinear road that runs diagonally from the northwest corner to the eastern edge of the image is quite smooth and characterizes the classified image well. The grid pattern in the southwestern corner is very good. The behavior of the tracing algorithm is predominately as expected, although some anomalies could be seen. In the southwest corner of Figure 16.18, there are two intersections that are represented as a set of two 90° bends. In these instances, the ridge in the magnitude image was more dominant towards the entering road rather than the continuing road. Another problem also seen in the southwest corner of the image is the tracing of roundabouts. There are several regions that appear to be quite noisy, in particular, the area in the southeast of the image. This apparent noise is due to the presence of many car parks in the area. As far as the bridge highlighted Figure 16.14 is concerned, Figure 16.17 shows that it still yielded a continuous ridge, albeit reduced in magnitude, so that the tracing algorithm correctly vectorized this area, and information that was lost in the initial classification has been recovered. The edges of the roads are also displayed in Figure 16.18. In areas of relatively low noise, the road edges have been calculated consistently. Smooth edges are obtained along areas of clearly defined roads. In areas where roads are less clearly defined, for example, with car parks or road junctions, the resultant magnitude is less representative of the road due to a saturation effect within the PCD.

To quantify the vectorization results, the evaluation methods described in (Wiedemann, 2003) were adopted. First, the reference centerlines were estimated from the reference road image. After that, both extracted and reference data are compared in vector format, in which the first step is to match the two datasets, whereas the second is to calculate the quality measures. We use the buffer width for the simple matching technique described in (Wiedemann, 2003). Completeness describes how well reference data are explained by the extracted data and is determined by the ratio of the length of the matched reference data to the total length of the reference data. For the results shown in Figure 16.18, completeness was 84%. Correctness represents the percentage of correctly extracted road data and is determined by the ratio of the length of the matched extraction to the total length of the matched extraction. Correctness was 75% in our test. The centerline RMS of all points expresses

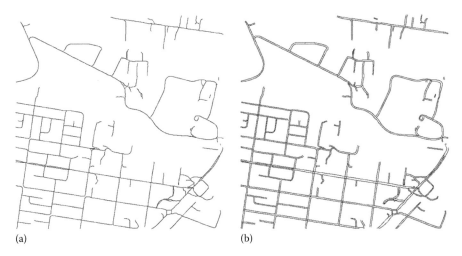

(a)   (b)

FIGURE 16.18 (a) Vectorized road centerlines and (b) vectorized road edges as calculated from the road centerline and the detected road width. (From Clode, S. et al., *Photogram. Eng. Rem. Sens.*, 73, 5, 517, 2007. With permission.)

the quadratic mean distance between the matched extracted and the matched reference points. It was determined to be ±1.70 m. The centerline RMS of all road segments considers the RMS values for each individual road segment and expresses the quadratic mean distance of these results, and it was ±1.56 m. The width RMS of all points and road segments are calculated in a similar manner to the centerline quality values but consider the difference in extracted and reference widths as opposed to the centerlines. The value for all points was ±1.66 and ±1.48 m for the segments. Detected intersections were manually classified against the reference data into TP, FP, and FN detections. Topological completeness and correctness are then calculated using Equations 16.8 and 16.9. The algorithm achieved a topological completeness of 87% and a topological correctness of 73%. The final evaluation of the vectorized road network is promising, and the results resemble the classification results. This suggests that the vectorization process was performed well, and that improvements need to be made to the classification algorithm to achieve better final quality values. The topological evaluation of the network extraction suggests that improvement in the current network model can be made. The removal of car parks from the classified road image would greatly improve the classification results. The application of several PCDs with different radii within a hierarchical framework during the vectorization stage may improve the vectorization of the road edges and allow a more robust determination of the road width.

REFERENCES

Alharthy, A., Bethel, J. (2003). Automated road extraction from LIDAR data. *Proceedings of the ASPRS Annual Conference*, Anchorage, AK, unpaginated CD-ROM.

Anguelov, D., Taskar, B., Chatalbashev, V., Koller, D., Gupta, D., Heitz, G., Ng, A. (2005). Discriminative learning of MRFs for segmentation of 3D scan data. *Proceedings of the 2005 IEEE Conference on Computer Vision and Pattern Recognition (CVPR)*, IEEE Computer Society, Los Alamitos, CA, pp. 169–176.

Axelsson, P. (2000). DEM generation from laser scanner data using adaptive TIN models. *International Archives of the Photogrammetry, Remote Sensing and Spatial Information Sciences*, 33(B4), 110–117.

Azizi, Z., Najafi, A., Sadeghian, S. (2014). Forest road detection using LiDAR data. *Journal of Forestry Research*, 25(4), 975–980.

Bishop, C. M. (2006). *Pattern Recognition and Machine Learning*, 1st ed. Springer: New York.

Boykov, Y. Y., Jolly, M.-P. (2001). Interactive graph cuts for optimal boundary and region segmentation of objects in n-d images. *Proceedings of the IEEE International Conference on Computer Vision (ICCV)*, 1, 105–112.

Boykov, Y. Y., Veksler, O., Zabih, R. (2001). Fast approximate energy minimization via graph cuts. *IEEE Transactions on Pattern Analysis and Machine Intelligence*, 23(11), 1222–1239.

Breiman, L. (2001). Random forests. *Machine Learning*, 45(1), 5–32.

Brenner, C. (2005). Constraints for modelling complex objects. *International Archives of the Photogrammetry, Remote Sensing and Spatial Information Sciences*, 36(3/W24), 49.

Brunn, A., Weidner, U. (1997). Extracting buildings from digital surface models. *International Archives of Photogrammetry and Remote Sensing*, 32(3–4W2), 27.

Chehata, N., Guo, L., Mallet, C. (2009). Airborne LiDAR feature selection for urban classification using random forests. *International Archives of the Photogrammetry, Remote Sensing and Spatial Information Sciences*, 38(3/W8), 207–212.

Chen, C., Liaw, A., Breiman, L. (2004). Using random forest to learn imbalanced data. Technical Report. University of California, Berkeley.

Clode, S., Rottensteiner, F., Kotsookos, P., Zelniker, E. (2007). Detection and vectorization of roads from LiDAR data. *Photogrammetric Engineering and Remote Sensing*, 73(3), 517–536.

Cortes, C., Vapnik, V. (1995). Support-vector networks. *Machine Learning*, 20(3), 273–297.

Cramer, M. (2010). The DGPF test on digital aerial camera evaluation—overview and test design. *Photogrammetrie Fernerkundung Geoinformation*, 2(2010), 73–82.

Dorninger, P., Pfeifer, N. (2008). A comprehensive automated 3D approach for building extraction, reconstruction, and regularization from airborne laser scanning point clouds. *Sensors*, 8, 7323–7343.

Farabet, C., Couprie, C., Najman, L., LeCun, Y. (2013). Learning hierarchical features for scene labelling. *IEEE Transactions on Pattern Analysis and Machine Intelligence*, 35(8), 1915–1929.

Fischler, M. A., Bolles, R. C. (1981). Random sample consensus: A paradigm for model fitting with applications to image analysis and automated cartography. *Communications of the ACM*, 24(6), 381–395.

Frey, B. J., MacKay, D. J. C. (1998). A revolution: Belief propagation in graphs with cycles. *Proceedings of the Neural Information Processing Systems Conference*, (10), 479–485.

Geman, S., Geman, D. (1984). Stochastic relaxation, Gibbs distributions, and the Bayesian restoration of images. *IEEE Transactions on Pattern Analysis and Machine Intelligence*, 6(6), 721–741.

Gislason, P. O., Benediktsson, J. A., Sveinsson, J. R. (2006). Random forests for land cover classification. *Pattern Recognition Letters*, 27(4), 294–300.

Heipke, C., Mayer, H., Wiedemann, C., Jamet, O. (1997). Evaluation of automatic road extraction. *International Archives of Photogrammetry and Remote Sensing*, 32(3/2W3), 56.

Heuel, S. (2004). *Uncertain Projective Geometry. Statistical Reasoning for Polyhedral Object Reconstruction.* Springer-Verlag: Berlin, Germany.

Höfle, B., Hollaus, M., Hagenauer, J. (2012). Urban vegetation detection using radiometrically calibrated small-footprint full-waveform airborne LiDAR data. *ISPRS Journal of Photogrammetry and Remote Sensing*, 67(2012), 134–147.

Hu, X., Tao, C. V., Hu, Y. (2004). Automatic road extraction from dense urban area by integrated processing of high resolution imagery and LIDAR data. *International Archives of Photogrammetry, Remote Sensing and Spatial Information Science*, 35(B3), 288.

Huang, H., Brenner, C., Sester, M. (2013). A generative statistical approach to automatic 3D building roof reconstruction from laser scanning data. *ISPRS Journal of Photogrammetry and Remote Sensing*, 79(2013), 29–43.

Kada, M., McKinley, L. (2009). 3D building reconstruction from LiDAR based on a cell decomposition approach. *International Archives of the Photogrammetry, Remote Sensing and Spatial Information Sciences*, 38(3/W4), 47–52.

Kager, H. (2000). Adjustment of algebraic surfaces by least squared distances. *International Archives of Photogrammetry and Remote Sensing*, 33(B3), 472.

Kraus, K., Pfeifer, N. (1998). Determination of terrain models in wooded areas with airborne laser scanner data. *ISPRS Journal of Photogrammetry and Remote Sensing*, 53(4), 193–203.

Kumar, S., Hebert, M. (2006). Discriminative random fields. *International Journal of Computer Vision*, 68(2), 179–201.

Lafarge, F., Mallet, C. (2012). Creating large-scale city models from 3D-point clouds: A robust approach with hybrid representation. *International Journal of Computer Vision*, 99(1), 69–85.

Lim, E., Suter, D. (2007). Conditional random field for 3D point clouds with adaptive data reduction. *Proceedings of the International Conference on Cyberworlds*, 404–408. http://ieeexplore.ieee.org/stamp/stamp.jsp?tp=&arnumber=4390945

Lim, E., Suter, D. (2009). 3D terrestrial LIDAR classifications with super-voxels and multi-scale conditional random fields. *Computer Aided Design*, 41(10), 701–710.

Lu, W., Murphy, K., Little, J., Sheffer, A., Hongbo, F. (2009). A hybrid conditional random field for estimating the underlying ground surface from airborne LiDAR data. *IEEE Transactions on Geoscience and Remote Sensing*, 47(8), 2913–2922.

Maas, H.-G. (1999). Fast determination of parametric house models from dense airborne laserscanner data. *International Archives of Photogrammetry and Remote Sensing*, 32(2W1), 1.

Mallet, C. (2010). Analysis of full-waveform Lidar data for urban area mapping. PhD Thesis. Télécom ParisTech, France.

Mallet, C., Bretar, F. (2009). Full-waveform topographic LiDAR: State-of-the-art. *ISPRS Journal of Photogrammetry and Remote Sensing*, 64(1), 1–16.

Mäntylä, M. (1988). *An Introduction to Solid Modeling: Principles of Computer Science*. Computer Science Press: Rockville, MA.

Munoz, D., Vandapel, N., Hebert, M. (2008). Directional associative Markov network for 3-D point cloud classification. *International Symposium on 3D Data Processing, Visualization and Transmission (3DPVT)*, Georgia Institute of Technology, Atlanta, GA, pp. 1–8.

Niemeyer, J., Rottensteiner, F., Sörgel, U. (2014). Contextual classification of Lidar data and building object detection in urban areas. *ISPRS Journal of Photogrammetry and Remote Sensing*, 87(2014), 152–165.

Niemeyer, J., Wegner, J., Mallet, C., Rottensteiner, F., Sörgel, U. (2011). Conditional random fields for urban scene classification with full waveform LiDAR data. *Photogrammetric Image Analysis (PIA)*. LNCS 6952, Springer, Heidelberg, Germany, pp. 233–244.

Ortner, M., Descombe, X., Zerubia, J. (2008). A marked point process of rectangles and segments for automatic analysis of digital elevation models. *IEEE Transactions on Pattern Analysis and Machine Intelligence*, 30(1), 105–119.

Oude Elberink, S., Vosselman, G. (2009). Building reconstruction by target based graph matching on incomplete laser data: Analysis and limitations. *Sensors*, 9(8), 6101–6118.

Paisitkriangkrai, S., Sherrah, J., Janney, P., Van-den Hengel, A. (2015). Effective semantic pixel labelling with convolutional networks and conditional random fields. *Proceedings of the IEEE Conference on Computer Vision and Pattern Recognition Workshops (CVPRW)*, pp. 36–43.

Pauly, M., Keiser, R., Gross, M. (2003). Multi-scale feature extraction on point-sampled surfaces. *Computer Graphics Forum*, 22(3), 81–89.

Perera, S., Maas, H.-G. (2014). Cycle graph analysis for 2D roof structure modelling: Concepts and performance. *ISPRS Journal of Photogrammetry and Remote Sensing*, 93(2014), 213–226.

Pfeifer, N., Mandlburger, G. (2017). LiDAR data filtering and DTM generation. In Shan, J., Toth, C. (Eds.), *Topographic Laser Ranging and Scanning—Principles and Processing*, 2nd ed. CRC Press, Boca Raton, FL, 349–378.

Poullis, C. (2013). A framework for automatic modeling from point cloud data. *IEEE Transactions on Pattern Analysis and Machine Intelligence*, 35(11), 2563–2574.

Rottensteiner, F. (2003). Automatic generation of high-quality building models from LiDAR data. *IEEE Computer Graphics and Applications*, 23(6), 42.

Rottensteiner, F. (2006). Consistent estimation of building parameters considering geometric regularities by soft constraints. *International Archives of the Photogrammetry, Remote Sensing and Spatial Information Sciences*, 36(3), 13.

Rottensteiner, F., Briese, C. (2002). A new method for building extraction in urban areas from high-resolution LIDAR data. *International Archives of the Photogrammetry, Remote Sensing and Spatial Information Sciences*, 34(3A), 295.

Rottensteiner, F., Sohn, G., Gerke, M., Wegner, J. D., Breitkopf, U., Jung, J. (2014). Results of the ISPRS benchmark on urban object detection and 3D building reconstruction. *ISPRS Journal of Photogrammetry and Remote Sensing*, 93(2014), 256–271.

Rottensteiner, F., Trinder, J., Clode, S., Kubik, K. (2005a). Using the Dempster Shafer method for the fusion of LIDAR data and multi-spectral images for building detection. *Information Fusion*, 6(4), 283.

Rottensteiner, F., Trinder, J., Clode, S., Kubik, K. (2005b). Automated delineation of roof planes in Lidar data. *International Archives of the Photogrammetry, Remote Sensing and Spatial Information Sciences*, 36(3/W19), 221.

Rottensteiner, F., Trinder, J., Clode, S., Kubik, K. (2007). Building detection by fusion of airborne laserscanner data and multispectral images: Performance evaluation and sensitivity analysis. *ISPRS Journal of Photogrammetry and Remote Sensing*, 62(2), 135.

Rutzinger, M., Rottensteiner, F., Pfeifer, N. (2009). A comparison of evaluation techniques for building extraction from airborne laser scanning. *IEEE Journal of Selected Topics in Applied Earth Observation and Remote Sensing*, 2(1), 11–20.

Sampath, A., Shan, J., 2007. Building boundary tracing and regularization from airborne Lidar point clouds. *Photogrammetric Engineering and Remote Sensing*, 73(7), 805–812.

Schenk, T., Csatho, B. (2002). Fusion of LIDAR data and aerial imagery for a more complete surface description. *International Archives of the Photogrammetry, Remote Sensing and Spatial Information Sciences*, 34(3A), 310.

Schmidt, M. (2012). UGM: A Matlab toolbox for probabilistic undirected graphical models. http://www.cs.ubc.ca/~schmidtm/Software/UGM.html (accessed June 22, 2016).

Shapovalov, R., Velizhev, A., Barinova, O. (2010). Non-associative Markov networks for 3D point cloud classification. *International Archives of the Photogrammetry, Remote Sensing and Spatial Information Sciences*, 38(3A), 103–108.

Shotton, J., Winn, J., Rother, C., Criminisi, A. (2009). Textonboost for image understanding: Multi-class object recognition and segmentation by jointly modeling texture, layout, and context. *International Journal of Computer Vision*, 81(1), 2–23.

Sithole, G., Vosselman, G. (2004). Experimental comparison of filter algorithms for bare-earth extraction from airborne laser scanning point clouds. *ISPRS Journal of Photogrammetry and Remote Sensing*, 59(1), 85–101.

Tarsha-Kurdi, F., Landes, T., Grussenmeyer, P. (2008). Extended RANSAC algorithm for automatic detection of building roof planes from Lidar data. *The Photogrammetric Journal of Finland*, 21(1), 97–109.

Vögtle, T., Steinle, E. (2003). On the quality of object classification and automated building modeling based on laserscanning data. *International Archives of the Photogrammetry, Remote Sensing and Spatial Information Sciences*, 34(3W13), 149.

Vosselman, G. (2002). On the estimation of planimetric offsets in laser altimetry data. *International Archives of the Photogrammetry, Remote Sensing and Spatial Information Sciences*, 34(3A), 375.

Vosselman, G., Dijkman, S. (2001). 3D building model reconstruction from point clouds and ground plans. *International Archives of the Photogrammetry, Remote Sensing and Spatial Information Sciences*, 34(3W4), 37.

Vosselman, G., Liang, Z. (2009). Detection of curbstones in airborne laser scanning data. *International Archives of Photogrammetry, Remote Sensing and Spatial Information Science*, 38(3/W8), 111–116.

Wagner, W., Ullrich, A., Ducic, V., Melzer, T., Studnicka, N. (2006). Gaussian decomposition and calibration of a novel small-footprint full-waveform digitising airborne laser scanner. *ISPRS Journal of Photogrammetry and Remote Sensing*, 60(2), 100.

Weidner, U., Förstner, W. (1995). Towards automatic building reconstruction from high resolution digital elevation models. *ISPRS Journal of Photogrammetry and Remote Sensing*, 50(4), 38.

Weinmann, M., Schmidt, A., Mallet, C., Hinz, S., Rottensteiner, F., Jutzi, B. (2015). Contextual classification of point cloud data by exploiting individual 3D neighbourhoods. *ISPRS Annals of the Photogrammetry, Remote Sensing and Spatial Information Sciences*, 2(3/W4), 271–278.

Wiedemann, C. (2003). External evaluation of road networks. *International Archives of Photogrammetry and Remote Sensing*, 34(3/W8), 93.

Xiong, B., Jancosek, B., Oude Elberink, S., Vosselman, G. (2015). Flexible building primitives for 3D building modeling. *ISPRS Journal of Photogrammetry and Remote Sensing*, 101(2015), 275–290.

Xiong, X., Munoz, D., Bagnell, J. A., Hebert, M. (2011). 3-D scene analysis via sequenced predictions over points and regions. *Proceedings of IEEE International Conference on Robotics and Automation (ICRA11)*, 2609–2616.

Zhu, P., Lu, Z., Chen, X., Honda, K., Eiumnoh, A. (2004). Extraction of city roads through shadow path reconstruction using laser data. *Photogrammetric Engineering and Remote Sensing*, 70(12), 1433.

17 Progressive Modeling of 3D Building Rooftops from Airborne LiDAR and Imagery

Jaewook Jung and Gunho Sohn

CONTENTS

17.1　INTRODUCTION

Urbanization is an inevitable movement that is not merely a modern phenomenon, but a rapid and historic transformation of human social roots on a global scale. According to the United Nations, half of the world's population lived in urban areas at the end of 2008, and the number will increase to about 70% by 2050 (International Herald Tribune, 2008). The rapid urbanization has led the dramatic change of city environments and has presented an urgent need to construct, synthesize, and update environmental information for the purpose of planning, managing, and making various critical decisions that impact growing cities. To create useful and accurate representations of various dynamic city entities, researchers put forth numerous efforts in computer vision, photogrammetry, and remote sensing fields in the last few decades. Particularly, a building, a structure very closely connected with human life, is recognized as the most important object in generating of 3D virtual models of city environment. Since initial efforts on automatic building extraction from remotely sensed data in the early 1990s, a large number of research studies have been conducted to recognize, detect, reconstruct, and represent building objects. As a result, many applications for web mapping services and mobile use have been developed by major companies, including Google, Apple, HERE, and Uber, and are able to provide 3D building models for consumer use.

According to Skyscrperpage.com, in 2015, there were over 2000 high-rise buildings in the city of Toronto, and 139 high-rise and mid-rise buildings were under construction in January that year (Economic Dashboard-Annual Summary, 2015). With expansion of different types of building structures, even more changes are expected to take place in the cityscapes. A city is a dynamic entity as the environment continuously changes. Accordingly, its virtual models also need to be regularly updated. To address the continuous changes in the city environment, companies like HERE have been updating their maps on a bi or trimonthly basis (HERE 360, 2015). However, for a large-scale area, newly generating building models whenever new data is acquired is cost-inefficient and labor-intensive. Therefore, existing building models should be reused and appropriately updated in cost-effective and automatic manners to record changes. As such, continuous modeling of 3D cityscapes using remotely sensed multidata taken at different epochs is expected to play an important role in generating timely and accurate building models.

In the perspective of building reconstruction, raw data acquired from remotely sensed data are converted into *building models*. A large number of building reconstruction methods, which range widely in terms of levels of automation (automatic vs. semiautomatic), data source (single data vs. multidata), and data processing strategies (data-driven, model-driven, or hybrid), have been explored effectively represent a full description of buildings. However, in spite of constant efforts, developing a *universal* intelligent machine enabling the massive generation of highly accurate rooftop models in a fully-automated manner still remains a challenging task. Many researchers (Ameri, 2000; Sohn and Dowman, 2007) pointed out several reasons for the problem as follows:

- Scene complexity: Remotely sensed data from the urban scene contain a large amount of information of nonbuilding objects (e.g., ground, tree, car, and clutter) in addition to the building objects. Although some heuristic knowledge (e.g., building height, certain brightness, or nearby shadow) can be used to recognize building objects, detecting individual buildings is not easy because buildings are attached and form blocks. In terms of building interpretation, buildings in urban scenes have enormous variants in structure and shapes with multistory planes, the landmark buildings of the city in particular. The variety of shapes cannot be described by common types of building structures. Thus, a method to simplify complex building scenes is required for effective interpretation.

- Incomplete cues: There is always a significant loss of information in data. Occlusion of buildings or building parts by themselves or adjacent objects causes problems in data integrity. Moreover, shadow, noise, low contrast, and superstructures on building roofs cause redundant or spurious cues, bringing about ambiguity and confusion to the building reconstruction process.

- Sensor dependency: Sensors used for building modeling have unique characteristics related to the acquisition mechanism. This inherent property has a considerable influence over the reconstructed building models; for instance, Light Detection and Ranging (LiDAR) data provides accurate plane information, whereas the accuracy of building boundaries is less than that of image data due to its irregular point distribution. Thus, fully understanding sensor characteristics is one of the most important tasks in building reconstruction.

Even though many algorithms for reconstructing 3D building models using single data source have been introduced and can provide promising results (Rottensteiner, 2014), the methods still have some limitations due to inherent sensor dependent properties, levels of automation, model accuracies, and missing data problems. One promising approach to address these problems is to combine multisensor data which have different characteristics. In this regard, combining LiDAR point clouds and optical imagery for building reconstruction have been exploited by many researchers (Haala and Kada, 2010). This is due to the fact that the characteristics of the modeling cues from the two data are complementary. Compared with LiDAR point clouds, the optical imagery better provides semantically rich information, geometrically accurate step, and eave edges, whereas it has weakness in detecting roof edges and 3D information such as planar patches when single imagery is used. However, LiDAR has somewhat opposite characteristics to optical imagery.

Regardless of which fusion approach is applied, the registration between different sensor data is recognized as an essential and prerequisite process. The accuracy of registration has a substantial impact on the quality of results. The registration method should provide accurate and robust relations between datasets taken from different sensors or from different viewpoints at different epochs. In addition, a registration between existing models and newly taken sensor data should be addressed, particularly in continuous city modeling. However, although many registration methods that deal with correspondence problems between different sensor data have been studied, the registration between valuable 3D building models over a large-scale area and remotely sensed data has been studied relatively less. Therefore, more researches on development of registration methods using valuable 3D building models are required for continuous city modeling.

The current chapter presents a novel research framework for continuously reconstructing 3D building rooftops using multisensor data, which are acquired at different epochs. This is mainly based on research works conducted by Sohn et al. (2013), Jung (2016), and Jung et al. (2016, 2017). This chapter starts by presenting literature review on building modeling, registration, and data fusion that are related to continuous modeling. Next, data-driven method is presented to reconstruct 3D building rooftop using a popular single data source (i.e., airborne LiDAR data). Then, a context-based geometric hashing (CGH) method is presented to align newly acquired image data with existing building models as a prerequisite process of the subsequent building refinement application. The existing building models are refined by a sequential fusion method. Lastly, experimental results are presented for the methods to aid in the assessment and discussion of system performance.

17.2 BACKGROUND

17.2.1 BUILDING RECONSTRUCTION

Building reconstruction can be recognized as a huge process for the generation of digital representations of physical buildings where raw data without any structured information are converted into highly structured 3D building models with rich semantic information. As initial efforts for automatically generating 3D building models began in the early 1990s, numerous techniques using various remotely sensed data have been explored in computer vision, photogrammetry, and remote sensing fields. In this section, we review existing building reconstruction methods in terms of reconstruction strategy and regularization.

17.2.1.1 Model-Driven versus Data-Driven

In the model-driven approaches, 3D building models are reconstructed by fitting parameterized primitives to data. This is possible due to the fact that many buildings in rural and suburban area have common shapes in whole building or building roof parts. These common roof shapes such as flat, gable, and hip roof are considered as standard primitives for representing building rooftop structures. Simple buildings can be well represented as regularized building models using predefined parameterized primitives even with low density data and presence of missing data. However, complex buildings and arbitrarily shaped buildings are difficult to model using a basic set of primitives. Moreover, the selection of the proper primitives among a set of primitives is not an easy task. To address the limitations, Verma et al. (2006) presented a parametric modeling method to reconstruct relatively complex buildings by combining simple parametric roof shapes that are categorized into four types of simple primitives. In this study, the roof–topology graph is constructed to represent the relationships among the various planar patches of approximate roof geometry. The constructed roof–topology graph is decomposed into subgraphs, which represents simple parametric roof shapes, and then parameters of the primitives are determined by fitting LiDAR data. Although they decomposed complex buildings into simple building parts, many building parts cannot be still explained by their four simple shape primitives. Kada and McKinley (2009) decomposed the building's footprint into cells that provided the basic building blocks. Three types of roof shapes including basic, connecting, and manual shapes are defined. Basic shapes consist of flat, shed, gabled, hipped, and Berliner roofs, whereas connecting shapes are used to connect the roofs of the sections with specific junction shapes. The parameterized roof shapes of all cells are determined from the normal direction of LiDAR points. The entire 3D building model is represented by integrating the parameterized roof elements with the neighboring pieces. Although a high level of automation is achieved, the method still requires manual works to adjust cell parameters and to model more complex roof shapes like mansard, cupola, barrel, and even some detail elements. Lafarge et al. (2010) reconstructed building models from a digital surface model (DSM) by combining generic and parametric methods. Buildings are considered as assemblages of 3D parametric blocks from a library. After extracting 2D building supports, 3D parametric blocks are placed on the 2D supports using Gibbs model that controls both the block assemblage and the fitting to data. The optimal configuration of 3D blocks is determined using the Bayesian framework. They mentioned that the optimization step needs to be improved to achieve both higher precision and shorter computing time as future work. Based on a predefined primitive library, Huang et al. (2013) conducted a generative modeling to reconstruct roof models that fit the data. The library provides three groups including 11 types of roof primitives whose parameters consist of position parameters, contour parameters, and shape parameters. Building roofs are represented as one primitive or an assemblage of primitives allowing primitives overlaps. For combining primitives, they derived combination and merging rules that consider both vertical and horizontal intersections. Reversible Jump Markov Chain Monte Carlo with a specified jump mechanism is conducted for the selection of roof primitives, and the sampling of their parameters. Although they have shown potential and flexibility of their method, there are issues to be solved: (1) Uncertainty and instability of the reconstructed building model, (2) Influence of prior knowledge and scene complexity on completeness of the reconstruction, and (3) Heavy computation time.

In contrast with model-driven approaches, data-driven approaches do not make any assumptions regarding to the building shapes, thus they can theoretically handle all kinds of buildings. However, the approach may cause considerable deformations due to the sensitivity to surface fluctuations and outliers in the data. Moreover, it requires a regularization step during the reconstruction process. In general, the generic approach starts by extracting building modeling cues such as surface primitives, step lines, intersection lines, and outer boundary lines followed by reconstructing the 3D building model. The segmentation procedure for extracting surface primitives divides a given dataset into homogeneous regions. Classical segmentation algorithms such as region growing (Rottensteiner et al., 2005; Kada and Wichmann, 2012) and random sample consensus (RANSAC) (Tarsha-Kurdi et al., 2008) can be used for segmenting building roof planes. Moreover, Sampath and Shan (2010) conducted eigenanalysis for each roof point within its Voronoi neighborhood, and then adopted the fuzzy k-means approach to cluster the planar points into roof segments based on their surface normal. Then, they separated the clusters into parallel and coplanar segments based on their distance and connectivity. Lafarge and Mallet (2012) extracted geometric shapes such as planes, cylinders, spheres, or cones for identifying the roof sections by fitting points into various geometric shapes and then proposed a method for arranging both the geometric shapes and the other urban components by propagating point labels based on Markov Random Field (MRF). Yan et al. (2014) proposed a global solution for roof segmentation. Initial segmentation is optimized by minimizing a global energy function consisting of the distances of LiDAR points to initial planes, spatial smoothness between data points, and the number of planes. After segmenting points or extracting homogeneous surface primitives, modeling cues such as intersection lines and step lines can be extracted on the basis of geometrical and topological relationships of the segmented roof planes. Intersection lines are easily obtained by intersecting two adjacent planes or segmented points, whereas step lines are extracted at roof plane boundary with abrupt height discontinuity. To extract step lines, Rottensteiner et al. (2005) detected edge candidate points and then extracted step lines from an adjustment considering edge points within user-specified threshold. Moreover, Sohn et al. (2008) proposed a step line extractor, called Compass Line filter (CLF), for extracting straight lines from irregularly distributed LiDAR points. Although outer boundary is one type of step line, it is recognized as a separate process in many data-driven approaches. Some researchers delineated initial boundary lines from building boundary points using alpha shape (Dorninger and Pfeifer, 2008), ball-pivoting (Verma et al., 2006), and contouring algorithm (Zhou and Neumann, 2008). Then, the initial boundary was simplified or regularized. The detail reviews for simplification or regularization of boundary will be given in Section 17.2.1.2.

Once all building modeling cues are collected, 3D building models are reconstructed by aggregating the modeling cues. To reconstruct topologically and geometrically correct 3D building models, Sohn et al. (2008) proposed the (Binary Space Partitioning) BSP technique that progressively partitions a building region into homogeneous binary convex polygons. Rau and Lin (2011) proposed a line-based roof model reconstruction algorithm, namely (Triangulated Irregular Network) TIN-Merging and Reshaping, to reconstruct topology with geometric modeling. Oude and Vosselman (2009), and Perera and Maas (2014) used a roof topology graph to preserve roof topology. In the latter, roof corners are geometrically modeled using the shortest closed cycles and the outermost cycle derived from the roof topology graph.

17.2.1.2   Building Boundary Regularization

Detection of building boundary is an intermediate step for 3D building reconstruction although it is not required in all building reconstruction algorithms. Generally, the initial boundary extracted from irregular LiDAR points has jagged shape with large numbers of vertices. Thus, a simplification or regularization process is required to delineate plausible building boundaries with certain regularities such as orthogonality, parallelism, and symmetry. In most methods, the boundary detection process starts by extracting boundary points from segmented points. From extracted boundary points, initial building boundaries are generated by tracing boundary points followed by a simplification or regularization process that improves the initial boundary. The easiest method to improve initial boundary is to simplify the initial boundary by removing vertices but preserving relevant points. The well-known Douglas–Peucker algorithm (Douglas and Peucker, 1973) is widely recognized as the most visually effective line simplification algorithm. The algorithm starts by selecting two points that have the longest distance and recursively adding vertices whose distance from the line is less than a given threshold. However, the performance of the algorithm fully depends on the used threshold and is substantially affected by outliers. Another approach extracts straight lines from boundary points using the Hough Transform (Morgan and Habib, 2002) or RANSAC (Fischler and Bolles, 1981). The extracted lines are then connected by intersections of the extracted straight lines to generate closed outer boundary lines. However, Brenner (2010) pointed out that the methods require some additional steps due to missing small building edges.

On the other hand, the regularization process imposes certain regularities when the initial boundary is simplified. Vosselman (1999) assumed that building outlines are along or perpendicular to the main direction of a building. After defining the position of a line by the first two boundary points, the line is updated using the succeeding boundary points until the distance of a point to the line exceeds some bound. The next line starts from this point in a direction perpendicular to the previous line. A similar approach was proposed by Sampath and Shan (2007). They grouped points on consecutive edges with similar slopes and then applied a hierarchical least squares solution to fit parametric lines representing the building boundary.

Some methods are based on the model hypothesis and verification approach. Ameri (2000) introduced the Feature Based Model Verification for modification and refinement of polyhedral-like building objects. In their approach, they imposed the geometrical and topological model information to the Feature-Based Model Verification process as external and internal constraints that consider linearity for straightening consecutive lines, connectivity for establishing topology between adjacent lines, orthogonality, and coplanarity. Then, the weighted least squares minimization was adopted to produce a good regularized description of a building model. Weidner and Förstner (1995) adopted the Minimum Description Length (MDL) concept to regularize noisy building boundaries. For four local consecutive points, 10 different hypothetical models are generated with respect to regularization criteria. Then, MDL, which depends on the mutual fit of the data and model and on the complexity of the model, is used to find the optimal regularity of the local configuration. Jwa et al. (2008) extended the MDL-based regularization method by proposing new implicit hypothesis generation rules and by redesigning model complexity terms in which line directionality, inner angle, and number of vertices are considered as geometric parameters. Furthermore, Sohn et al. (2012) used the MDL-based concept to regularize topologies within rooftop model. Zhou and Neumann (2012) introduced global regularities in building modeling to reflect the orientation and placement similarities among 2.5D elements that consist of planar roof patches and roof boundary segments. In their method, roof–roof regularities, roof–boundary regularities, and boundary–boundary regularities are defined, and then, the regularities are integrated into a unified framework.

17.2.2 DATA FUSION

The integration of data and knowledge from several sources is known as data fusion. Hall and Llinas (1997) defined data fusion as follows: "Data fusion techniques combine data from multiple sensors and related information from associated databases to achieve improved accuracy and more specific inferences than could be achieved by the use of a single sensor alone." In the remote sensing community, data fusion combines multiple sources of data acquired with different spatial and spectral resolution to improve the potential values and interpretation performances of the source data and to produce a high-quality visible representation of data. Remote sensing fusion techniques can be classified into three different levels: (1) pixel/data level, (2) feature level, and (3) decision level (Pohl and van Genderen, 1998). Pixel level fusion combines raw data from multiple sources to yield a single resolution datum. The pixel level fusion of optical images is well known as the pan-sharpening technique that improves spatial resolution of panchromatic image by injecting structural and textual details of multispectral images or synthetic aperture radar (SAR) images. Feature level fusion combines features extracted from multiple data sources. As features are extracted from different characteristics of different sensors, the extracted features can provide additional valuable properties for various applications. Decision or interpretation level fusion combines the results, which are individually processed, to make a final decision. The decision level fusion methods contain voting methods, statistical methods, and fuzzy methods.

In the current chapter, our interest is feature level fusion, particularly the fusion of LiDAR and optical images for building reconstruction. Although many building reconstruction algorithms using single data provide some promising results, the integration of two complementary datasets can improve the quality of 3D building models with an increase of available information. In particular, combining LiDAR point clouds and optical images for rooftop modeling has been exploited by many researchers (Haala and Kada, 2010). In previous studies, image information in the fusion approach is mainly used for four different purposes in terms of building reconstruction: (1) Extraction of building points while removing nonbuilding points such as tree points, (2) improvement of segmentation, (3) improvement of building boundary, and (4) texture mapping. For building region extraction, Chen et al. (2005) used spectral information and texture of color images. Sohn and Dowman (2007) used Normalized Difference Vegetation Index to discriminate between buildings and trees. Demir and Baltsavias (2012) detected building regions by combining results of four different building detection methods that were respectively derived from combinations of spectral information and Normalized Difference Vegetation Index of image data and spatial distribution of LiDAR data.

Awrangjeb et al. (2013) proposed an image line guided technique to robustly segment building points into individual roof planes. Lines extracted from images were classified into ground, tree, roof edge, and roof ridge-lines using the ground mask, color, and texture information of the image. Lines classified as roof edge or roof ridge were used to define robust seed regions for region growing for roof plane segmentation. Cheng et al. (2013) used images to refine initial roof point segmentation derived from LiDAR data based on the Shrink-Expand technique. Spectral and texture information (entropy) of images were used as a criterion for judging the reliability of segmentation.

Rottensteiner and Briese (2003) proposed wire frame fitting to improve the geometric quality of the polyhedral models created from LiDAR data. Image edges were matched with LiDAR-driven edges, and then the matched image edges were considered in the estimation of model parameters. Hu et al. (2006) proposed a hybrid modeling system in which building boundaries and plane surfaces were extracted from image and LiDAR data, respectively. Lee et al. (2008) proposed a method to extract the boundaries of complex buildings from LiDAR and photogrammetric images. Coarse building boundaries generated by LiDAR are simply substituted with image edges to extract precise building boundaries by matching with some constraints such as length ratio, angle, and distance. Kim and Habib (2009) similarly replaced initial building boundaries by 3D lines that have the biggest spectral difference between two flanking regions. Sohn et al. (2013) proposed a sequential fusion method to improve the boundary quality of existing building models based on the hypothesize and test (HAT) framework. Image lines were used to propose possible hypotheses. Cheng et al. (2013) also used image data to extract building boundary and step lines. After establishing relationships between 2D image lines and 3D LiDAR points, 3D lines were determined from multiview images. Two rectangle boxes along orthogonal directions of a line segment were analyzed to separate step and nonstep line segments. 3D building models were reconstructed by segmented roof points, 3D step lines, 3D ridge lines, and 3D boundaries using the Split-Merge-Shape method.

To achieve photorealistic rendering, Frueh et al. (2004) proposed a way to texture-map a LiDAR-driven 3D building models with oblique aerial images. After registering the oblique image with the existing building model, an optimal image for each triangle of the model was selected for texture by taking into account occlusion, image resolution, surface normal orientation, and coherence with neighbor triangles.

17.2.3 REGISTRATION

Registration is an essential process when multidatasets are used for various applications such as object recognition, environmental monitoring, change detection, and data fusion. In computer vision, remote sensing, and photogrammetry, this includes registrations of the same source taken from different viewpoints at different times (e.g., image to image), between datasets collected with different sensors (e.g., image and LiDAR), and between an existing model and remotely sensed raw data (e.g., map and image). Numerous registration methods have been proposed to solve the registration problems for given environments and for different purposes (Brown, 1992; Fonseca and Manjunath, 1996; Zitova and Flusser, 2003). Regardless of data types and applications, the registration process can be recognized as a feature extraction, and correspondence problem (or matching problem) between datasets. Brown (1992) categorized the existing matching methods into area-based and feature-based methods according to their nature. Area-based matching methods use image intensity values extracted from image patches. They deal with images without attempting to detect salient objects. Correspondences between two image patches are determined with a moving kernel sliding across a specific size of image search window or across the entire other image using correlation-like methods, Fourier methods, mutual information methods, and others. In contrast, feature-based methods use salient objects such as points, lines, and polygons to establish relations between two different datasets. In feature matching processes, correspondences are determined by considering the attributions of the used features. In model-to-image registration, most of the existing registration methods adopt a feature-based method because many 3D building models have no texture information.

In terms of features, point features such as line intersections, corners, and centroids of regions can be easily extracted from both models and images. Thus, Wunsch and Hirzinger (1996) applied the Iterative Closest Point algorithm to register 3D Computer-aided design (CAD) models with images. The Iterative Closest Point algorithm iteratively revises the transformation with two subprocedures. First, all closest point pair correspondences are computed. Then, the current registration is updated using the least square minimization of the displacement of matched point pair correspondences. In a similar way, Avbelj et al. (2010) used point features to align 3D wire-frame building models with infrared video sequences using a subsequent closeness-based matching algorithm. Lamdan and Wolfson (1988) used a geometric hashing method to recognize 3D objects in occluded scenes from 2D gray scale images. However, Frueh et al. (2004) pointed out that point features extracted from images cause false correspondences due to a large number of outliers.

As building models or man-made objects are mainly described by linear structures, many researchers have used lines or line segments instead of points as features. Hsu et al. (2000) used line features to estimate 3D pose of a video in which coarse pose was refined by aligning projected 3D models of line segments to oriented image gradient energy pyramids. Frueh et al. (2004) proposed a model to image registration for texture mapping of 3D models with oblique aerial images. Correspondences between line segments are computed by a rating function that consists of slope and proximity. As an exhaustive search to find optimal pose parameters was conducted, the method is affected by the sampling size of the parameter space, and it is computationally expensive. Eugster and Nebiker (2009) also used line features for real-time georegistration of video streams from unmanned aircraft systems. They applied relational matching, which does not only consider the agreement between an image feature and a model feature but also takes the relations between features into account. Avbelj et al. (2015) matched boundary lines of building models derived from DSM and hyperspectral images using an accumulator. Yang and Chen (2015) proposed a method to register Unmanned Aerial Vehicle (UAV)-borne sequent images and LiDAR data. They compared building outlines derived from LiDAR data with tensor gradient magnitudes and orientation in images to estimate key frame-image exterior orientation parameters (EOPs). Persad et al. (2015) matched linear features between Pan–Tilt–Zoom video images with 3D wireframe models based on a hypothesis-verification optimization framework. However, Tian et al. (2008) pointed out several reasons that make the use of lines or edge segments for registration of a difficult problem. First, edges or lines are extracted incompletely, and inaccurately, so that ideal edges might be broken into two or more small segments that are not connected to each other. Second, there is no strong disambiguating geometric constraint, whereas building models are reconstructed with certain regularities such as orthogonality and parallelism.

Utilizing a prior knowledge of building structures can reduce the matching ambiguities and the search space. Thus, Ding et al. (2008) used 2D orthogonal corners (2DOC) as a feature to recover the camera pose for texture mapping of 3D building models. The coarse camera parameters were determined by vertical vanishing points that correspond to vertical lines in the 3D models. Correspondences between image 2DOC and DSM 2DOC were determined using Hough transform and generalized M-estimator sample consensus. However, they described their error source as too limited to correct 2DOCs matches, in particular for residential areas. Moreover, Wang and Neumann (2009) pointed out that 2DOC features are not very distinctive because the features can be extracted from only orthogonal corners. Instead of using 2DOC, they proposed 3 connected segments as a feature that is more distinctive and repeatable. For putative feature matches, they applied a two level RANSAC method that consists of a local, and a global RANSAC for robust matching.

17.3 IMPLICIT REGULARIZATION FOR RECONSTRUCTING 3D BUILDING MODELS

This section proposes a data-driven modeling approach to reconstruct 3D rooftop models at city-scale from airborne laser scanning data. The focus of the method is to implicitly impose building regularity on 3D building rooftop models by introducing flexible regularity constraints. Figure 17.1 shows the overall workflow for reconstructing 3D rooftop models. In the element clustering step, building-labeled point clouds are clustered into homogeneous groups by applying height similarity and plane similarity. Based on segmented clusters, linear modeling cues including outer boundaries, intersection lines, and step lines are extracted. Topology elements among the modeling cues are recovered by the BSP technique. The regularity of the building rooftop model is achieved by an implicit regularization process in the framework of MDL combined with (HAT).

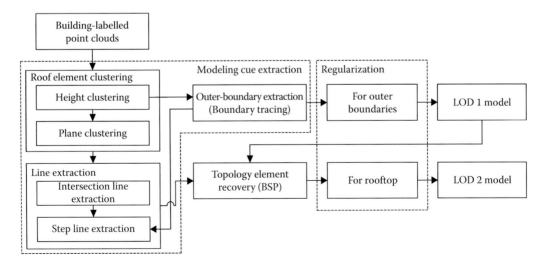

FIGURE 17.1   The overall workflow developed for reconstructing 3D rooftop models.

17.3.1   MODELING CUE EXTRACTION

The first step towards generating 3D building models using LiDAR data is to gather the evidence of building structures. Planes and lines are recognized as the most important evidence to interpret building structures due to the fact that 3D building rooftop models can be mainly represented by planar roof faces and edges. The two different modeling cues (planar and linear modeling cues) have different properties and can be separately extracted from LiDAR points.

17.3.1.1   Roof Element Clustering (Planar Modeling Cues)

Roof element clustering segments building-labeled points into homogeneous rooftop regions with a hierarchical structure. A building rooftop in an urban area is a combination of multiple stories, each of which consists of various shapes of flat and sloped planes. Directly extracting homogeneous regions from entire building points may result in difficulties due to a high degree of shape complexity. To reduce the complexity, the building-labeled points are decomposed into homogeneous clusters by sequentially applying height similarity and plane similarity in order.

In the height clustering step, the rooftop region $R = \{p_i \mid i = 1, 2, \ldots, m\}$ with n numbers of building-labeled points is divided into m height clusters $R = \{S_1, S_2, \ldots, S_m\}$. Height similarity at each point is measured over its adjacent neighboring points in TIN. A point with the maximum height is first selected as a seed point, and then a conventional region growing algorithm is applied to add neighbor points to a corresponding height cluster with a certain threshold δ_h. This process is repeated until all building rooftop points are assigned to one of the height clusters. As a result, the height clusters satisfy the property $R = \bigcup_{i=1}^{M} S_i, S_i \cap S_j = \{ \}, \forall i \neq j$. Note that each height cluster consists of one or more different roof planes. In the plane clustering step, each height cluster is decomposed into k plane clusters $\Pi = \{\pi_1, \pi_2, \ldots, \pi_k\}$ based on a plane similarity criterion. The well-known RANSAC algorithm is adopted to obtain reliable plane clusters. The process starts by randomly selecting three points as seed points to generate an initial plane. After a certain period of random sampling, a plane, which has the maximum number of inliers with a user defined tolerance distance ζ from the estimated plane, is selected as a best plane. Points, which are assigned in the previous iteration, are excluded in the next step. The process continues until all points of the height cluster are assigned into certain plane clusters. Figure 17.2b and c shows examples of height clusters and plane clusters, respectively, in which different colors represent different clusters.

17.3.1.2   Linear Modeling Cue Extraction

Once building-labeled points are segmented into homogeneous clusters with a hierarchical structure, linear modeling cues are extracted from the homogeneous clusters. We divide linear modeling cues into three different types to reduce the complexity in the modeling cue extraction process as follows: (1) outer boundaries of height clusters, (2) intersection lines, and (3) step lines within each height cluster.

In boundaries of height clusters, two adjacent planes have a large height discontinuity. Thus, outer boundaries of height clusters can be recognized as step lines. However, distinguishing between

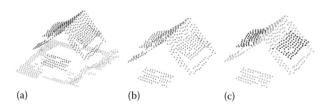

(a)   (b)   (c)

FIGURE 17.2   Roof element clustering: (a) building-labeled points (purple), (b) height clustering (pink and green), and (c) plane clustering (black, pink, blue, and purple).

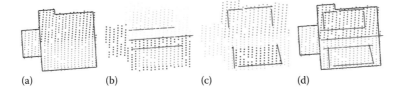

FIGURE 17.3   Modeling cues extraction: (a) outer boundaries (black), (b) intersection lines (red), (c) step lines (blue), and (d) combined modeling cues.

outer boundaries of height clusters and step lines within each height cluster can reduce ambiguity in the topology recovering process. Moreover, outer boundaries of height clusters can serve to generate the LOD1model. For these reasons, we separately extract outer boundaries of height clusters. The process starts by detecting boundary points of height clusters that share neighbor height clusters in a TIN structure. After selecting a starting boundary point, a next boundary point is determined by surveying neighbor boundary points, which are connected with the previous boundary point in TIN structure, and by selecting a boundary point that appears first in an anticlockwise direction. The process continues until the boundary is closed. Then, the closed boundary is regularized by the MDL-based regularization method that will be described in Section 17.3.3. An intersection line candidate is extracted by two adjacent roof planes. Candidates are accepted as valid intersection lines if they separate the point sets of the planes and if a sufficient number of points are close to the generated lines. For step lines, boundary points of plane clusters, which do not belong to outer boundaries or intersection lines, are considered as candidate points for step lines. Given a sequence $D = \{c_1, c_2, ..., c_l\}$ of l candidate points, step lines are extracted in a similar way to the Douglas–Peucker algorithm. The process starts with a straight line $(\overline{c_1c_2})$ connecting the first point and last point of the sequence and then recursively adding candidate points that have a distance larger than a user-defined tolerance. Each segment of the line segments is considered a step line. Figure 17.3 gives examples of each type of linear modeling cues.

17.3.2   BSP-Based Topology Construction

Once all modeling cues are collected, topological relations among the modeling cues are constructed by the BSP technique. In computer science, the BSP is a hierarchical partitioning method for recursively subdividing a space into convex sets with hyperlines. The BSP technique is used to recover a topologically and geometrically correct 3D building rooftop model from incomplete modeling cues. The topology recovery process consists of a partitioning step and plane merging step. In the partitioning step, a hierarchical binary tree is generated by dividing a parent region into two child regions with hyperlines (linear modeling cue). The partitioning optimum is achieved by maximizing partitioning score that consists of planar homogeneity, geometric regularity, and edge correspondence (Sohn et al., 2008). In plane merging step, the adjacent roof planes having similar normal vectors are merged. The merging process continues until no plane can be accepted by the coplanar similarity test. Once all polygons are merged together, 3D building rooftop model can be reconstructed by collecting final leave nodes in the BSP tree. Figure 17.4 shows results of partitioning step, merging step, and corresponding 3D rooftop model.

17.3.3   Implicit Regularization of Building Rooftop Model

Recovering error-free 3D rooftop models from erroneous modeling cues is a challenging task. Geometric constraints such as parallelism, symmetry, and orthogonality can be explicitly used as a prior knowledge on rooftop structures to compensate the limitations of erroneous modeling cues. However, explicitly imposing the constraints has limitations on describing complex buildings that

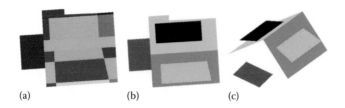

(a)   (b)   (c)

FIGURE 17.4 Binary space partitioning: (a) partitioning step, (b) merging step, and (c) reconstructed model.

appear in reality. In this section, we propose an implicit regularization in which regular patterns of building structures are not directly expressed, but implicitly imposed on reconstructed building models providing flexibility for describing more complex rooftop models. The proposed regularization process is conducted based on HAT optimization in MDL framework. Possible hypotheses are generated by incorporating regular patterns that are present in the given data. MDL is used as a criterion for selecting an optimal model out of the possible hypotheses.

17.3.3.1 *Minimum Description Length* Principles and Rooftop Modeling

The MDL proposed by Rissanen (1978) is a method for inductive inference that provides a generic solution to the model selection problem. The MDL is based on the idea of transmitting data as a coded message, in which the coding is based on some prearranged set of parametric statistical model. The full transmission has to include not only the encoded data values but also the coded model parameter values. Thus, the MDL consists of model complexity and model closeness as follows:

$$DL = \lambda\, L\big(D\,|\,H\big) + (1-\lambda)\, L\big(H\big) \tag{17.1}$$

where:
  $L(D|H)$ indicates a goodness-of-fit of observations D given a model H
  $L(H)$ represents how complex the model H is
  λ is a weight parameter for balancing the model closeness and the model complexity

Assuming that an optimal model representing the data has the minimal description length, the model selection process allows a model H to be converged to the optimal model H^* as follows:

$$H^* = \arg\min_{H \in \Phi}\{\lambda L(D\,|\,H) + (1-\lambda)L(H)\} \tag{17.2}$$

The first term in Equation 17.1 is optimized for good data attachment to the corresponding model. With an assumption that an irregular distribution of data $D = \{x_1,\ldots,x_n\}$ with n measurements caused by random errors follows a Gaussian distribution $x \sim N(\mu,\sigma^2)$ with expectation μ and variance σ^2, its density function can be represented as $P(x) = (1/\sigma\sqrt{2})e^{-((x-\mu)^2/(2\sigma^2))}$. By using a statistical model of the data, the degree of fit between a model and data can be measured by $L(D\,|\,\mu,\sigma^2)$, and then the term of model closeness can be rewritten in a logarithmic form as follows:

$$L\big(D\,|\,\mu,\sigma^2\big) = -\log_2 P(D) = -\left(\log_2 e^{-\dfrac{\sum(x-\mu)^2}{2\sigma^2}} + n\log_2 \dfrac{1}{\sigma\sqrt{2\pi}}\right) \tag{17.3}$$

$$= \frac{1}{2\ln 2}\sum\left(\frac{x-\mu}{\sigma}\right)^2 + n\log_2 \sigma + \frac{n}{2}\log_2 2\pi$$

In Equation 17.3, the last two terms can be ignored with an assumption that all the hypotheses have the same σ. Thus, the equation is simplified as follows:

$$L(D \mid H) = \frac{\Omega}{2\ln 2} \qquad (17.4)$$

where Ω is the weighted sum of the squared residuals between a model H and a set of observations D, that is $[D-H]^{\mathrm{T}}[D-H]$ in matrix form.

The second term in Equation 17.1 is designed to encode the model complexity. In this study, the model complexity is explained by three geometric factors: (1) the number of vertices N_v, (2) the number of identical line directions N_d, and (3) the inner angle transition $N_{\angle\theta}$. By using the three geometric factors, an optimal model is chosen if its polygon has a small number of vertices and a small number of the identical line directions, and if the inner angle transition is smoother or more orthogonal.

Suppose that N_v, N_d, and $N_{\angle\theta}$ are used for an initial model, whereas N_v', N_d', and $N_{\angle\theta}'$ are used for a hypothetical model generated from the initial model. To measure the description length for the number of vertices, we start by deriving the probability that a vertex is randomly selected from a given model, $P(v) = 1/N_v$. Then, it can be expressed in bits as $\log_2(N_v)$. As a hypothetic model generated by hypothesis generation process has N_v' vertices, its description length is $N_v' \log_2(N_v)$. Similarly, the probability for the number of identical line directions N_d is $P(d) = 1/N_d$ and can be expressed in bits as $\log_2(N_d)$. By considering the required number of line directions N_d', the description length for identical line direction is measured by $N_d' \log_2(N_d)$. To define line directions, we adopt CLF as shown in Figure 17.5. The CLF is determined by the whole set of eight filtering lines with different slopes $\{\theta_i: i=1,\ldots,8\}$ that is equally separated in steps of 22.5°. The representative angle for each slope, θ_i^{REP}, is calculated by a weighted averaging of angles that takes the summed line length of each CLF slope into account.

Lastly, the description length for inner angle transition is measured by assigning a certain penalty value to quantized inner angles. As depicted in Equation 17.5, the penalty values $\gamma_{i=0,1,2}$ are heuristically determined to have the minimum value of 0 (i.e., favor inner angle) if inner angle $\angle\theta$ is close to 90° or 180°, whereas the maximum value of 2 (i.e., unfavor inner angle) is assigned to very acute inner angles. This is because acute inner angle at two consecutive building vectors rarely appears in reality. Thus, the probability for $N_{\angle\theta}$ can be derived from an inner angle that is located in one of the quantized angles, $P(\angle\theta) = 1/N_{\angle\theta}$, and expressed in bits as $\log_2(N_{\angle\theta})$. In the optimal model, the cost imposed by penalty values is $\sum_{k=1}^{N_v'} \gamma_{i=0,1,2}$, and its description length is calculated by $N_{\angle\theta}' \log_2(N_{\angle\theta})$.

$$\gamma_{i=0,1,2} = \begin{cases} 0 & \text{if } 78.75° \le \angle\theta \le 101.25° \text{ or } 168.75° \le \angle\theta \le 180° \\ 1 & \text{if } 11.25° < \angle\theta < 78.75° \text{ or } 101.25° < \angle\theta < 168.75° \\ 3 & \text{if } 0° < \angle\theta \le 11.25° \end{cases} \qquad (17.5)$$

As a result, the description length for subterms of model complexity $L(H)$ is obtained by the summation of three geometric factors as follows:

$$L(H) = W_v N_v' \log_2(N_v) + W_d N_d' \log_2(N_d) + W_{\angle\theta} N_{\angle\theta}' \log_2(N_{\angle\theta}) \qquad (17.6)$$

where W_v, W_d, and $W_{\angle\theta}$ are weight values for each subfactor in the model complexity.

17.3.3.2 Hypothesis Generation

The hypothesis generation process proposes a set of possible hypotheses under certain configurations of a rooftop model (or building boundary). Suppose a rooftop model consists of a polygon $\Pi_A = \{v_1, v_2, v_3, v_4, v_5, v_6, v_7\}$ and a polygon $\Pi_B = \{v_3, v_4, v_5, v_8, v_9, v_{10}\}$, where v_3, v_4, and v_5 are common vertices in both polygons (Figure 17.6a). A task is to generate possible hypotheses at a certain vertex considering a given configuration of rooftop model. The hypothesis generation process starts

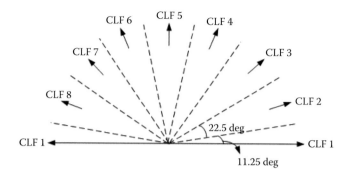

FIGURE 17.5 Compass line filter.

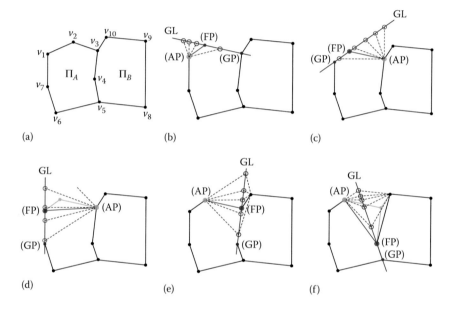

FIGURE 17.6 Examples of hypothesis generation (blue point: anchor point (AP), green point: removed point, purple point: guide point (GP), red point: floating point (FP), red circle: new possible positions of FP, red line: floating line (FL) and purple line: guide line (GL): (a) initial configuration, (b) case 1, (c) case 2, (d) case 3, (e) case 4, and (f) case 5.

by defining an Anchor Point (AP), Floating Point (FP), and Guide Point (GP) and then by deriving a Floating Line (FL=[AP, FP]) and Guiding Line (GL=[GP, FP]). The role of AP is to define the origin of a line to be changed (FL). FP is a point to be moved, whereas GP is used to generate GL that guides the movement of FP. Hypotheses are generated by moving FP along the GL with AP as an origin of FL. The orientation of FL is determined by representative angles of CLF that consists of eight directions as shown in Figure 17.5. There are different cases for hypothesis generation: (1) depending on a relative direction of AP and FP (forward [clockwise] and backward [anticlockwise]), (2) depending on whether a vertex is removed (removal or nonremoval), and (3) depending on whether FP is a common vertex in more than two adjacent polygons (common vertex or noncommon vertex). For the reader's understanding, some cases are explained as follows:

- Case 1 (forward, nonremoval, and noncommon vertex): As shown in Figure 17.6b, v_1 and v_2 are assigned as AP (blue circle) and FP (red point), respectively. Hypotheses are generated by moving FP along to the GL in which red circles represent new possible positions of v_2.

- Case 2 (backward, nonremoval, and noncommon vertex): As shown in Figure 17.6c, v_3 and v_2 are assigned as AP and FP, respectively. In contrast to case 1, FP is located in backward direction of AP.
- Case 3 (backward, removal, and noncommon vertex): As shown in Figure 17.6d, after removing v_2 (green point), v_3 and v_1 are assigned as AP and FP, respectively. New hypotheses are generated by moving v_1.
- Case 4 (forward, nonremoval, common vertex): As shown in Figure 17.6e, v_2 and v_3 are assigned as AP and FP, respectively. v_3 is a common vertex in Π_A and Π_B. As the position of v_3 changes, shapes of both polygons are changed.
- Case 5 (forward, removal, common vertex): As shown in Figure 17.6f, v_2 and v_4 are assigned as AP and FP, respectively. After v_3 is removed, v_4 is assigned as FP so that the position of v_4 is changed.

17.4 MATCHING AERIAL IMAGES TO 3D BUILDING MODELS

A concept of continuous city modeling is to progressively reconstruct city models by accommodating their changes recognized in spatiotemporal domain, while preserving unchanged structures. A first, critical step for continuous city modeling is to coherently register remotely sensed data taken at different epochs with existing building models. This section presents a new model-to-image registration method using a CGH method to align a single image with existing 3D building models. As shown in Figure 17.7, this model-to-image registration process consists of three steps: (1) feature extraction, (2) similarity measure, and (3) matching and estimating EOPs of a single image. For feature extraction, we propose two types of matching cues: edged corner features representing the saliency of building corner points with associated edges, and contextual relations among the edged corner features within an individual roof. A set of matched corners are found with given proximity measure through geometric hashing, and optimal matches are then finally determined by maximizing the matching cost encoding contextual similarity between matching candidates. Final matched

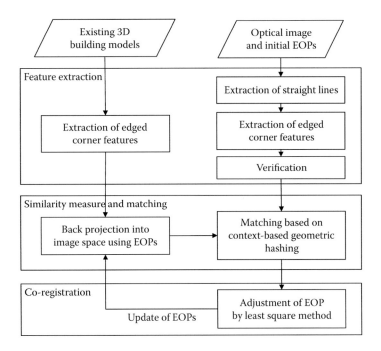

FIGURE 17.7   Flowchart of the proposed model-to-image registration method.

corners are used for adjusting EOPs of the single airborne image by the least square method based on collinearity equations.

17.4.1 FEATURE EXTRACTION

Feature extraction is the first step of the registration task. Feature selection should consider the properties of the given datasets, the application, and the required accuracy. In this study, we use two different types of features: edged corner features, and context features. An edged corner feature, which consists of a corner point and the two associated lines that potentially intersect at this point (*arms*), provides local structure information of a building. In the building models, it is relatively straightforward to extract this feature because each vertex of a building polygon can be treated as a corner and the connected lines as arms. In an image with rich texture information, various corner detectors and line detectors can be used to extract edged corner features. A context feature is defined as a characteristic spatial relation between two edged corner features selected within an individual roof. This context feature is used to represent global structure information so that more accurate and robust matching results can be achieved.

17.4.1.1 Edged Corner Feature Extraction from Image

Edged corner features from a single image are extracted by three separate steps; (1) extraction of straight lines, (2) extraction of corners and their arms, and (3) verification. The process starts with the extraction of straight lines from a single image by applying a straight line detector. We use Kovesi's algorithm, which relies on the calculation of phase congruency to localize, and link edges (Kovesi, 2011). Then, corners are extracted by estimating the intersection of the extracted straight lines, considering the proximity with a given distance threshold T_d. Afterwards, corner arms are determined by two straight lines used to extract the corner with fixed length. This procedure may produce incorrect corners because the proximity constraint is the only one considered. Thus, the verification process removes incorrectly extracted corners based on geometric and radiometric constraints. As a geometric constraint, the inner angle between two corner arms is calculated and investigated to remove corners with sharp inner angles. In general, many of building structures appears in regular shapes following orthogonality and parallelism in which small acute angles are found to be uncommon. Through this process, incorrectly extracted corners are filtered out by applying a user-defined inner angle threshold T_θ. For the radiometric constraint, we analyze the radiometric values (Digital Number [DN]) value or color value) of the left, and right flanking regions ($F_1^L, F_1^R, F_2^L, F_2^R$) of each corner arm with a flanking width (ε) as used in Ok et al. (2012). Figure 17.8a shows a configuration of a corner, its arms, and the concept of the flanking regions. In a correctly extracted corner, the average DN (or color) difference between F_1^L and F_2^R, $\|F_1^L - F_2^R\|$, or between F_1^R and F_2^L, $\|F_1^R - F_2^L\|$, is likely to be small, underlining the homogeneity of two regions, whereas average DN difference between F_1^L and F_2^L, $\|F_1^L - F_2^L\|$, or between F_1^R and F_2^R, $\|F_1^R - F_2^R\|$, should be large enough to underline the heterogeneity of two regions. Thus, we measure two radiometric properties:

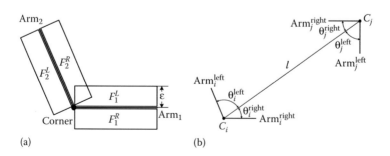

(a)   (b)

FIGURE 17.8 (a) Edged corner feature (corner and its arms) and flanking regions and (b) context feature.

the minimum average DN difference value of two neighbor flanking regions for homogeneity measurement, $D_{\min}^{\text{homo}} = \min\left(\left\|F_1^L - F_2^R\right\|, \left\|F_1^R - F_2^L\right\|\right)$, and the maximum DN difference value of two opposite flanking regions for heterogeneity measurement, $D_{\max}^{\text{hetero}} = \max\left(\left\|F_1^L - F_2^L\right\|, \left\|F_1^R - F_2^R\right\|\right)$. A corner is considered as an edged corner feature if the corner has a smaller $D_{\min}^{\text{homo}}$ than a threshold T_{homo} and if it has a larger $D_{\max}^{\text{hetero}}$ than a threshold T_{hetero}.

To determine thresholds for two radiometric properties, we assume that the intersection points are generated from both correct corners, and incorrect corners; and the two types of intersection points have different distributions with regards to their radiometric properties. As there are two cases (correct corner and incorrect corner) for the average DN difference values, we can use the Otsu's binarization method (Otsu, 1979) to automatically determine an appropriate threshold value. The method was originally designed to extract an object from its background for binary image segmentation based on histogram distribution. It calculates the optimum threshold by separating the two classes (foreground and background) in such a way that their intraclass variance is minimal. In our study, a histogram of homogeneity values (or heterogeneity values) for the entire selection of points is generated, and the optimal threshold for homogeneity (or heterogeneity) is automatically determined by Otsu's binarization method.

17.4.1.2 Context Features

Although an edged corner feature provides only local structure information about a building corner, context features partly impart global structure information related to the building configuration. Context features are set by selecting any two adjacent edged corner features, that is, four angles $(\theta_i^{\text{left}}, \theta_i^{\text{right}}, \theta_j^{\text{left}}, \theta_j^{\text{right}})$ between a line (l) connecting the two corners $(C_i$ and $C_j)$ and their arms $(\text{Arm}_i^{\text{left}}, \text{Arm}_i^{\text{right}}, \text{Arm}_j^{\text{left}}, \text{Arm}_j^{\text{right}})$ as shown in Figure 17.8b. Note that each angle is determined by the relative line connecting any two corners (l). The context feature, which is invariant under scale, translation, and rotation, is used to calculate contextual similarity in the proposed score function.

17.4.2 SIMILARITY MEASUREMENT AND PRIMITIVES MATCHING

Similarity measurement and matching process take place in the image space after the 3D building models are back-projected onto the image space using the collinearity equations with the initial EOPs (or updated EOPs). To find reliable and accurate correspondences between features extracted from a single image and building models, we introduce a CGH method in which the vote counting scheme of a standard geometric hashing is supplemented by a newly developed similarity score function. The similarity score function consists of a unary term and a contextual term. The unary term measures the similarity between edged corner features derived from the image and models, whereas the contextual term measures the geometric property of context features.

17.4.2.1 Geometric Hashing

Geometric hashing, a well-known indexing-based approach, is a model-based object recognition technique for retrieving objects in scenes from a constructed database (Wolfson and Rigoutsos, 1997). In geometric hashing, an object is represented as a set of geometric features such as points and lines, and by its geometric relations that are transformation-invariant under certain transformations. As only local invariant geometric features are used, geometric hashing can handle partly occluded objects. Geometric hashing consists of two main stages: the preprocessing stage, and the recognition stage. The preprocessing stage encodes the representation of the objects in a database and stores them in a hash table. Given a set of object points $(p_k; k = 0,\ldots,n)$, a pair of points $(p_i$ and $p_j)$ is selected as a base pair (Figure 17.9a). The base pair is scaled, rotated, and translated into the reference frame. In the reference frame, the magnitude of the base pair equals 1; the midpoint between p_i and p_j is placed at origin of the reference frame; the vector $\overrightarrow{p_i p_j}$ corresponds to a unit vector of the x axis. The remaining points of the model are located in the coordinate frame based on corresponding base pair (Figure 17.9b). The locations (to be used as index) are quantized by a proper bin size

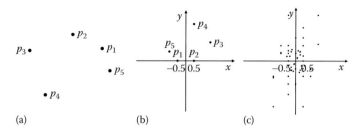

FIGURE 17.9 Geometric Hashing: (a) model points, (b) hashing table with base pair, and (c) all hashing table entries with all base pairs.

and recorded with the form (model ID, used base pair ID) in hash table. For all possible base pairs, all entries of points are similarly recorded in the hash table (Figure 17.9c).

In the subsequent recognition stage, the invariants, which are derived from geometric features in a scene, are used as indexing keys to assess the previously constructed hash table so that they can be matched with the stored models. In a similar way to the preprocessing stage, two points from a set of points in the scene are selected as the base pair. The remaining points are mapped to the hash table, and all entries in the corresponding hash table bin receive a vote. Correspondences are determined by a vote counting scheme, producing candidate matches.

Although geometric hashing can solve matching problems of rotated, translated, and partly occluded objects, it has some limitations. The first limitation is that the method is sensitive to the bin size used for quantization of the hash table. Although a large bin size in the hash table cannot separate between two close points, a small bin size cannot deal with the position error of the point. Second, geometric hashing can produce redundant solutions due to its vote counting scheme (Wolfson and Rigoutsos, 1997). Although it can significantly reduce candidate hypotheses, a verification step or additional fine matching step is required to find optimal matches. Third, geometric hashing has a weakness in cases in which the scene contains many features of similar shapes at different scales and rotations. Without any constraints (e.g., position, scale, and rotation) based on prior knowledge about the model, geometric hashing may produce incorrect matches due to the matching ambiguity. Fourth, the complexity of processing increases by the number of base pairs, and the number of features in the scene (Lamdan and Wolfson, 1988). To address these limitations, we enhance the standard geometric hashing by changing the vote counting scheme to a score function, and by adding several constraints such as scale difference of a base and specific selection of bases.

17.4.2.2 Context-Based Geometric Hashing

In CGH method, we describe the building model objects and the scene by sets of edged corner features. Edged corner features derived from input building models are used to construct the hash table in the preprocessing stage, whereas edged corner features derived from the single image are used in the recognition stage. Each given building model consists of several planes. Thus, in the preprocessing stage, we select two edged corner features that belong to the same plane of the building model as the base pair. It can reduce the complexity of the hashing table and ensures that the base pair retains the spatial information of the plane. The selected base pair is scaled, rotated, and translated to define the reference frame. The remaining edged corner features that belong to the whole building model are also transformed with the base pair. In contrast to the standard geometric hashing, our hashing table contains model IDs, feature IDs of the base pair, the scale of the base pair (the rate of real distance of base pair), an index for member edged corner features, and context features generated by combinations with edged corner features.

Once all possible base pairs are set, the recognition stage tries to retrieve corresponding features based on the designed score function. Two edged corner features from the image are selected as base pair with two constraints: (1) scale constraint, and (2) position constraint. As a constraint on

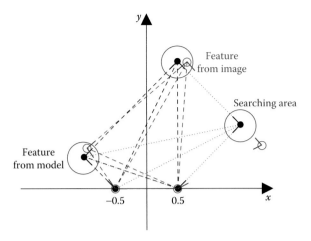

FIGURE 17.10   Context features to be used for calculating score function.

a scale, only those base pairs whose scale is similar to the scale of the base pair in the hash table are considered with an assumption that the initial EOPs provide an approximate scale of the image. Thus, if the scale ratio is smaller than a user defined threshold T_s, the base pair is excluded from the set of possible base pairs. In addition to scale constraint, the possible positions of a base pair can be also restricted with a proper searching space. This searching space can be determined by calculating error propagation with the amount of assumed errors (calculated by the iterative process) for initial EOPs (updated EOPs) of the image and the models. These two constraints reduce the matching ambiguity and the complexity of processing. After the selection of possible base pairs from the image, all remaining edged corner features in the image are transformed on the basis of a selected base pair. Afterwards, the optimal matches are determined by comparing a similarity score. The process starts by generating context features from the model, and the image in a reference frame. Given a model that consists of five edged corner features (black color), ten context features can be generated as shown in Figure 17.10. Note that all edged corner features derived from the model are not matched with edged corner features derived from the image (red color). Thus, only edged corner features, which have corresponding image edged corner features within the search area ($n = 4$ in Figure 17.10), and their corresponding context features ($m = 6$ in Figure 17.10 [red long-dash]) are considered in the calculation of the similarity score function.

The newly designed score function consists of a unary term, which measures the position differences of the matched points, and a contextual term, which measures length and angle differences of corresponding context features, as follows;

$$\text{Score} = \alpha \times \left[w \times \frac{\sum_{i=1}^{n} U(i)}{n} + (1-w) \times \frac{\sum_{i=1}^{n} \sum_{j=1}^{n} C(i,j)}{m} \right] \qquad (17.7)$$

where

$$\alpha = \begin{cases} 0 & \text{if } \dfrac{\# \text{ of matched features}}{\# \text{ of features in the model}} < T_c \\ 1 & \text{else} \end{cases} \qquad (17.8)$$

α is an indicator function in which the minimum number of features to be matched is determined depending on T_c ($T_c = 0.5$ in this study, at least 50% of corners in the model should be matched with corners from the image) so that all features of the model do not need to be detected in the image; n and m are the number of matched edged corner features and context features, respectively; w is a weight value which balances the unary term and the contextual term; in our case, $w = 0.5$ is heuristically selected.

- *Unary term*: The unary term $U(i)$ measures the position distance between edged corner features derived from the model, and the image in reference frame. The position difference $\left\| P_i^M - P_i^I \right\|$ between an edged corner feature in the model and its corresponding feature in the image is normalized by the distance N_i^P calculated by error propagation.

$$U(i) = \frac{N_i^P - \left\| P_i^M - P_i^I \right\|}{N_i} \tag{17.9}$$

- *Contextual term*: This term is designed to measure the similarity between context features in terms of length and four angles. The contextual term is calculated for all context features that are generated from matched edged corner features. For the length difference, $\left\| L_{ij}^M - L_{ij}^I \right\|$, the difference between lengths of context features in the model, and in the image is normalized by length N_{ij}^L of the context feature in the model. For angle differences, the angle difference $\left\| \theta_{ij}^{M_k} - \theta_{ij}^{I_k} \right\|$ between inner angles of a context feature is normalized by the N_{ij}^θ ($N_{ij}^\theta = \pi / 2$).

$$C(i, j) = \frac{N_{ij}^L - \left\| L_{ij}^M - L_{ij}^I \right\|}{N_{ij}^L} + \frac{\sum_{k=1}^4 \left(N_{ij}^\theta - \left\| \theta_{ij}^{M_k} - \theta_{ij}^{I_k} \right\| \right)}{4 \times N_{ij}^\theta} \tag{17.10}$$

For each model, a base pair and its corresponding corners that maximize the score function are selected as optimal matches. Note that if the maximum score is smaller than a certain threshold T_m, the matches are not considered as matched corners. Once all correspondences are determined, the EOPs of the image are adjusted through space resection using pairs of object coordinates of the existing building models, and newly derived image coordinates from the matching process. Values calculated from the similarity score function are used to weight matched pairs. The process continues until matched pairs do not change.

17.5   SEQUENTIAL MODELING OF BUILDING ROOFTOP

This section introduces a novel fusion method to sequentially reconstruct building rooftop models over time by integrating multisensor data. The proposed method integrates modeling cues from airborne imagery to refine the quality of rooftop models produced by an extant algorithm. In this study, a LiDAR-driven building rooftop model (L-Model), which was generated by the method proposed in Section 17.3, is used as an input vector (initial building rooftop model to be refined) of the sequential modeling chain. After extracting image features (I-Lines and I-Corners) from a single image, a set of new model hypotheses are generated by connecting the image feature and existing rooftop model. The modeling cue integration process is developed to progressively rectify geometric errors based on HAT optimization using MDL. A stochastic method, Markov Chain Monte Carlo (MCMC) coupled with Simulated Annealing (SA), is employed to generate model hypotheses and perform a global optimization to find the best solution. Figure 17.11 shows an overall workflow of the proposed method.

17.5.1   FEATURE EXTRACTION FROM OPTICAL IMAGE

As a man-made object that usually contains a certain amount of geometric regularity, the shape of building rooftops can be well described by lines and corners. Images are one of the most appropriate

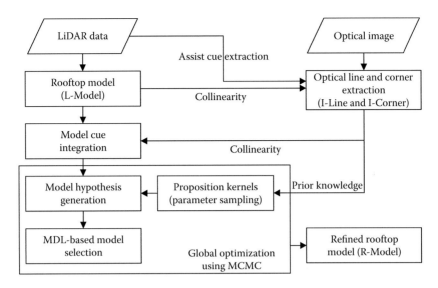

FIGURE 17.11   Flowchart for the proposed refinement algorithm.

data sources for acquiring the geometrically accurate lines and corners. Thus, we use lines and corners extracted from a single image to rectify modeling errors of LiDAR-driven building models. First, we extract modeling cues (lines and corners) from a single image. Then, the extracted 2D modeling cues in image space are transformed into 3D object space using collinearity equation.

17.5.1.1   Modeling Cue Extraction

From a single image, straight lines are extracted using Kovesi's algorithm (Kovesi, 2011) as shown in Figure 17.12a. Moreover, we adopted an algorithm proposed by Chabat et al. (1999) to extract corners. The corner detector is based on the analysis of local anisotropism and identifies corners as points with strong gradient without being oriented in a single dominant direction. The advantage of the adopted corner detector is its ability to detect true location of a corner and orientations of its arms (Figure 17.12b). The information of corners and their arms provide the local structure information of building shapes (*high-level* primitive geometric elements), facilitating a high-level interpretation of building structures. After extracting corners, each corner is assigned to corresponding lines through a greedy search based on its proximity and direction. The extracted modeling cues are used as a prior knowledge for generating rooftop model hypotheses.

17.5.1.2   Transformation between Image Space and Object Space

A transformation between image space and object space is an essential process in establishing a mapping relationship between 3D building models and 2D image features. Given interior orientation

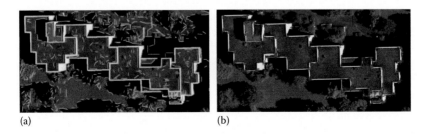

(a)   (b)

FIGURE 17.12   Modeling cue extraction: (a) straight lines and (b) corners and their arms.

parameters and EOPs of the image, the well-known collinearity condition between image space and object space is established. A transformation from object space in 3D to image space in 2D is straightforward with given interior orientation parameters and EOPs. However, its inverse conversion (from 2D to 3D) is known as an ill-posed problem due to the missing one dimension. One possible solution to address this problem is to use stereo images or multiple images. However, in our research framework, our method is limited to a single image due to constraints of multiple-view data availability. Thus, we use the height information of 3D building models to recover the missing third dimension. The process starts by back-projecting L-Model (LiDAR-driven rooftop model) and its associated LiDAR points with attributes including labels (building, nonbuilding) and plane segmentation IDs into the image space. I-Lines (line extracted from the imagery) corresponding to L-Model are determined using a proximity criterion; A I-Line is assigned to a line of L-Model as a conjugate line pair if the I-Line is found within a searching space (minimum bounding box) generated from the L-Model line projected onto the image space. Then, 3D coordinates of I-Lines are calculated by projecting 2D I-Lines onto their corresponding roof planes (L-Model) using the collinearity equation; starting and ending points of a I-Line is transformed following collinearity rays, each of which is intersected by the corresponding roof plane to calculate the parameters of 3D line in the object space. In a similar way, 2D coordinates of I-Corners (i.e., corners and their arms extracted from the imagery) are computed into 3D object space. It is noted that an I-Line can be shared by multiple roof planes. In this case, multiple 3D lines are calculated in the object space for a single I-Line, all of which will be considered as the modeling cues for generating the rooftop hypotheses.

17.5.2 Modeling Cue Integration

After all the image modeling cues (I-Lines and I-Corners) are extracted and transformed in the object space, the integration of I-Lines and I-Corners with existing rooftop models (L-Models) is conducted in the object space. This process establishes spatial relationships between modeling cues derived from two different data sources.

Suppose that L-Model, I-Lines, and I-Corners are denoted as a set of model lines $L^M = \{L_i^M; i = 0,...,l\}$, image lines $L^I = \{L_j^I; j = 0,...,m\}$, and image corners $C^I = \{C_k^I; k = 0,...,n\}$ where l, m and n represent the number of model lines, image lines, and image corners, respectively (Figure 17.13a). The first step of the cue integration is to identify their spatial relations by investigating a spatial proximity and geometric configuration among L^I, C^I, and L^M. Given a model line L_i^M, a set of image cues are determined as its conjugate cue pairs if they satisfy following spatial cue relations:

- *Image cues to model line relations*: An image cue (i.e., L_j^I or C_k^I) has assigned its membership to a model line L_i^M if a spatial proximity measured between the image cues and a model line in the object space is less than a prespecified threshold. A I-Line can belong to multiple model lines if it spans multiple model lines.
- *Image line to corner relations*: The image cues meeting the previous "image cues to model line relation are further filtered in order to determine an "image line to corner pair." An image line L_j^I is paired with an image corer C_k^I if the orientation of L_j^I is coaligned with the one of orientations of the edged arms C_k^I. Note that C_k^I is excluded for the hypothesis generation if it is not paired with any image line, whereas L_j^I is accepted even though its corner pair does not exist.

Once all the memberships of the image cues to each model line are found, the next cue integration process is to *physically* represent the spatial relations among the paired cues. This topological cue relation is conducted by generating *virtual* lines to connect the paired cues (Figure 17.13). When a

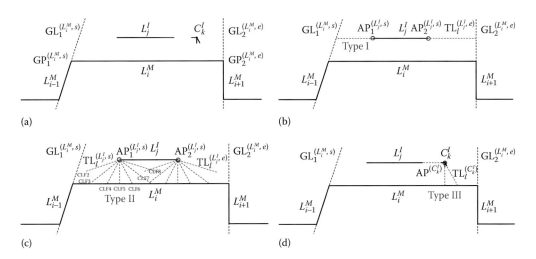

FIGURE 17.13 Topological lines (red) from a given configuration (a) (Type I: between guide lines and I-Lines (b), Type II: between I-Line and L-Model (c), and Type III: prior-guided (d)).

model line L_i^M is given and its paired image cues, L_j^I and C_k^I, are found, we define two different types of virtual lines as follows:

- *Guide line*: Two guide lines (GL), $GL_1^{(L_i^M, s)}$ and $GL_2^{(L_i^M, e)}$, are defined as the infinite lines that line parameters are identical to L_{i-1}^M and L_{i+1}^M (L_i^M's neighboring lines) and are generated from starting and ending points of L_i^M, respectively (Figure 17.13a). The starting (s) and ending point (e) of L_i^M is defined through guide points, $GP_1^{(L_i^M, s)}$ and $GP_2^{(L_i^M, e)}$.
- *Topological line*: A topological line (TL) TL_k is a virtual line to establish spatial relations between image cues and L_i^M. An AP is defined as a starting point of newly generated TL. $AP_1^{(L_j^I, s)}$ and $AP_2^{(L_j^I, e)}$ are the APs defined from the starting (s) and ending point (e) of an image line L_j^I, whereas $AP_1^{(C_k^I)}$ is defined from the image corner point C_k^I. Starting from these APs, a set of infinite lines are generated with CLF line slope angles determined. A line segment generated by intersecting the infinite lines with L_i^M or GL ($GL_1^{(L_i^M, s)}$ and $GL_2^{(L_i^M, s)}$) is considered as the TL to connect the image cues to its paired model line (Figure 17.13b–d).

The generation of TLs fully depends on a spatial configuration among modeling cues, L^M, L^I, and C^I. Three different types of TLs are defined as follows:

- *Type I*: This TLs, $TL_k^{(L_j^I, s)}$ and $TL_k^{(L_j^I, e)}$, are generated for connecting lines between an image line L_j^I and GL, $GL_1^{(L_i^M, s)}$ and $GL_2^{(L_i^M, e)}$, that are created from two APs of a model line L_i^M (Figure 17.13b). The direction of TLs is the same as one of L_j^I.
- *Type II*: Type II TLs are defined for connecting an image line L_j^I to a modeling line L_i^M (Figure 17.13c). A set of Type II TLs are generated from two APs, $AP_1^{(L_j^I, s)}$ and $AP_2^{(L_j^I, e)}$ of an image line L_j^I. Each TLs are generated using one of the representative angles of CLF. The angle is determined by quantizing the orientation distribution of lines extracted from a targeted building region captured in an airborne imagery.
- *Type III*: Type III TLs are generated for connecting an image corner C_k^I to its paired modeling line L_i^M. In Type III TL, the corner point C_k^I serves as an AP $AP_1^{(C_k^I)}$, and its arms are used as *a priori* knowledge to generate TL; each TL is generated following one of orientations of the edged arms C_k^I (Figure 17.13d).

17.5.3 Hypothesis Generation and Global Optimization

17.5.3.1 Model Hypothesis Generation

In the previous section, multiple topological relations integrating image features with existing model lines were established. The model hypothesis generation is a process to generate probable models reflecting the contribution of image features to improve existing models. Given L-Model lines and I-Lines, I-Corners, GL, and TLs, hypotheses are generated by sampling possible combinations. A hypothesis with one of possible combinations is generated by finding intersection points between TL and I-Line, between TL and L-Model line, and/or between guide line and I-Line. Figure 17.14 shows some examples generated with different combination sets for Figure 17.13a.

17.5.3.2 Global Optimization in MCMC Framework

Let $H = \{h_1,\ldots,h_m\}$ denote a set of all possible roof hypotheses. The optimal model H^* is selected through the direct comparison of DL values for all model candidates, in which H^* has the minimum DL (Section 17.3.3.1). However, it is not possible to explore a large hypothetical space to find the optimal solution. For example, the number of possible hypotheses at a certain configuration, even when new configurations are generated by a model line and an image line using Type II TLs, is 64 (8 directions for $AP_1^{(L_j^I,\,s)}$ and 8 directions for $AP_2^{(L_j^I,\,e)}$). Considering all model lines, image lines, and corners, it is too computationally expensive to compare all solutions. Thus, we employ a stochastic method, MCMC coupled with SA, to find a global optimization in the large hypothetical space. Compared with naive hypothesis generation (Section 17.3.3), the stochastic hypothesis generation can reduce computing time because it does not compare all possible hypotheses. Furthermore, the optimal model H^* is not sensitive to an initial building rooftop configuration because the optimization process is conducted by random sampling.

MCMC, first introduced by Metropolis et al. (1953), is a method for obtaining a sequence of random samples from a probability distribution for which direct sampling is difficult. As the name suggests, MCMC is a combination of concepts of Monte Carlo sampling and Markov Chain. The Monte Carlo sampling is a method to generate a set of samples from a target density to compute integrals. Markov Chain refers to a sequence of random variables generated by a Markov process whose transition probabilities between different values in the state space depend only on the random

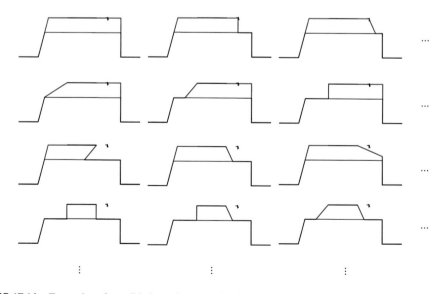

FIGURE 17.14 Examples of possible hypotheses (red) with respect to a given configuration (black).

variable's current state, $P(x_t \mid x_{t-1},...,x_0) = P(x_t \mid x_{t-1})$. By combining these two concepts, the MCMC sampler can effectively explore a configuration space and approximates a target density.

In this study, MCMC coupled with SA is used to solve our optimization problem by simulating a discrete Markov Chain (X_t), $t \in N$ on the configuration space Φ, which converges towards an invariant measure specified by the *DL*. The MCMC sampler performs transitions from one state to another, which can be managed by kernels Q_m (Section 17.5.3.3). If a rooftop configuration h transits to h' according to proposition kernels Q_m $(h \rightarrow h')$, the move between these configurations is accepted with the following probability:

$$A(h,h') = \min \left\{ 1, \frac{DL^{-1}(h')Q_m(h \mid h')}{DL^{-1}(h)Q_m(h' \mid h)} \right\} \tag{17.11}$$

If $A(h, h') = 1$, the new configuration h' is added to the Markov Chain. Otherwise, it remains at h.

A SA is then embedded in the MCMC to find the optimal configuration with the minimum global *DL* value. To perform the SA, the description length *DL* is replaced by $DL_{T_t} = DL(h)^{1/T_t}$, where T_t is the temperature parameter, which tends to zero as t approaches ∞. The acceptance rate is as follows:

$$A(h,h') = \min \left\{ 1, \frac{\left(DL^{-1}(h') \right)^{1/T_t} Q_m(h \mid h')}{\left(DL^{-1}(h) \right)^{1/T_t} Q_m(h' \mid h)} \right\} \tag{17.12}$$

A logarithmic decrease ensures the convergence to the global optimum for any initial configuration h_0. In practice, a geometric cooling scheme is preferred to accelerate the process and to give an approximate solution close to the optimal one as follows:

$$T_t = T_0 \alpha^t \tag{17.13}$$

where T_0, α, and t are the initial temperature, the decreasing coefficient, and the number of iterations, respectively. A slight adaption of the schedule was made in which the temperature is updated in every kth iteration of the algorithm.

17.5.3.3 Proposed Kernels

MCMC algorithms typically require the design of proposal mechanisms to propose candidate hypothesis. Appropriately designed proposal kernels let MCMC algorithm quickly converge by proposing configurations of interest. In our approach, three types of kernels (Q_1, Q_2, and, Q_3) are defined to perform moves between different configurations as follows:

- *Kernel Q_1 (with Type I TL)*: This kernel is designed to replace an L-Model line with an I-Line. The kernel does not add any vertex where an ending point of L-Model line moves to a point generated by intersecting the extension line of I-Line with GL. Thus, L-Model fully contains I-Line's properties (slope and position).
- *Kernel Q_2 (with Type II TL)*: This kernel is designed to change the shape at an ending point of an I-Line. This kernel adds vertices to a new configuration to represent the shape changes. The shape changes occur in two intersection points between I-Line and TL, and between TL and L-Model. The direction of the TL is determined by sampling one representative angle of CLF. Thus, a new configuration contains properties of I-Line, L-Model, and TL. In this Kernel, no prior knowledge for corners is used.
- *Kernel Q_3 (with Type III TL)*: This kernel is an advanced form of kernel Q_2, which adds vertices with prior knowledge for corners. A corner, evidence of a sudden change in building structure, serves to represent shape changes. Directions of the corner arms are used to guide the direction of TL. If there are more than two arms, one direction of the TLs is determined by randomly sampling the directions of the arms.

L-Model line L_i^M and I-Line L_j^I are randomly selected, and topological relations are established. The move from a configuration H to H' is realized by sampling Q_M with uniform distribution for each AP. Kernel Q_1 and Q_2 can be selected for all I-Lines, whereas Kernel Q_3 is only selected in the case that corners exist. After selecting the kernel type for each AP, a new building rooftop configuration is generated as explained in Section 17.5.3.1.

17.6   EXPERIMENTAL RESULTS

17.6.1   DATA AND EVALUATION METRICS

17.6.1.1   Data

The proposed methods were tested over the International Society for Photogrammetry and Remote Sensing (ISPRS) benchmark datasets provided by the ISPRS Commission III, WG3/4 (Rottensteiner et al., 2014). The ISPRS benchmark datasets consist of three subregions (Area 1, Area 2, and Area 3) of Vaihingen dataset, and two subregions (Area 4 and Area 5) of the Toronto dataset (Figure 17.15) with multisensor data, including aerial images and airborne laser scanning data. Moreover, the ISPRS provides reference datasets (building footprints and rooftop models in 3D) reconstructed by the manual stereo plotting method. The Vaihingen dataset contains typical European building types showing various degrees of shapes including gable, hip roof, and their mixture structures. The Toronto dataset contains representative scene characteristics of a modern mega city in North America including a mixture of low- and high-story building and a complex cluster of high-rise buildings. More detailed explanation on the acquisition setting and data characteristics can be found in Rottensteiner et al. (2014).

17.6.1.2   Evaluation Metrics

Performance evaluation is the process of analyzing the performance of a building reconstruction algorithm by comparing its results with the reference models or the results produced by other algorithms. In this study, we measure the performance characteristics of the proposed algorithms by combining multiple evaluation indices including confusion matrix, geometric accuracy (root mean square error [RMSE]), shape similarity, and angle difference.

A confusion matrix, also known as an error metric, has been often used for assessing the performance of an algorithm, typically spatial object detection, and supervised learning. As a quality measure for object reconstruction algorithms, each column of the matrix represents the instances in a reconstructed object (a predicted class in supervised learning), whereas each row represents the in a reference object (an actual class) or vice-versa. In the confusion matrix with two rows and two columns, we can compute the number of False Positive (FP), False Negative (FN), True Positive (TP),

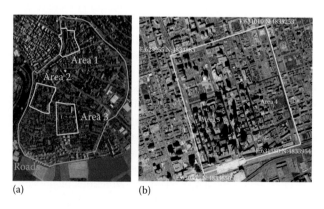

(a)   (b)

FIGURE 17.15   Test datasets: (a) Vaihingen and (b) downtown Toronto.

17.6.3 Results for Matching Aerial Image to 3D Building Models

The proposed CGH-based registration method was tested on Toronto dataset. Two different types of reference building models were prepared by: (1) a manual digitization process conducted by human operators, and (2) using a method proposed in Section 17.3 from airborne LiDAR point clouds. These two building models were used to investigate their respective effects on the performance of the CGH method. A total of 16 check points, which were evenly distributed throughout the images, were used to evaluate the accuracy of the EOPs.

From the aerial image, a total of 90,951 straight lines were extracted, and 258,486 intersection points were derived by intersecting any two straight lines found within 20 pixels of proximity constraint. Out of these, 57,767 intersection points were selected as edged corner features following the removal of 15%, and 60% of intersection points using geometric constraint, and radiometric constraints, respectively (Table 17.3). Figure 17.18b shows edged corner features extracted from the aerial image. As many of the intersection points are not likely to be corners, the majority of them were removed. The method correctly detected corners and arms in most cases even though some corners were visually difficult to detect due to their low contrasts.

After the existing building models were back-projected onto the image using error-contained EOPs, edged corner features were extracted from the vertices of the building models in the image space (Figure 17.18c and d). It should be noted that two different datasets were used as the existing building models. Some edged corner features extracted from both existing building models were not observed in the image due to occlusions caused by neighbor building planes. Moreover, some edged corner features, in particular those extracted from LiDAR-driven building models, do not

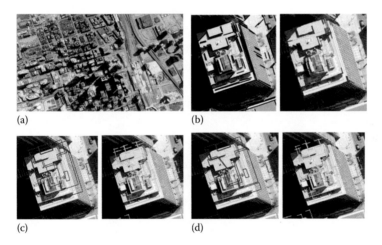

(a) (b)

(c) (d)

FIGURE 17.18 Features extraction: (a) image and existing building models, (b) features from image, features from (c) manually digitized models, and (d) LiDAR-generated models.

TABLE 17.3
Extracted Features and Matched Features

	Image		Existing Building Models	
	Intersections	Corners	Manually Digitized Building Models	LiDAR-Driven Building Models
# of features	258,486	57,767	8,895	7,757
# of matched features	–	–	693	381

TABLE 17.4
Parameter Setting

Feature Extraction				Geometric Hashing			
T_d	T_θ	T_{homo}	T_{hetero}	T_s	T_p	T_c	T_m
20 pixel	10°	Automatic	Automatic	0.98	Automatic	50%	0.6

match with the edged corner features extracted from the image due to modeling errors caused by irregular point distribution, occlusions, and the reconstruction mechanism. Thus, correspondences between edged corner features from the image and from the existing building models are likely to be partly established.

The proposed CGH method was applied to find correspondences between features derived from the image and from existing building models. When manually digitized building models are used as building models, a total of 693 edged corner features were matched using the parameters given in Table 17.4. It is noted that only models whose vertices were greater than T_c were considered to find possible building matches. For LiDAR-driven building models, only 381 edged corner features were matched (Table 17.3). The results show that more edged corner features are matched when manually digitized building models were used as the existing building models than when LiDAR-driven building models were used.

Based on matched edged corner features, EOPs for the image were calculated by applying the least square method based on collinearity equations. For qualitative assessment, the existing models were back-projected to the image with refined EOPs. Each column of Figures 17.19 and 17.20 shows back-projected building models with error-contained EOPs (a), matched edged corner features (b), and back-projected building models with refined EOPs (c). In the figures, boundaries of the existing building models are well matched to building boundaries in the image with refined EOPs.

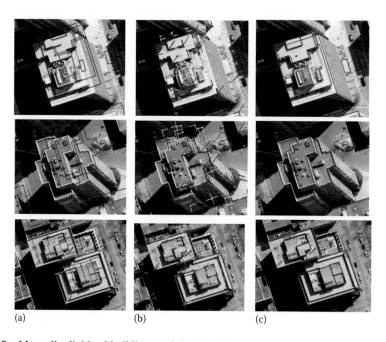

(a)   (b)   (c)

FIGURE 17.19 Manually digitized building models: (a) with error-contained EOPs, (b) matching relations (purple) between edged corner features extracted from the image (blue) and from the models (cyan), and (c) with refined EOPs.

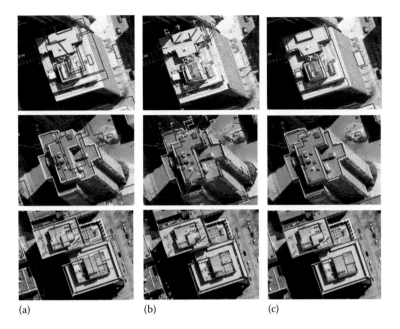

FIGURE 17.20 LiDAR-driven building models: (a) with error-contained EOPs, (b) matching relations (purple) between edged corner features extracted from the image (blue) and from the models (cyan), and (c) with refined EOPs.

In our quantitative evaluation, we assessed the RMSE of check points back-projected onto the image space using refined EOPs (Table 17.5). When reference building models were used as the existing building models, the results show that the average difference in x and y directions are -0.27 and 0.33 pixels, respectively, with RMSE of ±0.68 and ±0.71 pixels, respectively. The results with LiDAR-driven buildings models show that the average differences in x and y directions are -1.03 and 1.93 pixels, with RMSE of ±0.95 and ±0.89 pixels, respectively. Although LiDAR-driven building models are used, the accuracy of the EOPs is less than 2 pixels in image space (approximately 30 cm in ground sample distance). Considering that the point space (resolution) of the input airborne LiDAR dataset is larger than 0.3 m, the refined EOPs provide a greater accuracy for engineering applications.

To evaluate the effect on context feature, we set weight parameter w in score function (Equation 17.7) as 1 and 0.5, respectively, and then compared the results. When $w = 1$, the score function considers only the unary term without the effect of the contextual term so that the contextual force is ignored. As shown in Table 17.6, the results show that registration with only unary terms causes considerably low accuracy in both cases. In particular, with LiDAR-driven models, the accuracy is heavily affected. These results indicate that the use of context features has a positive effect on resolving the matching ambiguity, and thus improving the EOP accuracy by reinforcing contextual force.

TABLE 17.5

Quantitative Assessment with Check Points (Unit: Pixel)

Error-Contained Initial EOPs				Refined EOPs with Manually Digitized Building Models				Refined EOPs with LiDAR-Driven Building Models			
Ave.		RMSE		Ave.		RMSE		Ave.		RMSE	
x	y	x	y	x	y	x	y	x	y	x	y
20.51	−24.81	±6.64	±8.22	−0.27	0.33	±0.68	±0.71	−1.03	1.93	±0.95	±0.89

TABLE 17.6

Effect on Pair-Wise Feature (Unit: Pixel)

	Manually Digitized Building Models					LiDAR-Driven Building Models				
	No. of Matched Features	Ave.		RMSE		No. of Matched Features	Ave.		RMSE	
		x	y	x	y		x	y	x	y
Unary term only ($w = 1$)	542	−0.67	−0.39	±1.56	±1.84	361	5.98	1.17	±7.72	±5.31
Unary term and contextual term ($w = 0.5$)	693	−0.27	0.33	±0.68	±0.71	381	−1.03	1.93	±0.95	±0.89

17.6.4 RESULTS FOR SEQUENTIAL MODELING

The proposed sequential fusion method was tested on the Vaihingen and the Toronto datasets. Table 17.7 shows evaluation results using confusion matrix. For the initial models, the average of the completeness, correctness, and quality were 91.5%, 97.4%, and 89.2%, respectively. After applying the proposed sequential method, the average of the completeness, correctness, and quality were 93.9%, 96.8%, and 91.0%, respectively. The correctness was slightly deteriorated by small increase of false positives, whereas higher completeness and quality were achieved by a large decrease of *FN*s and an increase of TPs. The results imply that a loss of correctness is inevitable, but the proposed sequential fusion method can improve the overall quality of LiDAR-driven building models.

Angle-based and shape-based evaluations were conducted for 2D building boundaries (Table 17.8) and for 3D rooftop planes with 50% overlap (Table 17.9). For 2D building boundaries, the average orientation error reduced from 1.17° to 0.63°; the average Hausdorff distance decreased from 1.81 m to 0.48 m; the average turning function distance decreased from 0.042 to 0.007, respectively. For 3D rooftop planes, the average angle difference, Hausdorff distance, and turning function distance were improved by 0.3°, 0.3 m, and 0.004, respectively.

More specifically, Figure 17.21 shows distributions of angle difference, Hausdorff distance, and turning function distance for initial building models and for refined building models, respectively. The figures clearly show that the proposed sequential fusion method has positive effects for the corrections of orientation errors and shape deformations of LiDAR-driven building models.

TABLE 17.7

Assessment by Area-Based Confusion Matrix

		Between Reference and Initial Model			Between Reference and Refined Model		
Dataset	Area	Comp$_{Area}$ (%)	Corr$_{Area}$ (%)	Quality$_{Area}$ (%)	Comp$_{Area}$ (%)	Corr$_{Area}$ (%)	Quality$_{Area}$ (%)
Vaihingen	Area 1	90.6	98.8	89.6	92.2	98.6	91.0
	Area 2	91.3	99.7	91.0	95.8	98.6	94.5
	Area 3	88.6	99.7	88.4	92.0	98.5	90.8
Toronto	Area 4	93.7	96.9	90.9	95.3	96.4	92.0
	Area 5	93.1	92.0	86.1	94.2	91.8	86.9
Total		91.5	97.4	89.2	93.9	96.8	91.0

TABLE 17.8
Assessment by Angle-Based Index and Shape-Based Indices for 2D Building Boundary

Dataset	Area	Angle Difference (deg)	Hausdorff Distance (m)	Turning Function Distance	Angle Difference (deg)	Hausdorff Distance (m)	Turning Function Distance
		Between Reference and Initial Model			Between Reference and Refined Model		
Vaihingen	Area 1	1.32	1.33	0.049	0.65	1.32	0.043
	Area 2	1.62	1.26	0.040	0.82	0.83	0.029
	Area 3	0.59	0.93	0.031	0.70	0.72	0.028
Toronto	Area 4	1.30	2.44	0.046	0.61	2.14	0.039
	Area 5	1.04	3.10	0.046	0.35	2.37	0.035
Total		1.17	1.81	0.042	0.63	1.48	0.035

TABLE 17.9
Assessment by Angle-Based Index and Shape-Based Indices for 3D Rooftop Polygons

Dataset	Area	Angle Difference (deg)	Hausdorff Distance (m)	Turning Function Distance	Angle Difference (deg)	Hausdorff Distance (m)	Turning Function Distance
		Between Reference and Initial Model			Between Reference and Refined Model		
Vaihingen	Area 1	0.78	0.46	0.020	0.82	0.43	0.021
	Area 2	1.11	1.77	0.041	0.56	0.65	0.030
	Area 3	0.44	0.48	0.016	0.58	0.42	0.024
Toronto	Area 4	1.30	1.38	0.040	0.72	1.28	0.041
	Area 5	0.91	1.75	0.047	0.39	1.59	0.031
Total		0.91	1.17	0.033	0.61	0.87	0.029

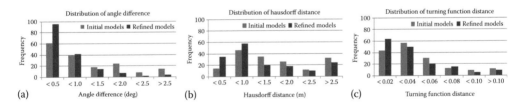

FIGURE 17.21 Distributions of (a) angle difference, (b) Hausdorff distance, and (c) turning function distance for initial models and refined models.

17.7 CONCLUSION

The current chapter aimed to continuously reconstruct 3D building rooftop models using multi-sensor data. To achieve the goal, we identified three critical steps and provided reasonable and promising solutions for each topic. The first step towards continuous city modeling was to devise a method to reconstruct robust and accurate regularized building rooftop models, regardless of scene complexity. The second step was to automatically register newly acquired image data with existing building models without any labor-intensive manual process, thus providing a reasonable solution for the sequential fusion process. The third step was to construct a method that can update the existing building models using multisensor data in a timely way. The following sections provide the conclusions of each critical topic toward continuous city modeling.

In section 17.3, we proposed an automatic building reconstruction method using LiDAR data that covered low-level modeling cue extraction to reconstruction of realistic 3D rooftop models. A hierarchical strategy for modeling cue extraction made it possible to effectively collect modeling evidence from complex building structures. The hypothesis generation method generated regular-shaped candidate hypotheses by implicitly designed rules from various configurations of building rooftop. MDL was used as a criterion for model selection to choose the best model among possible candidate hypotheses. In particular, the proposed implicit regularization provided flexibility for describing more complex rooftop models while preserving building regularity. The experimental results showed that the proposed building reconstruction method can robustly produce regularized building rooftop models regardless of building complexity. Area-based evaluations using confusion matrix were the average completeness of 91.5%, correctness of 97.4%, and quality of 89.2%, whereas object-based evaluation was the average completeness of 93.8%, correctness of 96.9%, and quality of 91.3% for roof planes larger than 10 m². However, results of object-based evaluations indicated that small size rooftops are not effectively extracted. Angle-based index showed that angle difference is approximately 1.17° compared with reference models. These results demonstrated that our proposed building reconstruction method is a reasonable solution. Three main modeling errors (shape deformation, boundary displacement, and orientation errors) were observed in the rooftop models reconstructed by our proposed method. The modeling errors of the proposed building rooftop modeling were caused by the inherent characteristics of LiDAR data. Thus, a fusion method to integrate complementary data (i.e., image data) with LiDAR data is required to rectify the modeling errors.

Section 17.4 proposed a new registration method, CGH, to align a single image with existing 3D building models. As an essential step for continuous city modeling, the newly acquired single image was aligned with large-scale existing building models without any labor-intensive manual processes. We solved the registration problems by adopting geometric hashing, a well-known model-to-image matching method. To compensate the limitations of standard geometric hashing methods, new features, namely the cornered edges and context feature, were proposed. The main contribution of this registration was the development of the new score function in the CGH-based registration method, which reinforce context forces to improve matching performance. Experimental results showed that the overall registration accuracy was under 2 pixels (under 30 cm in ground sample distance) over two different scenes that have different complexity. The amount of registration error is reasonable and acceptable for the sequential refinement process considering that airborne LiDAR point space is approximately 30 cm, and its position accuracy is approximately 15 cm. However, the quality of building models directly affected the accuracy of EOPs. In terms of feature used, the use of the context feature played a significant role in the matching process. It is due to that fact that contextual term made it possible to provide global information of building structures.

Section 17.5 proposed a sequential fusion method to refine existing LiDAR-driven 3D building models. The modeling errors of LiDAR-driven building rooftop models were progressively refined by image features derived from images in a HAT framework. A new method to connect the existing LiDAR-driven building model with the image cues was proposed to generate possible hypotheses. MDL was again used as a criterion for model selection. MCMC coupled with SA was employed for evaluation of global optimization. Experimental results on simulated data showed that the proposed fusion method can effectively recover the large amount of modeling errors that were often observed in LiDAR-driven building models. Confusion matrix-based evaluations showed that the completeness and quality were considerably improved, compared with those of initial LiDAR-driven models, whereas the correctness is slightly degenerated. Angle-based evaluation showed that orientation error, which accidently occurs in LiDAR-driven building models, can be refined by accurate orientations derived from image. Shape-based evaluations showed that partly deformed shapes in

LiDAR-driven models were improved by the sequential fusion method. The proposed refinement process provided a way that existing building models can be effectively reused by accommodating their changes recognized in temporal domain.

REFERENCES

Ameri, B., 2000. Automatic recognition and 3D reconstruction of buildings through computer vision and digital photogrammetry. PhD Thesis, University of Stuttgart, Germany.

Arkin, E., Chew, L. P., Huttenlocher, D. P., Kedem, K., Mitchell, J. S. B., 1991. An efficiently computable metric for comparing polygonal shapes. *IEEE Transactions on Pattern Analysis and Machine Intelligence*, 13(3):209–215.

Avbelj, J., Iwaszczuk, D., Müller, R., Reinartz, P., Stilla, U., 2015. Coregistration refinement of hyperspectral images and DSM: An object-based approach using spectral information. *ISPRS Journal of Photogrammetry and Remote Sensing*, 100:23–34.

Avbelj, J., Iwaszczuk, D., Stilla, U., 2010. Matching of 3D wire-frame building models with image features from infrared video sequences taken by helicopters or UAVs. *International Archives of the Photogrammetry, Remote Sensing and Spatial Information Sciences*, 38(3B):149–154.

Awrangjeb, M., Zhang, C., Fraser, C. S., 2013. Automatic extraction of building roofs using LIDAR data and multispectral imagery. *ISPRS Journal of Photogrammetry and Remote Sensing*, 83:1–18.

Brenner, C., 2010. Building extraction. In *Airborne and Terrestrial Laser Scanning*, George, V. and Hans-Gerd, M. (Eds.), Whittles Publishing, Dunbeath, UK, 2010. pp.169–212.

Brown, L. G., 1992. A survey of image registration techniques. *ACM Computing Surveys*, 24:326–376.

Chabat, F., Yang, G. Z., Hansell, D. M., 1999. A corner orientation detector. *Image and Vision Computing*, 17:761–769.

Chen, L., Teo, T., Rau, J., Liu, J., Hsu, W., 2005. Building reconstruction from LiDAR data and aerial imagery. In *Proceedings of the IEEE International Geoscience and Remote Sensing Symposium*, Seoul, South Korea, pp. 2846–2849.

Cheng, L., Tong, L., Chen, Y., Zhang, W., Shan, J., Liu, Y., Li, M., 2013. Integration of LiDAR data and optical multi-view images for 3D reconstruction of building roofs. *Optics and Lasers in Engineering*, 51(4):493–502.

Demir N., Baltsavias, E., 2012. Automated modeling of 3D building roofs using image and LiDAR data. In *ISPRS Annals of the Photogrammetry, Remote Sensing and Spatial Information Sciences*, XXII ISPRS Congress, Melbourne, Australia, I(4):35–40.

Ding, M., Lyngbaek, K., Zakhor, A., 2008. Automatic registration of aerial imagery with untextured 3D LiDAR models. In *IEEE Computer Society Conference on Computer Vision and Pattern Recognition (CVPR)*, Anchorage, Alaska.

Dorninger, P., Pfeifer, N., 2008. A comprehensive automated 3D approach for building extraction, reconstruction, and regularization from airborne laser scanning point clouds. *Sensors*, 2008(8):7323–7343.

Douglas, D., Peucker, T., 1973. Algorithms for the reduction of the number of points required to represent a digitized line or its caricature. *The Canadian Cartographer*, 10(2):112–122.

Economic Dashboard-Annual Summary, 2015. http://www.toronto.ca/legdocs/mmis/2015/ed/bgrd/backgroundfile-76322.pdf (accessed on July 26 , 2017).

Eugster, H., Neibiker, S., 2009. Real-time georegistration of video streams from mini or micro UAS using digital 3D city models. In *6th International Symposium on Mobile Mapping Technology*, Presidente Prudente, Sao Paulo, Brazil.

Fischler, M. A., Bolles, R. C., 1981. Random sample consensus: A paradigm for model fitting with applications to image analysis and automated cartography. *Communications of the ACM*, 24(6):381–395.

Fonseca, L. M. G., Manjunath B. S., 1996. Registration techniques for multisensor remotely sensed imagery. *Photogrammetric Engineering & Remote Sensing*, 62(9):1049–1056.

Frueh, C., Russell, S., Zakhor, A., 2004. Automated texture mapping of 3D city models with oblique aerial imagery. In *Proceeding of the 2nd International Symposium on 3D Data Processing, Visualization, and Transmission, (3DPVT'04)*, Thessaloniki, Greece.

Haala, N., Kada, M., 2010. An update on automatic 3D building reconstruction. *ISPRS Journal of Photogrammetry and Remote Sensing*, 65:570–580.

Hall, D. L., Llinas, J., 1997. An introduction to multisensor data fusion. *Proceedings of the IEEE*, 85(1):6–23.

HERE 360, 2015 (accessed on July 26, 2017). HERE updates maps worldwide for Android, iOS and Windows. http://360.here.com/2015/08/04/here-updates-maps-for-android-ios-and-windows/

Hsu, S., Samarasekera, S., Kumar, R., Sawhney, H., S., 2000. Pose estimation, model refinement, and enhanced visualization using video. In *Proceedings of IEEE International conference on Computer Vision and Pattern recognition*, Hilton Head, SC, pp. 488–495.

Hu, J., You, S., Neumann, U., 2006. Integrating LiDAR, aerial image and ground image for complete urban building model. In *Proceedings of the Third International Symposium on 3D Data Processing, Visualization and Transmission* (3DPVT'06), Chapel Hill, NC.

Huang, H., Brenner, C., Sester, M., 2013. A generative statistical approach to automatic 3D building roof reconstruction from laser scanning data. *ISPRS Journal of Photogrammetry and Remote Sensing*, 79:29–43.

Huttenlocher, D. P., Klanderman, G. A., Rucklidge, W. J., 1993. Comparing images using the Hausdorff distance. *IEEE Transaction on Pattern Analysis and Machine Intelligence*, 15(9):850–863.

International Herald Tribune, 2008. (accessed on July 26, 2017). UN says half the world's population will live in urban areas by end of 2008. https://web.archive.org/web/20090209221745/http://www.iht.com/articles/ap/2008/02/26/news/UN-GEN-UN-Growing-Cities.php

Jung, J., 2016. Continuous modeling of 3D building rooftops from airborne LiDAR and Imagery. PhD Thesis, York University, Ontario, Canada.

Jung, J., Sohn, G., Bang, K., Wichmann, A., Armenakis, C., Kada, M., 2016. Matching aerial images to 3D building models using context-based geometric hashing. *Sensors*, 16(6):932.

Jung, J., Jwa, Y., Sohn, G., 2017. Implicit regularization for reconstructing 3D building rooftop models using airborne LiDAR data. *Sensors*, 17(3):621.

Jwa, Y., Sohn, G., Cho, W., Tao, V., 2008. An Implicit geometric regularization of 3D building shape using airborne LiDAR data. *International Archives of the Photogrammetry, Remote Sensing and Spatial Information Sciences*, 37(B3A):69–76.

Kada, M., McKinley, L., 2009. 3D Building reconstruction from LiDAR based on a cell decomposition approach. *International Archives of the Photogrammetry, Remote Sensing and Spatial Information Sciences*, 38(3/W4):47–52.

Kada, M., Wichmann, A., 2012. Sub-surface growing and boundary generalization for 3D building reconstruction. *ISPRS Annals of the Photogrammetry, Remote Sensing and Spatial Information Sciences*, I(3):223–238.

Kim, C., Habib, A., 2009. Object-based integration of photogrammetric and LiDAR data for automated generation of complex polyhedral building models. *Sensors*, 9:5679–5701.

Kovesi, P. D., 2011. MATLAB and octave functions for computer vision and image processing. Centre for Exploration Targeting, School of Earth and Environment, The University of Western Australia.

Lafarge, F., Descombes, X., Zerubia, J., Pierrot-Deseilligny, M., 2010. Structural approach for building reconstruction from a single DSM. *IEEE Transactions on Pattern Analysis and Machine Intelligence*, 32(1):135–147.

Lafarge, F., Mallet C., 2012. Creating large-scale city models from 3d-point clouds: A robust approach with hybrid representation. *International Journal of Computer Vision*, 99(1):69–85.

Lamdan, Y., Wolfson, H. J., 1988. Geometric hashing: A general and efficient model-based recognition scheme. In *Proceedings of the IEEE 2nd International Conference on Computer Vision* (*ICCV*), Tampa, FL, pp. 238–249.

Lee, D. H., Lee, K. M., Lee, S. U., 2008. Fusion of LiDAR and imagery for reliable building extraction. *Photogrammetric Engineering and Remote Sensing*, 74(2):215–225.

Metropolis, N., Rosenbluth, A. W., Rosenbluth, M. N., Teller, A. H., Teller, E., 1953. Equations of state calculations by fast computing machines. *Journal of Chemical Physics*, 21:1087–1091.

Morgan, M., Habib, A., 2002. Interpolation of LiDAR data and automatic building extraction. In *ACSM-ASPRS 2002 Annual Conference Proceedings*, Denver, CO, November 12–14, 2002.

Ok, A. O., Wegner, J. D., Heipke, C., Rottensteiner, F., Soergel, U., Toprak, V., 2012. Matching of straight line segments from aerial stereo images of urban areas. *ISPRS Journal of Photogrammetry and Remote Sensing*, 74:133–152.

Otsu, N., 1979. A threshold selection method from gray-level histograms. *IEEE Transactions on Systems, Man, and Cybernetics*, 9(1):62–66.

Oude, E. S., Vosselman, G., 2009. Building reconstruction by target based graph matching on incomplete laser data: Analysis and limitations. *Sensors*, 9(8):6101–6118.

Perera, S., Mass, H. G., 2014. Cycle graph analysis for 3D roof structure modelling: Concepts and performance. *ISPRS Journal of Photogrammetry and Remote Sensing*, 93:213–226.

Persad, R. A., Armenakis, C., Sohn, G., 2015. Automatic co-registration of pan-tilt-zoom (PTZ) video images with 3D wireframe models. *Photogrammetric Engineering & Remote Sensing*, 81(11):847–859.

Pohl, C., van Genderen, J. L., 1998. Multisensor image fusion in remote sensing: Concepts, methods and applications. *International Journal of Remote Sensing*, 19(5):823–854.

Rau, J. Y., Lin, B. C., 2011. Automatic roof model reconstruction from ALS data and 2D ground plans based on side projection and the TMR algorithm. *ISPRS Journal of Photogrammetry and Remote Sensing*, 66(6):s13–s27.

Rissanen, J., 1978. Modeling by the shortest data description. *Automatica*, 14:465–471.

Rottensteiner, F., Briese, C., 2003. Automatic generation of building models from LiDAR data and the integration of aerial images. *International Archives of the Photogrammetry, Remote Sensing and Spatial Information Sciences*, 34(3/W13).

Rottensteiner, F., Sohn, G., Gerke, M., Wegner, J. D., Breitkopf, U., Jung, J., 2014. Results of the ISPRS benchmark on urban object detection and 3D building reconstruction. *ISPRS Journal of Photogrammetry and Remote Sensing*, 93:256–271.

Rottensteiner, F., Trinder, J., Clode, S., Kubik, K., 2005. Automated delineation of roof planes from LiDAR data. *International Archives of the Photogrammetry, Remote Sensing and Spatial Information Sciences*, 36(3/W4):221–226.

Sampath, A., Shan, J., 2007. Building boundary tracing and regularization from airborne LiDAR point clouds. *Photogrammetric Engineering and Remote Sensing*, 73(7):805–812.

Sampath, A., Shan J., 2010. Segmentation and reconstruction of polyhedral building roofs from aerial LiDAR point clouds. *IEEE Transactions on Geoscience and Remote Sensing*, 48(3):1554–1567.

Sohn, G., Dowman, I., 2007. Data fusion of high-resolution satellite imagery and LiDAR data for automatic building extraction. *ISPRS Journal of Photogrammetry and Remote Sensing*, 62(1):43–63.

Sohn, G., Huang, X., Tao, V., 2008. Using a binary space partitioning tree for reconstructing polyhedral building models from airborne LiDAR data. *Photogrammetric Engineering & Remote Sensing*, 74(11):1425–1438.

Sohn, G., Jung, J., Jwa, Y., Armenakis, C., 2013. Sequential modelling of building rooftops by integrating airborne LiDAR data and optical imagery: Preliminary results. *ISPRS Annals of the Photogrammetry, Remote Sensing and Spatial Information Sciences*, II-3(W1):27–33.

Sohn, G., Jwa, Y., Jung, J., Kim, H. B., 2012. An implicit regularization for 3D building rooftop modeling using airborne data. *ISPRS Annals of the Photogrammetry, Remote Sensing and Spatial Information Sciences*, I(3):305–310.

Tarsha-Kurdi, F., Landes, T., Grussenmeyer, P., 2008. Extended RANSAC algorithm for automatic detection of building roof planes from LiDAR data. *The Photogrammetric Journal of Finland*, 21(1):97–109.

Tian, Y., Gerke, M., Vosselman, G., Zhu, Q., 2008. Automatic edge matching across an image sequence based on reliable points. *International Archives of the Photogrammetry, Remote Sensing and Spatial Information Science*, 37(3B):657–662.

United Nations, 2014. World urbanization prospects: The 2014 revision. Available online: http://esa.un.org/unpd/wup/highlights/wup2014-highlights.pdf (assessed on March 12, 2016)

Verma, V., Kumar, R., Hsu, S., 2006. 3D building detection and modeling from aerial LiDAR data. In *The IEEE Computer Society Conference on Computer Vision and Pattern Recognition, (CVPR'06)*, New York, pp. 2213–2220.

Vosselman, G., 1999. Building reconstruction using planar faces in very high density height data. *International Archives of Photogrammetry, Remote Sensing and Spatial Information Sciences*, 32(3/2W5):87–92.

Wang, L., Neumann, U., 2009. A robust approach for automatic registration of aerial images with untextured aerial LiDAR data. In *Proceedings on 2009 IEEE Computer Society Conference, (CVPR 2009)*, Miami, FL, June 22–25, 2009, pp. 2623–2630.

Weidner, U., Förstner, W., 1995. Towards automatic building extraction from high resolution digital elevation models. *ISPRS Journal of Photogrammetry and Remote Sensing*, 50(4):38–49.

Wolfson, H. J., Rigoutsos, I., 1997. Geometric hashing: An overview. *IEEE Computational Science and Engineering*, 4(4):10–21.

Wunsch, P., Hirzinger, G., 1996. Registration of CAD-models to images by iterative inverse perspective matching. In *IEEE Proceedings of the 1996 International Conference on Pattern Recognition, (ICPR'96)*, Bavaria, Germany.

Yan, J., Shan, J., Jiang, W., 2014. A global optimization approach to roof segmentation from airborne LiDAR point clouds. *ISPRS Journal of Photogrammetry and Remote Sensing*, 94:183–193.

Yang, B., Chen, C., 2015. Automatic registration of UAV-borne sequent images and LiDAR data. *ISPRS Journal of Photogrammetry and Remote Sensing*, 101:262–274.

Zhou, Q. Y., Neumann, U., 2012. 2.5D building modeling by discovering global regularities. In *The IEEE Computer Society Conference on Computer Vision and Pattern Recognition*, Providence, RI, June 16–21, 2012, pp. 326–333.

Zhou, Q. Y., Neumann, U., 2008. Fast and extensible building modeling from airborne LiDAR data. In *Proceedings of the 16th ACM SIGSPATIAL International Conference on Advances in Geographic Information Systems*, (ACM GIS' 08), Irvine, CA, November 5–7, 2008 (on CD-ROM).

Zitova, B., Flusser, J., 2003. Image registration method: A survey. *Image and Vision Computing*, 21:977–1000.

18 A Framework for Automated Construction of Building Models from Airborne LiDAR Measurements

Keqi Zhang, Jianhua Yan, and Shu-Ching Chen

CONTENTS

18.1 INTRODUCTION

Building footprint, height, volume, and three-dimensional (3D) shape information can be used to estimate energy demand, quality of life, urban populations, property tax, and surface roughness (Jensen 2015). Building models are essential for 3D city or urban landscape models, urban flooding prediction, and assessment of urban heat island effects. Building models can be divided into two categories: simple (2.5D) and sophisticated (3D). Geometric attributes for a 2.5D building model consist of a footprint polygon and a height value. The geometric attributes for a 3D building model include not only a footprint polygon, but also planes or other types of surfaces for various parts of the roof as well as their projections (polygons) on the ground plane. Building heights of the 2.5D model are often constants, whereas building heights of the 3D model are variable. The advantage of the 2.5D model is that it requires fewer geometric attributes to delineate a building. The buildings are represented by various types of 3D *boxes*; therefore, the 3D rendering is fast. Many commercial geographic information system (GIS) software packages such as ArcGIS (www.esri.com) can display 2.5D building models effectively. The 2.5D building model is sufficient for applications such as numerical modeling of urban flooding and heat island effect, estimating urban population and energy demand, and large scale 3D visualization, all of which do not require the details of buildings. The key to extracting a 2.5D building model from airborne Light Detection and Ranging (LiDAR) measurements is to derive footprints. The building height value can be represented using statistical height values such as mean and median of LiDAR measurements within a footprint.

The disadvantage of the simple building model is the lack of detail and accuracy of building shapes. The 3D models overcome this limitation by offering more geometric information for 3D buildings. The 3D models are required by applications such as hurricane wind damage models for individual properties, property tax estimation, and detailed urban landscape modeling. The disadvantage of 3D models is that 3D rendering is slow. Most existing commercial GIS software cannot display 3D models efficiently due to compound geometric structures.

High-resolution data needed for extracting building models include aerial photographs, high-resolution satellite images, and airborne LiDAR measurements. Compared with aerial photographs and satellite images obtained by optical sensors, LiDAR measurements are not influenced by sun shadow and relief displacement, and provide direct measurements of building heights. However, airborne LiDAR systems generate voluminous and irregularly spaced 3D point measurements of objects, including ground, buildings, trees, and cars scanned by the laser beneath the aircraft. The sheer volumes of point data require dedicated algorithms for automated building reconstruction. This paper focuses on presenting a framework for construction of 2.5D and 3D building models from LiDAR measurements. The current chapter is arranged as follows. Section 18.2 reviews previous work. Section 18.3 describes the algorithms that derive building models. Section 18.4 describes the sample LiDAR dataset used in this study. Section 18.5 examines the results by applying the proposed framework to the sample dataset, and Section 18.6 includes conclusions.

18.2 LITERATURE REVIEW

18.2.1 Separation of Ground and Nonground Points

The critical step in building model construction is to identify building measurements from LiDAR data. Two ways are often utilized to identify building measurements from LiDAR data. One is to separate ground, buildings, trees, and other measurements from LiDAR data simultaneously using segmentation. Examples of this method can be found in Maas (1999) and Filin and Pfeifer (2006).

The other way is to separate the ground from nonground LiDAR measurements first and then identify the building points from nonground measurements. Numerous algorithms have been developed to identify ground measurements from LiDAR data (Meng et al. 2010, Pfeifer and Mandlburger 2017). For example, Vosselman (2000) proposed a slope-based filter to remove nonground measurements by comparing slopes between a LiDAR point and its neighbors. Shan and Sampath (2005) applied a slope based 1D bidirectional filter to LiDAR measurements along the cross track direction to label nonground points. Zhang et al. (2003) used mathematical morphology to identify ground measurements. The nonground measurements can be simply derived by removing identified ground data from a raw dataset. To facilitate the data processing, digital surface models (DSMs) interpolated from raw LiDAR measurements and digital terrain model (DTM) interpolated from ground measurements are often produced. The image for nonground objects is derived by subtracting DTM from DSM. Examples using this method to derive nonground objects can be found in Weidner and Forstner (1995), Morgan and Tempfli (2000), Rottensteiner and Jansa (2002), and Elaksher and Bethel (2002).

18.2.2 Building Measurement Identification

The next step is to extract building point measurements or pixels in the dataset for nonground objects that are dominated by trees and buildings. The distinct difference between buildings and trees is that the roof surfaces are approximately planar, whereas canopy surfaces are irregular. Several parameters based on this difference have been proposed for segmenting buildings and trees. For example, the first derivatives of heights are either a zero (flat roof) or constant (sloped roof) for a planar surface, and the second derivatives of a sloped planar surface are zero. The first and second derivatives of an irregular surface should be variable. Morgan and Tempfli (2000) applied Laplacian and Sobel operators to height surfaces to separate building and tree measurements. The problem with this method is that the derivatives from LiDAR measurements for roofs are not constant because of measurement errors. Small features such as chimneys, water tanks, and pipes on a roof surface can also produce abnormal derivative values. In addition, the derivatives at the edge of a building have a large variation, making it difficult to separate buildings from adjacent trees.

Another technique to separate building and tree measurements involves using a least squares method to estimate the parameters for a plane that fits a LiDAR point and its neighbors within a local window (Filin 2002, Morgan and Habib 2002, Zhang et al. 2006). It is expected that the deviations of roof points from their fitted planes will be small and the plane parameters will be similar and consistent, whereas deviations and plane parameters for tree points will be large and variable. Compared with derivatives, the plane parameters are less sensitive to individual outliers caused by chimneys and water tanks. The drawback of this method is that plane parameters are not robust at the boundary of the buildings because fewer points are available for parameter estimation.

Many airborne LiDAR systems are capable of deriving the first and last return measurements produced by multiple reflections of a laser pulse by the objects on the earth surface. The height difference between the first and last return measurements can be used to separate building and tree measurements (Alharthy and Bethel 2002). The height difference is usually large for tree measurements and close to zero for building measurements. A measurement is identified as a building measurement by comparing its height difference with a predefined threshold. However, this method does not work for areas covered by dense trees where laser pulses cannot penetrate. In addition, the elevation difference between first and last returns less than 3–4 m is not reliable because of the influence of the laser pulse width that is typically 10 nanoseconds (Carter et al. 2004).

Hough transform has also been used to separate building points from tree measurements or to identify them directly from a raw LiDAR dataset (Vosselman 2001, Rottensteiner and Jansa 2002, Overby et al. 2004). Data in the physical space are transformed to and analyzed in the parameter space. The advantage of the Hough transform is its tolerance of gaps in the feature boundary. However, it is difficult to define the optimum cell size for voting in the parameter space, which is influenced by the error range of LiDAR measurements, the sampling density, and local height changes of a roof surface. Unfortunately, local height changes of individual surfaces are different, thus it is difficult to quantify these changes using a single value. Usually, the cell size is set empirically, and if the cell size is too large, several real-world planes could be merged into one during building identification. In contrast, if the cell size is too small, one real-world plane could be split into several smaller planes. In addition, the adjacency of point measurements is not considered by the Hugh transform. Therefore, LiDAR measurements for separated but closely adjacent buildings with the same height could be mixed into the same category.

The LiDAR measurements for buildings and trees can be segmented on the basis of one or more of the above-mentioned parameters. There are two ways to perform the segmentation task. One is the point or pixel level classification. Maas (1999) employed raw elevation, Laplace filtered height, and maximum slope from LiDAR measurements to perform a supervised maximum likelihood classification. Filin (2002) separated LiDAR measurements for buildings, vegetation, ground, and other features using an unsupervised classification based on the position of a point, the parameters of the tangent plane to the point, and the relative height difference between the point and its neighbors. Elberink and Maas (2000) used unsupervised *k*-means classification in their texture-based segmentation. The problem with a point level classification is that the measurements for a building cannot be guaranteed to classify into the same category. Moreover, selection of training datasets for a supervised classification can be very time consuming. An alternative way is to find a building area using a region growing algorithm based on a seed point (Morgan and Habib 2002, Zhang et al. 2006, Yan et al. 2014). This method considers the adjacency of LiDAR measurements and the robustness of region growing processes (Zhang et al. 2006).

18.2.3 BUILDING MODEL CREATION

After measurements for a building are identified, a raw footprint can be derived by connecting boundary points of LiDAR measurements for a building. The raw footprint polygon has to be generalized for building models because the raw footprint includes considerable noise due to irregularly spaced LiDAR points. Alharthy and Bethel (2002) employed the histogram of boundary points to generalize the footprint edges by assuming that the buildings have only two dominant directions that are perpendicular with each other. Based on the similar assumption, Sampath and Shan (2006) used the least squares model to regularize the footprint edges. Zhang et al. (2006, 2009) published a method to refine a footprint iteratively based on estimated dominant directions. Footprints with oblique edges that are not perpendicular to dominant directions are allowed in this method as long as the total length of the oblique edges are less than the total length of the edges parallel or perpendicular to the dominant directions.

Simple building models can be derived by adding a uniform height value once a refined building footprint is derived. However, the process to derive 3D building models is more complicated. Schwalbe et al. (2005) categorized the methods to derive 3D building models into two categories: model-driven and data-driven. In the model-driven method, building models are identified by fitting predefined models into the LiDAR measurements. For example, Maas and Vosselman (1999) estimated parameters for primitive building models based on the invariant moment analysis. Brenner (2000, 2004) extended this method to complex buildings by splitting a building into simple primitives first and then fitting individual primitives using point clouds. However, building models for a study area are not always available in advance, which limits the application of the model-driven method.

In the data-driven method, building measurements are grouped first for different roof planes. For example, You and Lin (2011) employed the tensor voting algorithm and a region-growing method to extract building roof planes and boundary lines from LiDAR data. Sampath and Shan (2010) used the principle analysis to identify roof facets. Then, 2D topology of each building, which is represented by a set of connected planar roof surfaces projected onto a horizontal plane, is derived. Rottensteiner and Jansa (2002) approximated pixels on edges with line segments first and then intersected these line segments to derive the vertices of the 2D topology. Alharthy and Bethel (2002) derived the polygon for each roof plane by connecting the boundary points and simplifying the polygon edges using the Douglas–Peucker algorithm (Douglas and Peucker 1973). Polygons for each roof plane are then snapped into the 2D topology. As neighboring roof polygons may overlap or be separated from each other, it is not easy to snap the neighboring polygons together.

Raw 3D building models can be directly created from derived 2D topologies and identified roof planes. However, the quality of such building models is poor because a 2D topology is often noisy due to irregularly spaced LiDAR measurements. Uncertainty in estimated roof plane parameters due to errors in LiDAR measurements and segmentation also distort raw building models. A refinement of 2D topology and roof plane parameters is often needed to derive high quality building models. Many geometric constraints have been proposed to regularize and refine the 2D topology. Gruen and Wang (2001) proposed seven constraints formulated into weighted observation equations, and they enforced these constraints by using least squares adjustment. Before least-squares adjustment is applied, points, lines, or planes related to each constraint should be grouped together manually, a step that limits the application of this method. Other methods realized the difficulty in enforcing many constraints simultaneously and only enforced important constraints, especially parallelism. The parallelism constraint assumes that a building has two dominant directions perpendicular to each other, and building edges on the 2D topology should be parallel to either of the dominant directions (Vosselman 2001). However, this assumption is too strict and cannot be applied to buildings with edges oblique to both dominant directions.

The main objective of this chapter is to present a framework involving a series of algorithms for extraction of simple and sophisticated building models from LIDAR measurements (Figure 18.1). The framework consists of three major steps. First, the nonground and ground measurements

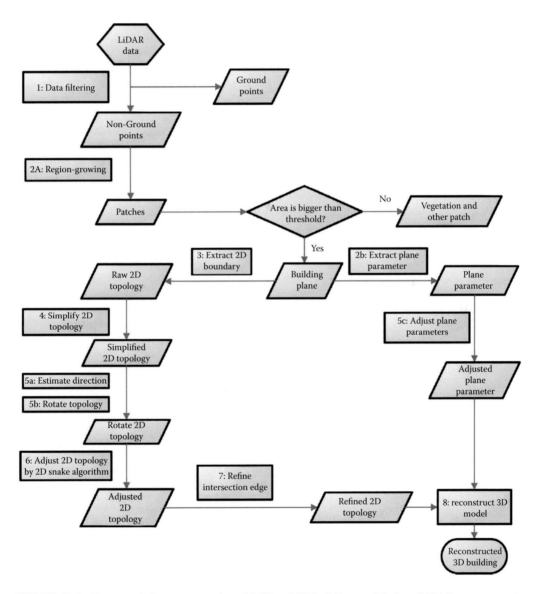

FIGURE 18.1 Framework for reconstruction of 2.5D and 3D building models from LiDAR measurements. Each rectangular block represents one step or a substep of the framework. The integer inside each rectangular block corresponds to the processing order of the step in the framework. (From Yan, J. et al., *IEEE Trans. Geosci. Remote Sens.*, 53, 1, 3, 2015. With Permission.)

are separated. Second, building measurements are identified by region growing using a local plane-fitting technique. Third, 2.5D building models are derived and adjusted on the basis of estimated dominant directions, and 3D building models are derived and refined on the basis of 2D topology of roof facets.

18.3   CONSTRUCTION OF BUILDING MODELS

18.3.1   Separating Ground and Nonground Measurements

The first step in the proposed framework for constructing building models is to separate ground and nonground measurements. We selected the progressive morphological filter for this task because this filter identified the ground and nonground measurements well for the sample dataset used in the current chapter (Zhang et al. 2003). Other filters can also be used in this step if they produce a good classification. The progressive morphological filter classifies ground and nonground measurements by comparing the elevation difference between interpolated and filtered surfaces. By gradually increasing the window size, the filter separates the measurements for different-sized nonground objects from ground data.

A 2D array, whose elements represent points falling in cells of a mesh overlaying the dataset, is employed to facilitate the filtering and building identification computations. The cell size (c_s) of the mesh is usually set to be less than the average spacing of the LiDAR points to reduce information loss. Each point measurement from the LiDAR dataset is assigned to a cell in terms of its horizontal x and y coordinates. If more than one point falls in the same cell, the point with the lowest elevation is selected as the array element. If no point exists in a cell, the array element for the cell is assigned as its nearest neighbor.

18.3.2   Segmenting Nonground Measurements and Identifying Building Patches

The second step of the framework is to separate building measurements from vegetation measurements using a region-growing algorithm (Adams and Bischof 1994) based on a plane-fitting technique. The rationale behind this algorithm is to group nonground measurements, which are located on the same planes, into patches. Building patches are much larger in size because measurements for a roof facet are almost always located in the same plane, whereas vegetation patch sizes are small due to large changes in elevations for an irregular canopy. Region-growing segmentation of nonground measurements starts first with the selection of the seed point for a group. Then, the neighbors of the seed point are recursively added into the group by comparing the elevations of the neighboring points with the plane generated by fitting points in the group using the least squares method.

Both triangulated irregular network (TIN) and grid data structures can be used to represent nonground measurements for segmentation. Only algorithms based on grid data structure are presented in detail in this chapter because the method based on TIN is similar to the method based on the grid. The only difference is that the region-growing algorithm based on a grid data structure recursively expands a building patch using the eight neighbors of a cell, whereas the region-growing algorithm based on a TIN structure expands a building patch using the three neighbors of a triangle. The region-growing algorithm for a grid data structure is presented as follows:

1. Create a set of areas for connected nonground measurements $A = \{S_1, S_2, .., S_N\}$. We start an area creation by selecting any nonground measurement (typically it is the most left nonground measurement) as a seed nonground measurement. An area is created by connecting recursively nonground measurements among the eight neighbors of the seed measurement. The creation of an area stops when no additional nonground neighbors can be connected. The area creation process is repeated until every nonground measurement is included in S_i.

2. Partition S_i into PI_i which represents a set of inside points and PB_i which represents a set of boundary points. If at least one of the eight neighbors of a point is a ground measurement, the point is defined as a boundary point. Otherwise, the point is an inside point.

3. Remove S_i from A if the area of $S_i <$ predefined min_surface or the number of inside points $= 0$.

4. For each S_i in A

   a. Given an inside point $p_k(x_0, y_0, z_0)$, a Cartesian coordinate system (x, y, and z) is established using p_k as the origin. In this coordinate system, a best fitting plane for p_k and its eight neighbors is derived by using the least-squares method. Assume that the plane is defined by:

$$z = ax + by + c \qquad (18.1)$$

   The parameters (a, b, c) can be derived by minimizing the sum of squares due to deviations (SSD) which is represented by ssd_k

$$ssd_k = \sum_{(p_k) \in K} (z_k - h_k)^2 \qquad (18.2)$$

   where:
   K is a set for p_k and its neighbors
   h_k and z_k are observed and plane fitted surface elevations, respectively

   We derive a set $SSD_i = \{ssd_1, sd_2, ..., sd_{Ni}\}$ which consist of minimized $ssds$ for all inside points in S_i.

   b. While $SSD_i \neq \emptyset$ (empty), Sort ssd in SSD_i in ascending order, set $P_j = \emptyset$, and set $j = 1$ for the first time.

   c. Select the point p_k with minimum ssd_k as a seed point for region-growing and add p_k to a set P_j.

   d. Label the neighbors of a seed point by examining whether they belong to the same set P_j through a plane-fitting technique. A plane is constructed based on the points in the category using a least-squares fit. If the elevation from a neighbor point to this plane is less than a predefined threshold Δh_T, the neighbor is added to P_j. Δh_T is determined by the elevation error of the LiDAR survey and is usually 15–30 cm. The process is continued until no point can be added to P_j.

   e. Remove points in P_j from S_i and corresponding $ssds$ from SSD_i, Go back to step b.

   f. Continue the above-mentioned process until $S_i = \emptyset$.

5. Let P represent a set of patch P_j derived from the above-mentioned process, we have $P = \{P_1, P_2, ..., P_M\}$ and $P_j = \{p_{1j}, p_{2j}, ..., p_{Nj}\}$, where p_{kj} is a kth point in patch j. Remove P_j from P if the area of P_j is less than a predefined threshold min_surface, and P_j is not completely enclosed by other patches.

6. Merge patches in P if they are connected and derive a new set PM. Through this process, adjacent roof surfaces from the same building, having different slopes, are merged into a large building patch. Remove PM_j from PM if the area of PM_j is less than a predefined threshold min_building. The remained patches in PM are building patches.

The above-mentioned algorithm indicates that the identification of building patches depends on the region growing process that, in turn, relies on the selection of seed points. Zhang et al. (2006) has demonstrated that the region growing process is robust when starting with a seed point with a minimum SSD.

18.3.3 Deriving 2.5D Building Models

Raw footprints are derived by connecting the boundary points for building patches. The boundary of a raw footprint contains noise because of the irregularly spaced LIDAR point measurements (Figure 18.2b). A refinement of the footprint boundary is needed to reduce the noise. First, the Douglas–Peucker algorithm is employed to generalize line segments in the footprint (Douglas and Peucker 1973). This algorithm works well in reducing the noise, but it removes critical corner vertices of the raw footprint in some cases, leading to an orientation distortion of the segments in the footprint (Figure 18.2c). To recover the removed critical vertices and distorted segments, the dominant directions of building footprints need to be estimated.

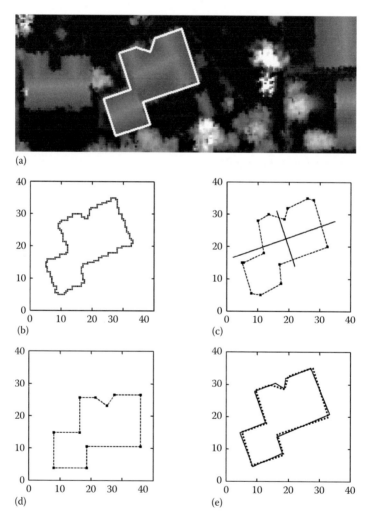

FIGURE 18.2 Example to illustrate the reconstruction process of a simple building model. The x and y coordinates are in meters. (a) The LiDAR image of a sample building with adjusted final footprint. (b) The raw footprint derived by connecting boundary points of identified building measurements through the region-growing algorithm. The raw footprint is noisy because of the interpolation of irregularly spaced LiDAR measurements. (c) The coarse footprint derived using the Douglas–Peucker algorithm. Two solid lines across the footprint are estimated dominant directions. (d) The adjusted footprint that recovered critical corner vertices removed by the Douglas–Peucker algorithm. The footprint was rotated clockwise according to the estimated dominant directions. (e) Comparison of the final footprint (solid) with corresponding known footprint (dash). (From Zhang, K. et al., *IEEE Trans. Geosci. Remote Sens.*, 44, 9, 2523, 2006. With Permission.)

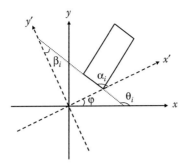

FIGURE 18.3 Relationship between angles β_i, θ_i, α_i, and φ. The coordinate system x' and y' is a counterclockwise rotation of the coordinate system x and y by an angle φ. θ_i is the counterclockwise intersection angle between a segment of a building footprint (solid square) and the axis x. α_i is the counterclockwise intersection angle between a segment and the axis x'. β_i is the minimum intersection angle between a segment and the axes x' and y'. (From Zhang, K. et al., *IEEE Trans. Geosci. Remote Sens.*, 44, 9, 2523, 2006. With Permission.)

A method based on weighted line segment lengths is used in this study to estimate the dominant directions of a building footprint. Let x' and y' represent possible dominant directions in a 2D coordinate system x and y (Figure 18.3). The dominant directions x' and y' are related to the coordinate system x and y through a counterclockwise rotation by an angle φ ($0°<=\varphi<90°$). Therefore, the key step to estimate the dominant directions is to find the rotation angle φ. Assuming that the counterclockwise intersection angle between a line segment and x axis is θ_i ($0°\leq\theta_i<180°$), we define

$$SL = \sum_{i=1}^{N-1} g(L_i) f\left(\beta_i(\theta_i,\phi)\right) \tag{18.3}$$

where:
N is the total number of vertices of a building footprint
L_i is the segment length
β_i ($0°\leq\beta_i<45°$) is the minimum intersection angle between a segment and the nearest axis in the coordinate system x' and y', and is determined by θ_i and φ, $g()$ is the weight function based on L_i, and $f()$ is the weight function based on θ_i and φ. The dominant building directions can be estimated by finding an optimum φ so that SL will reach a minimum.

A linear function is employed to represent $g()$

$$g(L_i) = \frac{L_i}{\sum_{i=1}^{N-1} L_i} \tag{18.4}$$

Finding the optimum φ depends heavily on $f()$ which can have many forms such as linear and exponential expressions. We would like to construct $f()$ such that the segments close to the dominant direction will have a small contribution to SL. Here a linear function is used to represent $f()$

$$f(\beta_i(\theta_i,\phi)) = \frac{\beta_i}{45} \tag{18.5}$$

Obviously, the closer the segment is to the dominant direction, the smaller the $f()$. β_i is determined by

$$\beta_i = \begin{cases} \min(\alpha_i, 90-\alpha_i) & \text{if } \alpha_i \leq 90° \\ \min(180-\alpha_i, \alpha_i - 90) & \text{if } \alpha_i > 90° \end{cases} \tag{18.6}$$

where α_i ($0° \leq \alpha_i < 180°$) is the counterclockwise intersection angle between a line segment and the axis x' and has the following relationship with θ_i and φ

$$\alpha_i = \begin{cases} \theta_i - \phi & \text{if } \theta_i \geq \phi \\ 180 + \theta_i - \phi & \text{if } \theta_i < \phi \end{cases} \tag{18.7}$$

Numerically, the optimum φ is found by comparing SL values for angles between $0°$ and $90°$. After φ is derived, the building footprint is rotated so that the x and y axes are aligned with the dominant directions of the buildings (Figure 18.2d).

It has been proven that the estimated dominant directions are the same as the directions of the parallel and perpendicular segments as long as the total length of the oblique lines is less than the total length of the parallel and perpendicular segments of a footprint (Zhang et al. 2006). After the dominant directions of a footprint are estimated, five operations including split, intersect, and three types of merge are employed to refine the building footprint. The selection of operations is mainly determined by the projections of a line segment in the two dominant directions (Zhang et al. 2006). Figure 18.2e displays an adjusted footprint using these five operations. The method works fine for buildings with flat roofs and large nonflat roofs. However, the robustness of the method is decreased for the buildings with many small nonflat roof facets and oblique edges when the point density of LiDAR measurements is less than one per square meter. To overcome this problem, the line segment weighted method was extended to use the areas of roof faces to estimate the dominant directions of the nonflat roof buildings (Zhang et al. 2009). Once a refined footprint is obtained, a 2.5D building model can be created by adding a uniform height value for the building.

18.3.4 Deriving 3D Building Models

18.3.4.1 2D Topology Extraction

In addition to the outline of the footprint, inside boundaries between adjacent roof facets and equations describing these roof facets need to be derived for a 3D building model. It is assumed that the roof facets follow plane equations in our 3D building model. The boundaries between roof facets are part of the 2D topology that is represented by a set of connected polygons that are the projections of roof facets on a horizontal plane. The edges and vertices that form the topology are derived from grouped LiDAR measurements for individual facets of a roof. To facilitate this discussion, measurements for each roof facet for a building are assigned a unique positive integer label starting from the label "1" (Figure 18.4a and b). Nonbuilding measurements surrounding the building are labeled as "0." To obtain the boundaries of a roof facet, the cell size of the mesh covering the dataset is reduced by half, and new points are inserted between the old cells.

The label of a newly inserted point p is determined by checking four pairs of neighbors of p as illustrated in Figure 18.5. If the label at the left bottom corner is the same as the label at the right upper corner, p is assigned the same label as these two neighbors (Figure 18.5a). If not, other pairs of neighbors (Figure 18.5b–d) are checked following the same procedure. If all four pairs of neighbors of p have different labels, p is identified as a boundary point separating two or more neighboring roof facets and is assigned a label value "–1" (Figure 18.4c).

A boundary point that joins at least three different roof facets (Figure 18.4d) is classified as a vertex of the 2D topology, whereas a boundary point that separates only two neighboring roof facets is defined as an edge point. An edge is derived by connecting all edge points between two vertices. The vertices and edges from boundary points constitute a raw 2D topology. The raw 2D topology is rotated clockwise so that the dominant directions of the footprint align with the x and y axes. The raw 2D topology is simplified using the Douglas–Peucker algorithm to reduce the noise due to the interpolation of the irregularly spaced LiDAR points. Parallel and perpendicular edges in a simplified topology can be distorted in some cases because no geometric constraints are applied during the simplification (Figure 18.4e). Therefore, a refinement of topology is needed to minimize this distortion.

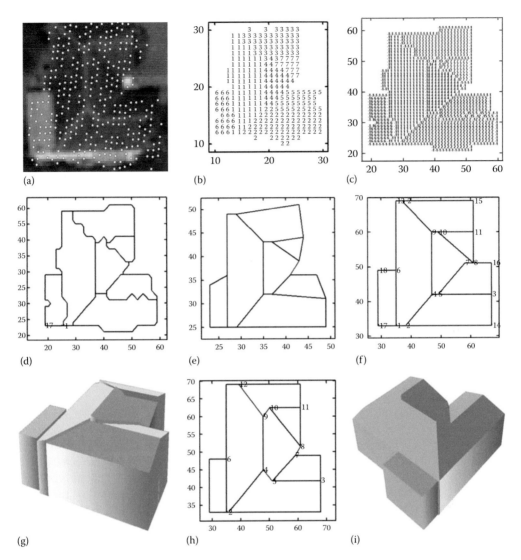

FIGURE 18.4 Example illustrating the reconstruction process of a 3D building model. The *x* and *y* coordinates are in meters. (a) Raw LiDAR points overlaid on the aerial photograph, (b) polygons for segmented roof facets, which are labeled by different positive integers, (c) roof facet polygons with labeled (−1) boundary, (d) raw 2D topology of the building, (e) simplified 2D topology using the Douglas–Peucker algorithm, (f) adjusted 2D topology by applying the snake algorithm, (g) 3D building model reconstructed on the basis of the topology from (f), (h) refined 2D topology by replacing intersection edges with the joint point of neighboring roof facets, and (i) the 3D building model reconstructed in ArcGIS based on the refined topology from (h). (From Yan, J. et al., *IEEE Trans. Geosci. Remote Sens.*, 53, 1, 3, 2015. With Permission.)

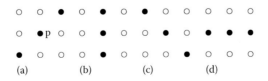

FIGURE 18.5 Patterns used to determine whether a point *p* is a boundary point. If four pairs of neighbors of *p* have different labels, *p* is a boundary point.

18.3.4.2 Roof Facet Adjustment

Before performing a topological adjustment, plane parameters for a roof facet are estimated using the points within a segmented patch for the roof facet. Parallel and perpendicular properties of a plane are also enforced using the variation of elevation values. Each roof plane is first examined whether it is flat by analyzing the variation (*Var*) in the elevations of LiDAR measurements for the plane using the following equation:

$$\mathrm{Var} = \sum \left(z_k - \bar{z} \right)^2, \quad \bar{z} = \frac{\sum z_k}{n} \qquad (18.8)$$

where:
   n is the number of measurements for a roof facet
   $\bar{z}$ is the average elevations of the LiDAR measurements z_k

If Var is less than the threshold T_Var^2, the plane is classified as horizontal, and the plane equation is adjusted to be $z = \bar{z}$. If not, the roof facet is classified as nonflat. Then, we examine whether the plane for a nonflat roof facet is perpendicular to *X-Z* plane. Each 3D measurement (x, y, z) on the plane is projected onto the *X-Z* plane to form a 2D point (x, z). The line segment with equation $z = ax + b$ which fits the projected 2D points are derived using the least squares method. The square root of SSD (*srssd*) is used to determine the perpendicular property of a roof facet

$$SSD = \sum (z'_k - z_k)^2, \; srssd = \sqrt{SSD} \qquad (18.9)$$

where z'_k and z_k are the estimated and observed elevations for point k, respectively. If *srssd* is less than the threshold T_Var, the plane is perpendicular to the *X-Z* plane and its plane equation is adjusted to be $z = ax + b$. If *srssd* is greater than T_Var, we then check whether the plane is perpendicular to *Y–Z* plane using the same procedure. Each 3D measurement (x, y, z) is projected onto the *Y–Z* plane to form a 2D point (y, z). The linear equation $z = a'y + b'$ fitting the 2D points is estimated using the least squares method. If *srssd* from observed and estimated z values is less than T_Var, the roof plane is perpendicular to *Y–Z* plane, and its plane equation is adjusted to be $z = a'y + b'$. Otherwise, the nonflat roof plane is neither perpendicular to *X–Z* nor *Y–Z*, and its plane equation is not adjusted.

18.3.4.3 Adjustment of 2D Topology

The snake algorithm is utilized to refine the 2D topology of a building. The snake algorithm was introduced by Kass et al. (1988) to locate features of interest in an image. First, an initial contour enclosing the feature of interest is selected, and then the contour is pulled toward the target feature by minimizing energy functions that represent constraint forces. The total energy E_{total} of a contour with a parametric representation $v(s) = (x(s), y(s))$ consists of the internal energy E_{int} and the external energy E_{ext} and can be written as

$$E_{\text{total}} = \int_0^1 E_{\text{int}}\left(v(s)\right) + E_{\text{ext}}\left(v(s)\right) ds \qquad (18.10)$$

The internal energy represents the forces which constrain the contour to be smooth and are formulated as:

$$E_{\text{int}}\left(v(s)\right) = \left(\alpha(s) \mid v_s(s) \mid^2 + \beta(s) \mid v_{ss}(s) \mid^2 \right) \qquad (18.11)$$

where the first-order energy term $v_s(s)$ moves the points closer to each other and the second-order energy term $v_{ss}(s)$ favors equidistant points. The external energy attracts the contour toward the feature of interest, such as the high intensity or high gradient areas. The minimization of energy can be implemented using finite difference, finite element, or dynamic programming methods (Kass et al. 1988, Amini et al. 1990, Cohen and Cohen 1993). Traditional snake algorithms only consider the outlines of components in an image. To adjust the 2D topology, both the deformation of the outline of a component and the inside structure of edges between polygons have to be considered.

A dynamic programming method for 1D snakes was extended to adjust 2D topology in our framework because the method guarantees a global minimum and is numerically stable (Yan et al. 2007, 2015). The dynamic programming method assumes that each vertex on the contour is adjustable to a position that belongs to a set of positions around the vertex. The set of positions determines the range of vertex variation. Each position of the vertex corresponds to one state of the vertex. A set of the vertices connecting contours corresponding to the minimal energy represent the vertices for an optimum 2D topology. The key for deriving the optimum 2D topology is to define appropriate energy functions for the snake algorithm. Traditional inner energy functions for a smooth contour shown in Equation 18.11 are not suitable for building topology adjustment because there are intersection angles between the two adjacent edges and the transition between two edges is not smooth. In addition, the distances between vertices of the topology are not equal in many cases.

The objective of building footprint adjustment is to enforce the requirement that the edges be parallel to the dominant directions in a 2D topology while keeping the adjusted footprint as close to its original location as possible. To meet this requirement, we are proposing new functions, direction energy E_{Dir}, to enforce the parallel constraint and, deviation energy E_{Dis}, to limit the deviation of the adjusted footprint from its original position. Given an edge $e' = (v', w')$ joining two vertices v' and w' on the footprint, and v' and w' are one of the possible states for vertices v and w, we define the direction energy E_{Dir} for the edge $e' = (v', w')$ as:

$$E_{Dir}\left(e' = (v', w')\right) =$$

$$\begin{cases} 0 & c(e) = 1, v'_x = w'_x \\ 0 & c(e) = 2, v'_y = w'_y \\ 0 & c(e) = 3, |v'_y - w'_y| \triangleright \text{T\_Projection} \quad \text{or } |v'_x - w'_x| \triangleright \text{T\_Projection} \\ |v' - w'| & \text{others} \end{cases} \tag{18.12}$$

where $c(e)$ with values of 1, 2, or 3 represents horizontal, vertical and nonadjustable oblique edges, respectively. A zero value is assigned to the direction energy function E_{Dir} for edge e' in the 2D topology if e' is either parallel to the dominant directions or e' is a nonadjustable oblique line. A nonadjustable oblique edge is the line whose projection on x or y axis is larger than a threshold T_Projection. The purpose in introducing nonadjustable lines is to preserve oblique edges in the 2D topology. All remained edges are classified as adjustable ones, and a penalty value proportional to the length of the edge is assigned to E_{Dir} for an adjustable edge e'. As a result of this energy function, the adjustable horizontal and vertical edges tend to deform and align with the dominant directions, and nonadjustable oblique edges tend to remain oblique.

The Energy function E_{Dis} is defined as the sum of the distance values between points on the adjusted edge (v', w') and the original edge (v, w):

$$E_{Dis}\left(e' = (v', w')\right) = \sum_{p \in v'w'} D_p(v, v', w, w') \tag{18.13}$$

where p is a point located on the edge connecting points v' and w'. The smaller the distance energy of the adjusted edge e', the closer e' is to the original edge e. Only the distance values of points in an area close to the original 2D topology need to be calculated as each vertex on the 2D topology is only allowed to move within a small window $W_v \times W_v$. To compute E_{Dis}, a distance transform is applied to the 2D topology to derive a gray scale image, the pixel intensity of which indicates shortest distances to boundaries. The edges between roof facets are rasterized using the same grid mesh for 2D topology extraction described in the previous section. Distance values of edge cells are initialized as 0 and distance values of their direct neighbors are assigned a value of 1. Distance values of direct neighbors of cells having values of 1 are assigned a value of 2 and so on. Figure 18.6 shows the distance value of the image after applying the distance transform to the 2D topology shown in Figure 18.4c. The distance value is calculated up to 3 cells. The total energy for each adjusted edge e' is determined as follows:

$$E\big(e' = (v', w')\big) = C_{\mathrm{Dir}} E_{\mathrm{Dir}}(e') + C_{\mathrm{Dis}} E_{\mathrm{Dis}}(e') \tag{18.14}$$

where C_{Dir} and C_{Dis} are weights for two energy terms. The snake algorithm transforms each adjustable vertex on the 2D topology and finds an optimum combination of vertices by minimizing the sum of the energy functions for all edges on the 2D topology. Figure 18.4f shows one example in which a contour was adjusted using the snake algorithm based on defined energy functions.

A 2D snake algorithm based on the energy functions described earlier is required to refine a 2D topology for a building. A brute force implementation of minimizing the energy specified in Equation 18.14 takes time that is exponential to the number of vertices and is not feasible for the snake problems with complicated topology. Unfortunately, the general 2D snake algorithm is a nondeterministic polynomial time (NP) completeness problem and the method to reduce the computational time significantly cannot be developed (Yan et al. 2007). However, a subset of 2D snake problems can be solved in polynomial time using an algorithm based on the graph theory

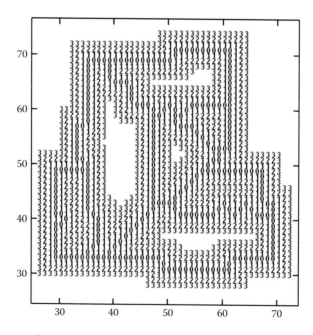

FIGURE 18.6 Distance values derived by applying distance transform to the 2D topology of a building. Label "0" indicates the edges of roof facets, and labels "1," "2," and "3" represent the distances from edges. The x and y coordinates are in meters. (From Yan, J. et al., *IEEE Trans. Geosci. Remote Sens.*, 53, 1, 3, 2015. With Permission.)

(Yan et al. 2015). This is accomplished by finding the minimum energy of a graph by progressively simplifying the graph using graph reduction operations. Each reduction operation simplifies the original graph while retaining the minimum energy of the graph. Four graph reduction operations are utilized in our 2D snake algorithm and more details about theoretical analysis and implementation can be found in Yan et al (2015). Figure 18.7 demonstrates the process of adjusting 2D topology using the 2D snake algorithm based on graph reduction. Our work demonstrates that over 98% of building topologies can be adjusted by the graph-based algorithm. Figure 18.8 shows an example adjusting the 2D topology of a complicated building using the graph-based algorithm. The building roof consists of 46 vertices and 67 boundaries, and all the roof planes are flat.

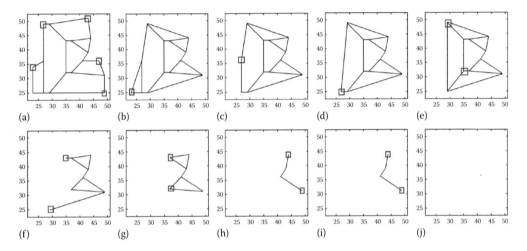

FIGURE 18.7 Example showing the graph reduction process for the 2D topology as shown in Figure 18.4e. (From Yan, J. et al., *IEEE Trans. Geosci. Remote Sens.*, 53, 1, 3, 2015. With Permission.)

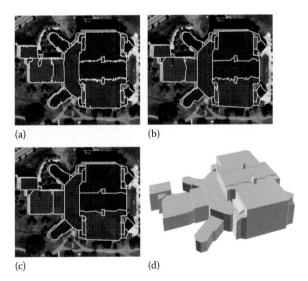

FIGURE 18.8 Derivation of a complex 3D building model from LiDAR measurements involves (a) raw 2D topology, (b) simplified 2D topology, (c) refined 2D topology through the 2D snake algorithm, and (d) reconstructed 3D building model. The background of (a), (b), and (c) is a black–white area photograph. The details of the roof surface of this building can also be found in Figure 18.12. (From Yan, J. et al., *IEEE Trans. Geosci. Remote Sens.*, 53, 1, 3, 2015. With Permission.)

After applying the 2D snake algorithm based on the proposed energy functions, the 2D topology is adjusted and shown in Figure 18.8c.

18.3.4.4 Adjustment of Intersection Edges

Some buildings, such as residential houses, mainly consist of nonhorizontal roof facets that form many intersection edges. The edges between building facets can be classified into two categories: intersection and step edges. A step edge separates either two parallel planes or two intersection planes with a height discontinuity (Figure 18.9a and b). An intersection edge separates two neighboring roof planes with height continuity (Figure 18.9c). Obviously, all edges of the footprint outline are step edges.

The height values at an intersection edge from two adjacent roof facets may be different because the 2D topology adjustment algorithm does not enforce the height continuity constraint. Figure 18.4g shows that extra walls are induced by the false height discontinuity between neighboring roof facets in the 3D model. To remove this inconsistency, the edges from the snake-based adjustment within the outline of the building footprints are replaced with the intersection segments of neighboring planes and the height continuity constraint for vertices on intersection edges is enforced. Before the adjustment operation is performed, intersection edges have to be identified. We determine an intersection edge using the following equation:

$$\text{DH}\big(e(v,w)\big) = \sum_{p \in vw} |h_1(p) - h_2(p)| / n \qquad (18.15)$$

where:

$\overline{vw}$ represents the set of grid cells containing the edge e, n is the total number of grid cells in the set

p is a grid cell in the set

$h_1(p)$ and $h_2(p)$ are the elevation of p on its two neighboring planes, respectively

If DH(e) is less than a predefined threshold T_Edge, edge e is classified as the intersection edge as shown in Figure 18.9c. If not, it is a step edge as shown in Figure 18.9b. The threshold T_Edge is determined by the error of LIDAR measurements. In our framework, T_Edge is set as $2\Delta h_T$. Δh_T is the elevation deviation threshold of a LIDAR point to a fitting plane and is used in the building patch segmentation algorithm in Section 18.3.2. As the height deviation of a point from the fitting plane error is assumed to be Δh_T in the worst situation, the height difference between two neighboring roof planes with height continuity can reach $2\Delta h_T$.

This step also checks each vertex if it is located on any intersection edge. If a vertex is located on an intersection edge or a step edge, it will be replaced by the joint point of the two roof facets

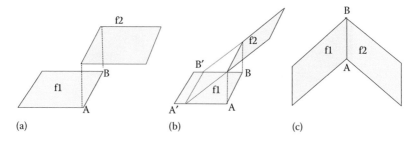

FIGURE 18.9 Types of edge (AB) between two roof facets: (a) step edge separating two paralleling planes, (b) step edge separating two intersection planes with height discontinuity, and (c) intersection edge separating two intersection planes with height continuity. (From Yan, J. et al., *IEEE Trans. Geosci. Remote Sens.*, 53, 1, 3, 2015. With Permission.)

forming the intersection edge or the wall where the step edge is located. If a vertex is located on two intersection edges, it will be replaced by the joint point of the three roof facets forming these two intersection edges. Figure 18.4h shows the 2D topology after refining several intersection edges in Figure 18.4f. Vertex 4 is replaced by the joint point of roof facets 1, 4, and 5 (see the facet labels in Figure 18.4c). Similar operations are applied to vertices 5, 9, and 10. It is evident that the refined 2D topology better approximates the real topology.

18.4   DATA PROCESSING

The study area is located at and around the campus of Florida International University (FIU), covering 6 km^2 of low relief topography. Surveyed features include residential houses, complex buildings, individual trees, forest stands, parking lots, open ground, ponds, roads, and canals. The data were collected on August 2003 with Optech ALTM 1233 systems operated by FIU. The Optech system recorded the coordinates (x, y, z) and intensity of the point measurements corresponding to the first and last laser returns. The dataset consists of five overlapping 340 m wide swaths of 13 cm diameter footprints spaced approximately 1 m apart. 2.5D and 3D building models were reconstructed for the FIU campus and an adjacent area to examine the effectiveness of the proposed framework. The thresholds used in our experiments for building reconstruction are listed in Tables 18.1 and 18.2.

TABLE 18.1
Parameters for Extracting 2.5D Building Models

Parameters	Values
Cell size (c_s) for progressive morphological filter	0.5 m
Height difference to aggregate a point (Δh_T)	0.2 m
Minimal surface on the roof (Min_Surface)	5 m^2 (20 c_s^2)
Minimal building area (Min_Building)	60 m^2 (240 c_s^2)
Douglas distance (T_Douglas)	1.5 m (3 c_s)
Threshold for footprint classification (T_SL)	0.3
Threshold for split adjustment (T_Projection_Final)	2 m (4 c_s)
Threshold for deviation (T_Deviation)	2 m (4 c_s)
Threshold for triangle area ratio (T_Ratio)	0.1
Threshold for footprint evaluation (T_Footprint)	0.85

Source: Zhang, K. et al., *IEEE Trans. Geosci. Remote Sens.*, 44, 9, 2523, 2006. With Permission.

TABLE 18.2
Parameters for Reconstructing 3D Building Models

Parameters	Values
Cell size (c_s) for snake algorithm	1 m
Height difference to aggregate a point (Δh_T)	0.15 m
Douglas distance (T_Douglas)	1.5 m
Threshold for split adjustment (T_Projection)	2 m
Threshold for plane adjusting T_Var	0.09 m
Weight for direction energy C_{Dir}	1
Weight for deviation energy C_{Dis}	5
Threshold for edge classification T_Edge	0.3 m
Window size for vertex adjusting W_v	5 m

A sensitivity analysis showed that small changes in most thresholds have little impact on the final results. It took 9, 2, 0.7, and 2 min for a personal computer with a 2.8 GHz processor and a 2 GB RAM to perform morphological filtering, building measurement identification, and 2.5D and 3D building reconstruction for the FIU campus dataset. A 2D array with about 7.2 million elements was employed to represent raw and interpolated points covering an area of 1.8 km². Aerial photographs and field investigation were used to help evaluate the reconstructed building models. The aerial photographs were collected in 1999 at a resolution of 0.3 m.

18.5 RESULTS

18.5.1 2.5D BUILDING MODEL

Both qualitative and quantitative methods were employed to measure the errors committed in extracting simple building models in this study. A qualitative method checked the quality of estimated dominant directions and derived footprints by visually comparing the extracted footprints with those in maps and aerial photographs. The quantitative method examines the accuracy of extracted footprints using count-based and area-based metric methods (Zhang et al. 2006). The count-based metric method quantifies commission and omission errors in the number of footprints identified, whereas the area-based metric method measures the area differences between identified and known building footprints. Figure 18.10 shows a 2.5D building map for the FIU campus based on refined building footprints and heights. The refined footprints were derived by applying footprint adjustment algorithms from Section 18.3.3 to raw footprints that were obtained by connecting boundary points of an identified building patch. The heights of the buildings were derived by averaging the elevation differences between building measurements and the DTM interpolated from ground measurements.

Comparison of 62 adjusted footprints with footprints from the map provided by the FIU Planning and Facility Management Department showed that all buildings were identified correctly in terms of number of extracted buildings. However, 10% of the building footprints were mistakenly removed, and 2% of the footprints were incorrectly included into the final output

FIGURE 18.10 2.5D building models for FIU campus. Each building model was created using the final footprint and average building height derived from LiDAR measurements. The DTM for building bases was derived by interpolating ground measurements identified by the progressive morphological filter. (From Zhang, K. et al., *IEEE Trans. Geosci. Remote Sens.*, 44, 9, 2523, 2006. With Permission.)

in terms of the total area of extracted buildings. These building models have been used to construct a 3D synthetic visual environment to animate hurricane-induced fresh water flooding at the FIU campus. The simple building model extraction algorithm was also applied to residential areas adjacent to the FIU campus. The commission and omission errors of extracted building footprints are both around 6% (Zhang et al. 2006).

The effectiveness of the footprint adjustment algorithm is well illustrated in Figures 18.11 and 18.12. Figure 18.11a shows that the raw building footprints have noise on their boundaries. Most of the noise was removed in the adjusted footprints as shown in Figure 18.11b, and the smoothness

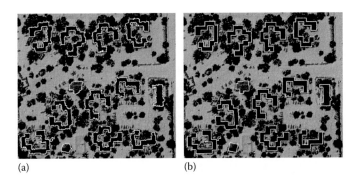

(a)   (b)

FIGURE 18.11 Comparison of raw (a) and adjusted (b) footprints for 2.5D building models. The small *zig-zag* noise in the raw footprints was removed in the adjusted footprints, making the adjusted footprints look more realistic. The background images were produced by interpolating point LiDAR measurements.

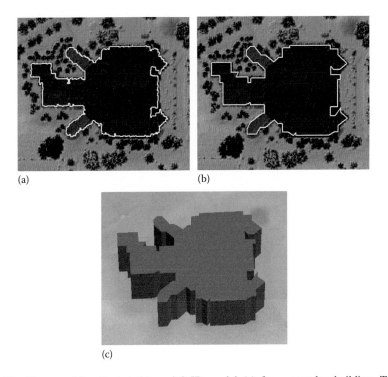

(a)   (b)

(c)

FIGURE 18.12 The raw (a), adjusted (b), and 2.5D model (c) for a complex building. Two dominant directions of the footprint are nearly horizontal (x) and vertical (y). Note that oblique portions of the building that are not aligned with the dominant directions were well preserved in the simple building model.

of the building outline was greatly improved. Figure 18.12 shows a complex building that consists of footprint segments parallel and oblique to the dominant directions. The portion of the building whose direction is different from the dominant directions was also adjusted appropriately.

18.5.2 3D BUILDING MODEL

Figure 18.7 shows an example of 3D building models at the FIU campus. The 2D topology of the building roof consists of 46 vertices and 67 boundaries. Although the topology looks very complicated, the noise in the edges of the building footprint was reduced greatly after the adjustment was performed using the snake-based algorithm (Figure 18.7c). A 3D building model was created on the basis of the adjusted roof facets, roof edges, and associated height data (Figure 18.7d). Compared with a 2.5D building model (Figure 18.12), a 3D model includes much more details of roof facets and looks more realistic. Figure 18.13 shows a 2.5D residential building with nonflat roof facets. Various nonflat roof facets were reconstructed well by the proposed algorithm.

Figure 18.14a and b shows extracted 2D topologies and 3D models for commercial and residential buildings next to the FIU campus. It is very difficult to quantify the errors of the algorithms for reconstructing 3D building models as no ground-truth data with higher accuracy

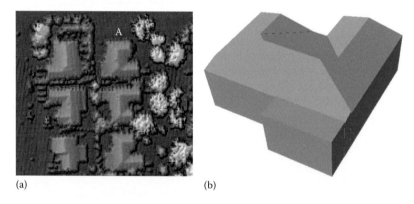

(a) (b)

FIGURE 18.13 Building and tree images interpolated from LiDAR measurements for a residential area (a) and a 3D building model for a residential building (b). The 3D building model was derived by applying the proposed building reconstruction framework to LiDAR measurements. The reconstructed model is for a building indicated by the letter A in (a).

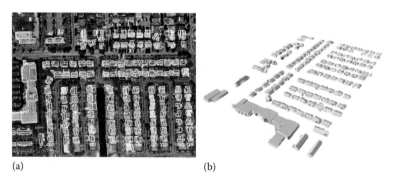

(a) (b)

FIGURE 18.14 2D topology (a) and reconstructed 3D models (b) for commercial and residential buildings next to the FIU campus. (From Yan, J. et al., *IEEE Trans. Geosci. Remote Sens.*, 53, 1, 3, 2015. With Permission.)

are available. Digitizing 3D building models manually from LiDAR measurements or DSM is impractical. Derivation of 3D building models from aerial photographs for the study areas is also impossible because no overlapped photographs are available. We qualitatively examined the quality of extracted 3D building models by comparing the models with raw LiDAR measurements, LiDAR DSMs, and aerial photographs in ArcGIS. The results indicated that most extracted 2D topologies represent the boundaries of connected roof facets well. Most buildings (196/211) were reconstructed properly after applying the building refinement algorithms.

18.6 DISCUSSION AND CONCLUSIONS

A framework including a series of algorithms has been developed to automatically reconstruct 2.5D and 3D building models from LiDAR measurements. The framework includes five major components: (1) the progressive morphological filter for separating the ground and nonground measurements, (2) a region-growing algorithm based on a local plane-fitting technique for segmenting building measurements, (3) an algorithm for estimating the dominant direction of a building, (4) a method for extracting 2D topology, and (5) a snake-based algorithm for adjusting the 2D topology. The entire process is highly automatic and requires little human input, which is very useful for processing voluminous LiDAR measurements.

The application of the framework to the FIU campus and adjacent residential areas shows that the region growing algorithm identified building patches well, the snake algorithm adjusted most buildings properly, and 2.5D and 3D building models were reconstructed effectively. The quantitative accuracy analysis indicates that all buildings were extracted and about 12% errors in the total area of footprints were committed by the proposed algorithms in reconstructing 2.5D building models, despite the fact that there are several complex buildings on the FIU campus.

Accurate segmentation of roof facets is critical for extraction of 2D topology and reconstruction of 3D building models. Numerical experiments demonstrated that the segmentation accuracy was limited by the density of LiDAR measurements. For example, the 3D model in Figure 18.14, did not generate a good adjustment of the edges between small roof facets due to the low density of LiDAR measurements. Segmentation is also sensitive to the errors in LiDAR measurements; therefore, the segmentation based on a single strip of LiDAR measurements is more robust than overlapped strips of LiDAR measurements because relatively large errors are introduced by the data from overlapped strips. This poses a serious challenge for using multistrip data for reconstruction of 3D building models. Further improvement in segmentation is needed for reconstruction of better 3D building models. Changes of boundaries between roof facets within a building footprint have little effect on the extraction of the outline of the footprint. The derivation of the footprint outline is robust because of a distinctive difference in elevations between building and adjacent ground measurements. Therefore, the performance of the region growing algorithm has much less effect on reconstruction of 2.5D building models than on reconstruction of 3D building models.

ACKNOWLEDGMENTS

The authors would like to thank the anonymous reviewers for valuable comments. This work was supported by the Florida Hurricane Alliance Research Program sponsored by the National Oceanic and Atmospheric Administration and the Florida Public Hurricane Loss Model Project sponsored by Florida Office of Insurance Regulation.

REFERENCES

Adams, R. and L. Bischof, 1994. Seeded region growing. *IEEE Transactions on Pattern Analysis and Machine Intelligence*, 16, 641–647.

Alharthy, A. and J. Bethel, 2002. Heuristic filtering and 3D feature extraction from LiDAR data. In *Proceedings of ISPRS Commission III, Symposium*, International Society for Photogrammetry and Remote Sensing (ISPRS), Graz, Austria, pp. 23–28.

Amini, A. A., T. E. Weymouth, and R. C. Jain, 1990. Using dynamic programming for solving variational problem in vision. *IEEE Transactions on Pattern Analysis and Machine Intelligence*, 12, 855–867.

Brenner, C., 2000. Towards fully automatic generation of city models. *International Archives of Photogrammetry and Remote Sensing*, 33(B3/1), 84–92.

Brenner, C., 2004. Modeling 3-D objects using weak CSG primitives. *International Archives of Photogrammetry, Remote Sensing and Spatial Information Sciences*, 35(2004), 1085–1090.

Carter, W. E., R. L. Shrestha, and K. C. Slatton, 2004. Photon counting airborne laser swath mapping (PC-ALSM). In *SPIE Proceedings*: *Remote Sensing Applications of the Global Position System*, SPIE-International Society for Optical Engine. Hawaii, HI, pp. 78–85.

Cohen, L. D. and I. Cohen, 1993. Finite-element methods for active contour models and balloons for 2-D and 3-D images. *IEEE Transactions on Pattern Analysis and Machine Learning*, 15, 1131–1147.

Douglas, D. H. and T. K. Peucker, 1973. Algorithms for the reduction of the number of points required to represent a digitized line or its caricature. *The Canadian Cartographer*, 10, 112–122.

Elaksher, A. F. and J. S. Bethel, 2002. Reconstructing 3D Buildings from LIDAR Data. In *ISPRS Commission III Symposium, Photogrammetric and Computer Vision*, ISPRS, Graz, Austria, pp. 102–107.

Elberink, S. O. and H. G. Maas, 2000. The use of anisotropic height texture measures for the segmentation of airborne laser scanner data. *International Archive of Photogrammetry & Remote Sensing*, 33(B3/2), 678–684.

Filin, S., 2002. Surface clustering from airborne laser scanning data. In *ISPRS Commission III Symposium, Photogrammetric and Computer Vision*, ISPRS, Graz, Austria, pp. 119–124.

Filin, S. and N. Pfeifer, 2006. Segmentation of airborne laser scanning data using a slope adaptive neighborhood. *ISPRS Journal of Photogrammetry and Remote Sensing*, 60, 71–80.

Gruen, A. and X. Wang, 2001. News from cybercity-modeler. In *Automatic Extraction of Man-Made Objects from Aerial and Space Images (III)*, Swets & Zeitlinger, Lisse, the Netherlands, pp. 93–101.

Jensen, J. R., 2015. *Remote Sensing of the Environment*, 4th ed. Prentice Hall: Upper Saddle River, NJ.

Kass, M., A. Witkin, and D. Terzopoulos, 1988. Snakes: Active contour models. *International Journal of Computer Vision*, 1, 321–331.

Maas, H. G., 1999. Fast determination of parametric house models from dense airborne laser scanner data. In *ISPRS Workshop on Mobile Mapping Technology*, Bangkok, Thailand.

Maas, H. G. and G. Vosselman, 1999. Two algorithms for extracting building models from raw laser altimetry data. *ISPRS Journal of Photogrammetry and Remote Sensing*, 54, 153–163.

Meng, X., N. Currit, and K. Zhao, 2010. Ground filtering algorithms for airborne LiDAR data: A review of critical issues. *Remote Sensing*, 2, 833–860.

Morgan, M. and A. Habib, 2002. Interpolation of LIDAR data and automatic building extraction. In *ACSM-ASPRS 2002 Annual Conference Proceedings*. Washington, DC.

Morgan, M. and K. Tempfli, 2000. Automatic building extraction from airborne laser scanning data. In *Proceeding of the 19th ISPRS Congress*, ASPRS, Amsterdam, the Netherlands, pp. 616–623.

Overby, J., L. Bodum, E. Kjems, and P. M. Ilsøe, 2004. Automatic 3D building reconstruction from airborne laser scanning and cadastral data using Hough transform. In *ISPRS Proceedings*, Organising Committee of the XXth International Congress for Photogrammetry and Remote Sensing. pp. 1–6.

Pfeifer, N. and G. Mandlburger, 2017. LiDAR data filtering and DTM generation. In Shan, J., and Toth, C.K., (Ed.), *Topographic Laser Ranging and Scanning*: *Principles and Processing*. CRC Press: Boca Raton, FL.

Rottensteiner, F. and J. Jansa, 2002. Automatic extraction of buildings from LIDAR data and aerial images. *International Archives of Photogrammetry and Remote Sensing*, 34(4), 569–574.

Sampath, A. and J. Shan, 2006. Clustering based planar roof extraction from LiDAR data. In *American Society for Photogrammetry and Remote Sensing Annual Conference*, ASPRS, Reno, NV. pp. 1–6.

Sampath, A. and J. Shan, 2010. Segmentation and reconstruction of polyhedral building roofs from aerial LiDAR point clouds. *IEEE Transactions on Geoscience and Remote Sensing*, 48, 1554–1567.

Schwalbe, E., H. G. Maas, and F. Seidel, 2005. 3-D building model generation from airborne laser scanner data using 2-D data and orthogonal point cloud projections. In *ISPRS WG III/3, III/4, V/3 Workshop Laser scanning 2005*, Enschede, the Netherlands.

Shan, J. and A. Sampath, 2005. Urban DEM generation from raw LIDAR data: A labeling algorithm and its performance. *Photogrammetric Engineering and Remote Sensing*, 71, 217–226.

Vosselman, G., 2000. Slope based filtering of laser altimetry data. *International Archives of Photogrammetry and Remote Sensing*, 33(B4), 958–964.

Vosselman, G., 2001. Building reconstruction using planar faces in very high density height data. *International Archive of Photogrammetry & Remote Sensing*, 34(3/W4), 211–218.

Weidner, U. and W. Forstner, 1995. Towards automatic building reconstruction from high resolution digital elevation models. *ISPRS Journal of Photogrammetry & Remote Sensing*, 50, 38–49.

Yan, J., J. Shan, and W. Jiang, 2014. A global optimization approach to roof segmentation from airborne LiDAR point clouds. *ISPRS Journal of Photogrammetry and Remote Sensing*, 94, 183–193.

Yan, J., K. Zhang, C. Zhang, S. C. Chen, and G. Narasimhan, 2007. A graph reduction method for 2D snake problems. In *IEEE Computer Society Conference on Computer Vision and Pattern Recognition*, Minneapolis, MN, pp. 1–6.

Yan, J., K. Zhang, C. Zhang, S. C. Chen, and G. Narasimhan, 2015. Automatic construction of 3-D building model from airborne LiDAR data through 2-D snake algorithm. *IEEE Transactions on Geoscience and Remote Sensing*, 53, 3–14.

You, R.-J. and B.-C. Lin, 2011. A quality prediction method for building model reconstruction using LiDAR data and topographic maps. *IEEE Transactions on Geoscience and Remote Sensing*, 49, 3471–3480.

Zhang, K., S. C. Chen, D. Whitman, M. L. Shyu, J. Yan, and C. Zhang, 2003. A progressive morphological filter for removing non-ground measurements from airborne LIDAR data. *IEEE Transactions on Geoscience and Remote Sensing*, 41, 872–882.

Zhang, K., J. Yan, and S. C. Chen, 2006. Automatic construction of building footprints from airborne LIDAR data. *IEEE Transactions on Geoscience and Remote Sensing*, 44, 2523–2533.

Zhang, K., J. Yan, and S. C. Chen, 2009. Automatic 3D building reconstruction from airborne LiDAR measurements. In *2009 Urban Remote Sensing Joint Event*, IEEE, Shanghai, China, pp. 483–487.

19 Quality of Buildings Extracted from Airborne Laser Scanning Data—Results of an Empirical Investigation on 3D Building Reconstruction

Eberhard Gülch, Harri Kaartinen, and Juha Hyyppä

CONTENTS

19.1   INTRODUCTION

Three-dimensional geographical information systems have been since more than two decades of increasing importance in urban areas in various applications such as urban planning, visualization, environmental studies and simulation (pollution, noise), tourism, facility management, telecommunication network planning, 3D cadastre, and vehicle or pedestrian navigation. In the late 1990s, European Organization for Experimental Photogrammetric Research (OEEPE) conducted a survey on 3D-city models. The scope of that study was to find out the state of the art of generating and using 3D city data. The study was based on a questionnaire sent out to about 200 European institutions (Fuchs et al., 1998). Fifty-five responded to the questionnaire from seventeen countries (curiously, 30% of the replies came from Germany, which were certainly influenced by the activities of three major mobile phone companies in 3D city modeling). The most important objects of interest according to the users were buildings (95%), traffic network (76%), and vegetation (71%). It was stated that there was a definite lack of economical techniques for producing 3D-city data. The lack of knowledge of information sources as well as high costs for building and vegetation acquisition hindered broader use at that time. Data sources used by the producers were aerial imagery (76%), map data (54%), classical survey data (46%), and aerial range data/laser scanning data (20%). At that time, range data were regarded by many producers as too expensive. In addition, the usual point density acquired at that time in airborne laser scanning (ALS) did not really allow for a broader application of object extraction in urban areas.

Semiautomatic and automatic methods for 3D city models are aimed at reducing the costs of providing these data with reasonable level of detail. Today, aerial photogrammetry is still one of the major techniques for obtaining accurate and reliable 3D building information. By applying digital aerial photogrammetry, the accurate measurement of points and structures can be supported, which are usually defined by a human operator (Brenner, 2005). Despite significant research efforts in the past, the low degree of automation achieved has remained for a long time the major problem. Thus, the majority of development work has in the last 20 years focused on semiautomatic systems, in which, for example, recognition and interpretation tasks are performed by the human operator, whereas modeling and precise measurement are supported by automation.

Due to the development of scanning systems and improvements in the accuracy of direct georeferencing, ALS became a feasible technology to provide range data in the early 1990s. At the end of that decade, ALS was already considered as a mature technology (Baltsavias, 1999). ALS provides since more than one decade, dense point clouds (more than 10 returns/m² are possible today) with 3D coordinates. This makes range data segmentation relatively easy and feasible for the modeling of buildings (Brenner, 2005).

A new research field was opened with the integration of laser point clouds and photogrammetric processes with aerial photos. This provided also new technological solutions. By combining the expected good height assessment accuracy of laser scanner and expected good planimetric accuracy of aerial images, both high accuracy and higher automation can in theory be obtained. However, despite the progress that has been made with integrating laser scanning systems and digital images, automated processing of the resulting datasets remained for long time at an early research stage (Brenner, 2005).

A short summary of the state of the art in building extraction methods a decade ago can be obtained from Baltsavias (2004), Brenner (2001, 2005), Gülch (2004), and Haala (2004). This is important to understand the basic conditions for the tests conducted, that are described in this chapter. More details on building extraction methods with emphasis on achievements or impact on the development can be obtained from Alharthy and Bethel (2004), Ameri and Fritsch (2000), Baillard and Dissard (2000), Baillard and Maître (1999), Braun et al. (1995), Brenner (2003), Brunn and Weidner (1997, 1998), Brunn et al. (1998), Centeno and Miqueles (2004), Chen et al. (2004), Cho et al. (2004), Cord and Declerq (2001), Cord et al. (2001), Dash et al. (2004), Dold and Brenner (2004), Elaksher and Bethel (2002), Elaksher et al. (2003), Fischer et al. (1998), Forlani et al. (2003, 2006), Fraser et al. (2002), Fuchs and Le-Men (2000), Fujii and Arikawa (2002), Förstner (1999), Grün (1997, 1998), Grün and Wang (1998a, b, 1999a, b), Grün et al. (2003), Haala et al. (1998), Haala and Brenner (1999), Haithcoat et al. (2001), Hofmann (2004), Hofmann et al. (2002, 2003), Jaynes et al. (2003), Jutzi and Stilla (2004), Jülge and Brenner (2004), Khoshelham (2004), Kim and Muller (1998), Lee and Choi (2004), Li et al. (2004), Maas and Vosselman (1999), Maas (2001), Madhavan et al. (2004), Mayer (1999), Morgan and Habib (2001, 2002), Neidhart and Sester (2003), Niederöst (2003), Noronha and Nevatia (2001), Oda et al. (2004), Oriot and Michel (2004), Overby et al. (2003), Paparoditis et al. (1998), Peternell and Steiner (2004), Rottensteiner (2000, 2003), Rottensteiner and Briese (2002), Rottensteiner and Jansa (2002), Rottensteiner et al. (2003, 2004, 2005), Sahar and Krupnik (1999), Schwalbe (2004), Sequeira et al. (1999), Shufelt (1999), Sinning-Meister et al. (1996), Sohn (2004), Stilla and Jurkiewicz (1999), Süveg and Vosselman (2004), Söderman et al. (2004), Taillandier and Deriche (2004), Takase et al. (2004), Tan and Shibasaki (2002), Tsay (2001), Tseng and Wang (2003), Vosselman (2002), Vosselman and Dijkman (2001), Vosselman and Süveg (2001), Wang and Grün (2003), Weidner and Förstner (1995), Zhan et al. (2002), Zhang and Wang (2004), and Zhao and Shibasaki (2003). Already a decade ago, one could observe a period of mutual exchange of experiences from laserscanner data analysis and multiimage matching to derive (dense) 3D point clouds, and thus we avoided to sort the example references into fixed categories.

All methods in the cited references are usually evaluated by the producer of the software only. Very rarely, methods are compared with several other competing methods. Even less comparisons were done for comparing photogrammetric methods based on digital imagery with methods based on ALS data.

Due to the rapid development of sensors and methods at the beginning of the millennium, it was proposed in 2003 and accepted that under the EuroSDR Commission III "Production Systems and Processes," headed by President Eberhard Gülch, a joint test would be undertaken to compare various methods. Juha Hyyppä was assigned the project leader. The major part of the investigations was carried out by Harri Kaartinen. The former Finnish Geodetic Institute (FGI), now Finnish Geospatial Research Institute, acted as pilot center.

The objectives of the EuroSDR Building Extraction project were to evaluate the quality, accuracy, feasibility, and economical aspects of the following

1. Semiautomatic building extraction based on photogrammetric techniques. Here, with the emphasis on commercial and/or operative systems.
2. Semiautomatic and automatic building extraction techniques based on high density laser scanner data. Here, with the emphasis on commercial and research systems.
3. Semiautomatic and automatic building extraction techniques based on integration of laser scanner data and aerial images. Here, mainly research systems were expected.

This book's chapter summarizes the results obtained in the comparison. We first describe the test data and the software used. Then we present the analysis of the test. We will focus on the ALS aspect, but we will include the digital photogrammetric techniques based on image data and also the results of hybrid systems, as they seemed to have a high potential in future developments, which is clearly confirmed by the current research and development in the field.

We will also focus on buildings, as this is still the major object type in 3D city modeling. EuroSDR has also performed a test on tree extraction from ALS data, headed by former EuroSDR Commission 2 president Juha Hyyppä, which is, however, not part of the chapter here.

In a final outlook part, we try to give recommendations for further development in the methods applied and the ways how results are evaluated on the basis of different available reference data.

It is, however, not possible to include current software in the test results, but we briefly try to relate the outcome reached to current trends in research and development.

19.2 TEST DATASETS

19.2.1 OVERVIEW

The project consisted of three sites provided by the FGI, namely Espoonlahti, Hermanni, and Senaatti and one site, Amiens, provided by the Institut Geographique National (IGN). The test sites were selected not only according to various criteria, such as specific characteristics, namely density, type and complexity of buildings, and vegetation and undulating terrain aspects, but also due to availability of various input data sources and the possibility to generate ground truth data.

For each test site, the following data were available and could be downloaded from the FGI ftp-site:

- Aerial images
- Camera calibration and image orientation information
- Ground control point coordinates and jpg-images of point locations (with the exception of Amiens)
- Airborne laser scanner data
- Cadastral map vectors of selected buildings (vector ground plans)
- Espoonlahti: 11 buildings
- Hermanni: 9 buildings
- Senaatti: 6 buildings/blocks
- Amiens: 7 buildings

Participants were requested to create the vectors of the 3D building models from the given areas in all four test sites using the materials described earlier. Participants could use any methods and data combinations they wished. The 3D-model should consist of permanent structures of the test area: buildings modeled in as much detail as possible (this mainly concerns roof structures) and terrain so that it would be possible to measure building heights using the model. The provided vector ground plans could be used in methods that require this information. They were also used by the pilot center to check the quality of the results.

19.2.2 FINNISH GEODETIC INSTITUTE TEST SITES ESPOONLAHTI, HERMANNI, AND SENAATTI

Each of the test sites had their own characteristics:

- Espoonlahti: With a significant variety of houses, partly row houses, undulating terrain and a large number of trees, Espoonlahti is located in Espoo, about 15 km west of Helsinki with high-rise buildings and terraced houses. Laser scanner data and one aerial image are shown in Figures 19.1 and 19.2.
- Hermanni: A large, simple block of flat or gabled houses with low vegetation, Hermanni is a residential area about 3 km from the main city center with four to six storey houses built mainly in the 1950s. Laserscanner data and one aerial image are shown in Figures 19.3 and 19.4.
- Senaatti: A typical European city center with some complex, historic buildings and no vegetation, Senaatti includes the area around the Senate Square in Helsinki main city centre, three to six storey houses, and Lutheran Cathedral built mainly in the nineteenth century. Laserscanner data and an aerial image are shown in Figures 19.5 and 19.6.

FIGURE 19.1   Airborne laser scanning data of FGI test site Espoonlahti. (Courtesy of Finnish Geodetic Institute, Finland. With Permission.)

FIGURE 19.2   Digital aerial image data of FGI test site Espoonlahti. (Courtesy of Finnish Geodetic Institute, Finland. With Permission.)

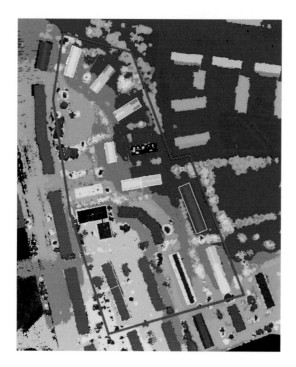

FIGURE 19.3 Airborne laser scanning data of FGI test site Hermanni. (Courtesy of Terrasolid Ltd and Helsinki City Survey Division, Finland. With Permission.)

FIGURE 19.4 Airborne laser scanning data of FGI test site Hermanni. (Courtesy of Helsinki City Survey Division, Finland. With Permission.)

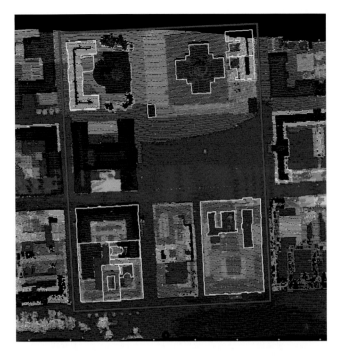

FIGURE 19.5   Airborne laser scanning data of FGI test site Senaatti. (Courtesy of Blom Kartta Oy, Finland. With Permission.)

FIGURE 19.6   Digital aerial image data of FGI test site Senaatti. (Courtesy of Helsinki City Survey Division, Finland. With Permission.)

TABLE 19.1

Features of Aerial Images of FGI Test Sites

	Espoonlahti	Hermanni	Senaatti
Photos	Stereo pair	Stereo pair	Stereo pair
Date	June 26, 2003	May 4, 2001	April 24, 2002
Camera	RC-30	RC-30	RC-30
Lens	15/4 UAG-S, no 13355	15/4 UAG-S, no 13260	15/4 UAG-S, no 13260
Calibration date	November 22, 2002	January 18, 2000	April 14, 2002
Flying height, scale	860 m, 1:5300	670 m, 1:4000	660 m, 1:4000
Pixel size	14 microns	15 microns	14 microns
Pixel size on the ground	7.5 cm	6 cm	5.5 cm

TABLE 19.2

Features of Laser Scanner Data of FGI Test Sites

	Espoonlahti	Hermanni	Senaatti
Acquisition	May 14, 2003	End of June 2002	June 14, 2000
Instrument	TopoSys Falcon	TopEye	TopoSys-1, pulse modulated
Flight altitude	400 m	200 m	800 m
Pulse frequency	83,000 Hz	7,000 Hz	83,000 Hz
Field of view	$\pm 7.15°$	$\pm 20°$	$\pm 7.1°$
Measurement density	10–20 per m^2	7–9 per m^2 on the average	1.6 per m^2 on the average
Swath width	100 m	Ab. 130 m	Ab. 200m
Mode	First pulse	2 pulses	First pulse

All FGI test sites were flown with an aerial film camera in large image scales and digitized with a photogrammetric scanner yielding ground pixel sizes of 5.5–7.5 cm (cf. Table 19.1).

In addition to the construction type, that differs from site to site, the test sites have also been flown with different laser scanners (TopEye, TopoSys-I, TopoSys-Falcon) and with different pulse densities (from 1.6 to about 20 pulses per m^2) as shown in Table 19.2. Last pulse data were not available for all the test sites, and therefore, first pulse data were considered as the major data.

Airborne images were in Tagged Image File Format (TIFF)-format, with orientation parameters and ground point coordinates that could be used as control or check points. Laser scanner data were in American Standard Code for Information Interchange (ASCII) format (XYZ point data). Cadastral map vectors were in Drawing Interchange File Format (DXF)-format.

In the preprocessing of Senaatti and Hermanni data, the laser scanner data were matched visually by the pilot center to map vectors in the XY-directions, and height was shifted using known height points. These shifts were done after the original laser scanner observations were transformed from WGS84-coordinates to a local rectangular coordinate system.

19.2.3 INSTITUT GEOGRAPHIQUE NATIONAL TEST SITE AMIENS

This study site, with a high density of small buildings, is located in the city of Amiens in Northern France. Laser acquisition with a density of 4 points/m^2 came from the TopoSys system (cf. Figure 19.7). Airborne images were part of a digital acquisition (June 23, 2001) using IGN's France digital camera. The images were in TIFF-format, and the ground pixel size is around 25 cm (cf. also Figure 19.8). The main characteristic of these acquisitions is the large overlap rate (around 80%) between the different images: each point of the terrain appears on a large number of images. Amiens consists of four strips with a total of 11 images. A digitized cadastral map, with 2D description of buildings, was provided in DXF-format.

FIGURE 19.7   Airborne laser scanning data of Amiens test site. (Courtesy of IGN, France. With Permission.)

FIGURE 19.8   Digital aerial image data of Amiens test site. (Courtesy of IGN, France. With Permission.)

19.3   ACCURACY EVALUATION

We decided to focus on two aspects of accuracy evaluation: (i) the analysis using reference points and (ii) the analysis using reference raster ground plans. When using reference points, we can derive information on the quality of location, building height and building length parameters, and roof inclination. With the comparison with ground plan reference, we can derive information on shape similarity or dissimilarity.

19.3.1   REFERENCE DATA

On test sites, Espoonlahti, Hermanni, and Senaatti reference data were collected from November–December 2003 using a Trimble 5602 DR200+ tacheometer. Measured targets included corners of walls, roofs, chimneys and equivalent constructions, and ground points next to building corners. In this survey altogether, about 980 points were measured in Espoonlahti, 400 points in Hermanni, and 200 points in Senaatti. Known points were used to orientate the tacheometer to the test site's coordinate system.

On all three FGI test sites, repeated observations to the same targets from different station setups were made to control the uniformity and accuracy of reference measurements. The differences in these repeated measurements were on average 4.7 cm in planimetry (maximum 8.3 cm) and 1.2 cm in height (maximum 3.5 cm) based on 19 control observations in total. It is a fact that distance measurement directly to the surface of, for example, a building wall is affected by the angle between the wall and the measuring beam, especially at long distances (as the laser footprint expands). This effect can also be seen in those control measurements in which measured distances were somewhat longer than those mainly used in this test, thus, giving a slightly more pessimistic value of the total accuracy expected.

Reference data for the Amiens test site included thirty-two roof points measured manually from aerial images. The given accuracy (standard deviation) of these points was 25 cm for X, Y, and Z. Reference data were measured and delivered by IGN, France.

After some minor updating and refinements on the basis of reference measurements in the field, the cadastral maps of Helsinki and Espoo City Survey Division were used to produce raster images of building ground plans with a pixel size of 10×10 cm^2 for the test sites Hermanni and Espoonlahti.

19.3.2   ACCURACY EVALUATION METHODS

Analysis Using Reference Points—Reference points were used to analyze the accuracy of the location (single point measurement), length (distance between two points), and roof inclination (slope between two points) of the modeled buildings. Single points were analyzed separately for planimetric and height errors. If a digital elevation model (DEM) was included, building heights were measured to the DEM using bilinear interpolation. If not, wall vectors were used. On some models, the same ground height was used for whole buildings resulting in large errors in building height determination. These cases were marked at the building height analysis stage and left out of the figures presenting the building height results.

Root mean squared error (abbreviated to RMSE, Equation 19.1) was calculated for building length, height, and roof inclination:

$$\text{RMSE} = \sqrt{\frac{\sum_{i=1}^{n}(e_{1i} - e_{2i})^2}{n}} \tag{19.1}$$

where:

e_{1i} is the result obtained with the described retrieved model
e_{2i} is the corresponding reference measured value
n is the number of samples

In addition, minimum, maximum, medium, mean, standard deviation, and interquartile range (IQR, Equation 19.2) values were calculated. IQR values represent the range between the 25th and 75th quartiles:

$$\text{IQR} = p_{75th} - p_{25th} \tag{19.2}$$

where:

p_{75th} is the value at the 75th quartile
p_{25th} is the value at the 25th quartile

For example, if the IQR is 20 cm and the median value is 0, then 50% of the errors are within ±10 cm. Significantly, deviating measurements (outliers) were detected using threshold levels: the lower bound at the 25th quartile minus 1.5*IQR and the upper bound at the 75th quartile plus 1.5*IQR. IQR is not as sensitive to large deviations (gross errors) as standard deviation/error. In addition, the coefficient of determination R^2 was calculated to help one to separate cases between low variability of the reference data and high estimation accuracy, and high variability of the reference data and low estimation accuracy. Descriptive statistics are presented before and after outliers have been removed.

Analysis Using Reference Raster Ground Plans—In the Espoonlahti and Hermanni test sites, reference raster ground plans were used to compute the total relative building area and total relative shape dissimilarity (Henricsson and Baltsavias, 1997). Total relative building area gives the difference between modeled building area and reference building area. Total relative shape dissimilarity is the sum of the area difference and the remaining overlap error, that is, the sum of missing area and extra area divided by the reference area. The vector models delivered by participants were rasterized (pixel size 10 cm) and compared with reference raster ground plans. For those participants that modeled only buildings with given ground plans, the reference raster ground plans consist only of these buildings. Total relative building area and shape dissimilarity were computed in two ways. First, all buildings in the test sites were included, giving the values for the whole test site. Second, only those buildings that exist on both reference and delivered model were included, giving the values for modeled buildings.

The eaves of the roofs are not included in reference raster ground plans, thus making the reference buildings somewhat smaller than extracted buildings in the Hermanni test site (in Espoonlahti, there were no eaves, just straight walls from the ground to the roof).

19.4   PARTICIPANTS' RESULTS

3D-models were obtained from 10 participants coming from 11 organizations (Table 19.3). In this table, we give the used formats of the received building vectors, and we have indicated the cases where we also received DEM information by the participants.

19.5   APPLIED METHODS AND SOFTWARE BY PARTNERS

19.5.1   OVERVIEW

The participants have applied quite different methods, and Table 19.4 summarizes the data use, degree of automation, and time use of those applied methods. The participants and their specific method are sorted for the used data. First, we give the purely image based methods, then some hybrid methods, and finally the methods relying on laser scanning data. Among those, we present at the end the methods that require ground plan data. The level of automation is indicated as given by the participants and ranges from low to high. Here, we can already see a tendency for high automation with the purely laser scanning methods using given ground plan information, whereas the methods relying on image data are mainly classified as low to medium. The time use was also given by the participants in one of the three categories: low, medium, and high. It was not able to receive absolute figures in hours or minutes and partly no answers were given, or they were unclear. Significantly, in the expected difficult sites Senaatti and Espoonlahti, we can observe also low time usage, whereas in the easy to moderate case of Hermanni, we can find also medium-to-high time usage. We have added also the experience of the operator as a further feature, showing that some of the software can be handled without extensive training.

In the following, we describe the single methods following the above-mentioned order. We base the descriptions on the provided description by the participants and references in literature. Please note that some of the methods may be not described in a comparable level of detail, due to confidentiality reasons.

TABLE 19.3

Participant-Generated 3D-Models and Data Formats

| | Test Site | | | | | | | |
| | Espoonlahti | | Hermanni | | Senaatti | | Amiens | |
Participant	Building Vectors	DEM	Building Vectors	DEM	Building Vectors	DEM	Building Vectors	DEM
CyberCity AG, Switzerland	DXF	ArcView	DXF	ArcView	DXF	ArcView		
Delft University of Technology, the Netherlands			DXF and VRML	VRML	DXF and VRML	VRML		
Hamburg University of Applied Sciences and Nebel + Partner, Germany	DXF		DXF		DXF		DXF	
Institut Geographique National, France (IGN)	DXF and VRML	DXF and VRML	DXF and VRML	DXF and VRML	DXF and VRML	DXF and VRML	DXF and VRML	DXF and VRML
Swedish Defence Research Agency (FOI)	DXF		DXF		DXF		DXF	
University of Aalborg, Denmark			VRML					
C+B Technik, Germany	DXF and ASCII		DXF and ASCII		DXF and ASCII		DXF and ASCII	
Institut Cartografic de Catalunya, Spain (ICC)	DGN		DGN		DGN		(DGN)	
Dresden University of Technology, Germany			VRML		VRML			
University of Applied Sciences, Stuttgart, Germany	DXF		DXF		DXF			

19.5.2 CyberCity AG

CyberCity used aerial images, camera calibration, and exterior orientation information. Work was done by experienced personnel. CyberCity methods are based on concepts reported by Grün and Wang (1998a, b; 1999a, b) and commercialization of the methods in CyberCity AG. The modeling is based on two steps. First, manual photogrammetric stereo measurements are carried out using an in-house developed digital photogrammetric workstation Visual Star. The point measurements are taken in a certain order, and points are attached certain codes using CyberCity's Point Cloud Coding System that will help the more automatic process. Then, the resulting point clouds are imported into the main module CC-Modeler that automatically triangulates the roof points and creates the 3D building models. The building footprints (i.e., derived from the roof outline) and walls are generated by projecting and intersecting the boundary roof points with the digital terrain model (DTM), if that exists. Alternatively, cadastral building footprints can be projected back to the roof, which results in realistic roof overhangs and fulfills the consistency between 2D cadastral

TABLE 19.4
Summary of Used Data, Level of Automation, and Time Use of Building Extraction

Participant	Operator	Laser Data	Aerial Images	Ground Plan	Level of Automation	Espoonlahti	Hermanni	Senaatti	Amiens
Cybercity	Exp.		100%		Low	Medium	Medium	Medium	
Hamburg	Inexp.		100%		Low	Medium		High	
Stuttgart	Inexp.		100%		Low-high	Low	Low	?	
IGN (Amiens)	Exp.		100%		Medium				Low
IGN (Espl, Her, Sen)	Exp.	50%	50%		Medium	Low	Low	Low	
ICC laser + aerial	Inexp.	80%	20%		Low-medium	High	Medium	High	
Nebel + Partner	Inexp.	90%	10%		Medium	High	High	Medium	High
ICC laser	Inexp.	100%			Low-medium	High	High	High	?
FOI	Exp.	100%			High	Low	Low	Low	Low
FOI outlines	Exp.	100%		X	High	Low	Low	Low	Low
C+B Technik	Exp.	100%		X	Medium-high	?	?	?	?
Delft	Exp.	100%		X	Medium-high		?	Medium	
Aalborg	Inexp.	100%		X	High		?		
Dresden	Exp.	100%		X	High		Low	Low	

? = not available

maps and 3D city models. With CC-Edit geometric corrections can be carried out to get accurate and correct 3D city models with planar faces and without overlaps or gaps between neighboring buildings and so on. This is, for example, required to derive orthogonal and parallel lines of buildings from the manual stereo measurements.

19.5.3 Hamburg University of Applied Sciences

Hamburg University of Applied Sciences used a digital photogrammetric workstation DPW770 with SocetSet from BAE Systems/Leica Geosystems (BAE System, 2000) to extract buildings by manual stereo measurements using aerial images, camera calibration, and exterior orientation information. Photogrammetrically measured data were corrected and improved in an AutoCAD 2000 system. All measurements and modeling was performed by an inexperienced person. After import of the digital aerial images in the digital station using the given camera calibration and exterior orientation data, an automatic interior orientation of each image was performed. The quality of the exterior orientation of each stereo pair was checked for existing y-parallaxes and by comparing the measured with given control points. If the exterior orientation was accepted, each building was measured manually by the operator using the software module Feature Extraction of SOCET Set. The operator could select three different roof types (flat, peaked, and gabled), and the sequence of photogrammetric measurements depends on the roof type. Feature Extraction (Auto Create Mode) automatically creates the walls of the buildings based on the value specified for the ground. Complex buildings were broken down into simpler components, which were later combined in AutoCAD. Finally, the measured data were transferred to AutoCAD 2000 for the correction of some measurements and for modeling of the complex buildings. This approach can be regarded as highly manual.

19.5.4   STUTTGART UNIVERSITY OF APPLIED SCIENCES

Stuttgart used inJECT1.9 (prerelease) software to semiautomatically extract buildings using aerial images, camera calibration, and exterior orientation information. For a description of the used automated tools see Gülch and Müller (2001). An inexperienced student performed the building extraction as a part of a diploma thesis, supervised by E. Gülch.

The general workflow was as follows:

- Preparation of orientation and image data and direct import to inJECT1.9 (prerelease).
- Measurement of buildings without stereo-viewing using parametric (Figure 19.9) or polyhedral (Figure 19.10) 3D building models with one common ground height for a building part or for a building composite. Partly, the building models were measured as Constructive Solid Geometry (CSG) structure forming composite buildings. These were not merged inside inJECT but were merged externally. The results presented here give the pure measurements inside inJECT. Snapping function, rectangular enforcement for polygonal measurements, and automatic enforcement of planarity were frequently used for the polyhedral

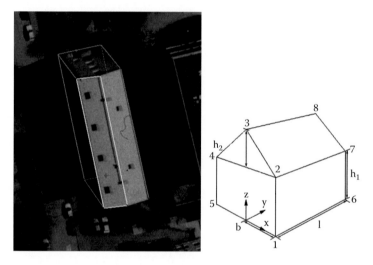

FIGURE 19.9 Parametric building models (here saddleback-roof building in test site Hermanni). (Courtesy of University of Applied Sciences, Stuttgart, Germany. With Permission.)

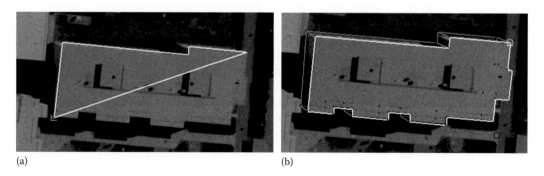

FIGURE 19.10 Polyhedral flat-roof building. (a) First five corners are measured. (b) The whole roof outline is measured and with one matched ground point the walls are derived. (Courtesy of University of Applied Sciences, Stuttgart, Germany. With Permission.)

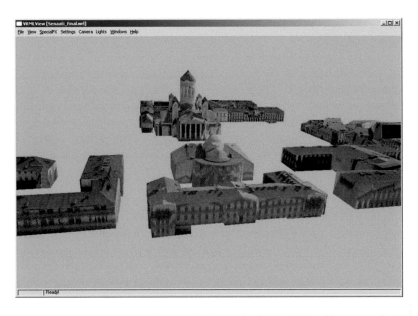

FIGURE 19.11   Building extraction in FGI test site Senaatti with inJECT with parametric and polyhedral building models and textured 3D view in a VRML Browser. (Student exercise by A. Novacheva and S.H. Foo, supervisor E. Gülch. With Permission.)

buildings. Basically, all ground height measurements were done by image matching. The rooftop height and some shape features of saddleback and hip-roof buildings were measured by area-based and feature-based image matching.
- Storage in geography markup language (GML3) and export to DXF and partly virtual reality modeling language (VRML) (Figure 19.11).

In addition to the derived GML3 and DXF files, a visualization using VRML was automatically derived with automatically extracted texture (see example in Figure 19.11).

19.5.5 Institut Géographique National

IGN used calibrated aerial images in multiview context in the Amiens test site and calibrated aerial images and laser digital surface model (DSM) in the Espoonlahti, Hermanni, and Senaatti test sites. All work was carried out by experienced personnel, one person working with Amiens, Espoonlahti, and Hermanni and one with Senaatti. For each test site, IGN created a pseudocadastral map manually using aerial images.

The workflow was divided into preparation (fully automatic procedures such as DSM and true orthoprocessing), cadastral map edition and pruning, 3D reconstruction of buildings, and quality control. Materials and methods varied between test sites.

Amiens test site—First a DSM and the true ortho image were processed by correlation using the multiview context. Then, two methods of reconstruction were used. For prismatic models, the 2D shape was edited, and the median height in the DSM was measured. For other models, the skeleton of the central ridge was edited manually in one image, and the system automatically reconstructed the 3D shape. Finally, the quality was controlled with a difference image of 3D polygons and DSM.

Espoonlahti test site—Pseudocadastral maps were edited on a single image with height adjustment using the roll-button of the mouse. For each 2D polygon, a median height was measured on the laser DSM. Finally, the quality was controlled with a difference image of 3D polygons and DSM.

Hermanni test site—Pseudocadastral maps were edited on a single image with height adjustment using the roll-button of the mouse. Reconstruction was fully automatic using these 2D polygons and laser DSM.

Senaatti test site—On this test site, several modules were used. Sometimes, pseudocadastral maps were edited on a single image with height adjustment using the roll-button of the mouse. After this, reconstruction was carried out using a model driven approach (Flamanc et al. 2003). Moreover, manual tools were used for complex buildings. Finally, special tools such as dome edition mode were used to extract specific structures.

19.5.6 Institut Cartografic de Catalunya

Institut Cartografic de Catalunya (ICC) used TerraScan, TerraPhoto, and TerraModeler software by Terrasolid (2003) to extract buildings using laser scan data with aerial images (marked with *ICC laser + aerial* in the results) and without aerial images (marked with *ICC laser* in the results). All work was carried out by one person. The major difference of the applied approach with the Nebel + Partner, who also used TerraScan for the test, was that in the ICC process, the orientation information of aerial images was applied. Thus, the building outlines could be derived using the image data.

19.5.7 Nebel + Partner GmbH

Nebel + Partner used TerraScan software by Terrasolid (2003) to extract buildings using laser scan data. All measurements and modeling were performed by an inexperienced person. Aerial images were not used for any measurements. Image crops were only used as superimposed images for better visual interpretation of the laser point clouds during the measurement of the roofs. Before manual measurements with the point clouds, each laser point set was automatically classified as buildings, ground elevation, and high, medium, and low vegetation by TerraScan. In the TerraScan software tool—Construct Building—an algorithm automatically finds roofs based on laser hits on planar surfaces of the roofs, which results in vectorized planes of each roof. Roof boundaries can also be created or modified manually. In TerraScan, the operator can select three different building boundary types (rectangle, rectangular, and polygon). A *Rectangle* boundary consists of exactly four edges, whereas a *Rectangular* boundary has more than four edges, a polygon is a free form polygon with angles also different of 90°.

19.5.8 FOI

FOI used their own in-house software and methods to extract buildings using laser data without given ground plan (marked with *FOI* in the results) and with given ground plan (marked *FOI outlines* in the results). All work was carried out by an experienced person.

Without a ground plan, the three preprocessing steps of the building extraction algorithm were:

1. Rasterize the data to obtain digital surface models (DSMzmax and DSMzmin)
2. Estimate the terrain surface (DTM)
3. Classify the data above the ground surface (vegetation/buildings)

Each group of connected pixels classified as buildings were used as the ground plan for the building extraction algorithm.

When using the given ground plan, instead of classifying the data (earlier step 3), the outlines were used to create a classification image with buildings. The classification image is used to specify the ground plans of the buildings. The outlines were not used when estimating the roof polygons along the contour of the building.

The raw laser scanner data were resampled into a regular grid (0.25 × 0.25 m²). Two grids were saved, one containing the lowest height value in each cell (DSMzmin), and one containing the highest z-value (DSMzmax). The building reconstruction methods are based on DSMzmin to reduce noise from antennas for example. The ground surface (DTM) is then estimated (Elmqvist, 2002). Having estimated the ground surface, the remaining pixels are further classified as buildings and vegetation. Each group of connected pixels classified as buildings is used as ground plan for the building extraction algorithm. If the given ground plan is used for the building extraction algorithm, the classification of buildings and vegetation is not needed.

For each ground plan detected in the classification or obtained otherwise, elevation data are used to extract planar roof faces. The planar faces of the roof are detected using clustering of surface normals. The algorithm works on gridded but not interpolated data. For each pixel within the building ground plan, a window is formed. The surface normal is estimated by fitting a plane to the elevation data within the window. The plane parameters are estimated using a Least Squares adjustment.

A clustering of the parameters of the surface normals is then performed. In the clustering space, the cell with the largest number of points is used as initial parameters of a plane, and an initial estimation of which pixels belong to the plane is made using these parameter values. Finally, an enhanced estimation of the plane parameters using a reweighted Least Squares adjustment is performed, and a final estimation of the pixels that belong to the plane is obtained. These pixels form a roof segment. This clustering process is repeated on the remaining nonsegmented pixels within the ground plan. Only roof segments having a certain minimum area are kept. After the clustering process, region growing is performed on the roof segments until all pixels within the ground plan belong to a roof segment.

Having segmented the roof faces of a building, the relationship between the faces is defined. By following the outlines of roof faces, so-called topological points are first added. Topological points are defined as vertices where a roof face's neighbor changes. Next, each section of an outline between any two topological points is defined either as an intersection line, height jump section or both. For sections defined as both an intersection line and jump section, a new topological point is added to split the section.

The jump sections are often noisy and difficult to model. Therefore, for each jump section, lines are estimated using the 2D Hough transform. The slopes of the estimated lines are then adjusted according to the orientation of the plane. The intersection points of the estimated lines are used as vertices along the jump sections. These vertices along the jump sections are used together with the topological points to define the roof polygons.

After having defined the roof polygons, a 3D model of the building can be created. The roof of the model is created using the defined roof polygons. The height value of the vertices is obtained from the z-value of the plane at the location of a vertex. Wall segments are inserted between any two vertices along jump sections.

19.5.9 C+B TECHNIK

C+B Technik used their own software to extract buildings using laser scan data and given ground plans. In a first step, a triangle net is created from the laser scanner data, which builds the basic data structure for the modeling process. The basic computation method selects the laser scanner points within a building polygon. Triangles are combined to create surfaces, and triangle sides are classified as edge lines. The resulting surfaces, edges, and corners are analyzed and edited to achieve the typical building objects that are, for example, inner vertical walls, horizontal or tilted roof planes, and horizontal ridges. The surfaces are also adapted to the building polygon, so that, for example, an edge line and a house corner fit together. For each situation, a complete automatic solution is not possible, so that a complete check and possible editing of the result must be performed. The extraction of buildings from laser scanner data is split first into an automatic computation step and second into an interactive check and editing step. The results for each building have the following logical structure: the ground polygon, the outer walls, the inner walls, and the roof surfaces. The results can be output as ASCII- or DXF-data.

19.5.10   Delft University of Technology

Delft used their own software and methods to extract buildings using laser data and ground plans based on studies by Vosselman and Dijkman (2001), and Vosselman and Süveg (2001). All work was carried out by an experienced person. If building outlines were not available, they were manually drawn in a display of the laser points with color-coded heights. In the Hermanni test site, only buildings with given ground plan were modeled.

If the point cloud within a building polygon can be represented by a simple roof shape (flat, shed, gable, hip, gambrel, spherical, or cylindrical roof), the model of this roof is fitted to the points with a robust Least Squares estimation. Often, building models have to be decomposed interactively such that all parts correspond to the above-mentioned shape primitives. If the point cloud is such that all roof planes can be detected automatically, an automatic reconstruction is attempted based on the intersection of detected neighboring roof faces and the detection of height jump edges between roof faces.

If the above-mentioned two situations do not apply, the building polygon is split into two or more parts until each part fulfills one of the aforementioned two conditions. Optionally, point clouds can be edited to remove outlier points that would disable an automatic roof reconstruction.

19.5.11   University of Aalborg

The modeling of buildings was carried out by three university students as part of their studies (Frederiksen et al., 2004). The applied method used laser scanner and 2D vector map data to extract and model the buildings. The applied method can be used to model rectangular buildings with gable roofs in different levels.

A 2D vector map is not only used to extract the data within the buildings but also as building outline.

After the extraction of the data, the data are interpolated into a rectangular grid. For every point in the grid, a surface normal based on the point and its neighboring points are calculated (Burrough and McDonnell, 1998, p. 190–192). Now, all points belonging to a roof plane have to be found. An assumption is made: the two longest parallel sides of the building each have one roof plane orientated towards the side of the building, meaning a surface normal on the roof plane is perpendicular in the xy-plane to the side of the building. This means that every point is examined regarding its surface normal, and surface normals perpendicular to the side of the building and with approximately the same orientation are grouped. The result is two groups of surface normals—each group representing a roof plane.

The points of each roof plane are now adjusted by the principles of Least Squares adjustment and a method called *the Danish method* (Juhl, 1980). The Danish method is a method used in adjustments and is used to automatically weight down points with large residuals compared with the determined roof plane from the Least Squares adjustment. This means that points not belonging to the roof plane, for example, points on chimneys, are sorted out of the group automatically and will not have influence on the final determination of the parameters of the roof planes (Juhl, 1980).

Gable roofs at different levels will also show only two groups of points with approximately similar surface normals, because the method cannot distinguish between roof planes at different levels. To find the border between the buildings, a gradient filter is applied to the grid, and large gradients are grouped by means of a connected components analysis. Groups are adjusted by Least Squares adjustment to a line, and thus a building with gable roofs at different levels is divided into rectangular buildings with two roof planes.

19.5.12   Dresden University of Technology

The Dresden method (Hofmann, 2004) only uses point clouds obtained by a presegmentation of airborne laser scanner data. All works were carried out by an experienced person. It is a plane-based approach that presumes that buildings are characterized by planes. It utilizes a triangulated irregular network (TIN)-structure that is calculated into the point cloud. The method only uses point clouds of the laser scanner data that contain one building. To get such point clouds, polygons coarsely framing the building can be used to extract the points (e.g., in ArcGIS). The polygons can be created manually, or map or ground plan information can be used.

The parameters of every TIN-mesh, which define its position in space uniquely, are mapped into a 3D triangle-mesh parameter space. As all triangles of a roof face have similar parameters, they form clusters in the parameter space.

Those clusters that represent roof faces are detected with a cluster analysis technique. By analyzing the clusters, significant roof planes are derived from the 3D triangle parameter space while taking common knowledge of roofs into account. However, no prior knowledge of the roof as, for example, the number of roof faces is required. The obtained roof planes are intersected in accordance to their position in space. Analyzing the intersected roof faces determines the roof outlines, and the ground plan is derived.

19.5.13   DISCUSSION

From the earlier descriptions, we can identify not only commercial systems but also in-house systems used by various organizations. We have only two commercial systems (CyberCity, inJECT) dedicated for some semiautomatic extraction from digital imagery. We have at least one in-house system (IGN) for it as well. Others base on standard digital photogrammetric Workstations. We have a majority of systems using ALS data, but we can identify really only few commercial systems like Terrascan and C+B, one in-house system (FOI), and several very advanced research approaches. Several participants based their results on inexperienced personnel, which seems to be an indication that those software packages can be handled without extensive training, such as earlier stereo plotter measurements that required highly trained personnel. We can also observe the need for some approaches to have ground plan information available, which certainly has some effect on the overall applicability and costs, which is restricted or influenced then by the availability and costs of such additional data. From the hybrid approaches, we can find various ways how the second data source (images, respectively laser data) is actually used. There is so far no clear trend visible for a best strategy to involve both.

19.6   RESULTS OF THE EMPIRICAL TEST

We want first to present the results of the single approaches, and then we want to compare laser scanning approaches with photogrammetric approaches based on image data. Here, we group also the different degrees of hybrid approaches using both input sources.

We want in each case to classify the results in the following categories:

- Building location, height, and length
- Roof inclination

In a last major step, we focus on the effects of automation for the used methods.

We present the major results for different datasets and give comments to their analysis. For more details on the full evaluation, we refer to Kaartinen et al. (2005a, b) and Kaartinen and Hyyppä (2006).

19.7   RESULTS OF SINGLE APPROACHES

19.7.1   BUILDING LOCATION, HEIGHT, AND LENGTH

Figure 19.12 shows the differences between various models with respect to elevation and planimetric errors for single point targets defined by IQR quality measure. The quality measures are given separately for north–south and east–west directions due to the different point spacing in along and

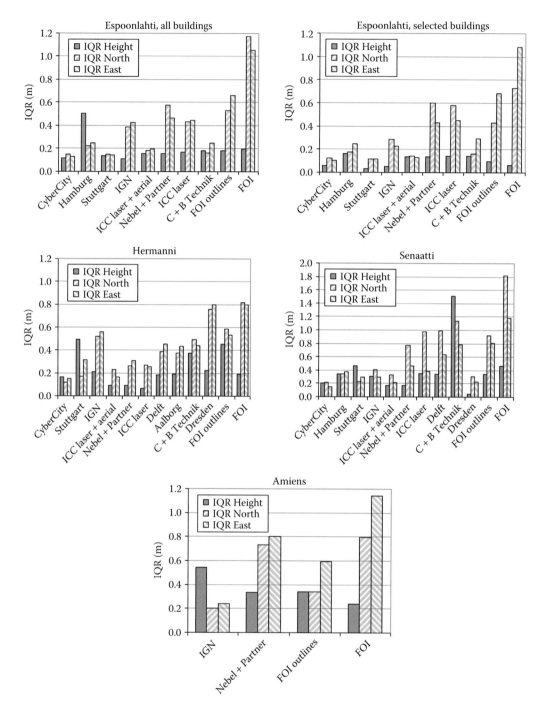

FIGURE 19.12 Location accuracy (IQR) of the models with respect to single points. The selected buildings (four) in Espoonlahti had been measured by all participants. (Courtesy to Finnish Geospatial Research Institute, Finland. With Permission.)

across track directions of TopoSys systems. Building location accuracies for all modeled buildings in Espoonlahti are given for all and for selected buildings (same four buildings modeled by all participants).

Figure 19.13 gives information about building height defined by IQR and RMSE quality measures. Some methods do not use DEM information; in these cases, no values are given.

Figure 19.14 gives information about building length defined by IQR and RMSE measures.

CyberCity achieved a good quality in all three Finnish test sites. However, it should be remembered that measurements in CyberCity are done manually from stereoscopic images, and automation is mainly used for connecting points to derive 3D models. Thus, the manual part of the process is verified in this analysis. The quality improved when image quality was better, but in general, the accuracy variation within the test sites was low.

The Hamburg and Stuttgart models were affected more by the site and site-wise data characteristics. As the Hermanni test site was relatively easy, the Stuttgart model used high automation in Hermanni resulting in decreased performance. With the most difficult test site, Espoonlahti, the Stuttgart model was almost as good with all buildings as the CyberCity model and even better with selected buildings (same buildings modeled by all partners). It has to be remembered that the Stuttgart process was done by a student, and with CyberCity, the process was done by an expert.

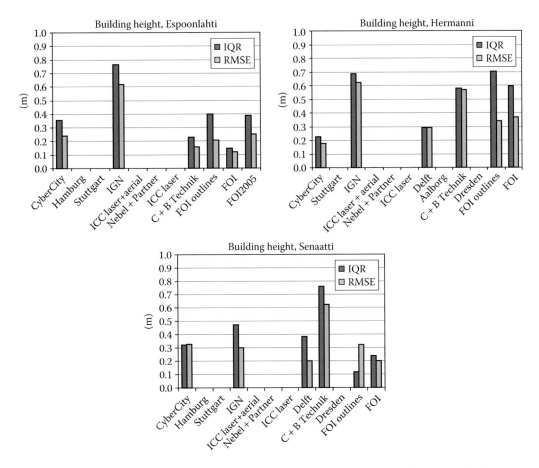

FIGURE 19.13 Building height accuracy. IQR: all observations, RMSE: outliers removed. For some approaches, that did not use DEM information, no results are given. (Courtesy of Finnish Geospatial Research Institute, Finland. With Permission.)

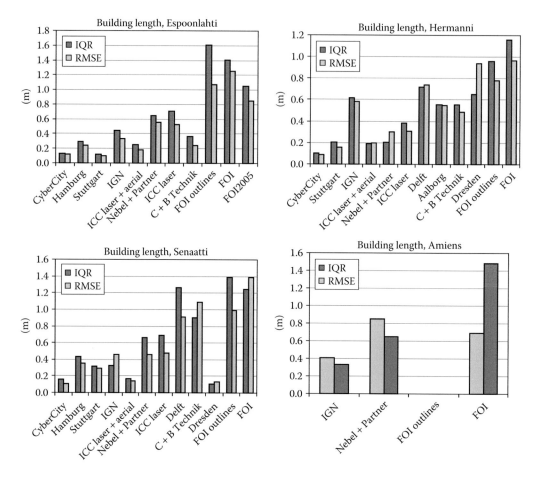

FIGURE 19.14 Building length accuracy. IQR: all observations, RMSE: outliers removed. (Courtesy to Finnish Geospatial Research Institute, Finland. With Permission.)

The Hamburg model showed lower performance than these two other photogrammetric techniques even though it was based on a more standard photogrammetric process but with an inexperienced operator.

Hybrid techniques using TerraScan (see ICC laser + aerial) showed a quality almost comparable with the best photogrammetric methods. The same method (i.e., TerraScan) used by Nebel and Partner showed typically lower accuracy that can be explained by the lower amount of time spent and the lower amount of aerial image information used in the process. In ICC, aerial images, and image orientations were used to measure the outlines, and in Nebel and Partner, images were used for visual interpretation purposes. It seems that good quality results can be obtained when building outlines are measured using image information and laser-derived planes.

In Hermanni, where all participants using laser data provided a model, the ICC model (based on TerraScan) showed the best accuracy. Moreover, the Delft and Aalborg models (the latter developed by undergraduate students within the project) resulted in reasonably good accuracy. Delft, Aalborg, and C+B Technik mainly used given ground plans to determine building location, so planimetric accuracy is directly affected by the ground plans that do not include roof eaves. This can also be seen in the results: the planimetric accuracy is almost the same for these three, and also point deviations in Hermanni seem to have *common* points. It should be noted that the generation of the FOI model was fully automatic, and their newest extraction algorithm was based on the use of first and last pulse data, and last pulse data were not provided in the test, as it was not available from all test sites. C+B did exceptionally well in Espoonlahti, which was the most difficult area, but they provided only the buildings where outlines were given. One possible explanation for the variability of the C+B performance in various test sites is that as it is based on TINs, it works better than plane-based models (laser points defining a plane) with more complex and with small structures. Moreover, the Dresden model performed extremely well in Senaatti, but not in Hermanni. The number of points used in the analysis was extremely low for the Dresden model.

19.7.2   ROOF INCLINATION

Figure 19.15 gives the roof inclination accuracy for each model, defined with IQR and RMSE quality measures.

In Hermanni, the best result was obtained with the Dresden model, and basically, all laser based methods resulted in less than 1° error (RMSE). The C+B approach had a larger error than the photogrammetric processes. Most probably, the TIN based principle of C+B does not give as good results as using all the points hitting one surface and defining a plane using all of them. In photogrammetry, the roof inclination is obtained from two measurements, and therefore, the same accuracy as laser was clearly not achieved. When the building size was smaller and/or the pulse density was lower, this difference between photogrammetry and laser scanning was reduced or even disappeared (Figure 19.15).

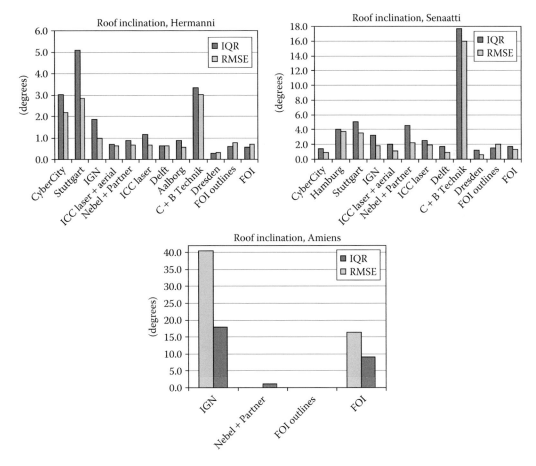

FIGURE 19.15   Roof inclination accuracy of the models. IQR: all observations, RMSE: outliers removed. (Courtesy of Finnish Geospatial Research Institute, Finland. With Permission.)

19.8   COMPARISON OF LASER SCANNING AND PHOTOGRAMMETRY

The purpose here is to give an evaluation of the performance of the applied methods based on laser scanning data and those using aerial image data (photogrammetric methods) and all hybrid approaches. We group the results accordingly to derive conclusions for the two major competing approaches currently used.

19.8.1   BUILDING OUTLINES, HEIGHT, AND LENGTH

In general, photogrammetric methods were more accurate in determining building outlines (mean of IQR North and IQR East), as may be seen in Figure 19.16. Taking all test sites into account, the IQR value of photogrammetric methods ranged from 14 to 36 cm (average 21 cm, median 22 cm, and standard deviation (std.) 7.2 cm of IQR values). The corresponding values for aerial image assisted laser scanning ranged from 20 to 76 cm (mean 44 cm, median 46 cm, std. 18.5 cm). Laser scanning based building outline errors ranged from 20 to 150 cm (mean 66 cm, median 60 cm, std. 33.2 cm). We display also linear regression line where applicable in the following figures.

Point density, shadowing of trees, and complexity of the structure were the major reasons for site-wise variation of the laser scanner-based results. The lowest accuracy was obtained with the lowest pulse density (Senaatti). Moreover, in Amiens, the complexity deteriorated the performance. It was almost impossible to reveal the transition from one house to another using DSM data in Amiens. The small number of trees, simple building structure, and relatively high pulse density resulted in the highest accuracy at the Hermanni test site.

The effect of the point density on the achieved average accuracy (planimetric, mean of IQR North and IQR East, and height errors) is depicted in Figure 19.17. The figure clearly shows how point density improvement can slowly reduce both height and planimetric errors. Roughly speaking, we can state that an increase of pulse density to 10 times higher will result in reduction of errors by 50%.

In building length determination (Figure 19.18), laser-based methods were not as accurate as photogrammetric methods, as could be expected from the above. The photogrammetrically derived lengths varied from 14 to 51 cm (RMSE, mean 26 cm, median 22 cm, std. 12.6 cm). Lengths

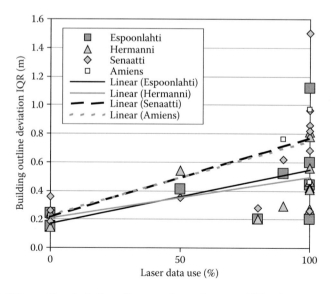

FIGURE 19.16 Building outline deviation, all targets. Laser data use of 0% refers to photogrammetric methods, 100% to fully laser scanning based techniques and intermediate values to hybrid techniques. (Courtesy of Finnish Geospatial Research Institute, Finland. With Permission.)

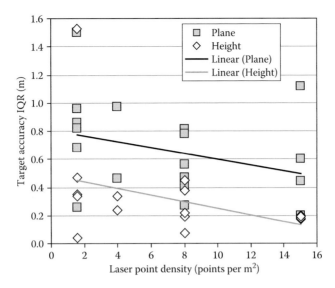

FIGURE 19.17   Average accuracy versus laser point density. (Courtesy of Finnish Geospatial Research Institute, Finland. With Permission.)

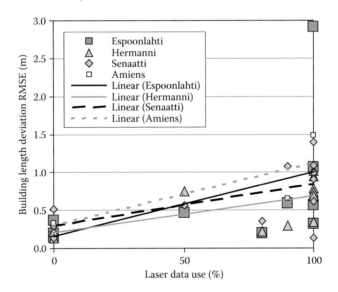

FIGURE 19.18   Building length deviation (all observations). (Courtesy to Finnish Geospatial Research Institute, Finland. With Permission.)

obtained with aerial image assisted laser scanning varied from 19 to 108 cm (mean 59.4 cm, median 57 cm, std. 31.2 cm). The laser scanning based lengths varied from 13 to 292 cm (mean 93 cm, median 84.5 cm, std. 60.9 cm). With laser scanning, the complexity of the buildings was the major cause for site-wise variation rather than the point density.

The IQR value for laser scanning height determination ranged from 4 to 153 cm (mean 32 cm, median 22 cm, std. 31.5 cm). One fully automatic method caused high errors modifying the mean value. Laser scanning assisted by aerial images resulted in IQR values between 9 and 34 cm (mean 18 cm, median 16.5 cm, std. 8.5 cm). Photogrammetric height determination ranged from 14 to 54 cm (mean 33 cm, median 35 cm, std. 18 cm). The accuracy of height determination exactly followed the laser scanning point density. With the high-density data in Espoonlahti, all participants were able to provide average heights with a better accuracy than 20 cm IQR value.

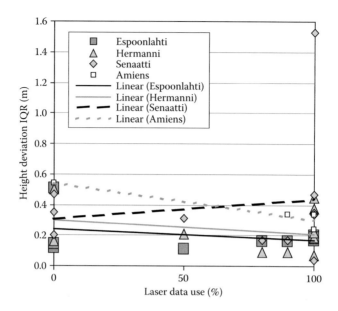

FIGURE 19.19 Target height deviation. (Courtesy of Finnish Geospatial Research Institute, Finland. With Permission.)

The comparison of various laser point densities and methods for height determination and with photogrammetry is given in Figure 19.19. Photogrammetric methods are on the left side (laser data use 0%), and laser scanning methods on the right (laser data use 100%). In the cases of Hermanni, Amiens, and Espoonlahti, the regression line is decreasing indicating that on the average, height determination is slightly more accurate using laser scanner than photogrammetric techniques. In Senaatti, the opposite conclusion can be drawn due to the low pulse density and to an unsuccessful fully automatic building extraction case.

In Figure 19.20, the target height and plane deviations are shown separately for eaves and ridges for Hermanni and Senaatti. As expected, for laser scanning methods, the accuracy is better for ridges, which is determined by the intersection of two planes. In theory, the height determination of ridges is extremely accurate if the number of hits defining the planes is high enough. The planimetric accuracy of ridges and eaves was more accurate with photogrammetry than with laser scanning.

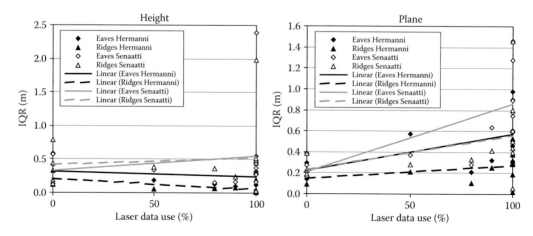

FIGURE 19.20 Target height and plane (mean of north/east) deviations for different targets. (Courtesy to Finnish Geospatial Research Institute, Finland. With Permission.)

19.8.2 ROOF INCLINATION

Roof inclination determination was more accurate when using laser data than photogrammetry, but a large variation in quality exists due to methods and test sites (i.e., complex buildings). The RMSE (all observations included) using laser scanning for roof inclination varied from 0.3° to 15.9° (mean 2.7°, median 0.85°, std. 4.4°). The corresponding values for aerial image assisted laser scanning ranged from 0.6 to 2.3° (mean 1.3°, median 1.1°, std. 0.6°) and for photogrammetry ranged from 1.0° to 17.9° (mean 5.2°, median 3.2°, std. 6.3°). In Senaatti and Amiens, the roof inclinations are steep and roofs short, so even small errors in target height determination lead to large errors in inclination angle. The Hermanni test site is relatively easy for both methods. In Hermanni, the accuracy of roof inclination determination was about 2.5° for photogrammetric methods and about 1° (RMSE, all observations included) for laser-based methods.

19.8.3 SHAPE SIMILARITY

There are buildings or parts of buildings missing on many models. The reasons for this are not always known, but at least, the shadowing of big trees and the complexity of buildings can cause problems. Participants may also strive for different goals in building extraction, some want to model the buildings as detailed as possible, and some are aiming to get only the basic form of buildings. For automatic methods the vegetation causes problems also as trees are classified as buildings. On other methods, the extra buildings are mainly temporary constructions, such as site huts for construction sites, which are usually impossible to separate out from *real* buildings. Images showing the differences in 2D between reference raster and delivered models are given in the following. Figure 19.21 shows the similarity in the FGI test site Hermanni for two photogrammetric approaches. Figure 19.22 shows the similarity in the FGI test site Hermanni for three laser scanning approaches. In the Hermanni site, the extracted buildings are large in size; thus, all buildings has been modeled by partners in example, and thus, the shape dissimilarity reveals quality difference of the models.

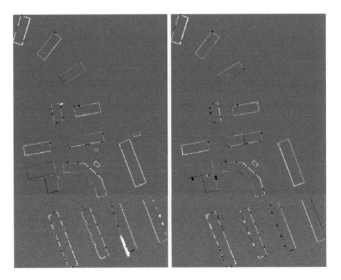

FIGURE 19.21   Image data only: CyberCity, Stuttgart (white = extra area, black = missing area). (Courtesy to Finnish Geospatial Research Institute, Finland. With Permission.)

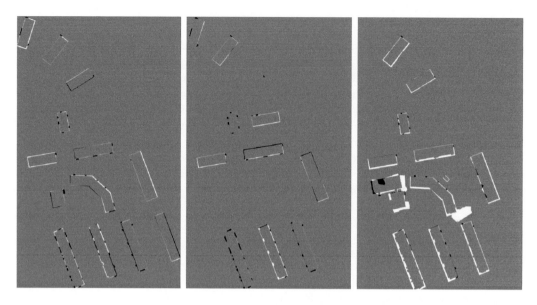

FIGURE 19.22 Laser data only: ICC laser, Aalborg, FOI (white = extra area, black = missing area). (Courtesy to Finnish Geospatial Research Institute, Finland. With Permission.)

19.9 EFFECT OF AUTOMATION

We want now to analyze the effects of automation in the applied methods. We therefore group methods according to their degree of automation and analyze overall planimetric and height accuracies.

The degree of automation varied significantly among the participants of this test. In general, the laser data allow higher automation in creating the models. Editing of the complex building models is needed for slowing down the process. Even though some laser-based processes are relatively automatic, the processes are still under development. In general, the planar target accuracy is affected by the degree of automation (and method, Figure 19.23). The accuracy of low automation methods

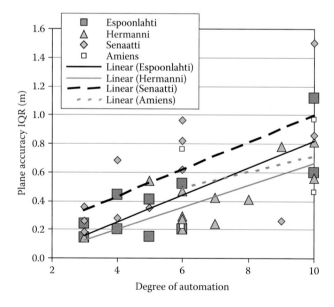

FIGURE 19.23 Obtained overall planimetric accuracy (mean of north/east) as a function of automation. (Courtesy to Finnish Geospatial Research Institute, Finland. With Permission.)

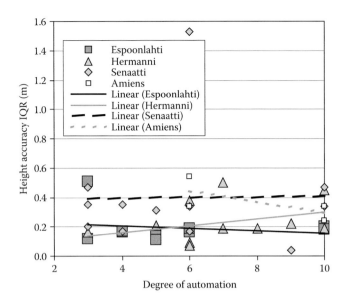

FIGURE 19.24 Obtained overall height accuracy as a function of automation. (Courtesy to Finnish Geospatial Research Institute, Finland. With Permission.)

is about 20–30 cm, whereas on high automation methods, it is about 60–100 cm (IQR). The target height accuracy seems to be almost independent of the degree of automation (Figure 19.24). The degree of automation was estimated as a value from 0 to 10 based on the workflow charts and the procedure descriptions and comments provided by participants.

In general, the methods using aerial images and interactive processes are capable of producing more detail in building models, but only some providers modeled more detailed structures such as chimneys and ventilation equipment on roofs.

19.10 DISCUSSION AND CONCLUSIONS

It can be concluded that the applied photogrammetric techniques and hybrid techniques using both laser scanning and image data provide the highest accuracy and level of detail in 3D city reconstruction. Despite the high amount of research during the last two decades, the level of automation for building extraction in a high level-of-detail is still relatively low.

Improvement in automation can be achieved most significantly by utilizing the synergy of laser and photogrammetry. The photogrammetric techniques are powerful for visual interpretation of the area, measurement of the building outlines, and of small details (e.g., chimneys), whereas laser scanning gives height, roof planes, and ridge information at its best as well as better estimations of the DTM surface in urban areas. When the advantages of both methods are implemented in a single system well, high accuracy and a relatively high level of automation can be achieved. In this test, there was only one real hybrid system. However, several participants incorporated both major data sources in various ways. This only shows that there was no commonly agreed strategy available, and this is still valid today. We still strongly recommend to focus on this aspect to derive commonly accepted and proofed methods for multisensor fusion in this field. Both International Society for Photogrammetry and Remote Sensing (ISPRS) and EuroSDR working groups have started to deal with this subject and continue with tests incorporating also the usage of dense image matching point clouds.

Most airborne laser scanners today can be equipped with medium-format or even large-format digital aerial cameras with RGB and Near infrared channels. This means that usually at least the swath width of the scanner is covered by image information. This must not necessarily allow for stereo, as single overlapping images would be sufficient to produce orthophotos, as the DTM and the Digital Surface Model is derived from simultaneously recorded laser scanning data. In summary, we can see also today increasing R&D efforts in the multisensor fusion field not only from academia but also from companies.

The quality of the laser scanning data has a significant impact on the accuracy of the fully laser based models. As pulse density is increasing substantially and there are new possibilities to use pulse information (multiple pulse, waveforms), it is visible that the accuracy of the laser-based object extraction techniques follow this technical development. The rapid rate of implementation development of laser scanning data can also be seen from the improvement of FOI's results during the test. On the other hand, when the study was initiated, the implementation level of the majority of laser-based techniques was low. It can be seen from Table 19.4 (time spent), that thanks to the good implementation of photogrammetric methods, the time spent on these models is not much higher than those of laser scanning.

We see, based on the presented results and development of the last 20 years, that the following:

- There is still a need to further integrate laser scanning and photogrammetric techniques in a more advanced way. The accuracy of single height measurements with the laser should be used in the photogrammetric process to increase the automation. We can see very promising results reported by companies on automated building extraction based on laser point clouds and without the usage of footprint information, but using the additional aerial images as source for building outline detection.
- The quality improvements of digital camera data offer an increase in the quality of photogrammetric techniques and will also allow higher automation by using remote sensing methods in the image processing. This is supported by the RGB and Near infrared recording capabilities of medium- and large-format digital aerial cameras, but not really fully exploited.
- The rapid technological development in laser scanning will allow higher accuracies to be obtained by techniques relying only on laser scanning.
- Laser scanning allows higher automation needed in the building extraction process. There is still a need to implement laser-based techniques into practical working processes. On the other hand, a clear trend since about 2009 towards dense image matching offers new potential for photogrammetric techniques in building extraction, that could compete, at least in some aspects, with ALS. We can see commercial software developments such as Trimble's Building Generator, using point clouds from laser scanning and/or dense image matching and footprint information to derive Level-of-Detail 1 and 2 building models in a highly automated fashion, applied in many state mapping agencies in Germany.
- New algorithmic innovations will emerge in many areas, thanks to digital camera data, dense image matching, and laser scanning developments that will allow a further increase in automation. However, there is still quite some room for reaching higher reliability and a greater independency from given data as for example, footprint.

In the project, first pulse data were used for the building extraction. It is known already that the use of last pulse and combined use of first and last pulse would lead to improvements.

In practice, the availability of the material (aerial images, laser point clouds, or matching point clouds) determines significantly what kind of techniques can be used. The costs of laser scanning are usually higher than aerial imagery, although in many cities, the material has already been collected. In many cities, building outline information exists as well as some kind of existing 3D models. So there is not always a general recommendation or solution for all specific problems. The availability of the sources and their up-to-dateness definitely affect their potential usage.

To be able to separate methodological development from development due to improvement of the data, it is recommended to use the same test sites, with old and new laser scanning and aerial data, for future verifications. The value of such empirical investigations is rather high and has been confirmed not only by the high number of participants of this test but also by many fruitful discussions on the results of this investigation presented at various conferences and meetings. Still, it is the first and to our knowledge comprehensive empirical comparison of this type.

In future, an even bigger challenge is to find the answers to the following: how existing models are to be automatically updated, how changes are reliably verified, and how existing information can be used optimally in an even more automatic process.

ACKNOWLEDGMENTS

All participants, i.e. George Vosselman (International Institute for Geo-Information Science and Earth Observation), Alexandra Hofmann (Dresden University of Technology, presently Definiens), Urs Mäder (CyberCity AG), Åsa Persson, Ulf Söderman and Magnus Elmqvist (Swedish Defence Research Agency), Antonio Ruiz and Martina Dragoja (Institut Cartogràfic de Catalunya), Sylvain Airault, David Flamanc and Gregoire Maillet (MATIS IGN), Thomas Kersten (Hamburg University of Applied Sciences), Jennifer Carl and Robert Hau (Nebel + Partner), Emil Wild (C+B Technik GmbH) and Lise Lausten Frederiksen, Jane Holmgaard and Kristian Vester (University of Aalborg) and Eberhard Gülch (Stuttgart University of Applied Sciences) are gratefully acknowledged for cooperation in the project. Hannu Hyyppä from Helsinki University of Technology and Leena Matikainen from FGI also contributed significantly. Support of Espoo and Helsinki City Survey Divisions, FM-Kartta Oy (Blom Kartta Oy) and Terrasolid Oy in providing data for the project is gratefully acknowledged.

REFERENCES

Alharthy, A. and Bethel, J., 2004. Detailed building reconstruction from airborne laser data using a moving surface method, *Proceedings of the XXth ISPRS Congress, International Archives of Photogrammetry, Remote Sensing and Spatial Information Sciences*, Istanbul, Turkey, 35(B3/III), 237–242.

Ameri, B. and Fritsch, D., 2000. Automatic 3D building reconstruction using plane roof structures, In *Proceedings of the ASPRS Congress*, Washington. Available at www.itp.uni-stuttgart.de/publications/2000/pub2000.html.

BAE Systems, 2000. SOCET Set—User's Manual Version 4.3.

Baillard, C. and Dissard, O., 2000. A stereo matching algorithm for urban digital elevation models, *Photogrammetric Engineering and Remote Sensing*, 66(9), 1119–1128.

Baillard, C. and Maître, H., 1999. 3-D reconstruction of urban scenes from aerial stereo imagery: A focusing strategy, *Computer Vision and Image Understanding*, 76(3), 244–258.

Baltsavias, E.P., 1999. Airborne laser scanning: Existing systems and firms and other resources, *ISPRS Journal of Photogrammetry and Remote Sensing*, 54(1999), 164–198.

Baltsavias, E.P., 2004. Object extraction and revision by image analysis using existing geodata and knowledge: Current status and steps towards operational systems, *ISPRS Journal of Photogrammetry and Remote Sensing*, 58(3–4), 129–151.

Braun, C. et al., 1995. Models for photogrammetric building reconstruction, *Computers and Graphics*, 19(1), 109–118.

Brenner, C., 2001. City models—automation in research and practice, *Photogrammetric Week'01*, 2001, 149–158.

Brenner, C., 2003. Building reconstruction from laser scanning and images, In *Proceedings ITC Earth Observation Science Department Workshop on Data Quality in Earth Observation Techniques*, Enschede, the Netherlands, November 2003, 8 pp.

Brenner, C., 2005. Building reconstruction from images and laser scanning, *International Journal of Applied Earth Observation and Geoinformation*, 6, 186–198.

Brunn, A. et al., 1998. A hybrid concept for 3D building acquisition, *ISPRS Journal of Photogrammetry and Remote Sensing*, 53(2), 119–129.

Brunn, A. and Weidner, U., 1997. Extracting buildings from digital surface models, *International Archives of Photogrammetry, Remote Sensing and Spatial Information Sciences*, 32(3–4W2), 27–34.

Brunn, A. and Weidner, U., 1998. Hierarchical bayesian nets for building extraction using dense digital surface models, *ISPRS Journal of Photogrammetry and Remote Sensing*, 53(5), 296–307.

Burrough, P. and McDonnell, R., 1998. *Principles of Geographical Information Systems*, Oxford University Press, Oxford, UK.

Centeno, J.A.S. and Miqueles, J., 2004. Extraction of buildings in Brazilian urban environments using high resolution remote sensing imagery and laser scanner data, *International Archives of Photogrammetry, Remote Sensing and Spatial Information Sciences*, 35(B4), 4.

Chen, L.-C., Teo, T.-A., Shao, Y.-C., Lai, Y.-C. and Rau, J-Y., 2004. Fusion of LiDAR data and optical imagery for building modelling, *International Archives of Photogrammetry, Remote Sensing and Spatial Information Sciences*, 35(B4), 6.

Cho, W. et al., 2004. Pseudo-grid based building extraction using airborne LiDAR data, *International Archives of Photogrammetry, Remote Sensing and Spatial Information Sciences*, 35(B3/III), 378–381.

Cord, M. and Declerq, D., 2001. Three-dimensional building detection and modeling using a statistical approach, *IEEE Transactions on Image Processing*, 10(5), 715–723.

Cord, M., Jordan, M. and Cocquerez, J.P., 2001. Accurate building structure recovery from high-resolution aerial imagery, *Computer Vision and Image Understanding*, 82(2), 138–173.

Dash, J., Steinle, E., Singh, R.P. and Bähr, H.P., 2004. Automatic building extraction from laser scanning data: An input tool for disaster management, *Advances in Space Research*, 33, 317–322.

Dold, C. and Brenner C., 2004. Automatic matching of terrestrial scan data as a basis for the generation of detailed 3D city models, *International Archives of Photogrammetry, Remote Sensing and Spatial Information Sciences*, 35(B3/III), 1091–1096.

Elaksher, A.F. and Bethel, J.S., 2002. Reconstructing 3D buildings from LiDAR data. *International Archives of Photogrammetry, Remote Sensing and Spatial Information Sciences*, 34(3A), 102–107.

Elaksher, A.F., Bethel, J.S. and Mikhail, E.M., 2003. Roof boundary extraction using multiple images, *Photogrammetric Record*, 18(101), 27–40.

Elmqvist, M., 2002. Ground surface estimation from airborne laser scanner data using active shape models, In *Proceedings of the ISPRS Technical Commission III Symposium "Photogrammetric Computer Vision" PCV'02*, Graz, Austria, September 2002, *International Archives of Photogrammetry, Remote Sensing and Spatial Information Sciences*, Vol. 34, Part 3A, 114–118.

Fischer, A., Kolbe, T.H., Lang, F., Cremers, A.B. Förstner, W., Plümer, L. and Steinhage, V., 1998. Extracting buildings from aerial images using hierarchical aggregation in 2D and 3D, *Computer Vision and Image Understanding*, 72(2), 185–203.

Flamanc, D., Maillet, G. and Jibrini, H., 2003. 3D city models: An operational approach using aerial images and cadastral maps, *International Archives of Photogrammetry, Remote Sensing and Spatial Information Sciences*, 34, 3/W8, 6.

Forlani, G., Nardinocchi, C., Scaioni, M. and Zingaretti, P., 2003. Building reconstruction and visualization from LiDAR data, *International Archives of Photogrammetry, Remote Sensing and Spatial Information Sciences*, 34(5W12), 151–156.

Forlani, G., Nardinocchi, C., Scaioni, M. and Zingaretti, P., 2006. Complete classification of raw LiDAR data and 3D reconstruction of buildings, *Pattern Analysis and Applications*, 8(4), 357–374.

Fraser, C.S., Baltsavias, E. and Grün, A., 2002. Processing of Ikonos imagery for submetre 3D positioning and building extraction, *ISPRS Journal of Photogrammetry and Remote Sensing*, 56(3), 177–194.

Frederiksen, L., Holmgaard, J. and Vester, K., 2004. Automatiseret 3D-bygningsmodellering – fran laserscanning til bygningsmodel, University of Aalborg (in Danish).

Fuchs, C., Gülch, E. and Förstner, W., 1998. OEEPE Survey on 3D-City Models, Bundesamt für Kartographie und Geodäsie, Frankfurt, Germany, OEEPE Publication, 35, 9–123.

Fuchs, F. and Le-Men, H., 2000. Efficient subgraph isomorphism with 'a priori' knowledge – Application to 3D reconstruction of buildings for cartography, *Lecture Notes in Computer Science*, 1876, 427–436.

Fujii, K. and Arikawa, T., 2002. Urban object reconstruction using airborne laser elevation image and aerial image, *IEEE Transactions on Geoscience and Remote Sensing*, 40(10), 2234–2240.

Förstner, W., 1999. 3D-city models: Automatic and semiautomatic acquisition methods, In Fritsch, D. and Spiller, R. (Eds.), *Photogrammetric Week'99*, Wichman Verlag, Heidelberg, Germany, CD-ROM, pp. 291–303.

Grün, A., 1997. Automation in building extraction, *Photogrammetric Week'97*, 175–186.

Grün, A., 1998. TOBAGO – A semi-automated approach for the generation of 3-D building models, *ISPRS Journal of Photogrammetry and Remote Sensing*, 53(2), 108–118.

Grün, A., Li, Z. and Wang, X. 2003. Generation of 3D city models with linear array CCD-sensors. Available at http://carstad.gsfc.nasa.gov/topics/JBRESEARCH/STARLABO/Opt3D_citymod_new.pdf.

Grün, A. and Wang, X., 1998a. CC-Modeler: A topology generator for 3-D city models, *ISPRS Journal of Photogrammetry and Remote Sensing*, 53(5), 286–295.

Grün, A. and Wang, X., 1998b. Acquisition and management of urban 3D data using cybercity modeler. Available at www.cybercity.tv/pub/ccm_3D_data.pdf.

Grün, A. and Wang, X., 1999a. CyberCity modeller, a tool for interactive 3-D city model generation, *Photogrammetric Week'99*, 317–327.

Grün, A. and Wang, X., 1999b. CyberCity spatial information system (CC-SIS): A new concept for the management of 3-D city models in a hybrid GIS, In *Proceedings of the 20th Asian Conference on Remote Sensing*, Hong Kong, China, Asian Association on Remote Sensing, November 22–25, 1999, 8 pp.

Gülch, E. 2004. Erfassung von 3D-geodaten mit digitaler photogrammetrie, In Coors, V. and Zipf, A. (Eds.), *3D-Geoinformationssysteme – Grundlagen und Anwendungen*, Herbert Wichmann Verlag, Heidelberg, Germany, pp. 4–25.

Gülch, E. and Müller, H., 2001. New applications of semi-automatic building extraction, In *Proceedings of the third International Workshop on Automatic Extraction of Man-Made Objects from Aerial and Space Images*, Ascona, Switzerland, Swets & Zeitlinger/Balkema, the Netherlands.

Haala, N. 2004. Laserscanning zur dreidimensionalen erfassung von stadtgebieten, In Coors, V. and Zipf, A. (Eds.), *3D-Geoinformationssysteme – Grundlagen und Anwendungen*, Herbert Wichmann Verlag, Heidelberg, Germany, pp. 26–38.

Haala, N. and Brenner, C., 1999. Extraction of buildings and trees in urban environments, *ISPRS Journal of Photogrammetry and Remote Sensing*, 54, 123–129.

Haala, N., Brenner, C. and Anders, K.-H., 1998. 3D urban GIS from laser altimeter and 2D map data, *International Archives of Photogrammetry, Remote Sensing and Spatial Information Sciences*, 32(3/1), 339–346.

Haithcoat, T.L., Song, W. and Hipple, J.D., 2001. Building footprint extraction and 3-D reconstruction from LiDAR data, *IEEE/ISPRS Joint Workshop on Remote Sensing and Data Fusion over Urban Areas*, CD-ROM, 5 pp.

Henricsson, O. and Baltsavias, E., 1997. 3-D building reconstruction with ARUBA: A qualitative and quantitative evaluation, In *Proceedings of the International Workshop Automatic Extraction of Man-Made Objects from Aerial and Space Images (II)*, Ascona, Switzerland, http://e-collection.ethbib.ethz.ch/show?type=inkonf&nr=96.

Hofmann, A.D., 2004. Analysis of tin-structure parameter spaces in airborne laser scanner data for 3-D building model generation, *International Archives of Photogrammetry, Remote Sensing and Spatial Information Sciences*, 35(B3/III), 6 pp.

Hofmann, A., Maas, H.-G. and Streilein, A., 2002. Knowledge-based building detection based on laser scanner data and topographic map information, In *Proceedings of the ISPRS Commission III, Symposium 2002, Photogrammetric Computer Vision, PCV02*, Graz, Austria, ISPRS Archives—Volume XXXIV Part 3 A+B, 2002, 6 pp.

Hofmann, A., Maas, H.-G. and Streilein, A., 2003. Derivation of roof types by cluster analysis in parameter spaces of airborne laser scanner point clouds, In *Proceedings of the ISPRS Working Group III/3 Workshop '3-D Reconstruction from Airborne Laserscanner and InSAR data'*, Dresden, Germany, 2003. Available at http://www.isprs.org/commission3/wg3/workshop_laserscanning/.

Jaynes, C., Riseman, E. and Hanson, A., 2003. Recognition and reconstruction of buildings from multiple aerial images, *Computer Vision and Image Understanding*, 90(1), 68–98.

Juhl, J., 1980. Det analytiske stråleudjævningsprogram sANA, Laboratoriet for fotogrammetri og landmåling, University of Aalborg (in Danish).

Jutzi, B. and Stilla, U., 2004. Extraction of features from objects in urban areas using spacetime analysis of recorded laser pulses, *International Archives of Photogrammetry, Remote Sensing and Spatial Information Sciences*, 35(B2), 6.

Jülge K. and Brenner, C., 2004. Object extraction from terrestrial laser scan data for a detailed building description, *International Archives of Photogrammetry, Remote Sensing and Spatial Information Sciences*, 35(B3/III), 1079–1084.

Kaartinen, H. et al., 2005a. EuroSDR building extraction comparison, In *Proceedings of the ISPRS Hannover Workshop High-Resolution Earth Imaging for Geospatial Information*, May 17–20, 2005, CD-ROM, 6 pp.

Kaartinen, H. et al., 2005b. Accuracy of 3D city models: EuroSDR comparison, *International Archives of Photogrammetry, Remote Sensing and Spatial Information Sciences*, 36(Part 3/W19), CD-ROM, 227–232.

Kaartinen, H. and Hyyppä, J., 2006. EuroSDR-Project Commission 3 "Evaluation of Building Extraction", Final Report, *EuroSDR – European Spatial Data Research, Official Publication* No 50, 9–77.

Khoshelham K., 2004. Building extraction from multiple data sources: A data fusion framework for reconstruction of generic models, *International Archives of Photogrammetry, Remote Sensing and Spatial Information Sciences*, 35(B3/III), 980–986.

Kim, T. and Muller, J.P., 1998. A technique for 3D building reconstruction, *Photogrammetric Engineering and Remote Sensing*, 64(9), 923–939.

Lee, I. and Choi, Y., 2004. Fusion of terrestrial laser scanner data and images for building reconstruction, *International Archives of Photogrammetry, Remote Sensing and Spatial Information Sciences*, 35(B5), 6.

Li, B.J., Li, Q.Q., Shi, W.Z. and Wu, F.F., 2004. Feature extraction and modeling of urban building from vehicle-borne laser scanning data, *International Archives of Photogrammetry, Remote Sensing and Spatial Information Sciences*, 35(B5), 6.

Maas, H.-G., 2001. The suitability of airborne laser scanner data for automatic 3D object reconstruction, In Baltsavias, E. (Ed.), *Automatic Extraction of Man-Made Objects from Aerial and Space Images III*, Swets & Zeitlinger B.V., Lissie, the Netherlands, pp. 291–296.

Maas, H.-G. and Vosselman, G., 1999. Two algorithms for extracting building models from raw laser altimetry data, *ISPRS Journal of Photogrammetry & Remote Sensing*, 54(2–3), 153–163.

Madhavan, B.B., TachibanaK., Sasagawa T., Okada H. and Shimozuma Y., 2004. Automatic extraction of shadow regions in high-resolution ADS40, *International Archives of Photogrammetry, Remote Sensing and Spatial Information Sciences*, 35(B3/III), 808–810.

Mayer, H., 1999. Automatic object extraction from aerial imagery—A survey focusing on buildings, *Computer Vision and Image Understanding*, 74(2), 138–149.

Morgan, M. and Habib, A., 2001. 3D TIN for automatic building extraction from airborne laser scanner data, In *Proceedings of the American Society of Photogrammetry and Remote Sensing* (ASPRS), St. Louis, MO.

Morgan, M. and Habib, A., 2002. Interpolation of LiDAR data and automatic building extraction, In *Proceedings of the ACSM-ASPRS Annual Conference*, Washington D.C. CD-ROM.

Neidhart, H. and Sester, M., 2003. Identifying building types and building clusters using 3D-laser scanning and GIS-data, *International Archives of Photogrammetry, Remote Sensing and Spatial Information Sciences*, 35(B4), 6.

Niederöst, M., 2003. Detection and reconstruction of buildings for automated map updating, Diss., Technische Wissenschaften ETH Zurich. (Mitteilungen/Institut für Geodäsie und Photogrammetrie/ETH), 139 pp.

Noronha, S. and Nevatia, R., 2001. Detection and modeling of buildings from multiple aerial images, *IEEE Transactions on Pattern Analysis and Machine Intelligence*, 23(5), 501–518.

Oda, K., Takano, T., Doihara, T. and Shibasaki, R., 2004. Automatic building extraction and 3D city modeling from LiDAR data based on Hough transformation, *International Archives of Photogrammetry, Remote Sensing and Spatial Information Sciences*, 35(B3:III), 277–280.

Oriot, H. and Michel, A., 2004. Building extraction from stereoscopic aerial images, *Applied Optics*, 43(2), 218–226.

Overby, J., Bodum, L., Kjems, E. and Ilsøe, P.M., 2003. Automatic 3D building reconstruction from airborne laser scanning and cadastral data using Hough transform, *International Archives of Photogrammetry, Remote Sensing and Spatial Information Sciences*, 35(B3), 6.

Paparoditis, N., Cord, M., Jordan, M. and Cocquerez, J.P., 1998. Building detection and reconstruction from mid- and high-resolution aerial imagery, *Computer Vision and Image Understanding*, 72(2), 122–142.

Peternell, M. and Steiner, T., 2004. Reconstruction of piecewise planar objects from point clouds, *Computer-Aided Design*, 36(4), 333–342.

Rottensteiner, F., 2000. Semi-automatic building reconstruction integrated in strict bundle block adjustment, *International Archives of Photogrammetry and Remote Sensing*, 33(B2/2), 461–468.

Rottensteiner, F., 2003. Automatic generation of high-quality building models from LiDAR data, *IEEE Computer Graphics and Applications*, 23(6), 42–50.

Rottensteiner, F. and Briese, Ch., 2002. A new method for building extraction in urban areas from high-resolution LiDAR data, *International Archives of Photogrammetry, Remote Sensing and Spatial Information Sciences*, 34(3A), 295–301.

Rottensteiner, F. and Jansa, J., 2002. Automatic extraction of buildings from LiDAR data and aerial images. Available at http://www.ipf.tuwien.ac.at/publications/fr_jj_ottawa_02.pdf.

Rottensteiner, F., Trinder, J., Clode, S. and Kubik, K., 2003. Detecting buildings and roof segments by combining LIDAR data and multispectral images. Available at http://sprg.massey.ac.nz/ivcnz/Proceedings/IVCNZ_12.pdf.

Rottensteiner, F., Trinder, J., Clode, S. and Kubik, K., 2004. Fusing airborne laser scanner data and aerial imagery for the automatic extraction of buildings in densely built-up areas, *International Archives of Photogrammetry, Remote Sensing and Spatial Information Sciences*, 35(B3:III), 512–517.

Rottensteiner, F., Trinder, J., Clode, S. and Kubik, K., 2005. Using the dempster-shafer method for the fusion of LIDAR data and multi-spectral images for building detection, *Information Fusion*, 6(4), 283–300.

Sahar, L. and Krupnik, A., 1999. Semiautomatic extraction of building outlines from large-scale aerial images, *Photogrammetric Engineering and Remote Sensing*, 65(4), 459–465.

Schwalbe, E., 2004. 3d Building model generation from airborne laserscanner data by straight line detection in specific orthogonal projections, *International Archives of Photogrammetry, Remote Sensing and Spatial Information Sciences*, 35(B3/III), 249–254.

Sequeira, V., Ng, K., Wolfart, E., Goncalves, J.G.M. and Hogg, D., 1999. Automated reconstruction of 3D models from real environments, *ISPRS Journal of Photogrammetry and Remote Sensing*, 54(1), 1–22.

Shufelt, J.A., 1999. Performance evaluation and analysis of monocular building extraction from aerial imagery, *IEEE Transactions on Pattern Analysis and Machine Intelligence*, 21(4), 311–326.

Sinning-Meister, M., Grün, A. and Dan, H., 1996. 3D City models for CAAD-supported analysis and design for urban areas, *ISPRS Journal of Photogrammetry and Remote Sensing*, 51, 196–208.

Söderman, U., Ahlberg, S., Elmqvist, M. and Persson, Å., 2004. Three-dimensional environment models from airborne laser radar data, In *Proceedings of SPIE Defense and Security Symposium*, 5412, 12 pp. Available at http://www.sne.foi.se/documents/SNE_SPIE_DS_04.pdf.

Sohn G., 2004. Extraction of buildings from high-resolution satellite data and LiDAR, *International Archives of Photogrammetry, Remote Sensing and Spatial Information Sciences*, 35(B3/III), 1036–1042.

Stilla, U. and Jurkiewicz, K., 1999. Reconstruction of building models from maps and laser altimeter data, *Lecture Notes on Computer Science*, 1737, 34–46.

Süveg, I. and Vosselman, G., 2004. Reconstruction of 3D building models from aerial images and maps, *ISPRS Journal of Photogrammetry and Remote Sensing*, 58(3–4), 202–224.

Taillandier, F. and Deriche, R., 2004. Automatic building extraction from aerial images: A generic bayesian framework, *International Archives of Photogrammetry, Remote Sensing and Spatial Information Sciences*, 35(B3:III), 343–348.

Takase, Y. et al., 2004. Reconstruction and visualization of "virtual time-space of Kyoto", A 4D-GIS of the city, *International Archives of Photogrammetry, Remote Sensing and Spatial Information Sciences*, 35(B5), 6.

Tan, G. and Shibasaki, R., 2002. A research for the extraction of 3D urban building by using airborne laser scanner data, In *Proceedings of the 23rd Asian Conference on Remote Sensing*, Kathmandu, Nepal, November 25–29, 2002, 5 pp.

TerraSolid, 2003. TerraScan User's Guide.

Tsay, J.-R., 2001. A concept and algorithm for 3D city surface modelling, In *Proceedings of the ISPRS Commission III/WGIII/2*, 4 pp. Available at http://www.isprs.org/commission3/proceedings02/papers/paper092.pdf.

Tseng, Y.H. and Wang, S.D., 2003. Semiautomated building extraction based on CSG model-image fitting, *Photogrammetric Engineering and Remote Sensing*, 69(2), 171–180.

Vosselman, G., 2002. Fusion of laser scanning data, maps, and aerial photographs for building reconstruction. Available at http://www.geo.tudelft.nl/frs/papers/2002/vosselman.igarss02.pdf.

Vosselman, G. and Dijkman, S., 2001. 3D building model reconstruction from point clouds and ground plans, *International Archives of Photogrammetry, Remote Sensing and Spatial Information Sciences*, 34(3W4), 37–43.

Vosselman, G. and Süveg, I., 2001. Map based building reconstruction from laser data and images. *Automatic Extraction of Man-Made Objects from Aerial and Space Images (III)*, Ascona, Switzerland, June 11–15, Balkema, the Netherlands, 231–239.

Wang, X. and Grün, A., 2003. A hybrid GIS for 3D city models, *International Archives of Photogrammetry, Remote Sensing and Spatial Information Sciences*, 33(4/3), 1165–1172.

Weidner, U. and Förstner, W., 1995. Towards automatic building extraction from high-resolution digital elevation models, *ISPRS Journal of Photogrammetry and Remote Sensing*, 50(4), 38–49.

Zhan, Q., Molenaar, M. and Tempfli, K., 2002. Building extraction from laser data by reasoning on image segments in elevation slices, *International Archives of Photogrammetry and Remote Sensing*, 34(3B), 305–308.

Zhang, Y. and Wang, R., 2004. Multi-resolution and multi-spectral image fusion for urban object extraction, *International Archives of Photogrammetry, Remote Sensing and Spatial Information Sciences*, 35(B3/III), 960–966.

Zhao, H.J. and Shibasaki, R., 2003. Reconstructing a textured CAD model of an urban environment using vehicle-born range scanners and line cameras, *Machine Vision and Applications*, 14(1), 35–41.

Index

Note: Page numbers followed by f and t refer to figures and tables respectively.